ADVANCED GEOMETRY

(*MATHEMATICAL ANALYSIS OF UNIFIED ARTICLES*)

notionpress.com

ADVANCED GEOMETRY

(*MATHEMATICAL ANALYSIS OF UNIFIED ARTICLES*)

Harish Chandra Rajpoot

notionpress.com

Notion Press

5 Muthu Kalathy Street, Triplicane, Chennai - 600 005

First Published by Notion Press 2014

ISBN: 978-93-83808-15-1

This book is dedicated to my loving parents, elder brother Mr. Chandra Shekhar Rajpoot & respected Guru Mr. Upendra Sir for their love, affection & invaluable sacrifice &

To The Learners

With hope that this book will develop interest & skill in

Advanced Geometry

Providing advanced knowledge & wide applications of solid angle

Preface

Although the concept of solid angle & its brief mathematical formations are existing from decades but its concept & practical applications in the field of engineering are not much enhanced. More over this article, so far, has not been included as a chapter even as a topic in any of the academic books relevant to UP, CBSE, ICSE or other boards.

Concept of solid angle still needs to be developed & improved to make its practical applications in the field of engineering such as Photometry, Radiometry & Space science. Mathematical analysis of solid angle becomes essential for its applications in basic & applied physics so in the field of engineering. Geometry, calculus, trigonometry have been used to analyse solid angle. Students of High school & Intermediate even higher education such as B Tech, BSc etc. have basic concept of solid angle incomprehensible. Although a number of academic books with no or little description of solid angle are available as reference books.

This book is although an outcome of a deep & long studies up to more than two years but including all the basic & applied concepts & wide applications of solid angle in Physics so in the engineering. The objective of writing this book is mainly to provide the advanced concept of solid angle & apply it in Mathematics & Physics. It includes all the concepts & mathematical formulae of solid angle in easy-to-comprehend way for learners, educators & academicians. Most of the articles of this book are created & rest are modified on the basis of new academic research.

The text extends into two units first one has six chapters & second has two chapters. Unit 1 includes analysis of solid angle. Chapter 1 deals with the introduction of solid angle which has definitions & Fundamental Theorem of Solid Angle. Chapter 2 deals with analysis of plane figures such as triangle, rectangle, rhombus, regular polygon & circle with neat line diagrams. Chapter 3 deals with the approximate analysis of solid angle & approximation theorem for symmetrical plane & its applications. Chapter 4 deals with the analysis of 3D figures such as sphere, circular cylinder, right cone, right pyramid & right prism along with graphs. Chapter 5 deals with the significance & applications of solid angle in Solid state (chemistry), Photometry/Radiometry & Space science. Chapter 6 deals with the Graphical analysis of solid angle Unit 2 has been included as an additional knowledge which deals with the miscellaneous analysis of new articles. Chapter 1 deals with the analysis of frustums of sphere, circular cylinder & right cone. Chapter 2 deals with the analysis of axioms of vectors & trigonometry, analysis of star-like regular-polygon & right pyramid and regular tetrahedron.

Numerical examples, easy concepts & chapter wise summary are salient features of the entire book. I think that the present edition will be very useful to intermediate, graduate & undergraduate learners to develop new skill in this subject. Although the best efforts have been made by the author to make this book error free still some errors might have crept in. Therefore, all the students, readers are requested to give their feedback to the author. It is suggested to UP, CBSE, ICSE & other institutions to adapt useful articles of this book to their respective academic syllabus to enhance the skill of students/learners & make practical applications of solid angle in Photometry/Radiometry.

Harish Chandra Rajpoot

Acknowledgements

After a long & deep study, I could complete this book (created as a new academic research) written on my interesting subject "Advanced Geometry". I had to study & work constantly more than two years on this book for new findings in this subject. I offer my great sense of gratitude to my parents & elder brother who always, loved, inspired, encouraged, helped & have been showering their blessings on me. Also, I am thankful to all my teachers specially M r. Upendra Sir, an experienced teacher of physics at Oxford Model I.C. Syam Nagar, Kanpur, who not only taught & guided but also helped me with all the aspects in my studies & inspired me to do something new in the field of education. He had been my best guide & directed me for such a work. The author is deeply thankful to ALMIGHTY GOD who kept him physically fit during the completion of this academic research based book.

Hoping my work will be useful for all the learners & wishing them a profitable & enjoyable journey through this book.

Harish Chandra Rajpoot
B Tech. (Mech. Engg.)
M.M.M. Govt. Engg. College
Gorakhpur, Uttar Pradesh, India
rajpootharishchandra@gmail.com

Note: I hereby claim that most of the articles of this book as '**Advanced Geometry**' are created & rest are modified by me.

Harish Chandra Rajpoot

Contents

Unit.1. Analysis of Solid Angle

This unit carries five chapters. Each of the chapters deals with the articles related to the analysis & estimation of the solid angle subtended by the plane-figures as well as by the solid-figures at the given point in the space. These chapters are as follows

1. Introduction to solid angle
2. Analysis of 2D-figures
3. Approximate analysis of solid angle
4. Analysis of 3D-figures
5. Significance & applications of solid angle.

Chapter.1. deals with 1. Introduction to solid angle2. Definition of solid angle

3. Fundamental theorem of solid angle 4. Applications of fundamental theorem

Chapter.2. deals with 1. Integral theorem of solid angle

2. Corollary of the plane-figure 3. Applications of the corollary

Chapter.3. deals withanalysis of the following 3D-figures 1. Sphere 2. Cylinder 3. Cone

4. Right pyramid & 5. Right prism

In this unit of analysis of solid angle, a given 'point' means a certain location of the 'viewer' observing a given object (2D or 3D) in the space.

In this unit, we are to study & estimate the solid angle subtended by different plane & solid figures at certain points in the space.

Numerical problems both solved and unsolved have been added in order to make their applications in geometry & photometry/radiometry. Some of the problems are relevant to the geometry & some of them are relevant to the photometry & radiometry. All the problems are quantitative.

Chapter 1
Introduction to Solid Angle

1.1 Concept of Solid Angle

As the name indicates that the solid angle is a tree dimensional angle. The solid angle is subtended by an object at a certain point in the space. It indicatesthat how an object appears to the observer standing at a certain point in the space. The solid angle is dependent particularly on the geometrical shape (i.e. the geometrical dimensions) of the object & its configuration w.r.t. the given point (i.e. the viewer) in the space.

We know that the angle subtended by a straight line at any point in the space is called plane-angle. The plane-angle shows that how closer the line is to the given point (viewer) in the space i.e. higher the angle subtended by the straight line at a given point, closer is the line to the given point in the space.

Similarly, the angle subtended by the object (2D or 3D-figures) at any point in the space is called solid angle. It shows that how closer the object is to the given point in the space i.e. higher the solid angle subtended by the object at a given point, closer is the object to the given point in the space.

The solid angle also indicates that how likely an object appears to a spherical surface appears to an observer located at the given point in the space.

If we place a large spherical surface of certain radius with centre at the given point in the space after the given objectw.r.t. the same point then we find that the object projects a certain area on the spherical surface.

The solid angle also indicates the measure of area of projection of the given object on a spherical surface with unit radius & centre at the given point in the space.

1.2 Definition of Solid Angle

It can generally be defined as "The three dimensional angle subtended by an object at a certain point in the space "

It is denoted by $'\omega'$ (omega). Its unit is steradian (in brief 'sr.').

Mathematically, it is given by the following expression

$$\omega = \frac{area\ of\ visible\ surface\ of\ the\ object\ (A)}{(\ distance(r)\ of\ any\ point\ on\ visible\ surface\ from\ the\ given\ point)^2}$$

Where, the distance 'r' is constant from all the points on the visible surface (of area 'A') of the object to the given point (i.e. observer) in the space.

It is clear that the above condition i.e. the distance 'r' is constant for all the points on the visible surface of an object, is true for an infinitesimal small circular area taken on the visible surface of the given object.

Or the distance 'r' is constant for all the points of a spherical surface from its centre.

Thus the solid angle can also be defined as "The angle subtended by any part of the spherical surface at its centre is called solid angle."

The solid angle can also be defined as "The measure of area of projection of a given object on a spherical surface with unit radius & centre at the given point in the space."

The solid angle can also be defined as "The degree of symmetrical & spherical distribution of an object w.r.t. a given point in the space which is similar to a spherical surface with centre at the given point & radius equal to the distance of any arbitrary point on the surface to the same point in the space."

Cone of vision: It is an imaginarysolid cone with vertex at the given point, obtained by joining all the points of an object to the given point in the space. This cone may be 1.Right or 2.Skew depending on geometrical shape of the object & its configuration w.r.t. the given point in the space.

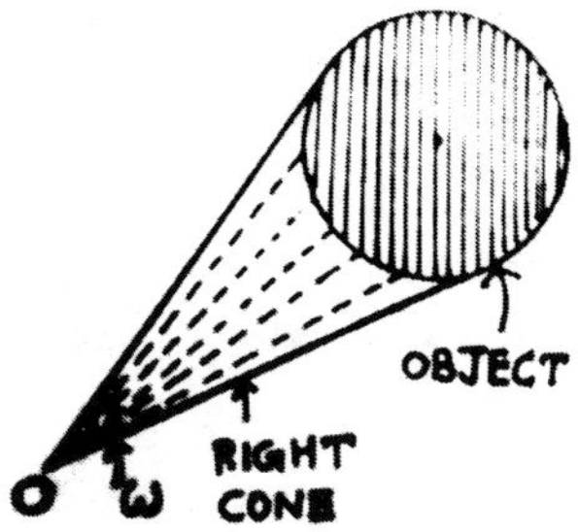

Fig(1) **Fig(2)**

Right cone of vision is obtained in case of the objects appearing circular to the observer located at the given point in the space while in other cases the cone of vision is always skew.

(see the figures (1) & (2) above)

Thus, the solid angle can also be defined as "The three dimensional angle subtended by the cone of vision at its vertex (i.e. at the given point)."

1.3 Fundamental Theorem of Solid Angle

Solid angle subtended by an object at a given point in the space is given by the following expression

$$\omega = \int_S \frac{dA}{r^2} \forall\ (0 \le \omega \le 4\pi)$$

Where 'r' is the distance of any arbitrary point on visible surface of the object from the given point in the space which is constant for all the points on elementary area 'dA' of visible surface 'S' of the object.

Proof: let there be an extended object with the visible surface 'S' and a given point say 'O'

in the space. (as shown in the figure(3) below)

Now, consider an infinitesimal small circular area 'dA' with the centre at any arbitrary point say 'P' on visible surface 'S' of the object at a distance say 'r' from the given point 'O'.

Since, the elementary circular area 'dA' is so small that all the points of area 'dA' may be considered to be at a constant distance 'r' from the point 'O' in the space.

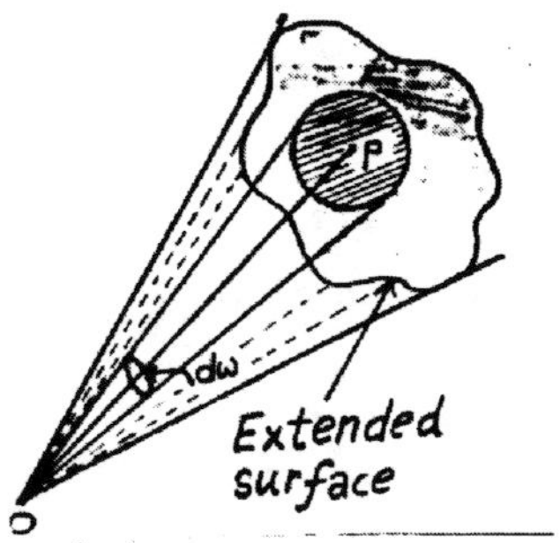

Fig(3)

Hence, by the definition of the solid angle

The solid angle subtended by the elementary area 'dA' at the point 'O' is given

According to the definition, mathematically it is given as follows

$$d\omega = \frac{elementary\ area\ (dA)}{(distance(r))^2} = \frac{dA}{r^2}$$

(Since, distance ‘r’ is constant for all the points on elementary area ‘dA’ of visible surface)

Thus, solid angle subtended by the extended surface ‘S’ of the object is obtained

By integrating above expression both the sides, we have total solid angle (ω) subtended by the object at the given point as follows

$$\omega = \int d\omega = \int_S \frac{dA}{r^2}$$

1.4 Applications of the Fundamental Theorem

Fundamental theorem is the most basic theorem of solid angle. All the theorems can be derived by this fundamental theorem. it has a no. of the applications.

1.4.1 Solid Angle of the Right Cone

Let there be a right cone having vertex at the point ‘P’ & the apex angle ‘$2\alpha'$

(as shown in the figure(4) below)

Now, consider an imaginary spherical surface with centre at the vertex ‘P’ & the radius ‘R’ such that it intercepts the right cone at a circular section with the centre ‘M’ & the diameter AB (see the fig.)

$$\Rightarrow PA = PB = R(radius\ of\ the\ spherical\ surface)$$

Now, consider an infinitesimal small circular strip with the thickness $'Rd\theta'$& the radius $'Rsin\theta$’ at an angle θ with the axis PM.

Hence, the area of the elementary circular strip

$$\Rightarrow dA = (periphery)(thichness) = [2\pi(Rsin\theta)][Rd\theta]$$

$$= 2\pi R^2 sin\theta d\theta$$

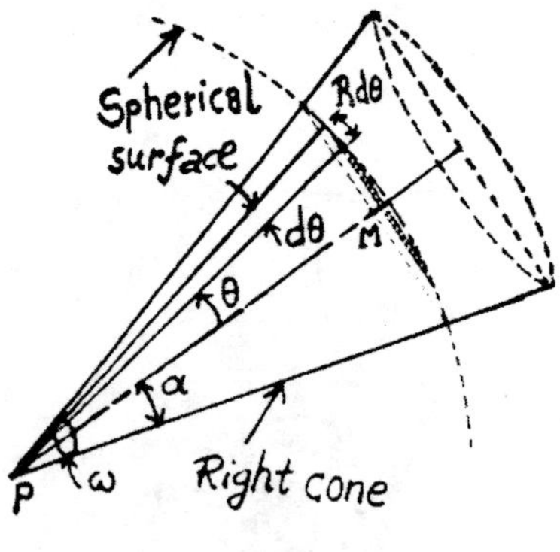

Fig(4)

∴ The total area of intercepted surface of the sphere by the right cone is obtained by integrating the area dA from ‘0’ to $'\alpha'$ as follows

$$\Rightarrow A = \int_0^\alpha dA = \int_0^\alpha 2\pi R^2 sin\theta d\theta = 2\pi R^2(-cos\theta)_0^\alpha = 2\pi R^2(cos0 - cos\alpha)$$
$$= 2\pi R^2(1 - cos\alpha)$$

Since, the right cone & the spherical surface are intercepting

Hence, the solid angle subtended by the right cone at the vertex ‘P’

$$= solid\ angle\ subtended\ by\ the\ intercepted\ surface\ of\ the\ sphere$$

$$\Rightarrow \omega = \int_S \frac{dA}{R^2} = \frac{1}{R^2}\int_S dA\ \ (since,\ 'R' is\ constant)$$

$$= \frac{A}{R^2} = \frac{2\pi R^2(1 - cos\alpha)}{R^2} = 2\pi(1 - cos\alpha)$$

$$\Rightarrow [\omega = 2\pi(1 - cos\alpha)] \quad \ldots\ldots\ldots\ldots\ldots\ldots\ldots..(1)$$

Note: In all the numerical problems, plane-angle istaken in 'radian'& solid angle in 'steradian'. A 'Point-source of light/energy' means 'Diffused point-source' i.e. it is emitting light/energy uniformly in all the directions.

1.4.2 Area of projection: If the solid angle subtended by a given object at a given point in the space is 'ω' then the area of projection (interception) of the object on a spherical surface with a radius 'R' & centre at the given point is given as

$$A_p = \omega R^2 \quad \ldots\ldots\ldots\ldots\ldots\ldots\ldots\ldots\ldots\ldots\ldots (1.1)$$

The above formula is also applicable to find the area of the spherical surface intercepted by a right cone with vertex at the centre of the spherical surface.

1.4.3 Luminous intensity: It can be defined as the luminous flux emitted by a uniform light source per unit solid angle. It is denoted by 'I'. Its unit is Candela (in brief 'cd'). Other unit is lumen per steradian (in brief 'lm/sr.').

1 cd = 1 1m/sr.

1.4.4 Luminous flux density: It can be defined as the luminous flux striking per unit area of a surface is called 'luminous flux density or luminosity'. It is denoted by 'E'. Its unit is lux orlm/m^2.

If the luminous intensity of a uniform point-source of light is 'I' & 'ω' is the solid angle subtended by a given surface with area 'A' at the point-source. Then the luminous flux density

$$\Rightarrow E = \frac{total\ flux\ striking\ the\ surface}{area\ of\ the\ surface} = \frac{F}{A} = \frac{I \times \omega}{A}$$

$$\Rightarrow luminous\ flux\ intercepted\ by\ the\ surface, F = I \times \omega$$

$$where,\ '\omega' is\ the\ solid\ angle\ subtended\ by\ the\ surface\ at\ the\ point\ source.$$

Note: In actual practice, value of luminous flux density 'E' varies from maximum at the foot of perpendicular (F.O.P.), drawn from point-source to the plane, to minimum at the points away from F.O.P. When a point-source is placed at any point on a given plane-surface, a brightest spot (point of maximum intensity) is obtained at the same point &the value of E is indeterminate (infinite) since total intercepted radiation is concentrated at that point. Although the value of total radiation (energy or luminous flux) intercepted by a given plane surface can be determined when a uniform point-source is placed at a certain point on the plane.

Solved Examples

Example 1.1: Find out solid angle subtended by a right cone with apex angle $\frac{\pi}{3}$ rad.

Sol. given that

$$2\alpha = \frac{\pi}{3}\ or\ \alpha = \frac{\pi}{6}$$

Hence, the solid angle subtended by the right cone at the apex

$$\Rightarrow \omega = 2\pi(1 - cos\alpha) = 2\pi\left(1 - cos\frac{\pi}{6}\right) = 2\pi\left(1 - \frac{\sqrt{3}}{2}\right) = 0.268\pi\ sr\ or\ 0.842\ sr\ \ Ans.$$

Example 1.2: Find out solid angle subtended by a right cone with apex angle $\frac{\pi}{2}$ rad.

Sol. given that

$$2\alpha = \frac{\pi}{2}\ or\ \alpha = \frac{\pi}{4}$$

Hence, the solid angle subtended by the right cone at the apex

$$\Rightarrow \omega = 2\pi(1 - cos\alpha) = 2\pi\left(1 - cos\frac{\pi}{4}\right) = 2\pi\left(1 - \frac{1}{\sqrt{2}}\right) = 0.586\pi\ sr\ or\ 1.84\ sr\ \ Ans.$$

Example 1.3: Find out solid angle subtended by a right cone at the apex, having a normal height 15cm. & a radius 6cm.

Sol. given that

$$h = 15\ cm.\ \ \&\ r = 6cm.$$

Now, the apex angle (2α) of the right cone can be found by the following formula

$$\Rightarrow tan\alpha = \frac{r}{h} = \frac{6}{15} = 0.4\ or\ \alpha = \tan^{-1} 0.4 = 0.38\ rad.$$

Hence, the solid angle subtended by the right cone at the apex

$$\Rightarrow \omega = 2\pi(1 - \cos(0.38)) = 2\pi(1 - 0.928) = 0.144\pi\ sr\ or\ 0.452\ sr\ Ans.$$

Example 1.4: Find out area of a spherical surface with a radius 10 cm. intercepted by a right cone with apex angle $\frac{\pi}{4}$ & apex point coincident with the centre of the spherical surface.

Sol. given that

$$2\alpha = \frac{\pi}{4}\ \ or\ \alpha = \frac{\pi}{8}$$

Now, the solid angle subtended by the right cone at the apex (i.e. at the centre of spherical surface)

$$\Rightarrow \omega = 2\pi(1 - cos\alpha) = 2\pi\left(1 - cos\frac{\pi}{8}\right) = 2\pi(1 - 0.924) = 0.152\pi\ sr\ or\ 0.478\ sr$$

∴ Solid angle subtended by the intercepted area of spherical surface at the centre

= solid angle subtended by the right cone at the apex (i.e. at the centre of spherical surface)

$$\Rightarrow \omega = 0.478\ sr$$

Hence, the area of spherical surface intercepted by the right cone is given from the eq(1.1)

$$\Rightarrow A_p = \omega R^2 \qquad where\ 'R' is\ the\ radius\ of\ the\ spherical\ surface$$

$$= 0.478\ \times (10)^2 = 47.8\ cm^2 \qquad Ans.$$

Example 1.5: Solid angle subtended by an object at a uniform point-source of light is 0.54sr. Find out the area of projection of a given object on a spherical surface with a radius 25 cm. & centre coincident with the given point-source.

Sol. given that

$$\omega = 0.54\ sr\ \ \&\ R = 25\ cm.$$

Hence, the area of projection of the object on the spherical surface is given from the eq(1.1)

$$\Rightarrow A_p = \omega R^2$$

$$where\ 'R' is\ the\ radius\ of\ the\ spherical\ surface$$

$$= 0.54\ \times (25)^2 = 337.5\ cm^2 \qquad Ans.$$

Example 1.6: A uniform point-source of light having luminous intensity 100cd (lm/sr.) located at the apex of a conical tower with an apex angle10^o. Find out the luminous flux intercepted by surface of the tower.

If the length of tower is increased by 5 cm. what will be the change in the flux striking the surface of the tower?

Sol. given that

$$luminous\ intensity, I = 100\ lm/sr.\ \&$$

$$\alpha = 10^o = 10 \times \frac{\pi}{180} = 0.174\, rad.$$

Now, the solid angle subtended by the conical tower at the point-source

$$\Rightarrow \omega = 2\pi(1 - \cos\,(0.174)) = 2\pi(1 - 0.985) = 0.03\pi\ sr\ or\ 0.095\ sr$$

Hence, the luminous flux intercepted by the surface of conical tower

$$\Rightarrow F = \mathrm{I} \times \omega$$

$$\therefore F = 100 \times 0.095 = 9.5\ lm \qquad Ans.$$

If, the length of the tower is increased then there is no. change in the flux until the apex angle of tower is changed.

Example 1.7: A point-emittermounted on a conical antenna with apex angle 15^ouniformly emitting the radiation-energy in all the directions at a rate 2kW/sr. Find out radiation-energy/secescaping the antenna.

If apex angle of antenna is decreased by 6° what will be the change in the radiation-energy/sec

1) Escaping the antennaor obstructed by the antenna& 2) how much power will be saved or lost?(Neglect the effect of reflection)

Sol. given that

$$Emissive\ power, P = 2kW/sr\ = 2000\ W/sr.\&$$

$$2\alpha = 15^o\ or\ \alpha = \frac{15}{2} = 7.5^o = 7.5 \times \frac{\pi}{180} = 0.131\, rad.$$

Now, the solid angle subtended by the conical antenna at the point-source

$$\Rightarrow \omega = 2\pi(1 - \cos\,(0.131)) = 2\pi(1 - 0.991) = 0.018\pi\ sr\ or\ 0.056\ sr$$

Thus, the radiation-energy obstructed by the surface of conical antenna,

$$E_1{}' = \mathrm{P} \times \omega = 2000 \times 0.056 = 112\, J/sec.$$

Total radiation-energy emitted by the source,

$$E_t = 4\pi \times P = 4\pi \times 2000 = 25132.741 J/sec.$$

Hence, radiation-energy escaping the antenna,

$$E_1 = (total\ energy\ emitted\ per\ \text{sec.}) - (energy\ obstructed\ per\ sec.)$$

$$= E_t - E_1{}' = 25132.741 - 112 = 25020.741\ J/sec. or\ 25.021\ kJ/s \qquad Ans.$$

If, the apex angle of antenna is decreasedby 6°

$$\mathbf{2\alpha' = 15^o - 6^o = 9^o\ or\ \alpha = \frac{9^o}{2} = 4.5^o = 4.5 \times \frac{\pi}{180} = 0.078\, rad.}$$

Now, the solid angle subtended by the conical antenna at the point-source

$$\mathbf{\Rightarrow \omega = 2\pi(1 - cos\,(0.078)) = 2\pi(1 - 0.997) = 0.019\ sr}$$

Thus, the radiation-energy obstructed by the surface of conical antenna,

$$\mathbf{E_2{}' = P \times \omega = 2000 \times 0.019 = 38\, J/sec.}$$

Now, radiation-energy escaping the antenna

$$\mathbf{E_2 = (total\ energy\ emitted\ per\ sec.) - (energy\ obstructed\ per\ sec.)}$$

$$\mathbf{= E_t - E_2' = 25132.741 - 38}$$

$$\mathbf{= 25094.741\, J/sec.\ or\ 25.095\ kJ/sec.}$$

1. Change in the radiation-energy escaping the antenna or obstructed by the antenna

$$\mathbf{= E_2 - E_1 = 25.095 - 25.021 = 0.074\ kJ/sec\ or\ 74\, J/sec}$$

2. Since, the radiation-energy obstructed by antenna is decreased, hence the power will be saved

$$\therefore power\ saved = E_2 - E_1$$

$$= E_1' - E_2' = 112 - 38 = 74\, J/sec\ or\ 74\, W \qquad Ans.$$

Example 1.8:A uniform point-source of light having luminous intensity 110 cd (lm/sr.) is located at the centre of a cube-box with each side 16cm. Now, if a small circular hole ofradius 2.5cm is made at the centre of one of the faces. Find out the luminous flux escaping the box.

(Neglect thickness of the box & the effect of reflection)

Sol. given that

$$luminous\ intensity, I = 110\ lm/sr$$

$$side\ of\ cube\ box, a = 16cm.\ \&\ radius\ of\ the\ hole, r = 2.5cm.$$

Now, the distance of point-source from the centre of one of the faces of the box

$$= \frac{a}{2} = \frac{16}{2} = 8cm.$$

Let the plane-angle subtended by the circular hole at the point-source be '2α'

Then, we have

$$\Rightarrow cos\alpha = \frac{\left(\frac{a}{2}\right)}{\sqrt{\left(\frac{a}{2}\right)^2 + r^2}} = \frac{8}{\sqrt{8^2 + 2.5^2}} = 0.954$$

If, we assume the circular hole to be the base of a right cone having apex at the point-source, the solid angle subtended by the circular hole at the point-source

$$\Rightarrow \omega = 2\pi(1 - \cos\ \alpha) = 2\pi(1 - 0.954) = 0.092\pi\ sr\ or\ 0.289\ sr$$

Hence, the luminous flux escaping the box,

$$F = I \times \omega = 110 \times 0.289 = 31.79\ lm \qquad Ans.$$

Example 1.9: A uniform point-source of light having luminous intensity 111.4 cd (lm/sr.) is located at the centre of a cuboidal box with dimensions$20cm. \times 18cm. \times 12cm.$ Now,a small circular hole of radius 1.6cmis made at the centre of a face of$20cm. \times 12cm$ find out the luminous flux

1) Escaping the box&2) retained in the bow.

(Neglect thickness of the box & the effect of reflection)

Sol. given that

$$luminous\ intensity, I = 111.4\ lm/sr$$

$$dimensions\ of\ cuboidal\ box, l = 20cm., b = 18cm.\ \&\ h = 12cm.$$

$$radius\ of\ the\ hole, r = 1.6cm.$$

Now, the distance of point-source from the centre of the face of $20cm \times 12cm$

$$= \frac{b}{2} = \frac{18}{2} = 9cm.$$

Let the plane-angle subtended by the circular hole at the point-source be '2α'

Then, we have

$$\Rightarrow cos\alpha = \frac{\left(\frac{b}{2}\right)}{\sqrt{\left(\frac{b}{2}\right)^2 + r^2}} = \frac{9}{\sqrt{9^2 + 1.6^2}} = 0.984$$

If, we assume the circular hole to be the base of a right cone having apex at the point-source, the solid angle subtended by the circular hole at the point-source

$$\Rightarrow \omega = 2\pi(1 - \cos\ \alpha) = 2\pi(1 - 0.984) = 0.032\pi\ sr\ or\ 0.1\ sr$$

1. The luminous flux escaping the box,

$$F = I \times \omega = 111.4 \times 0.1 = 11.14\ lm \qquad Ans.$$

2. We know that the total luminous flux emitted by the point-source

$$\Rightarrow F_t = 4\pi I = 4\pi \times 111.4 = 1399.893\ lm$$

Hence, the luminous flux retained in the box

$$= toatl\ flux\ emitted - flux\ escaping\ the\ box$$

$$= F_t - F = 1399.893 - 11.14 = 1388.753\ lm \qquad Ans.$$

Example 1.10: A uniform point-source of light having luminous intensity 120 cd (lm/sr.) is located at the centre of a cube-box with each side 25cm. Now, if a small circular hole of radius 1.5cm is made at the centre of each of the faces. Find out the luminous flux

1) Escaping the box. 2) Retained in the box.

(Neglect thickness of the box & the effect of reflection)

Sol. given that

$$luminous\ intensity, I = 120\ lm/sr$$

$$side\ of\ cube\ box, a = 25cm.\ \ \&\ radius\ of\ the\ hole, r = 1.5cm.$$

Now, the distance of point-source from the centre of each of the faces of the box

$$= \frac{a}{2} = \frac{25}{2} = 12.5cm.$$

Let the plane-angle subtended by the circular hole at the point-source be '2α'

Then, we have

$$\Rightarrow cos\alpha = \frac{\left(\frac{a}{2}\right)}{\sqrt{\left(\frac{a}{2}\right)^2 + r^2}} = \frac{12.5}{\sqrt{12.5^2 + 1.5^2}} = 0.993$$

If, we assume the circular hole to be the base of a right cone having apex at the point-source, the solid angle subtended by the circular hole at the point-source

$$\Rightarrow \omega = 2\pi(1 - cos\ \alpha) = 2\pi(1 - 0.993) = 0.014\pi\ sr\ or\ 0.044\ sr$$

1. The luminous flux escaping the box

$$F = 6(I \times \omega)$$

$$(since\ there\ are\ six\ holes\ each\ in\ the\ face\ of\ the\ box)$$

$$= 6(120 \times 0.044) = 6 \times 5.28 \ = 31.68\ lm \qquad Ans.$$

2. We know that the total luminous flux emitted by the point-source

$$\Rightarrow F_t = 4\pi I = 4\pi \times 120 = 1507.964\ lm$$

Hence, the luminous flux retained in the box

$$= toatl\ flux\ emitted - flux\ escaping\ the\ box$$

$$= F_t - F = 1507.964 - 31.68 = 1476.284\ lm \qquad Ans.$$

Example 1.11: A uniform point-source of light having luminous intensity 100 cd (lm/sr.) is located at the centre of a spherical shell with a radius 14cm. Now, if a small circular hole of radius 1.8cmis made in the shell find out the luminous flux 1) escaping the shell.& 2) retained in the shell

(Neglect the effect of reflection)

Sol. given that

$$luminous\ intensity, I = 100\ lm/sr$$

$$radius\ of\ the\ spherical\ shell, R = 14cm.\ \ \&\ radius\ of\ the\ hole, r = 1.8cm.$$

Let the plane-angle subtended by the circular hole at the point-source be '2α'

Then, we have

$$\Rightarrow cos\alpha = \frac{\sqrt{R^2 - r^2}}{R} = \frac{\sqrt{14^2 - 1.8^2}}{14} = 0.992$$

If, we assume the circular hole to be the base of a right cone having apex at the point-source, the solid angle subtended by the circular hole at the point-source located at the centre of the shell

$$\Rightarrow \omega = 2\pi(1 - cos\ \alpha) = 2\pi(1 - 0.992) = 0.016\pi\ sr\ or\ 0.05\ sr$$

1. The luminous fluxescaping the shell,

$$F = I \times \omega = 100 \times 0.05 = 5\ lm \qquad Ans.$$

2. We know that the total luminous flux emitted by the point-source

$$\Rightarrow F_t = 4\pi I = 4\pi \times 100 = 1256.637\ lm$$

Hence, the luminous flux retained in the shell

$$= total\ flux\ emitted - flux\ escaping\ the\ shell$$

$$= F_t - F = 1256.637 - 5 = 1251.637\ lm \qquad Ans.$$

Example 1.12: A uniform point-source of light having luminous intensity 111.4cd (lm/sr.) is located at the centre of a cube-box with internal dimensions $25cm. \times 25cm. \times 25cm.$&thickness of wall 2.5cm. Now, if a small circular hole of radius 20mm. is made at the centre of one of the faces. Find out the luminous flux

1.) Escaping the box & 2) retained in the box.(Neglect the effect of reflection)

Sol. given that

$$luminous\ intensity, I = 111.4\ lm/sr$$

$$since, internal\ length\ of\ sides\ of\ cube\ box = 25cm.$$

$$\therefore\ external\ length\ of\ sides\ of\ cube\ box, a = 25 + 2 \times 2.5 = 30cm.$$

$$thickness\ of\ the\ wall = 3\ cm\ \ \ \&\ \ \ radius\ of\ the\ hole, r = 20mm = 2cm.$$

Now, the distance of point-source from the centre of one of the faces of the box

$$= \frac{a}{2} = \frac{30}{2} = 15cm.$$

Let the plane-angle subtended by the circular hole at the point-source be '2α'

Then, we have

$$\Rightarrow cos\alpha = \frac{\left(\frac{a}{2}\right)}{\sqrt{\left(\frac{a}{2}\right)^2 + r^2}} = \frac{15}{\sqrt{15^2 + 2^2}} = 0.991$$

If, we assume the circular hole to be the base of a right cone having apex at the point-source, the solid angle subtended by the circular hole at the point-source

$$\Rightarrow \omega = 2\pi(1 - cos\ \alpha) = 2\pi(1 - 0.991) = 0.018\pi\ sr\ or\ 0.056\ sr$$

1. Luminous flux escaping the box

$$F = I \times \omega = 111.4 \times 0.056 = 6.238\ lm \qquad Ans.$$

2. We know that the total luminous flux emitted by the point-source

$$\Rightarrow F_t = 4\pi I = 4\pi \times 111.4 = 1399.894\ lm$$

Hence, luminous flux retained in the box

$$= \boldsymbol{total\ flux\ emitted - flux\ escaping\ the\ box}$$

$$= \boldsymbol{F_t - F = 1399.894 - 6.238 = 1393.656\ lm} \qquad \textbf{Ans.}$$

Example 1.13: A small bead of radioactive element with a radius 0.5mm is placed at the centre of bottom of a lead tube with a length 16cm. & a radius 1cm. Assume that the element is uniformly emitting radiation-energy in all the directions at a rate of 2.5 mW/sr.

Find out the radiation-energyescaping the tube per second.

(Neglect the effect of reflection)

Sol. given that

$$\boldsymbol{Emissive\ power, P = 2.5\ mW/sr = 2.5 \times 10^{-3}\ W/sr}$$

$$\boldsymbol{radius\ of\ the\ bead, r = 0.5mm}$$

$$\boldsymbol{length\ of\ tube, l = 160mm.} \&\ radius\ of\ tube, R = 10mm.$$

Let the plane-angle subtended by circular opening of the tube at the centre of the bottom be '2α'

Then, we have

$$\boldsymbol{cos\alpha = \frac{(l-r)}{\sqrt{(l-r)^2 + R^2}} = \frac{l}{\sqrt{l^2 + R^2}}} \quad \boldsymbol{(since, radius\ of\ the\ bead\ is\ negligible)}$$

$$= \boldsymbol{\frac{160}{\sqrt{160^2 + 10^2}} = \frac{160}{160.31} = 0.998}$$

Now, solid angle subtended by the opening at the bead

$$\boldsymbol{\Rightarrow \omega = 2\pi(1 - cos\alpha) = 2\pi(1 - 0.998) = 0.0126\ sr}$$

Hence, radiation-energy escaping the tube

$$\boldsymbol{= P \times \omega = 2.5 \times 10^{-3} \times 0.0126 = 3.15 \times 10^{-5} j/sec} \qquad \textbf{Ans.}$$

Example 1.14: A uniform point-source of light having luminous intensity 111.4 lm/sr, is located at a height 140mm. on the vertical axis of a circular disc with a radius 15mm. What will be the luminous flux striking the disc? Also find the luminous flux density on the disc.

Sol. given that

$$\boldsymbol{luminous\ intensity, I = 111.4\ lm/sr}$$

$$\boldsymbol{height, h = 140mm} \&\ radius\ of\ the\ disc, R = 15mm$$

If the plane angle subtended by the disc at the point-source is '2α'

Then we have,

$$\boldsymbol{\Rightarrow cos\alpha = \frac{h}{\sqrt{h^2 + R^2}} = \frac{140}{\sqrt{140^2 + 15^2}} = 0.994}$$

If, we assume the circular disc to be the base of a right cone having apex at the point-source, the solid angle subtended by the circular disc at the point-source

$$\boldsymbol{\Rightarrow \omega = 2\pi(1 - cos\ \alpha) = 2\pi(1 - 0.994) = 0.038\ sr}$$

Hence, the luminous flux striking the disc

$$\boldsymbol{\Rightarrow F = I \times \omega = 111.4 \times 0.038 = 4.233\ lm} \qquad \textbf{Ans.}$$

Now, we know that the luminous flux density

$$\boldsymbol{\Rightarrow E = \frac{F}{A} = \frac{F}{\pi R^2} = \frac{4.233}{\pi(15)^2} = 5.998 \times 10^{-3}\ lm/mm^2}$$

$$\boldsymbol{= 5.998 \times 10^3\ lm/m^2\ or\ 5.998 \times 10^3 lux} \qquad \textbf{Ans.}$$

Example 1.15: Find out the luminous intensity of a uniform point-source of light which is located at a height 165mm on the vertical of a circular disc with a diameter 25.4mm if the luminous flux intercepted by the disc is 2.4 lm. Also, find out the luminous flux density.

Sol. given that

$$height, h = 165\ mm, \qquad radius\ of\ the\ disc, R = \frac{25.4}{2} = 12.7\ mm.\ \&$$

$$luminous\ flux, F = 2.4\ lm.$$

If the plane angle subtended by the disc at the point-source is '2α'

Then, we have

$$\Rightarrow cos\alpha = \frac{h}{\sqrt{h^2 + R^2}} = \frac{165}{\sqrt{165^2 + 12.7^2}} = 0.997$$

If, we assume the circular disc to be the base of a right cone having apex at the point-source, the solid angle subtended by the circular disc at the point-source

$$\Rightarrow \omega = 2\pi(1 - cos\ \alpha) = 2\pi(1 - 0.997) = 0.019\ sr$$

Hence, the luminous flux striking the disc

$$\Rightarrow F = I \times \omega = I \times 0.019 \Rightarrow 2.4 = I \times 0.019$$

$$\Rightarrow I = \frac{2.4}{0.019} = 126.316\ lm/sr\ or\ 126.316\ cd \qquad Ans.$$

Now, luminous flux density

$$\Rightarrow E = \frac{F}{A} = \frac{F}{\pi R^2} = \frac{2.4}{\pi(12.7)^2} = 4.736 \times 10^{-3}\ lm/mm^2$$

$$= 4.736 \times 10^3\ lm/m^2\ or\ 4.736 \times 10^3\ lux \qquad Ans.$$

Example 1.16: A uniform point-source of light having luminous intensity100 cd is located at a height 120mm on the vertical axis of circular disc such that the luminous flux density on the disc is 2.5×10^3 lux. Find out the radius of the disc.

Sol. given that

$$luminous\ intensity, I = 100\ lm/sr,$$

$$luminous\ flux\ density, E = 2.5 \times 10^3 = 2500\ lux\ \&$$

$$height, h = 120mm = 0.12m$$

$$now.\ let\ radius\ of\ the\ disc = R$$

If the plane angle subtended by the disc at the point-source is '2α'

Then, we have

$$\Rightarrow cos\alpha = \frac{h}{\sqrt{h^2 + R^2}} = \frac{0.12}{\sqrt{0.12^2 + R^2}} = \frac{0.12}{\sqrt{0.0144 + R^2}}$$

If, we assume the circular disc to be the base of a right cone having apex at the point-source, the solid angle subtended by the circular disc at the point-source

$$\Rightarrow \omega = 2\pi(1 - cos\ \alpha) = 2\pi\left(1 - \frac{0.12}{\sqrt{0.0144 + R^2}}\right)$$

Now, luminous flux density

$$\Rightarrow E = \frac{I \times \omega}{A} \Rightarrow 2500 = \frac{100 \times \omega}{\pi R^2} = \frac{100}{\pi R^2}\left[2\pi\left(1 - \frac{0.12}{\sqrt{0.0144 + R^2}}\right)\right]$$

$$\Rightarrow 12.5 = \frac{1}{R^2}\left(1 - \frac{0.12}{\sqrt{0.0144 + R^2}}\right) \Rightarrow 1 - \frac{0.12}{\sqrt{0.0144 + R^2}} = 12.5R^2$$

$$\Rightarrow \frac{0.12}{\sqrt{0.0144 + R^2}} = 1 - 12.5R^2 \ \ or \ \ \sqrt{0.0144 + R^2} = \frac{0.12}{(1 - 12.5R^2)}$$

On, taking square of both the sides, we get

$$\Rightarrow 0.0144 + R^2 = \frac{0.0144}{(1 - 12.5R^2)^2} \Rightarrow (0.0144 + R^2)(1 + 156.25R^4 - 25R^2) = 0.0144$$

$$\Rightarrow 0.0144 + R^2 + 2.25R^4 + 156.25R^6 - 0.36R^2 - 25R^4 = 0.0144$$

$$\Rightarrow R^2(156.25R^4 - 22.75R^2 + 0.64) = 0$$

On solving the above equation, we get

$$R^2 = 0, R^2 = 0.107 \ \& \ R^2 = 0.038 \ \ or \ R = 0, R = 0.327 \ \& \ R = 0.195$$

$$Hence, the\ radius\ of\ the\ disc, R = 0.327m\ or\ 0.195m \qquad Ans.$$

Thus, we find that two values of the radius (R) of the circular disc are possible.

Example 1.17: A uniform point-source of light emitting 1400 lm luminous-flux is located at a height 150mm on the vertical axis of circular disc such that the luminous flux intercepted by the disc is 175 lm. What is the luminous intensity of the point-source? Also find out the radius of the disc& luminous flux density.

Sol. given that

$$total\ luminous\ flux, F_t = 1400\ lm,$$

$$luminous\ flux\ intercepted\ by\ the\ disc, F = 175\ lm\ \&$$

$$height, h = 150mm = 0.15m$$

$$now, let\ radius\ of\ the\ disc = R$$

We know that the total luminous flux emitted by a point source is given as

$$F_t = 4\pi \times I \ \Rightarrow I = \frac{F_t}{4\pi} = \frac{1400}{4\pi} = 111.408\ lm/sr \qquad Ans.$$

If the plane angle subtended by the disc at the point-source is '2α'

Then, we have

$$\Rightarrow cos\alpha = \frac{h}{\sqrt{h^2 + R^2}} = \frac{0.15}{\sqrt{0.15^2 + R^2}} = \frac{0.15}{\sqrt{0.0225 + R^2}}$$

If, we assume the circular disc to be the base of a right cone having apex at the point-source, the solid angle subtended by the circular disc at the point-source

$$\Rightarrow \omega = 2\pi(1 - cos\ \alpha) = 2\pi\left(1 - \frac{0.15}{\sqrt{0.0225 + R^2}}\right)$$

Hence, luminous flux intercepted by the disc,

$$F = I \times \omega \Rightarrow 175 = 111.408 \times 2\pi\left(1 - \frac{0.15}{\sqrt{0.0225 + R^2}}\right)$$

$$\Rightarrow 1 - \frac{0.15}{\sqrt{0.0225 + R^2}} = 0.25 \ \ or \ \frac{0.15}{\sqrt{0.0225 + R^2}} = 1 - 0.25 = 0.75$$

$$\Rightarrow \sqrt{0.0225 + R^2} = \frac{0.15}{0.75} = \frac{1}{5} \ \ or\ 0.0225 + R^2 = 0.04 \qquad (on\ squaring\ both\ the\ sides)$$

$$\Rightarrow R^2 = 0.04 - 0.0225 = 0.0175\ or$$

$$R = 0.132\ m\ or\ 132mm. \qquad Ans.$$

Now, luminous flux density

$$\Rightarrow E = \frac{F}{A} = \frac{175}{\pi(0.132)^2} = \frac{175}{0.0547} = 3196.983\ lm/m^2\ or\ 3.197 \times 10^3\ lux. \qquad Ans.$$

Example 1.18: A uniform point-source of light having luminous intensity 150cd is located on the vertical axis of circular disc with a radius 100mm. such that the luminous flux density on the disc is 9.6×10^3 lux. Find out the height of the point-source from the centre of disc.

Sol. given that

$$luminous\ intensity, I = 150\ lm/sr,$$

$$luminous\ flux\ density, E = 9.6 \times 10^3 = 9600\ lux\ \&$$

$$radius, R = 100mm = 0.1m$$

$$now.\ let\ the\ height = h$$

If the plane angle subtended by the disc at the point-source is '2α'

Then, we have

$$\Rightarrow cos\alpha = \frac{h}{\sqrt{h^2 + R^2}} = \frac{h}{\sqrt{h^2 + 0.1^2}} = \frac{h}{\sqrt{h^2 + 0.01}}$$

If, we assume the circular disc to be the base of a right cone having apex at the point-source, the solid angle subtended by the circular disc at the point-source

$$\Rightarrow \omega = 2\pi(1 - cos\ \alpha) = 2\pi\left(1 - \frac{h}{\sqrt{h^2 + 0.01}}\right)$$

Now, luminous flux density

$$\Rightarrow E = \frac{I \times \omega}{A} \Rightarrow 9600 = \frac{150 \times \omega}{\pi(0.1)^2} = \frac{15000}{\pi}\left[2\pi\left(1 - \frac{h}{\sqrt{h^2 + 0.01}}\right)\right]$$

$$\Rightarrow 0.32 = \left(1 - \frac{h}{\sqrt{h^2 + 0.01}}\right) \Rightarrow \frac{h}{\sqrt{h^2 + 0.01}} = 1 - 0.32 = 0.68$$

$$\Rightarrow \sqrt{h^2 + 0.01} = \frac{h}{0.68} = 1.47h$$

Now, on taking square of both the sides, we get

$$\Rightarrow h^2 + 0.01 = 2.161h^2\ or\ 1.161h^2 = 0.01\ or\ h = \sqrt{\frac{0.01}{1.161}}$$

$$h = 0.0928\ m\ or\ 9.28\ cm \qquad Ans.$$

Example 1.19: A lamp is mounted at a normal height 35 cm. from the centre of a circular table with a radius 450mm. Find out the luminous flux striking the table & the flux density.

(Assume lamp as a point-source emitting the luminous flux 1400 lm. uniformly in all the directions.)

Sol. given that

$$total\ luminous\ flux, F_t = 1400\ lm,$$

$$height, h = 35cm = 0.35\ m\ \&\ radius\ of\ the\ table = 450mm = 0.45\ m$$

We know that the total luminous flux emitted by a point source is given as

$$F_t = 4\pi \times I \ \Rightarrow I = \frac{F_t}{4\pi} = \frac{1400}{4\pi} = 111.408\ lm/sr$$

If the plane angle subtended by the circular table at the point-source is '2α'

Then, we have

$$\Rightarrow cos\alpha = \frac{h}{\sqrt{h^2 + R^2}} = \frac{0.35}{\sqrt{0.35^2 + 0.45^2}} = 0.614$$

If, we assume the circular table to be the base of a right cone having apex at the point-source, then solid angle subtended by the circular disc at the point-source

$$\Rightarrow \omega = 2\pi(1 - \cos\ \alpha) = 2\pi(1 - 0.614) = 2.425\ sr$$

Hence, luminous flux intercepted by the table,

$$F = I \times \omega \qquad \Rightarrow F = 111.408 \times 2.425 = 270.164\ lm. \qquad Ans.$$

Now, luminous flux density

$$E = \frac{luminous\ flux\ intercepted\ by\ the\ table}{area\ of\ the\ table} = \frac{F}{A}$$

$$\Rightarrow E = \frac{270.164}{\pi(0.45)^2} = \frac{270.164}{0.636} = 424.671\ lm/m^2 or\ 424.671\ lux. \qquad Ans.$$

Example 1.20: Find out the solid angle subtended by the circular plane$x^2 + y^2 = 25$ at a point$(0, 0, 7)$.

Sol. Equation of the circle: $x^2 + y^2 = 25$

Compare the given equation with the standard equation of the circle: $x^2 + y^2 = R^2$

∴ Radius of the circle,

$$R = 5\ \&\ centre\ at\ the\ origin\ (0, 0)$$

If, we assume the given circle as the base of a right cone having apex at the given point$(0, 0, 7)$ (lying on the Z-axis) & apex angle $'2\alpha'$.

We have,

$$\Rightarrow cos\alpha = \frac{h}{\sqrt{R^2 + h^2}} = \frac{7}{\sqrt{5^2 + 7^2}} = 0.814$$

Hence, the solid angle subtended by the circular plane at the point$(0, 0, 7)$

$$\Rightarrow \omega = 2\pi(1 - \cos\ \alpha) = 2\pi(1 - 0.814) = 1.169\ sr \qquad Ans.$$

Example 1.21: Find out the solid angle subtended by the circular plane$x^2 + y^2 = 169$ at a point$(0, 0, -11)$.

Sol. Equation of the circle: $x^2 + y^2 = 169$

Compare the given equation with the standard equation of the circle: $x^2 + y^2 = R^2$

∴ Radius of the circle,

$$R = 13\ \&\ centre\ at\ the\ origin\ (0, 0)$$

If, we assume the given circle as the base of a right cone having apex at the given point$(0, 0, -11)$ (lying on the Z'-axis)& apex angle 2α.

($-ve$ sign of z co-ordinate indicates that the given point$(-6, -9, -25)$ is lying on $-ve$ side of Z-axis)

We have,

$$\Rightarrow cos\alpha = \frac{h}{\sqrt{R^2 + h^2}} = \frac{11}{\sqrt{13^2 + 11^2}} = 0.646$$

Hence, the solid angle subtended by the circular plane at the point$(0, 0, -11)$

$$\Rightarrow \omega = 2\pi(1 - \cos\ \alpha) = 2\pi(1 - 0.646) = 2.224\ sr \qquad Ans.$$

Example 1.22: Determine the value of solid angle subtended by the circular plane$x^2 + y^2 - 10x - 4y - 20 = 0$ at a point $(5, 2, 10)$.

Sol. Equation of the circle: $x^2 + y^2 - 10x - 4y - 20 = 0$

Compare the given equation with the standard equation of the circle:

$$x^2 + y^2 + 2gx + 2fy + c = 0$$

We have,

$$g = -5, f = -2 \;\&\; c = -20$$

$$\therefore \; centre\ of\ the\ circle, (-g, -f) \equiv (5, 2)\&$$

$$radius, R = \sqrt{g^2 + f^2 - c} = \sqrt{(-5)^2 + (-2)^2 - (-20)} = \sqrt{25 + 4 + 20} = \sqrt{49} = 7$$

If, we assume the given circle as the base of a right cone having apex at the given point$(5, 2, 10)$ (lying on a axis parallel to Z-axis & passing through the centre $(5, 2)$of the circle) & apex angle 2α.

We have,

$$\Rightarrow cos\alpha = \frac{h}{\sqrt{R^2 + h^2}} = \frac{10}{\sqrt{7^2 + 10^2}} = 0.819$$

Hence, the solid angle subtended by the circular plane at the point$(5, 2, 10)$

$$\Rightarrow \omega = 2\pi(1 - \cos\alpha) = 2\pi(1 - 0.819) = 1.137\ sr \qquad Ans.$$

Example 1.23: Find out the value of solid angle subtended by the circular plane $x^2 + y^2 + 12x + 18y - 724 = 0$ at a point $(-6, -9, -25)$.

Sol. Equation of the circle: $x^2 + y^2 + 12x + 18y - 724 = 0$

Compare the given equation with the standard equation of the circle:

$$x^2 + y^2 + 2gx + 2fy + c = 0$$

We have,

$$g = 6, f = 9 \;\&\; c = -724$$

$$\therefore \; centre\ of\ the\ circle, (-g, -f) \equiv (-6, -9)\&$$

$$radius, R = \sqrt{g^2 + f^2 - c} = \sqrt{6^2 + 9^2 - (-724)} = \sqrt{36 + 81 + 724} = \sqrt{841} = 29$$

If, we assume the given circle as the base of a right cone having apex at the given point$(-6, -9, -25)$ (lying on a axis parallel to Z-axis & passing through the centre$(-6, -9)$ of the circle) & apex angle 2α.

($-ve$ sign of z co-ordinate indicates that the given point$(-6, -9, -25)$ is lying on $-ve$ side of Z-axisi.e. in the seventh octant)

We have,

$$\Rightarrow cos\alpha = \frac{h}{\sqrt{R^2 + h^2}} = \frac{25}{\sqrt{29^2 + 25^2}} = 0.653$$

Hence, the solid angle subtended by the circular plane at the point $(-6, -9, -25)$

$$\Rightarrow \omega = 2\pi(1 - \cos\alpha) = 2\pi(1 - 0.653) = 2.18\ sr \qquad Ans.$$

Example 1.24: Find out the value of solid angle subtended by the circular plane $x^2 + y^2 + 16x - 20y - 197 = 0$ at a point $(-8, 10, 18)$.

Sol. Equation of the circle: $x^2 + y^2 + 16x - 20y - 197 = 0$

Compare the given equation with the standard equation of the circle:

$$x^2 + y^2 + 2gx + 2fy + c = 0$$

We have,

$$g = 8, f = -10 \;\&\; c = -197$$

$$\therefore \; centre\ of\ the\ circle, (-g, -f) \equiv (-8, 10)\&$$

$$radius, R = \sqrt{g^2 + f^2 - c} = \sqrt{8^2 + (-10)^2 - (-197)} = \sqrt{64 + 100 + 197} = \sqrt{361} = 19$$

If, we assume the given circle as the base of a right cone having apex at the given point$(-8, 10, 18)$ (lying on a axis parallel to Z-axis & passing through the centre $(-8, 10)$ of the circle) & apex angle 2α.

(The given point$(-8, 10, 18)$ is lying on $-ve$ side of Z-axis i.e. in the second octant)

We have,

$$\Rightarrow \boldsymbol{cos\alpha} = \frac{\boldsymbol{h}}{\sqrt{\boldsymbol{R^2 + h^2}}} = \frac{\mathbf{18}}{\sqrt{\mathbf{19^2 + 18^2}}} = \mathbf{0.688}$$

Hence, the solid angle subtended by the circular plane at the point$(-8, 10, 18)$

$$\Rightarrow \boldsymbol{\omega = 2\pi(1 - cos\ \alpha) = 2\pi(1 - 0.688) = 1.96\ sr} \qquad \textit{Ans.}$$

Example 1.25: Find out the total flux intercepted by the circular plane $x^2 + y^2 - 6x + 14y - 23 = 0$ when a uniform point-source of 250 cd is located at a point$(3, -7, -8)$. If the intercepted flux is uniformly distributed over the entire plane then what is flux density?

Sol. Equation of the circle: $x^2 + y^2 - 6x + 14y - 23 = 0$

Compare the given equation with the standard equation of the circle:

$$\boldsymbol{x^2 + y^2 + 2gx + 2fy + c = 0}$$

We have,

$$g = -3, f = 7\ \&\ c = -23$$

$$\therefore \boldsymbol{centre\ of\ the\ circle, (-g, -f) \equiv (3, -7)\&}$$

$$\boldsymbol{radius, R = \sqrt{g^2 + f^2 - c} = \sqrt{(-3)^2 + 7^2 - (-23)} = \sqrt{9 + 49 + 23} = \sqrt{81} = 9}$$

If, we assume the given circle as the base of a right cone having apex at the given point$(3, -7, -8)$ (lying on a axis parallel to Z-axis & passing through the centre $(3, -7)$ of the circle) & apex angle 2α.

(The given point$(3, -7, -8)$ is lying on $-ve$ side of Z-axis i.e. in the eighth octant)

We have,

$$\Rightarrow \boldsymbol{cos\alpha} = \frac{\boldsymbol{h}}{\sqrt{\boldsymbol{R^2 + h^2}}} = \frac{\mathbf{8}}{\sqrt{\mathbf{9^2 + 8^2}}} = \mathbf{0.664}$$

Hence, the solid angle subtended by the circular plane at the point$(3, -7, -8)$

$$\Rightarrow \boldsymbol{\omega = 2\pi(1 - cos\ \alpha) = 2\pi(1 - 0.664) = 2.111\ sr}$$

Hence, the total flux intercepted by the plane is given as

$$\Rightarrow \boldsymbol{flux, F = I \times \omega = 250 \times 2.111 = 527.75\ lm} \qquad \textit{Ans.}$$

$$\Rightarrow \boldsymbol{flux\ density, F.D.} = \frac{\boldsymbol{F}}{\boldsymbol{A}} = \frac{\boldsymbol{F}}{\boldsymbol{\pi R^2}} = \frac{\mathbf{527.75}}{\boldsymbol{\pi \times 9^2}} = \boldsymbol{2.074\ lm\ per\ sq.unit.} \qquad \textit{Ans.}$$

Example 1.26: A circular plane with a radius 'R' subtends $\pi\ sr$ solid angle at a point lying at a height 'h' on a vertical axis passing the centre of the plane. Find out the ratio $h:R$. If the radius of the circle is $5\sqrt{3}\ unit$ find out the height of the point from the centre.

Sol. assume the circular plane to be the base of a right cone with apex at the given point & apex angle 2α then we have

$$\Rightarrow \boldsymbol{cos\alpha} = \frac{\boldsymbol{h}}{\sqrt{\boldsymbol{R^2 + h^2}}}$$

Hence, the solid angle subtended by the plane at the given point

$$\Rightarrow \omega = 2\pi(1 - \cos\alpha) \Rightarrow \pi = 2\pi\left(1 - \frac{h}{\sqrt{R^2 + h^2}}\right)$$

$$\Rightarrow \left(\mathbf{1} - \frac{\left(\frac{\boldsymbol{h}}{\boldsymbol{R}}\right)}{\sqrt{\mathbf{1} + \left(\frac{\boldsymbol{h}}{\boldsymbol{R}}\right)^2}}\right) = \frac{\boldsymbol{\pi}}{\boldsymbol{2\pi}} = \frac{\mathbf{1}}{\mathbf{2}} \Rightarrow \frac{\left(\frac{\boldsymbol{h}}{\boldsymbol{R}}\right)}{\sqrt{\mathbf{1} + \left(\frac{\boldsymbol{h}}{\boldsymbol{R}}\right)^2}} = \mathbf{1} - \frac{\mathbf{1}}{\mathbf{2}} = \frac{\mathbf{1}}{\mathbf{2}}$$

On squaring both the sides, we have

$$\Rightarrow \frac{\left(\frac{h}{R}\right)^2}{1+\left(\frac{h}{R}\right)^2} = \frac{1}{4} \Rightarrow 4\left(\frac{h}{R}\right)^2 = 1 + \left(\frac{h}{R}\right)^2 \Rightarrow \left(\frac{h}{R}\right)^2 = \frac{1}{3} \text{ or } \frac{h}{R} = \frac{1}{\sqrt{3}} \qquad \textit{Ans.}$$

Hence, the height of the point from the centre of the circle

$$\Rightarrow h = \frac{R}{\sqrt{3}} = \frac{5\sqrt{3}}{\sqrt{3}} = 5\ unit \qquad \textit{Ans.}$$

Summary

1. Solid angle: "The three dimensional angle subtended by an object at a certain point in the space"
   It is denoted by $'\omega'$ (omega). Its unit is steradian (in brief 'sr').
   Mathematically, it is given by the following expression
   $$\omega = \frac{area\ of\ visible\ surface\ of\ the\ object}{(distance\ (r) of\ any\ point\ on\ visible\ surfeace\ from\ the\ given\ point)^2}$$
   Where, distance 'r' from the given point (observer) to all the points on visible surface of the object is constant.
   Although the distance 'r' from the given point to all the points of visible surface is no more constant except in case of a spherical surface. But still the distance 'r' may be assumed to be constant for an infinitesimal small area 'dA' of on visible surface of the given object.
   This concept derives the fundamental theorem which is the integration of solid angle subtended by infinitesimal small area 'dA' on visible surface at the given point in the space.
2. The solid angle subtended by an object at a given point (observer) in the space is given by fundamental theorem
   $$\omega = \int_s \frac{dA}{r^2} \qquad \forall\ (0 \le \omega \le 4\pi)$$
   Where, distance 'r' is constant within limit i.e. distance (r) from the given point is constant for all the points of infinitesimal small area 'dA' of visible surface of the object.
3. Solid angle subtended by a right cone with apex angle $'2\alpha'$ at the apex point is given as
   $$\omega = 2\pi(1 - cos\alpha) \qquad \ldots\ldots\ldots\ldots\ldots\ldots \quad (I)$$
   If normal height of the right cone is 'H' & radius of the base is 'R' then the apex angle $'2\alpha'$ is related as
   $$cos\alpha = \frac{H}{\sqrt{H^2 + R^2}}$$
   $$\therefore\ \omega = 2\pi\left(1 - \frac{H}{\sqrt{H^2 + R^2}}\right) \qquad \ldots\ldots\ldots\ldots\ldots\ldots \quad (II)$$

Exercise 1

Q.1.1. Find out the solid angle subtended by a right cone with apex angle 55°.

Q.1.2. Find out the solid angle subtended by a right cone with apex angle $\frac{\pi}{5}$ rad.

Q.1.3. Find out the apex angle of a right cone subtending a solid angle $\frac{\pi}{2}$ sr. at the apex point.

Q.1.4. Find out the solid angle subtended by a right cone at the apex, having a normal height 18cm. & a radius 5cm.

Q.1.5. Find out the normal height of a right cone if solid angle subtended by it at the apex is $\frac{\pi}{4}$ sr. (take radius of the base = 6cm.)

Q.1.6. If the area of a spherical surface with a radius 12 cm. intercepted by a right cone with apex point coincident with the centre of the spherical surface is 60.67 cm^2. What is the apex angle of right cone?

Q.1.7. A uniform point-source of light having luminous intensity 111.4 lm/sr, is located at a height 150 mm. on the vertical axis of a circular disc with a radius 1.25cm. What will be the luminous flux striking the disc?

Q.1.8. Find out the luminous intensity of a uniform point-source of light which is located at a height 165mm on the vertical of a circular disc with a diameter 25.4mm. If the luminous flux intercepted by the disc is 200 lm.

Q.1.9. A uniform point-source of light is located on the axis of a circular disc with a diameter 3cm such that the area of projection of disc on a spherical surface with a radius 50cm & centre coincident with the point-source is 256 cm^2. Find out the distance of the point-source from the centre of disc.

If the luminous flux striking the disc is 120 lm then what is the luminous intensity of the point-source?

Q.1.10. A uniform point-source of light having luminous intensity 100 cd is located at a height 120mm on the vertical axis of circular disc such that the luminous flux density on the disc is 60 lux. Find out the radius of the disc.

Q.1.11. A uniform point-source of light having luminous intensity 120 cd (lm/sr.) is located at the centre of a cube-box with each side 250cm. Now, if a small circular hole of diameter 80cm. is made at the centre of one of the faces. Find out the luminous flux retained in the box.

(Neglect thickness of the box & the effect of reflection)

Q.1.12. A uniform point-source of light having luminous intensity 100cd (lm/sr) is located at the centre of a cube-box with external dimensions $25cm.\times 25cm.\times 25cm.$& thickness of wall 25mm. Now, if a small circular hole of radius 25mm. is made at the centre of one of the faces. Find out the luminous flux

1) Escaping the box & 2) retained in the box.(Neglect the effect of reflection)

Q.1.13. A uniform point-source of light having luminous intensity 111.4 cd (lm/sr.) is located at the centre of a cuboidal box with dimensions $30cm.\times 20cm.\times 10cm.$ Now, if a small circular hole of radius 24mm. is made at the centre of a face of $20cm.\times 10cm$. Find out the luminous flux

1) Escaping the box & 2) retained in the bow.

(Neglect thickness of the box & the effect of reflection)

Q.1.14. A lamp is mounted at a normal height 40 cm. from the centre of a circular table with a radius 400 mm. Find out the luminous intensity of the lamp if luminous flux striking the table is 250 lm. & the luminous flux density.

(Assume lamp as a point-source emitting the luminous flux uniformly in all the directions.)

Q.1.15. Find out the solid angle subtended by the circular plane $x^2 + y^2 = 289$ at a point$(0, 0, -7)$.

Q.1.16. Find out the total flux intercepted by the circular plane $x^2 + y^2 - 8x + 24y - 129 = 0$ when a uniform point-source of 250 cd is located at a point$(4, -12, -8)$. If the intercepted flux is uniformly distributed over the entire plane then what is flux density?

Q.1.17. A circular plane with a radius 'R' intercepts one-eighth of the total luminous flux emitted from a uniform point-source located a height 'h' on the vertical axis passing through the centre of the plane. Determine the ratio h:R.

Chapter 2
Analysis of 2D-Figures

This chapter deals with the analysis of solid angle subtended by the plane figures only.

2.1 Integral Theorem of Solid Angle

The solid angle subtended by a given object at the given point in the space is given by the following expression (in the form of spherical co-ordinates)

$$\omega = \int_{\varphi_1}^{\varphi_2} \left| \int_{\theta_1}^{\theta_2} sin\theta d\theta \right| d\varphi \qquad \forall (0 \le \theta \le \pi \,\&\, -\pi \le \varphi \le \pi)$$

Where $'\theta'\&'\varphi'$ are the parameters(variables) depending on the

1. Geometrical shape of the object &
2. Configuration(orientation) of the shape w.r.t. the given point in the space.

Derivation:

Let there be an object with extended surface & a given point say 'O' in the space (as shown in the figure(5) below)

Now, consider an imaginary spherical surface with centre at the given point say 'O'& the radius 'R' such that the area of the projection of the given object on the spherical surface is 'A'.

Now, consider an elementary area of projection 'dA' on the spherical surface having $arc\ legth\ 'Rsin\theta d\varphi'\&\ width\ \ 'Rd\theta'$ (see the fig (5) above)

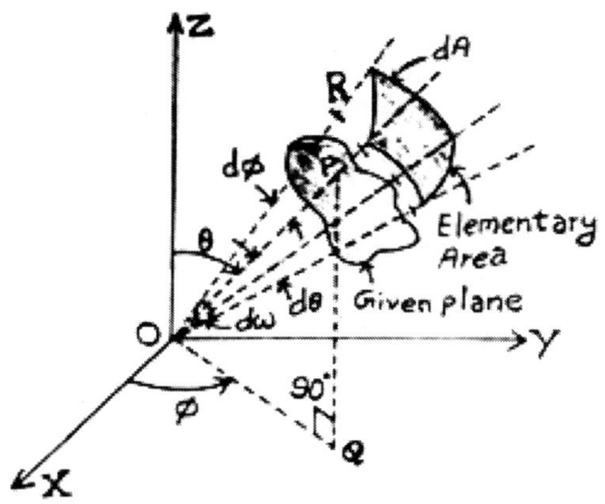

Fig (5)

Elementary area of projection

$$\Rightarrow dA = (length)(width) = (Rsin\theta d\varphi)(Rd\theta)$$
$$= R^2 sin\theta d\theta d\varphi$$

Hence, the total area of projection is obtained by double integration of the above expression by applying the limits of the parameters $'\theta'\ \&\ '\varphi' from\ \ \theta_1\ to\ \theta_2 \& \varphi_1\ to\ \varphi_2$, we get

$$A = \int_{\varphi_1}^{\varphi_2} \left(\int_{\theta_1}^{\theta_2} R^2 sin\theta d\theta \right) d\varphi = R^2 \int_{\varphi_1}^{\varphi_2} \left(\int_{\theta_1}^{\theta_2} sin\theta d\theta \right) d\varphi$$

Hence the solid angle subtended by the area of projection 'dA' at the given point 'O' using fundamental theorem of solid angle, we get

$$\omega = \int_s \frac{dA}{R^2} \Rightarrow \omega = \frac{1}{R^2} \int_s dA = \frac{A}{R^2} (since, radius\ 'R' is\ constant)$$

$$\omega = \frac{R^2}{R^2} \int_{\varphi_1}^{\varphi_2} \left(\int_{\theta_1}^{\theta_2} sin\theta d\theta \right) d\varphi = \int_{\varphi_1}^{\varphi_2} \left(\int_{\theta_1}^{\theta_2} sin\theta d\theta \right) d\varphi$$

Applying the limits of variables $\theta\ \&\ \varphi$, we have

$$\Rightarrow \left[\omega = \int_{\varphi_1}^{\varphi_2} \left| \int_{\theta_1}^{\theta_2} sin\theta d\theta \right| d\varphi \right] \quad \forall (0 \le \theta \le \pi \,\&\, -\pi \le \varphi \le \pi)$$

$$(here\ '|\quad|'\ is\ used\ for\ positive\ result)$$

*This theorem is very significant for finding the solid angle subtended any object at any point in the space.

But the double integration is solvable only for certain geometrical figures & the locations of the point in the space.

Integral theorem of solid angle derives a corollary for the plane-figures which is extremely useful for analysis of the plane-figures i.e. for finding the solid angle subtended by different plane-figures at certain locations of the given point in the space.

2.2 Corollary of Plane-figure:

If the foot of perpendicular drawn from a given point to the given plane lies on the plane withinits boundary then the solid angle subtended by the plane in the first quadrant w.r.t. the foot of perpendicular assumed as the origin is given by the following expression

$$\left[\omega = \int_0^{\varphi} \frac{f(htan\varphi)d\varphi}{\sqrt{\left(f(htan\varphi)\right)^2 + h^2sec^2\varphi}}\right] \quad \forall \left(0 \leq \varphi \leq \frac{\pi}{2}\right)$$

Where 'h' is the perpendicular height of a given point from the plane in the space,

$y = f(x) = f(htan\varphi) \neq 0$ is the equation of plane boundary in the first quadrant w.r.t. the foot of perpendicular assuming as the origin point.

$'\varphi'$ is the parameter (variable) & $tan\varphi = \frac{x}{h}$

Derivation:

Let there be a given point say 'P' at a height 'h' lying on the normal axis to the given plane & the point 'O' be the foot of the perpendicular drawn from the point 'P'

(see the figure(6) bellow, showing the top-view & the point 'P' is lying on the axis normal to the plane of the paper)

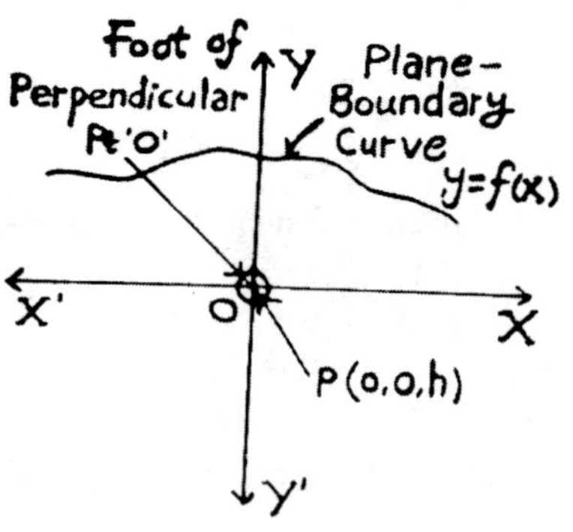

Fig (6)

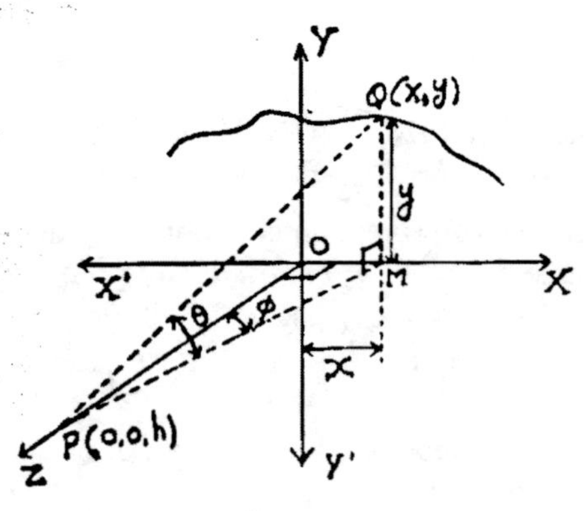

Fig (7)

Now, let the equation of boundary-curve of the plane be $y = f(x)$ w.r.t. the foot ofperpendicular 'O' & consider a point say Q(x, y) on the curve $y = f(x)$in the first quadrant (as shown in the figure(7) below)

In right ΔPOM

$$\Rightarrow tan\varphi = \frac{OM}{OP} \Rightarrow tan\varphi = \frac{x}{h} \Rightarrow x = h\, tan\varphi$$

$$\Rightarrow y = f(x) = f(\, htan\varphi)$$

In right ΔQMP

$$\Rightarrow tan\theta = \frac{QM}{PM} = \frac{QM}{\sqrt{OP^2 + OM^2}} = \frac{y}{\sqrt{x^2 + h^2}} = \frac{f(htan\varphi)}{\sqrt{(f(htan\varphi))^2 + h^2}}$$

$$\Rightarrow tan\theta = \frac{f(htan\varphi)}{hsec\varphi}$$

$$\Rightarrow sin\theta = \frac{tan\theta}{\sqrt{1 + tan^2\theta}} = \frac{f(htan\varphi)}{\sqrt{(f(htan\varphi))^2 + h^2sec^2\varphi}}$$

Now, applying Integral-Theorem of Solid Angle,

Solid angle subtended by the plane in the first quadrant at the given point 'P'

$$\omega = \int_{\varphi_1}^{\varphi_2} \left| \int_{\theta_1}^{\theta_2} sin\theta d\theta \right| d\varphi$$

Applying the limits of parameters $'\theta'$ & $'\varphi'$ in the first quadrant (see fig(8) below), we have

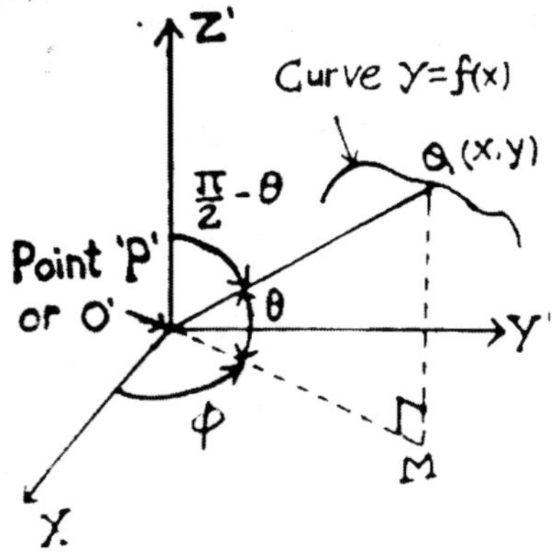

Fig (8)

$$\omega = \int_{\varphi_1=0}^{\varphi_2=\varphi} \left| \int_{\theta_1=\frac{\pi}{2}-\theta}^{\theta_2=\frac{\pi}{2}} sin\theta d\theta \right| d\varphi = \int_{\varphi_1=0}^{\varphi_2=\varphi} \left| [-cos\theta]_{\frac{\pi}{2}-\theta}^{\frac{\pi}{2}} \right| d\varphi$$

$$= \int_0^{\varphi} |sin\theta| d\varphi = \int_0^{\varphi} sin\theta d\varphi \left(since, \quad 0 \le \varphi \le \frac{\pi}{2}\right)$$

$$= \int_0^{\varphi} \frac{f(htan\varphi)d\varphi}{\sqrt{\left(f(htan\varphi)\right)^2 + h^2sec^2\varphi}}$$

On setting the value of $sin\theta$in the above expression, we get

$$\Rightarrow \left[\omega = \int_0^{\varphi} \frac{f(htan\varphi)d\varphi}{\sqrt{(f(htan\varphi))^2 + h^2sec^2\varphi}}\right] \quad \forall \left(h \ge 0 \ \& \ 0 \le \varphi \le \frac{\pi}{2}\right)$$

*The equation of the plane-boundary $y = f(x) \neq 0$in the first quadrant depends on the location of the point w.r.t. the given plane or on the profile of plane-boundary w.r.t. the given point in the space &$y = f(x)$is the equation of single-boundary of the plane i.e. it is differentiable at each point of the curve in the first quadrant.

Note: The above corollary is applicable, for a plane-figure only when the foot of perpendicular drawn from a given point to the planelies within the boundary of the plane, to find out the solid angle subtended by the plane at the given point in the first quadrant w.r.t. the foot of perpendicular assuming as the origin.

However, the corollary of plane figure is very significant for analysis of the plane-figures along with some limitations due to insolvable integrals obtained from the corollary.

Complex functions of boundary curve of the plane-figures are obtained corresponding to the location of a given point in the space that brings about the complexity of calculations. Hence, the use of this corollary is limited for ease of calculations.

The limitations of corollary can be summarised as follows

Limitations of Corollary:

1. The corollary is applicable only for the plane-figures.
2. Foot of the perpendicular drawn from a given point to the plane must lie within boundary of the plane.
3. Complex & insolvable integrals are obtained in case of the a) Scalene triangular plane b) Elliptical plane c) Inclined circular plane d) Parallelogram-like plane etc.
4. Complex & insolvable integrals are also obtained for different locations of a given point in the space however the foot of perpendicular, drawn from the point to the plane, lies within the boundary (of plane).
5. The corollary brings about the complex & insolvable integrals even in case of the linear functions of plane-boundary-curves.

*There are some key points related to the analysis of the plane-figures which simplifies the calculations involved in the analysis of plane-figures.

KEY POINTS:

1. If a given plane consists of 'n' no. of the elementary planes then the solid angle subtended by a given plane is equal to the sum of the solid angles subtended by elementary planes at the same point in the space.

   Mathematically, the solid angle $'\omega'$ subtended by a given plane having area 'A' at a certain point in the space is given as follows

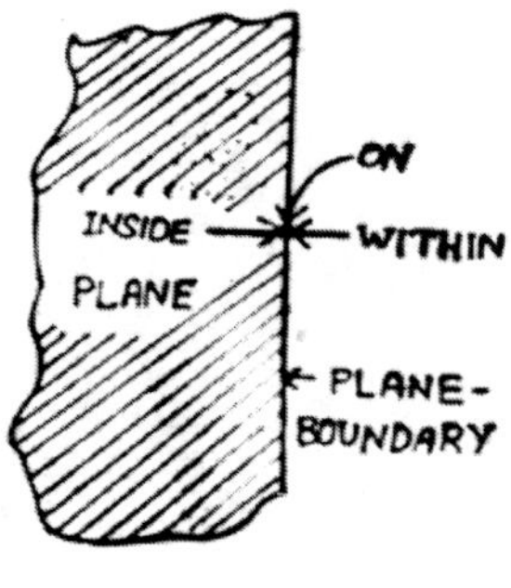

Fig (9)

$$\Rightarrow \omega = \omega_1 + \omega_2 + \omega_3 + \dots\dots\dots\dots + \omega_{n-1} + \omega_n$$

$$\Rightarrow \left[\omega = \sum_{i=1}^{i=n} (\omega_i) \quad \forall A = A_1 + A_2 + A_3 + \dots\dots\dots + A_{n-1} + A_n = \sum_{i=1}^{i=n} (A_i)\right]$$

Where $\omega_1, \omega_2, \omega_3, \dots\dots\dots \omega_{n-1}, \omega_n$ *are the solid angles subtended by the elementary plane* areas$A_1, A_2, A_3 \dots\dots\dots A_{n-1}, A_n$ *respectively.*

2. The solid angle subtended by a given plane at a certain point remains constant if the plane is rotated about the foot of the perpendicular drawn from a given point to the plane.

3. The solid angle subtended by a plane at any point lying on the plane inside the boundary in the first quadrant is given from the corollary on putting h = 0 in the expression of the corollary (see the figure (9) below)

   Hence, we get

$$\omega = \int_0^{\varphi} \frac{f((0)tan\varphi)d\varphi}{\sqrt{(f((0)tan\varphi))^2 + (0)sec^2\varphi}}$$

$$\Rightarrow \omega = \int_0^{\varphi} \frac{f(0)d\varphi}{\sqrt{(f(0))^2 + 0}} = \int_0^{\varphi} \frac{f(0)d\varphi}{f(0)} = \int_0^{\varphi} d\varphi = \varphi$$

$$= (plane\ angle\ subtended\ by\ X - axis\ at\ the\ point\ tending\ to\ origin\ 'O')$$

$$= \frac{\pi}{2}$$

$\varphi = \frac{\pi}{2}$ $sr$is the solid angle subtended by a plane in the first quadrant only but the plane is divided into four quadrants w.r.t. any point lying on the plane inside the boundary.

Hence, the solid subtended by the plane at any point lying on the plane inside the boundary

$\Rightarrow \omega = 4\varphi = 4.\frac{\pi}{2} = 2\pi\ sr$ (Irrespective of area & boundary (geometry) of the plane)

$$since\ for\ each\ point\ lying\ on\ the\ plane\ inside\ the\ boundary,$$
$$the\ apex\ angle\ of\ the\ cone\ of\ vision\ is\ 2\alpha = \pi$$

Hence, the solid angle subtended by the plane is also obtained from equation (1)

$$\Rightarrow \omega = 2\pi(1 - cos\alpha) = 2\pi\left(1 - cos\frac{\pi}{2}\right) = 2\pi\ sr$$

4. The solid angle subtended by any plane at any point lying on the plane-boundary is obtained by putting h = 0 in the expression of the corollary, we get

$$\boldsymbol{\omega} = \int_0^{\varphi} \frac{\boldsymbol{f((0)tan\varphi)d\varphi}}{\sqrt{(\boldsymbol{f((0)tan\varphi)})^2 + \boldsymbol{(0)sec^2\varphi}}} \Rightarrow \boldsymbol{\omega} = \int_0^{\varphi} \frac{\boldsymbol{f(0)d\varphi}}{\sqrt{(\boldsymbol{f(0)})^2 + \boldsymbol{0}}} = \int_0^{\varphi} \frac{\boldsymbol{f(0)d\varphi}}{\boldsymbol{f(0)}}$$

$$= \int_0^{\varphi} d\varphi = \varphi$$

$$= (plane - angle\ subtended\ by\ X - axis\ at\ the\ point\ tending\ to\ origin\ 'O')$$

$$= \frac{\pi}{2}$$

$\varphi = \frac{\pi}{2}$ $sr$is the solid angle subtended by the plane in the first quadrant only but the plane is divided into two quadrants w.r.t. any point lying on the plan-boundary.

Hence, the solid subtended by the plane at any point lying on the plane within the boundary

$\Rightarrow \omega = 2\varphi = 2.\frac{\pi}{2} = \pi\ sr$ (Irrespective area & boundary (geometry) of the plane)

5. The solid angle subtended by an infinitely large plane at any point lying at a finite normal distance from the plane is given from the equation (1)

$$\Rightarrow \omega = 2\pi(1 - cos\alpha) = 2\pi\left(1 - cos\frac{\pi}{2}\right) = 2\pi\ sr$$

$$(since\ for\ any\ point\ lying\ at\ a\ finite\ normal\ distance\ from\ the\ plane,$$
$$the\ apex\ angle\ of\ the\ cone\ of\ vision\ is\ 2\alpha = \pi\ )$$

6. If a given plane is divided into 'n' no. of the equiangular segments (a part (segment) of the plane whose boundary is bounded by two straight lines & a plane-curve) w.r.t. any point lying on the plane inside the boundary (as shown in the figure(10) below)

Then the solid angle subtended by each of the segments at the same point say 'P' inside the boundary

$$\omega = \frac{1}{n}(solid\ angle\ subtended\ by\ the\ whole\ plane\ at\ the\ point\ 'P')$$

$= \frac{1}{n}(2\pi) = \frac{2\pi}{n}\ sr\ \ (irrespective\ of\ area\ \&\ geometry\ of\ the\ plane)$ ss

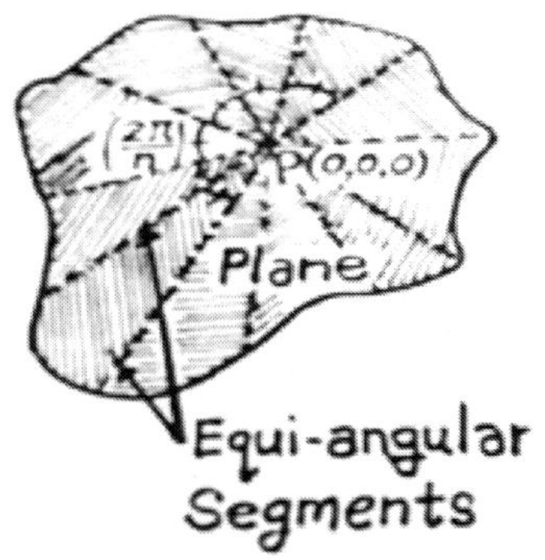

Fig (10)

7. Since, the apex angle of the cone of vision of a finite plane at the infinity is $2\alpha = 0$

Hence, the solid angle subtended by any finite plane at the infinity is obtained by putting $\alpha = 0$ in equation (1), we get

$$\Rightarrow \omega = 2\pi(1 - cos\alpha) = 2\pi(1 - cos0) = 2\pi = 0$$

2.3 Applications of Corollary

There are basic applications of the corollary for finding the solid angle subtended by the

1. Right triangular plane 2.Rectangular plane & 3. Circular plane

at any point lying on the normal axis passing through the certain point of the plane.

While, other equations (formulae) have been derived from the basic equations of planes mentioned above. Although, equations of all the planes can be derived from the corollary of plane-figure but for ease of calculations basic equations are applied to find out the solid angle subtended at the certain point by some particular plane-figures like

1. Isosceles triangle 2.Isosceles right triangle 3.Equilateral triangle 4.Square 5.Rhombus

6. Right trapezium 7. Regularpolygon

2.3.1 Solid angle subtended by a right-triangular plane at any point lying on the axis normal to the plane & passing through one of the acute angled vertices

Method 1: Using Fundamental Theorem

Let there be a right triangular plane ONM having perpendicular ON = p & the base MN = b and a given point say P(0, 0, h) at a height 'h' lying on the axis (i.e. Z-axis) passing through the acute angled vertex say 'O' (as shown in the figure (11) below)

In rightΔPOR

$$\Rightarrow \boldsymbol{sec\varphi = \frac{PR}{PO} \Rightarrow PR = POsec\varphi = hsec\varphi} \ \&$$

$$\boldsymbol{tan\varphi = \frac{OR}{PO} \Rightarrow OR = POtan\varphi = htan\varphi = x}$$

Now, the equation of the straight line OM passing through the origin 'O'

$$\Rightarrow y = mx = \frac{b}{p}x \quad \left(since, slope\ of\ the\ line\ OM = \frac{b}{p}\right)$$

$$\Rightarrow y = \frac{b}{p}x = \frac{b}{p}htan\varphi$$

Now, consider a point Q(x, y) on the straight line OM

$$\Rightarrow QR = y = \frac{b}{p}htan\varphi$$

In rightΔPRQ

$$\Rightarrow sin\theta = \frac{QR}{PQ} = \frac{QR}{\sqrt{QR^2 + PR^2}} = \frac{\left(\frac{b}{p}htan\varphi\right)}{\sqrt{\left(\frac{b}{p}htan\varphi\right)^2 + h^2sec^2\varphi}}$$

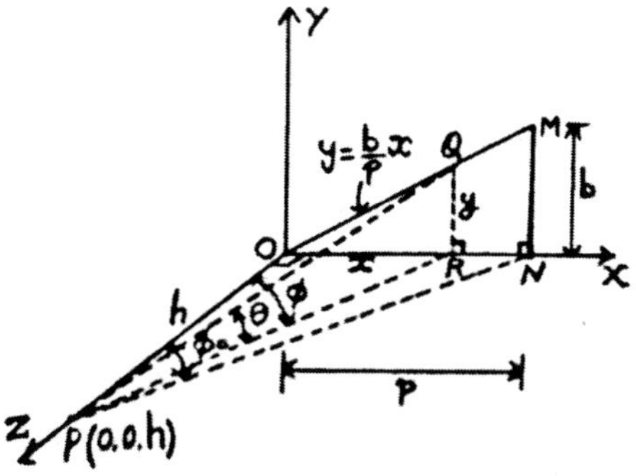

Fig(11)

$$sin\theta = \frac{bsin\varphi}{\sqrt{b^2sin^2\varphi + p^2}}$$

In rightΔPON

$\Rightarrow cos\varphi_0 = \frac{OP}{PN} = \frac{OP}{\sqrt{PO^2+ON^2}} = \frac{h}{\sqrt{h^2+p^2}}$

Let's consider an imaginary sphical surface having radius 'R' & centre at the given point 'P' such that the area of projection of the given plane ONM on the spherical surface w.r.t. the given point 'P' is 'A'

Now, consider an elementary area of projection 'dA' of the plane ONM on the spherical surface in the first quadrant YOX (see the figure(12) below)

Now, elementary area of projection in the first quadrant

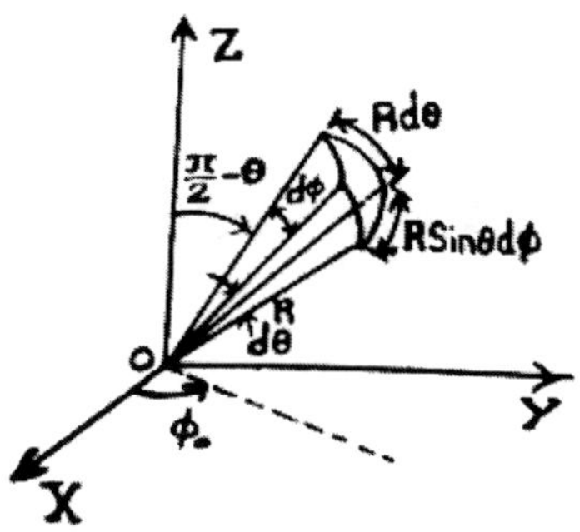

Fig (12)

$\Rightarrow dA = (length)(width) = (Rsin\theta d\varphi)(Rd\theta) = R^2sin\theta d\theta d\varphi$

Hence, area of projection of the plane ONM on the spherical surface in the first quadrant is obtained by integrating the above expression in the first quadrant & applying the limits of

$'\theta' from\ \theta_1 = \frac{\pi}{2} - \theta\ to\ \theta_2 = \frac{\pi}{2} \&' \varphi' from\ \varphi_1 = 0\ to\ \varphi_2 = \varphi_0$

we get

$$\Rightarrow A = \int_0^{\varphi_0} \left[\int_{\frac{\pi}{2}-\theta}^{\frac{\pi}{2}} R^2 sin\theta d\theta \right] d\varphi = R^2 \int_0^{\varphi_0} \left[\int_{\frac{\pi}{2}-\theta}^{\frac{\pi}{2}} sin\theta d\theta \right] d\varphi$$

$$= R^2 \int_0^{\varphi_0} d\varphi \, [-cos\theta]_{\frac{\pi}{2}-\theta}^{\frac{\pi}{2}} = R^2 \int_0^{\varphi_0} sin\theta d\varphi = R^2 \int_0^{\varphi_0} \frac{bsin\varphi}{\sqrt{b^2sin^2\varphi + p^2}} d\varphi$$

$$\Rightarrow A = R^2 \int_0^{\varphi_0} \frac{bsin\varphi}{\sqrt{b^2 - b^2cos^2\varphi + p^2}} d\varphi \,(since,\ sin^2\alpha = 1 - cos^2\alpha)$$

$$= R^2 \int_0^{\varphi_0} \frac{bsin\varphi}{b\sqrt{\frac{b^2+p^2}{b^2} - cos^2\varphi}} d\varphi = R^2 \int_0^{\varphi_0} \frac{sin\varphi}{\sqrt{K^2 - cos^2\varphi}} d\varphi$$

$$where\ K = \sqrt{\frac{b^2+p^2}{b^2}} = \frac{\sqrt{b^2+p^2}}{b}$$

$$\Rightarrow A = R^2 \left[-\sin^{-1}\left\{\frac{cos\varphi}{K}\right\} \right]_0^{\varphi_0} = R^2 \left[-\sin^{-1}\left\{\frac{cos\varphi_0}{K}\right\} + \sin^{-1}\left\{\frac{cos0}{K}\right\} \right]$$

$$= R^2\left[sin^{-1}\left\{\frac{1}{K}\right\} - sin^{-1}\left\{\frac{cos\varphi_0}{K}\right\}\right]$$

Now, on putting the values of 'K' & $'cos\varphi_0'$, we get

$$\Rightarrow A = R^2\left[sin^{-1}\left\{\frac{1}{\frac{\sqrt{b^2+p^2}}{b}}\right\} - sin^{-1}\left\{\frac{\frac{h}{\sqrt{h^2+p^2}}}{\frac{\sqrt{b^2+p^2}}{b}}\right\}\right]$$

$$= R^2\left[\sin^{-1}\left\{\frac{b}{\sqrt{b^2+p^2}}\right\} - \sin^{-1}\left\{\left(\frac{b}{\sqrt{b^2+p^2}}\right)\left(\frac{h}{\sqrt{h^2+p^2}}\right)\right\}\right]$$

Now, the solid angle subtended by the right triangular plane ONM at the given point 'P'

= solid angle subtended by the area of projection of triangle ONM on the spherical surface at the same point 'P'

Using Fundamental Theorem

$$\Rightarrow \omega = \int_s \frac{dA}{r^2} = \int_s \frac{dA}{R^2} = \frac{1}{R^2}\int_s = \frac{A}{R^2}$$

$(since, R\ is\ constant\ for\ each\ point\ on\ the\ area\ of\ projection\ on\ the\ spherical\ surface)$

$$\Rightarrow \omega = \frac{1}{R^2}\left[R^2\ sin^{-1}\left\{\frac{b}{\sqrt{b^2+p^2}}\right\} - sin^{-1}\left\{\left(\frac{b}{\sqrt{b^2+p^2}}\right)\left(\frac{h}{\sqrt{h^2+p^2}}\right)\right\}\right]$$

$$= sin^{-1}\left\{\frac{b}{\sqrt{b^2+p^2}}\right\} - sin^{-1}\left\{\left(\frac{b}{\sqrt{b^2+p^2}}\right)\left(\frac{h}{\sqrt{h^2+p^2}}\right)\right\}$$

$$\Rightarrow \left[\omega = \sin^{-1}\left\{\frac{b}{\sqrt{b^2+p^2}}\right\} - \sin^{-1}\left\{\left(\frac{b}{\sqrt{b^2+p^2}}\right)\left(\frac{h}{\sqrt{h^2+p^2}}\right)\right\}\right] \quad \forall(h \geq 0) \ldots\ldots (2)$$

Method 2: Using Integral Theorem

We know that the solid angle subtended by an object at a given point in the space is given by Integral-Theorem as follows

$$\omega = \int_{\varphi_1}^{\varphi_2}\left|\int_{\theta_1}^{\theta_2} sin\theta d\theta\right| d\varphi$$

Hence, the solid angle subtended by the right triangular plane ONM is obtained by applying the limits of the parameters $'\theta'$ & $'\varphi'$ for whole the plane ONM (see the figure (12) above)

Now, applying limits of $'\theta'\ from\ \theta_1 = \frac{\pi}{2} - \theta\ to\ \theta_2 = \frac{\pi}{2}$&

$$'\varphi'\ from\ \varphi_1 = 0\ to\ \varphi_2 = \varphi_0$$

$$\Rightarrow \omega = \int_{\varphi_1=0}^{\varphi_2=\varphi_0}\left|\int_{\theta_1=\frac{\pi}{2}-\theta}^{\theta_2=\frac{\pi}{2}} sin\theta d\theta\right| d\varphi = \int_{\varphi_1=0}^{\varphi_2=\varphi_0}\left|[-cos\theta]_{\frac{\pi}{2}-\theta}^{\frac{\pi}{2}} d\varphi\right|$$

$$= \int_{\varphi_1=0}^{\varphi_2=\varphi_0}\left|\left[-cos\left(\frac{\pi}{2}\right) + cos\left(\frac{\pi}{2} - \theta\right)\right]\right| d\varphi = \int_{\varphi_1=0}^{\varphi_2=\varphi_0} |sin\theta| d\varphi$$

$$\left[\text{On setting the value of } sin\theta = \frac{bsin\varphi}{\sqrt{b^2sin^2\varphi + p^2}}\right]$$

$$= \int_{\varphi_1=0}^{\varphi_2=\varphi_0} \left| \frac{bsin\varphi}{\sqrt{b^2sin^2\varphi + p^2}} \right| d\varphi$$

$$\Rightarrow \omega = \int_0^{\varphi_0} \frac{bsin\varphi}{\sqrt{b^2sin^2\varphi + p^2}} d\varphi \left(since, 0 \le \varphi \le \frac{\pi}{2} \& |f(x)| = f(x) \ \forall f(x) > 0\right)$$

$$\Rightarrow \omega = \int_0^{\varphi_0} \frac{bsin\varphi}{\sqrt{b^2 - b^2cos^2\varphi + p^2}} d\varphi \ (since, \ \ sin^2\alpha = 1 - cos^2\alpha)$$

$$= \int_0^{\varphi_0} \frac{bsin\varphi}{b\sqrt{\frac{b^2 + p^2}{b^2} - cos^2\varphi}} d\varphi = \int_0^{\varphi_0} \frac{sin\varphi}{\sqrt{K^2 - cos^2\varphi}} d\varphi$$

$$where \ K = \sqrt{\frac{b^2 + p^2}{b^2}} = \frac{\sqrt{b^2 + p^2}}{b}$$

$$\Rightarrow \omega = \left[-\sin^{-1}\left\{\frac{cos\varphi}{K}\right\}\right]_0^{\varphi_0} = \left[-\sin^{-1}\left\{\frac{cos\varphi_0}{K}\right\} + \sin^{-1}\left\{\frac{cos0}{K}\right\}\right]$$

$$= \left[\sin^{-1}\left\{\frac{1}{K}\right\} - \sin^{-1}\left\{\frac{cos\varphi_0}{K}\right\}\right]$$

Now, on putting the values of 'K' & $'cos\varphi_0'$, we get

$$\Rightarrow \omega = \left[\sin^{-1}\left\{\frac{1}{\frac{\sqrt{b^2 + p^2}}{b}}\right\} - \sin^{-1}\left\{\frac{\frac{h}{\sqrt{h^2 + p^2}}}{\frac{\sqrt{b^2 + p^2}}{b}}\right\}\right]$$

$$= \sin^{-1}\left\{\frac{b}{\sqrt{b^2 + p^2}}\right\} - \sin^{-1}\left\{\left(\frac{b}{\sqrt{b^2 + p^2}}\right)\left(\frac{h}{\sqrt{h^2 + p^2}}\right)\right\}$$

$$\Rightarrow \left[\omega = \sin^{-1}\left\{\frac{b}{\sqrt{b^2 + p^2}}\right\} - \sin^{-1}\left\{\left(\frac{b}{\sqrt{b^2 + p^2}}\right)\left(\frac{h}{\sqrt{h^2 + p^2}}\right)\right\}\right] \quad \forall (h \ge 0) \ldots \ldots (2)$$

Method 3: Using Corollary

Let there be a right triangular plane ONM having perpendicular ON = p & the base MN = b and a given point say P(0, 0, h) at a height 'h' lying on the axis (i.e. Z-axis) passing through the acute angled vertex say 'O' (as shown in the figure (13) below)

Since, the plane-boundary is bounded by the straight lines, $x = b, y = 0 \ \& \ y = \frac{b}{p}x$

But $y = f(x) \ne 0$

Hence, the equation of the boundary-curve $y = f(x) = \frac{b}{p}x \ \forall (0 \le x \le p)$

Also the vertex 'O' is the foot of the perpendicular PO drawn from the given point 'P'

Let's consider a point Q(x, y) on the boundary curve (i.e. straight line OM)$y = \frac{b}{p}x$

Now, in right $\Delta POM'$

$$\Rightarrow \ tan\varphi = \frac{OM'}{OP} = \frac{x}{h} \Rightarrow x = htan\varphi$$

From the right ΔPON

$$\Rightarrow\ cos\varphi_0 = \frac{OP}{PN} = \frac{OP}{\sqrt{OP^2 + ON^2}} = \frac{h}{\sqrt{h^2 + p^2}}$$

Now, the equation of the boundary line OM

$$\Rightarrow y = f(x) = \frac{b}{p}x$$

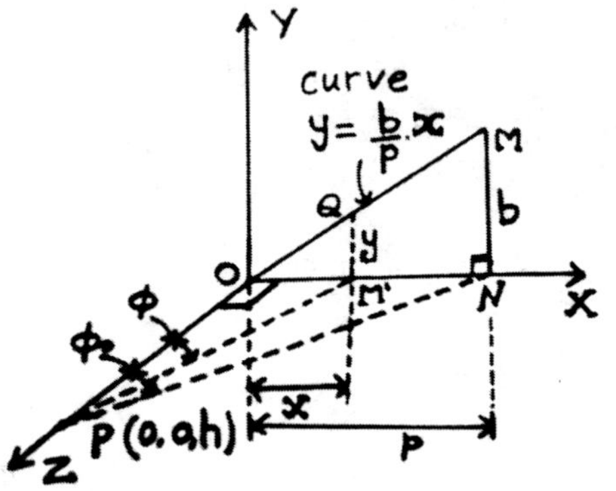

Fig(13)

$$\Rightarrow f(htan\varphi) = \frac{b}{p}(htan\varphi) = \frac{bh}{p}tan\varphi$$

Now, using Corollary of the plane-figure

Solid angle subtended by the right triangular plane ONM in the first quadrant YOX

$$\Rightarrow \omega = \int_0^{\varphi} \frac{f(htan\varphi)d\varphi}{\sqrt{(f(htan\varphi))^2 + h^2sec^2\varphi}} = \int_0^{\varphi} \frac{\frac{bh}{p}tan\varphi d\varphi}{\sqrt{\left(\frac{bh}{p}tan\varphi\right)^2 + h^2sec^2\varphi}}$$

Now, applying the limits of the parameter $'\varphi'$, we get

$$\omega = \frac{b}{p}\int_0^{\varphi=\varphi_0} \frac{tan\varphi d\varphi}{\sqrt{\left(\frac{b}{p}tan\varphi\right)^2 + sec^2\varphi}} = \frac{b}{p}\int_0^{\varphi=\varphi_0} \frac{\frac{sin\varphi}{cos\varphi}d\varphi}{\sqrt{\frac{b^2sin^2\varphi}{p^2cos^2\varphi} + \frac{1}{cos^2\varphi}}}$$

$$= \frac{b}{p}\int_0^{\varphi=\varphi_0} \frac{sin\varphi d\varphi}{\sqrt{\frac{b^2}{p^2}sin^2\varphi + 1}} = \frac{b}{p}\int_0^{\varphi=\varphi_0} \frac{sin\varphi d\varphi}{\sqrt{\frac{b^2}{p^2} - \frac{b^2}{p^2}cos^2\varphi + 1}}$$

$$= \frac{b}{p}\int_0^{\varphi=\varphi_0} \frac{sin\varphi d\varphi}{\frac{b}{p}\sqrt{\frac{b^2 + p^2}{b^2} - cos^2\varphi}} = \int_0^{\varphi=\varphi_0} \frac{sin\varphi d\varphi}{\sqrt{\frac{b^2 + p^2}{b^2} - cos^2\varphi}}$$

$$= \int_0^{\varphi=\varphi_0} \frac{sin\varphi d\varphi}{\sqrt{K^2 - cos^2\varphi}}$$

Where

$$K^2 = \frac{b^2 + p^2}{b^2} \ or \ K = \frac{\sqrt{b^2 + p^2}}{b}$$

Now, to find the integral of the above integration

$$I = \int \frac{sin\varphi d\varphi}{\sqrt{K^2 - cos^2\varphi}}$$

let $t = cos\varphi \Rightarrow dt = -sin\varphi d\varphi$

$$I = \int \frac{-dt}{\sqrt{K^2 - t^2}} = -sin^{-1}\left(\frac{t}{K}\right) = -sin^{-1}\left(\frac{cos\varphi}{K}\right)$$

On putting the value, we get

$$\omega = \left[-sin^{-1}\left(\frac{cos\varphi}{K}\right)\right]_0^{\varphi=\varphi_0} = \left[-sin^{-1}\left(\frac{cos\varphi_0}{K}\right) + sin^{-1}\left(\frac{1}{K}\right)\right]$$

Now, putting the value of 'K' &$cos\varphi_0$in the above expression, we have

$$\omega = \left[\sin^{-1}\left(\frac{1}{\frac{\sqrt{b^2+p^2}}{b}}\right) - \sin^{-1}\left(\frac{\frac{h}{\sqrt{h^2+p^2}}}{\frac{\sqrt{b^2+p^2}}{b}}\right)\right]$$

$$= \left[\sin^{-1}\left\{\frac{b}{\sqrt{b^2+p^2}}\right\} - \sin^{-1}\left\{\left(\frac{b}{\sqrt{b^2+p^2}}\right)\left(\frac{h}{\sqrt{h^2+p^2}}\right)\right\}\right]$$

$$\Rightarrow \left[\omega = \sin^{-1}\left\{\frac{b}{\sqrt{b^2+p^2}}\right\} - \sin^{-1}\left\{\left(\frac{b}{\sqrt{b^2+p^2}}\right)\left(\frac{h}{\sqrt{h^2+p^2}}\right)\right\}\right] \quad \forall(h \geq 0) \ldots\ldots\ldots (2)$$

Note: The perpendicular 'p' & the base 'b' are justified according to the location of the point 'P' with the foot of perpendicular at the acute angled vertex 'O' or 'M'

If the given point P(0, 0, h) is lying on the axis normal to the plane & passing through the acute angled vertex 'M' then the solid angle subtended by the right triangular plane ONM is obtained by replacing 'b' by 'p' & 'p' by 'b' in the equation(2)

Thus, we have

$$\left[\boldsymbol{\omega = \sin^{-1}\left\{\frac{p}{\sqrt{p^2+b^2}}\right\} - \sin^{-1}\left\{\left(\frac{p}{\sqrt{p^2+b^2}}\right)\left(\frac{h}{\sqrt{h^2+b^2}}\right)\right\}}\right] \quad \boldsymbol{\forall(h \geq 0) \ldots\ldots\ldots (3)}$$

Important deductions:

1. The solid angle subtended by the right triangular plane at a point lying on the acute angled vertex say 'O' is obtained by putting h = 0 in the eq(2), we get

$$\omega = sin^{-1}\left\{\frac{b}{\sqrt{b^2+p^2}}\right\} - \sin^{-1}\left\{\left(\frac{b}{\sqrt{b^2+p^2}}\right)\left(\frac{0}{\sqrt{0+p^2}}\right)\right\}$$

$$= sin^{-1}\left\{\frac{b}{\sqrt{b^2+p^2}}\right\} - \sin^{-1}\{0\}$$

$$\Rightarrow \left[\omega = sin^{-1}\left\{\frac{b}{\sqrt{b^2+p^2}}\right\}\right] \quad \ldots\ldots\ldots\ldots (4)$$

2. If the point 'P' is lying on the acute angled vertex 'M' (i.e. for which h = 0) then the solid angle subtended by the right triangular plane is obtained by putting h = 0 in the eq(3)

   Thus, we have

$$\omega = sin^{-1}\left\{\frac{p}{\sqrt{p^2+b^2}}\right\} - sin^{-1}\left\{\left(\frac{p}{\sqrt{p^2+b^2}}\right)\left(\frac{0}{\sqrt{0+b^2}}\right)\right\}$$

$$= sin^{-1}\left\{\frac{p}{\sqrt{p^2+b^2}}\right\} - sin^{-1}\{0\}$$

$$\Rightarrow \left[\omega = sin^{-1}\left\{\frac{p}{\sqrt{p^2+b^2}}\right\}\right] \quad \ldots\ldots\ldots\ldots\ldots\ldots\ldots\ldots (5)$$

*if p = b then the solid angle subtended by the right triangular plane at the given point 'P'

$$\omega = sin^{-1}\left\{\frac{b}{\sqrt{b^2+b^2}}\right\} = sin^{-1}\left\{\frac{b}{b\sqrt{2}}\right\} = sin^{-1}\left\{\frac{1}{\sqrt{2}}\right\} = \frac{\pi}{4}\ sr$$

2.3.2 Solid angle subtended by a right triangular plane at any point lying on the axis normal to the plane & passing through the right angled vertex

Let there be a right triangular plane ABC having orthogonal sides AB = a & BC = b& a given point say P(0, 0, h) at a height lying on the axis normal to the plane & passing through the right angled vertex 'B' (as shown in the figure (14) below)

Now, draw a perpendicular BN from the vertex 'B' to the hypotenuse AC of the right ΔABC

We know that in a right triangle, the length of the perpendicular drawn from the right angled vertex to the hypotenuse is given as

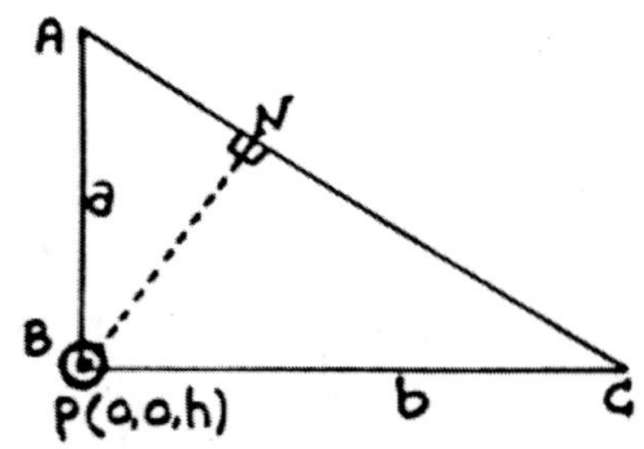

Fig (14)

$$= \frac{(height)(base)}{\sqrt{(height)^2+(base)^2}}$$

$$\therefore in\ right\ \Delta ABC$$

$$\Rightarrow BN = \frac{(AB)(BC)}{\sqrt{(AB)^2+(BC)^2}} = \frac{ab}{\sqrt{a^2+b^2}}$$

In right ΔBNA

$$\Rightarrow AN = \sqrt{AB^2 - BN^2} = \sqrt{a^2 - \left(\frac{ab}{\sqrt{a^2+b^2}}\right)^2} = \sqrt{\frac{a^4 + a^2b^2 - a^2b^2}{a^2+b^2}}$$

$$\Rightarrow AN = \frac{a^2}{\sqrt{a^2+b^2}}$$

Similarly, in right ΔBNC

$$\Rightarrow CN = \sqrt{BC^2 - BN^2} = \sqrt{b^2 - \left(\frac{ab}{\sqrt{a^2+b^2}}\right)^2}$$

$$= \sqrt{\frac{a^2b^2 + b^4 - a^2b^2}{a^2+b^2}} \Rightarrow CN = \frac{b^2}{\sqrt{a^2+b^2}}$$

Now, we have two right triangles $\Delta BNA \& \Delta BNC$ which the right ΔABC is consisting of.

Hence, the solid angle subtended by the right triangular plane ABC at the point 'P' will be equal to the sum of the solid angles subtended by the right $\Delta BNA \& \Delta BNC$

$$\therefore\ \omega = \omega_{\Delta BNA} + \omega_{\Delta BNC}$$

Using equation (2) for a right triangular plane, we have

$$\Rightarrow \omega_{\Delta BNA} = \sin^{-1}\left\{\frac{AN}{\sqrt{AN^2+BN^2}}\right\} - \sin^{-1}\left\{\left(\frac{AN}{\sqrt{AN^2+BN^2}}\right)\left(\frac{h}{\sqrt{h^2+BN^2}}\right)\right\}$$

$$= \sin^{-1}\left\{\frac{\frac{a^2}{\sqrt{a^2+b^2}}}{\sqrt{\left(\frac{a^2}{\sqrt{a^2+b^2}}\right)^2+\left(\frac{ab}{\sqrt{a^2+b^2}}\right)^2}}\right\} - \sin^{-1}\left\{\left(\frac{\frac{a^2}{\sqrt{a^2+b^2}}}{\sqrt{\left(\frac{a^2}{\sqrt{a^2+b^2}}\right)^2+\left(\frac{ab}{\sqrt{a^2+b^2}}\right)^2}}\right)\left(\frac{h}{\sqrt{h^2+\left(\frac{ab}{\sqrt{a^2+b^2}}\right)^2}}\right)\right\}$$

$$= \sin^{-1}\left\{\frac{a^2}{a\sqrt{a^2+b^2}}\right\} - \sin^{-1}\left\{\left(\frac{a^2}{a\sqrt{a^2+b^2}}\right)\left(\frac{h\sqrt{a^2+b^2}}{\sqrt{h^2(a^2+b^2)+a^2b^2}}\right)\right\}$$

$$\Rightarrow \omega_{\Delta BNA} = \sin^{-1}\left\{\frac{a}{\sqrt{a^2+b^2}}\right\} - \sin^{-1}\left\{\frac{ah}{\sqrt{h^2(a^2+b^2)+a^2b^2}}\right\}$$

Similarly, $\Rightarrow \omega_{\Delta BNC} = \sin^{-1}\left\{\frac{CN}{\sqrt{CN^2+BN^2}}\right\} - \sin^{-1}\left\{\left(\frac{CN}{\sqrt{CN^2+BN^2}}\right)\left(\frac{h}{\sqrt{h^2+BN^2}}\right)\right\}$

$$= \sin^{-1}\left\{\frac{\frac{b^2}{\sqrt{a^2+b^2}}}{\sqrt{\left(\frac{b^2}{\sqrt{a^2+b^2}}\right)^2+\left(\frac{ab}{\sqrt{a^2+b^2}}\right)^2}}\right\} - \sin^{-1}\left\{\left(\frac{\frac{b^2}{\sqrt{a^2+b^2}}}{\sqrt{\left(\frac{b^2}{\sqrt{a^2+b^2}}\right)^2+\left(\frac{ab}{\sqrt{a^2+b^2}}\right)^2}}\right)\left(\frac{h}{\sqrt{h^2+\left(\frac{ab}{\sqrt{a^2+b^2}}\right)^2}}\right)\right\}$$

$$= \sin^{-1}\left\{\frac{b^2}{b\sqrt{a^2+b^2}}\right\} - \sin^{-1}\left\{\left(\frac{b^2}{b\sqrt{a^2+b^2}}\right)\left(\frac{h\sqrt{a^2+b^2}}{\sqrt{h^2(a^2+b^2)+a^2b^2}}\right)\right\}$$

$$\Rightarrow \omega_{\Delta BNC} = \sin^{-1}\left\{\frac{b}{\sqrt{a^2+b^2}}\right\} - \sin^{-1}\left\{\frac{bh}{\sqrt{h^2(a^2+b^2)+a^2b^2}}\right\}$$

$$\Rightarrow \omega = \omega_{\Delta BNA} + \omega_{\Delta BNC} = \sin^{-1}\left\{\frac{a}{\sqrt{a^2+b^2}}\right\} - \sin^{-1}\left\{\frac{ah}{\sqrt{h^2(a^2+b^2)+a^2b^2}}\right\}$$

$$+\sin^{-1}\left\{\frac{b}{\sqrt{a^2+b^2}}\right\} - \sin^{-1}\left\{\frac{bh}{\sqrt{h^2(a^2+b^2)+a^2b^2}}\right\}$$

$$\Rightarrow \omega = \left[\sin^{-1}\left\{\frac{a}{\sqrt{a^2+b^2}}\right\} + \sin^{-1}\left\{\frac{b}{\sqrt{a^2+b^2}}\right\}\right]$$

$$-\left[\sin^{-1}\left\{\frac{ah}{\sqrt{h^2(a^2+b^2)+a^2b^2}}\right\} + \sin^{-1}\left\{\frac{bh}{\sqrt{h^2(a^2+b^2)+a^2b^2}}\right\}\right]$$

$$\Rightarrow \omega = \left[\sin^{-1}\left\{\frac{a}{\sqrt{a^2+b^2}}\frac{a}{\sqrt{a^2+b^2}} + \frac{b}{\sqrt{a^2+b^2}}\frac{b}{\sqrt{a^2+b^2}}\right\}\right]$$

$$-\left[\sin^{-1}\left\{\frac{ah}{\sqrt{h^2(a^2+b^2)+a^2b^2}}\frac{a\sqrt{h^2+b^2}}{\sqrt{h^2(a^2+b^2)+a^2b^2}} + \frac{bh}{\sqrt{h^2(a^2+b^2)+a^2b^2}}\frac{b\sqrt{h^2+a^2}}{\sqrt{h^2(a^2+b^2)+a^2b^2}}\right\}\right]$$

$$\left(since,\ \ \sin^{-1}x + \sin^{-1}y = \sin^{-1}\left(x\sqrt{1-y^2}+y\sqrt{1-x^2}\right)\ \forall(-1 \le (x,y) \le 1)\right)$$

$$\Rightarrow \omega = \sin^{-1}\left\{\frac{a^2+b^2}{a^2+b^2}\right\} - \sin^{-1}\left\{\frac{h\left(a^2\sqrt{h^2+b^2} + b^2\sqrt{h^2+a^2}\right)}{h^2(a^2+b^2)+a^2b^2}\right\}$$

$$\Rightarrow \omega = \sin^{-1}\{1\} - \sin^{-1}\left\{\frac{h\left(a^2\sqrt{h^2+b^2} + b^2\sqrt{h^2+a^2}\right)}{h^2(a^2+b^2)+a^2b^2}\right\}$$

$$\Rightarrow \omega = \frac{\pi}{2} - \sin^{-1}\left\{\frac{h\left(a^2\sqrt{h^2+b^2} + b^2\sqrt{h^2+a^2}\right)}{h^2(a^2+b^2)+a^2b^2}\right\} = \cos^{-1}\left\{\frac{h\left(a^2\sqrt{h^2+b^2} + b^2\sqrt{h^2+a^2}\right)}{h^2(a^2+b^2)+a^2b^2}\right\}$$

$$\left(since,\ \frac{\pi}{2} - \cos^{-1}x = \sin^{-1}x \quad \forall(-1 \le x \le 1)\right)$$

$$\Rightarrow \left[\omega = \cos^{-1}\left\{\frac{h\left(a^2\sqrt{h^2+b^2} + b^2\sqrt{h^2+a^2}\right)}{h^2(a^2+b^2)+a^2b^2}\right\}\right] \forall\ (h \ge 0) \ldots\ldots\ldots\ldots\ldots\ldots (6)$$

Important deduction:

The solid angle subtended by the right triangular plane at a point lying on the right angled vertex (i.e. for which h = 0) is obtained by putting h = 0 in the above equation (6)

$$\Rightarrow \omega = \cos^{-1}\left\{\frac{h\left(a^2\sqrt{h^2+b^2} + b^2\sqrt{h^2+a^2}\right)}{h^2(a^2+b^2)+a^2b^2}\right\} = \cos^{-1}\left\{\frac{(0)\left(a^2\sqrt{0+b^2} + b^2\sqrt{0+a^2}\right)}{(0)(a^2+b^2)+a^2b^2}\right\}$$

$$= \cos^{-1}\{0\} = \frac{\pi}{2}\ sr$$

2.3.3 Solid angle subtended by an isosceles triangular plane at any point lying on the axis normal to the plane & passing through the common vertex of the equal sides

Let there be an isosceles triangular plane ABC having base BC = b & the perpendicular AM = p drawn from the common vertex 'A' to the base BC and a given point say P(0, 0, h) at a height 'h' lying on the axis normal to the plane & passing through the common vertex 'A' of the equal sides AB & AC. (as shown in the figure(15) bellow)

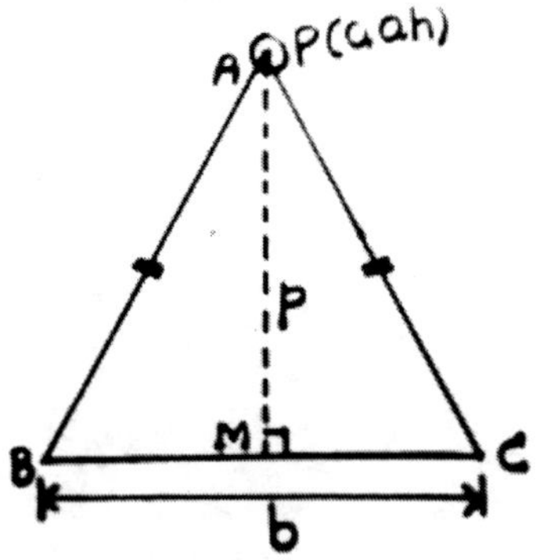

Fig (15)

In isosceles ΔABC

$$\Rightarrow AB = AC,\ BM = CM = \frac{b}{2} \& AM = AM = p$$

Thus, the right triangles ΔAMB & ΔAMC are congruent triangles which the isosceles ΔABC is consisting of.

Hence, the solid angle subtended by the isosceles ΔABC at the given point 'P'

$$\Rightarrow \omega = \omega_{\Delta AMB} + \omega_{\Delta AMC} = 2(\omega_{\Delta AMB})$$

Using formula from equation (2), we have

$$\Rightarrow \omega_{\Delta AMB} = \sin^{-1}\left\{\frac{BM}{\sqrt{BM^2+AM^2}}\right\} - \sin^{-1}\left\{\left(\frac{BM}{\sqrt{BM^2+AM^2}}\right)\left(\frac{h}{\sqrt{h^2+AM^2}}\right)\right\}$$

$$= \sin^{-1}\left\{\frac{\left(\frac{b}{2}\right)}{\sqrt{\left(\frac{b}{2}\right)^2+p^2}}\right\} - \sin^{-1}\left\{\left(\frac{\left(\frac{b}{2}\right)}{\sqrt{\left(\frac{b}{2}\right)^2+p^2}}\right)\left(\frac{h}{\sqrt{h^2+p^2}}\right)\right\}$$

$$= \sin^{-1}\left\{\frac{b}{\sqrt{b^2+4p^2}}\right\} - \sin^{-1}\left\{\left(\frac{b}{\sqrt{b^2+4p^2}}\right)\left(\frac{h}{\sqrt{h^2+p^2}}\right)\right\}$$

Hence, we get

$$\Rightarrow \omega = 2(\omega_{\Delta AMB}) = 2\left[\sin^{-1}\left\{\frac{b}{\sqrt{b^2+4p^2}}\right\} - \sin^{-1}\left\{\left(\frac{b}{\sqrt{b^2+4p^2}}\right)\left(\frac{h}{\sqrt{h^2+p^2}}\right)\right\}\right]$$

$$\omega = 2\left[\sin^{-1}\left\{\frac{b}{\sqrt{b^2+4p^2}}\right\} - \sin^{-1}\left\{\left(\frac{b}{\sqrt{b^2+4p^2}}\right)\left(\frac{h}{\sqrt{h^2+p^2}}\right)\right\}\right] \quad \forall(h \geq 0) \dots\dots (7)$$

Important deduction:

The solid angle subtended by the isosceles triangular plane at a point lying on the common vertex 'A' of equal sides (i.e. for which h = 0) is obtained by putting h = 0 in the above eq(7)

Thus, we have

$$\omega = 2\left[\sin^{-1}\left\{\frac{b}{\sqrt{b^2+4p^2}}\right\} - \sin^{-1}\left\{\left(\frac{b}{\sqrt{b^2+4p^2}}\right)\left(\frac{0}{\sqrt{0+p^2}}\right)\right\}\right]$$

$$= 2\left[\sin^{-1}\left\{\frac{b}{\sqrt{b^2+4p^2}}\right\} - \sin^{-1}\{0\}\right] = 2\sin^{-1}\left\{\frac{b}{\sqrt{b^2+4p^2}}\right\}$$

$$\Rightarrow \boxed{\omega = 2\sin^{-1}\left\{\frac{b}{\sqrt{b^2+4p^2}}\right\}} \quad \dots\dots\dots\dots\dots\dots\dots (8)$$

* if b = p, then the solid angle subtended by the isosceles triangular plane at the given point 'P'

$$\omega = 2\sin^{-1}\left\{\frac{p}{\sqrt{p^2+4p^2}}\right\} = 2\sin^{-1}\left\{\frac{p}{p\sqrt{5}}\right\} = 2\sin^{-1}\left\{\frac{1}{\sqrt{5}}\right\} = 0.927\ sr$$

2.3.4 Solid angle subtended by an isosceles triangular plane at any point lying on the axis normal to the plane & passing through the mid-point of the unequal side

Let there be an isosceles triangular plane ABC having the base BC = b & the perpendicular OA = p and a given point say P(0, 0, h) at a height 'h' lying on the axis normal to the plane &passing through the mid-point 'O' of the base BC (as shown in the figure (16) bellow)

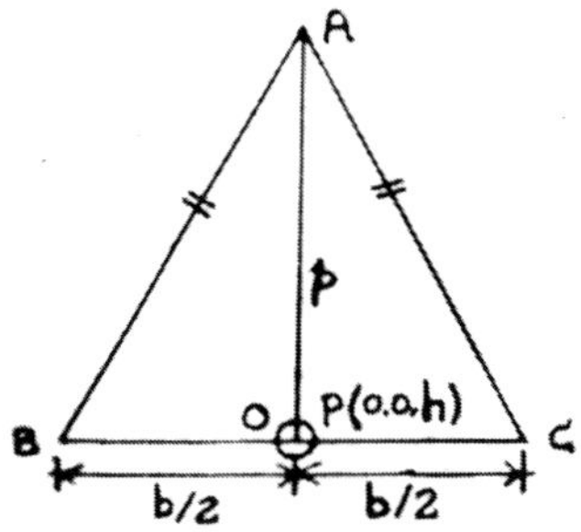

Fig (16)

In isosceles ΔABC

$$\Rightarrow AB = AC, OB = OC = \frac{b}{2} \& \ OA = OA = p$$

Thus, ΔAOB & ΔAOC are congruent right triangles

Hence, the solid angle subtended by the isosceles ΔABCat the given point 'P'

$$\Rightarrow \omega = \omega_{\Delta AOB} + \omega_{\Delta AOC} = 2(\omega_{\Delta AOB})$$

Using formula from the eq(6)& putting the values, we get

$$\Rightarrow \omega_{\Delta AOB} = \cos^{-1}\left\{\frac{h\left(OA^2\sqrt{h^2+OB^2} + OB^2\sqrt{h^2+OA^2}\right)}{h^2(OA^2+OB^2)+OA^2OB^2}\right\}$$

$$= \cos^{-1}\left\{\frac{h\left(p^2\sqrt{h^2+\left(\frac{b}{2}\right)^2} + \left(\frac{b}{2}\right)^2\sqrt{h^2+p^2}\right)}{h^2\left(p^2+\left(\frac{b}{2}\right)^2\right)+p^2\left(\frac{b}{2}\right)^2}\right\} = \cos^{-1}\left\{\frac{h\left(2p^2\sqrt{4h^2+b^2} + b^2\sqrt{h^2+p^2}\right)}{h^2(4p^2+b^2)+p^2b^2}\right\}$$

$$\Rightarrow \omega = 2(\omega_{\Delta AOB}) = 2\cos^{-1}\left\{\frac{h\left(2p^2\sqrt{4h^2+b^2} + b^2\sqrt{h^2+p^2}\right)}{h^2(4p^2+b^2)+p^2b^2}\right\}$$

$$\Rightarrow \left[\omega = 2\cos^{-1}\left\{\frac{h\left(2p^2\sqrt{4h^2+b^2}\ +\ b^2\sqrt{h^2+p^2}\right)}{h^2(4p^2+b^2)+p^2b^2}\right\}\right] \quad \forall(h \geq 0) \ldots\ldots\ldots\ldots\ldots\ldots (9)$$

Important deduction:

The solid angle subtended by the isosceles triangular plane at a point lying on the mid-point 'O' of the unequal side BC of the plane (i.e. for which h = 0)(see the fig(13) above) is obtained by putting h = 0 in the eq(9)

Thus, we get

$$\omega = 2\cos^{-1}\left\{\frac{(0)\left(2p^2\sqrt{4h^2+b^2}\ +\ b^2\sqrt{h^2+p^2}\right)}{(0)(4p^2+b^2)+p^2b^2}\right\} = 2\cos^{-1}\{0\} = 2\frac{\pi}{2} = \pi\ sr$$

Note: It is also true for any point lying on the boundary of the plane, $\omega = \pi\ sr$ while for any point lying on the plane inside the boundary, $\omega = 2\pi\ sr$ (see the fig (9))

2.3.5 Solid angle subtended by an isosceles right triangular plane at any point lying on the axis normal to the plane & passing through the right angled vertex

Method 1:

Let there be an isosceles right triangular plane AOB having equal sides OA & OB each of length 'a'& a given point say P(0, 0, h) at a height 'h' lying on the axis normal to the plane & passing through the right angled vertex 'O' (as shown in the figure(17) bellow)

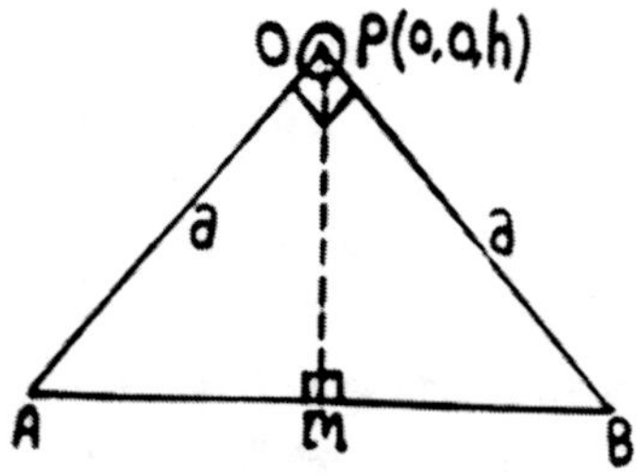

Fig (17)

We know that the solid angle subtended by the right triangular plane at any point lying on the axis normal to the plane & passing through the right angled vertex is given by eq(6) as follows

$$\omega = \cos^{-1}\left\{\frac{h\left(a^2\sqrt{h^2+b^2}\ +\ b^2\sqrt{h^2+a^2}\right)}{h^2(a^2+b^2)+a^2b^2}\right\}$$

Hence, on putting b = a in the above equation, we get

$$\omega = \cos^{-1}\left\{\frac{h\left(a^2\sqrt{h^2+a^2}\ +\ a^2\sqrt{h^2+a^2}\right)}{h^2(a^2+a^2)+a^2a^2}\right\} = \cos^{-1}\left\{\frac{2ha^2\sqrt{h^2+a^2}}{2h^2a^2+a^4}\right\}$$

$$= \cos^{-1}\left\{\frac{2ha^2\sqrt{h^2+a^2}}{a^2(2h^2+a^2)}\right\} = \cos^{-1}\left\{\frac{2h\sqrt{h^2+a^2}}{(2h^2+a^2)}\right\}$$

$$\Rightarrow \left[\omega = \cos^{-1}\left\{\frac{2h\sqrt{h^2+a^2}}{(2h^2+a^2)}\right\}\right] \quad \forall(h \geq 0) \ldots\ldots\ldots\ldots\ldots\ldots (10)$$

Method 2:

As shown in the above figure (17), draw a perpendicular OM from the right angled vertex 'O' to the hypotenuse AB

Now, in isosceles right $\Delta AOB \Rightarrow AB = \sqrt{OA^2+OB^2} = \sqrt{a^2+a^2} = a\sqrt{2}$

$$\Rightarrow AM = BM = \frac{AB}{2} = \frac{a\sqrt{2}}{2} = \frac{a}{\sqrt{2}} \&$$

$$in\ right\ \Delta OMA \Rightarrow OM = \sqrt{OA^2 - AM^2} = \sqrt{a^2 - \left(\frac{a}{\sqrt{2}}\right)^2} = \sqrt{\frac{a^2}{2}} = \frac{a}{\sqrt{2}}$$

Now, in isosceles right ΔAOB

$$\Rightarrow OA = OB = a, AM = BM = \frac{a}{\sqrt{2}} \& OM = OM = \frac{a}{\sqrt{2}}$$

Thus, ΔOMA & $\Delta OMB$are congruent right triangles which the isosceles right ΔAOB is consisting of. Hence, the solid angle subtended by the isosceles right triangular plane AOB

$$\Rightarrow \omega = \omega_{\Delta OMA} + \omega_{\Delta OMB} = 2(\omega_{\Delta OMA})$$

Now, using the formula from eq(2), we get

$$\Rightarrow \omega_{\Delta OMA} = \sin^{-1}\left\{\frac{AM}{\sqrt{AM^2 + OM^2}}\right\} - \sin^{-1}\left\{\left(\frac{AM}{\sqrt{AM^2 + OM^2}}\right)\left(\frac{h}{\sqrt{h^2 + OM^2}}\right)\right\}$$

$$= \sin^{-1}\left\{\frac{\left(\frac{a}{\sqrt{2}}\right)}{\sqrt{\left(\frac{a}{\sqrt{2}}\right)^2 + \left(\frac{a}{\sqrt{2}}\right)^2}}\right\} - \sin^{-1}\left\{\left(\frac{\left(\frac{a}{\sqrt{2}}\right)}{\sqrt{\left(\frac{a}{\sqrt{2}}\right)^2 + \left(\frac{a}{\sqrt{2}}\right)^2}}\right)\left(\frac{h}{\sqrt{h^2 + \left(\frac{a}{\sqrt{2}}\right)^2}}\right)\right\}$$

$$= \sin^{-1}\left\{\frac{a}{a\sqrt{2}}\right\} - \sin^{-1}\left\{\left(\frac{a}{a\sqrt{2}}\right)\left(\frac{h\sqrt{2}}{\sqrt{2h^2 + a^2}}\right)\right\} = \sin^{-1}\left\{\frac{1}{\sqrt{2}}\right\} - \sin^{-1}\left\{\frac{h}{\sqrt{2h^2 + a^2}}\right\}$$

$= \frac{\pi}{4} - \sin^{-1}\left\{\frac{h}{\sqrt{2h^2+a^2}}\right\}$, Hence, we get

$$\omega = 2(\omega_{\Delta OMA}) = 2\left[\frac{\pi}{4} - \sin^{-1}\left\{\frac{h}{\sqrt{2h^2 + a^2}}\right\}\right] = \frac{\pi}{2} - 2\sin^{-1}\left\{\frac{h}{\sqrt{2h^2 + a^2}}\right\}$$

$$= \frac{\pi}{2} - \sin^{-1}\left\{2\left(\frac{h}{\sqrt{2h^2 + a^2}}\right)\left(\frac{\sqrt{h^2 + a^2}}{\sqrt{2h^2 + a^2}}\right)\right\}$$

$$\left(since,\ 2\sin^{-1} x = \sin^{-1}\left(2x\sqrt{1 - x^2}\right)\ \forall(-1 \le x \le 1)\right)$$

$$= \frac{\pi}{2} - \sin^{-1}\left\{\frac{2h\sqrt{h^2 + a^2}}{2h^2 + a^2}\right\} = \cos^{-1}\left\{\frac{2h\sqrt{h^2 + a^2}}{2h^2 + a^2}\right\}$$

$$\left(since,\ \frac{\pi}{2} - \sin^{-1} x = \cos^{-1} x\ \ \forall(-1 \le x \le 1)\right)$$

$$\Rightarrow \left[\omega = \cos^{-1}\left\{\frac{2h\sqrt{h^2 + a^2}}{2h^2 + a^2}\right\}\right] \quad \forall(h \ge 0) \ldots\ldots\ldots\ldots\ldots (10)$$

Method 3:

in isosceles right $\Delta AOB \Rightarrow AB = \sqrt{OA^2 + OB^2} = \sqrt{a^2 + a^2} = a\sqrt{2}$

in right ΔOMA

$$\Rightarrow OM = \sqrt{OA^2 - AM^2} = \sqrt{a^2 - \left(\frac{a}{\sqrt{2}}\right)^2} = \sqrt{\frac{a^2}{2}} = \frac{a}{\sqrt{2}}$$

We know that the solid angle subtended by the isosceles triangular plane at any point lying on the axis normal to the plane & passing through the common vertex of the equal sides is given by eq(7) as follows

$$\omega = 2\left[\sin^{-1}\left\{\frac{b}{\sqrt{b^2 + 4p^2}}\right\} - \sin^{-1}\left\{\left(\frac{b}{\sqrt{b^2 + 4p^2}}\right)\left(\frac{h}{\sqrt{h^2 + p^2}}\right)\right\}\right]$$

Now, on putting the values of b = base = AB = $a\sqrt{2}$& p = perpendicular = OM = $\frac{a}{\sqrt{2}}$, we get

$$\omega = 2\left[\sin^{-1}\left\{\frac{a\sqrt{2}}{\sqrt{\left(a\sqrt{2}\right)^2+4\left(\frac{a}{\sqrt{2}}\right)^2}}\right\} - \sin^{-1}\left\{\left(\frac{a\sqrt{2}}{\sqrt{\left(a\sqrt{2}\right)^2+4\left(\frac{a}{\sqrt{2}}\right)^2}}\right)\left(\frac{h}{\sqrt{h^2+\left(\frac{a}{\sqrt{2}}\right)^2}}\right)\right\}\right]$$

$$= 2\left[\sin^{-1}\left\{\frac{a}{a\sqrt{2}}\right\} - \sin^{-1}\left\{\left(\frac{a}{a\sqrt{2}}\right)\left(\frac{h\sqrt{2}}{\sqrt{2h^2+a^2}}\right)\right\}\right] = 2\left[\sin^{-1}\left\{\frac{1}{\sqrt{2}}\right\} - \sin^{-1}\left\{\frac{h}{\sqrt{2h^2+a^2}}\right\}\right]$$

$$= 2\left[\frac{\pi}{4} - \sin^{-1}\left\{\frac{h}{\sqrt{2h^2+a^2}}\right\}\right] = \frac{\pi}{2} - 2\sin^{-1}\left\{\frac{h}{\sqrt{2h^2+a^2}}\right\}$$

$$= \frac{\pi}{2} - \sin^{-1}\left\{2\left(\frac{h}{\sqrt{2h^2+a^2}}\right)\left(\frac{\sqrt{h^2+a^2}}{\sqrt{2h^2+a^2}}\right)\right\}$$

$$\left(since,\ 2\sin^{-1}x = \sin^{-1}\left(2x\sqrt{1-x^2}\right)\ \forall(-1 \le x \le 1)\right)$$

$$= \frac{\pi}{2} - \sin^{-1}\left\{\frac{2h\sqrt{h^2+a^2}}{2h^2+a^2}\right\} = \cos^{-1}\left\{\frac{2h\sqrt{h^2+a^2}}{2h^2+a^2}\right\}$$

$$\left(since,\ \frac{\pi}{2} - \sin^{-1}x = \cos^{-1}x\ \ \forall(-1 \le x \le 1)\right)$$

$$\Rightarrow \left[\omega = \cos^{-1}\left\{\frac{2h\sqrt{h^2+a^2}}{2h^2+a^2}\right\}\right] \quad \forall(h \ge 0) \ldots\ldots\ldots\ldots\ldots\ldots\ldots (10)$$

Important deduction:

The solid angle subtended by the isosceles right triangular plane at a point lying on the right angled vertex is obtained by putting h = 0 in the above equation, we get

$$\boldsymbol{\omega = \cos^{-1}\left\{\frac{2(0)\sqrt{h^2+a^2}}{2(0)^2+a^2}\right\} = \cos^{-1}\{0\} = \frac{\pi}{2}\ sr}$$

2.3.6 Solid angle subtended by an isosceles right triangular plane at any point lying on the axis normal to the plane & passing through the acute angled vertex

Let there be an isosceles right triangular plane OBC having equal sides OB & BC each of the length 'a' and a given point sat P(0, 0, h) at a height 'h' lying on the axis normal to the plane & passing through the acute angled vertex say 'O' (as shown in the figure (18) below)

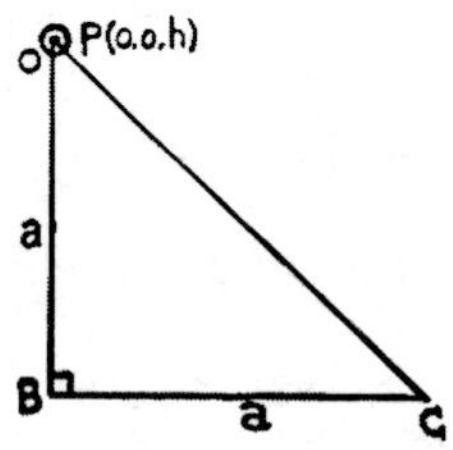

Fig (18)

We know that the solid angle subtended by the right triangular plane at any point lying on the axis normal to the plane & passing through the acute angled vertex is given from the eq(2) as follows

$$\omega = \sin^{-1}\left\{\frac{b}{\sqrt{b^2+p^2}}\right\} - \sin^{-1}\left\{\left(\frac{b}{\sqrt{b^2+p^2}}\right)\left(\frac{h}{\sqrt{h^2+p^2}}\right)\right\}$$

Now, on putting the values b = base = BC & p = perpendicular = OB, we have

$$\Rightarrow \omega = \sin^{-1}\left\{\frac{BC}{\sqrt{BC^2+OB^2}}\right\} - \sin^{-1}\left\{\left(\frac{BC}{\sqrt{BC^2+OB^2}}\right)\left(\frac{h}{\sqrt{h^2+OB^2}}\right)\right\}$$

$$\omega = \sin^{-1}\left\{\frac{a}{\sqrt{a^2+a^2}}\right\} - \sin^{-1}\left\{\left(\frac{a}{\sqrt{a^2+a^2}}\right)\left(\frac{h}{\sqrt{h^2+a^2}}\right)\right\} (since,\ OB = BC = a)$$

$$= \sin^{-1}\left\{\frac{1}{\sqrt{2}}\right\} - \sin^{-1}\left\{\left(\frac{1}{\sqrt{2}}\right)\left(\frac{h}{\sqrt{h^2+a^2}}\right)\right\} = \frac{\pi}{4} - \sin^{-1}\left\{\frac{h}{\sqrt{2(h^2+a^2)}}\right\}$$

$$\Rightarrow \left[\omega = \frac{\pi}{4} - \sin^{-1}\left\{\frac{h}{\sqrt{2(h^2 + a^2)}}\right\}\right] \quad \forall(h \geq 0) \ldots\ldots\ldots\ldots\ldots\ldots..(11)$$

Important deduction:

The solid angle subtended by the isosceles right triangular plane at a point lying on the acute angled vertex of the plane is obtained by putting h = 0 in the above eq(11), we get

$$\omega = \frac{\pi}{4} - \sin^{-1}\left\{\frac{0}{\sqrt{2(0 + a^2)}}\right\} = \frac{\pi}{4} - \sin^{-1}\{0\} = \frac{\pi}{4}\ sr$$

Also, we know that the solid angle subtended by the plane at any point lying on the plane inside the boundary is $\omega = 2\pi\ sr$(irrespective of area & boundary of the plane). Now, if the plane is divided in eight equiangular parts w.r.t. the same point on the plane then the point can be treated as the acute angled vertex for each of eight segments. Hence, the solid angle subtended by each segment (right triangular plane) at the same point(acute angled vertex) is

$$\omega = \frac{1}{8}.2\pi = \frac{\pi}{4}\ sr$$

2.3.7 Solid angle subtended by an isosceles right triangular plane at any point lying on the axis normal to the plane & passing through the mid-point of hypotenuse

Method 1:

Let there be an isosceles right triangular plane ACB having equal sides AC & BC each of the length 'a' and a given point say P(0, 0, h) at a height 'h' lying on the axis normal to the plane & passing through the mid-point 'O' of the hypotenuse AB (as shown in the figure (19) below)

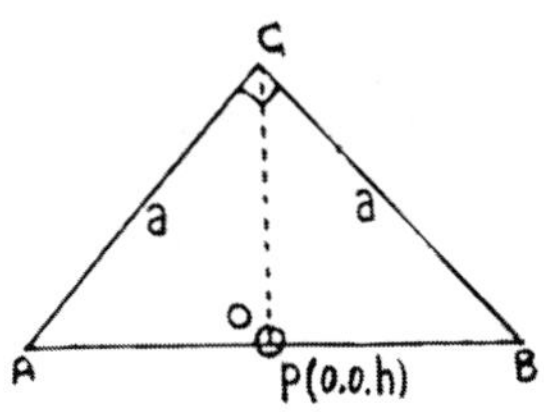

Fig (19)

Now, join the mid-point 'O' to the acute angled vertex 'C',

Now, in isosceles right $\Delta ACB \Rightarrow AB = \sqrt{AC^2 + BC^2} = \sqrt{a^2 + a^2} = a\sqrt{2}$

$$\Rightarrow OA = OB = \frac{AB}{2} = \frac{a\sqrt{2}}{2} = \frac{a}{\sqrt{2}} \&$$

$$\Rightarrow CO = \sqrt{AC^2 - OA^2} = \sqrt{a^2 - \left(\frac{a}{\sqrt{2}}\right)^2} = \frac{a}{\sqrt{2}}$$

Since, we know that the solid angle subtended by the isosceles triangular plane at any point lying on the axis normal to the plane & passing through the mid-point of the unequal side is given by the eq(9) as follows

$$\omega = 2\cos^{-1}\left\{\frac{h\left(2p^2\sqrt{4h^2 + b^2} + b^2\sqrt{h^2 + p^2}\right)}{h^2(4p^2 + b^2) + p^2b^2}\right\}$$

Now, putting the values of b = base = AB = $a\sqrt{2}$& p = perpendicular = CO = $\frac{a}{\sqrt{2}}$, we get

$$\omega = 2\cos^{-1}\left\{\frac{h\left(2\left(\frac{a}{\sqrt{2}}\right)^2\sqrt{4h^2 + \left(a\sqrt{2}\right)^2} + \left(a\sqrt{2}\right)^2\sqrt{h^2 + \left(\frac{a}{\sqrt{2}}\right)^2}\right)}{h^2\left(4\left(\frac{a}{\sqrt{2}}\right)^2 + \left(a\sqrt{2}\right)^2\right) + \left(\frac{a}{\sqrt{2}}\right)^2\left(a\sqrt{2}\right)^2}\right\}$$

$$= 2\cos^{-1}\left\{\frac{h\left(\sqrt{2}a^2\sqrt{2h^2 + a^2} + \sqrt{2}a^2\sqrt{2h^2 + a^2}\right)}{4h^2a^2 + a^4}\right\} = 2\cos^{-1}\left\{\frac{2\sqrt{2}ha^2\sqrt{2h^2 + a^2}}{4h^2a^2 + a^4}\right\}$$

$$\Rightarrow \left[\omega = 2\cos^{-1}\left\{\frac{2\sqrt{2}h\sqrt{2h^2+a^2}}{4h^2+a^2}\right\}\right] \quad \forall(h \geq 0) \quad \ldots\ldots\ldots\ldots(12)$$

Or

$$\omega = 2\left[\frac{\pi}{2} - \sin^{-1}\left\{\frac{2\sqrt{2}h\sqrt{2h^2+a^2}}{4h^2+a^2}\right\}\right] \text{(since, } \cos^{-1}x = \frac{\pi}{2} - \sin^{-1}x\text{)}$$

$$= \pi - 2\sin^{-1}\left\{\frac{2\sqrt{2}h\sqrt{2h^2+a^2}}{4h^2+a^2}\right\}$$

$$\omega = \pi - 4\sin^{-1}\left\{\frac{h\sqrt{2}}{\sqrt{4h^2+a^2}}\right\} \quad \forall(h \geq 0) \ldots\ldots\ldots\ldots(12)$$

$$\left(\text{since, } \sin^{-1}\left(2x\sqrt{1-x^2}\right) = 2\sin^{-1}x \ \forall(-1 \leq x \leq 1) \ \& \ here \ x = \frac{h\sqrt{2}}{\sqrt{4h^2+a^2}}\right)$$

Method 2:

In the isosceles right $\Delta ACB \Rightarrow AC = BC, OA = OB = \frac{a}{\sqrt{2}} \ \& \ CO = CO = \frac{a}{\sqrt{2}}$

Thus, the triangles ΔCOA & ΔCOB are congruent isosceles right triangles

Hence, the solid angle subtended by the isosceles right triangular plane ACB

$$\Rightarrow \omega = \omega_{\Delta COA} + \omega_{\Delta COB} = 2(\omega_{\Delta COA})$$

As we know that the solid angle subtended by the isosceles right triangular plane at any point lying on the axis normal to the plane & passing through the right angled vertex is given by the eq(10) as follows

$\omega = \cos^{-1}\left\{\frac{2h\sqrt{h^2+a^2}}{(2h^2+a^2)}\right\}$ (on putting the values of a = OA = CO = $\frac{a}{\sqrt{2}}$)

$$\Rightarrow \omega_{\Delta COA} = \cos^{-1}\left\{\frac{2h\sqrt{h^2+\left(\frac{a}{\sqrt{2}}\right)^2}}{\left(2h^2+\left(\frac{a}{\sqrt{2}}\right)^2\right)}\right\} = \cos^{-1}\left\{\frac{2\sqrt{2}h\sqrt{2h^2+a^2}}{(4h^2+a^2)}\right\}$$

$$\Rightarrow \left[\omega = 2(\omega_{\Delta COA}) = 2\cos^{-1}\left\{\frac{2\sqrt{2}h\sqrt{2h^2+a^2}}{(4h^2+a^2)}\right\} = \pi - 4\sin^{-1}\left\{\frac{h\sqrt{2}}{\sqrt{4h^2+a^2}}\right\}\right]$$

$$\forall(h \geq 0) \ldots\ldots\ldots\ldots\ldots\ldots\ldots\ldots(12)$$

Important deduction:

The solid angle subtended by the isosceles right triangular plane at a point lying on the mid-point of the hypotenuse is obtained by putting h = 0 in the above eq(12), we get

$$\omega = 2\cos^{-1}\left\{\frac{2\sqrt{2}(0)\sqrt{2h^2+a^2}}{(4(0)+a^2)}\right\} = 2\cos^{-1}\{0\} = 2\frac{\pi}{2} = \pi \ sr$$

*It is also true for each point lying on the boundary of the plane.

2.3.8 Solid angle subtended by an equilateral triangular plane at any point lying on the axis normal to the plane & passing through the vertex

Method 1:

Let there be an equilateral triangular plane ABC having each side equal to 'a' &a given point say P(0, 0, h) at a height 'h' lying on the axis normal to the plane & passing through the vertex say 'A' (as shown in the figure(20) below)

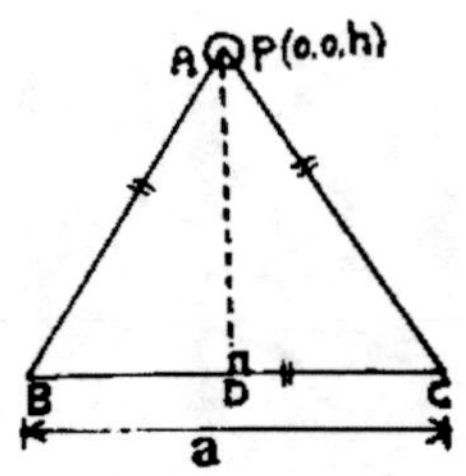

Fig (20)

Now, draw a perpendicular AD from the vertex 'A' to the side BC

In rightΔADB

$$\Rightarrow AD = \sqrt{AB^2 - BD^2} = \sqrt{a^2 - \left(\frac{a}{2}\right)^2} = \frac{\sqrt{3}a}{2}$$

In equilateralΔABC

$$\Rightarrow AB = AC = a, BD = CD = \frac{a}{2} \& AD = AD = \frac{\sqrt{3}a}{2}$$

Thus, ΔADB & ΔADC are congruent right triangles

Hence, the solid angle subtended by the equilateral triangular plane ABC at the given point 'P'

$$\Rightarrow \omega = \omega_{\Delta ADB} + \omega_{\Delta ADC} = 2(\omega_{\Delta ADB})$$

We know that the solid angle subtended by the right triangular plane at any point lying on the axis normal to the plane & passing through the acute angled vertex is given by the eq(2) as follows

$$\omega = \sin^{-1}\left\{\frac{b}{\sqrt{b^2+p^2}}\right\} - \sin^{-1}\left\{\left(\frac{b}{\sqrt{b^2+p^2}}\right)\left(\frac{h}{\sqrt{h^2+p^2}}\right)\right\}$$

Now, putting the values of b = base = BD $= \frac{a}{2}$ & p = perpendicular = AD $= \frac{\sqrt{3}a}{2}$, we get

$$\omega_{\Delta ADB} = \sin^{-1}\left\{\frac{\frac{a}{2}}{\sqrt{\left(\frac{a}{2}\right)^2 + \left(\frac{\sqrt{3}a}{2}\right)^2}}\right\} - \sin^{-1}\left\{\left(\frac{\frac{a}{2}}{\sqrt{\left(\frac{a}{2}\right)^2 + \left(\frac{\sqrt{3}a}{2}\right)^2}}\right)\left(\frac{h}{\sqrt{h^2 + \left(\frac{\sqrt{3}a}{2}\right)^2}}\right)\right\}$$

$$\omega_{\Delta ADB} = \sin^{-1}\left\{\frac{a}{2a}\right\} - \sin^{-1}\left\{\left(\frac{a}{2a}\right)\left(\frac{2h}{\sqrt{4h^2+3a^2}}\right)\right\} = \sin^{-1}\left\{\frac{1}{2}\right\} - \sin^{-1}\left\{\frac{h}{\sqrt{4h^2+3a^2}}\right\}$$

$$= \frac{\pi}{6} - \sin^{-1}\left\{\frac{h}{\sqrt{4h^2+3a^2}}\right\}$$

Hence, we have

$$\Rightarrow \omega = 2(\omega_{\Delta ADB}) = 2\left[\frac{\pi}{6} - \sin^{-1}\left\{\frac{h}{\sqrt{4h^2+3a^2}}\right\}\right]$$

$$\Rightarrow \left[\omega = \frac{\pi}{3} - 2\sin^{-1}\left\{\frac{h}{\sqrt{4h^2+3a^2}}\right\}\right] \quad \forall (h \geq 0) \ldots\ldots\ldots\ldots\ldots\ldots\ldots\ldots (13)$$

Method 2:

In equilateral $\Delta ABC \Rightarrow AB = BC = AC = a \,\&\, AD = \frac{\sqrt{3}a}{2}$

We know that the solid angle subtended by the isosceles triangular plane at any point lying on the axis normal to the plane & passing through the common vertex of the equal sidesis given by the eq(7) as follows

$$\omega = 2\left[\sin^{-1}\left\{\frac{b}{\sqrt{b^2+4p^2}}\right\} - \sin^{-1}\left\{\left(\frac{b}{\sqrt{b^2+4p^2}}\right)\left(\frac{h}{\sqrt{h^2+p^2}}\right)\right\}\right]$$

Now, putting the values of b = base = BC = a & p = perpendicular = AD $= \frac{\sqrt{3}a}{2}$ in the above expression

Hence, the solid angle subtended by the equilateral plane ABC at the given point 'P'

$$\omega = 2\left[\sin^{-1}\left\{\frac{a}{\sqrt{a^2+4\left(\frac{\sqrt{3}a}{2}\right)^2}}\right\} - \omega \sin^{-1}\left\{\left(\frac{a}{\sqrt{a^2+4\left(\frac{\sqrt{3}a}{2}\right)^2}}\right)\left(\frac{h}{\sqrt{h^2+\left(\frac{\sqrt{3}a}{2}\right)^2}}\right)\right\}\right]$$

$$= 2\left[\sin^{-1}\left\{\frac{a}{2a}\right\} - \omega \sin^{-1}\left\{\left(\frac{a}{2a}\right)\left(\frac{2h}{\sqrt{4h^2+3a^2}}\right)\right\}\right]$$

$$= 2\left[\sin^{-1}\left\{\frac{1}{2}\right\} - \sin^{-1}\left\{\left(\frac{h}{\sqrt{4h^2+3a^2}}\right)\right\}\right] = 2\left[\frac{\pi}{6} - \sin^{-1}\left\{\left(\frac{h}{\sqrt{4h^2+3a^2}}\right)\right\}\right]$$

$$\Rightarrow \left[\omega = \frac{\pi}{3} - 2\sin^{-1}\left\{\frac{h}{\sqrt{4h^2+3a^2}}\right\}\right] \quad \forall (h \geq 0) \qquad \ldots\ldots\ldots\ldots\ldots\ldots\ldots\ldots (13)$$

Important deduction:

The solid angle subtended by the equilateral triangular plane at a point lying on the vertex (i.e. for which h= 0) is obtained by putting h = 0 in the above eq(13)

Thus, we get

$$\omega = \frac{\pi}{3} - 2\sin^{-1}\left\{\frac{0}{\sqrt{4(0)+3a^2}}\right\} = \frac{\pi}{3} - 2\sin^{-1}\{0\} = \frac{\pi}{3} sr$$

Also, we know from the key-point-6if a plane is divided into 'n' no. of the equiangular segments w.r.t. any point lying on the plane inside the boundary then the solid angle subtended by the each segment at the same point is given as

$$\omega = \frac{2\pi}{n} sr$$

If the plane is divided into six equiangular segments then each segment can be treated as the equilateral triangle (irrespective of area & geometry) for the given point (as the vertex) on the plane then the solid angle subtended by each of the segment(equilateral triangle) at the same point(vertex) is

$$\omega = \frac{2\pi}{6} = \frac{\pi}{3} sr$$

2.3.9 Solid angle subtended by an equilateral triangular plane at any point lying on the axis normal to the plane & passing through the mid-point of one of its sides

Let there be a equilateral triangular plane ABC having each side equal to 'a' & a given point say P(0, 0, h) at a height 'h' lying on the axis normal to the plane & passing through the mid-point 'O' of the side say BC. (as shown in the figure(21) below)

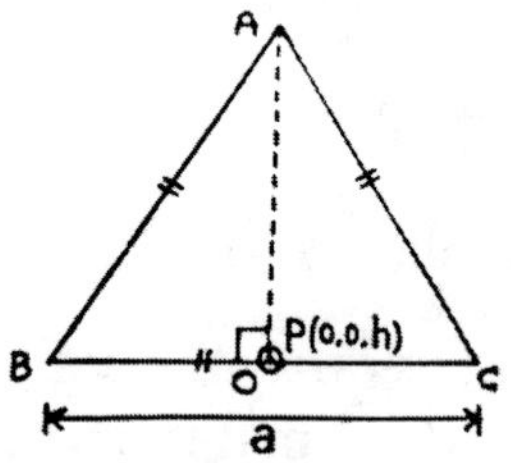

Fig (21)

Now, join the mid-point 'O' of the side BC to the opposite vertex 'A' thus, we have line AO normal to BC

In right ΔAOB

$$\Rightarrow AO = \sqrt{AB^2 - OB^2} = \sqrt{a^2 - \left(\frac{a}{2}\right)^2} = \frac{a\sqrt{3}}{2}$$

Now, in equilateral $\Delta ABC \Rightarrow AB = BC = AC = a, OB = OC = \frac{a}{2} \& AO = AO = \frac{a\sqrt{3}}{2}$

Thus, $\Delta AOB \& \Delta AOC$are congruent right triangles.

Hence, the solid angle subtended by the equilateral triangular plane ABC at the given point 'P'

$$\Rightarrow \omega = \omega_{\Delta AOB} + \omega_{\Delta AOC} = 2(\omega_{\Delta AOB})$$

We know that the solid angle subtended by the right triangular plane at any point lying on the axis normal to the plane passing through the right angled vertex is given by eq(6) as follows

$$\omega = \cos^{-1}\left\{\frac{h\left(a^2\sqrt{h^2+b^2} + b^2\sqrt{h^2+a^2}\right)}{h^2(a^2+b^2)+a^2b^2}\right\}$$

Hence, on putting the values of a = AO $=\frac{a\sqrt{3}}{2}$ & b = OB $=\frac{a}{2}$, we get

$$\omega_{\Delta AOB} = \cos^{-1}\left\{\frac{h\left(\left(\frac{a\sqrt{3}}{2}\right)^2\sqrt{h^2+\left(\frac{a}{2}\right)^2} + \left(\frac{a}{2}\right)^2\sqrt{h^2+\left(\frac{a\sqrt{3}}{2}\right)^2}\right)}{h^2\left(\left(\frac{a\sqrt{3}}{2}\right)^2+\left(\frac{a}{2}\right)^2\right)+\left(\frac{a\sqrt{3}}{2}\right)^2\left(\frac{a}{2}\right)^2}\right\}$$

$$= \cos^{-1}\left\{\frac{2h\left(3a^2\sqrt{4h^2+a^2} + a^2\sqrt{4h^2+3a^2}\right)}{16h^2a^2+3a^4}\right\} = \cos^{-1}\left\{\frac{2h\left(3\sqrt{4h^2+a^2} + \sqrt{4h^2+3a^2}\right)}{16h^2+3a^2}\right\}$$

$$\Rightarrow \omega = 2(\omega_{\Delta AOB}) = 2\cos^{-1}\left\{\frac{2h\left(3\sqrt{4h^2+a^2} + \sqrt{4h^2+3a^2}\right)}{16h^2+3a^2}\right\} \quad \forall (h \geq 0) \ldots\ldots\ldots\ldots\ldots (14)$$

Method 2:

In equilateral ΔABC , we have $BC = a$ & $AO = \frac{a\sqrt{3}}{2}$

Since, we know that the solid angle subtended by the isosceles triangular plane at any point lying on the axis normal to the plane & passing through the mid-point of the unequal side is given by the eq(9) as follows

$$\omega = 2\cos^{-1}\left\{\frac{h\left(2p^2\sqrt{4h^2+b^2} + b^2\sqrt{h^2+p^2}\right)}{h^2(4p^2+b^2)+p^2b^2}\right\}$$

Now, putting the values of b = base = BC = a& p = perpendicular = AO= $\frac{a\sqrt{3}}{2}$, we get

$$\omega = 2\cos^{-1}\left\{\frac{h\left(2\left(\frac{a\sqrt{3}}{2}\right)^2\sqrt{4h^2+a^2} + a^2\sqrt{h^2+\left(\frac{a\sqrt{3}}{2}\right)^2}\right)}{h^2\left(4\left(\frac{a\sqrt{3}}{2}\right)^2+a^2\right)+\left(\frac{a\sqrt{3}}{2}\right)^2a^2}\right\}$$

$$= 2\cos^{-1}\left\{\frac{2h\left(3a^2\sqrt{4h^2+a^2} + a^2\sqrt{4h^2+3a^2}\right)}{16h^2a^2+3a^4}\right\} = 2\cos^{-1}\left\{\frac{2h\left(3\sqrt{4h^2+a^2} + \sqrt{4h^2+3a^2}\right)}{16h^2+3a^2}\right\}$$

$$\Rightarrow \omega = 2\cos^{-1}\left\{\frac{2h\left(3\sqrt{4h^2+a^2} + \sqrt{4h^2+3a^2}\right)}{16h^2+3a^2}\right\} \quad \forall (h \geq 0) \ldots\ldots\ldots\ldots\ldots (14)$$

Important deduction:

The solid angle subtended by the equilateral triangular plane at a point lying on the mid-point of any of the equal sides is obtained by putting h = 0 in the above eq(14)

Thus, we get

$$\omega = 2\cos^{-1}\left\{\frac{2(0)\left(3\sqrt{4(0)+a^2} + \sqrt{4(0)+3a^2}\right)}{16(0)+3a^2}\right\} = 2\cos^{-1}\{0\} = 2\frac{\pi}{2} = \pi\ sr$$

*it's also true for any point lying on boundary of the plane.

2.3.10 Solid angle subtended by an equilateral triangular plane at any point lying on the axis normal to the plane & passing through the centre

Let there be a equilateral triangular plane ABC having each side equal to 'a' & a given point say P(0, 0, h) at a height 'h' lying on the axis normal to the plane & passing through the centre 'O' of the plane. (as shown in the figure(22) below)

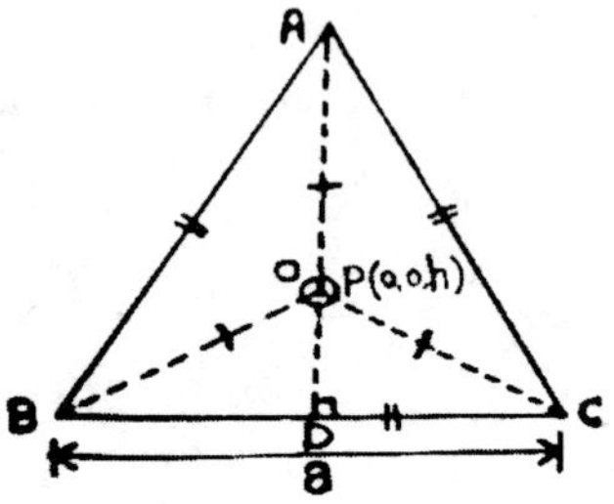

Fig (22)

We know that the area of the equilateral triangle is given as

$$\Delta = \frac{\sqrt{3}}{4}(side)^2 \Rightarrow \Delta = \frac{\sqrt{3}}{4}a^2 \ \& \ semiperemeter = \frac{sum\ of\ the\ sides}{2} \Rightarrow s = \frac{3a}{2}$$

$$\Rightarrow OD = inscribed\ radius = \frac{\Delta}{s} = \frac{\frac{\sqrt{3}}{4}a^2}{\frac{3a}{2}} = \frac{a}{2\sqrt{3}}$$

In equilateral $\Delta ABC, we\ have\ AB = BC = AC\ \&\ OA = OB = OC$

Thus, $\Delta AOB, \Delta BOC\ \&\ \Delta AOC$ are congruent isosceles triangles.

Hence, the solid angle subtended by the equilateral triangular plane ABC at the given point 'P'

$$\Rightarrow \omega = \omega_{\Delta AOB} + \omega_{\Delta BOC} + \omega_{\Delta AOC} = 3(\omega_{\Delta BOC})$$

We know that the solid angle subtended by the isosceles triangular plane at any point lying on the axis normal to the plane & passing through the common vertex of the equal sides is given by eq(7) as follows

$$\omega = 2\left[\sin^{-1}\left\{\frac{b}{\sqrt{b^2+4p^2}}\right\} - \sin^{-1}\left\{\left(\frac{b}{\sqrt{b^2+4p^2}}\right)\left(\frac{h}{\sqrt{h^2+p^2}}\right)\right\}\right]$$

Now, on putting the values of $b = base = BC = a\ \&\ p = perpendicular = OD = \frac{a}{2\sqrt{3}}$,

we get

$$\omega_{\Delta BOC} = 2\left[\sin^{-1}\left\{\frac{a}{\sqrt{a^2+4\left(\frac{a}{2\sqrt{3}}\right)^2}}\right\} - \sin^{-1}\left\{\left(\frac{a}{\sqrt{a^2+4\left(\frac{a}{2\sqrt{3}}\right)^2}}\right)\left(\frac{h}{\sqrt{h^2+\left(\frac{a}{2\sqrt{3}}\right)^2}}\right)\right\}\right]$$

$$= 2\left[\sin^{-1}\left\{\frac{a\sqrt{3}}{2a}\right\} - \sin^{-1}\left\{\left(\frac{a\sqrt{3}}{2a}\right)\left(\frac{2h\sqrt{3}}{\sqrt{12h^2+a^2}}\right)\right\}\right]$$

$$= 2\left[\sin^{-1}\left\{\frac{\sqrt{3}}{2}\right\} - \sin^{-1}\left\{\left(\frac{3h}{\sqrt{12h^2+a^2}}\right)\right\}\right] = 2\left[\frac{\pi}{3} - \sin^{-1}\left\{\left(\frac{3h}{\sqrt{12h^2+a^2}}\right)\right\}\right]$$

$$\Rightarrow \left[\omega = 3(\omega_{\Delta BOC}) = 2\pi - 6\sin^{-1}\left(\frac{3h}{\sqrt{12h^2+a^2}}\right)\right] \quad \forall (h \ge 0)\ \ldots..(15)$$

Important deduction:

The solid angle subtended by the equilateral triangular plane at a point lying on the centre of the plane is obtained by putting h = 0 in the above eq(15)

Thus, we get

$$\omega = 2\pi - 6\sin^{-1}\left(\frac{3(0)}{\sqrt{12(0)+a^2}}\right) = 2\pi - 2\sin^{-1}\{0\} = 2\pi \ sr$$

2.3.11 Solid angle subtended by a rectangular plane at any point lying on the axis normal to the plane & passing through the centre

Method 1: Using Fundamental Theorem

Let there be a rectangular plane ABCD having length 'l' & width 'b'& a given point say P(0, 0, h) at a height 'h' lying on the axis normal to the plane & passing through the centre 'O' of the plane. (i.e. lying on the Z-axis) (as shown in the figure(23) below)

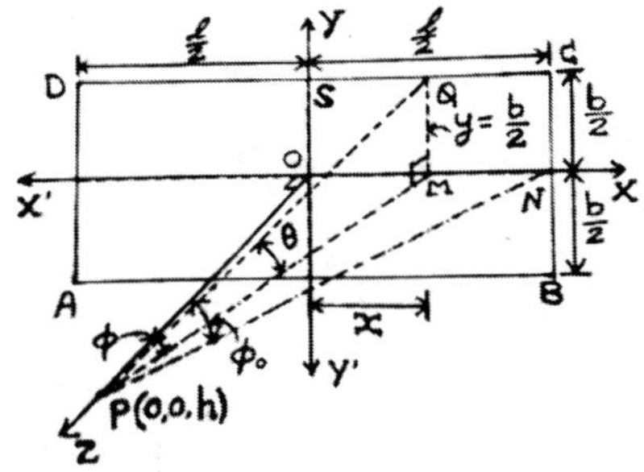

Fig (23)

In right ΔPOM

$$\Rightarrow sec\varphi = \frac{PM}{PO} \Rightarrow PM = POsec\varphi = hsec\varphi$$

In right ΔPMQ

$$\Rightarrow sin\theta = \frac{QM}{PQ} = \frac{QM}{\sqrt{QM^2+PM^2}} = \frac{\left(\frac{b}{2}\right)}{\sqrt{\left(\frac{b}{2}\right)^2 + h^2sec^2\varphi}}$$

$$sin\theta = \frac{bcos\varphi}{\sqrt{b^2cos^2\varphi + 4h^2}}$$

In right ΔPON

$$\Rightarrow sin\varphi_0 = \frac{ON}{PN} = \frac{ON}{\sqrt{ON^2+PO^2}} = \frac{\left(\frac{l}{2}\right)}{\sqrt{\left(\frac{l}{2}\right)^2 + h^2}} = \frac{l}{\sqrt{l^2+4h^2}}$$

Let's consider an imaginary spherical surface having radius 'R' & centre at the given point 'P' such that the area of projection of the given plane ABCD on the spherical surface w.r.t. the given point 'P' is 'A'

Now, consider an elementary area of projection 'dA' of the plane ONCS on the spherical surface in the first quadrant YOX (see the figure(24) below)

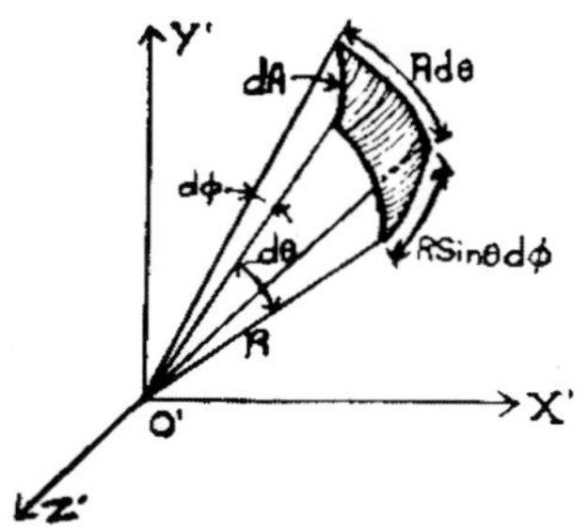

Fig (24)

Now, elementary area of projection in the first quadrant

$$\Rightarrow dA = (length)(width) = (Rsin\theta d\varphi)(Rd\theta) = R^2sin\theta d\theta d\varphi$$

Hence, area of projection of the plane ONCS on the spherical surface in the first quadrant is obtained by integrating the above expression in the first quadrant & applying the limits of $'\theta' from\ \theta_1 = \frac{\pi}{2} - \theta\ to\ \theta_2 = \frac{\pi}{2}$&$'\varphi' from\ \varphi_1 = 0\ to\ \varphi_2 = \ \varphi_0$ (see the figure (25) below)

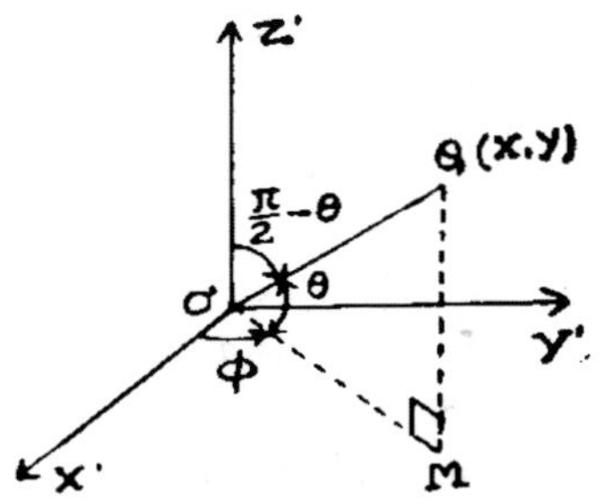

Fig (25)

$$\Rightarrow A' = \int_0^{\varphi_0}\left[\int_{\frac{\pi}{2}-\theta}^{\frac{\pi}{2}} R^2 sin\theta d\theta\right]d\varphi = R^2\int_0^{\varphi_0}\left[\int_{\frac{\pi}{2}-\theta}^{\frac{\pi}{2}} sin\theta d\theta\right]d\varphi$$

$$= R^2\int_0^{\varphi_0} d\varphi\,[-cos\theta]_{\frac{\pi}{2}-\theta}^{\frac{\pi}{2}} = R^2\int_0^{\varphi_0} sin\theta d\varphi = R^2\int_0^{\varphi_0}\frac{bcos\varphi}{\sqrt{b^2 - b^2sin^2\varphi + 4h^2}}d\varphi$$

$$\Rightarrow A' = R^2\int_0^{\varphi_0}\frac{bcos\varphi}{\sqrt{b^2 - b^2sin^2\varphi + 4h^2}}d\varphi\,(since,\ \ cos^2\alpha = 1 - sin^2\alpha)$$

$$= R^2\int_0^{\varphi_0}\frac{bcos\varphi}{b\sqrt{\frac{b^2+4h^2}{b^2} - sin^2\varphi}}d\varphi = R^2\int_0^{\varphi_0}\frac{cos\varphi}{\sqrt{K^2 - sin^2\varphi}}d\varphi$$

$$where\ K = \sqrt{\frac{b^2+4h^2}{b^2}} = \frac{\sqrt{b^2+4h^2}}{b}$$

$$\Rightarrow A' = R^2\left[\sin^{-1}\left\{\frac{sin\varphi}{K}\right\}\right]_0^{\varphi_0} = R^2\left[\sin^{-1}\left\{\frac{sin\varphi_0}{K}\right\} - 0\right] = R^2\sin^{-1}\left\{\frac{sin\varphi_0}{K}\right\}$$

Now, on putting the values of 'K' & $'sin\varphi_0'$, we get

$$\Rightarrow A' = R^2\sin^{-1}\left\{\frac{\frac{l}{\sqrt{l^2+4h^2}}}{\frac{\sqrt{b^2+4h^2}}{b}}\right\} = R^2\sin^{-1}\left\{\frac{lb}{\sqrt{(l^2+4h^2)(b^2+4h^2)}}\right\}$$

Hence, the total area of the projection of the rectangular plane ABCD

A = 4A' $(since, area\ of\ rectangle\ ABCD = 4(area\ of\ rectangle\ ONCS))$

$$\Rightarrow A = 4R^2\sin^{-1}\left\{\frac{lb}{\sqrt{(l^2+4h^2)(b^2+4h^2)}}\right\}$$

Now, the solid angle subtended by the rectangular plane ABCD at the given point 'P'

= solid angle subtended by the area of projection of rectangle ABCD on the spherical surface at the same point 'P'

$$\Rightarrow \omega = \int_s\frac{dA}{r^2} = \int_s\frac{dA}{R^2} = \frac{1}{R^2}\int_s dA = \frac{A}{R^2}$$

$(since, R\ is\ constant\ for\ each\ point\ on\ the\ area\ of\ projection\ on\ the\ spherical\ surface)$

$$\Rightarrow \omega = \frac{1}{R^2}\left[4R^2\sin^{-1}\left\{\frac{lb}{\sqrt{(l^2+4h^2)(b^2+4h^2)}}\right\}\right] = 4\sin^{-1}\left\{\frac{lb}{\sqrt{(l^2+4h^2)(b^2+4h^2)}}\right\}$$

$$\Rightarrow \left[\omega = 4\sin^{-1}\left\{\frac{lb}{\sqrt{(l^2+4h^2)(b^2+4h^2)}}\right\}\right]\quad \forall(h \geq 0) \ldots\ldots\ldots\ldots\ldots (16)$$

Method 2: Using Integral Theorem

We know that the solid angle subtended by an object at a given point in the space is given by Integral-Theorem as follows

$$\omega = \int_{\varphi_1}^{\varphi_2}\left|\int_{\theta_1}^{\theta_2} sin\theta d\theta\right|d\varphi$$

Hence, the solid angle subtended by the rectangular plane ABCD is obtained by applying the limits of the parameters $'\theta'$ & $'\varphi'$ for whole the plane ABCD divided into four symmetrical parts about the centre 'O' (also origin) (see the figure (25) above)

Now, applying limits of $'\theta'$ $from\ \theta_1 = \frac{\pi}{2} - \theta\ to\ \theta_2 = \frac{\pi}{2} + \theta \&$

$$'\varphi'\ from\ \varphi_1 = -\varphi_0\ to\ \varphi_2 = \varphi_0$$

$$\Rightarrow \omega = \int_{\varphi_1=-\varphi_0}^{\varphi_2=\varphi_0} \left| \int_{\theta_1=\frac{\pi}{2}-\theta}^{\theta_2=\frac{\pi}{2}+\theta} sin\theta d\theta \right| d\varphi = \int_{\varphi_1=-\varphi_0}^{\varphi_2=\varphi_0} \left| [-cos\theta]_{\frac{\pi}{2}-\theta}^{\frac{\pi}{2}+\theta} d\varphi \right|$$

$$= \int_{\varphi_1=-\varphi_0}^{\varphi_2=\varphi_0} \left| \left[-cos\left(\frac{\pi}{2}+\theta\right) + cos\left(\frac{\pi}{2}-\theta\right) \right] \right| d\varphi = 2\int_{\varphi_1=-\varphi_0}^{\varphi_2=\varphi_0} |sin\theta| d\varphi$$

on putting the value of $sin\theta$,we get

$$\omega = 2\int_{\varphi_1=-\varphi_0}^{\varphi_2=\varphi_0} \left| \frac{bcos\varphi}{\sqrt{b^2cos^2\varphi + 4h^2}} \right| d\varphi = 4\int_{\varphi_1=0}^{\varphi_2=\varphi_0} \left| \frac{bcos\varphi}{\sqrt{b^2cos^2\varphi + 4h^2}} \right| d\varphi$$

$$\left(since,\ \int_{-a}^{a} f(x)dx = 2\int_{0}^{a} f(x)dx\ iff\ f(-x) = f(x) \right)$$

$$\Rightarrow \omega = 4\int_{0}^{\varphi_0} \frac{bcos\varphi}{\sqrt{b^2cos^2\varphi + 4h^2}} d\varphi \left(since, 0 \le \varphi \le \frac{\pi}{2} \& |f(x)| = f(x)\ \ \forall f(x) > 0 \right)$$

$$= 4\int_{0}^{\varphi_0} \frac{bcos\varphi}{\sqrt{b^2 - b^2sin^2\varphi + 4h^2}} d\varphi = 4\int_{0}^{\varphi_0} \frac{bcos\varphi}{b\sqrt{\frac{b^2+4h^2}{b^2} - sin^2\varphi}} d\varphi$$

$$\Rightarrow \omega = 4\int_{0}^{\varphi_0} \frac{cos\varphi}{\sqrt{K^2 - sin^2\varphi}} d\varphi$$

$$where\ K = \sqrt{\frac{b^2+4h^2}{b^2}} = \frac{\sqrt{b^2+4h^2}}{b}$$

On solving the integration as follows

$$I = \int \frac{cos\varphi}{\sqrt{K^2 - sin^2\varphi}} d\varphi\ \ let\ sin\varphi = t \Rightarrow cos\varphi d\varphi = dt$$

$$\Rightarrow I = \int \frac{dt}{\sqrt{K^2 - t^2}} = \sin^{-1}\left(\frac{t}{K}\right) + C = \sin^{-1}\left(\frac{\sin\varphi}{K}\right) + C$$

Where, ‘C’ is the constant of integration.

Hence, we get

$$\Rightarrow \omega = 4\int_{0}^{\varphi_0} \frac{cos\varphi}{\sqrt{K^2 - sin^2\varphi}} d\varphi = 4[I]_0^{\varphi_0}$$

$$\Rightarrow \omega = 4\left[\sin^{-1}\left\{\frac{sin\varphi}{K}\right\} + C\right]_0^{\varphi_0} = 4\left[\sin^{-1}\left\{\frac{sin\varphi_0}{K}\right\} - 0\right] = 4\sin^{-1}\left\{\frac{sin\varphi_0}{K}\right\}$$

Now, on putting the values of ‘K’ & $'sin\varphi_0'$, we get

$$\Rightarrow \omega = 4\sin^{-1}\left\{\frac{\frac{l}{\sqrt{l^2+4h^2}}}{\frac{\sqrt{b^2+4h^2}}{b}}\right\} = 4\sin^{-1}\left\{\frac{lb}{\sqrt{(l^2+4h^2)(b^2+4h^2)}}\right\}\ \ \forall(h \ge 0) \dots\dots\dots\dots(16)$$

Method 3: Using Corollary

Let there be a rectangular plane ABCD having length ‘l’ & width ‘b’ & a given point say P(0, 0, h) at a height ‘h’ lying on the axis normal to the plane & passing through the centre ‘O’ of the plane. (i.e. lying on the Z-axis) (as shown in the figure (26) below)

The rectangular plane ABCD is divided into four identical rectangular planes w.r.t. the origin 'O' (also the centre of the plane ABCD)

Since, the rectangle ONCE is bounded by the straight lines x = 0, y = 0, x = $\frac{l}{2}$ & y = $\frac{b}{2}$

But $y = f(x) \neq 0$

Hence, the equation of the boundary-curve $y = f(x) = \frac{b}{2}\ \forall\left(0 \leq x \leq \frac{l}{2}\right)$

Let's consider a point Q(x, y) on the boundary curve (i.e. straight line OM) $y = \frac{b}{2}$

Now, in rightΔPOM

$$\Rightarrow tan\varphi = \frac{OM}{OP} = \frac{x}{h} \Rightarrow x = htan\varphi$$

$$\therefore f(htan\varphi) = \frac{b}{2}$$

in right ΔPON

$$\Rightarrow sin\varphi_0 = \frac{ON}{PN} = \frac{ON}{\sqrt{ON^2 + PO^2}} = \frac{\left(\frac{l}{2}\right)}{\sqrt{\left(\frac{l}{2}\right)^2 + h^2}} = \frac{l}{\sqrt{l^2 + h^2}}$$

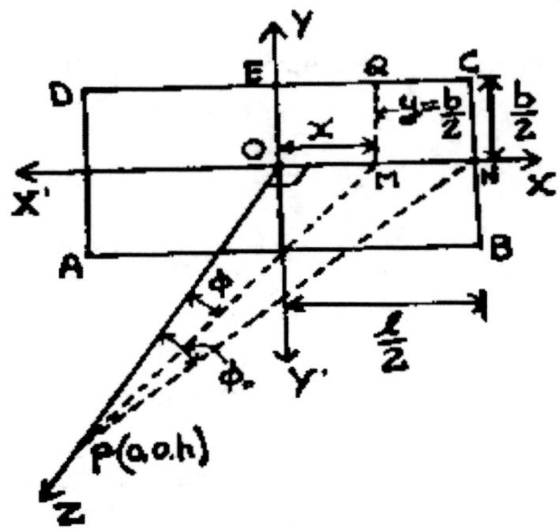

Fig (26)

Now, the solid angle subtended by the given rectangle ABCD at the given point 'P'

$$\Rightarrow \omega = 4\omega_{ONCB}\ (since, area\ of\ rectangle\ ABCD = 4(area\ of\ rectangle\ ONCE))$$

Hence, using Corollary

$$\Rightarrow \omega = \int_0^{\varphi} \frac{f(htan\varphi)d\varphi}{\sqrt{(f(htan\varphi))^2 + h^2sec^2\varphi}}$$

now, on putting the value of $f(htan\varphi)$& applying the limit of φ, we get

$$\omega_{ONCB} = \int_0^{\varphi=\varphi_0} \frac{\left(\frac{b}{2}\right)d\varphi}{\sqrt{\left(\frac{b}{2}\right)^2 + h^2sec^2\varphi}} = \int_0^{\varphi=\varphi_0} \frac{bd\varphi}{\sqrt{b^2 + \frac{4h^2}{cos^2\varphi}}} = \int_0^{\varphi=\varphi_0} \frac{bcos\varphi d\varphi}{\sqrt{b^2cos^2\varphi + 4h^2}}$$

$$= \int_0^{\varphi_0} \frac{bcos\varphi}{\sqrt{b^2 - b^2sin^2\varphi + 4h^2}} d\varphi$$

$$= \int_0^{\varphi_0} \frac{bcos\varphi}{b\sqrt{\frac{b^2 + 4h^2}{b^2} - sin^2\varphi}} d\varphi = \int_0^{\varphi_0} \frac{cos\varphi}{\sqrt{K^2 - sin^2\varphi}} d\varphi$$

$$where\ K = \sqrt{\frac{b^2 + 4h^2}{b^2}} = \frac{\sqrt{b^2 + 4h^2}}{b}$$

$$\Rightarrow \omega = \left[\sin^{-1}\left\{\frac{sin\varphi}{K}\right\}\right]_0^{\varphi_0} = \left[\sin^{-1}\left\{\frac{sin\varphi_0}{K}\right\} - 0\right] = \sin^{-1}\left\{\frac{sin\varphi_0}{K}\right\}$$

Now, on putting the values of 'K' & $'sin\varphi_0'$, we get

$$\Rightarrow \omega_{ONCB} = \sin^{-1}\left\{\frac{\frac{l}{\sqrt{l^2 + 4h^2}}}{\frac{\sqrt{b^2 + 4h^2}}{b}}\right\} = \sin^{-1}\left\{\frac{lb}{\sqrt{(l^2 + 4h^2)(b^2 + 4h^2)}}\right\}$$

Hence, the solid angle subtended by the rectangular plane ABCD at the given point 'P'

$$\Rightarrow \omega = 4\omega_{ONCB} = 4\sin^{-1}\left\{\frac{lb}{\sqrt{(l^2+4h^2)(b^2+4h^2)}}\right\} \quad \forall(h \geq 0) \ldots\ldots\ldots\ldots\ldots\ldots\ldots(16)$$

Method 4: Alternative

Let there be a rectangular plane ABCD having length 'l' & width 'b' & a given point say P(0, 0, h) at a height 'h' lying on the axis normal to the plane & passing through the centre 'O' of the plane. (i.e. lying on the Z-axis) (as shown in the figure (27) below)

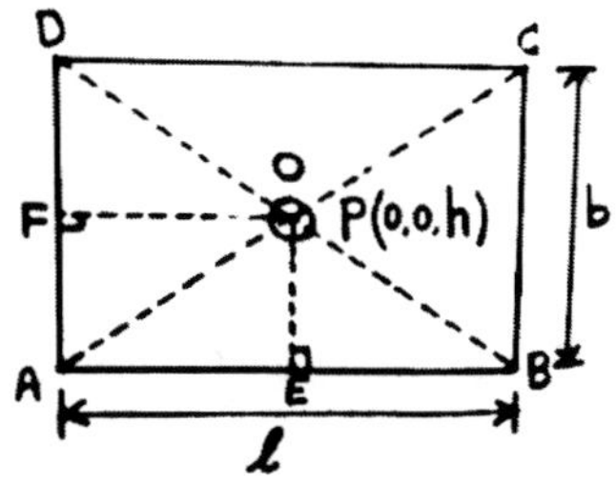

Fig (27)

Now, draw the diagonals AC & BD and the perpendiculars OE & OF from the centre 'O' to the sides AB & AD respectively which are bisecting to each-other at the centre 'O'.

In rectangle ABCD we have

$$OA = OB = OC = OD = \frac{\sqrt{l^2+b^2}}{2}, \qquad AB = CD = l \ \& \ AD = BC = b$$

Thus, ΔAOB & ΔCOD *as well as* ΔAOD & ΔBOC are congruent isosceles triangles which the rectangle ABCD consists of.

Hence, the solid angle subtended by the rectangular plane ABCD at the given point 'P'

$$\Rightarrow \omega = \omega_{\Delta AOB} + \omega_{\Delta COD} + \omega_{\Delta AOD} + \omega_{\Delta BOC} = 2(\omega_{\Delta AOB} + \omega_{\Delta AOD})$$

We know that the solid angle subtended by the isosceles triangular plane at any point lying on the axis normal to the plane & passing through the common vertex of the equal sides is given by eq(7) as follows

$$\omega = 2\left[\sin^{-1}\left\{\frac{b}{\sqrt{b^2+4p^2}}\right\} - \sin^{-1}\left\{\left(\frac{b}{\sqrt{b^2+4p^2}}\right)\left(\frac{h}{\sqrt{h^2+p^2}}\right)\right\}\right]$$

Now, on putting the values of b = base = AB = l & p = perpendicular = OE = $\frac{b}{2}$, we get

$$\omega_{\Delta AOB} = 2\left[\sin^{-1}\left\{\frac{l}{\sqrt{l^2+4\left(\frac{b}{2}\right)^2}}\right\} - \sin^{-1}\left\{\left(\frac{l}{\sqrt{l^2+4\left(\frac{b}{2}\right)^2}}\right)\left(\frac{h}{\sqrt{h^2+\left(\frac{b}{2}\right)^2}}\right)\right\}\right]$$

$$= 2\left[\sin^{-1}\left\{\frac{l}{\sqrt{l^2+b^2}}\right\} - \sin^{-1}\left\{\left(\frac{l}{\sqrt{l^2+b^2}}\right)\left(\frac{2h}{\sqrt{b^2+4h^2}}\right)\right\}\right]$$

Similarly, on putting the values of b = base = AD = b & p = perpendicular = OF = $\frac{l}{2}$, we get

$$\omega_{\Delta AOD} = 2\left[\sin^{-1}\left\{\frac{b}{\sqrt{b^2+4\left(\frac{l}{2}\right)^2}}\right\} - \sin^{-1}\left\{\left(\frac{b}{\sqrt{b^2+4\left(\frac{l}{2}\right)^2}}\right)\left(\frac{h}{\sqrt{h^2+\left(\frac{l}{2}\right)^2}}\right)\right\}\right]$$

$$= 2\left[\sin^{-1}\left\{\frac{b}{\sqrt{l^2+b^2}}\right\} - \sin^{-1}\left\{\left(\frac{b}{\sqrt{l^2+b^2}}\right)\left(\frac{2h}{\sqrt{l^2+4h^2}}\right)\right\}\right]$$

$$\therefore \omega = 2(\omega_{\Delta AOB} + \omega_{\Delta AOD})$$

Now, on putting the values, we have

$$\omega = 4\left[\sin^{-1}\left\{\frac{l}{\sqrt{l^2+b^2}}\right\} - \sin^{-1}\left\{\left(\frac{l}{\sqrt{l^2+b^2}}\right)\left(\frac{2h}{\sqrt{b^2+4h^2}}\right)\right\}\right]$$
$$+4\left[\sin^{-1}\left\{\frac{b}{\sqrt{l^2+b^2}}\right\} - \sin^{-1}\left\{\left(\frac{b}{\sqrt{l^2+b^2}}\right)\left(\frac{2h}{\sqrt{l^2+4h^2}}\right)\right\}\right]$$

$$\Rightarrow \omega = 4\left[\sin^{-1}\left\{\frac{l}{\sqrt{l^2+b^2}}\right\} + \sin^{-1}\left\{\frac{b}{\sqrt{l^2+b^2}}\right\}\right]$$

$$-4\left[\sin^{-1}\left\{\left(\frac{l}{\sqrt{l^2+b^2}}\right)\left(\frac{2h}{\sqrt{b^2+4h^2}}\right)\right\} + \sin^{-1}\left\{\left(\frac{b}{\sqrt{l^2+b^2}}\right)\left(\frac{2h}{\sqrt{l^2+4h^2}}\right)\right\}\right]$$

$$\Rightarrow \omega = 4\sin^{-1}\left\{\frac{l}{\sqrt{l^2+b^2}}\frac{l}{\sqrt{l^2+b^2}} + \frac{b}{\sqrt{l^2+b^2}}\frac{b}{\sqrt{l^2+b^2}}\right\}$$

$$-4\sin^{-1}\left\{\left(\frac{l}{\sqrt{l^2+b^2}}\right)\left(\frac{2h}{\sqrt{b^2+4h^2}}\right)\frac{l\sqrt{l^2+b^2+4h^2}}{\sqrt{l^2+b^2}\sqrt{l^2+4h^2}} + \left(\frac{b}{\sqrt{l^2+b^2}}\right)\left(\frac{2h}{\sqrt{l^2+4h^2}}\right)\frac{b\sqrt{l^2+b^2+4h^2}}{\sqrt{l^2+b^2}\sqrt{b^2+4h^2}}\right\}$$

$$\left(since,\ \ \sin^{-1}x + \sin^{-1}y = \sin^{-1}\left(x\sqrt{1-y^2} + y\sqrt{1-x^2}\right)\right)\quad \forall(-1 \le (x,y) \le 1)$$

$$\Rightarrow \omega = 4\left[\sin^{-1}\left\{\frac{l^2+b^2}{l^2+b^2}\right\} - \sin^{-1}\left\{\frac{2h(l^2+b^2)\sqrt{l^2+b^2+4h^2}}{(l^2+b^2)\sqrt{l^2+4h^2}\sqrt{b^2+4h^2}}\right\}\right]$$

$$\Rightarrow \omega = 4\left[\sin^{-1}\{1\} - \sin^{-1}\left\{\frac{2h\sqrt{l^2+b^2+4h^2}}{\sqrt{l^2+4h^2}\sqrt{b^2+4h^2}}\right\}\right]$$

$$= 4\left[\frac{\pi}{2} - \sin^{-1}\left\{\frac{2h\sqrt{l^2+b^2+4h^2}}{\sqrt{l^2+4h^2}\sqrt{b^2+4h^2}}\right\}\right] = 4\cos^{-1}\left\{\frac{2h\sqrt{l^2+b^2+4h^2}}{\sqrt{l^2+4h^2}\sqrt{b^2+4h^2}}\right\}$$

$$\left(since,\ \frac{\pi}{2} - \sin^{-1}x = \cos^{-1}x \qquad \forall(-1 \le x \le 1)\right)$$

$$\omega = 4\sin^{-1}\left\{\frac{lb}{\sqrt{l^2+4h^2}\sqrt{b^2+4h^2}}\right\}$$

$$\left(since, \cos^{-1}x = \sin^{-1}\left(\sqrt{1-x^2}\right) \qquad \forall(-1 \le x \le 1)\right)$$

$$\Rightarrow \left[\omega = 4\sin^{-1}\left\{\frac{lb}{\sqrt{(l^2+4h^2)(b^2+4h^2)}}\right\}\right] \quad \forall(h \ge 0) \quad \ldots\ldots\ldots\ldots\ldots\ldots\ldots (16)$$

Important deduction:

The solid angle subtended by the rectangular plane at a point lying on the centre of the plane is obtained by putting h = 0 in the above eq(16), hence we get

$$\omega = 4\sin^{-1}\left\{\frac{lb}{\sqrt{(l^2+4(0))(b^2+4(0))}}\right\} = 4\sin^{-1}\left\{\frac{lb}{lb}\right\} = 4\sin^{-1}\{1\} = 4\frac{\pi}{2} = 2\pi\ sr$$

*It's also true for any point lying inside the boundary of the plane.

2.3.12 Solid angle subtended by a rectangular plane at any point lying on the axis normal to the plane & passing through one of the vertices

Let there be a rectangular plane ABCD having length 'l' & width 'b' & a given point say P(0, 0, h) at a height 'h' lying on the axis normal to the plane & passing through the vertex say 'A' of the plane. (as shown in the figure (28) below)

Now, draw the diagonal AC to obtain two right triangles ΔABC & ΔADC which the rectangle ABCD consists of.

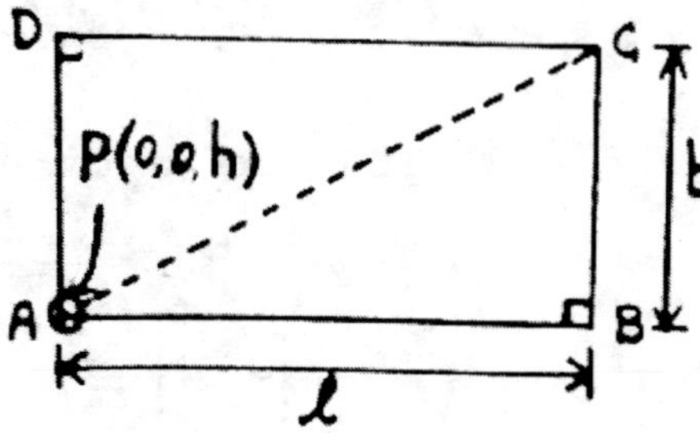

Fig (28)

Thus, the solid angle subtended by the rectangular plane ABCD at the given point 'P'

$$\Rightarrow \omega = \omega_{\Delta ABC} + \omega_{\Delta ADC}$$

We know that the solid angle subtended by the right triangular plane at any point lying on the axis normal to the plane & passing through the acute angled vertex is given by eq(2) as follows

$$\omega = \sin^{-1}\left\{\frac{b}{\sqrt{b^2+p^2}}\right\} - \sin^{-1}\left\{\left(\frac{b}{\sqrt{b^2+p^2}}\right)\left(\frac{h}{\sqrt{h^2+p^2}}\right)\right\}$$

Now, on putting the values of b = base = BC = b & p = perpendicular = AB = l, we get

$$\omega_{\Delta ABC} = \sin^{-1}\left\{\frac{b}{\sqrt{b^2+l^2}}\right\} - \sin^{-1}\left\{\left(\frac{b}{\sqrt{b^2+l^2}}\right)\left(\frac{h}{\sqrt{h^2+l^2}}\right)\right\}$$

$$\Rightarrow \omega_{\Delta ABC} = \sin^{-1}\left\{\frac{b}{\sqrt{l^2+b^2}}\right\} - \sin^{-1}\left\{\left(\frac{b}{\sqrt{l^2+b^2}}\right)\left(\frac{h}{\sqrt{l^2+h^2}}\right)\right\}$$

Similarly, on putting the values of b = base = BC = b & p = perpendicular = AB = l, we get

$$\omega_{\Delta ADC} = \sin^{-1}\left\{\frac{l}{\sqrt{l^2+b^2}}\right\} - \sin^{-1}\left\{\left(\frac{l}{\sqrt{l^2+b^2}}\right)\left(\frac{h}{\sqrt{b^2+h^2}}\right)\right\}$$

$$\Rightarrow \omega = \omega_{\Delta ABC} + \omega_{\Delta ADC} = \sin^{-1}\left\{\frac{b}{\sqrt{l^2+b^2}}\right\} - \sin^{-1}\left\{\left(\frac{b}{\sqrt{l^2+b^2}}\right)\left(\frac{h}{\sqrt{l^2+h^2}}\right)\right\}$$

$$+ \sin^{-1}\left\{\frac{l}{\sqrt{l^2+b^2}}\right\} - \sin^{-1}\left\{\left(\frac{l}{\sqrt{l^2+b^2}}\right)\left(\frac{h}{\sqrt{b^2+h^2}}\right)\right\}$$

$$\Rightarrow \omega = \left[\sin^{-1}\left\{\frac{l}{\sqrt{l^2+b^2}}\right\} + \sin^{-1}\left\{\frac{b}{\sqrt{l^2+b^2}}\right\}\right]$$

$$-\left[\sin^{-1}\left\{\left(\frac{l}{\sqrt{l^2+b^2}}\right)\left(\frac{h}{\sqrt{b^2+h^2}}\right)\right\} + \sin^{-1}\left\{\left(\frac{b}{\sqrt{l^2+b^2}}\right)\left(\frac{h}{\sqrt{l^2+h^2}}\right)\right\}\right]$$

$$\Rightarrow \omega = \left[\sin^{-1}\left\{\frac{l}{\sqrt{l^2+b^2}}\frac{l}{\sqrt{l^2+b^2}} + \frac{b}{\sqrt{l^2+b^2}}\frac{b}{\sqrt{l^2+b^2}}\right\}\right]$$

$$-\left[\sin^{-1}\left\{\left(\frac{l}{\sqrt{l^2+b^2}}\right)\left(\frac{h}{\sqrt{b^2+h^2}}\right)\frac{l\sqrt{l^2+b^2+h^2}}{\sqrt{l^2+b^2}\sqrt{l^2+h^2}} + \left(\frac{b}{\sqrt{l^2+b^2}}\right)\left(\frac{h}{\sqrt{l^2+h^2}}\right)\frac{b\sqrt{l^2+b^2+h^2}}{\sqrt{l^2+b^2}\sqrt{b^2+h^2}}\right\}\right]$$

$$\left(since,\ \ \sin^{-1}x + \sin^{-1}y = \sin^{-1}\left(x\sqrt{1-y^2} + y\sqrt{1-x^2}\right)\right) \quad \forall(-1 \le (x,y) \le 1)$$

$$\Rightarrow \omega = \left[\sin^{-1}\left\{\frac{l^2+b^2}{l^2+b^2}\right\}\right] - \left[\sin^{-1}\left\{\frac{h(l^2+b^2)\sqrt{l^2+b^2+h^2}}{(l^2+b^2)\sqrt{l^2+h^2}\sqrt{b^2+h^2}}\right\}\right]$$

$$\Rightarrow \omega = \sin^{-1}\{1\} - \sin^{-1}\left\{\frac{h\sqrt{l^2+b^2+h^2}}{\sqrt{l^2+h^2}\sqrt{b^2+h^2}}\right\} = \frac{\pi}{2} - \sin^{-1}\left\{\frac{h\sqrt{l^2+b^2+h^2}}{\sqrt{l^2+h^2}\sqrt{b^2+h^2}}\right\}$$

$$\left(since,\ \frac{\pi}{2} - \sin^{-1}x = \cos^{-1}x \quad \forall(-1 \le x \le 1)\right)$$

$$= \cos^{-1}\left\{\frac{h\sqrt{l^2+b^2+h^2}}{\sqrt{l^2+h^2}\sqrt{b^2+h^2}}\right\} = \sin^{-1}\left\{\frac{lb}{\sqrt{l^2+h^2}\sqrt{b^2+h^2}}\right\}$$

$$\left(since, \cos^{-1}x = \sin^{-1}\left(\sqrt{1-x^2}\right) \quad \forall(-1 \le x \le 1)\right)$$

$$\Rightarrow \left[\omega = sin^{-1}\left\{\frac{lb}{\sqrt{(l^2+h^2)(b^2+h^2)}}\right\}\right] \quad \forall(h \ge 0) \ldots\ldots\ldots\ldots\ldots (17)$$

Method 2:

Consider, a large rectangular plane CEFG having length '2l' , width '2b' & centre at the vertex 'A' of the given rectangular plane ABCD such that its sides are parallel to the corresponding sides of the given plane ABCD. (as shown in the figure (29) below)

In this case, area of the given rectangle ABCD is one-fourth of that of the rectangle CEFG

Hence, the solid angle subtended by the given rectangular plane ABCD at the given point 'P'

$$\Rightarrow \omega_{ABCD} = \frac{1}{4}\omega_{CEFG}$$

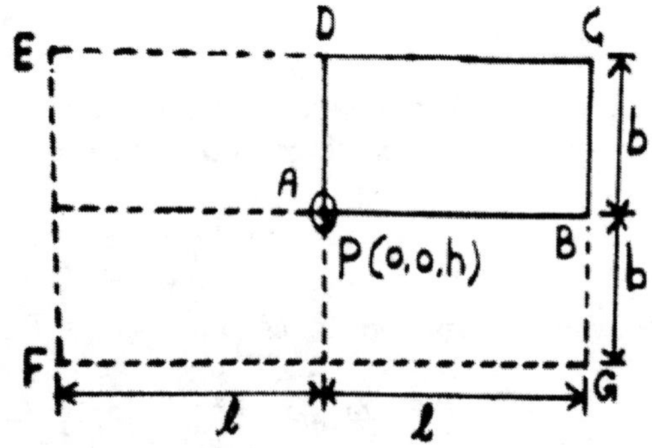

Fig (29)

We know that the solid angle subtended by the rectangular plane at any point lying on the axis normal to the plane & passing through the centre is given by eq(16) as follows

$$\omega = 4\sin^{-1}\left\{\frac{lb}{\sqrt{(l^2+4h^2)(b^2+4h^2)}}\right\}$$

Now, on putting the values of l = length = EC=FG = 2l & b = width = EF = CG = 2b, we get

$$\omega_{CEFG} = 4\sin^{-1}\left\{\frac{(2l)(2b)}{\sqrt{((2l)^2+4h^2)((2b)^2+4h^2)}}\right\} = 4\sin^{-1}\left\{\frac{lb}{\sqrt{(l^2+h^2)(b^2+h^2)}}\right\}$$

Hence,

$$\Rightarrow \omega_{ABCD} = \frac{1}{4}\omega_{CEFG} = \frac{1}{4}\left[4\sin^{-1}\left\{\frac{lb}{\sqrt{(l^2+h^2)(b^2+h^2)}}\right\}\right]$$

$$= \sin^{-1}\left\{\frac{lb}{\sqrt{(l^2+h^2)(b^2+h^2)}}\right\}$$

$$\Rightarrow \left[\omega = \sin^{-1}\left\{\frac{lb}{\sqrt{(l^2+h^2)(b^2+h^2)}}\right\}\right] \quad \forall(h \geq 0) \ldots\ldots\ldots\ldots\ldots\ldots\ldots(17)$$

The above eq(17) is standard formula used for solving the numerous problems related to the analysis of the rectangular plane.

Important deduction:

The solid angle subtended by the rectangular plane at a point lying on the vertex of the plane is obtained by putting h = 0 in the above eq(17), hence we have

$$\omega = \sin^{-1}\left\{\frac{lb}{\sqrt{(l^2+0)(b^2+0)}}\right\} = \sin^{-1}\left\{\frac{lb}{lb}\right\} = \sin^{-1}\{1\} = \frac{\pi}{2}\ sr$$

Also, if we divide a plane into four equiangular segments w.r.t. any point lying on the plane inside the boundary then each segment can be treated as the rectangular plane for the same point lying on the plane and the solid angle subtended by each segment (rectangular plane) at the same point (vertex) is given as follows

$$\omega = \frac{2\pi}{n} = \frac{2\pi}{4} = \frac{\pi}{2}\ sr$$

2.3.13 Solid angle subtended by a rectangular plane at any point lying on the axis normal to the plane & passing through mid-point of the side

Let there be a rectangular plane ABCD having length 'l' & width 'b' & a given point say P(0, 0, h) at a height 'h' lying on the axis normal to the plane & passing through the mid-point 'O' of the side say AB. (as shown in the figure (30) below)

Now, draw a perpendicular OE from the mid-point 'O' to the opposite side CD

In rectangle ABCD, we have

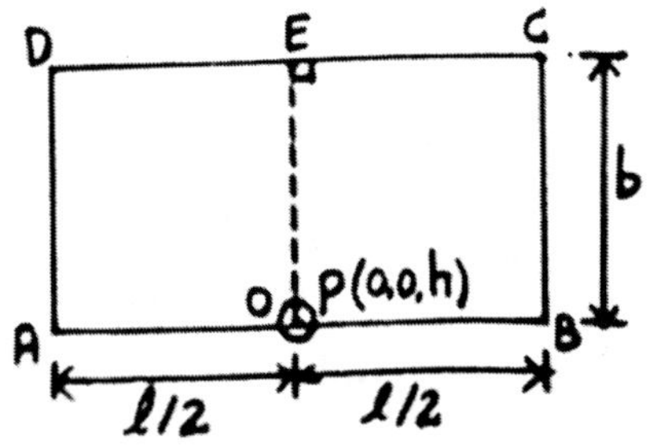

Fig (30)

$$OA = OB = EC = ED = \frac{l}{2} \ \& \ AD = OE = BC = b$$

Thus, we have two congruent rectangles OADE & OBCE which the given rectangle ABCD consists of.

Hence, the solid angle subtended by the given rectangular plane ABCD at the given point 'P'

$$\Rightarrow \omega = \omega_{OADE} + \omega_{OBDE} = 2(\omega_{OADE})$$

We know that the solid angle subtended by the rectangular plane at any point lying on the axis normal to the plane & passing through the vertex is given by the eq(17) as follows

$$\omega = \sin^{-1}\left\{\frac{lb}{\sqrt{(l^2+h^2)(b^2+h^2)}}\right\}$$

Now, on putting the values ofl = length = OA $= \frac{l}{2}$ & b = width = OE = b, we get

$$\omega_{OADE} = \sin^{-1}\left\{\frac{\left(\frac{l}{2}\right)b}{\sqrt{\left(\left(\frac{l}{2}\right)^2+h^2\right)(b^2+h^2)}}\right\} = \sin^{-1}\left\{\frac{lb}{\sqrt{(l^2+4h^2)(b^2+h^2)}}\right\}$$

$$\Rightarrow \omega = 2(\omega_{OADE}) = 2\sin^{-1}\left\{\frac{lb}{\sqrt{(l^2+4h^2)(b^2+h^2)}}\right\}$$

$$\Rightarrow \left[\omega = 2\sin^{-1}\left\{\frac{lb}{\sqrt{(l^2+4h^2)(b^2+h^2)}}\right\}\right] \quad \forall (h \geq 0) \ldots\ldots\ldots\ldots\ldots\ldots (18)$$

Important deduction 1:

If the given point P(0, 0, h) is lying on the axis normal to the plane & passing through the mid-point 'O' of the side say AD (as shown in the figure (31) below)

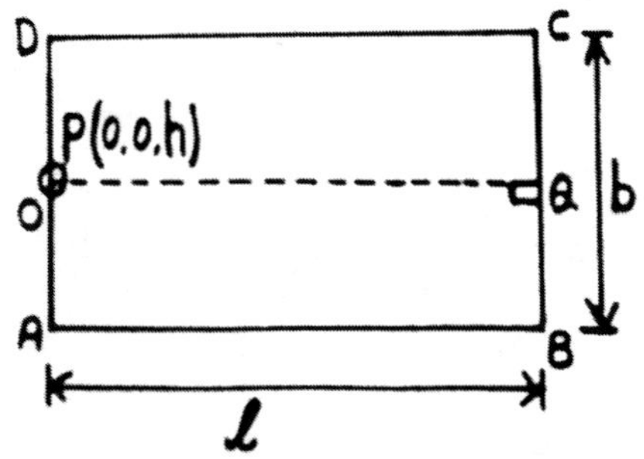

Fig (31)

In this case, the solid angle subtended by the rectangular plane ABCD at the given point 'P' is obtained by putting l = b& b = l in the eq(18) as follows

$$\omega = 2\sin^{-1}\left\{\frac{bl}{\sqrt{(b^2+4h^2)(l^2+h^2)}}\right\} = 2\sin^{-1}\left\{\frac{lb}{\sqrt{(l^2+h^2)(b^2+4h^2)}}\right\}$$

$$\Rightarrow \left[\omega = 2\sin^{-1}\left\{\frac{lb}{\sqrt{(l^2+h^2)(b^2+4h^2)}}\right\}\right] \quad \forall(h \geq 0) \ldots\ldots\ldots\ldots\ldots\ldots\ldots\ldots\ldots (19)$$

Important deduction 2:

The solid angle subtended by the rectangular plane at any point lying on the boundary of the plane is obtained by putting h = 0 in the above eq(18) or (19) as follows

$$\omega = 2\sin^{-1}\left\{\frac{lb}{\sqrt{(l^2+0)(b^2+0)}}\right\} = 2\sin^{-1}\left\{\frac{lb}{lb}\right\} = 2\sin^{-1}\{1\} = 2\frac{\pi}{2} = \pi\ sr$$

2.3.14 Solid angle subtended by a rectangular plane at any point lying on the axis normal to the plane & passing through any point within the boundary

Let there be a rectangular plane ABCD having length 'l' & width 'b' & a given point say P(0, 0, h) at a height 'h' lying on the axis normal to the plane & passing through any point say 'O' within the boundary of the plane. (as shown in the figure (32) below)

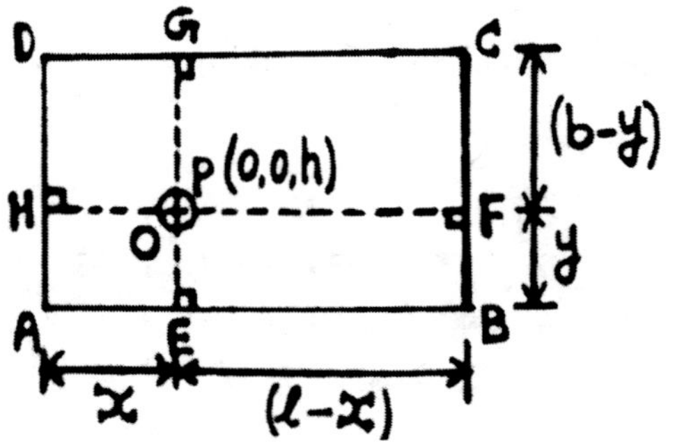

Fig (32)

Now, draw the perpendiculars OE, OF, OG & OH from the point 'O' to the sides AB, BC, CD & AD respectively

Thus, the rectangle ABCD is divided into four unequal rectangles AEOH, BEOF, OGDH & OGCF such that

$$\text{AE} = \text{HO} = \text{DG} = x, \text{EB} = \text{OF} = \text{GC} = (l-x), HA = OE = FB = y\ \&$$
$$HD = OG = FC = (b-y)$$

$$\Rightarrow \omega = \omega_{AEOH} + \omega_{BEOF} + \omega_{OGDH} + \omega_{OGCF}$$

We know that the solid angle subtended by the rectangular plane at any point lying on the axis normal to the plane & passing through the vertex is given by the eq(17) as follows

$$\omega = \sin^{-1}\left\{\frac{lb}{\sqrt{(l^2+h^2)(b^2+h^2)}}\right\}$$

Now, on putting the values of length 'l' & the width 'b', we get

$$\Rightarrow \omega = \sin^{-1}\left\{\frac{xy}{\sqrt{(x^2+h^2)(y^2+h^2)}}\right\} + \sin^{-1}\left\{\frac{(l-x)y}{\sqrt{((l-x)^2+h^2)(y^2+h^2)}}\right\}$$
$$+\sin^{-1}\left\{\frac{x(b-y)}{\sqrt{(x^2+h^2)((b-y)^2+h^2)}}\right\} + \sin^{-1}\left\{\frac{(l-x)(b-y)}{\sqrt{((l-x)^2+h^2)((b-y)^2+h^2)}}\right\}$$
$$\forall(h \geq 0) \ldots\ldots\ldots\ldots\ldots (20)$$

Note: Above formula is in most generalised form for a rectangular plane-figure. It is very significant for the solid angle analysis of the rectangular planes. All the formulae can be obtained from this formula by applying the different conditions as follows

Case 1: The solid angle subtended by the rectangular plane at any point lying on the axis normal to the plane & passing through the centre of the plane is obtained by putting $x = \frac{l}{2}$ & $y = \frac{b}{2}$ in the above eq(20), hence we get

$$\Rightarrow \omega = \sin^{-1}\left\{\frac{\left(\frac{l}{2}\right)\left(\frac{b}{2}\right)}{\sqrt{\left(\left(\frac{l}{2}\right)^2 + h^2\right)\left(\left(\frac{b}{2}\right)^2 + h^2\right)}}\right\} + \sin^{-1}\left\{\frac{\left(l-\frac{l}{2}\right)\left(\frac{b}{2}\right)}{\sqrt{\left(\left(l-\frac{l}{2}\right)^2 + h^2\right)\left(\left(\frac{b}{2}\right)^2 + h^2\right)}}\right\}$$

$$+\sin^{-1}\left\{\frac{\left(\frac{l}{2}\right)\left(b-\frac{b}{2}\right)}{\sqrt{\left(\left(\frac{l}{2}\right)^2 + h^2\right)\left(\left(b-\frac{b}{2}\right)^2 + h^2\right)}}\right\} + \sin^{-1}\left\{\frac{\left(l-\frac{l}{2}\right)\left(b-\frac{b}{2}\right)}{\sqrt{\left(\left(l-\frac{l}{2}\right)^2 + h^2\right)\left(\left(b-\frac{b}{2}\right)^2 + h^2\right)}}\right\}$$

$$\Rightarrow \omega = \sin^{-1}\left\{\frac{lb}{\sqrt{(l^2+4h^2)(b^2+4h^2)}}\right\} + \sin^{-1}\left\{\frac{lb}{\sqrt{(l^2+4h^2)(b^2+4h^2)}}\right\}$$

$$+\sin^{-1}\left\{\frac{lb}{\sqrt{(l^2+4h^2)(b^2+4h^2)}}\right\} + \sin^{-1}\left\{\frac{lb}{\sqrt{(l^2+4h^2)(b^2+4h^2)}}\right\}$$

$$\Rightarrow \omega = 4\sin^{-1}\left\{\frac{lb}{\sqrt{(l^2+4h^2)(b^2+4h^2)}}\right\}$$

Thus, result obtained is same to the eq(16).

Case 2: The solid angle subtended by the rectangular plane at any point lying on the axis normal to the plane & passing through the vertex of the plane is obtained by putting $x = 0$ & $y = 0$ in the above eq(20), hence we get

$$\Rightarrow \omega = \sin^{-1}\left\{\frac{(0)(0)}{\sqrt{((0)^2+h^2)((0)^2+h^2)}}\right\} + \sin^{-1}\left\{\frac{(l-0)(0)}{\sqrt{((l-0)^2+h^2)((0)^2+h^2)}}\right\}$$

$$+\sin^{-1}\left\{\frac{(0)(b-0)}{\sqrt{((0)^2+h^2)((b-0)^2+h^2)}}\right\} + \sin^{-1}\left\{\frac{(l-0)(b-0)}{\sqrt{((l-0)^2+h^2)((b-0)^2+h^2)}}\right\}$$

$$\Rightarrow \omega = \sin^{-1}\{0\} + \sin^{-1}\{0\}$$

$$+\sin^{-1}\{0\} + \sin^{-1}\left\{\frac{lb}{\sqrt{(l^2+h^2)(b^2+h^2)}}\right\}$$

$$\Rightarrow \omega = \sin^{-1}\left\{\frac{lb}{\sqrt{(l^2+h^2)(b^2+h^2)}}\right\}$$

Thus, result obtained is same to the eq(17).

Case 3: The solid angle subtended by the rectangular plane at any point lying on the axis normal to the plane & passing through the mid-point of longer side is obtained by putting

$x = \frac{l}{2}$ & $y = 0$ in the above eq(20), hence we get

$$\Rightarrow \omega = \sin^{-1}\left\{\frac{\left(\frac{l}{2}\right)(0)}{\sqrt{\left(\left(\frac{l}{2}\right)^2 + h^2\right)((0)^2 + h^2)}}\right\} + \sin^{-1}\left\{\frac{\left(l - \frac{l}{2}\right)(0)}{\sqrt{\left(\left(l - \frac{l}{2}\right)^2 + h^2\right)((0)^2 + h^2)}}\right\}$$

$$+\sin^{-1}\left\{\frac{\left(\frac{l}{2}\right)(b - 0)}{\sqrt{\left(\left(\frac{l}{2}\right)^2 + h^2\right)((b - 0)^2 + h^2)}}\right\} + \sin^{-1}\left\{\frac{\left(l - \frac{l}{2}\right)(b - 0)}{\sqrt{\left(\left(l - \frac{l}{2}\right)^2 + h^2\right)((b - 0)^2 + h^2)}}\right\}$$

$$\Rightarrow \omega = \sin^{-1}\{0\} + \sin^{-1}\{0\}$$

$$+\sin^{-1}\left\{\frac{lb}{\sqrt{(l^2 + 4h^2)(b^2 + h^2)}}\right\} + \sin^{-1}\left\{\frac{lb}{\sqrt{(l^2 + 4h^2)(b^2 + h^2)}}\right\}$$

$$\Rightarrow \omega = 2\sin^{-1}\left\{\frac{lb}{\sqrt{(l^2 + 4h^2)(b^2 + h^2)}}\right\}$$

Thus, result obtained is same to the eq(18)

Case 4: The solid angle subtended by the rectangular plane at any point lying on the axis normal to the plane & passing through the mid-point of smaller side is obtained by putting $x = 0 \;\&\; y = \frac{b}{2}$ in the above eq(20), hence we get

$$\Rightarrow \omega = \sin^{-1}\left\{\frac{(0)\left(\frac{b}{2}\right)}{\sqrt{((0)^2 + h^2)\left(\left(\frac{b}{2}\right)^2 + h^2\right)}}\right\} + \sin^{-1}\left\{\frac{(l - 0)\left(\frac{b}{2}\right)}{\sqrt{((l - 0)^2 + h^2)\left(\left(\frac{b}{2}\right)^2 + h^2\right)}}\right\}$$

$$+\sin^{-1}\left\{\frac{(0)\left(b - \frac{b}{2}\right)}{\sqrt{((0)^2 + h^2)\left(\left(b - \frac{b}{2}\right)^2 + h^2\right)}}\right\} + \sin^{-1}\left\{\frac{(l - 0)\left(b - \frac{b}{2}\right)}{\sqrt{((l - 0)^2 + h^2)\left(\left(b - \frac{b}{2}\right)^2 + h^2\right)}}\right\}$$

$$\Rightarrow \omega = \sin^{-1}\{0\} + \sin^{-1}\left\{\frac{lb}{\sqrt{(l^2 + h^2)(b^2 + 4h^2)}}\right\}$$

$$+\sin^{-1}\{0\} + \sin^{-1}\left\{\frac{lb}{\sqrt{(l^2 + h^2)(b^2 + 4h^2)}}\right\}$$

$$\Rightarrow \omega = 2\sin^{-1}\left\{\frac{lb}{\sqrt{(l^2 + h^2)(b^2 + 4h^2)}}\right\}$$

Thus, result obtained is same to the eq(19)

2.3.15 Solid angle subtended by a rectangular plane at any point in the space

Let there be a rectangular plane ABCD having length 'l' & width 'b' ($\forall\, l \geq b$)and a given point say 'P' at a distance 'r' from the centre 'O' of the plane. (as shown in the figure (33) below)

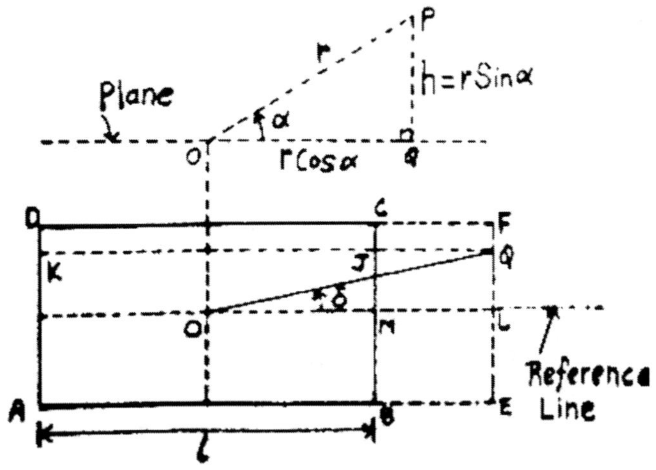

Fig (33)

Now, draw a perpendicular PQ from the given point 'P' to the plane of rectangle ABCD & join the given point 'P' & the foot of perpendicular 'Q' to the centre 'O' such that

$'\alpha' \rightarrow$ is the angle of inclination of the line OP with OQ (or with the given plane ABCD)

(also called 'angle of elevation' of the given point 'P' in the space)

$'\delta' \rightarrow$is the angle between the line OQ & the reference line*

(also called 'angle of deviation' of the given point 'P' in the space)

(*Reference line: The line passing through the centre & parallel to the longer side of the given rectangular plane.)

Now, extend the sides AB & CD and draw the lines QK & FE passing through the point 'Q' & parallel to the sides AB & BC respectively. Extended lines AB, CD, QK & the reference line intersecting the line FE at the points 'E', 'F', 'Q' & 'L' respectively.

In right $\Delta OQP \Rightarrow cos\alpha = \frac{OQ}{OP} = \frac{OQ}{r} \Rightarrow OQ = rcos\alpha$ &

$$sin\alpha = \frac{PQ}{OP} = \frac{PQ}{r} \Rightarrow PQ = rsin\alpha = h(let)$$

Where 'h' is the normal height of the given point 'P' from the rectangular plane ABCD

In right $\Delta OLQ \Rightarrow cos\delta = \frac{OL}{OQ} = \frac{OL}{rcos\alpha} \Rightarrow OL = rcos\alpha cos\delta$ &

$$sin\delta = \frac{QL}{OQ} = \frac{QL}{rcos\alpha} \Rightarrow QL = rcos\alpha sin\delta$$

$$\Rightarrow ML = OL - OM = -\frac{l}{2} + rcos\alpha cos\delta$$

$$\Rightarrow FQ = FL - QL = \frac{b}{2} - rcos\alpha sin\delta$$

$$\Rightarrow DF = DC + CF = DC + ML = l + rcos\alpha cos\delta - \frac{l}{2} = \frac{l}{2} + rcos\alpha cos\delta \&$$

$$\Rightarrow QE = QL + LE = rcos\alpha sin\delta + \frac{b}{2} = \frac{b}{2} + rcos\alpha sin\delta$$

It is clear from the figure (33) that the solid angle subtended by the rectangular plane ABCD at the given point lying on the axis PQ at a normal height h = $PQ = rsin\alpha$

$$\Rightarrow \omega = \omega_{QKAE} + \omega_{QKDF} - \omega_{QJBE} - \omega_{QJCF}$$

Now, for ease of calculation let's assume

$$L_1 = DF = \frac{l}{2} + rcos\alpha cos\delta \qquad L_2 = ML = -\frac{l}{2} + rcos\alpha cos\delta$$

$$B_1 = QE = \frac{b}{2} + rcos\alpha sin\delta \qquad B_2 = FQ = \frac{b}{2} - rcos\alpha sin\delta$$

$\&H = PQ = rsin\alpha$ and

$$\omega_{11} = \omega_{QKAE} = \sin^{-1}\left\{\frac{L_1B_1}{\sqrt{({L_1}^2 + H^2)({B_1}^2 + H^2)}}\right\} \qquad \omega_{12} = \omega_{QKDF} = \sin^{-1}\left\{\frac{L_1B_2}{\sqrt{({L_1}^2 + H^2)({B_2}^2 + H^2)}}\right\}$$

$$\omega_{21} = \omega_{QJBE} = \sin^{-1}\left\{\frac{L_2B_1}{\sqrt{({L_2}^2 + H^2)({B_1}^2 + H^2)}}\right\} \qquad \omega_{22} = \omega_{QJCF} = \sin^{-1}\left\{\frac{L_2B_2}{\sqrt{({L_2}^2 + H^2)({B_2}^2 + H^2)}}\right\}$$

Now, on putting the corresponding values we can find the solid angle subtended by the given rectangular plane ABCD at the point 'P' as follows

$$\omega = \omega_{11} + \omega_{12} - \omega_{21} - \omega_{22} \qquad \text{.........................(21)}$$

We find that the value of ω depends on the following variables

$$'\alpha',\ '\delta',\ 'l',\ 'b'\&'r'$$

Case 1: The solid angle subtended by the square plane having each side of length 'a' at any point in the space can be obtained by putting l = b = a in the above expressions of eq(21) as follows

$$L_1 = \frac{a}{2} + rcos\alpha cos\delta \qquad L_2 = -\frac{a}{2} + rcos\alpha cos\delta$$

$$B_1 = \frac{a}{2} + rcos\alpha sin\delta \qquad B_2 = \frac{a}{2} - rcos\alpha sin\delta$$

$\&H = rsin\alpha$ and

$$\omega_{11} = \sin^{-1}\left\{\frac{L_1B_1}{\sqrt{({L_1}^2 + H^2)({B_1}^2 + H^2)}}\right\} \qquad \omega_{12} = \sin^{-1}\left\{\frac{L_1B_2}{\sqrt{({L_1}^2 + H^2)({B_2}^2 + H^2)}}\right\}$$

$$\omega_{21} = \sin^{-1}\left\{\frac{L_2B_1}{\sqrt{({L_2}^2 + H^2)({B_1}^2 + H^2)}}\right\} \qquad \omega_{22} = \sin^{-1}\left\{\frac{L_2B_2}{\sqrt{({L_2}^2 + H^2)({B_2}^2 + H^2)}}\right\}$$

Now, on putting the corresponding values we can find the solid angle subtended by the given square plane at the given point as follows

$$\omega = \omega_{11} + \omega_{12} - \omega_{21} - \omega_{22}$$

Case 2: The solid angle subtended by the rectangular plane at any point lying on the axis normal to the plane & passing through the centre 'O' is obtained by putting $\alpha = \frac{\pi}{2} \& \delta = 0$ in the above expressions of eq(21) as follows

$$L_1 = \frac{l}{2} + rcos\frac{\pi}{2}cos0 = \frac{l}{2} \qquad L_2 = -\frac{l}{2} + rcos\frac{\pi}{2}cos0 = -\frac{l}{2}$$

$$B_1 = \frac{b}{2} + rcos\frac{\pi}{2}sin0 = \frac{b}{2} \qquad B_2 = \frac{b}{2} - rcos\frac{\pi}{2}sin0 = \frac{b}{2}$$

$\&H = rsin\frac{\pi}{2} = r = h$ (height) and

$$\omega_{11} = \sin^{-1}\left\{\frac{\left(\frac{l}{2}\right)\left(\frac{b}{2}\right)}{\sqrt{\left(\left(\frac{l}{2}\right)^2 + h^2\right)\left(\left(\frac{b}{2}\right)^2 + h^2\right)}}\right\} \quad \omega_{12} = \sin^{-1}\left\{\frac{\left(\frac{l}{2}\right)\left(\frac{b}{2}\right)}{\sqrt{\left(\left(\frac{l}{2}\right)^2 + h^2\right)\left(\left(\frac{b}{2}\right)^2 + h^2\right)}}\right\}$$

$$\omega_{21} = \sin^{-1}\left\{\frac{\left(-\frac{l}{2}\right)\left(\frac{b}{2}\right)}{\sqrt{\left(\left(\frac{l}{2}\right)^2 + h^2\right)\left(\left(\frac{b}{2}\right)^2 + h^2\right)}}\right\} \quad \omega_{22} = \sin^{-1}\left\{\frac{\left(-\frac{l}{2}\right)\left(\frac{b}{2}\right)}{\sqrt{\left(\left(\frac{l}{2}\right)^2 + h^2\right)\left(\left(\frac{b}{2}\right)^2 + h^2\right)}}\right\}$$

$$\omega_{11} = \sin^{-1}\left\{\frac{lb}{\sqrt{(l^2 + 4h^2)(b^2 + 4h^2)}}\right\} \quad \omega_{12} = \sin^{-1}\left\{\frac{lb}{\sqrt{(l^2 + 4h^2)(b^2 + 4h^2)}}\right\}$$

$$\omega_{21} = -\sin^{-1}\left\{\frac{lb}{\sqrt{(l^2 + 4h^2)(b^2 + 4h^2)}}\right\} \quad \omega_{22} = -\sin^{-1}\left\{\frac{lb}{\sqrt{(l^2 + 4h^2)(b^2 + 4h^2)}}\right\}$$

Now, on putting the corresponding values we can find the solid angle subtended by the given rectangular plane at any point lying on the axis normal to the plane & passing through the centre as follows

$$\omega = \omega_{11} + \omega_{12} - \omega_{21} - \omega_{22}$$

$$\Rightarrow \omega = \sin^{-1}\left\{\frac{lb}{\sqrt{(l^2 + 4h^2)(b^2 + 4h^2)}}\right\} + \sin^{-1}\left\{\frac{lb}{\sqrt{(l^2 + 4h^2)(b^2 + 4h^2)}}\right\}$$

$$+\sin^{-1}\left\{\frac{lb}{\sqrt{(l^2 + 4h^2)(b^2 + 4h^2)}}\right\} + \sin^{-1}\left\{\frac{lb}{\sqrt{(l^2 + 4h^2)(b^2 + 4h^2)}}\right\}$$

$$\Rightarrow \omega = 4\sin^{-1}\left\{\frac{lb}{\sqrt{(l^2 + 4h^2)(b^2 + 4h^2)}}\right\}$$

Thus, we get the same result as eq(16).

Case 3: The solid angle subtended by the rectangular plane at a point lying on the centre of the plane is obtained by putting $\alpha = 0$ & $\delta = 0$ & $r = 0$ in the above expressions of the eq(21) as follows

$$L_1 = \frac{l}{2} + (0)cos0cos0 = \frac{l}{2} \quad L_2 = -\frac{l}{2} + (0)cos0cos0 = -\frac{l}{2}$$

$$B_1 = \frac{b}{2} + (0)cos0sin0 = \frac{b}{2} \quad B_2 = \frac{b}{2} - (0)cos0sin0 = \frac{b}{2}$$

$\& H = (0)sin0 = 0 =$ height and

$$\omega_{11} = \sin^{-1}\left\{\frac{\left(\frac{l}{2}\right)\left(\frac{b}{2}\right)}{\sqrt{\left(\left(\frac{l}{2}\right)^2 + (0)^2\right)\left(\left(\frac{b}{2}\right)^2 + (0)^2\right)}}\right\} \quad \omega_{12} = \sin^{-1}\left\{\frac{\left(\frac{l}{2}\right)\left(\frac{b}{2}\right)}{\sqrt{\left(\left(\frac{l}{2}\right)^2 + (0)^2\right)\left(\left(\frac{b}{2}\right)^2 + (0)^2\right)}}\right\}$$

$$\omega_{21} = \sin^{-1}\left\{\frac{\left(-\frac{l}{2}\right)\left(\frac{b}{2}\right)}{\sqrt{\left(\left(\frac{l}{2}\right)^2 + (0)^2\right)\left(\left(\frac{b}{2}\right)^2 + (0)^2\right)}}\right\} \quad \omega_{22} = \sin^{-1}\left\{\frac{\left(-\frac{l}{2}\right)\left(\frac{b}{2}\right)}{\sqrt{\left(\left(\frac{l}{2}\right)^2 + (0)^2\right)\left(\left(\frac{b}{2}\right)^2 + (0)^2\right)}}\right\}$$

$$\omega_{11} = \sin^{-1}\left\{\frac{lb}{lb}\right\} = \sin^{-1}\{1\} = \frac{\pi}{2} \qquad \omega_{12} = \sin^{-1}\left\{\frac{lb}{lb}\right\} = \sin^{-1}\{1\} = \frac{\pi}{2}$$

$$\omega_{21} = -\sin^{-1}\left\{\frac{lb}{lb}\right\} = -\sin^{-1}\{1\} = -\frac{\pi}{2} \quad \omega_{22} = -\sin^{-1}\left\{\frac{lb}{lb}\right\} = -\sin^{-1}\{1\} = -\frac{\pi}{2}$$

Now, on putting the corresponding values we can find the solid angle subtended by the given rectangular plane at a point lying on the centre of the plane as follows

$$\omega = \omega_{11} + \omega_{12} - \omega_{21} - \omega_{22}$$

$$\Rightarrow \omega = \frac{\pi}{2} + \frac{\pi}{2} + \frac{\pi}{2} + \frac{\pi}{2} = 2\pi \ sr$$

*It's also true for any point lying on the plane inside the boundary.

Thus, we find that this formula is very significant for the analysis of the solid angle subtended by a given rectangular plane at any point in the space.

2.3.16 Solid angle subtended bya bounded orthogonal plane at any point in 3D-space

Let there be a given point say P(x, y, z) in 3D-space of X, Y & Z-orthogonal axes. (as shown in the figure (34) below)

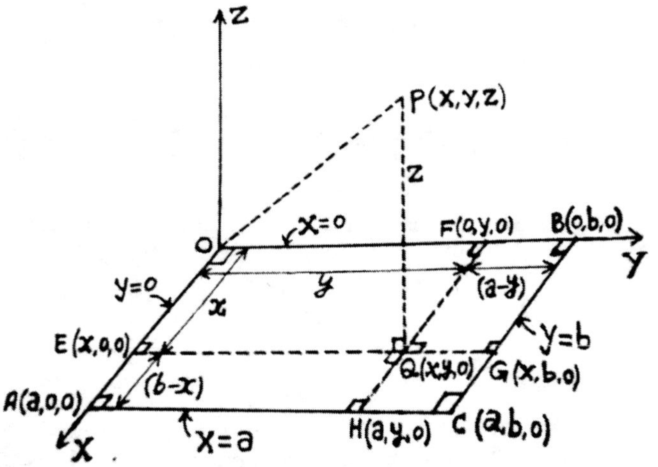

Fig(34)

Now, consider a region on the XY-plane bounded by the straight lines y = 0 (X-axis), x = 0 (Y-axis) y = b & x = a intersecting one another at the points O(0, 0, 0), B(0, b, 0), C(a, b, 0) & A(a, 0, 0) consecutively.

Draw a perpendicular PQ from the point 'P' to the XY-plane & then draw the perpendiculars QE, QF, QG & QH from the foot of the perpendicular 'Q' to the straight lines y = 0 (X-axis), x = 0 (Y-axis) y = b & x = a respectively.

Thus, the rectangle AOBC is divided into four rectangles QEOF, QEAH, QFBG & QHCG

Hence, the solid angle subtended by the bounded region on the XY-plane at the given point P(x, y, z)

= solid angle subtended by the rectangular plane AOBC at the same point p(x, y, z)

$$\Rightarrow \omega = \omega_{AOBC} = \omega_{QEOF} + \omega_{QEAH} + \omega_{QFBG} + \omega_{QHCG}$$

We know that the solid angle subtended by the rectangular plane at any point lying on the axis normal to the plane & passing through the vertex is given by the eq(17) as follows

$$\omega = \sin^{-1}\left\{\frac{lb}{\sqrt{(l^2+h^2)(b^2+h^2)}}\right\}$$

On setting the corresponding values

$$\Rightarrow \omega_{QEOF} = \sin^{-1}\left\{\frac{xy}{\sqrt{(x^2+z^2)(y^2+z^2)}}\right\}, \omega_{QEAH} = \sin^{-1}\left\{\frac{(b-x)y}{\sqrt{((b-x)^2+z^2)(y^2+z^2)}}\right\}$$

$$\Rightarrow \omega_{QFBG} = \sin^{-1}\left\{\frac{x(a-y)}{\sqrt{(x^2+z^2)((a-y)^2+z^2)}}\right\} \& \omega_{QHCG} = \sin^{-1}\left\{\frac{(b-x)(a-y)}{\sqrt{((b-x)^2+z^2)((a-y)^2+z^2)}}\right\}$$

Hence, the solid angle subtended by bounded region on the XY-plane at the point(x, y, z)

$$\Rightarrow \omega = \sin^{-1}\left\{\frac{xy}{\sqrt{(x^2+z^2)(y^2+z^2)}}\right\} + \sin^{-1}\left\{\frac{(b-x)y}{\sqrt{((b-x)^2+z^2)(y^2+z^2)}}\right\}$$

$$+ \sin^{-1}\left\{\frac{x(a-y)}{\sqrt{(x^2+z^2)((a-y)^2+z^2)}}\right\} + \sin^{-1}\left\{\frac{(b-x)(a-y)}{\sqrt{((b-x)^2+z^2)((a-y)^2+z^2)}}\right\} \ldots\ldots\ldots (22)$$

Case 1: The solid angle subtended by the YZ-plane bounded by the straight lines z = 0 (Y-axis), y = 0 (Z-axis), z = b & y = a is obtained by putting x = y, y = z & z = x in the eq(22)

Thus, we have

$$\Rightarrow \omega = \sin^{-1}\left\{\frac{yz}{\sqrt{(y^2+x^2)(z^2+x^2)}}\right\} + \sin^{-1}\left\{\frac{(b-y)z}{\sqrt{((b-y)^2+x^2)(z^2+x^2)}}\right\}$$

$$+ \sin^{-1}\left\{\frac{y(a-z)}{\sqrt{(y^2+x^2)((a-z)^2+x^2)}}\right\} + \sin^{-1}\left\{\frac{(b-y)(a-z)}{\sqrt{((b-y)^2+x^2)((a-z)^2+x^2)}}\right\} \ldots\ldots\ldots (23)$$

Case 2: The solid angle subtended by the ZX-plane bounded by the straight lines x = 0 (Z-axis), z = 0 (X-axis), x = b & z = a is obtained by putting x = z, y = x & z = y in the eq(22)

Thus, we have

$$\Rightarrow \omega = \sin^{-1}\left\{\frac{zx}{\sqrt{(z^2+y^2)(x^2+y^2)}}\right\} + \sin^{-1}\left\{\frac{(b-z)x}{\sqrt{((b-z)^2+y^2)(x^2+y^2)}}\right\}$$

$$+ \sin^{-1}\left\{\frac{z(a-x)}{\sqrt{(z^2+y^2)((a-x)^2+y^2)}}\right\} + \sin^{-1}\left\{\frac{(b-z)(a-x)}{\sqrt{((b-z)^2+y^2)((a-x)^2+y^2)}}\right\} \ldots\ldots\ldots (24)$$

Case 3: The solid angle subtended by the XY-plane at a point tending to the origin (i.e. $z \to 0$) is obtained by taking the limit of ω at $x = 0$, $y = 0$ & $z \to 0$ from eq(22) thus, we have

$$\omega = \lim_{z\to 0}\left[\sin^{-1}\left\{\frac{xy}{\sqrt{(x^2+z^2)(y^2+z^2)}}\right\} + \sin^{-1}\left\{\frac{(b-x)y}{\sqrt{((b-x)^2+z^2)(y^2+z^2)}}\right\} + \sin^{-1}\left\{\frac{x(a-y)}{\sqrt{(x^2+z^2)((a-y)^2+z^2)}}\right\} + \sin^{-1}\left\{\frac{(b-x)(a-y)}{\sqrt{((b-x)^2+z^2)((a-y)^2+z^2)}}\right\}\right]_{(x=0,y=0)}$$

$$\Rightarrow \omega = \lim_{z\to 0}\left[\sin^{-1}\left\{\frac{(0)(0)}{\sqrt{((0)^2+z^2)((0)^2+z^2)}}\right\} + \sin^{-1}\left\{\frac{(b-0)(0)}{\sqrt{((b-0)^2+z^2)((0)^2+z^2)}}\right\} + \sin^{-1}\left\{\frac{(0)(a-0)}{\sqrt{((0)^2+z^2)((a-0)^2+z^2)}}\right\} + \sin^{-1}\left\{\frac{(b-0)(a-0)}{\sqrt{((b-0)^2+z^2)((a-0)^2+z^2)}}\right\}\right]$$

$$\Rightarrow \omega = \lim_{z\to 0}\left[\sin^{-1}\{0\} + \sin^{-1}\{0\} + \sin^{-1}\{0\} + \sin^{-1}\left\{\frac{ab}{\sqrt{(b^2+z^2)(a^2+z^2)}}\right\}\right]$$

$$\Rightarrow \omega = \lim_{z\to 0}\left[\sin^{-1}\left\{\frac{ab}{\sqrt{(b^2+z^2)(a^2+z^2)}}\right\}\right] = \sin^{-1}\left\{\frac{ab}{\sqrt{(b^2+0)(a^2+0)}}\right\}$$

$$= \sin^{-1}\left\{\frac{ab}{ab}\right\} = \sin^{-1}\{1\} = \frac{\pi}{2}\ sr$$

*It is true that the solid angle included at the origin by each quadrant out of eight quadrants in 3D space

$$= \frac{1}{8}(complete\ solid\ angle) = \frac{1}{8}(4\pi) = \frac{\pi}{2}\ sr$$

*The same result is obtained in chapter 4.Analysis of 3D-figure.

2.3.17 Solid angle subtended by a square plane at any point lying on the axis normal to the plane & passing through the centre

Method 1:

Let there be a square plane ABCD having each side of length 'a' & a given point say P(0, 0, h) at a height 'h' lying on the axis normal to the plane & passing through the centre 'O' of the plane. (as shown in the figure (35) below)

Now, draw the diagonals AC & BD bisecting each other at the right angle at the centre 'O' & a perpendicular OE from the centre 'O' to the side AB

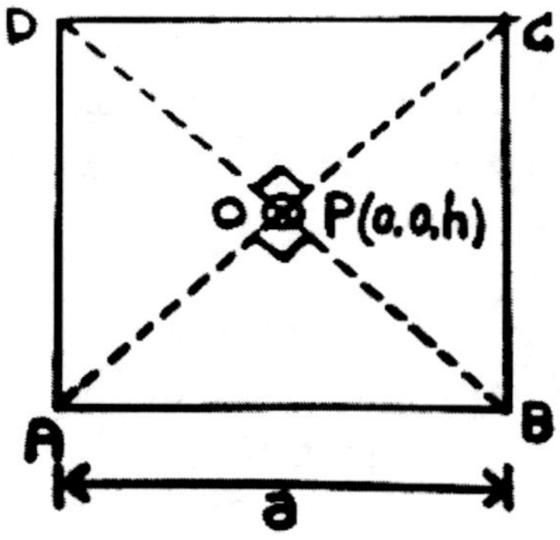

Fig (35)

In right $\Delta ABC \Rightarrow AC = \sqrt{AB^2+BC^2} = \sqrt{a^2+a^2} = a\sqrt{2}$

$$\Rightarrow OA = \frac{AC}{2} = \frac{a\sqrt{2}}{2} = \frac{a}{\sqrt{2}}$$

In the square ABCD, we have OA = OB = OC = OD $= \frac{a}{\sqrt{2}}$ & AB = BC = CD = AD = a

Thus $\Delta AOB, \Delta BOC, \Delta COD$ & ΔAOD are congruent isosceles right triangles which the square ABCD consists of.

Hence, the solid angle subtended by the square plane ABCD at the given point 'P'

$$\Rightarrow \omega = \omega_{\Delta AOB} + \omega_{\Delta BOC} + \omega_{\Delta COD} + \omega_{\Delta AOD} = 4(\omega_{\Delta AOB})$$

We know that the solid angle subtended by the isosceles right triangular plane at any point lying on the axis normal to the plane & passing through the right angled vertex is given by eq(10) as follows

$$\omega = \cos^{-1}\left\{\frac{2h\sqrt{h^2+a^2}}{2h^2+a^2}\right\}$$

Now, on putting the value of a = OA= OB $= \frac{a}{\sqrt{2}}$

$$\Rightarrow \omega_{\Delta AOB} = \cos^{-1}\left\{\frac{2h\sqrt{h^2+\left(\frac{a}{\sqrt{2}}\right)^2}}{2h^2+\left(\frac{a}{\sqrt{2}}\right)^2}\right\} = \cos^{-1}\left\{\frac{2h\sqrt{2}\sqrt{2h^2+a^2}}{4h^2+a^2}\right\}$$

$$\left(since,\ \frac{\pi}{2} - \sin^{-1}x = \cos^{-1}x \qquad \forall(-1 \le x \le 1)\right)$$

$$\Rightarrow \omega_{\Delta AOB} = \sin^{-1}\left\{\frac{a^2}{a^2+4h^2}\right\}\left(since,\ \cos^{-1}x = \sin^{-1}\sqrt{1-x^2} \qquad \forall(-1 \le x \le 1)\right)$$

Hence, $\Rightarrow \omega = 4(\omega_{\Delta AOB})$

$$\Rightarrow \left[\omega = 4\sin^{-1}\left\{\frac{a^2}{a^2+4h^2}\right\}\right] \quad \forall(h \geq 0) \dots\dots\dots\dots\dots\dots\dots\dots\dots\dots\dots (25)$$

Method 2:

Let there be a square plane ABCD having each side of length 'a' & a given point say P(0, 0, h) at a height 'h' lying on the axis normal to the plane & passing through the centre 'O' of the plane. (as shown in the figure (36) below)

Now, draw the diagonals AC & BD bisecting each other at the centre 'O' & a perpendicular OE from the centre 'O' to the side AB

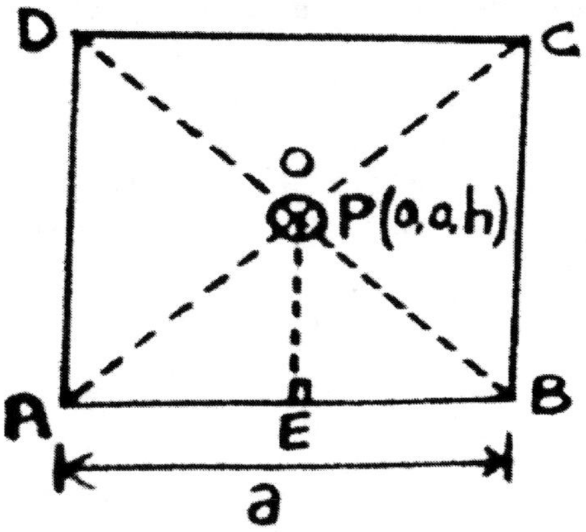

Fig (36)

In right $\Delta ABC \Rightarrow AC = \sqrt{AB^2+BC^2} = \sqrt{a^2+a^2} = a\sqrt{2}$

$$\Rightarrow OA = \frac{AC}{2} = \frac{a\sqrt{2}}{2} = \frac{a}{\sqrt{2}}$$

In the square ABCD, we have OA = OB = OC = OD $=\frac{a}{\sqrt{2}}$ & AB = BC = CD = AD = a

Thus $\Delta AOB, \Delta BOC, \Delta COD$ & ΔAOD are congruent isosceles triangles which the square ABCD consists of.

Hence, the solid angle subtended by the square plane ABCD at the given point 'P'

$$\Rightarrow \omega = \omega_{\Delta AOB} + \omega_{\Delta BOC} + \omega_{\Delta COD} + \omega_{\Delta AOD} = 4(\omega_{\Delta AOB})$$

We know that the solid angle subtended by the isosceles triangular plane at any point lying on the axis normal to the plane & passing through the common vertex of the equal sides is given by eq(7) as follows

$$\omega = 2\left[\sin^{-1}\left\{\frac{b}{\sqrt{b^2+4p^2}}\right\} - \sin^{-1}\left\{\left(\frac{b}{\sqrt{b^2+4p^2}}\right)\left(\frac{h}{\sqrt{h^2+p^2}}\right)\right\}\right]$$

Now, on putting the values of b = base = AB = a & p = perpendicular = OE = $\frac{a}{2}$, we get

$$\Rightarrow \omega_{\Delta AOB} = 2\left[\sin^{-1}\left\{\frac{a}{\sqrt{a^2+4\left(\frac{a}{2}\right)^2}}\right\} - \sin^{-1}\left\{\left(\frac{a}{\sqrt{a^2+4\left(\frac{a}{2}\right)^2}}\right)\left(\frac{h}{\sqrt{h^2+\left(\frac{a}{2}\right)^2}}\right)\right\}\right]$$

$$\omega_{\Delta AOB} = 2\left[\sin^{-1}\left\{\frac{a}{a\sqrt{2}}\right\} - \sin^{-1}\left\{\left(\frac{a}{a\sqrt{2}}\right)\left(\frac{2h}{\sqrt{4h^2+a^2}}\right)\right\}\right]$$

$$= 2\left[\sin^{-1}\left\{\frac{1}{\sqrt{2}}\right\} - \sin^{-1}\left\{\left(\frac{h\sqrt{2}}{\sqrt{4h^2+a^2}}\right)\right\}\right] = 2\left[\frac{\pi}{4} - \sin^{-1}\left\{\left(\frac{h\sqrt{2}}{\sqrt{4h^2+a^2}}\right)\right\}\right]$$

$$= \frac{\pi}{2} - 2\sin^{-1}\left\{\frac{h\sqrt{2}}{\sqrt{4h^2+a^2}}\right\} = \frac{\pi}{2} - \sin^{-1}\left\{2\frac{h\sqrt{2}}{\sqrt{4h^2+a^2}}\frac{\sqrt{2h^2+a^2}}{\sqrt{4h^2+a^2}}\right\}$$

$$\left(since,\ 2\sin^{-1}x = \sin^{-1}(2x\sqrt{1-x^2}) \quad \forall(-1 \leq x \leq 1)\right)$$

$$= \frac{\pi}{2} - \sin^{-1}\left\{\frac{2h\sqrt{2}\sqrt{2h^2+a^2}}{4h^2+a^2}\right\} = \cos^{-1}\left\{\frac{2h\sqrt{2}\sqrt{2h^2+a^2}}{4h^2+a^2}\right\}$$

$$\left(since,\ \frac{\pi}{2} - \sin^{-1}x = \cos^{-1}x \quad \forall(-1 \leq x \leq 1)\right)$$

$$\Rightarrow \omega_{\Delta AOB} = \sin^{-1}\left\{\frac{a^2}{a^2+4h^2}\right\}\left(since,\ \cos^{-1}x = \sin^{-1}\sqrt{1-x^2}\qquad \forall(-1\leq x\leq 1)\right)$$

Hence, $\Rightarrow \omega = 4(\omega_{\Delta AOB})$

$$\Rightarrow \boxed{\omega = 4\sin^{-1}\left\{\frac{a^2}{a^2+4h^2}\right\}}\qquad \forall(h\geq 0)\ \ldots\ldots\ldots\ldots\ldots\ldots (25)$$

Method 3: Alternative

We know that the solid angle subtended by the rectangular plane at any point lying on the axis normal to the plane & passing through the centre is given by the eq(16) as follows

$$\omega = 4\sin^{-1}\left\{\frac{lb}{\sqrt{(l^2+4h^2)(b^2+4h^2)}}\right\}$$

Now, on putting l = b = a in the above expression, we get the solid angle subtended by the square plane at any point lying on the axis normal to the plane & passing through the centre

$$\omega = 4\sin^{-1}\left\{\frac{a.a}{\sqrt{(a^2+4h^2)(a^2+4h^2)}}\right\} = 4\sin^{-1}\left\{\frac{a^2}{\sqrt{(a^2+4h^2)^2}}\right\} = 4\sin^{-1}\left\{\frac{a^2}{a^2+4h^2}\right\}$$

$$\Rightarrow \boxed{\omega = 4\sin^{-1}\left\{\frac{a^2}{a^2+4h^2}\right\}}\qquad \forall(h\geq 0)\ \ldots\ldots\ldots\ldots\ldots\ldots (25)$$

Important deduction: the solid angle subtended by the square plane at any point lying on the plane completely inside the boundary is obtained by putting h = 0 in the above eq(25) as follows

$$\omega = 4\sin^{-1}\left\{\frac{a^2}{a^2+4(0)^2}\right\} = 4\sin^{-1}\left\{\frac{a^2}{a^2}\right\} = 4\sin^{-1}\{1\} = 4\frac{\pi}{2} = 2\pi\ sr$$

2.3.18 Solid angle subtended by a square plane at any point lying on the axis normal to the plane & passing through one of the vertices

Let there be a square plane ABCD having each side of length 'a' & a given point sayP(0, 0, h) at a height 'h' lying on the axis normal to the plane & passing through the vertex say 'A' of the plane. (as shown in the figure (37) below)

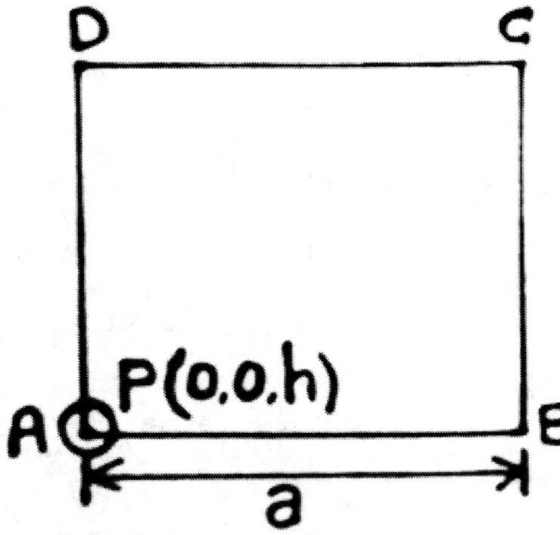

Fig (37)

We know that the solid angle subtended by the rectangular plane at any point lying on the axis normal to the plane & passing through the centre is given by the eq(17) as follows

$$\omega = \sin^{-1}\left\{\frac{lb}{\sqrt{(l^2+h^2)(b^2+h^2)}}\right\}$$

Now, on putting l = b = a in the above expression, we get the solid angle subtended by the square plane at any point lying on the axis normal to the plane & passing through the vertex

$$\omega = \sin^{-1}\left\{\frac{a.a}{\sqrt{(a^2+h^2)(a^2+h^2)}}\right\} = \sin^{-1}\left\{\frac{a^2}{\sqrt{(a^2+h^2)^2}}\right\}$$

$$= \sin^{-1}\left\{\frac{a^2}{a^2+h^2}\right\}$$

$$\Rightarrow \boxed{\omega = \sin^{-1}\left\{\frac{a^2}{a^2+h^2}\right\}}\qquad \forall(h\geq 0)\ \ldots\ldots\ldots\ldots\ldots\ldots (26)$$

Important deduction: The solid angle subtended by the square plane at any point lying on the plane completely inside the boundary is obtained by putting h = 0 in the above eq(26) as follows

$$\omega = \sin^{-1}\left\{\frac{a^2}{a^2+(0)^2}\right\} = \sin^{-1}\left\{\frac{a^2}{a^2}\right\} = \sin^{-1}\{1\} = \frac{\pi}{2}\ sr$$

2.3.19 Solid angle subtended by a square plane at any point lying on the axis normal to the plane & passing through the mid-point of one of the sides

Method 1:

Let there be a square plane ABCD having each side of length 'a' & a given point say P(0, 0, h) at a height 'h' lying on the axis normal to the plane& passing through the mid-point 'O' of the side say CD. (as shown in the figure (38) below)

Draw a perpendicular OE from the point 'O' to the side CD, thus the square ABCD is divided into two rectangles OADE & OBCE.

In the rectangle ABCD, we have OA = OB = EC = ED = $\frac{a}{2}$ & AB = BC = OE = a

Thus, the rectangles OADE & OBCE are congruent rectangles which the given rectangle ABCD consists of.

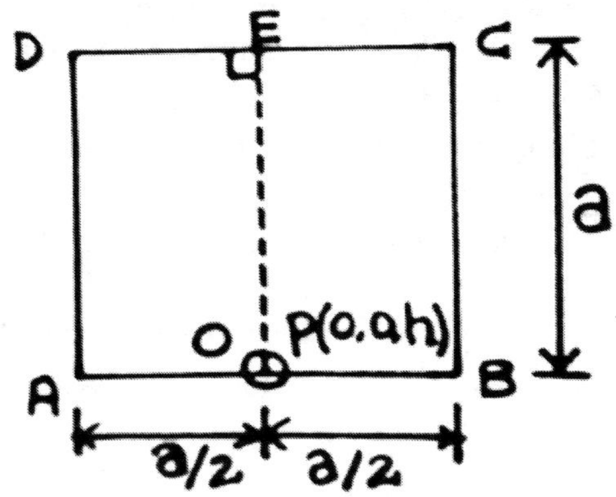

Fig (38)

Hence, the solid angle subtended by the given square plane ABCD at the given point 'P'

$$\Rightarrow \omega = \omega_{OADE} + \omega_{OBCE} = 2(\omega_{OADE})$$

We know that the solid angle subtended by the rectangular plane at any point lying on the axis normal to the plane & passing through one of the vertices is given by the eq(17) as follows

$$\omega = \sin^{-1}\left\{\frac{lb}{\sqrt{(l^2+h^2)(b^2+h^2)}}\right\}$$

Now, on putting the values of $l = legth = AD = a\ \&\ b = width = OA = \frac{a}{2}$ in the above expression we have

$$\omega_{OADE} = \sin^{-1}\left\{\frac{a\left(\frac{a}{2}\right)}{\sqrt{(a^2+h^2)\left(\left(\frac{a}{2}\right)^2+h^2\right)}}\right\} = \sin^{-1}\left\{\frac{a^2}{\sqrt{(a^2+h^2)(a^2+4h^2)}}\right\}$$

Hence, $\Rightarrow \omega = 2(\omega_{OADE})$

$$\Rightarrow \left[\omega = 2\sin^{-1}\left\{\frac{a^2}{\sqrt{(a^2+h^2)(a^2+4h^2)}}\right\}\right] \quad \forall (h \geq 0) \ldots\ldots\ldots\ldots\ldots\ldots (27)$$

Method 2:

We know that the solid angle subtended by the rectangular plane at any point lying on the axis normal to the plane & passing through mid-point of the side is given by the eq(18) as follows

$$\omega = 2\sin^{-1}\left\{\frac{lb}{\sqrt{(l^2+4h^2)(b^2+h^2)}}\right\}$$

Now, on putting l = b = a in the above expression, we get

$$\omega = 2\sin^{-1}\left\{\frac{a.a}{\sqrt{(a^2+4h^2)(a^2+h^2)}}\right\} = 2\sin^{-1}\left\{\frac{a^2}{\sqrt{(a^2+h^2)(a^2+4h^2)}}\right\} \text{ or}$$

$$\Rightarrow \left[\omega = 2\sin^{-1}\left\{\frac{a^2}{\sqrt{(a^2+h^2)(a^2+4h^2)}}\right\}\right] \quad \forall(h \geq 0) \ldots\ldots\ldots\ldots (27)$$

2.3.20 Solid angle subtended by a rhombus-like plane at any point lying on the axis normal to the plane & passing through the centre

Method 1:

Let there be a rhombus-like plane ABCD having diagonals $AC = 2d_1 \& BD = 2d_2$ bisecting each other at right angle at the centre 'O' of the plane ABCD and a given point say P(0, 0, h) at a height 'h' lying on the axis normal to the plane & passing through the centre 'O' (as shown in the figure(39) below)

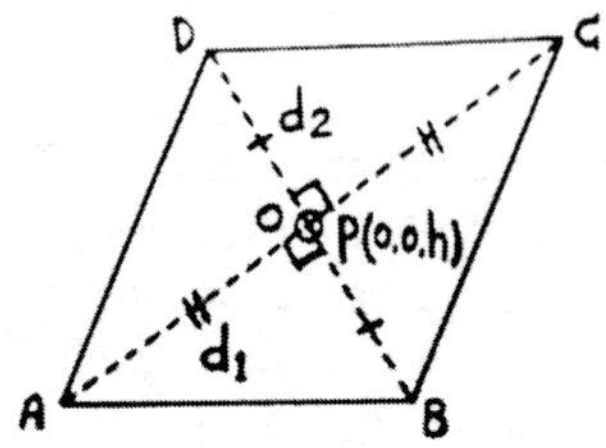

Fig (39)

Since, the diagonals are bisecting each other at right angle at the centre 'O' of the plane.

Thus, in rhombus ABCD we have

$$AB = BC = CD = AD = \sqrt{{d_1}^2 + {d_2}^2}, OA = OC = d_1 \& OB = OD = d_2$$

Hence, $\Delta AOB, \Delta BOC, \Delta COD \;\&\; \Delta AOD$ are congruent right triangles which the rhombus ABCD consists of.

∴ The solid angle subtended by the rhombus-like plane ABCD at the given point 'P'

$$\Rightarrow \omega = \omega_{\Delta AOB} + \omega_{\Delta BOC} + \omega_{\Delta COD} + \omega_{\Delta AOD} = 4(\omega_{\Delta AOB})$$

We know that the solid angle subtended by the right triangular plane at any point lying on the axis normal to the plane & passing through the centre of the plane is given by the eq(6) as follows

$$\omega = \cos^{-1}\left\{\frac{h\left(a^2\sqrt{h^2+b^2} + b^2\sqrt{h^2+a^2}\right)}{h^2(a^2+b^2)+a^2b^2}\right\}$$

now, on setting $a = d_1 \& b = d_2$ in the above expression, we get

$$\omega_{\Delta AOB} = \cos^{-1}\left\{\frac{h\left({d_1}^2\sqrt{h^2+{d_2}^2} + {d_2}^2\sqrt{h^2+{d_1}^2}\right)}{h^2({d_1}^2+{d_2}^2)+{d_1}^2{d_2}^2}\right\}$$

$$\Rightarrow \left[\omega = 4\cos^{-1}\left\{\frac{h\left({d_1}^2\sqrt{h^2+{d_2}^2} + {d_2}^2\sqrt{h^2+{d_1}^2}\right)}{h^2({d_1}^2+{d_2}^2)+{d_1}^2{d_2}^2}\right\}\right] \quad \forall(h \geq 0) \ldots\ldots\ldots (28)$$

Method 2:

Since, in rhombus ABCD we have

$$AB = BC = CD = AD = \sqrt{{d_1}^2 + {d_2}^2}, \qquad AC = AC = 2d_1 \& OB = OD = d_2$$

Thus, $\Delta ABC \;\&\; \Delta ADC$ are congruent isosceles triangles which the rhombus ABCD consists of.

∴ The solid angle subtended by the rhombus-like plane ABCD at the given point 'P'

$$\Rightarrow \omega = \omega_{\Delta ABC} + \omega_{\Delta ADC} = 2(\omega_{\Delta ADC})$$

We know that the solid angle subtended by the right triangular plane at any point lying on the axis normal to the plane & passing through the mid-point of the unequal side is given by the eq(9) as follows

$$\omega = 2\cos^{-1}\left\{\frac{h\left(2p^2\sqrt{4h^2+b^2} + b^2\sqrt{h^2+p^2}\right)}{h^2(4p^2+b^2)+p^2b^2}\right\}$$

Now, on setting b = base = AC = $2d_1$& p = perpendicular = OD = d_2 in the above expression, we have

$$\omega = 2\cos^{-1}\left\{\frac{h\left(2{d_2}^2\sqrt{4h^2+(2d_1)^2} + (2d_1)^2\sqrt{h^2+{d_2}^2}\right)}{h^2\left(4{d_2}^2+(2d_1)^2\right)+{d_2}^2(2d_1)^2}\right\}$$

$$\omega = 2\cos^{-1}\left\{\frac{h\left(2{d_2}^2\sqrt{4h^2+(2d_1)^2} + (2d_1)^2\sqrt{h^2+{d_2}^2}\right)}{h^2\left(4{d_2}^2+(2d_1)^2\right)+{d_2}^2(2d_1)^2}\right\}$$

$$= 2\cos^{-1}\left\{\frac{4h\left({d_2}^2\sqrt{h^2+{d_1}^2} + {d_1}^2\sqrt{h^2+{d_2}^2}\right)}{4\left(h^2\left({d_2}^2+{d_1}^2\right)+{d_2}^2{d_1}^2\right)}\right\}$$

$$= 2\cos^{-1}\left\{\frac{h\left({d_1}^2\sqrt{h^2+{d_2}^2} + {d_2}^2\sqrt{h^2+{d_1}^2}\right)}{h^2\left({d_1}^2+{d_2}^2\right)+{d_1}^2{d_2}^2}\right\}$$

$$\Rightarrow \omega = 2(\omega_{\Delta ADC})$$

$$\Rightarrow \left[\omega = 4\cos^{-1}\left\{\frac{h\left({d_1}^2\sqrt{h^2+{d_2}^2} + {d_2}^2\sqrt{h^2+{d_1}^2}\right)}{h^2\left({d_1}^2+{d_2}^2\right)+{d_1}^2{d_2}^2}\right\}\right] \quad \forall(h \geq 0) \ldots\ldots\ldots\ldots\ldots\ldots (28)$$

2.3.21 Solid angle subtended by a right trapezoidal plane at any point lying on the axis normal to the plane

Let there be a right trapezoidal plane ABCD having parallel sides AD = a& BC = b $(such\ that\ b \geq a)$at a perpendicular distance 'p' and a given point say P(0, 0, h) at a height 'h' lying on the axis normal to the plane.

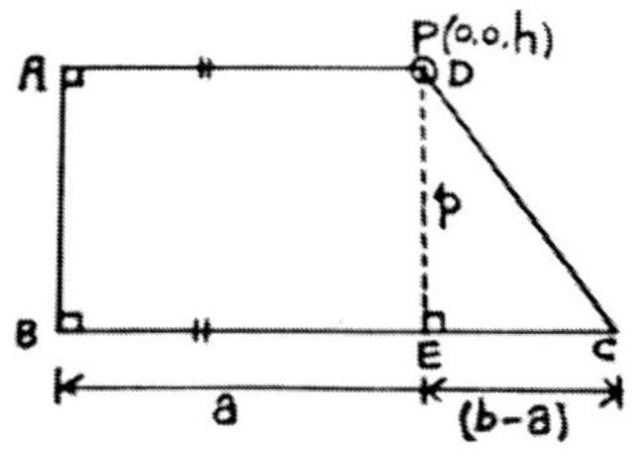

Fig (40)

Now, let's consider the different positions of the normal axis to the plane as follows

Case 1: Normal axis passing through the vertex 'D'

(as shown in the figure (40) below)

It is obvious from the above figure (40) that the right trapezoidal consists of the rectangle ABED & a right ΔDEC

Hence, the solid angle subtended by the given right trapezoidal plane ABCD at the given point 'P'

$$\Rightarrow \omega = \omega_{ABED} + \omega_{\Delta DEC}$$

We know that the solid angle subtended by the rectangular plane at any point lying on the axis normal to the plane & passing through the vertex is given by the eq(17) as follows

$$\omega = \sin^{-1}\left\{\frac{lb}{\sqrt{(l^2+h^2)(b^2+h^2)}}\right\}$$

Now, on setting l = length = AD = a & b = width = AB = p in the above expression, we get

$$\Rightarrow \omega_{ABED} = \sin^{-1}\left\{\frac{ap}{\sqrt{(a^2+h^2)(p^2+h^2)}}\right\}$$

We know that the solid angle subtended by the right triangular plane at any point lying on the axis normal to the plane & passing through the acute angled vertex is given by the eq(2) as follows

$$\omega = \sin^{-1}\left\{\frac{b}{\sqrt{b^2+p^2}}\right\} - \sin^{-1}\left\{\left(\frac{b}{\sqrt{b^2+p^2}}\right)\left(\frac{h}{\sqrt{h^2+p^2}}\right)\right\}$$

Now, on setting the values of b = base = EC = $(b-a)$& p = perpendicular = DE = p in the above expression, we get

$$\Rightarrow \omega_{\Delta DEC} = \sin^{-1}\left\{\frac{(b-a)}{\sqrt{(b-a)^2+p^2}}\right\} - \sin^{-1}\left\{\left(\frac{(b-a)}{\sqrt{(b-a)^2+p^2}}\right)\left(\frac{h}{\sqrt{h^2+p^2}}\right)\right\}$$

Hence, the solid angle subtended by the given right trapezoidal plane ABCD at the given point 'P'

$\Rightarrow \omega = \omega_{ABED} + \omega_{\Delta DEC}$(on setting the values)

$$\Rightarrow \left[\omega = \sin^{-1}\left\{\frac{ap}{\sqrt{(a^2+h^2)(p^2+h^2)}}\right\} + \sin^{-1}\left\{\frac{(b-a)}{\sqrt{(b-a)^2+p^2}}\right\} - \sin^{-1}\left\{\left(\frac{(b-a)}{\sqrt{(b-a)^2+p^2}}\right)\left(\frac{h}{\sqrt{h^2+p^2}}\right)\right\}\right]$$

$$\forall (h \geq 0) \ldots\ldots\ldots\ldots\ldots (29)$$

Important deduction: The solid angle subtended by the right trapezoidal plane ABCD at a point lying on the vertex 'D' is obtained by setting h = 0 in the above eq(29) as follows

$$\omega = \sin^{-1}\left\{\frac{ap}{\sqrt{(a^2+(0)^2)(p^2+(0)^2)}}\right\} + \sin^{-1}\left\{\frac{(b-a)}{\sqrt{(b-a)^2+p^2}}\right\} - \sin^{-1}\left\{\left(\frac{(b-a)}{\sqrt{(b-a)^2+p^2}}\right)\left(\frac{(0)}{\sqrt{(0)^2+p^2}}\right)\right\}$$

$$\Rightarrow \omega = \sin^{-1}\left\{\frac{ap}{ap}\right\} + \sin^{-1}\left\{\frac{(b-a)}{\sqrt{(b-a)^2+p^2}}\right\} - \sin^{-1}\{0\} = \sin^{-1}\{1\} + \sin^{-1}\left\{\frac{(b-a)}{\sqrt{(b-a)^2+p^2}}\right\}$$

$$\Rightarrow \omega = \frac{\pi}{2} + \sin^{-1}\left\{\frac{(b-a)}{\sqrt{(b-a)^2+p^2}}\right\} \quad \ldots\ldots\ldots\ldots\ldots\ldots\ldots\ldots\ldots (30)$$

*if b = a then, we get

$$\Rightarrow \omega = \frac{\pi}{2} + \sin^{-1}\left\{\frac{(a-a)}{\sqrt{(a-a)^2+p^2}}\right\} = \frac{\pi}{2} + \sin^{-1}\left\{\frac{(0)}{\sqrt{(0)^2+p^2}}\right\} = \frac{\pi}{2} + 0 = \frac{\pi}{2}\ sr$$

Case 2: normal axis passing through the vertex 'C'

(as shown in the figure (41) below)

Now, draw a perpendicular CF to the extended side AD

It is obvious from the above figure (41) that the rectangle ABCF consists of the right trapezium ABCD & a right ΔCFD

Hence, the solid angle subtended by the given rectangular plane ABCF at the given point 'P'

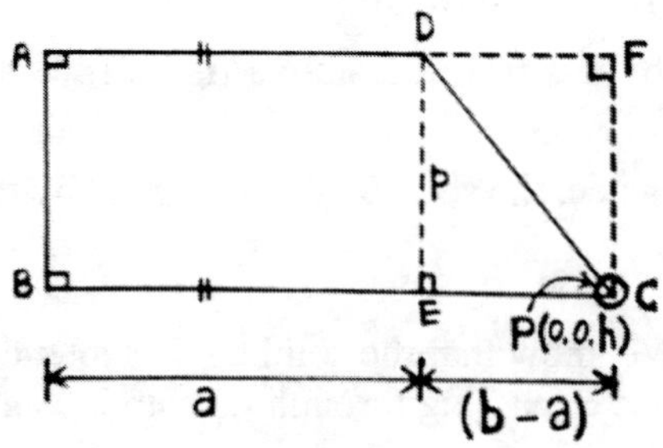

$\Rightarrow \omega_{ABCF} = \omega_{ABCD} + \omega_{\Delta DEC} \Rightarrow \omega_{ABCD} = \omega_{ABCF} - \omega_{\Delta CFD}$

Fig (41)

We know that the solid angle subtended by the rectangular plane at any point lying on the axis normal to the plane & passing through the vertex is given by the eq(17) as follows

$$\omega = \sin^{-1}\left\{\frac{lb}{\sqrt{(l^2+h^2)(b^2+h^2)}}\right\}$$

Now, on setting l = length = BC = b & b = width = AB = p in the above expression, we get

$$\Rightarrow \omega_{ABCF} = \sin^{-1}\left\{\frac{bp}{\sqrt{(b^2+h^2)(p^2+h^2)}}\right\}$$

We know that the solid angle subtended by the right triangular plane at any point lying on the axis normal to the plane & passing through the acute angled vertex is given by the eq(2) as follows

$$\omega = \sin^{-1}\left\{\frac{b}{\sqrt{b^2+p^2}}\right\} - \sin^{-1}\left\{\left(\frac{b}{\sqrt{b^2+p^2}}\right)\left(\frac{h}{\sqrt{h^2+p^2}}\right)\right\}$$

Now, on setting the values of b = base = DF = $(b-a)$& p = perpendicular = CF = DE = p in the above expression, we get

$$\Rightarrow \omega_{\Delta CFD} = \sin^{-1}\left\{\frac{(b-a)}{\sqrt{(b-a)^2+p^2}}\right\} - \sin^{-1}\left\{\left(\frac{(b-a)}{\sqrt{(b-a)^2+p^2}}\right)\left(\frac{h}{\sqrt{h^2+p^2}}\right)\right\}$$

Hence, the solid angle subtended by the given right trapezoidal plane ABCD at the given point 'P'

$$\Rightarrow \omega = \omega_{ABCF} - \omega_{\Delta CFD} \text{ (on setting the values)}$$

$$\Rightarrow \left[\omega = \sin^{-1}\left\{\frac{ap}{\sqrt{(a^2+h^2)(p^2+h^2)}}\right\} + \sin^{-1}\left\{\left(\frac{(b-a)}{\sqrt{(b-a)^2+p^2}}\right)\left(\frac{h}{\sqrt{h^2+p^2}}\right)\right\} - \sin^{-1}\left\{\frac{(b-a)}{\sqrt{(b-a)^2+p^2}}\right\}\right]$$

$$\forall (h \geq 0) \ldots\ldots\ldots\ldots\ldots (31)$$

Important deduction: The solid angle subtended by the right trapezoidal plane ABCD at a point lying on the vertex 'D' is obtained by setting h = 0 in the above eq(31) as follows

$$\omega = \sin^{-1}\left\{\frac{ap}{\sqrt{(a^2+(0)^2)(p^2+(0)^2)}}\right\} + \sin^{-1}\left\{\left(\frac{(b-a)}{\sqrt{(b-a)^2+p^2}}\right)\left(\frac{(0)}{\sqrt{(0)^2+p^2}}\right)\right\} - \sin^{-1}\left\{\frac{(b-a)}{\sqrt{(b-a)^2+p^2}}\right\}$$

$$\Rightarrow \omega = \sin^{-1}\left\{\frac{ap}{ap}\right\} + \sin^{-1}\{0\} - \sin^{-1}\left\{\frac{(b-a)}{\sqrt{(b-a)^2+p^2}}\right\}$$

$$\Rightarrow \omega = \frac{\pi}{2} - \sin^{-1}\left\{\frac{(b-a)}{\sqrt{(b-a)^2+p^2}}\right\} \quad \ldots\ldots\ldots\ldots\ldots\ldots\ldots (32)$$

Case 3: normal axis passing through the point 'O'

The point 'O' is the foot of the perpendicular DO drawn from the vertex 'D' to the side BC.

(as shown in the figure (42) below)

It is obvious from the above figure (42) that the right trapezium ABCD consists of a rectangle ABOD & a right ΔCOD

Hence, the solid angle subtended by the given right trapezoidal plane ABCD at the given point 'P'

$$\Rightarrow \omega_{ABCD} = \omega_{ABOD} + \omega_{\Delta COD}$$

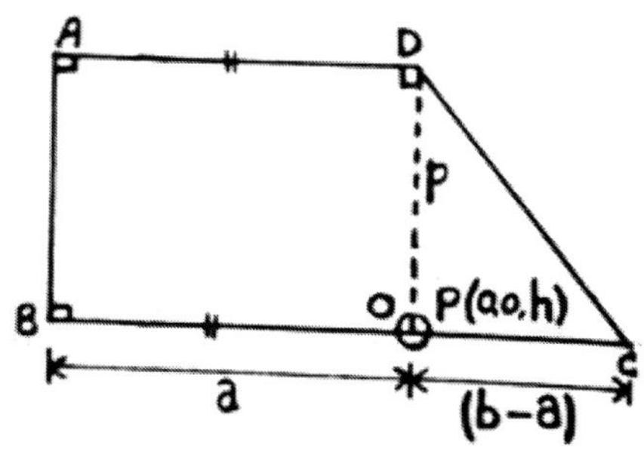

Fig (42)

We know that the solid angle subtended by the rectangular plane at any point lying on the axis normal to the plane & passing through the vertex is given by the eq(17) as follows

$$\omega = \sin^{-1}\left\{\frac{lb}{\sqrt{(l^2+h^2)(b^2+h^2)}}\right\}$$

Now, on setting l = length = AD = a & b = width = AB = p in the above expression, we get

$$\Rightarrow \omega_{ABOD} = \sin^{-1}\left\{\frac{ap}{\sqrt{(a^2+h^2)(p^2+h^2)}}\right\}$$

We know that the solid angle subtended by the right triangular plane at any point lying on the axis normal to the plane & passing through the right angled vertex is given by the eq(6) as follows

$$\omega = \cos^{-1}\left\{\frac{h\left(a^2\sqrt{h^2+b^2} + b^2\sqrt{h^2+a^2}\right)}{h^2(a^2+b^2)+a^2b^2}\right\}$$

Now, on setting the values of a = OD = p& $b = OC = (b-a)$ in the above expression, we get

$$\Rightarrow \omega_{\Delta COD} = \cos^{-1}\left\{\frac{h\left(p^2\sqrt{h^2+(b-a)^2} + (b-a)^2\sqrt{h^2+p^2}\right)}{h^2(p^2+(b-a)^2)+p^2(b-a)^2}\right\}$$

Hence, the solid angle subtended by the given right trapezoidal plane ABCD at the given point 'P'

$$\Rightarrow \omega = \omega_{ABOD} + \omega_{\Delta COD} \quad \text{(on setting the values)}$$

$$\Rightarrow \left[\omega = \sin^{-1}\left\{\frac{ap}{\sqrt{(a^2+h^2)(p^2+h^2)}}\right\} + \cos^{-1}\left\{\frac{h\left(p^2\sqrt{h^2+(b-a)^2} + (b-a)^2\sqrt{h^2+p^2}\right)}{h^2(p^2+(b-a)^2)+p^2(b-a)^2}\right\}\right]$$

$$\forall(h \geq 0) \ldots\ldots\ldots\ldots\ldots (33)$$

Important deduction: The solid angle subtended by the right trapezoidal plane ABCD at a point lying on the point 'O' is obtained by setting h = 0 in the above eq(33) as follows

$$\omega = \sin^{-1}\left\{\frac{ap}{\sqrt{(a^2+(0)^2)(p^2+(0)^2)}}\right\} + \cos^{-1}\left\{\frac{(0)\left(p^2\sqrt{(0)^2+(b-a)^2} + (b-a)^2\sqrt{(0)^2+p^2}\right)}{(0)^2(p^2+(b-a)^2)+p^2(b-a)^2}\right\}$$

$$\omega = \sin^{-1}\left\{\frac{ap}{ap}\right\} + \cos^{-1}\{0\} = \sin^{-1}\{1\} + \cos^{-1}\{0\} = \frac{\pi}{2} + \frac{\pi}{2} = \pi\ sr$$

2.3.22 Solid angle subtended by a regular polygonal plane at any point lying on the axis normal to the plane & passing through the centre

Let there be a regular polygonal plane$A_1A_2A_3 \ldots\ldots A_n$ having 'n' no. of the sides each equal to the length 'a' & a given point say P(0, 0, h) at a height 'h' lying on the axis normal to the plane & passing through the centre 'O'. (as shown in the figure(43) below)

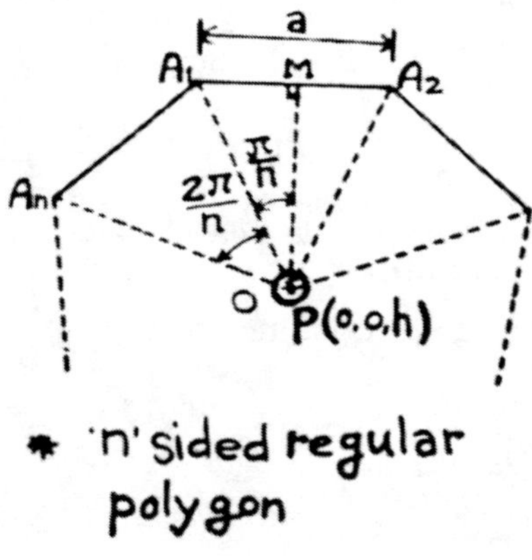

Fig (43)

Now, join the centre 'O' to all the vertices $A_1, A_2, A_3, \ldots\ldots \& A_n$

In regular polygon$A_1A_2A_3 \ldots\ldots A_n$, we have

$$A_1A_2 = A_2A_3 = \ \ldots = A_{n-1}A_n = a \ \& \ OA_1 = OA_2 = \ \ldots\ldots = OA_n = cicumscribed\ radius$$

Thus, $\Delta OA_1A_2, \Delta OA_2A_3, \ldots\ldots\ldots\ldots\ldots, \Delta OA_{n-1}A_n$ are congruent isosceles triangles which the regular polygon consists of.

Now, consider an isosceles ΔOA_1A_2& draw a perpendicular OM from the centre 'O' to the side A_1A_2

In right ΔOMA_1

$$\Rightarrow tan\left(\frac{\pi}{n}\right) = \frac{A_1M}{OM} = \frac{\left(\frac{a}{2}\right)}{OM} \Rightarrow OM = \frac{a}{2}cot\frac{\pi}{n}$$

Hence, the solid angle subtended by the regular polygon at the given point 'P'

$$\Rightarrow \omega = \omega_{\Delta OA_1A_2} + \omega_{\Delta OA_2A_3} + \omega_{\Delta OA_3A_4} + \ \ldots\ldots\ldots + \omega_{\Delta OA_nA_1} = n\left(\omega_{\Delta OA_1A_2}\right)$$

We know that the solid angle subtended by the isosceles triangular plane at any point lying on the axis normal to the plane & passing through the common vertex of the equal sides is given by the eq(7) as follows

$$\omega = 2\left[\sin^{-1}\left\{\frac{b}{\sqrt{b^2+4p^2}}\right\} - \sin^{-1}\left\{\left(\frac{b}{\sqrt{b^2+4p^2}}\right)\left(\frac{h}{\sqrt{h^2+p^2}}\right)\right\}\right]$$

Now, on setting the values of

$b = base = A_1A_2 = a \ \& \ p = perpendicular = OM = \frac{a}{2}cot\frac{\pi}{n}$in the above expression,

we get

$$\omega_{\Delta OA_1A_2} = 2\left[\sin^{-1}\left\{\frac{a}{\sqrt{a^2+4\left(\frac{a}{2}cot\frac{\pi}{n}\right)^2}}\right\} - \sin^{-1}\left\{\left(\frac{a}{\sqrt{a^2+4\left(\frac{a}{2}cot\frac{\pi}{n}\right)^2}}\right)\left(\frac{h}{\sqrt{h^2+\left(\frac{a}{2}cot\frac{\pi}{n}\right)^2}}\right)\right\}\right]$$

$$\Rightarrow \omega_{\Delta OA_1A_2} = 2\left[\sin^{-1}\left\{\frac{a}{a\sqrt{1+cot^2\frac{\pi}{n}}}\right\} - \sin^{-1}\left\{\left(\frac{a}{a\sqrt{1+cot^2\frac{\pi}{n}}}\right)\left(\frac{2h}{\sqrt{4h^2+a^2cot^2\frac{\pi}{n}}}\right)\right\}\right]$$

$$= 2\left[\sin^{-1}\left\{\frac{1}{cosec\frac{\pi}{n}}\right\} - \sin^{-1}\left\{\left(\frac{1}{cosec\frac{\pi}{n}}\right)\left(\frac{2h}{\sqrt{4h^2+a^2cot^2\frac{\pi}{n}}}\right)\right\}\right]$$

$$= 2\left[\sin^{-1}\left\{sin\frac{\pi}{n}\right\} - \sin^{-1}\left\{\left(sin\frac{\pi}{n}\right)\left(\frac{2h}{\sqrt{4h^2+a^2cot^2\frac{\pi}{n}}}\right)\right\}\right]$$

$$= 2\left[\frac{\pi}{n} - \sin^{-1}\left\{\frac{2hsin\frac{\pi}{n}}{\sqrt{4h^2+a^2cot^2\frac{\pi}{n}}}\right\}\right]$$

Hence, we get

$$\Rightarrow \omega = n\left(\omega_{\Delta OA_1A_2}\right) = 2n\left[\frac{\pi}{\mathrm{n} - \sin^{-1}\left\{\dfrac{2hsin\frac{\pi}{n}}{\sqrt{4h^2 + a^2cot^2\frac{\pi}{n}}}\right\}}\right]$$

$$\Rightarrow \boxed{\omega = 2\pi - 2n\sin^{-1}\left\{\frac{2hsin\frac{\pi}{n}}{\sqrt{4h^2 + a^2cot^2\frac{\pi}{n}}}\right\}} \quad \forall(h \geq 0\,, n \in N\ \&n \geq 3) \ldots\ldots\ldots(34)$$

Important deduction1: The formulae for the different regular polygonal plane can be derived by setting the different values of 'n' in the eq(34) as follows

Case 1: Solid angle subtended by the equilateral triangular plane at any point lying on the axis normal to the plane & passing through the centre

Is obtained by setting n = 3 in the eq(34), we get

$$\omega = 2\pi - 2(3)\sin^{-1}\left\{\frac{2hsin\frac{\pi}{3}}{\sqrt{4h^2 + a^2cot^2\frac{\pi}{3}}}\right\} = 2\pi - 6\sin^{-1}\left\{\frac{2h\frac{\sqrt{3}}{2}}{\sqrt{4h^2 + a^2\left(\frac{1}{\sqrt{3}}\right)^2}}\right\}$$

$$\Rightarrow \omega = 2\pi - 6\sin^{-1}\left\{\frac{3h}{\sqrt{12h^2 + a^2}}\right\}$$

Above result is same to that of the eq(15).

Case 2: Solid angle subtended by the square plane at any point lying on the axis normal to the plane & passing through the centre

Is obtained by setting n = 4 in the eq(34), we get

$$\omega = 2\pi - 2(4)\sin^{-1}\left\{\frac{2hsin\frac{\pi}{4}}{\sqrt{4h^2 + a^2cot^2\frac{\pi}{4}}}\right\} = 2\pi - 8\sin^{-1}\left\{\frac{2h\frac{1}{\sqrt{2}}}{\sqrt{4h^2 + a^2(1)^2}}\right\}$$

$$\Rightarrow \omega = 2\pi - 8\sin^{-1}\left\{\frac{h\sqrt{2}}{\sqrt{4h^2 + a^2}}\right\} = 4\left[\frac{\pi}{2} - 2\sin^{-1}\left\{\frac{h\sqrt{2}}{\sqrt{4h^2 + a^2}}\right\}\right]$$

$$= 4\left[\frac{\pi}{2} - \sin^{-1}\left\{2\left(\frac{h\sqrt{2}}{\sqrt{4h^2 + a^2}}\right)\left(\frac{\sqrt{2h^2 + a^2}}{\sqrt{4h^2 + a^2}}\right)\right\}\right]$$

$$\left(since,\ 2\sin^{-1}x = \sin^{-1}\left(2x\sqrt{1 - x^2}\right) \quad \forall(-1 \leq x \leq 1)\right)$$

$$= 4\left[\frac{\pi}{2} - \sin^{-1}\left\{\frac{2h\sqrt{2}\sqrt{2h^2 + a^2}}{4h^2 + a^2}\right\}\right] = 4\cos^{-1}\left\{\frac{2h\sqrt{2}\sqrt{2h^2 + a^2}}{4h^2 + a^2}\right\}$$

$$\left(since,\ \frac{\pi}{2} - \sin^{-1}x = \cos^{-1}x \quad \forall(-1 \leq x \leq 1)\right)$$

$$= 4\sin^{-1}\left\{\frac{a^2}{4h^2 + a^2}\right\}$$

$$\left(since,\ \cos^{-1}x = \sin^{-1}\sqrt{1 - x^2} \quad \forall(-1 \leq x \leq 1)\right)$$

The result obtained is same to that of the eq(25)

Case 3: Solid angle subtended by the regular pentagonal plane at any point lying on the axis normal to the plane & passing through the centre

Is obtained by setting n = 5 in the eq(34), we get

$$\omega = 2\pi - 2(5)\sin^{-1}\left\{\frac{2hsin\frac{\pi}{5}}{\sqrt{4h^2 + a^2cot^2\frac{\pi}{5}}}\right\} = 2\pi - 10\sin^{-1}\left\{\frac{2h(0.587)}{\sqrt{4h^2 + a^2(1.376)^2}}\right\}$$

$$\Rightarrow \omega = 2\pi - 10\sin^{-1}\left\{\frac{1.175h}{\sqrt{4h^2 + 1.893a^2}}\right\} \quad \forall(h \geq 0) \dots\dots\dots\dots\dots\dots\dots\dots\dots\dots (35)$$

Case 4: Solid angle subtended by the regular hexagonal plane at any point lying on the axis normal to the plane & passing through the centre

Is obtained by setting n = 6 in the eq(34), we get

$$\omega = 2\pi - 2(6)\sin^{-1}\left\{\frac{2hsin\frac{\pi}{6}}{\sqrt{4h^2 + a^2cot^2\frac{\pi}{6}}}\right\} = 2\pi - 12\sin^{-1}\left\{\frac{2h\frac{1}{2}}{\sqrt{4h^2 + a^2(\sqrt{3})^2}}\right\}$$

$$\Rightarrow \omega = 2\pi - 12\sin^{-1}\left\{\frac{h}{\sqrt{4h^2 + 3a^2}}\right\} \quad \forall(h \geq 0) \dots\dots\dots\dots\dots\dots\dots\dots\dots\dots (36)$$

Thus, further by setting n = 7, 8, 9,…….upto n, different formulae can be derived.

Important deduction 2: The solid angle subtended by any plane at any point lying on the plane completely inside the boundary is obtained by setting h = 0 in the eq(34) as follows

$$\omega = \left[2\pi - 2n\sin^{-1}\left\{\frac{2hsin\frac{\pi}{n}}{\sqrt{4h^2 + a^2cot^2\frac{\pi}{n}}}\right\}\right]_{h=0}$$

$$\omega = 2\pi - 2n\sin^{-1}\left\{\frac{2(0)sin\frac{\pi}{n}}{\sqrt{4(0)^2 + a^2cot^2\frac{\pi}{n}}}\right\} = 2\pi - 2n\sin^{-1}\{0\} = 2\pi\ sr$$

Important deduction 3: The solid angle subtended by an infinitely large plane at any point lying at a finite distance from the plane is obtained by taking limit of $'\omega'$ at $a \to \infty$ from the eq(34) as follows

$$\Rightarrow \omega = \lim_{a\to\infty}\left[2\pi - 2n\sin^{-1}\left\{\frac{2hsin\frac{\pi}{n}}{\sqrt{4h^2 + a^2cot^2\frac{\pi}{n}}}\right\}\right]$$

$$= \lim_{a\to\infty}\left[2\pi - 2n\sin^{-1}\left\{\frac{2\left(\frac{h}{a}\right)sin\frac{\pi}{n}}{\sqrt{4\left(\frac{h}{a}\right)^2 + cot^2\frac{\pi}{n}}}\right\}\right] = 2\pi - 2n\sin^{-1}\left\{\frac{2(0)sin\frac{\pi}{n}}{\sqrt{4(0)^2 + cot^2\frac{\pi}{n}}}\right\}$$

$$= 2\pi - 2n\sin^{-1}\{0\} = 2\pi\ sr$$

Important deduction 4: The solid angle subtended by an finite plane at any point lying at infinity from the plane is obtained by taking limit of $'\omega'$ at $h \to \infty$ from the eq(34) as follows

$$\Rightarrow \omega = \lim_{h\to\infty}\left[2\pi - 2n\sin^{-1}\left\{\frac{2hsin\frac{\pi}{n}}{\sqrt{4h^2 + a^2cot^2\frac{\pi}{n}}}\right\}\right]$$

$$= 2\pi - 2n\sin^{-1}\left\{\frac{2sin\frac{\pi}{n}}{2}\right\} = 2\pi - 2n\sin^{-1}\left\{sin\frac{\pi}{n}\right\} = 2\pi - 2n\frac{\pi}{n} = 2\pi - 2\pi = 0$$

2.3.23 Solid angle subtended by a circular plane at any point lying on the axis normal to the plane & passing through the centre

Method 1: Using Fundamental Theorem

Let there be a circular plane having radius equal to 'R' & centre 'O' and a given point say P(0, 0, h) at a height 'h' lying on the axis normal to the plane & passing through the centre 'O' (as shown in the figure (44) below)

In right $\Delta POM \Rightarrow sec\varphi = \frac{PM}{PO} \Rightarrow PM = POsec\varphi = hsec\varphi$

$$\&tan\varphi = \frac{OM}{PO} = \frac{x}{h} \Rightarrow x = htan\varphi$$

Now, the equation of the circle with radius 'R' & centre at the origin

$$\Rightarrow x^2 + y^2 = R^2$$

$$\Rightarrow y = \sqrt{R^2 - x^2} \Rightarrow y = \sqrt{R^2 - (htan\varphi)^2} = \sqrt{R^2 - h^2tan^2\varphi}$$

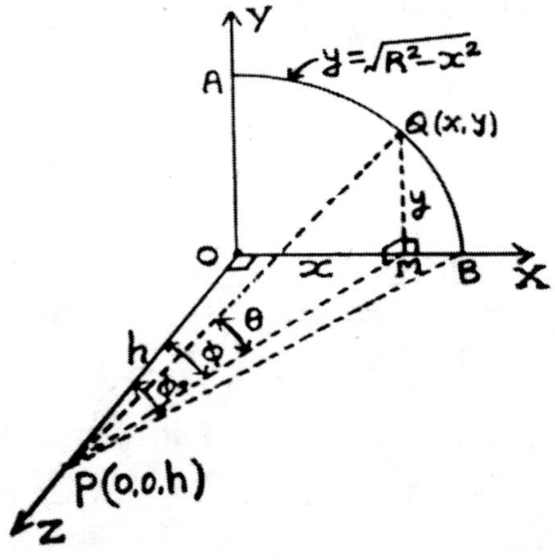

Fig (44)

Now, consider a point Q(x, y) on the circle

$$\Rightarrow QM = y = \sqrt{R^2 - h^2tan^2\varphi}$$

In right ΔPMQ

$$\Rightarrow sin\theta = \frac{QM}{PQ} = \frac{QM}{\sqrt{QM^2 + PM^2}} = \frac{\sqrt{R^2 - h^2tan^2\varphi}}{\sqrt{\left(\sqrt{R^2 - h^2tan^2\varphi}\right)^2 + h^2sec^2\varphi}}$$

$$sin\theta = \frac{\sqrt{R^2 - h^2tan^2\varphi}}{\sqrt{R^2 + h^2(sec^2\varphi - tan^2\varphi)}} = \frac{\sqrt{R^2 - h^2tan^2\varphi}}{\sqrt{h^2 + R^2}} \quad (since, sec^2\varphi - tan^2\varphi = 1)$$

rightΔPOB

$$\Rightarrow tan\varphi_0 = \frac{OB}{OP} = \frac{R}{h}$$

Let's consider an imaginary spherical surface having radius 'R' & centre at the given point 'P' such that the area of projection of the given circular plane on the spherical surface w.r.t. the given point 'P' is 'A'

Now, consider an elementary area of projection 'dA' of the circular plane OAB on the spherical surface in the first quadrant YOX (see the figure(45) below)

Now, elementary area of projection in the first quadrant

$$\Rightarrow dA = (length)(width) = (Rsin\theta d\varphi)(Rd\theta) = R^2 sin\theta d\theta d\varphi$$

Hence, area of projection(A_1) of the circular plane OAB on the spherical surface in the first quadrant is obtained by integrating the above expression in the first quadrant & applying the limits of $'\theta' from\ \theta_1 = \frac{\pi}{2} - \theta\ to\ \theta_2 = \frac{\pi}{2} \& '\varphi' from\ \varphi_1 = 0\ to\ \varphi_2 = \varphi_0$

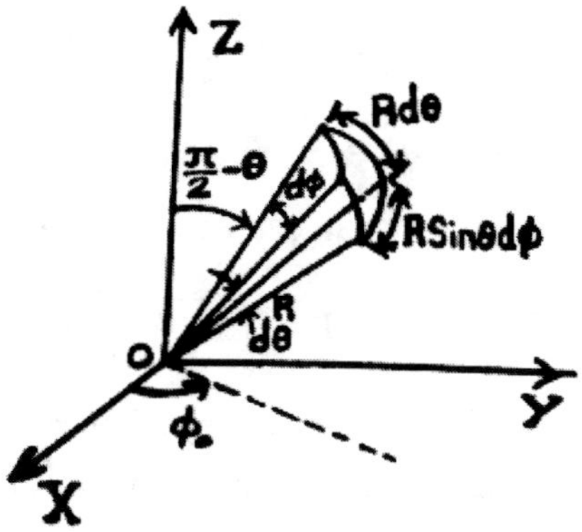

Fig (45)

we get

$$\Rightarrow A_1 = \int_0^{\varphi_0}\left[\int_{\frac{\pi}{2}-\theta}^{\frac{\pi}{2}} R^2 sin\theta d\theta\right] d\varphi = R^2\int_0^{\varphi_0}\left[\int_{\frac{\pi}{2}-\theta}^{\frac{\pi}{2}} sin\theta d\theta\right] d\varphi$$

$$= R^2\int_0^{\varphi_0} d\varphi\,[-cos\theta]_{\frac{\pi}{2}-\theta}^{\frac{\pi}{2}} = R^2\int_0^{\varphi_0} d\varphi\left[-cos\frac{\pi}{2} - cos\left(\frac{\pi}{2}-\theta\right)\right] = R^2\int_0^{\varphi_0} sin\theta d\varphi$$

$$= R^2\int_0^{\varphi_0}\frac{\sqrt{R^2 - h^2tan^2\varphi}}{\sqrt{h^2+R^2}}d\varphi = \frac{R^2h}{\sqrt{h^2+R^2}}\int_0^{\varphi=\varphi_0}\sqrt{\left(\frac{R}{h}\right)^2 - h^2tan^2\varphi\, d\varphi}$$

$$\Rightarrow A_1 = \frac{R^2h}{\sqrt{h^2+R^2}}\int_0^{\varphi=\varphi_0}\sqrt{K^2 - tan^2\varphi}\, d\varphi$$

Where $K = \frac{R}{h} = tan\varphi_0$ ....(I)

Now, solving the integral as follows

$$\Rightarrow I = \int\sqrt{K^2 - tan^2\varphi}\, d\varphi \quad let\ tan\varphi = Ksin\alpha \Rightarrow sec^2\varphi d\varphi = Kcos\alpha d\alpha\ \&$$

$$tan\alpha = \frac{sin\alpha}{cos\alpha} = \frac{sin\alpha}{\sqrt{1 - sin^2\alpha}} = \frac{tan\varphi}{\sqrt{K^2 - tan^2\varphi}}$$

$$\Rightarrow d\varphi = \frac{Kcos\alpha}{(1 + tan^2\varphi)}d\alpha = \frac{Kcos\alpha}{(1 + K^2sin^2\alpha)}d\alpha$$

$$\Rightarrow I = \int\sqrt{K^2 - K^2sin^2\alpha}\frac{Kcos\alpha}{(1 + K^2sin^2\alpha)}d\alpha = \int\frac{K^2cos^2\alpha}{(1 + K^2sin^2\alpha)}d\alpha$$

$$= \int\frac{K^2(1 - sin^2\alpha)}{(1 + K^2sin^2\alpha)}d\alpha = \int\frac{1 + K^2 - (1 + K^2sin^2\alpha)}{(1 + K^2sin^2\alpha)}d\alpha$$

$$= (1 + K^2)\int\frac{d\alpha}{(1 + K^2sin^2\alpha)} - \int d\alpha = (1 + K^2)\int\frac{sec^2\alpha d\alpha}{(sec^2\alpha + K^2tan^2\alpha)} - \alpha$$

$$= (1 + K^2)\int\frac{sec^2\alpha d\alpha}{(1 + (1 + K^2)tan^2\alpha)} - \alpha = \frac{(1 + K^2)}{(1 + K^2)}\int\frac{sec^2\alpha d\alpha}{\left(\left(\frac{1}{\sqrt{1 + K^2}}\right)^2 + tan^2\alpha\right)} - \alpha$$

$$= \int \frac{sec^2\alpha d\alpha}{\left(\left(\frac{1}{\sqrt{1+K^2}}\right)^2 + tan^2\alpha\right)} - \alpha \quad let\ tan\alpha = t \Rightarrow sec^2\alpha d\alpha = dt$$

$$= \int \frac{dt}{\left(\left(\frac{1}{\sqrt{1+K^2}}\right)^2 + t^2\right)} - \tan^{-1} t = \frac{1}{\left(\frac{1}{\sqrt{1+K^2}}\right)} \tan^{-1}\left(\frac{t}{\left(\frac{1}{\sqrt{1+K^2}}\right)}\right) - \tan^{-1} t$$

$$= \sqrt{1+K^2}\tan^{-1}\left(t\sqrt{1+K^2}\right) - \tan^{-1} t = \sqrt{1+K^2}\tan^{-1}\left(tan\alpha\sqrt{1+K^2}\right) - \alpha$$

On setting the value of $tan\alpha$, we get

$$\Rightarrow I = \sqrt{1+K^2}\tan^{-1}\left(\frac{tan\varphi\sqrt{1+K^2}}{\sqrt{K^2 - tan^2\varphi}}\right) - \tan^{-1}\left(\frac{tan\varphi}{\sqrt{K^2 - tan^2\varphi}}\right) + C$$

Where, 'C' is the constant of integration.

Thus, the area of projection of the circular plane OAB in the first quadrant YOX

$$\Rightarrow A_1 = \frac{R^2 h}{\sqrt{h^2+R^2}} \int_0^{\varphi=\varphi_0} \sqrt{K^2 - tan^2\varphi}\, d\varphi = \frac{R^2 h}{\sqrt{h^2+R^2}} [I]_0^{\varphi_0}$$

$$= \frac{R^2 h}{\sqrt{h^2+R^2}} \left[\sqrt{1+K^2}\tan^{-1}\left(\frac{tan\varphi\sqrt{1+K^2}}{\sqrt{K^2 - tan^2\varphi}}\right) - \tan^{-1}\left(\frac{tan\varphi}{\sqrt{K^2 - tan^2\varphi}}\right) + C\right]_0^{\varphi_0}$$

$$= \frac{R^2 h}{\sqrt{h^2+R^2}} \left[\sqrt{1+K^2}\tan^{-1}\left(\frac{tan\varphi_0\sqrt{1+K^2}}{\sqrt{K^2 - tan^2\varphi_0}}\right) - \tan^{-1}\left(\frac{tan\varphi_0}{\sqrt{K^2 - tan^2\varphi_0}}\right) + C\right]$$

$$- \frac{R^2 h}{\sqrt{h^2+R^2}} \left[\sqrt{1+K^2}\tan^{-1}\left(\frac{\tan(0)\sqrt{1+K^2}}{\sqrt{K^2 - tan^2(0)}}\right) - \tan^{-1}\left(\frac{\tan(0)}{\sqrt{K^2 - tan^2(0)}}\right) + C\right]$$

$$= \frac{R^2 h}{\sqrt{h^2+R^2}} \left[\sqrt{1+K^2}\tan^{-1}\left(\frac{\sqrt{1+K^2}}{\sqrt{K^2 - K^2}}\right) - \tan^{-1}\left(\frac{K}{\sqrt{K^2 - K^2}}\right)\right] - \frac{h}{\sqrt{h^2+R^2}}(0)$$

From eq(I) $\Rightarrow K = \frac{R}{h} = tan\varphi_0$

$$= \frac{R^2 h}{\sqrt{h^2+R^2}} \left[\sqrt{1+K^2}\tan^{-1}(\infty) - \tan^{-1}(\infty)\right] = \frac{R^2 h}{\sqrt{h^2+R^2}} \left[\sqrt{1+K^2}\frac{\pi}{2} - \frac{\pi}{2}\right]$$

$$= \frac{\pi R^2}{2}\left[\frac{h}{\sqrt{h^2+R^2}}\sqrt{1+\left(\frac{R}{h}\right)^2} - \frac{h}{\sqrt{h^2+R^2}}\right] = \frac{\pi R^2}{2}\left[\frac{h}{\sqrt{h^2+R^2}}\frac{\sqrt{h^2+R^2}}{h} - \frac{h}{\sqrt{h^2+R^2}}\right]$$

$$\Rightarrow A_1 = \frac{\pi R^2}{2}\left[1 - \frac{h}{\sqrt{h^2+R^2}}\right]$$

Now, the area of projection (A) of the entire circular plane

$$A = 4A_1 = 4\frac{\pi R^2}{2}\left[1 - \frac{h}{\sqrt{h^2+R^2}}\right] \quad (since\ the\ plane\ is\ symmetrical\ in\ all\ four\ quadrants)$$

$$\Rightarrow A = 2\pi R^2\left[1 - \frac{h}{\sqrt{h^2+R^2}}\right]$$

Hence, the solid angle subtended by the given circular plane at the given point 'P'

Using Fundamental Theorem of Solid Angle as follows

$$\omega = \int d\omega = \int_s \frac{dA}{r^2} = \int_s \frac{dA}{R^2} = \frac{1}{R^2}\int_s dA \qquad (since,\ 'R' is\ constant\ for\ each\ point\ )$$

$$\Rightarrow \omega = \frac{A}{R^2} = \frac{2\pi R^2\left[1 - \frac{h}{\sqrt{h^2+R^2}}\right]}{R^2} = 2\pi\left[1 - \frac{h}{\sqrt{h^2+R^2}}\right]$$

$$\Rightarrow \omega = 2\pi\left[1 - \frac{h}{\sqrt{h^2+R^2}}\right] \qquad \forall(h \geq 0) \qquad \ldots\ldots\ldots\ldots\ldots\ldots\ldots\ldots\ldots\ldots..(37)$$

Method 2: Using Integral Theorem

We know that the solid angle subtended by an object at a given point in the space is given by Integral-Theorem as follows

$$\omega = \int_{\varphi_1}^{\varphi_2}\left|\int_{\theta_1}^{\theta_2} sin\theta d\theta\right| d\varphi$$

Hence, the solid angle subtended by the rectangular plane ABCD is obtained by applying the limits of the parameters $'\theta'$ & $'\varphi'$ for whole the plane ABCD divided into four symmetrical parts about the centre 'O' (also origin)

(see the the figure (21) above)

Now, applying limits of $'\theta'\ from\ \theta_1 = \frac{\pi}{2} - \theta\ to\ \theta_2 = \frac{\pi}{2} + \theta$ &

$$'\varphi'\ from\ \varphi_1 = -\varphi_0\ to\ \varphi_2 = \varphi_0$$

$$\Rightarrow \omega = \int_{\varphi_1=-\varphi_0}^{\varphi_2=\varphi_0}\left|\int_{\theta_1=\frac{\pi}{2}-\theta}^{\theta_2=\frac{\pi}{2}+\theta} sin\theta d\theta\right| d\varphi = \int_{\varphi_1=-\varphi_0}^{\varphi_2=\varphi_0}\left|[-cos\theta]_{\frac{\pi}{2}-\theta}^{\frac{\pi}{2}+\theta} d\varphi\right|$$

$$= \int_{-\varphi_0}^{\varphi_0}\left|\left[-cos\left(\frac{\pi}{2}+\theta\right) + cos\left(\frac{\pi}{2}-\theta\right)\right]\right| d\varphi = 2\int_{-\varphi_0}^{\varphi_0}|sin\theta| d\varphi$$

Now, on setting the value of $sin\theta$ in the above expression, we get

$$\omega = 2\int_{-\varphi_0}^{\varphi_0}\left|\frac{\sqrt{R^2 - h^2tan^2\varphi}}{\sqrt{h^2+R^2}}\right| d\varphi = 4\int_0^{\varphi_0}\left|\frac{\sqrt{R^2 - h^2tan^2\varphi}}{\sqrt{h^2+R^2}}\right| d\varphi$$

$$\left(since,\ \int_{-a}^{a} f(x)dx = 2\int_0^a f(x)dx\ iff\ f(-x) = f(x)\right)$$

$$= 4\int_0^{\varphi_0}\frac{\sqrt{R^2 - h^2tan^2\varphi}}{\sqrt{h^2+R^2}} d\varphi \quad \left(since, 0 \leq \varphi \leq \frac{\pi}{2} \& |f(x)| = f(x)\ \forall\ f(x) > 0\right)$$

$$= \frac{4h}{\sqrt{h^2+R^2}}\int_0^{\varphi_0}\sqrt{\left(\frac{R}{h}\right)^2 - tan^2\varphi d\varphi} = \frac{4h}{\sqrt{h^2+R^2}}\int_0^{\varphi_0}\sqrt{K^2 - tan^2\varphi} d\varphi$$

Where $K = \frac{R}{h} = tan\varphi_0$ ....(I)

Now, solving the integral as follows

$$\Rightarrow I = \int\sqrt{K^2 - tan^2\varphi}\, d\varphi \quad let\ tan\varphi = Ksin\alpha \Rightarrow sec^2\varphi d\varphi = Kcos\alpha d\alpha\ \&$$

$$tan\alpha = \frac{sin\alpha}{cos\alpha} = \frac{sin\alpha}{\sqrt{1 - sin^2\alpha}} = \frac{tan\varphi}{\sqrt{K^2 - tan^2\varphi}}$$

$$\Rightarrow d\varphi = \frac{Kcos\alpha}{(1 + tan^2\varphi)} d\alpha = \frac{Kcos\alpha}{(1 + K^2sin^2\alpha)} d\alpha$$

$$\Rightarrow I = \int \sqrt{K^2 - K^2 sin^2\alpha}\frac{Kcos\alpha}{(1 + K^2 sin^2\alpha)}d\alpha = \int \frac{K^2 cos^2\alpha}{(1 + K^2 sin^2\alpha)}d\alpha$$

$$= \int \frac{K^2(1 - sin^2\alpha)}{(1 + K^2 sin^2\alpha)}d\alpha = \int \frac{1 + K^2 - (1 + K^2 sin^2\alpha)}{(1 + K^2 sin^2\alpha)}d\alpha$$

$$= (1 + K^2)\int \frac{d\alpha}{(1 + K^2 sin^2\alpha)} - \int d\alpha = (1 + K^2)\int \frac{sec^2\alpha d\alpha}{(sec^2\alpha + K^2 tan^2\alpha)} - \alpha$$

$$= (1 + K^2)\int \frac{sec^2\alpha d\alpha}{(1 + (1 + K^2)tan^2\alpha)} - \alpha = \frac{(1 + K^2)}{(1 + K^2)}\int \frac{sec^2\alpha d\alpha}{\left(\left(\frac{1}{\sqrt{1 + K^2}}\right)^2 + tan^2\alpha\right)} - \alpha$$

$$= \int \frac{sec^2\alpha d\alpha}{\left(\left(\frac{1}{\sqrt{1 + K^2}}\right)^2 + tan^2\alpha\right)} - \alpha \quad let\ tan\alpha = t \Rightarrow sec^2\alpha d\alpha = dt$$

$$= \int \frac{dt}{\left(\left(\frac{1}{\sqrt{1 + K^2}}\right)^2 + t^2\right)} - \tan^{-1} t = \frac{1}{\left(\frac{1}{\sqrt{1 + K^2}}\right)}\tan^{-1}\left(\frac{t}{\left(\frac{1}{\sqrt{1 + K^2}}\right)}\right) - \tan^{-1} t$$

$$= \sqrt{1 + K^2}\tan^{-1}\left(t\sqrt{1 + K^2}\right) - \tan^{-1} t = \sqrt{1 + K^2}\tan^{-1}\left(tan\alpha\sqrt{1 + K^2}\right) - \alpha$$

On setting the value of $tan\alpha$, we get

$$\Rightarrow I = \sqrt{1 + K^2}\tan^{-1}\left(\frac{tan\varphi\sqrt{1 + K^2}}{\sqrt{K^2 - tan^2\varphi}}\right) - \tan^{-1}\left(\frac{tan\varphi}{\sqrt{K^2 - tan^2\varphi}}\right) + C$$

Where 'C' is the constant of integration

Thus, the area of projection of the circular plane OAB in the first quadrant YOX

$$\Rightarrow \omega = \frac{4h}{\sqrt{h^2 + R^2}}\int_0^{\varphi=\varphi_0} \sqrt{K^2 - tan^2\varphi}\, d\varphi = \frac{4h}{\sqrt{h^2 + R^2}}[I]_0^{\varphi_0}$$

$$= \frac{4h}{\sqrt{h^2 + R^2}}\left[\sqrt{1 + K^2}\tan^{-1}\left(\frac{tan\varphi\sqrt{1 + K^2}}{\sqrt{K^2 - tan^2\varphi}}\right) - \tan^{-1}\left(\frac{tan\varphi}{\sqrt{K^2 - tan^2\varphi}}\right) + C\right]_0^{\varphi_0}$$

$$= \frac{4h}{\sqrt{h^2 + R^2}}\left[\sqrt{1 + K^2}\tan^{-1}\left(\frac{tan\varphi_0\sqrt{1 + K^2}}{\sqrt{K^2 - tan^2\varphi_0}}\right) - \tan^{-1}\left(\frac{tan\varphi_0}{\sqrt{K^2 - tan^2\varphi_0}}\right) + C\right]$$

$$- \frac{4h}{\sqrt{h^2 + R^2}}\left[\sqrt{1 + K^2}\tan^{-1}\left(\frac{\tan(0)\sqrt{1 + K^2}}{\sqrt{K^2 - tan^2(0)}}\right) - \tan^{-1}\left(\frac{\tan(0)}{\sqrt{K^2 - tan^2(0)}}\right) + C\right]$$

$$= \frac{4h}{\sqrt{h^2 + R^2}}\left[\sqrt{1 + K^2}\tan^{-1}\left(\frac{\sqrt{1 + K^2}}{\sqrt{K^2 - K^2}}\right) - \tan^{-1}\left(\frac{K}{\sqrt{K^2 - K^2}}\right)\right] - \frac{h}{\sqrt{h^2 + R^2}}[0]$$

From eq(I) $\Rightarrow K = \frac{R}{h} = tan\varphi_0$

$$\omega = \frac{4h}{\sqrt{h^2 + R^2}}\left[\sqrt{1 + K^2}\tan^{-1}(\infty) - \tan^{-1}(\infty)\right] = \frac{4h}{\sqrt{h^2 + R^2}}\left[\sqrt{1 + K^2}\frac{\pi}{2} - \frac{\pi}{2}\right]$$

$$= \frac{4\pi}{2}\left[\frac{h}{\sqrt{h^2 + R^2}}\sqrt{1 + \left(\frac{R}{h}\right)^2} - \frac{h}{\sqrt{h^2 + R^2}}\right] = 2\pi\left[\frac{h}{\sqrt{h^2 + R^2}}\frac{\sqrt{h^2 + R^2}}{h} - \frac{h}{\sqrt{h^2 + R^2}}\right]$$

$$= 2\pi\left[1 - \frac{h}{\sqrt{h^2 + R^2}}\right]$$

$$\Rightarrow \omega = 2\pi\left[1 - \frac{h}{\sqrt{h^2 + R^2}}\right] \qquad \forall(h \geq 0) \qquad \ldots\ldots\ldots\ldots\ldots\ldots\ldots..(37)$$

Method 3: Using Corollary

Let there be a circular plane having radius equal to 'R' & centre 'O' and a given point say P(0, 0, h) at a height 'h' lying on the axis normal to the plane & passing through the centre 'O' (as shown in the figure (46) below)

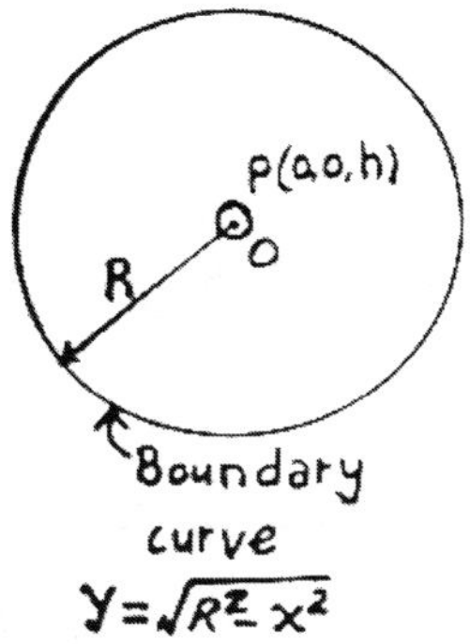

Fig (46)

We know that the equation of the circle with radius 'R' &centre at the origin 'O' is given as

$$x^2 + y^2 = R^2 \Rightarrow y = \sqrt{R^2 - x^2} \quad \forall(-R \leq x \leq R)$$

Since, the circular plane is symmetrical in all the four quadrants about the origin 'O' (also the centre of the plane)& the plane is bounded by the following curves in the first quadrant

$$x = 0 \text{ (Y-axis)}, y = 0 \text{ (X-axis) } \& y = \sqrt{R^2 - x^2} \quad \forall(0 \leq x \leq R)$$

(see the figure (47) below)

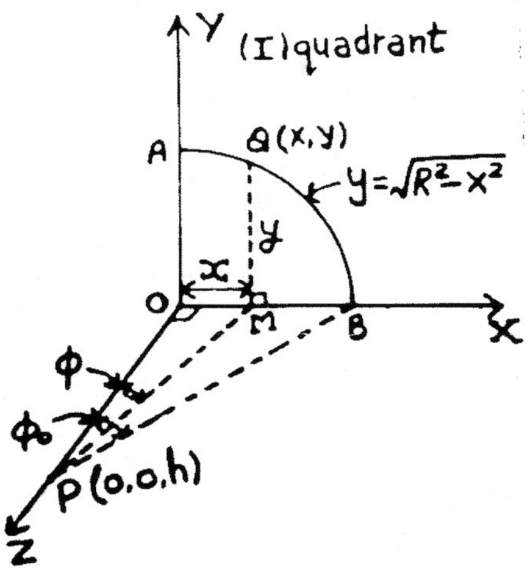

Fig (47)

Hence, the equation of the boundary curve of the circular plane in the first quadrant

$$y = \sqrt{R^2 - x^2} \quad \forall(0 \leq x \leq R)(since, \; y = f(x) \neq 0)$$

$$\therefore f(x) = \sqrt{R^2 - x^2}$$

Now, consider an arbitrary point Q(x, y) on the boundary curve $f(x) = \sqrt{R^2 - x^2}$

In right ΔPOM

$$\Rightarrow tan\varphi = \frac{OM}{OP} = \frac{x}{h} \Rightarrow x = htan\varphi$$

$$\Rightarrow f(htan\varphi) = \sqrt{R^2 - (htan\varphi)^2} = \sqrt{R^2 - h^2tan^2\varphi}$$

In right ΔPOB

$$\Rightarrow tan\varphi_0 = \frac{OB}{OP} = \frac{R}{h}$$

Hence, the solid angle subtended by the given circular plane at the given point 'P' in the first quadrant, using Corollary as follows

$$\Rightarrow \omega_1 = \int_0^{\varphi} \frac{f(htan\varphi)d\varphi}{\sqrt{(f(htan\varphi))^2 + h^2sec^2\varphi}} = \int_0^{\varphi} \frac{\sqrt{R^2 - h^2tan^2\varphi}\ d\varphi}{\sqrt{\left(\sqrt{R^2 - h^2tan^2\varphi}\right)^2 + h^2sec^2\varphi}}$$

On applying the limit of φ from 0 to φ_0, we get

$$\omega_1 = h\int_0^{\varphi=\varphi_0} \frac{\sqrt{\left(\frac{R}{h}\right)^2 - tan^2\varphi}\ d\varphi}{\sqrt{R^2 + h^2(sec^2\varphi - tan^2\varphi)}} = h\int_0^{\varphi=\varphi_0} \frac{\sqrt{K^2 - tan^2\varphi}\ d\varphi}{\sqrt{h^2 + R^2}}$$

$$(since, sec^2\varphi - tan^2\varphi = 1)$$

Where $K = \frac{R}{h} = tan\varphi_0$ …………… (I)

$$\Rightarrow \omega_1 = \frac{h}{\sqrt{h^2 + R^2}}\int_0^{\varphi=\varphi_0} \sqrt{K^2 - tan^2\varphi}\, d\varphi$$

Now, solving the integral as follows

$$\Rightarrow I = \int \sqrt{K^2 - tan^2\varphi}\, d\varphi \quad let\ tan\varphi = Ksin\theta \Rightarrow sec^2\varphi d\varphi = Kcos\theta d\theta\ \&$$

$$tan\theta = \frac{sin\theta}{cos\theta} = \frac{sin\theta}{\sqrt{1 - sin^2\theta}} = \frac{tan\varphi}{\sqrt{K^2 - tan^2\varphi}}$$

$$\Rightarrow d\varphi = \frac{Kcos\theta}{(1 + tan^2\varphi)}d\theta = \frac{Kcos\theta}{(1 + K^2sin^2\theta)}d\theta$$

$$\Rightarrow I = \int \sqrt{K^2 - K^2sin^2\theta}\frac{Kcos\theta}{(1 + K^2sin^2\theta)}d\theta = \int \frac{K^2cos^2\theta}{(1 + K^2sin^2\theta)}d\theta$$

$$= \int \frac{K^2(1 - sin^2\theta)}{(1 + K^2sin^2\theta)}d\theta = \int \frac{1 + K^2 - (1 + K^2sin^2\theta)}{(1 + K^2sin^2\theta)}d\theta$$

$$= (1 + K^2)\int \frac{d\theta}{(1 + K^2sin^2\theta)} - \int d\theta = (1 + K^2)\int \frac{sec^2\theta d\theta}{(sec^2\theta + K^2tan^2\theta)} - \theta$$

$$= (1 + K^2)\int \frac{sec^2\theta d\theta}{(1 + (1 + K^2)tan^2\theta)} - \theta = \frac{(1 + K^2)}{(1 + K^2)}\int \frac{sec^2\theta d\theta}{\left(\left(\frac{1}{\sqrt{1 + K^2}}\right)^2 + tan^2\theta\right)} - \theta$$

$$= \int \frac{sec^2\theta d\theta}{\left(\left(\frac{1}{\sqrt{1 + K^2}}\right)^2 + tan^2\theta\right)} - \theta \quad let\ tan\theta = t \Rightarrow sec^2\theta d\theta = dt$$

$$= \int \frac{dt}{\left(\left(\frac{1}{\sqrt{1 + K^2}}\right)^2 + t^2\right)} - \tan^{-1} t = \frac{1}{\left(\frac{1}{\sqrt{1 + K^2}}\right)}\tan^{-1}\left(\frac{t}{\left(\frac{1}{\sqrt{1 + K^2}}\right)}\right) - \tan^{-1} t$$

$$= \sqrt{1 + K^2}\tan^{-1}\left(t\sqrt{1 + K^2}\right) - \tan^{-1} t = \sqrt{1 + K^2}\tan^{-1}\left(tan\theta\sqrt{1 + K^2}\right) - \theta$$

On setting the value of $tan\theta$, we get

$$\Rightarrow I = \sqrt{1 + K^2}\tan^{-1}\left(\frac{tan\varphi\sqrt{1 + K^2}}{\sqrt{K^2 - tan^2\varphi}}\right) - \tan^{-1}\left(\frac{tan\varphi}{\sqrt{K^2 - tan^2\varphi}}\right) + C$$

Where 'C' is the constant of integration

$$\Rightarrow \omega_1 = \frac{h}{\sqrt{h^2+R^2}}\int_0^{\varphi=\varphi_0}\sqrt{K^2-tan^2\varphi}\,d\varphi = \frac{h}{\sqrt{h^2+R^2}}[I]_0^{\varphi_0}$$

$$= \frac{h}{\sqrt{h^2+R^2}}\left[\sqrt{1+K^2}\tan^{-1}\left(\frac{tan\varphi\sqrt{1+K^2}}{\sqrt{K^2-tan^2\varphi}}\right)-\tan^{-1}\left(\frac{tan\varphi}{\sqrt{K^2-tan^2\varphi}}\right)+C\right]_0^{\varphi_0}$$

$$= \frac{h}{\sqrt{h^2+R^2}}\left[\sqrt{1+K^2}\tan^{-1}\left(\frac{tan\varphi_0\sqrt{1+K^2}}{\sqrt{K^2-tan^2\varphi_0}}\right)-\tan^{-1}\left(\frac{tan\varphi_0}{\sqrt{K^2-tan^2\varphi_0}}\right)+C\right]$$

$$-\frac{h}{\sqrt{h^2+R^2}}\left[\sqrt{1+K^2}\tan^{-1}\left(\frac{\tan(0)\sqrt{1+K^2}}{\sqrt{K^2-tan^2(0)}}\right)-\tan^{-1}\left(\frac{\tan(0)}{\sqrt{K^2-tan^2(0)}}\right)+C\right]$$

$$= \frac{h}{\sqrt{h^2+R^2}}\left[\sqrt{1+K^2}\tan^{-1}\left(\frac{\sqrt{1+K^2}}{\sqrt{K^2-K^2}}\right)-\tan^{-1}\left(\frac{K}{\sqrt{K^2-K^2}}\right)\right]-\frac{h}{\sqrt{h^2+R^2}}[0]$$

From eq(I) $\Rightarrow K = \frac{R}{h} = tan\varphi_0$

$$= \frac{h}{\sqrt{h^2+R^2}}\left[\sqrt{1+K^2}\tan^{-1}(\infty)-\tan^{-1}(\infty)\right] = \frac{h}{\sqrt{h^2+R^2}}\left[\sqrt{1+K^2}\frac{\pi}{2}-\frac{\pi}{2}\right]$$

$$= \frac{\pi}{2}\left[\frac{h}{\sqrt{h^2+R^2}}\sqrt{1+\left(\frac{R}{h}\right)^2}-\frac{h}{\sqrt{h^2+R^2}}\right] = \frac{\pi}{2}\left[\frac{h}{\sqrt{h^2+R^2}}\frac{\sqrt{h^2+R^2}}{h}-\frac{h}{\sqrt{h^2+R^2}}\right]$$

$$\Rightarrow \omega_1 = \frac{\pi}{2}\left[1-\frac{h}{\sqrt{h^2+R^2}}\right]$$

Hence, the solid angle subtended by the entire circular plane at the given point 'P'

$$\Rightarrow \omega = 4\omega_1 = 4\frac{\pi}{2}\left[1-\frac{h}{\sqrt{h^2+R^2}}\right] \quad (since\ the\ plane\ is\ symmetrical\ in\ all\ four\ quadrants)$$

$$\Rightarrow \omega = 2\pi\left[1-\frac{h}{\sqrt{h^2+R^2}}\right] \qquad \forall(h\geq 0) \qquad \ldots\ldots\ldots\ldots(37)$$

Method 4: Alternative

Let there be a circular plane having radius equal to 'R' & centre 'O' and a given point say P(0, 0, h) at a height 'h' lying on the axis normal to the plane & passing through the centre 'O' (as shown in the figure (48) below)

in right $\Delta POB \Rightarrow cos\alpha = \frac{OP}{PB} = \frac{OP}{\sqrt{OP^2+OB^2}} = \frac{h}{\sqrt{h^2+R^2}}$ ....(I)

Now, consider an imaginary right cone with the apex at the given point P(0, 0, h) & apex angle $'2\alpha'$ such that its lateral face is coincided with the circular boundary of the given plane

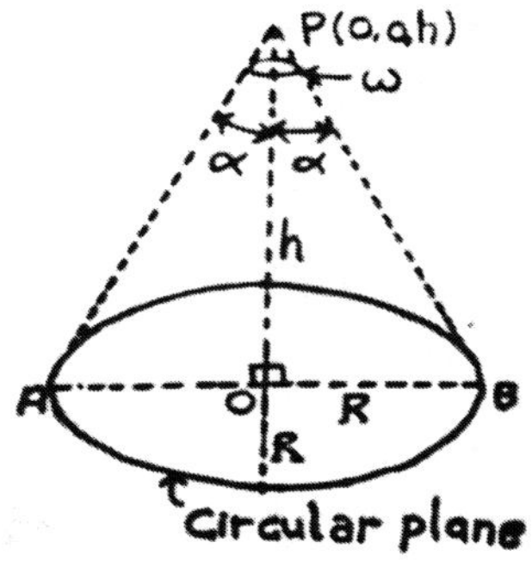

Fig (48)

In this case, the solid angle subtended by the circular plane at the point 'P'

= solid angle subtended by the right cone at the apex 'P'

We know that the solid angle subtended by the right cone at the apex is given by the eq(1) as follows

$\omega = 2\pi(1-cos\alpha)$where $'2\alpha'$is the apex angle

Hence,the solid angle subtended by the circular plane at the point 'P'

Now, on setting the value of $cos\alpha$from the eq(I), we get

$$\omega = 2\pi\left[1-\frac{h}{\sqrt{h^2+R^2}}\right] \forall(h\geq 0) \qquad \ldots\ldots\ldots\ldots(37)$$

2.3.24 Solid angle subtended by a radial segment of the circular plane at any point lying on the vertical axis passing through the centre

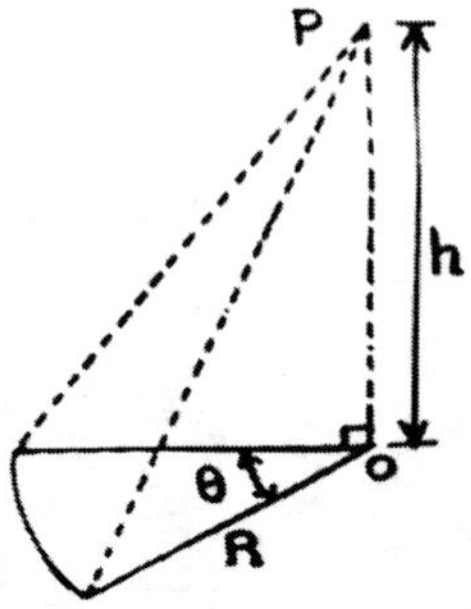

Fig (49)

Let there be a radial segment of the circular plane with radius 'R' & centre 'O' and a given point say 'P' at a height 'h' lying on the vertical axis passing through the centre 'O'. (as shown in the figure)

Let $'\theta'$ be the angle subtended by the radial segment at the centre 'O'

We know that the solid angle subtended by the complete circular plane at any point lying on the vertical axis passing through the centre is given by eq(37) as follows

$$\omega = 2\pi\left[1 - \frac{h}{\sqrt{h^2 + R^2}}\right]$$

A complete circular plane is symmetrical about its centre & subtends an angle $'2\pi'$ at its centre

Also, the radial segment is symmetrical about the centre 'O' with angle θ i.e. it is a symmetrical part of complete circular plane with angle 2π

Hence, using symmetry the solid angle subtended by the given radial segment at the same point 'P'

$$\Rightarrow \omega = \frac{\theta}{2\pi}\left(2\pi\left[1 - \frac{h}{\sqrt{h^2 + R^2}}\right]\right) = \theta\left[1 - \frac{h}{\sqrt{h^2 + R^2}}\right] \ \forall (h \geq 0) \dots\dots\dots\dots\dots.. (38)$$

2.3.25 Infinitely long rectangular plane ($l \to \infty$)

Here, we are to analyse a rectangular plane having infinite length ($l \to \infty$)& afinite width (b)in order to find out the solid angle subtended by the same plane at any point lying on the vertical axis passing through the certain points within the boundary of the plane. The rectangular plane having one finite dimension while other infinite subtends a finite solid angle at any point in the space

It is to be noted that the value of solid angle is finite subtended by the infinitely long cylinder having the finite radius. Let us consider some particular locations of the vertical axis on which the given point is lying.

2.3.26 Solid angle subtended by an infinitely long rectangular plane with finite width at any point lying on the vertical axis passing through the vertex of the plane

There are different methods of finding the value of solid angle. Let's discuss them consequently as follows

Method 1: Using Fundamental Theorem

Let there be an infinitely long rectangular plane having finite width 'b' & a given point say

P(0, 0, h) at a height 'h' lying on the vertical axis passing through the vertex 'O' of the plane. (i.e. lying on the Z-axis) (as shown in the figure (50) below)

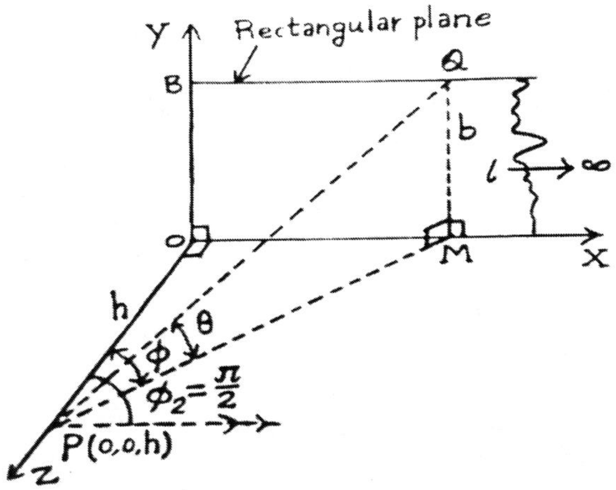

Fig (50)

In rightΔPOM

$$\Rightarrow sec\varphi = \frac{PM}{PO} \Rightarrow PM = POsec\varphi = hsec\varphi$$

In rightΔPMQ

$$\Rightarrow sin\theta = \frac{QM}{PQ} = \frac{QM}{\sqrt{QM^2 + PM^2}} = \frac{b}{\sqrt{b^2 + h^2sec^2\varphi}}$$

$$sin\theta = \frac{bcos\varphi}{\sqrt{b^2cos^2\varphi + h^2}}$$

Let's consider an imaginary spherical surface having radius 'R' & centre at the given point 'P' such that the area of projection of the given infinitely long rectangular plane on the spherical surface w.r.t. the given point 'P' is 'A'

Now, consider an elementary area of projection 'dA' of the plane on the spherical surface in the first quadrant YOX (see the figure(51) below)

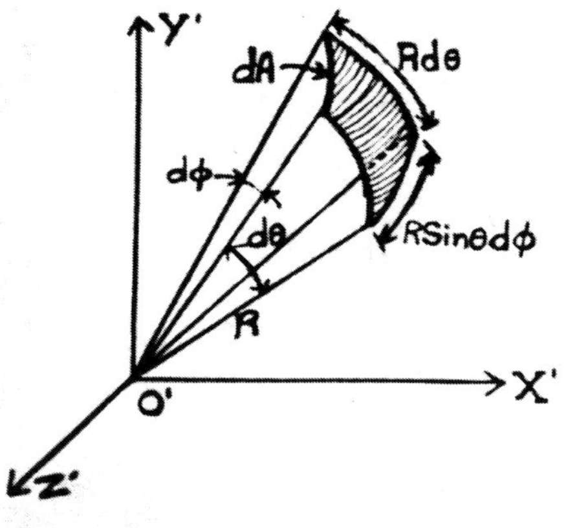

Fig (51)

Now, elementary area of projection in the first quadrant (in form of spherical co-ordinates)

$$\Rightarrow dA = (length)(width) = (Rsin\theta d\varphi)(Rd\theta) = R^2sin\theta d\theta d\varphi$$

Hence, area of projection of the plane on the spherical surface is obtained by integrating the above expression in the first quadrant & applying the limits of $'\theta' from\ \theta_1 = \frac{\pi}{2} - \theta\ to\ \theta_2 = \frac{\pi}{2}$&$'\varphi' from\ \varphi_1 = 0\ to\ \varphi_2 = \frac{\pi}{2}$

$$\Rightarrow A=\int_{0}^{\frac{\pi}{2}}\left[\int_{\frac{\pi}{2}-\theta}^{\frac{\pi}{2}}R^2 sin\theta d\theta\right]d\varphi=R^2\int_{0}^{\frac{\pi}{2}}\left[\int_{\frac{\pi}{2}-\theta}^{\frac{\pi}{2}}sin\theta d\theta\right]d\varphi$$

$$=R^2\int_{0}^{\frac{\pi}{2}}d\varphi\,[-cos\theta]_{\frac{\pi}{2}-\theta}^{\frac{\pi}{2}}=R^2\int_{0}^{\frac{\pi}{2}}sin\theta d\varphi=R^2\int_{0}^{\frac{\pi}{2}}\frac{bcos\varphi}{\sqrt{b^2cos^2\varphi+h^2}}d\varphi$$

$$\Rightarrow A=R^2\int_{0}^{\frac{\pi}{2}}\frac{bcos\varphi}{\sqrt{b^2-b^2sin^2\varphi+h^2}}d\varphi\,(since,\ cos^2\alpha=1-sin^2\alpha)$$

$$=R^2\int_{0}^{\frac{\pi}{2}}\frac{bcos\varphi}{b\sqrt{\frac{b^2+h^2}{b^2}-sin^2\varphi}}d\varphi=R^2\int_{0}^{\frac{\pi}{2}}\frac{cos\varphi}{\sqrt{K^2-sin^2\varphi}}d\varphi$$

$$where\ K=\sqrt{\frac{b^2+h^2}{b^2}}=\frac{\sqrt{b^2+h^2}}{b}$$

$$\Rightarrow A=R^2\left[\sin^{-1}\left\{\frac{sin\varphi}{K}\right\}\right]_0^{\frac{\pi}{2}}=R^2\left[\sin^{-1}\left\{\frac{sin\frac{\pi}{2}}{K}\right\}-\sin^{-1}\left\{\frac{\sin(0)}{K}\right\}\right]=R^2\sin^{-1}\left\{\frac{1}{K}\right\}$$

Now, on putting the values of K,we get

$$\Rightarrow A=R^2\sin^{-1}\left\{\frac{1}{\frac{\sqrt{b^2+h^2}}{b}}\right\}=R^2\sin^{-1}\left\{\frac{b}{\sqrt{b^2+h^2}}\right\}$$

Now, the solid angle subtended by the infinitely long rectangular plane at the given point 'P' = solid angle subtended by the area of projection (A) of the plane on the spherical surface at the same point 'P'

$$\Rightarrow \omega=\int_s\frac{dA}{r^2}=\int_s\frac{dA}{R^2}=\frac{1}{R^2}\int_s dA=\frac{A}{R^2}(where, A\ is\ area\ of\ projection)$$

$$(since, R\ is\ constant\ for\ each\ point\ on\ the\ area\ of\ projection\ on\ the\ spherical\ surface)$$

$$\Rightarrow \omega=\frac{1}{R^2}\left[R^2\sin^{-1}\left\{\frac{b}{\sqrt{(b^2+h^2)}}\right\}\right]=\sin^{-1}\left\{\frac{b}{\sqrt{(b^2+4h^2)}}\right\}$$

$$\Rightarrow\left[\omega=\sin^{-1}\left\{\frac{b}{\sqrt{(b^2+4h^2)}}\right\}\right]\quad\forall(h\geq 0)\ldots\ldots\ldots\ldots\ldots\ldots\ldots(2.4)$$

Method 2: Using Integral Theorem

We know that the solid angle subtended by an object at a given point in the space is given by Integral-Theorem as follows

$$\omega=\int_{\varphi_1}^{\varphi_2}\left|\int_{\theta_1}^{\theta_2}sin\theta d\theta\right|d\varphi$$

Hence, the solid angle subtended by the infinitely long rectangular plane is obtained by applying the limits of the parameters $'\theta'$ & $'\varphi'$ for whole the plane (see the figure (52) below)

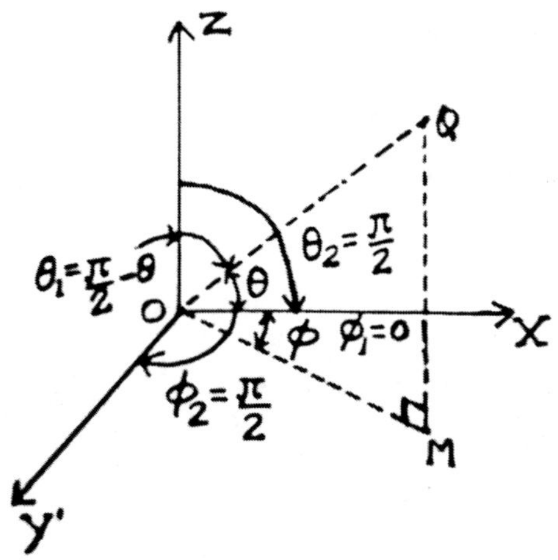

Fig (52)

Now, applying limits of $'\theta'$ $from\ \theta_1 = \frac{\pi}{2} - \theta\ to\ \theta_2 = \frac{\pi}{2}$&

$$'\varphi'\ from\ \varphi_1 = 0\ to\ \varphi_2 = \frac{\pi}{2}$$

$$\Rightarrow \omega = \int_{\varphi_1=0}^{\varphi_2=\frac{\pi}{2}} \left| \int_{\theta_1=\frac{\pi}{2}-\theta}^{\theta_2=\frac{\pi}{2}} sin\theta d\theta \right| d\varphi = \int_{\varphi_1=0}^{\varphi_2=\frac{\pi}{2}} \left| [-cos\theta]_{\frac{\pi}{2}-\theta}^{\frac{\pi}{2}} d\varphi \right|$$

$$= \int_{\varphi_1=0}^{\varphi_2=\frac{\pi}{2}} \left| \left[-cos\frac{\pi}{2} + cos\left(\frac{\pi}{2} - \theta\right)\right] \right| d\varphi = \int_{\varphi_1=0}^{\varphi_2=\frac{\pi}{2}} |[0 + sin\theta]| d\varphi = \int_0^{\frac{\pi}{2}} |sin\theta| d\varphi$$

on putting the value of $sin\theta$,we get

$$\Rightarrow \omega = \int_0^{\frac{\pi}{2}} \left| \frac{bcos\varphi}{\sqrt{b^2cos^2\varphi + h^2}} \right| d\varphi = \int_0^{\frac{\pi}{2}} \left| \frac{bcos\varphi}{\sqrt{b^2cos^2\varphi + h^2}} \right| d\varphi$$

$$\Rightarrow \omega = \int_0^{\frac{\pi}{2}} \frac{bcos\varphi}{\sqrt{b^2cos^2\varphi + 4h^2}} d\varphi \left(since, 0 \le \varphi \le \frac{\pi}{2} \& |f(x)| = f(x)\ \forall f(x) > 0\right)$$

$$= \int_0^{\frac{\pi}{2}} \frac{bcos\varphi}{\sqrt{b^2 - b^2sin^2\varphi + 4h^2}} d\varphi = \int_0^{\frac{\pi}{2}} \frac{bcos\varphi}{b\sqrt{\frac{b^2+4h^2}{b^2} - sin^2\varphi}} d\varphi$$

$$\Rightarrow \omega = \int_0^{\frac{\pi}{2}} \frac{cos\varphi}{\sqrt{K^2 - sin^2\varphi}} d\varphi$$

$$where\ K = \sqrt{\frac{b^2 + 4h^2}{b^2}} = \frac{\sqrt{b^2 + 4h^2}}{b}$$

On solving the integration as follows

$$I = \int \frac{cos\varphi}{\sqrt{K^2 - sin^2\varphi}} d\varphi\ \ let\ sin\varphi = t \Rightarrow cos\varphi d\varphi = dt$$

$$\Rightarrow I = \int \frac{dt}{\sqrt{K^2 - t^2}} = \sin^{-1}\left(\frac{t}{K}\right) + C = \sin^{-1}\left(\frac{\sin\varphi}{K}\right) + C$$

Where, 'C' is the constant of integration.

Hence, we get

$$\Rightarrow \omega = \int_0^{\frac{\pi}{2}} \frac{cos\varphi}{\sqrt{K^2 - sin^2\varphi}} d\varphi = [I]_0^{\frac{\pi}{2}}$$

$$\Rightarrow \omega = \left[\sin^{-1}\left\{\frac{sin\varphi}{K}\right\} + C\right]_0^{\frac{\pi}{2}} = \left[\sin^{-1}\left\{\frac{sin\frac{\pi}{2}}{K}\right\} - \sin^{-1}\left\{\frac{\sin(0)}{K}\right\} - 0\right] = \sin^{-1}\left\{\frac{1}{K}\right\}$$

Now, on putting the values of K, we get

$$\Rightarrow \omega = \sin^{-1}\left\{\frac{1}{\frac{\sqrt{b^2+4h^2}}{b}}\right\} = \sin^{-1}\left\{\frac{b}{\sqrt{(b^2+4h^2)}}\right\} \quad \forall(h \geq 0) \ldots\ldots\ldots\ldots\ldots (2.4)$$

Method 3: Using Corollary

Let there be an infinitely long rectangular planehaving finite width 'b' & a given point say

P(0, 0, h) at a height 'h' lying on the vertical axis passing through the vertex 'O' of the plane. (i.e. lying on the Z-axis) (as shown in the figure (53) below)

Since, the rectangular plane is bounded by the straight lines x = 0, y = 0& y =b

But $y = f(x) \neq 0$

Hence, the equation of the boundary-curve $y = f(x) = b \ \forall(0 \leq x < \infty)$

Let's consider a point Q(x, y) on the boundary curve (i.e. straight line BQ) $y = b$

Now, in rightΔPOM

$$\Rightarrow tan\varphi = \frac{OM}{OP} = \frac{x}{h} \Rightarrow x = htan\varphi$$

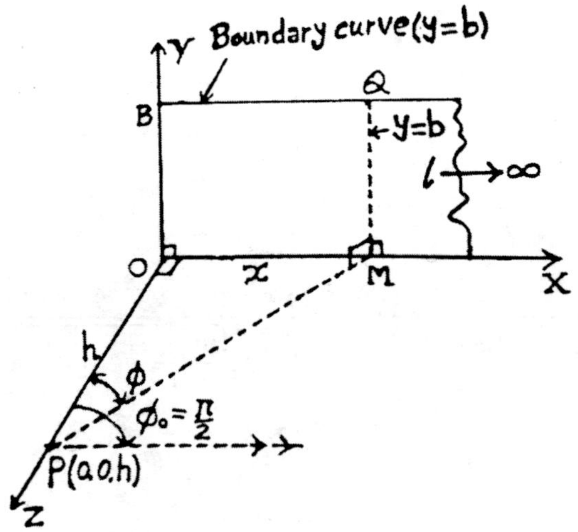

Fig(53)

$$\therefore f(htan\varphi) = b$$

Now, the solid angle subtended by the given rectangular plane at the given point 'P'

By using Corollary

$$\Rightarrow \omega = \int_0^{\varphi} \frac{f(htan\varphi)d\varphi}{\sqrt{(f(htan\varphi))^2 + h^2sec^2\varphi}}$$

Now, on putting the value of $f(htan\varphi)$& applying the limit of φ, we get

$$\omega = \int_0^{\varphi=\varphi_0} \frac{bd\varphi}{\sqrt{b^2 + h^2sec^2\varphi}} = \int_0^{\varphi=\frac{\pi}{2}} \frac{bd\varphi}{\sqrt{b^2 + h^2sec^2\varphi}} = \int_0^{\frac{\pi}{2}} \frac{bcos\varphi d\varphi}{\sqrt{b^2cos^2\varphi + h^2}}$$

$$= \int_0^{\frac{\pi}{2}} \frac{bcos\varphi}{\sqrt{b^2 - b^2sin^2\varphi + h^2}} d\varphi$$

$$= \int_0^{\frac{\pi}{2}} \frac{bcos\varphi}{b\sqrt{\frac{b^2+h^2}{b^2} - sin^2\varphi}} d\varphi = \int_0^{\frac{\pi}{2}} \frac{cos\varphi}{\sqrt{K^2 - sin^2\varphi}} d\varphi$$

$$where\ K = \sqrt{\frac{b^2+h^2}{b^2}} = \frac{\sqrt{b^2+h^2}}{b}$$

$$\Rightarrow \omega = \left[\sin^{-1}\left\{\frac{sin\varphi}{K}\right\}\right]_0^{\frac{\pi}{2}} = \left[\sin^{-1}\left\{\frac{sin\frac{\pi}{2}}{K}\right\} - \sin^{-1}\left\{\frac{\sin(0)}{K}\right\} - 0\right] == \sin^{-1}\left\{\frac{1}{K}\right\}$$

Now, on putting the values of K, we get

$$\Rightarrow \omega = \sin^{-1}\left\{\frac{1}{\frac{\sqrt{b^2+h^2}}{b}}\right\} = \sin^{-1}\left\{\frac{b}{\sqrt{b^2+h^2}}\right\} \ \forall(h \geq 0) \ \ldots\ldots\ldots\ldots\ldots\ldots\ldots (2.4)$$

Method 4: Alternative method

We know that the solid angle subtended a rectangular plane at any point lying on the vertical axis passing through the vertex is given by eq(17) as follows

$$\omega = \sin^{-1}\left\{\frac{lb}{\sqrt{(l^2+h^2)(b^2+h^2)}}\right\}$$

Now taking the limit of ω *when lenth*, $l \to \infty$ from the above expression as follows

$$\Rightarrow \omega = \lim_{l\to\infty} \sin^{-1}\left\{\frac{lb}{\sqrt{(l^2+h^2)(b^2+h^2)}}\right\} = \lim_{l\to\infty} \sin^{-1}\left\{\frac{b}{\sqrt{\frac{(l^2+h^2)}{l^2}(b^2+h^2)}}\right\}$$

$$= \lim_{l\to\infty} \sin^{-1}\left\{\frac{b}{\sqrt{\left(1+\frac{h^2}{l^2}\right)(b^2+h^2)}}\right\} = \sin^{-1}\left\{\frac{b}{\sqrt{\left(1+\frac{h^2}{(\infty)}\right)(b^2+h^2)}}\right\}$$

$$= \sin^{-1}\left\{\frac{b}{\sqrt{(1+0)(b^2+h^2)}}\right\} = \sin^{-1}\left\{\frac{b}{\sqrt{(b^2+h^2)}}\right\}$$

$$\Rightarrow \omega = \sin^{-1}\left\{\frac{b}{\sqrt{b^2+h^2}}\right\} \ \forall(h \geq 0) \ \ldots\ldots\ldots\ldots\ldots\ldots (2.4)$$

The above formula is very important to find the value of solid angle subtended by an infinitely long rectangular plane at any point lying on the vertical axis passing through any point of the given plane.

Important deduction: The solid angle subtended by the infinitely long rectangular plane at a point lying on the vertex of the plane (i.e. h = 0) is obtained by setting h = 0 in the above expression as follows

$$\omega = \sin^{-1}\left\{\frac{b}{\sqrt{b^2+(0)^2}}\right\} = \sin^{-1}\left\{\frac{b}{b}\right\} = \sin^{-1}\{1\} = \frac{\pi}{2} sr$$

It's true for the vertex of a rectangular plane irrespective of the geometrical dimensions.

Now, let us consider the following cases of the location of vertical axis

2.3.27 Solid angle subtended by an infinitely long rectangular plane at any point lying on the vertical axis passing through any point of finite side of the plane

Let there be an infinitely long rectangular plane having finite width 'b' & a given point say

P(0, 0, h) at a height 'h' lying on the vertical axis passing through the point 'O' of finite side AB = b of the given plane.

Let 'y' be the distance of the point 'O' (i.e. foot of perpendicular drawn from the point 'P') from the vertex 'A' (as shown in the figure (54) below)

Let's divide the given plane into two elementary planes indicated as '1' & '2'

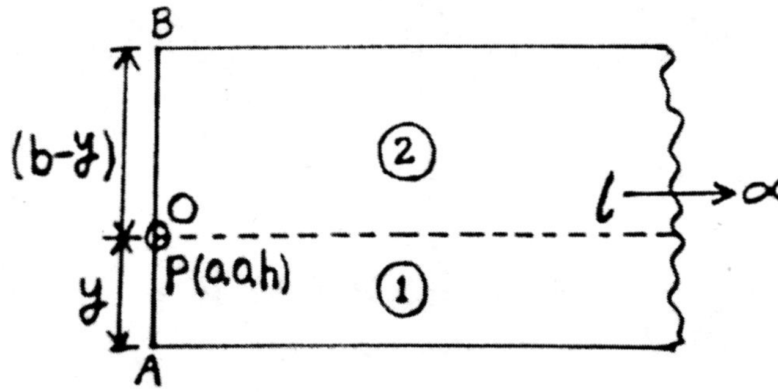

Fig (54)

It is obvious that solid angle subtended by the given plane at the point 'P'

= (solid angle subtended by plane-1) + (solid angle subtended by plane-2) at the same point 'P'

$$\Rightarrow \omega = \omega_1 + \omega_2$$

We know that the solid angle subtended by an infinitely long rectangular plane with finite width at any point lying on the vertical axis passing through the vertex of the planeis given from eq(2.4) as follows

$$\omega = \sin^{-1}\left\{\frac{b}{\sqrt{b^2+h^2}}\right\}$$

On setting b = y, we get

$$\Rightarrow \omega_1 = \sin^{-1}\left\{\frac{y}{\sqrt{y^2+h^2}}\right\}$$

Similarly, on setting $b = (b-y)$, we get

$$\Rightarrow \omega_2 = \sin^{-1}\left\{\frac{(b-y)}{\sqrt{(b-y)^2+h^2}}\right\}$$

$$\therefore \omega = \sin^{-1}\left\{\frac{y}{\sqrt{y^2+h^2}}\right\} + \sin^{-1}\left\{\frac{(b-y)}{\sqrt{(b-y)^2+h^2}}\right\} \quad \forall (h \geq 0) \ldots\ldots\ldots\ldots (2.5)$$

Case: When the vertical axis passes through the mid-point of the finite side AB of an infinitely long rectangular plane

(as shown in the figure (55) below)

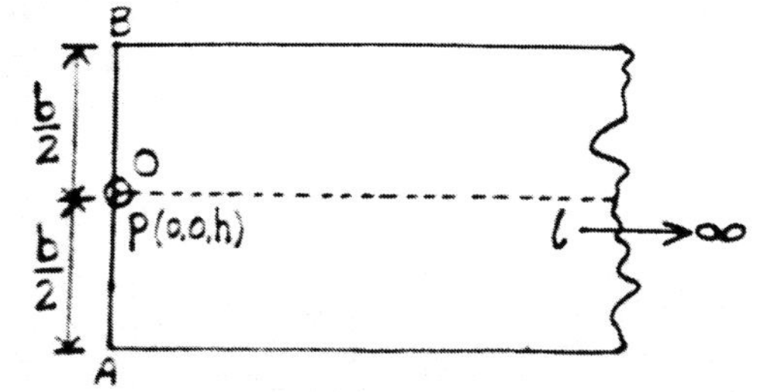

Fig (55)

In this case, the solid angle by an infinitely long rectangular plane with finite width (AB = b) at any point lying on the vertical axis passing through the mid-point of finite side AB of the planeis obtained by setting y= $\frac{b}{2}$ in the eq(2.5) as follows

$$\omega = \sin^{-1}\left\{\frac{\left(\frac{b}{2}\right)}{\sqrt{\left(\frac{b}{2}\right)^2 + h^2}}\right\} + \sin^{-1}\left\{\frac{\left(b-\frac{b}{2}\right)}{\sqrt{\left(b-\frac{b}{2}\right)^2 + h^2}}\right\}$$

$$= \sin^{-1}\left\{\frac{b}{\sqrt{b^2+4h^2}}\right\} + \sin^{-1}\left\{\frac{b}{\sqrt{b^2+4h^2}}\right\} = 2\sin^{-1}\left\{\frac{b}{\sqrt{b^2+4h^2}}\right\}$$

$$\Rightarrow \omega = 2\sin^{-1}\left\{\frac{b}{\sqrt{b^2+4h^2}}\right\} \quad \forall(h \geq 0) \quad \dots\dots\dots\dots\dots\dots\dots.(2.6)$$

2.3.28 Solid angle subtended by a rectangular plane extended infinitely to both the sides of finite centre-line at any point lying on the vertical axis passing through any point of centre-line of the plane

Let there be a rectangular plane having finite width AB = b & the length is extended to infinity about both the sides of finite central line AB (or about both the sides of foot of perpendicular 'O' drawn from the given point P(0, 0, h) lying at a height 'h' on a vertical axis) & 'y' be the distance of the point 'O' (i.e. foot of perpendicular 'O' drawn from the point 'P') from the end point 'A' of central line AB (as shown in the figure (56) below)

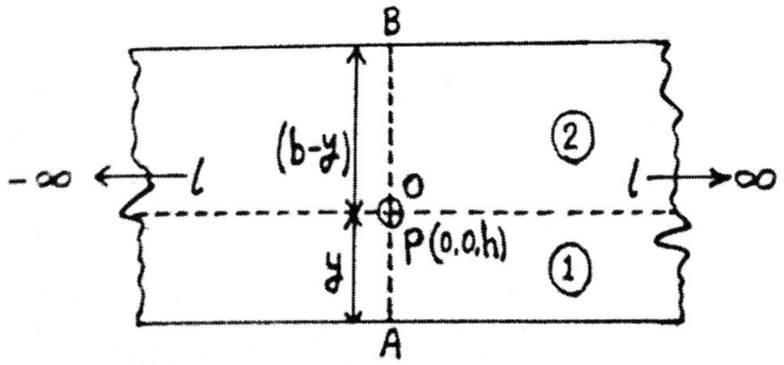

Fig (56)

In this case, we divide the given rectangular plane into two identical elementary planes one L.H.S.(left hand side) tending to $(-\infty)$& other R.H.S. (right hand side) tending to $(+\infty)$about the finite centre-line AB

Hence, the solid angle subtended by the given plane at the point 'P'

$$\Rightarrow \omega = \omega_{LHS} + \omega_{RHS} = 2\omega_{RHS}$$

Now, divide the R.H.S. elementary sub-planes as planes '1' & '2'

$$\omega_{RHS} = \omega_1 + \omega_2 \Rightarrow \omega = 2(\omega_1 + \omega_2)$$

We know that the solid angle subtended by an infinitely long rectangular plane with finite width at any point lying on the vertical axis passing through the vertex of the planeis given from eq(2.4) as follows

$$\omega = \sin^{-1}\left\{\frac{b}{\sqrt{b^2 + h^2}}\right\}$$

On setting b = y, we get

$$\Rightarrow \omega_1 = \sin^{-1}\left\{\frac{y}{\sqrt{y^2 + h^2}}\right\}$$

Similarly, on setting $b = (b - y)$, we get

$$\Rightarrow \omega_2 = \sin^{-1}\left\{\frac{(b-y)}{\sqrt{(b-y)^2 + h^2}}\right\}$$

$$\therefore \omega = 2\left[\sin^{-1}\left\{\frac{y}{\sqrt{y^2 + h^2}}\right\} + \sin^{-1}\left\{\frac{(b-y)}{\sqrt{(b-y)^2 + h^2}}\right\}\right] \quad \forall (h \geq 0) \ldots\ldots\ldots\ldots\ldots (2.7)$$

Let's consider the following cases of the location of the vertical axis passing through the given plane as follows

Case 1: The vertical axis passes through the mid-point of finite centre-line AB of the plane (as shown in the figure (57) below)

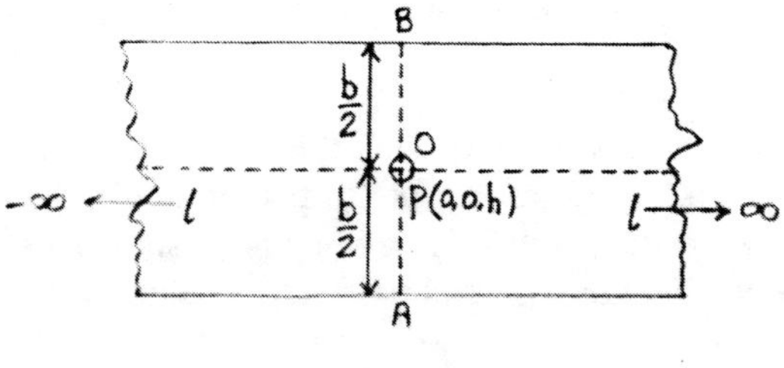

Fig (57)

In this case, the solid angle subtended by the rectangular plane extended infinitely about both the sides of centre-line of the plane at any point say 'P' lying on the vertical axis passing through the mid-point of finite centre line AB of the plane is obtained by setting y= $\frac{b}{2}$in above eq(2.7) as follows

$$\omega = 2\left[\sin^{-1}\left\{\frac{\left(\frac{b}{2}\right)}{\sqrt{\left(\frac{b}{2}\right)^2 + h^2}}\right\} + \sin^{-1}\left\{\frac{\left(b - \frac{b}{2}\right)}{\sqrt{\left(b - \frac{b}{2}\right)^2 + h^2}}\right\}\right]$$

$$= 2\left[\sin^{-1}\left\{\frac{b}{\sqrt{b^2 + 4h^2}}\right\} + \sin^{-1}\left\{\frac{b}{\sqrt{b^2 + 4h^2}}\right\}\right] = 4\sin^{-1}\left\{\frac{b}{\sqrt{b^2 + 4h^2}}\right\}$$

$$\Rightarrow \omega = 4\sin^{-1}\left\{\frac{b}{\sqrt{b^2+4h^2}}\right\} \quad \forall(h \geq 0) \quad \dots\dots\dots\dots\dots\dots\dots (2.8)$$

Case 2: The vertical axis passes through the mid-point of infinitely extended side (as shown in the figure (58) below)

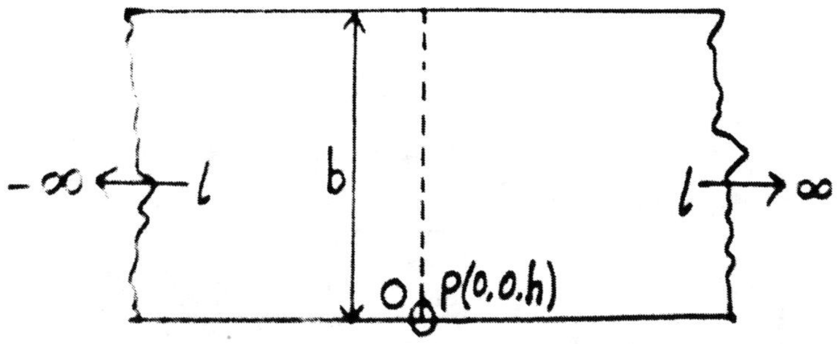

Fig (58)

In this case, the solid angle subtended by the rectangular plane extended infinitely about both the sides of centre-line AB of the plane at any point say 'P' lying on the vertical axis passing through the mid-point of infinitely extended side is obtained by setting $y = 0$in above eq(2.7) as follows

$$\omega = 2\left[\sin^{-1}\left\{\frac{(0)}{\sqrt{(0)^2+h^2}}\right\} + \sin^{-1}\left\{\frac{(b-0)}{\sqrt{(b-0)^2+h^2}}\right\}\right]$$

$$= 2\left[\sin^{-1}\{0\} + \sin^{-1}\left\{\frac{b}{\sqrt{b^2+h^2}}\right\}\right] = 2\sin^{-1}\left\{\frac{b}{\sqrt{b^2+h^2}}\right\}$$

$$\Rightarrow \omega = 2\sin^{-1}\left\{\frac{b}{\sqrt{b^2+h^2}}\right\} \quad \forall(h \geq 0) \quad \dots\dots\dots\dots\dots\dots\dots (2.9)$$

Important deductions:

1. Solid angle subtended by the rectangular plane extended infinitely about both the sides of finite centre-line AB of the plane at any point lying on the plane completely inside the boundary is obtained by setting h = 0 in the eq(2.8) as follows

$$\omega = 4\sin^{-1}\left\{\frac{b}{\sqrt{b^2+4(0)^2}}\right\} = 4\sin^{-1}\left\{\frac{b}{b}\right\} = 4\sin^{-1}\{1\} = 4\frac{\pi}{2} = 2\pi\ sr$$

2. Solid angle subtended by the rectangular plane extended infinitely about both the sides of finite centre-line AB of the plane at any point lying on the infinitely extended side between its end-points is obtained by setting h = 0 in the eq(2.9) as follows

$$\omega = 2\sin^{-1}\left\{\frac{b}{\sqrt{b^2+(0)^2}}\right\} = 2\sin^{-1}\left\{\frac{b}{b}\right\} = 2\sin^{-1}\{1\} = 2\frac{\pi}{2} = \pi\ sr$$

3. Solid angle subtended by an infinitely long rectangular plane at the mid-point of the finite side is obtained by setting h = 0 in the eq(2.6) as follows

$$\omega = 2\sin^{-1}\left\{\frac{b}{\sqrt{b^2+4(0)^2}}\right\} = 2\sin^{-1}\left\{\frac{b}{b}\right\} = 2\sin^{-1}\{1\} = 2\frac{\pi}{2} = \pi\ sr$$

2.3.29Solid angle subtended by an isosceles triangular plane with infinite perpendicular at any point lying on the vertical axis passing through the mid-point of the finite base

Let there be an isosceles triangular plane with infinite perpendicular and finite base with mid-point 'O' & length 'b' and a given point say 'P' lying on the vertical axis passing through the mid-point 'O' of the base (as shown in the figure (59) below)

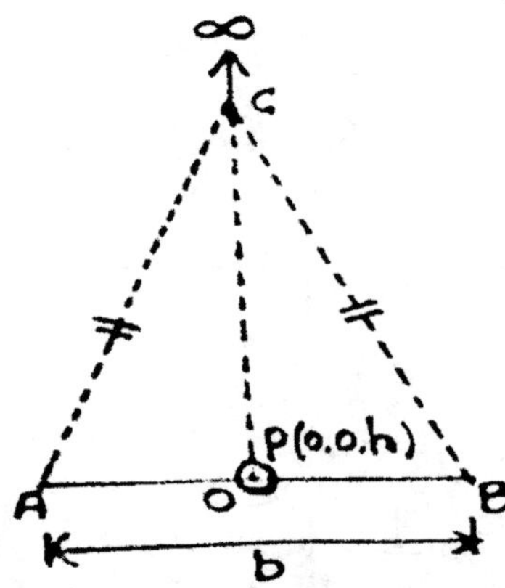

Fig (59)

In this case, the solid angle subtended by the isosceles triangular plane with infinite perpendicular at point the given point 'P' is obtained by taking the limit of ω *at* $p \to \infty$ from the eq(9) as follows

$$\Rightarrow \omega = \lim_{p\to\infty} 2\cos^{-1}\left\{\frac{h\left(2p^2\sqrt{4h^2+b^2} + b^2\sqrt{h^2+p^2}\right)}{h^2(4p^2+b^2)+p^2b^2}\right\}$$

$$= 2\lim_{p\to\infty} \cos^{-1}\left\{\frac{h\left(2\sqrt{4h^2+b^2} + \frac{b^2}{p}\sqrt{\left(\frac{h}{p}\right)^2+1}\right)}{h^2\left(4+\left(\frac{b}{p}\right)^2\right)+b^2}\right\}$$

$$= 2\cos^{-1}\left\{\frac{h\left(2\sqrt{4h^2+b^2} + \frac{b^2}{(\infty)}\sqrt{\left(\frac{h}{\infty}\right)^2+1}\right)}{h^2\left(4+\left(\frac{b}{\infty}\right)^2\right)+b^2}\right\}$$

$$= 2\cos^{-1}\left\{\frac{h\left(2\sqrt{4h^2+b^2} + (0)\sqrt{0+1}\right)}{h^2(4+0)+b^2}\right\} = 2\cos^{-1}\left\{\frac{2h\sqrt{4h^2+b^2}}{4h^2+b^2}\right\}$$

$$\Rightarrow \omega = 2\cos^{-1}\left\{\frac{2h}{\sqrt{4h^2+b^2}}\right\}(h \ge 0) \quad \dots\dots\dots\dots\dots\dots\dots (2.10)$$

The above result is comparable with the equation (2.6) i.e. the result obtained in the above equation can be manipulated to obtain the expression of eq(2.6). Thus, an isosceles triangular plane having infinite height will as a rectangular plane having infinite length. The above result can be reduced to that of eq(2.6) as follows

$$\omega = 2\cos^{-1}\left\{\frac{2h}{\sqrt{4h^2+b^2}}\right\} = 2\sin^{-1}\left\{\sqrt{1-\left(\frac{2h}{\sqrt{4h^2+b^2}}\right)^2}\right\}$$

$$\left(since, \cos^{-1}x = \sin^{-1}\sqrt{1-x^2} \quad \forall\, x \in [-1,1]\right)$$

$$= 2\sin^{-1}\left\{\sqrt{\frac{4h^2+b^2-4h^2}{4h^2+b^2}}\right\} = 2\sin^{-1}\left\{\sqrt{\frac{b^2}{4h^2+b^2}}\right\} = 2\sin^{-1}\left\{\frac{b}{\sqrt{b^2+4h^2}}\right\}$$

Thus, the above result is same as that of the eq(2.6).

Thus we have,

$$\Rightarrow \omega = 2\cos^{-1}\left\{\frac{2h}{\sqrt{4h^2+b^2}}\right\} = 2\sin^{-1}\left\{\frac{b}{\sqrt{b^2+4h^2}}\right\} (h \geq 0) \quad \ldots\ldots\ldots\ldots\ldots\ldots (2.10)$$

Solved Examples

Example 2.1: Find out the solid angle subtended by a right triangular plane with orthogonal sides 12cm & 15cm at a point lying at a height 10cm on the vertical axis passing through the vertex common to the

1. Hypotenuse & smaller orthogonal side, & 2.hypotenuse & greater orthogonal side.

Sol. given that

Orthogonal sides are 12cm & 15cm & normal height, h = 10cm.

We know that the solid angle subtended by a right triangular plane at any point lying on the vertical axis passing through the acute angled vertex is given from the eq(2) as follows

$$\omega = \sin^{-1}\left\{\frac{b}{\sqrt{b^2+p^2}}\right\} - \sin^{-1}\left\{\left(\frac{b}{\sqrt{b^2+p^2}}\right)\left(\frac{h}{\sqrt{h^2+p^2}}\right)\right\}$$

1. When the axis passes through the vertex common to the hypotenuse & smaller orthogonal side. We have,

b = 15cm, p = 12cm & h = 10cm

Now, on setting the values we have

$$\omega = \sin^{-1}\left\{\frac{15}{\sqrt{15^2+12^2}}\right\} - \sin^{-1}\left\{\left(\frac{15}{\sqrt{15^2+12^2}}\right)\left(\frac{10}{\sqrt{10^2+12^2}}\right)\right\}$$

$$= \sin^{-1}\left\{\frac{15}{\sqrt{369}}\right\} - \sin^{-1}\left\{\left(\frac{15}{\sqrt{369}}\right)\left(\frac{10}{\sqrt{244}}\right)\right\} = 0.896 - 0.523 = 0.373\ sr \qquad Ans.$$

2. When the axis passes through the vertex common to the hypotenuse & greater orthogonal side. We have,

b = 12cm, p = 15cm & h = 10cm

Now, on setting the values we have

$$\omega = \sin^{-1}\left\{\frac{12}{\sqrt{12^2+15^2}}\right\} - \sin^{-1}\left\{\left(\frac{12}{\sqrt{12^2+15^2}}\right)\left(\frac{10}{\sqrt{10^2+15^2}}\right)\right\}$$

$$= \sin^{-1}\left\{\frac{12}{\sqrt{369}}\right\} - \sin^{-1}\left\{\left(\frac{12}{\sqrt{369}}\right)\left(\frac{10}{\sqrt{325}}\right)\right\} = 0.675 - 0.354 = 0.321\ sr \qquad Ans.$$

Example 2.2: Find out the solid angle subtended by a right triangular plane with orthogonal sides 5cm & 12cm at a point lying at a height 3cm at the vertex common to the

1. Hypotenuse & smaller orthogonal side & 2. Hypotenuse & larger orthogonal side

Sol. given that

Orthogonal sides are 5cm & 12cm

We know that the solid angle subtended by a right triangular plane at the acute angled vertex is given from the eq(4) as follows

$$\omega = \sin^{-1}\left\{\frac{b}{\sqrt{b^2+p^2}}\right\}$$

1. for the vertex common to the hypotenuse & smaller orthogonal side. We have,

b = 12cm, p = 5cm

Now, on setting the values we have

$$\omega = \sin^{-1}\left\{\frac{12}{\sqrt{12^2+5^2}}\right\} = \sin^{-1}\left\{\frac{12}{13}\right\} = 1.176\ sr \qquad Ans.$$

2. for the vertex common to the hypotenuse & larger orthogonal side. We have,

   b = 5cm, p = 12cm

   Now, on setting the values we have

$$\omega = \sin^{-1}\left\{\frac{5}{\sqrt{5^2+12^2}}\right\} = \sin^{-1}\left\{\frac{5}{13}\right\} = 0.395\ sr \qquad Ans.$$

Example 2.3: Find out the solid angle subtended by a right triangular plane with orthogonal sides 9cm & 13cmat a point lying at a height 6cm on the vertical axis passing through the right angled vertex.

Sol. given that

Orthogonal sides are 9cm & 13cm & normal height, h = 6cm

We know that the solid angle subtended by a right triangular plane at any point lying on the vertical axis passing through the right angled vertex is given from the eq(6) as follows

$$\omega = \cos^{-1}\left\{\frac{h\left(a^2\sqrt{h^2+b^2}+b^2\sqrt{h^2+a^2}\right)}{h^2(a^2+b^2)+a^2b^2}\right\}$$

On setting $a = 9cm, b = 13cm$ & $h = 6cm$ as follows, we have

$$\omega = \cos^{-1}\left\{\frac{6\left(9^2\sqrt{6^2+13^2}+13^2\sqrt{6^2+9^2}\right)}{6^2(9^2+13^2)+9^2(13^2)}\right\} = \cos^{-1}\left\{\frac{6\left(81\sqrt{205}+169\sqrt{117}\right)}{36(250)+81(169)}\right\}$$

$$= \cos^{-1}\left\{\frac{486\sqrt{205}+1014\sqrt{117}}{22689}\right\} = 0.66\ sr \qquad Ans$$

Example 2.4: The solid angle subtended by a right triangular plane with orthogonal sides 14cm & 17cm, at a point lying on the vertical axis passing through the common vertex of hypotenuse & smaller orthogonal side, is 0.35sr. Find out height of the point from the plane.

Sol. let the height of given point be 'h'

We know that the solid angle subtended by a right triangular plane at any point lying on the vertical axis passing through the acute angled vertex is given from the eq(2) as follows

$$\omega = \sin^{-1}\left\{\frac{b}{\sqrt{b^2+p^2}}\right\} - \sin^{-1}\left\{\left(\frac{b}{\sqrt{b^2+p^2}}\right)\left(\frac{h}{\sqrt{h^2+p^2}}\right)\right\}$$

Since, the axis passes through the vertex common to the hypotenuse & smaller orthogonal side hence on setting the values b = 17cm & p = 14cm, we have

$$\omega = \sin^{-1}\left\{\frac{17}{\sqrt{17^2+14^2}}\right\} - \sin^{-1}\left\{\left(\frac{17}{\sqrt{17^2+14^2}}\right)\left(\frac{h}{\sqrt{h^2+14^2}}\right)\right\}$$

$$\Rightarrow 0.35 = \sin^{-1}\left\{\frac{17}{\sqrt{485}}\right\} - \sin^{-1}\left\{\left(\frac{17}{\sqrt{485}}\right)\left(\frac{h}{\sqrt{h^2+196}}\right)\right\}$$

$$= 0.882 - \sin^{-1}\left\{\frac{17h}{\sqrt{485}\sqrt{h^2+196}}\right\}$$

$$\sin^{-1}\left\{\frac{17h}{\sqrt{485}\sqrt{h^2+196}}\right\} = 0.882 - 0.35 = 0.532$$

$$\frac{17h}{\sqrt{485}\sqrt{h^2+196}} = \sin(0.532) = 0.507\ or\ \frac{h^2}{h^2+196} = \frac{485\times(0.507)^2}{289} = 0.431$$

$$2.318h^2 = h^2 + 196 \Rightarrow h^2 = \frac{196}{1.318} = 148.71\ or\ h = 12.195cm \qquad Ans.$$

Example 2.5: Find out the area of a right triangular plane with a orthogonal side 15cm, which subtends a solid angle 0.983sr at a point lying at a height 5cm on the vertical axis passing through the right angled vertex.

Sol. given that

Orthogonal side is 15cm, solid angle, $\omega = 0.983sr$& normal height, h = 11cm

Let other orthogonal side be, x

Now, on setting the values in eq(6) as follows

$$\omega = 0.983 = \cos^{-1}\left\{\frac{5\left(x^2\sqrt{5^2+15^2}+15^2\sqrt{5^2+x^2}\right)}{5^2(x^2+15^2)+(15^2)x^2}\right\}$$

$$\frac{5\left(15.811x^2+225\sqrt{25+x^2}\right)}{25x^2+5625+225x^2} = \cos(0.983) = 0.554$$

$$138.5x^2+3093.75 = 79.055x^2+1125\sqrt{25+x^2}$$

$$59.445x^2+3093.75 = 1125\sqrt{25+x^2} \; or \; 0.0528x^2+2.75 = \sqrt{25+x^2}$$

Now, taking square of both the sides, we have

$$(0.0528x^2+2.75)^2 = 25+x^2$$

$$0.00279x^4+7.562+0.29x^2 = 25+x^2 \; or \; 0.00279x^4-0.71x^2-17.438 = 0$$

Now, solving the biquadratic eq. for the value of x, we have

$$\Rightarrow x^2 = \frac{-(-0.71)\pm\sqrt{(-0.71)^2-4\times(0.00279)\times(-17.438)}}{2(0.00279)} = \frac{0.71\pm 0.836}{0.00558}$$

$$\Rightarrow x^2 = -22.58 \; or \; 277.061$$

Since, length of side (x) is positive hence the negative value of x^2 is not acceptable, we have

$$x^2 = 277.061 \; or \; x = 16.645cm$$

$$hence, the\ area\ of\ right\ \Delta = \frac{1}{2}(base)\times(normal\ height)$$

$$\therefore\ area\ of\ the\ plane\ = \frac{1}{2}(15\times 16.645) = 124.837\ cm^2\ or\ 0.0125\ m^2 \qquad Ans.$$

Example 2.6: Calculate the area of an isosceles right-triangular plane subtending a solid angle 0.464sr at a point lying at a height 8cm on the vertical axis passing through one of the acute angled vertices.

Sol. given that

Solid angle, $\omega = 0.464sr$& normal height, h = 8cm

Let the length of orthogonal sides be x.

Method.1. Using equation (2) for a right triangular plane

Now, on setting $\omega = 0.464sr,\ a = b = x\ \&\ h = 8cm$ in the eq(2), we have

$$\omega = 0.464 = \sin^{-1}\left\{\frac{x}{\sqrt{x^2+x^2}}\right\} - \sin^{-1}\left\{\left(\frac{x}{\sqrt{x^2+x^2}}\right)\left(\frac{8}{\sqrt{8^2+x^2}}\right)\right\}$$

$$0.464 = \sin^{-1}\left\{\frac{1}{\sqrt{2}}\right\} - \sin^{-1}\left\{\left(\frac{1}{\sqrt{2}}\right)\left(\frac{8}{\sqrt{64+x^2}}\right)\right\} = \frac{\pi}{4} - \sin^{-1}\left(\frac{4\sqrt{2}}{\sqrt{64+x^2}}\right)$$

$$\sin^{-1}\left(\frac{4\sqrt{2}}{\sqrt{64+x^2}}\right) = 0.785-0.464 = 0.321 \; or \; \frac{4\sqrt{2}}{\sqrt{64+x^2}} = \sin(0.321) = 0.315$$

$$\sqrt{64+x^2} = \frac{4\sqrt{2}}{0.315} = 17.958 \; or \; 64+x^2 = 322.49$$

$$\Rightarrow x^2 = 258.49 \; or \; x = 16.077cm$$

$$\therefore \; area \; of \; isoceles \; righttriangular \; plane \; = \frac{1}{2}(16.077)\times(16.077)$$

$$= 129.235 \; cm^2$$

Method.2. Using equation (11) for an isosceles right-triangular plane

On setting $\omega = 0.464sr, \; a = b = x \;\&\; h = 8cm$ in the eq(11), we have

$$\omega = 0.464 = \frac{\pi}{4} - \sin^{-1}\left(\frac{8}{\sqrt{2(8^2+x^2)}}\right)$$

$$\sin^{-1}\left(\frac{4\sqrt{2}}{\sqrt{64+x^2}}\right) = 0.785 - 0.464 = 0.321 \; or \; \frac{4\sqrt{2}}{\sqrt{64+x^2}} = \sin(0.321) = 0.315$$

$$\sqrt{64+x^2} = \frac{4\sqrt{2}}{0.315} = 17.958 \; or \; 64+x^2 = 322.49$$

$$\Rightarrow x^2 = 258.49 \; or \; x = 16.077cm$$

$$\therefore \; area \; of \; isoceles \; righttriangular \; plane \; = \frac{1}{2}(16.077)\times(16.077)$$

$$= 129.235 \; cm^2 \; or \; 0.0129 \; m^2 \qquad Ans.$$

Example 2.7: Find out the solid angle subtended by a isosceles triangular plane with one side 24cm & each of the equal sides 13cm, at a point lying at a height 5cm on the vertical axis passing through the common vertex of the equal sides.

Sol. given that one side (base) is 24cm & each of the equal sides is 13cm & height, h = 5cm

The perpendicular drawn from the common vertex of equal sides, intersects the base at the mid-point, hence the length of perpendicular to the base

$$\Rightarrow p = \sqrt{13^2 - \left(\frac{24}{2}\right)^2} = \sqrt{13^2 - 12^2} = 5cm$$

The isosceles triangular plane can be divided into two right triangular planes

Thus, the solid angle subtended by a right triangular plane at the given point is obtained by setting b = 12cm, p = 5cm & h = 5cm in the eq(2) as follows

$$\omega = \sin^{-1}\left\{\frac{12}{\sqrt{12^2+5^2}}\right\} - \sin^{-1}\left\{\left(\frac{12}{\sqrt{12^2+5^2}}\right)\left(\frac{5}{\sqrt{5^2+5^2}}\right)\right\}$$

$$= \sin^{-1}\left\{\frac{12}{13}\right\} - \sin^{-1}\left\{\left(\frac{12}{13}\right)\left(\frac{5}{5\sqrt{2}}\right)\right\} = 1.176 - 0.711 = 0.465 \; sr$$

$$hence, solid \; angle \; subtended \; by \; the \; isosceles \; triangular \; plane \; = 2\omega$$

$$= 2\times 0.465 = 0.93 \; sr \qquad Ans.$$

Example 2.8: Calculate the solid angle subtended by an equilateral triangular plane with each side 30cm at appoint lying at a height 10cm on the vertical axis passing through the centre of the plane.

Sol. given that

Length of each side, $a = 30cm \;\&\; height, h = 10cm$

We know that the solid angle subtended by an equilateral triangular plane at any point lying on the vertical axis passing through the centre of the plane is given by the eq(15) as follows

$$\omega = 2\pi - 6\sin^{-1}\left(\frac{3h}{\sqrt{12h^2 + a^2}}\right)$$

On setting the values in the eq(15), we get

$$\omega = 2\pi - 6\sin^{-1}\left(\frac{3(10)}{\sqrt{12(10^2) + 30^2}}\right) = 2\pi - 6\sin^{-1}\left(\frac{30}{\sqrt{2100}}\right)$$

$$= 2\pi - 6 \times 0.714 = 2\pi - 4.284 = 2\ sr \qquad Ans.$$

Example 2.9: Determine the total flux intercepted by a right triangular plane with orthogonal sides 20cm & 28cm if a uniform point-source of intensity 110 cd is located at a height 47cm on the vertical axis passing through the right angled vertex. If the flux is uniformly distributed over the entire plane find out the flux density.

Sol. given that

Orthogonal sides 20cm & 28cm, height, h = 47cm & intensity, I = 110 cd

Hence on setting $a = 20cm, b = 28cm \; \& \; h = 47cm$ in the eq(6), we get

Solid angle subtended by the plane at the point-source,

$$\omega = \cos^{-1}\left\{\frac{47\left(20^2\sqrt{47^2 + 28^2} + 28^2\sqrt{47^2 + 20^2}\right)}{47^2(20^2 + 28^2) + (20^2)(28^2)}\right\}$$

$$= \cos^{-1}\left\{\frac{47(21883.327 + 40045.443)}{2615456 + 313600}\right\} = \cos^{-1}\left\{\frac{2910652.19}{2929056}\right\}$$

$$= \cos^{-1}\{0.9937\} = 0.112\ sr$$

Hence, flux intercepted by the plane

$$F = I \times \omega = 110 \times 0.112 = 12.32\ lm \qquad Ans.$$

Area of the plane,

$$\Rightarrow A = \frac{1}{2}(base) \times (normal\ height) = \frac{1}{2}(20 \times 28) = 280\ cm^2\ or\ 0.028\ m^2$$

Assuming uniform distribution of the flux over the entire plane, the flux density

$$F.D. = \frac{Total\ flux\ intercepted\ by\ the\ plane}{Area\ of\ the\ plane}$$

$$= \frac{F}{A} = \frac{12.32}{0.028} = 440\ lux \qquad Ans.$$

Example 2.10: Three uniform point-sources, each of 200 W, are separately located at an equal height 4m on the vertical axes passing through the vertices of a right triangular plane with orthogonal sides 28m & 35m. Find out the total radiation energy striking the plane.

Sol. given that

Orthogonal sides 28m & 35cm, height, h = 4 m &

Luminosity of the point-sources, L = 200 W

1. Energy intercepted by the plane emitted by the point-source-1 lying on the vertical axis passing through the vertex common to the hypotenuse & smaller orthogonal side

   On setting b = 35 m, p = 28 m & height, h = 4 m in the eq(2), we get

   Solid angle subtended by the plane at the point-source-1

$$\omega = \sin^{-1}\left\{\frac{35}{\sqrt{35^2 + 28^2}}\right\} - \sin^{-1}\left\{\left(\frac{35}{\sqrt{35^2 + 28^2}}\right)\left(\frac{4}{\sqrt{4^2 + 28^2}}\right)\right\}$$

$$= \sin^{-1}\left\{\frac{35}{\sqrt{2009}}\right\} - \sin^{-1}\left\{\left(\frac{35}{\sqrt{2009}}\right)\left(\frac{4}{\sqrt{800}}\right)\right\} = 0.896 - 0.111 = 0.785\ sr$$

$$intercepted\ energy, E_1 = \left(\frac{L}{4\pi}\right) \times \omega = \left(\frac{200}{4\pi}\right) \times 0.785 = 12.494\ W$$

2 Energy intercepted by the plane emitted by the point-source-2 lying on the vertical axis passing through the vertex common to the hypotenuse & larger orthogonal side

On setting b = 28 m, p = 35 m & height, h = 4 m in the eq(2), we get

Solid angle subtended by the plane at the point-source-2

$$\omega = \sin^{-1}\left\{\frac{28}{\sqrt{28^2+35^2}}\right\} - \sin^{-1}\left\{\left(\frac{28}{\sqrt{28^2+35^2}}\right)\left(\frac{4}{\sqrt{4^2+35^2}}\right)\right\}$$

$$= \sin^{-1}\left\{\frac{28}{\sqrt{2009}}\right\} - \sin^{-1}\left\{\left(\frac{28}{\sqrt{2009}}\right)\left(\frac{4}{\sqrt{1241}}\right)\right\} = 0.675 - 0.071 = 0.604\ sr$$

$$intercepted\ energy, E_2 = \left(\frac{L}{4\pi}\right) \times \omega = \left(\frac{200}{4\pi}\right) \times 0.604 = 9.613\ W$$

3. Energy intercepted by the plane emitted by the point-source-3 lying on the vertical axis passing through the right angled vertex

On setting a = 28 m, b = 35 m & height, h = 4 m in the eq(6), we get

Solid angle subtended by the plane at the point-source-3

$$\omega = \cos^{-1}\left\{\frac{4\left(28^2\sqrt{4^2+35^2}+35^2\sqrt{4^2+28^2}\right)}{4^2(28^2+35^2)+(28^2)(35^2)}\right\} = \cos^{-1}\left\{\frac{4(27618.619+34648.232)}{32144+960400}\right\}$$

$$= \cos^{-1}\left\{\frac{249067.404}{992544}\right\} = 1.317\ sr$$

$$intercepted\ energy, E_3 = \left(\frac{L}{4\pi}\right) \times \omega = \left(\frac{200}{4\pi}\right) \times 1.317 = 20.96\ W$$

Hence the total radiation energy intercepted by the plane,

$$\Rightarrow E = E_1 + E_2 + E_3$$

$$= 12.494 + 9.613 + 20.96 = 43.067\ W\ or\ 43.067\ j/sec \qquad Ans.$$

Example 2.11: Find out the area of a right triangular plane with one of the orthogonal sides 19cm, intercepts $^1/_{16}$ of the total radiation energy emitted by a uniform point-source located at a height 7cm on the vertical axis passing though the right angled vertex.

Sol. given that

One of the orthogonal sides, 19cm, height, h = 7cm

Let the solid angle subtended by the plane at the point-source with luminosity L, be ω then

We know that the radiation energy intercepted by the plane is given as

$$E = \left(\frac{L}{4\pi}\right) \times \omega$$

$$\Rightarrow \left(\frac{L}{4\pi}\right) \times \omega = \frac{1}{16}(total\ radiation\ energy\ (L)) = \frac{L}{16}$$

$$\omega = \frac{4\pi}{16} = \frac{\pi}{4}\ sr$$

Let another orthogonal side of the plane be 'x' hence

On setting$a = x, b = 19cm\ \&\ h = 7cm$ in the eq(6), we have

$$\omega = \frac{\pi}{4} = \cos^{-1}\left\{\frac{7\left(x^2\sqrt{7^2+19^2}+19^2\sqrt{7^2+x^2}\right)}{7^2(x^2+19^2)+(x^2)(19^2)}\right\}$$

$$\frac{7\left(20.248x^2+361\sqrt{49+x^2}\right)}{49x^2+17689+361x^2}=\cos\frac{\pi}{4}=\frac{1}{\sqrt{2}}=0.707$$

$$289.87x^2+12506.123=141.736x^2+2527\sqrt{49+x^2}$$

$$148.134x^2+12506.123=2527\sqrt{49+x^2}\ or\ \ 0.0586x^2+4.949=\sqrt{49+x^2}$$

Now, taking square of both the sides, we have

$$(0.0586x^2+4.949)^2=49+x^2$$

$$0.00343x^4+24.493+0.58x^2=49+x^2\ \ or\ \ 0.00343x^4-0.42x^2-24.507=0$$

Now, solving the biquadratic eq. for the value of x, we have

$$\Rightarrow x^2=\frac{-(-0.42)\pm\sqrt{(-0.42)^2-4\times(0.00343)\times(-24.507)}}{2(0.00343)}=\frac{0.42\pm0.716}{0.00686}$$

$$\Rightarrow x^2=-43.149\ or\ 165.598$$

Since, length of side (x) is positive hence the negative value of x^2 is not acceptable, we have

$$x^2=165.598\ or\ x=12.868cm$$

$$hence, the\ area\ of\ right\ \Delta=\frac{1}{2}(base)\times(normal\ height)$$

$$\therefore\ area\ of\ the\ plane\ =\frac{1}{2}(19\times12.868)=122.246\ cm^2\ or\ 0.0122\ m^2\qquad Ans.$$

Example 2.12: An equilateral triangular plane intercepts 300 lm flux emitted by a uniform point-source of 200 cd located at a height 16cm on the vertical axis passing though the centre of the plane. If the flux is uniformly distributed over the entire plane what will the flux density?

Sol. given that

Intercepted flux, F = 300 lm, intensity, I = 200 cd & height, h = 16cm

Let the solid angle subtended by the plane at the point-source be ω then flux intercepted by the plane

$$F=I\times\omega$$

$$\Rightarrow 300=200\times\omega\quad\Rightarrow\omega=\frac{300}{200}=1.5\ sr$$

On setting h = 16cm in the eq(15), we get

$$\omega=1.5=2\pi-6\sin^{-1}\left(\frac{3(16)}{\sqrt{12(16^2)+a^2}}\right)$$

$$6\sin^{-1}\left(\frac{48}{\sqrt{3072+a^2}}\right)=2\pi-1.5=4.783\ \ or\ \ \frac{48}{\sqrt{3072+a^2}}=sin\left(\frac{4.783}{6}\right)=0.715$$

$$3072+a^2=4506.822\ or\ a^2=1434.822\ or\ a=37.879cm$$

Now, area of the plane,

$$\Rightarrow A=\frac{\sqrt{3}}{4}\times(side)^2=\frac{\sqrt{3}}{4}\times(37.879)^2=621.296\ cm^2\ or\ 0.0621\ m^2$$

Hence, the flux density over the plane,

$$\Rightarrow F.D.=\frac{F}{A}=\frac{300}{0.0621}=4830.918\ lux\ \ or\ 4.83\ k\ lux\qquad Ans.$$

Example 2.13: An equilateral triangular plane with each side 'a' intercepts x fraction of the total radiation emitted by a uniform point-source located at a height 'h' on the vertical axis passing through the centre of plane. Prove that

$$\frac{a}{h}=\sqrt{9cosec^2\left\{\frac{\pi}{3}(1-2x)\right\}-12}\quad \forall\ x<\frac{1}{2}.$$

Sol. let the luminosity of the point-source be 'L' &

Solid angle subtended by the plane at the point-source be ω then the radiation energy intercepted by the plane

$$E=\left(\frac{L}{4\pi}\right)\times\omega$$

But it is given that the intercepted radiation

$$=x\times(total\ radiation\ enrgy\ (L))=xL$$

$$\Rightarrow\left(\frac{L}{4\pi}\right)\times\omega=xL\quad or\quad \omega=4\pi x$$

But the solid angle subtended by an equilateral triangular plane at any point lying on the vertical axis passing through the centre is given by the eq(15) as follows

$$\omega=2\pi-6\sin^{-1}\left(\frac{3h}{\sqrt{12h^2+a^2}}\right)$$

On setting the value of solid angle (ω), we have

$$4\pi x=2\pi-6\sin^{-1}\left(\frac{3h}{\sqrt{12h^2+a^2}}\right)$$

$$6\sin^{-1}\left(\frac{3h}{\sqrt{12h^2+a^2}}\right)=2\pi-4\pi x\qquad or\qquad \sin^{-1}\left(\frac{3h}{\sqrt{12h^2+a^2}}\right)=\frac{\pi-2\pi x}{3}$$

$$\Rightarrow\frac{3h}{\sqrt{12h^2+a^2}}=sin\frac{\pi}{3}(1-2x)$$

Taking square on both the sides, we get

$$\frac{9h^2}{12h^2+a^2}=sin^2\left\{\frac{\pi}{3}(1-2x)\right\}\ or\ \frac{12h^2+a^2}{9h^2}=cosec^2\left\{\frac{\pi}{3}(1-2x)\right\}$$

$$\frac{12h^2+a^2}{12h^2}=\frac{3}{4}cosec^2\left\{\frac{\pi}{3}(1-2x)\right\}$$

Now, on subtracting 1 from both the sides, we have

$$\frac{12h^2+a^2}{12h^2}-1=\frac{3}{4}cosec^2\left\{\frac{\pi}{3}(1-2x)\right\}-1$$

$$or\quad \frac{12h^2+a^2-12h^2}{12h^2}=\frac{a^2}{12h^2}=\frac{3cosec^2\left\{\frac{\pi}{3}(1-2x)\right\}-4}{4}$$

$$or\quad \frac{a^2}{h^2}=3\left(3cosec^2\left\{\frac{\pi}{3}(1-2x)\right\}-4\right)$$

$$or\quad \frac{a^2}{h^2}=9cosec^2\left\{\frac{\pi}{3}(1-2x)\right\}-12\quad or\quad \frac{a}{h}=\sqrt{9cosec^2\left\{\frac{\pi}{3}(1-2x)\right\}-12}$$

Now, on applying the condition of the value of variable 'x' as follows

$$9cosec^2\left\{\frac{\pi}{3}(1-2x)\right\}-12>0$$

$$cosec^2\left\{\frac{\pi}{3}(1-2x)\right\}>\frac{12}{9}\qquad or\qquad cosec^2\left\{\frac{\pi}{3}(1-2x)\right\}>\left(\frac{2}{\sqrt{3}}\right)^2$$

$$or\ cosec^2\left\{\frac{\pi}{3}(1-2x)\right\}>cosec^2\frac{\pi}{3}\qquad or\qquad \frac{\pi}{3}(1-2x)<\frac{\pi}{3}\ or\ 0<(1-2x)<1$$

$$or\ 2x < 1 < (1 + 2x) \Rightarrow \qquad 2x < 1\ \ or\ x < \frac{1}{2}\&$$

$$\Rightarrow 1 < (1 + 2x) \quad or \quad x > 0\ (i.e. fration\ is\ always\ positive) \qquad hence, we\ get$$

$$\frac{a}{h} = \sqrt{9cosec^2\left\{\frac{\pi}{3}(1 - 2x)\right\} - 12} \quad \forall\, x < \frac{1}{2}. \qquad proved.$$

Example 2.14: What is the solid angle subtended by a rectangular plane having dimensions 14cm & 17cm at a point lying at a height 5cm on a vertical axis passing through the vertex?

Sol. given that

$$length, l = 17cm, width, b = 14cm\ \&\ normal\ height, h = 5cm$$

We know that solid angle subtended by a rectangular plane with dimensions 'l' & 'b' at any point lying at a normal height 'h' on the axis passing through the centre is given by the eq(16) as follows

$$\omega = 4\sin^{-1}\left(\frac{lb}{\sqrt{(l^2 + 4h^2)(b^2 + 4h^2)}}\right)$$

On setting the values, we get

$$\omega = 4\sin^{-1}\left(\frac{17 \times 14}{\sqrt{((17)^2 + 4(5)^2)((14)^2 + 4(5)^2)}}\right)$$

$$= 4\sin^{-1}\left(\frac{238}{\sqrt{115144}}\right) = 4\sin^{-1}(0.701)$$

$$= 3.107\ sr \qquad Ans.$$

Example 2.15: Find out the solid angle subtended by a rectangular plane of $10cm \times 24cm$ at a point lying at a height 4cm on a vertical axis passing through the mid-point of

1. Smaller side & 2.larger side

Sol. given that

$$length, l = 24cm, width, b = 10cm\ \&\ normal\ height, h = 4cm$$

1. When the axis passes through the mid-point of larger side

We know that solid angle subtended by a rectangular plane with dimensions 'l' & 'b' at any point lying at a normal height 'h' on the axis passing through the mid-point of side 'l' is given by the eq(18) as follows

$$\omega = 2\sin^{-1}\left(\frac{lb}{\sqrt{(l^2 + 4h^2)(b^2 + h^2)}}\right)$$

On setting the values, we get

$$\omega = 2\sin^{-1}\left(\frac{24 \times 10}{\sqrt{((24)^2 + 4(4)^2)((10)^2 + (4)^2)}}\right)$$

$$= 2\sin^{-1}\left(\frac{240}{\sqrt{74240}}\right) = 2\sin^{-1}(0.881)$$

$$= 2.156\ sr \qquad Ans.$$

2. When the axis passes through the mid-point of smaller side

We know that solid angle subtended by a rectangular plane with dimensions 'l' & 'b' at any point lying at a normal height 'h' on the axis passing through the mid-point of side 'b' is given by the eq(18) as follows

$$\omega = 2\sin^{-1}\left(\frac{lb}{\sqrt{(l^2 + h^2)(b^2 + 4h^2)}}\right)$$

On setting the values, we get

$$\omega = 2\sin^{-1}\left(\frac{24\times 10}{\sqrt{((24)^2+(4)^2)((10)^2+4(4)^2)}}\right)$$

$$= 2\sin^{-1}\left(\frac{240}{\sqrt{97088}}\right) = 2\sin^{-1}(0.77)$$

$$= 1.758\ sr \qquad\qquad Ans.$$

Example 2.16: If the solid angle subtended by a rectangular plane of $9cm \times 16cm$ at a point lying on a vertical axis passing through the centre is 2.566sr. What is the normal height of the point from the plane?

Sol. given that

$$length, l = 24cm, width, b = 10cm \ \&\ solid\ angle, \omega = 2.566\ sr$$

Let the normal height of point from the centre of plane be 'h' then we have

$$\omega = 4\sin^{-1}\left(\frac{lb}{\sqrt{(l^2+4h^2)(b^2+4h^2)}}\right)$$

On setting the values, we get

$$\omega = 2.566 = 4\sin^{-1}\left(\frac{16\times 9}{\sqrt{((16)^2+4h^2)((9)^2+4h^2)}}\right)$$

$$\Rightarrow \frac{144}{\sqrt{(256+4h^2)(81+4h^2)}} = sin\left(\frac{2.566}{4}\times\frac{180^o}{\pi}\right) = 0.598$$

$$\Rightarrow 20736 + 324h^2 + 1024h^2 + 16h^4 = 57985.928$$

$$16h^4 + 1348h^2 - 37249.928 = 0$$

On solving, we get

$$h^2 = 21.927, -106.177$$

$$\Rightarrow h^2 = 21.927 \ or\ h = 4.682 \quad (since,\ \ h \geq 0)$$

Hence, the normal height of the point, $h = 4.682$ *Ans.*

Example 2.17: A rectangular plane of $l \times b$ intercepts 30% of the total radiation emitted by a uniform point-source located at a normal height 'h' from one of the vertices. If $h : l = {}^2/_3$ find out the value of $h : b$.

Sol. given that

$$length = l,\ \ width = b,\ \ \frac{h}{l} = \frac{2}{3}$$

Let the luminosity of point-source be 'L' then the radiation energy intercepted by the plane

$$\Rightarrow E_i = \left(\frac{L}{4\pi}\right)\times\omega = \left(\frac{L}{4\pi}\right)\times\sin^{-1}\left(\frac{lb}{\sqrt{(l^2+h^2)(b^2+h^2)}}\right)$$

But, the energy intercepted by the plane is 30% of the total energy, hence we get

$$E_i = L\times\frac{30}{100} = 0.3L$$

Now, equating both the results

$$0.3L = \left(\frac{L}{4\pi}\right)\times\sin^{-1}\left(\frac{lb}{\sqrt{(l^2+h^2)(b^2+h^2)}}\right)$$

$$\Rightarrow \frac{lb}{\sqrt{(l^2+h^2)(b^2+h^2)}} = sin(1.2\pi) = sin\left(1.2\pi\times\frac{180^o}{\pi}\right) = sin216^o$$

$$\Rightarrow \frac{1}{\sqrt{\left(1+\left(\frac{h}{l}\right)^2\right)\left(1+\left(\frac{h}{b}\right)^2\right)}} = sin216^o$$

On setting the value of $h/_l$

$$\left(1+\left(\frac{2}{3}\right)^2\right)\left(1+\left(\frac{h}{b}\right)^2\right) = cosec^2 216^o = 2.894$$

$$\Rightarrow 1+\left(\frac{h}{b}\right)^2 = \frac{2.894}{1.444} = 2 \ or \ \left(\frac{h}{b}\right)^2 = 1$$

$$\Rightarrow \frac{h}{b} = 1 \qquad\qquad Ans.$$

Example 2.18: Find out the flux density over a square plane with each side 34cm if four uniform point-sources each of 200 cd are located at the same normal height 50cm from the vertices. (Assume that the flux is uniformly distributed over the entire plane)

Sol. given that

$$side\ of\ the\ square\ plane, a = 34cm,\ \ normal\ height, h = 50cm\ \&$$
$$luminous\ intensity, I = 200\ cd$$

Solid angle subtended by a square plane at any point lying at a normal height 'h' from the one of the vertices is given from the eq(26) as follows

$$\omega = \sin^{-1}\left(\frac{a^2}{a^2+h^2}\right)$$

On setting the values, we get

$$\omega = \sin^{-1}\left(\frac{(34)^2}{(34)^2+(50)^2}\right)$$
$$= \sin^{-1}(0.316) = 0.321\ sr$$

Luminous flux intercepted by the square plane emitted by all the point-sources

$$\Rightarrow F = 4 \times I \times \omega$$
$$= 4 \times 200 \times 0.321 = 256.8\ lm$$

Hence, the flux density over the plane

$$\Rightarrow F.D. = \frac{F}{A} = \frac{256.8}{(0.34)^2} = 2221.453\ lux \qquad\qquad Ans.$$

Example 2.19: A square plane intercepts x fraction of the total radiation emitted by a uniform point-source located at a normal height 'h' from the centre. Prove that area (A) of the plane is given as

$$A = \frac{4h^2}{cosec\pi x - 1} \quad \forall \left(x \le \frac{1}{2}\right).$$

Sol. let each side of the square plane be 'a'& the luminosity of the point-source be 'L'

Solid angle subtended by the square plane at the point-source

$$\omega = 4\sin^{-1}\left(\frac{a^2}{a^2+4h^2}\right) \qquad\qquad (from\ the\ eq(25))$$

Now, the radiation energy intercepted by the plane

$$\Rightarrow E_i = \left(\frac{L}{4\pi}\right) \times \omega = \left(\frac{L}{4\pi}\right) \times 4\sin^{-1}\left(\frac{a^2}{a^2+4h^2}\right)$$

But, fraction of total radiation energy emitted by the source, intercepted by the plane, is x

$$\Rightarrow E_i = x \times (total\ energy\ emitted\ by\ the\ point - source) = xL$$

On equating both the results, we get

$$\left(\frac{L}{4\pi}\right) \times 4\sin^{-1}\left(\frac{a^2}{a^2+4h^2}\right) = xL$$

$$\Rightarrow \frac{a^2}{a^2+4h^2} = sin\pi x$$

$$or \quad \frac{A}{A+4h^2} = sin\pi x \qquad (since,\ \ area, A = a^2)$$

$$\Rightarrow \frac{A+4h^2}{A} = cosec\pi x \ \ or \ \ \frac{A+4h^2}{A} - 1 = cosec\pi x - 1$$

$$or \quad \frac{A+4h^2-A}{A} = cosec\pi x - 1$$

$$or \quad \frac{4h^2}{A} = cosec\pi x - 1$$

$$\Rightarrow \quad A = \frac{4h^2}{cosec\pi x - 1}$$

It is obvious that the value of solid angle varies as follows

$$\Rightarrow 0 < \sin^{-1}\left(\frac{a^2}{a^2+4h^2}\right) \leq \frac{\pi}{2} (\forall\ a > 0)$$

$$or \ \ 0 < 4\sin^{-1}\left(\frac{a^2}{a^2+4h^2}\right) \leq 2\pi$$

$$\Rightarrow 0 < \omega \leq 2\pi$$

Thus, the value of energy intercepted by plane varies as follows

$$\Rightarrow 0 < (E_i = xL) \leq \left(\frac{L}{4\pi}\right) \times \omega_{max}$$

$$\Rightarrow 0 < xL \leq \left(\frac{L}{4\pi}\right) \times 2\pi \ \ or \ \ 0 < x \leq \frac{1}{2}$$

$$\Rightarrow \ \ x \leq \frac{1}{2} (since, fraction\ (x) is\ always\ positive)$$

Hence, applying the limit, we have

$$A = \frac{4h^2}{cosec\pi x - 1} \quad \forall \left(x \leq \frac{1}{2}\right) \qquad\qquad proved.$$

Example 2.20: Find out the energy density over a rhombus like plane with diagonals 12cm & 22cm if a uniform point-source of 500W is located at a normal height 60cm from the centre. (Assume that the flux is uniformly distributed over the entire plane).

Sol. given that

$$diagonals, 2d_1 = 12cm\ \&\ 2d_2 = 22cm, \qquad normal\ height, h = 60cm\ \&$$

$$luminosity\ of\ point\ source, L = 500\ W$$

Solid angle subtended by a rhombus like plane with diagonals $2d_1$& $2d_2$ at any point lying at a normal height ‘h’ from the centre of the plane is given by the eq(28) as follows

$$\omega = 4\cos^{-1}\left\{\frac{h\left(d_1{}^2\sqrt{h^2+d_2{}^2}+d_2{}^2\sqrt{h^2+d_1{}^2}\right)}{h^2(d_1{}^2+d_2{}^2)+d_1{}^2d_2{}^2}\right\}$$

On setting the values, we get

$$\omega = 4\cos^{-1}\left\{\frac{60\left((6)^2\sqrt{(60)^2+(11)^2}+(11)^2\sqrt{(60)^2+(6)^2}\right)}{(60)^2((6)^2+(11)^2)+(6)^2(11)^2}\right\}$$

$$= 4\cos^{-1}\left\{\frac{569532.582}{569556}\right\}$$

$$= 4\cos^{-1}\{0.999\} = 0.179\ sr$$

$$\therefore\ radiation\ energy\ intercepted\ by\ the\ plane, \qquad E_i = \left(\frac{L}{4\pi}\right)\times\omega$$

$$E_i = \left(\frac{500}{4\pi}\right)\times 0.179 = 7.122\ W$$

Area of the rhombus

$$\Rightarrow A = 2d_1d_2 = 2\times 6\times 11 = 132\ cm^2 = 0.0132\ m^2$$

$$hence,\ \ energy\ density = \frac{E_i}{A}$$

$$= \frac{7.122}{0.0132} = 539.545\ Wm^{-2} \qquad Ans.$$

Example 2.21: Find out the energy density over a regular pentagonal plane with each side 12cm if a uniform point-source of 200W is located at a normal height 20cm from the centre. (Assume that the flux is uniformly distributed over the entire plane).

Sol. given that

$$number\ of\ sides, n = 5,\ \ length\ of\ each\ side, a = 12cm,$$

$$normal\ height, h = 20cm\ \&$$

$$luminosity\ of\ point\ source, L = 200\ W$$

We know that solid angle subtended by a regular pentagonal plane with 'n' number of the sides each of length 'a' at any point lying at a normal height 'h' from the centre of the plane is given by eq(34) as follows

$$\omega = 2\pi - 2n\sin^{-1}\left\{\frac{2h sin\frac{\pi}{n}}{\sqrt{4h^2+a^2cot^2\frac{\pi}{n}}}\right\}$$

On setting the values, we get

$$\omega = 2\pi - 2\times 5\sin^{-1}\left\{\frac{2(20)sin\frac{\pi}{5}}{\sqrt{4(20)^2+(12)^2cot^2\frac{\pi}{5}}}\right\}$$

$$= 2\pi - 10\sin^{-1}\left\{\frac{23.511}{\sqrt{1872.797}}\right\} = 2\pi - 10\sin^{-1}\{0.543\}$$

$$= 2\pi - 5.74 = 0.543\ sr$$

$$\therefore\ radiation\ energy\ intercepted\ by\ the\ plane, \qquad E_i = \left(\frac{L}{4\pi}\right)\times\omega$$

$$E_i = \left(\frac{200}{4\pi}\right) \times 0.543 = 8.642\ W$$

We know that the area of a regular pentagon is given as

$$\Rightarrow A = \frac{1}{4} na^2 cot \frac{\pi}{n}$$

On setting the values we get

$$A = \frac{1}{4} \times 5(12)^2 cot \frac{\pi}{5} = 247.749\ cm^2 = 0.0248\ m^2$$

$$hence,\ energy\ density = \frac{E_i}{A}$$

$$= \frac{8.642}{0.0248} = 348.468\ Wm^{-2} \qquad Ans.$$

Example 2.22: A regular hexagonal plane intercepts 240 lm flux when a uniform point source of 200 cd is located at a normal height 10cm from the centre of the plane. What will be area of the plane?

Sol. given that

$$number\ of\ sides, n = 6,\ \ normal\ height, h = 10cm,$$
$$luminous\ flux\ intercepted\ by\ the\ plane, F = 240\ lm\ \&$$
$$intensity\ of\ point\ source, I = 200\ cd$$

Let 'a' be the length of each side of the plane

We know that solid angle subtended by a regular pentagonal plane with 'n' number of the sides each of length 'a' at any point lying at a normal height 'h' from the centre of the plane is given by eq(34) as follows

$$\omega = 2\pi - 2n\sin^{-1}\left\{\frac{2h sin\frac{\pi}{n}}{\sqrt{4h^2 + a^2 cot^2\frac{\pi}{n}}}\right\}$$

On setting the values, we get

$$\omega = 2\pi - 2 \times 6\sin^{-1}\left\{\frac{2(10) sin\frac{\pi}{6}}{\sqrt{4(10)^2 + a^2 cot^2\frac{\pi}{6}}}\right\}$$

$$= 2\pi - 12\sin^{-1}\left\{\frac{10}{\sqrt{400 + 3a^2}}\right\}$$

Hence, flux intercepted by the plane is given as follows

$$F = I \times \omega$$

$$\Rightarrow 240 = 200 \times \left[2\pi - 12\sin^{-1}\left\{\frac{10}{\sqrt{400 + 3a^2}}\right\}\right]$$

$$or\ \ 12\sin^{-1}\left\{\frac{10}{\sqrt{400 + 3a^2}}\right\} = 2\pi - 1.2 = 5.083$$

$$or\ \ \frac{10}{\sqrt{400 + 3a^2}} = sin\left(\frac{5.083}{12} \times \frac{180^o}{\pi}\right) = 0.411$$

$$or\ \ \ 400 + 3a^2 = 591.993$$

$$or\ a^2 = \frac{591.993 - 400}{3} = 63.998$$

We know that the area of a regular pentagon is given as

$$\Rightarrow A = \frac{1}{4}na^2 cot\frac{\pi}{n}$$

On setting the values we get

$$A = \frac{1}{4} \times 6(63.998)cot\frac{\pi}{6} = 166.272\ cm^2 \qquad \textit{Ans.}$$

Example 2.23: Find out the flux density over a circular plane with a radius 35cm if a uniform point-source emitting 1400 lm flux is located at a normal height 40cm from the centre. (Assume that the flux is uniformly distributed over the entire plane).

Sol. given that

$$radius\ of\ the\ plane, R = 35cm,\ \ normal\ height, h = 40cm\ \&$$

$$total\ flux\ emitted\ by\ the\ point\ source, F_t = 1400\ lm$$

Intensity of the source,

$$\Rightarrow I = \frac{F_t}{4\pi} = \frac{1400}{4\pi} = 111.408\ cd$$

Solid angle subtended by the circular plane at the point-source

$$\omega = 2\pi\left(1 - \frac{h}{\sqrt{h^2 + R^2}}\right)$$

On setting the values, we get

$$\omega = 2\pi\left(1 - \frac{40}{\sqrt{(40)^2 + (35)^2}}\right) = 2\pi(1 - 0.752) = 1.558\ sr$$

Area of the circular plane

$$\Rightarrow A = \pi R^2 = \pi(35)^2 = 3848.451 cm^2\ or\ 0.385\ m^2$$

Hence, the flux density

$$\Rightarrow \quad F.D. = \frac{I \times \omega}{A} = \frac{111.408 \times 1.558}{0.385} = 450.84\ lux \qquad \textit{Ans.}$$

Example 2.24: Find out the total radiation energy intercepted by an infinitely long rectangular strip having width 325mm if a uniform point-source of 300 W is mounted at a normal height 1.5m from the centre of strip.

Sol. given that

$$width\ of\ strip, b = 325mm = 0.325m, \qquad normal\ height, h = 1.5m\ \&$$

$$luminosity\ of\ the\ point\ source, L = 300\ W$$

We know that the solid angle subtended by an infinitely long rectangular plane having width 'b' at any point lying at a normal height 'h' from the centre is given by eq(2.8) as follows

$$\omega = 4\sin^{-1}\left(\frac{b}{\sqrt{b^2 + 4h^2}}\right)$$

On setting the values, we get

$$\omega = 4\sin^{-1}\left(\frac{0.325}{\sqrt{(0.325)^2 + 4(1.5)^2}}\right)$$

$$= 4\sin^{-1}\left(\frac{0.325}{\sqrt{9.106}}\right) = 0.432\ sr$$

Hence, the total energy intercepted by the strip

$$\Rightarrow E_i = \left(\frac{L}{4\pi}\right) \times \omega$$

$$= \left(\frac{300}{4\pi}\right) \times 0.432 = 10.313\ W \qquad Ans.$$

Example 2.25: A regular polygonal plane, having 'n' number of the sides each of length 'a', intercepts x fraction of the total radiation emitted by a uniform point-source located at a normal height 'h' from the centre of plane. Prove that

$$\frac{h}{a} = \frac{sin\left\{\frac{\pi}{n}(1-2x)\right\}}{2tan\frac{\pi}{n}\sqrt{sin^2\frac{\pi}{n} - sin^2\left\{\frac{\pi}{n}(1-2x)\right\}}} \quad \forall \left(x \le \frac{1}{2}\right)$$

Sol. given that

$$number\ of\ the\ sides = n, length\ of\ each\ side = a\ \ \&$$

$$fraction\ of\ the\ radaition\ intercepted\ by\ the\ plane = x$$

Let the 'L' be the luminosity of the point-source

We know that solid angle subtended by a regular pentagonal plane with 'n' number of the sides each of length 'a' at any point lying at a normal height 'h' from the centre of the plane is given by eq(34) as follows

$$\omega = 2\pi - 2n\sin^{-1}\left\{\frac{2hsin\frac{\pi}{n}}{\sqrt{4h^2 + a^2cot^2\frac{\pi}{n}}}\right\}$$

Energy intercepted by the plane

$$\Rightarrow E_i = \left(\frac{L}{4\pi}\right) \times \omega$$

$$= \left(\frac{L}{4\pi}\right) \times \left[2\pi - 2n\sin^{-1}\left\{\frac{2hsin\frac{\pi}{n}}{\sqrt{4h^2 + a^2cot^2\frac{\pi}{n}}}\right\}\right]$$

But the fraction of the total radiation, emitted by point-source, intercepted by the plane is x

Hence, we have

$$\Rightarrow E_i = x \times (total\ energy\ emitted\ by\ the\ source) = xL$$

On equating both the results, we have

$$\left(\frac{L}{4\pi}\right) \times \left[2\pi - 2n\sin^{-1}\left\{\frac{2hsin\frac{\pi}{n}}{\sqrt{4h^2 + a^2cot^2\frac{\pi}{n}}}\right\}\right] = xL$$

$$\Rightarrow 2n\sin^{-1}\left\{\frac{2hsin\frac{\pi}{n}}{\sqrt{4h^2 + a^2cot^2\frac{\pi}{n}}}\right\} = 2\pi - 4x\pi$$

$$\Rightarrow \frac{2hsin\frac{\pi}{n}}{\sqrt{4h^2 + a^2cot^2\frac{\pi}{n}}} = sin\left\{\frac{\pi}{n}(1-2x)\right\}$$

On squaring both the sides as follows

$$\Rightarrow \frac{4h^2sin^2\frac{\pi}{n}}{4h^2 + a^2cot^2\frac{\pi}{n}} = sin^2\left\{\frac{\pi}{n}(1-2x)\right\}$$

$$\Rightarrow \frac{4h^2 + a^2cot^2\frac{\pi}{n}}{4h^2sin^2\frac{\pi}{n}} = \frac{1}{sin^2\left\{\frac{\pi}{n}(1-2x)\right\}}$$

$$\Rightarrow \frac{4h^2 + a^2cot^2\frac{\pi}{n}}{4h^2} - 1 = \frac{sin^2\frac{\pi}{n}}{sin^2\left\{\frac{\pi}{n}(1-2x)\right\}} - 1$$

$$\Rightarrow \frac{4h^2 + a^2cot^2\frac{\pi}{n} - 4h^2}{4h^2} = \frac{sin^2\frac{\pi}{n} - sin^2\left\{\frac{\pi}{n}(1-2x)\right\}}{sin^2\left\{\frac{\pi}{n}(1-2x)\right\}}$$

$$\Rightarrow \frac{a^2}{4h^2} = \frac{sin^2\frac{\pi}{n} - sin^2\left\{\frac{\pi}{n}(1-2x)\right\}}{cot^2\frac{\pi}{n}sin^2\left\{\frac{\pi}{n}(1-2x)\right\}}$$

$$\Rightarrow \frac{a^2}{h^2} = \frac{4\left[sin^2\frac{\pi}{n} - sin^2\left\{\frac{\pi}{n}(1-2x)\right\}\right]}{cot^2\frac{\pi}{n}sin^2\left\{\frac{\pi}{n}(1-2x)\right\}} = \frac{4tan^2\frac{\pi}{n}\left[sin^2\frac{\pi}{n} - sin^2\left\{\frac{\pi}{n}(1-2x)\right\}\right]}{sin^2\left\{\frac{\pi}{n}(1-2x)\right\}}$$

$$\Rightarrow \frac{h^2}{a^2} = \frac{sin^2\left\{\frac{\pi}{n}(1-2x)\right\}}{4tan^2\frac{\pi}{n}\left[sin^2\frac{\pi}{n} - sin^2\left\{\frac{\pi}{n}(1-2x)\right\}\right]}$$

$$or\ \frac{h}{a} = \frac{sin\left\{\frac{\pi}{n}(1-2x)\right\}}{2tan\frac{\pi}{n}\sqrt{sin^2\frac{\pi}{n} - sin^2\left\{\frac{\pi}{n}(1-2x)\right\}}}$$

It is obvious that the value of the solid angle subtended by a regular polygonal plane varies as follows

$$0 < \left[2\pi - 2n\sin^{-1}\left\{\frac{2hsin\frac{\pi}{n}}{\sqrt{4h^2 + a^2cot^2\frac{\pi}{n}}}\right\}\right] \leq 2\pi \qquad \forall\,(a > 0)$$

$$\Rightarrow 0 < \omega \leq 2\pi \qquad \Rightarrow \omega_{max} = 2\pi\ sr$$

Hence, the value of the radiation energy intercepted by the plane varies as follows

$$0 < (E_i = xL) \leq \left(\frac{L}{4\pi}\right) \times \omega_{max}$$

$$\Rightarrow\ 0 < x \leq \frac{1}{2}\ \ or \qquad x \leq \frac{1}{2}(since,\ \ fraction\ (x)\ is\ always + ve)$$

Hence, on applying the limit we have

$$\frac{h}{a} = \frac{sin\left\{\frac{\pi}{n}(1-2x)\right\}}{2tan\frac{\pi}{n}\sqrt{sin^2\frac{\pi}{n} - sin^2\left\{\frac{\pi}{n}(1-2x)\right\}}} \qquad \forall\left(x \leq \frac{1}{2}\right) \qquad proved.$$

Miscellaneous questions:

Note: All the dimensions have same units in the examples below.

Example 2.26: Find out the solid angle subtended by plane lamina ABCD at the point 'P' (indicated in the fig(2.1) below)

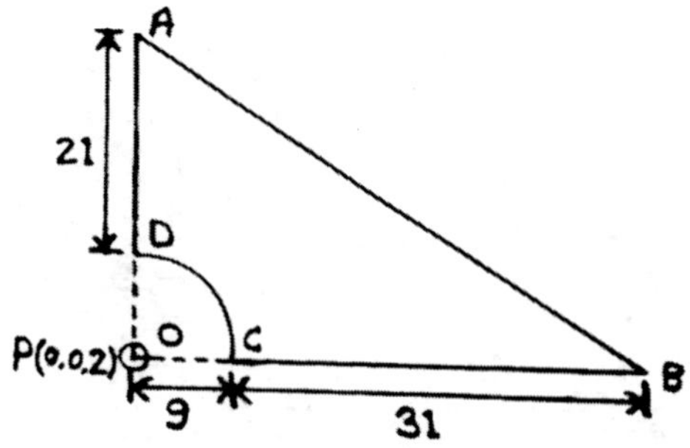

Fig (2.1)

Sol. solid angle subtended by the right ΔAOB at the point P (0, 0, 2)

(From eq(6))

$$\omega_{\Delta AOB} = \cos^{-1}\left\{\frac{(OP)\left((OA)^2\sqrt{(OP)^2+(OB)^2}+(OB)^2\sqrt{(OP)^2+(OA)^2}\right)}{(OP)^2((OA)^2+(OB)^2)+(OA)^2(OB)^2}\right\}$$

$$= \cos^{-1}\left\{\frac{(2)\left((30)^2\sqrt{(2)^2+(40)^2}+(40)^2\sqrt{(2)^2+(30)^2}\right)}{(2)^2((30)^2+(40)^2)+(30)^2(40)^2}\right\}$$

$$= \cos^{-1}\left\{\frac{168303.041}{1450000}\right\} = \cos^{-1}\{0.116\} = 1.454\ sr$$

Now, the solid angle subtended by circular lamina OCD at the point 'P'

(From eq(37))

$$\Rightarrow \omega_{OCD} = \frac{1}{4}\times 2\pi\left(1-\frac{h}{\sqrt{h^2+R^2}}\right)$$

$$= \frac{1}{4}\times 2\pi\left(1-\frac{2}{\sqrt{2^2+9^2}}\right) = 1.23\ sr$$

$\therefore$ *solid angle subtended by the lamina ABCD at the 'P'*

$$\Rightarrow \omega_{ABCD} = \omega_{\Delta AOB} - \omega_{OCD}$$

$$= 1.454 - 1.23 = 0.224\ sr$$

Ans.

Example 2.27: Find out the solid angle subtended by rectangular lamina ABCD (having an equilateral triangular hole at the centre 'O')at the point 'P' (indicated in the fig(2.2) below

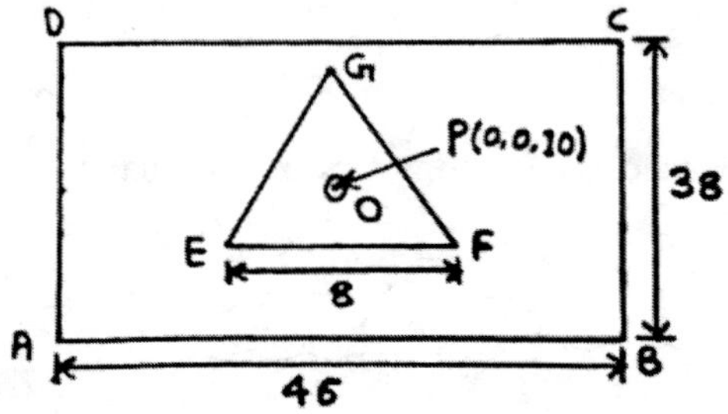

Fig 2.2

Sol. solid angle subtended by rectangular lamina ABCD (without hole) at the point 'P'

(From the eq(16))

$$\omega_{ABCD} = 4\sin^{-1}\left(\frac{lb}{\sqrt{(l^2+4h^2)(b^2+4h^2)}}\right)$$

$$= 4\sin^{-1}\left(\frac{46\times 38}{\sqrt{((46)^2+4(10)^2)((38)^2+4(10)^2)}}\right)$$

$$= 4\sin^{-1}\left(\frac{1748}{2153.951}\right)$$

$$= 4\sin^{-1}(0.811) = 3.783\ sr$$

Solid angle subtended by equilateral triangular lamina EFG at the point 'P'

(From the eq(15))

$$\Rightarrow \omega_{\Delta EFG} = 2\pi - 6\sin^{-1}\left(\frac{3h}{\sqrt{12h^2 + a^2}}\right)$$

$$= 2\pi - 6\sin^{-1}\left(\frac{3 \times 10}{\sqrt{12(10)^2 + (8)^2}}\right) = 2\pi - 6\sin^{-1}\left(\frac{30}{35.553}\right)$$

$$= 2\pi - 6\sin^{-1}(0.844) = 2\pi - 6.028 = 0.255\ sr$$

$\therefore$ *solid angle subtended by the lamina ABCD (with hole) at the 'P'*

$$\Rightarrow \omega'_{ABCD} = \omega_{ABCD} - \omega_{\Delta EFG}$$

$$= 3.783 - 0.255 = 3.528\ sr \qquad Ans.$$

Example 2.28: Find out the solid angle subtended by plane lamina ABCDEO at the point 'P' (indicated in the fig(2.3) below)

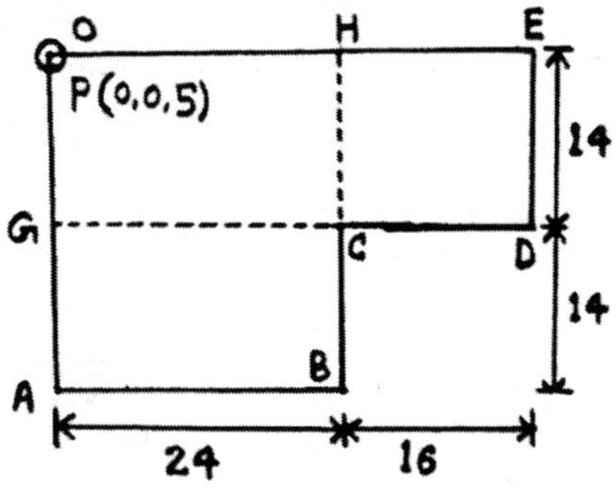

Fig (2.3)

Sol. solid angle subtended by rectangular lamina OABH at the point 'P'
(From the eq(17))

$$\omega_{OABH} = \sin^{-1}\left(\frac{lb}{\sqrt{(l^2 + h^2)(b^2 + h^2)}}\right)$$

$$= \sin^{-1}\left(\frac{28 \times 24}{\sqrt{((28)^2 + (5)^2)((24)^2 + (5)^2)}}\right)$$

$$= \sin^{-1}\left(\frac{672}{697.287}\right) = \sin^{-1}(0.964) = 1.302\ sr$$

Similarly, solid angle subtended by rectangular lamina OEDG at the point 'P'

$$\omega_{OEDG} = \sin^{-1}\left(\frac{40 \times 14}{\sqrt{((40)^2 + (5)^2)((14)^2 + (5)^2)}}\right)$$

$$= \sin^{-1}\left(\frac{560}{599.27}\right) = \sin^{-1}(0.934) = 1.205\ sr$$

Similarly, solid angle subtended by rectangular lamina OHCG at the point 'P'

$$\omega_{OHCG} = \sin^{-1}\left(\frac{24 \times 14}{\sqrt{((24)^2 + (5)^2)((14)^2 + (5)^2)}}\right)$$

$$= \sin^{-1}\left(\frac{336}{364.446}\right) = \sin^{-1}(0.922) = 1.173\ sr$$

Hence, solid angle subtended by the lamina ABCDEO at the point 'P'

$$\Rightarrow \omega_{ABCDEO} = \omega_{OABH} + \omega_{OEDG} - \omega_{OHCG}$$

$$= 1.302 + 1.205 - 1.173 = 1.334\ sr \qquad Ans.$$

Example 2.29: Find out the solid angle subtended by plane lamina ABCD (having a square hole at the centre 'O') at the point 'P' (indicated in the fig(2.4) below)

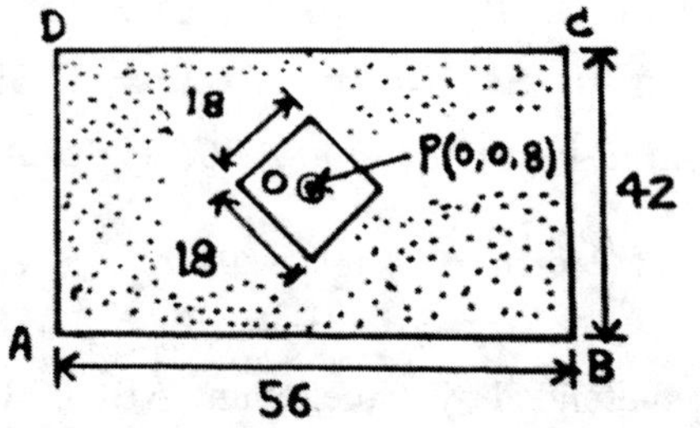

Fig (2.4)

Sol. solid angle subtended by rectangular lamina ABCD (without hole) at the point 'P'

(From the eq(16))

$$\omega_{ABCD} = 4\sin^{-1}\left(\frac{lb}{\sqrt{(l^2+4h^2)(b^2+4h^2)}}\right)$$

$$= 4\sin^{-1}\left(\frac{56\times 42}{\sqrt{((56)^2+4(8)^2)((42)^2+4(8)^2)}}\right)$$

$$= 4\sin^{-1}\left(\frac{2352}{2617.602}\right)$$

$$= 4\sin^{-1}(0.898) = 4.461\ sr$$

Now, solid angle subtended by square hole at the point 'P'

(From the eq(25))

$$\omega' = 4\sin^{-1}\left(\frac{a^2}{a^2+4h^2}\right)$$

$$= 4\sin^{-1}\left(\frac{(18)^2}{(18)^2+4(8)^2}\right)$$

$$= 4\sin^{-1}(0.559) = 2.373\ sr$$

Hence, solid angle subtended by the rectangular lamina ABCD (with hole) at the point 'P'

$$\Rightarrow \omega'_{ABCD} = \omega_{ABCD} - \omega'$$

$$= 4.461 - 2.373 = 2.088\ sr$$ *Ans.*

Example 2.30: Find out the solid angle subtended by square lamina ABCD (having a rectangular hole at the centre 'O') at the point 'P' (indicated in the fig(2.5) below)

Sol. solid angle subtended by square lamina ABCD at the point 'P'

(From the eq(25))

$$\omega_{ABCD} = 4\sin^{-1}\left(\frac{a^2}{a^2+4h^2}\right)$$

$$= 4\sin^{-1}\left(\frac{(26)^2}{(26)^2+4(12)^2}\right)$$

$$= 4\sin^{-1}(0.54) = 2.282\ sr$$

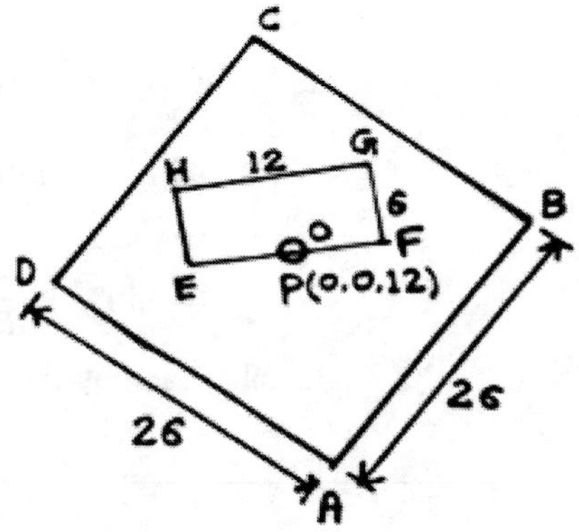

Fig 2.5

Now, solid angle subtended by rectangular hole EFGH at the point 'P'
(From the eq(18))

$$\omega_{EFGH} = 2\sin^{-1}\left(\frac{lb}{\sqrt{(l^2+4h^2)(b^2+h^2)}}\right)$$

$$= 2\sin^{-1}\left(\frac{12\times 6}{\sqrt{((12)^2+4(12)^2)((6)^2+(12)^2)}}\right)$$

$$= 2\sin^{-1}\left(\frac{72}{360}\right)$$

$$= 2\sin^{-1}(0.2) = 0.403\ sr$$

Hence, solid angle subtended by the square lamina ABCD (with hole) at the point 'P'

$$\Rightarrow \omega'_{ABCD} = \omega_{ABCD} - \omega_{EFGH}$$

$$= 2.282 - 0.403 = 1.879\ sr$$ *Ans.*

Example 2.31: Find out the solid angle subtended by regular hexagonal lamina ABCDEF (having a circular hole at the centre 'O') at the point 'P' (indicated in the fig(2.6) below)

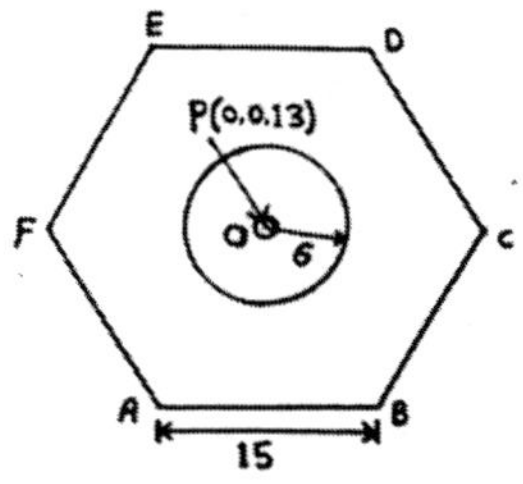

Fig 2.6

Sol. solid angle subtended by regular hexagonal lamina ABCDEF (without hole) at the point 'P'

(From the eq(34))

$$\Rightarrow \omega_{ABCDEF} = 2\pi - 2n\sin^{-1}\left(\frac{2h sin\frac{\pi}{n}}{\sqrt{4h^2 + a^2 cot^2\frac{\pi}{n}}}\right)$$

On setting the values, we get

$$\omega_{ABCDEF} = 2\pi - 2\times 6\sin^{-1}\left(\frac{2\times 13 sin\frac{\pi}{6}}{\sqrt{4(13)^2 + (15)^2 cot^2\frac{\pi}{6}}}\right)$$

$$= 2\pi - 12\sin^{-1}\left(\frac{13}{\sqrt{1351}}\right) = 2\pi - 12\sin^{-1}(0.354)$$

$$= 2\pi - 4.342 = 1.941\ sr$$

Now, the solid angle subtended by circular hole at the point 'P'
(From eq(37))

$$\Rightarrow \omega' = 2\pi\left(1 - \frac{h}{\sqrt{h^2+R^2}}\right)$$

$$= 2\pi\left(1 - \frac{13}{\sqrt{(13)^2+(6)^2}}\right) = 0.578\ sr$$

Hence, solid angle subtended by the square lamina ABCDEF (with hole) at the point 'P'

$$\Rightarrow \omega'_{ABCDEF} = \omega_{ABCDEF} - \omega'$$

$$= 1.941 - 0.578 = 1.363\ sr$$ *Ans.*

Example 2.32: A uniform point-source of 100 cd is located at point P (0, 0, 12) Find out the total flux intercepted by plane lamina ABCDEFFGHJKLM (having a circular hole at the centre 'O') (indicated in the fig(2.7) below)

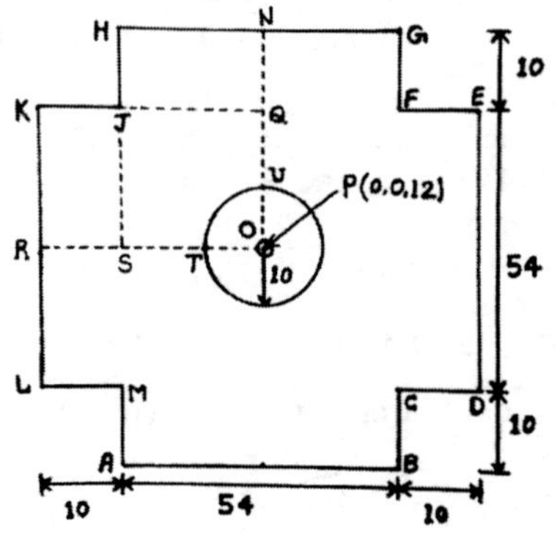

Fig (2.7)

Sol. solid angle subtended by rectangular lamina ONHS (without hole)

(From the eq(17))

$$\omega_{ONHS} = \sin^{-1}\left(\frac{lb}{\sqrt{(l^2+h^2)(b^2+h^2)}}\right)$$

$$= \sin^{-1}\left(\frac{37\times 27}{\sqrt{((37)^2+(12)^2)((27)^2+(12)^2)}}\right)$$

$$= \sin^{-1}\left(\frac{999}{1149.282}\right) = \sin^{-1}(0.869) = 1.053\ sr$$

Solid angle subtended by square lamina OQJS (without hole) at the point 'P'

(From the eq(26))

$$\omega_{OQJS} = \sin^{-1}\left(\frac{a^2}{a^2+h^2}\right)$$

$$= \sin^{-1}\left(\frac{(27)^2}{(27)^2+(12)^2}\right)$$

$$= \sin^{-1}(0.835) = 0.988\ sr$$

Now, solid angle subtended by plane lamina ONHJKR (without hole) at the point 'P'

$$\omega_{ONHJKR} = \omega_{ONHS} + \omega_{OQKR} - \omega_{OQJS}$$

$$= 2\omega_{ONHS} - \omega_{OQJS}(since,\ \ \omega_{OQKR} = \omega_{ONHS})$$

$$= 2\times 1.053 - 0.988 = 1.118\ sr$$

Solid angle subtended by plane lamina ABCDEFGHJKLM (without hole) at the point 'P'

$$\Rightarrow \omega_{ABCDEFGHJKLM} = 4\times \omega_{ONHJKR} \qquad (by\ symmetry)$$

$$= 4\times 1.118 = 4.472\ sr$$

Now, solid angle subtended by the circular hole at the point 'P'

(From eq(37))

$$\Rightarrow \omega' = 2\pi\left(1-\frac{h}{\sqrt{h^2+R^2}}\right)$$

$$= 2\pi\left(1 - \frac{12}{\sqrt{(12)^2 + (10)^2}}\right) = 1.456\ sr$$

Thus, solid angle subtended by plane lamina ABCDEFGHJKLM (with hole) at the point 'P'

$$\Rightarrow \omega'_{ABCDEFGHJKLM} = \omega_{ABCDEFGJKLM} - \omega'$$

$$= 4.472 - 1.456 = 3.016\ sr$$

Hence, flux intercepted by the plane lamina ABCDEFGHJKLM (with hole)

$$\Rightarrow F = I \times \omega = 100 \times 3.016 = 301.6\ lm \qquad Ans.$$

Example 2.33: A uniform point-source of 1400 lm is located at point P (0, 0, 8) Find out the total flux intercepted by I-sectionABCDEFGHJKLM (indicated in the fig(2.8) below)

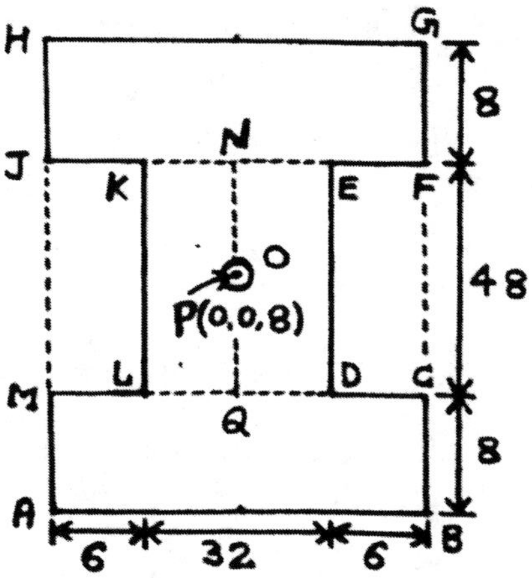

Fig (2.8)

Sol.Solid angle subtended by rectangular lamina JNQM at the point 'P'

(From the eq(18))

$$\omega_{JNQM} = 2\sin^{-1}\left(\frac{lb}{\sqrt{(l^2 + 4h^2)(b^2 + h^2)}}\right)$$

$$= 2\sin^{-1}\left(\frac{48 \times 22}{\sqrt{((48)^2 + 4(8)^2)((22)^2 + (8)^2)}}\right)$$

$$= 2\sin^{-1}\left(\frac{1056}{1184.432}\right)$$

$$= 2\sin^{-1}(0.891) = 2.199\ sr$$

Similarly, Solid angle subtended by rectangular lamina KNQL at the point 'P'

$$\omega_{KNQL} = 2\sin^{-1}\left(\frac{48 \times 16}{\sqrt{((48)^2 + 4(8)^2)((16)^2 + (8)^2)}}\right)$$

$$= 2\sin^{-1}\left(\frac{768}{905.097}\right)$$

$$= 2\sin^{-1}(0.848) = 2.024\ sr$$

Solid angle subtended by plane lamina JKLM (i.e. void) at the point 'P'

$$\omega_{JKLM} = \omega_{JNQM} - \omega_{KNQL}$$

$$= 2.199 - 2.024 = 0.175\ sr$$

Now, solid angle subtended by rectangular lamina ABGH (without voids) at the point 'P' (From the eq(16))

$$\omega_{ABGH} = 4\sin^{-1}\left(\frac{lb}{\sqrt{(l^2+4h^2)(b^2+4h^2)}}\right)$$

$$= 4\sin^{-1}\left(\frac{64\times 44}{\sqrt{((64)^2+4(8)^2)((44)^2+4(8)^2)}}\right)$$

$$= 4\sin^{-1}\left(\frac{2816}{3088.622}\right)$$

$$= 4\sin^{-1}(0.912) = 4.592\ sr$$

Now, solid angle subtended by I-section ABCDEFGHJKLM

$$\omega_o = \omega_{ABGH} - \omega_{JKLM} - \omega_{EFCD}$$

$$= \omega_{ABGH} - 2\omega_{JKLM}\ (since,\ \omega_{JKLM} = \omega_{EFCD})$$

$$= 4.592 - 2\times 0.175 = 4.242\ sr$$

Hence, flux intercepted by I-section

$$\Rightarrow F = I\times\omega_o = \left(\frac{F_t}{4\pi}\right)\times\omega_o$$

$$= \left(\frac{1400}{4\pi}\right)\times 4.242 = 472.595\ lm \qquad Ans.$$

Example 2.34: A uniform point-source of 1600 lm is located at point P (0, 0, 4) Find out the total flux intercepted by Channel-section ABCOEFGH (indicated in the fig(2.9) below)

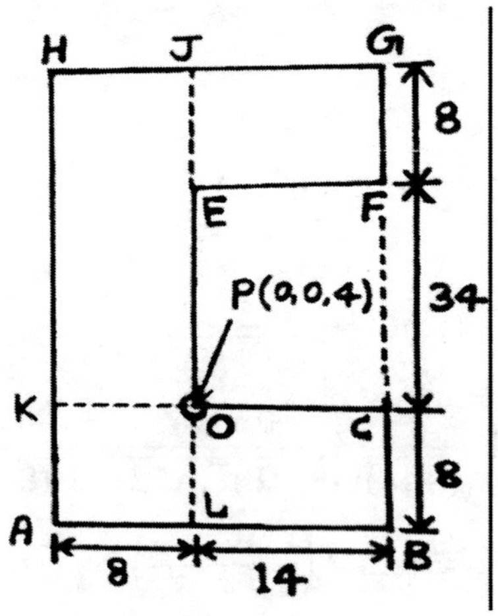

Fig (2.9)

Sol. solid angle subtended by rectangular lamina OCGJ (without void) at the point 'P' (From the eq(17))

$$\omega_{OCGJ} = \sin^{-1}\left(\frac{lb}{\sqrt{(l^2+h^2)(b^2+h^2)}}\right)$$

$$= \sin^{-1}\left(\frac{42\times 14}{\sqrt{((42)^2+(4)^2)((14)^2+(4)^2)}}\right)$$

$$= \sin^{-1}\left(\frac{588}{614.296}\right) = \sin^{-1}(0.957) = 1.276\ sr$$

Similarly, solid angle subtended by rectangular lamina OJHK at the same point

$$\omega_{OJHK} = \sin^{-1}\left(\frac{42 \times 8}{\sqrt{((42)^2+(4)^2)((8)^2+(4)^2)}}\right)$$

$$= \sin^{-1}\left(\frac{336}{377.359}\right) = \sin^{-1}(0.89) = 1.097\ sr$$

Similarly, solid angle subtended by rectangular lamina OCBL at the same point

$$\omega_{OCBL} = \sin^{-1}\left(\frac{14 \times 8}{\sqrt{((14)^2+(4)^2)((8)^2+(4)^2)}}\right)$$

$$= \sin^{-1}\left(\frac{112}{130.23}\right) = \sin^{-1}(0.86) = 1.035\ sr$$

Similarly, solid angle subtended by rectangular lamina EFCO (i.e. void) at the same point 'P'

$$\omega_{EFCO} = \sin^{-1}\left(\frac{34 \times 14}{\sqrt{((34)^2+(4)^2)((14)^2+(4)^2)}}\right)$$

$$= \sin^{-1}\left(\frac{476}{498.462}\right) = \sin^{-1}(0.955) = 1.27\ sr$$

And solid angle subtended by square lamina OLAK at the point 'P'

(From the eq(26))

$$\omega_{OLAK} = \sin^{-1}\left(\frac{a^2}{a^2+h^2}\right)$$

$$= \sin^{-1}\left(\frac{8^2}{(8)^2+(4)^2}\right)$$

$$= \sin^{-1}(0.8) = 0.927\ sr$$

Now, solid angle subtended by rectangular lamina ABGH (without void) at the point 'P'

$$\Rightarrow \omega_{ABGH} = \omega_{OCGJ} + \omega_{OJHK} + \omega_{OCBL} + \omega_{OLAK}$$

$$= 1.276 + 1.097 + 1.035 + 0.927 = 4.335\ sr$$

Solid angle subtended by Channel-section ABCOEFGH at the point 'P'

$$\Rightarrow \omega_o = \omega_{ABGH} - \omega_{EFCO}$$

$$= 4.335 - 1.27 = 3.065\ sr$$

Hence, flux intercepted by Channel-section

$$\Rightarrow F = I \times \omega_o = \left(\frac{F_t}{4\pi}\right) \times \omega_o$$

$$= \left(\frac{1600}{4\pi}\right) \times 3.065 = 390.248\ lm \qquad Ans.$$

Example 2.35: A uniform point-source of 100 W is located at point P (0, 0, 13) Find out total radiation energy intercepted by infinitely long rectangular lamina (having a rectangular hole of 32 × 14at the centre 'O') (indicated in the fig(2.10) below)

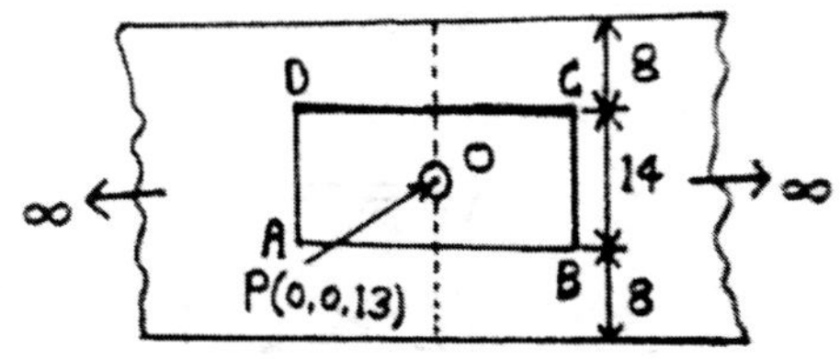

Fig (2.10)

Sol. solid angle subtended by infinitely long rectangular lamina (without hole) at the point 'P'
(From eq(2.8))

$$\omega_\infty = 4\sin^{-1}\left(\frac{b}{\sqrt{b^2+4h^2}}\right)$$
$$= 4\sin^{-1}\left(\frac{30}{\sqrt{(30)^2+4(13)^2}}\right)$$
$$= 4\sin^{-1}(0.756) = 3.429\ sr$$

Now, solid angle subtended by rectangular hole ABCD at the point 'P'
(From the eq(16))

$$\omega_{ABGH} = 4\sin^{-1}\left(\frac{lb}{\sqrt{(l^2+4h^2)(b^2+4h^2)}}\right)$$
$$= 4\sin^{-1}\left(\frac{32\times 14}{\sqrt{((32)^2+4(13)^2)((14)^2+4(13)^2)}}\right)$$
$$= 4\sin^{-1}\left(\frac{448}{1217.538}\right)$$
$$= 4\sin^{-1}(0.368) = 1.507\ sr$$

Solid angle subtended by infinitely long rectangular lamina (with hole) at the point 'P'

$$\Rightarrow \omega'_\infty = \omega_\infty - \omega_{ABGH}$$
$$= 3.429 - 1.507 = 1.922\ sr$$

Hence, the total radiation energy intercepted by the infinitely long rectangular lamina (with hole)

$$\Rightarrow E_i = \left(\frac{L}{4\pi}\right)\times\omega \quad (given\ that\ luminosity, L = 100\ W)$$
$$= \left(\frac{100}{4\pi}\right)\times 1.922 = 15.295\ J/\sec \qquad Ans.$$

Example 2.36: A uniform point-source of 500 W is located at point P (0, 0, 9) Find out total radiation energy intercepted by infinitely long rectangular lamina (having a semi-circular cut centred at the point 'O') (indicated in the fig(2.11) below)

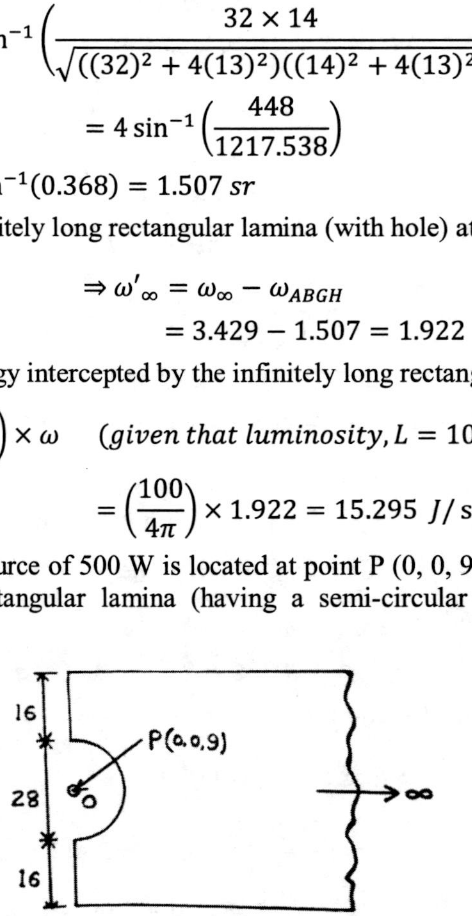

Fig (2.11)

Sol. solid angle subtended by infinitely long rectangular lamina (without cut) at the point 'P'
(From eq(2.6))

$$\omega_\infty = 2\sin^{-1}\left(\frac{b}{\sqrt{b^2+4h^2}}\right)$$
$$= 2\sin^{-1}\left(\frac{60}{\sqrt{(60)^2+4(9)^2}}\right)$$
$$= 2\sin^{-1}(0.958) = 2.56\ sr$$

Now, solid angle subtended by the semi-circular cut at the point ‘P’

(From eq(37))

$$\Rightarrow \omega' = \frac{1}{2} \times 2\pi\left(1 - \frac{h}{\sqrt{h^2 + R^2}}\right)$$

$$= \pi\left(1 - \frac{9}{\sqrt{(9)^2 + (14)^2}}\right) = 1.443\ sr$$

Solid angle subtended by infinitely long rectangular lamina (with void) at the point ‘P’

$$\Rightarrow \omega'_{\infty} = \omega_{\infty} - \omega'$$

$$= 2.56 - 1.443 = 1.117\ sr$$

Hence, the total radiation energy intercepted by the infinitely long rectangular lamina (with hole)

$$\Rightarrow E_i = \left(\frac{L}{4\pi}\right) \times \omega \quad (given\ that\ luminosity, L = 500\ W)$$

$$= \left(\frac{500}{4\pi}\right) \times 1.117 = 44.444\ J/\sec \qquad Ans.$$

Example 2.37: A uniform point-source of 300 W is located at point P (0, 0, 11) Find out total radiation energy intercepted by right-triangular lamina (having a semi-circular cut centred at the mid-point ‘O’ of hypotenuse) (indicated in the fig(2.12) below)

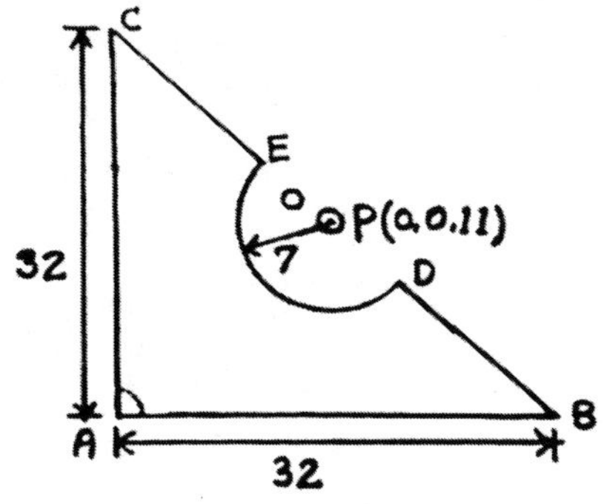

Fig (2.12)

Sol. Method 1:

Solid angle subtended by isosceles right-triangular lamina ABC (without cut) at the point ‘P’ (from eq(12))

$$\omega_{\Delta ABC} = \pi - 4\sin^{-1}\left\{\frac{h\sqrt{2}}{\sqrt{4h^2 + a^2}}\right\}$$

$$= \pi - 4\sin^{-1}\left\{\frac{11\sqrt{2}}{\sqrt{4(11)^2 + (32)^2}}\right\}$$

$$= \pi - 4\sin^{-1}\{0.4\} = \pi - 1.646 = 1.495\ sr$$

Now, solid angle subtended by the semi-circular cut at the point ‘P’

(From eq(37))

$$\Rightarrow \omega' = \frac{1}{2} \times 2\pi\left(1 - \frac{h}{\sqrt{h^2 + R^2}}\right)$$

$$= \pi\left(1 - \frac{11}{\sqrt{(11)^2 + (7)^2}}\right) = 0.491\ sr$$

Solid angle subtended by isosceles right-triangular lamina ABC (with cut) at the point ‘P’

$$\Rightarrow \omega'_{\Delta ABC} = \omega_{\Delta ABC} - \omega'$$
$$= 1.495 - 0.491 = 1.004\ sr$$

Hence, the total radiation energy intercepted by isosceles right-triangular lamina (with cut)

$$\Rightarrow E_i = \left(\frac{L}{4\pi}\right) \times \omega \quad (given\ that\ luminosity, L = 300\ W)$$

$$= \left(\frac{300}{4\pi}\right) \times 1.004 = 23.969\ J/\sec \qquad Ans.$$

Method 2:

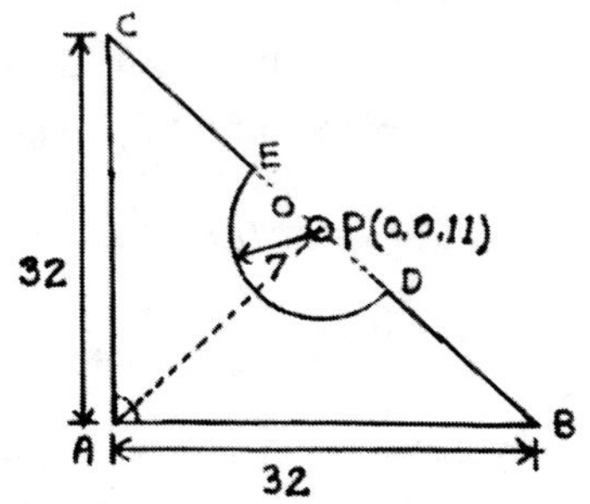

Join the vertex 'A' to the mid-point 'O' of hypotenuse BC. Thus ΔAOB & ΔAOC are congruent right triangles

$$\Rightarrow OA = OB = OC = \frac{BC}{2} = \frac{32\sqrt{2}}{2} = 16\sqrt{2}$$

Solid angle subtended by the right triangular lamina AOB (without cut) at the point 'P'

(From eq(6))

$$\omega_{\Delta AOB} = \cos^{-1}\left\{\frac{(OP)\left((OA)^2\sqrt{(OP)^2 + (OB)^2} + (OB)^2\sqrt{(OP)^2 + (OA)^2}\right)}{(OP)^2((OA)^2 + (OB)^2) + (OA)^2(OB)^2}\right\}$$

$$= \cos^{-1}\left\{\frac{(11)\left(\left(16\sqrt{2}\right)^2\sqrt{(11)^2 + \left(16\sqrt{2}\right)^2} + \left(16\sqrt{2}\right)^2\sqrt{(11)^2 + \left(16\sqrt{2}\right)^2}\right)}{(11)^2\left(\left(16\sqrt{2}\right)^2 + \left(16\sqrt{2}\right)^2\right) + \left(16\sqrt{2}\right)^2\left(16\sqrt{2}\right)^2}\right\}$$

$$= \cos^{-1}\left\{\frac{283396.509}{386048}\right\} = \cos^{-1}\{0.734\} = 0.747\ sr$$

Solid angle subtended by isosceles right-triangular lamina ABC (without cut) at the point 'P'

$$\Rightarrow \omega_{\Delta ABC} = \omega_{\Delta AOB} + \omega_{\Delta AOC}$$
$$= 2 \times \omega_{\Delta AOB} (since,\ \ \omega_{\Delta AOB} = \omega_{\Delta AOB})$$
$$= 2 \times 0.747 = 1.494\ sr$$

Now, solid angle subtended by the semi-circular cut at the point 'P'

(From eq(37))

$$\Rightarrow \omega' = \frac{1}{2} \times 2\pi\left(1 - \frac{h}{\sqrt{h^2 + R^2}}\right)$$

$$= \pi\left(1 - \frac{11}{\sqrt{(11)^2 + (7)^2}}\right) = 0.491\ sr$$

Solid angle subtended by isosceles right-triangular lamina ABC (with cut) at the point 'P'

$$\Rightarrow \omega'_{\Delta ABC} = \omega_{\Delta ABC} - \omega'$$
$$= 1.494 - 0.491 = 1.003\ sr$$

Hence, the total radiation energy intercepted by isosceles right-triangular lamina (with cut)

$$\Rightarrow E_i = \left(\frac{L}{4\pi}\right) \times \omega \quad (given\ that\ luminosity, L = 300\ W)$$

$$= \left(\frac{300}{4\pi}\right) \times 1.003 = 23.945\ J/\sec \qquad Ans.$$

Note: In both the methods, the answers are approximately same in magnitude. Mathematical calculations may cause minor difference (usually negligible) in decimal places of the values.

Example 2.38: A uniform point-source of 1000 lm is located at point P (0, 0, 7) Find out the total flux intercepted by plane lamina ABCDEFG (indicated in the fig(2.13) below)

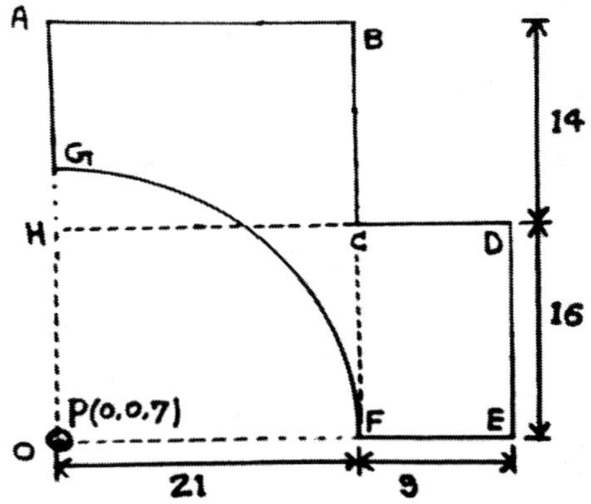

Fig (2.13)

Sol. solid angle subtended by rectangular lamina OFBA (without void) at the point 'P'

(From the eq(17))

$$\omega_{OFBA} = \sin^{-1}\left(\frac{lb}{\sqrt{(l^2+h^2)(b^2+h^2)}}\right)$$

$$= \sin^{-1}\left(\frac{30 \times 21}{\sqrt{((30)^2+(7)^2)((21)^2+(7)^2)}}\right)$$

$$= \sin^{-1}\left(\frac{630}{681.916}\right) = \sin^{-1}(0.924) = 1.178\ sr$$

Similarly, solid angle subtended by rectangular lamina OEDH at the same point

$$\omega_{OEDH} = \sin^{-1}\left(\frac{30 \times 16}{\sqrt{((30)^2+(7)^2)((16)^2+(7)^2)}}\right)$$

$$= \sin^{-1}\left(\frac{480}{538}\right) = \sin^{-1}(0.892) = 1.102\ sr$$

Similarly, solid angle subtended by rectangular lamina OHCF at the same point

$$\omega_{OHCF} = \sin^{-1}\left(\frac{21 \times 16}{\sqrt{((21)^2+(7)^2)((16)^2+(7)^2)}}\right)$$

$$= \sin^{-1}\left(\frac{336}{386.588}\right) = \sin^{-1}(0.869) = 1.053\ sr$$

Now, solid angle subtended by plane lamina ABCDEO (without void) at the point 'P'

$$\Rightarrow \omega_{ABCDEO} = \omega_{OFBA} + \omega_{OEDH} - \omega_{OHCF}$$
$$= 1.178 + 1.102 - 1.053 = 1.227\ sr$$

Now, solid angle subtended by the circular cut at the point 'P'

(From eq(37))

$$\Rightarrow \omega' = \frac{1}{4} \times 2\pi \left(1 - \frac{h}{\sqrt{h^2 + R^2}}\right)$$
$$= \frac{\pi}{2}\left(1 - \frac{7}{\sqrt{(7)^2 + (21)^2}}\right) = 1.074\ sr$$

Solid angle subtended by plane lamina ABCDEFG (with cut) at the point 'P'

$$\Rightarrow \omega'_{ABCDEFG} = \omega_{ABCDEO} - \omega'$$
$$= 1.227 - 1.074 = 0.153\ sr$$

Hence, flux intercepted by plane lamina

$$\Rightarrow F = I \times \omega'_{ABCDEFG} = \left(\frac{F_t}{4\pi}\right) \times \omega_o$$
$$= \left(\frac{1000}{4\pi}\right) \times 0.153 = 12.175\ lm \qquad Ans.$$

Planes bounded by curves:

Example 2.39: Find out solid angle subtended by the plane bounded by straight lines $x = 2, y = 5$ & $7x + 9y = 122$ at the point $P(11, 5, 3)$.

Sol. In order to identify the geometry of plane bounded by straight lines plot the straight lines on the XY-plane as follows

$$AB \equiv x = 2,\ \ BC \equiv y = 5\ \ \&\ \ AC \equiv 7x + 9y = 122$$

(as shown in the figure(2.14) below)

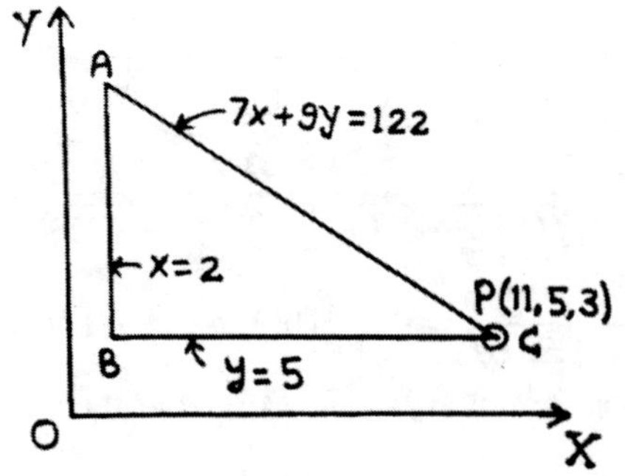

Fig (2.14)

It is obvious that the lines $AB \equiv x = 2,\ BC \equiv y = 5$ are intersecting normal to each other

By the plot, it is clear that the plane bounded is a right triangle ($i.e.\ \Delta ABC$)

On solving the equations of lines

$$x = 2\ \&\ 7x + 9y = 122$$
$$\Rightarrow 7(2) + 9y = 122$$
$$or\ 9y = 122 - 14 = 108\ \Rightarrow y = 12$$
$$\Rightarrow A \equiv (2, 12)$$

$$by\ solving, x = 2\ \&\ y = 5 \Rightarrow B \equiv (2, 5)$$

$$by\ solving, y = 5\ \&\ 7x + 9y = 122$$

$$7x + 9(5) = 122$$

$$or\ 7x = 122 - 45 = 77 \Rightarrow y = 11$$

$$\Rightarrow C \equiv (11, 5)$$

It is obvious that the point $P \equiv (11, 5, 3)$ is located at a normal height 3 unit from the vertex $C \equiv (11, 5)$ on the vertical axis

$$\Rightarrow AB = \sqrt{(2-2)^2 + (12-5)^2} = 7\ \&$$

$$BC = \sqrt{(11-2)^2 + (5-5)^2} = 9$$

Hence, the solid angle subtended by bounded plane ABC at the point 'P'

(From the eq(2))

$$\Rightarrow \omega_{\Delta ABC} = \sin^{-1}\left\{\frac{b}{\sqrt{b^2+p^2}}\right\} - \sin^{-1}\left\{\left(\frac{b}{\sqrt{b^2+p^2}}\right)\left(\frac{h}{\sqrt{h^2+p^2}}\right)\right\}$$

$$= \sin^{-1}\left\{\frac{7}{\sqrt{7^2+9^2}}\right\} - \sin^{-1}\left\{\left(\frac{7}{\sqrt{7^2+9^2}}\right)\left(\frac{3}{\sqrt{3^2+9^2}}\right)\right\}$$

$$= \sin^{-1}\{0.614\} - \sin^{-1}\{0.194\}$$

$$= 0.661 - 0.195 = 0.466\ sr \qquad Ans.$$

Example 2.40: Find out solid angle subtended by the plane bounded by straight lines $x - \sqrt{3}y = 0, \sqrt{3}x + y = 0\ \&(5 + 12\sqrt{3})x + (12 - 5\sqrt{3})y = 120$ at the point $P(0, 0, 3)$.

Sol. In order to identify the geometry of plane bounded by straight lines plot the straight lines on the XY-plane as follows

$$OA \equiv x - \sqrt{3}y = 0, \quad OB \equiv \sqrt{3}x + y = 0\ \&$$

$$AB \equiv (5 + 12\sqrt{3})x + (12 - 5\sqrt{3})y = 120$$

(as shown in the figure(2.15) below)

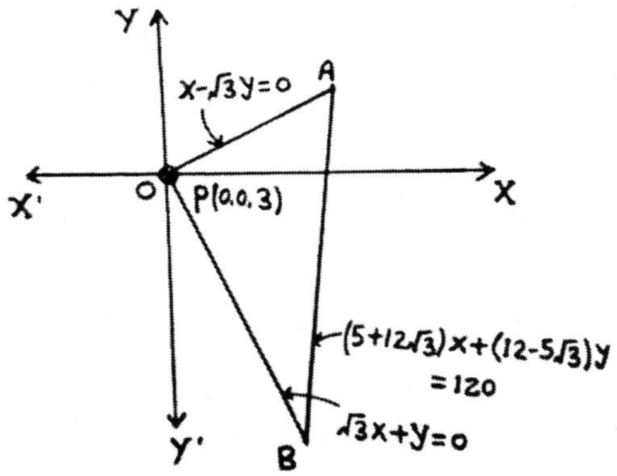

Fig (2.15)

$$slope\ of\ line\ OA, m_1 = -\left[\frac{1}{-\sqrt{3}}\right] = \frac{1}{\sqrt{3}}$$

$$slope\ of\ line\ OB, m_2 = -\left[\frac{\sqrt{3}}{1}\right] = -\sqrt{3}$$

$$\Rightarrow m_1m_2 = \left(\frac{1}{\sqrt{3}}\right)(-\sqrt{3}) = -1 \quad (condition\ of\ normalicy)$$

Thus, lines OA & OB are intersecting each other at right angle

By the plot, it is clear that the plane bounded is a right triangle ($i.e.\ \Delta ABC$)

On solving the equations of lines

$$x - \sqrt{3}y = 0\ \&(5 + 12\sqrt{3})x + (12 - 5\sqrt{3})y = 120$$

$$\Rightarrow (5 + 12\sqrt{3})\sqrt{3}y + (12 - 5\sqrt{3})y = 120$$

$$or\ (5\sqrt{3} + 36 + 12 - 5\sqrt{3})y = 120\ or\ y = \frac{5}{2}$$

$$\Rightarrow x = \frac{5\sqrt{3}}{2} \Rightarrow A \equiv \left(\frac{5\sqrt{3}}{2}, \frac{5}{2}\right)$$

On solving the equations of lines

$$\sqrt{3}x + y = 0\ \&(5 + 12\sqrt{3})x + (12 - 5\sqrt{3})y = 120$$

$$\Rightarrow (5 + 12\sqrt{3})x + (12 - 5\sqrt{3})(-\sqrt{3}x) = 120$$

$$or\ (5 + 12\sqrt{3} - 12\sqrt{3} + 15)x = 120\ or\ x = 6$$

$$\Rightarrow y = -6\sqrt{3} \Rightarrow A \equiv (6, -6\sqrt{3})$$

$$\Rightarrow OA = \sqrt{\left(\frac{5\sqrt{3}}{2} - 0\right)^2 + \left(\frac{5}{2} - 0\right)^2} = 5\ \&$$

$$BC = \sqrt{(6-0)^2 + (-6\sqrt{3} - 0)^2} = 12$$

It is obvious that the point $P \equiv (0, 0, 3)$ is located at a normal height 3 unit from right angled vertex $O \equiv (0, 0)$ on the Z-axis

Hence, the solid angle subtended by bounded plane ABC at the point 'P'

(From the eq(6))

$$\Rightarrow \omega_{\Delta AOB} = \cos^{-1}\left\{\frac{(OP)\left((OA)^2\sqrt{(OP)^2 + (OB)^2} + (OB)^2\sqrt{(OP)^2 + (OA)^2}\right)}{(OP)^2((OA)^2 + (OB)^2) + (OA)^2(OB)^2}\right\}$$

$$= \cos^{-1}\left\{\frac{(3)\left((5)^2\sqrt{(3)^2 + (12)^2} + (12)^2\sqrt{(3)^2 + (5)^2}\right)}{(3)^2((5)^2 + (12)^2) + (5)^2(12)^2}\right\}$$

$$= \cos^{-1}\left\{\frac{3446.67}{5121}\right\} = \cos^{-1}\{0.673\} = 0.832\ sr \qquad Ans.$$

Example 2.41: A uniform point-source of 200 cd is located at the point $P(0, 0, 5)$. Find out flux densityover the plane bounded by straight lines $x = \pm 24\ \&\ y = \pm 10$.(Assume that flux is uniformly distributed over the entire plane).

Sol. In order to identify the geometry of plane bounded by straight lines plot the straight lines on the XY-plane as follows

$$AB \equiv y = -10,\ BC \equiv x = 24,\ CD \equiv y = 10\ \&\ AD \equiv x = -24$$

(as shown in the figure(2.16) below)

It is obvious that pair of parallel straight lines BC & AD intersects the pair of parallel straight line AB & CD.

By the plot, it is clear that the plane bounded is a rectangle $(i.e.\ ABCD)$

On solving the equations, we get vertices of the rectangle ABCD

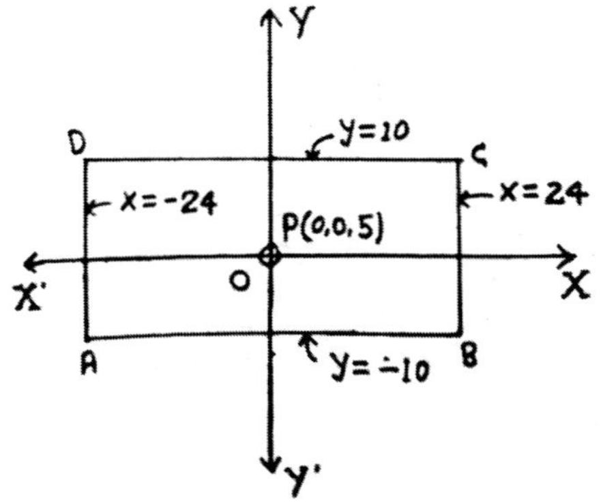

Fig (2.16)

$$\Rightarrow A \equiv (-24,-10),\quad B \equiv (24,-10),\quad C \equiv (24,10) \& \ D \equiv (-24,10)$$

$$\Rightarrow AB = CD = \sqrt{(-24-24)^2 + (-10+10)^2} = 48 \quad \&$$

$$BC = AD = \sqrt{(24-24)^2 + (-10-10)^2} = 20$$

The point $P \equiv (0,0,5)$ is located at a normal height 5 unit on Z-axis (normal to the plane of paper)(as indicated in the above figure)

Now, solid angle subtended by rectangular plane ABCD at the point 'P'

(From the eq(16))

$$\omega_{ABCD} = 4\sin^{-1}\left(\frac{lb}{\sqrt{(l^2+4h^2)(b^2+4h^2)}}\right)$$

$$= 4\sin^{-1}\left(\frac{48 \times 20}{\sqrt{((48)^2+4(5)^2)((20)^2+4(5)^2)}}\right)$$

$$= 4\sin^{-1}\left(\frac{960}{1096.358}\right)$$

$$= 4\sin^{-1}(0.876) = 4.27\ sr$$

Hence, the flux density over the rectangular plane

$$F.D. = \frac{I \times \omega}{A} = \frac{200 \times 4.27}{48 \times 20} = 0.889\ lm/sq.unit \qquad Ans.$$

Example 2.42: A uniform point-source of 111.4 cd is located at the point $P(0,0,4)$. Find out flux density over the plane bounded by straight lines $x - \sqrt{3}y = \pm 16$ &$\sqrt{3}x + y = \pm 16$. (Assume that flux is uniformly distributed over the entire plane).

Sol. In order to identify the geometry of plane bounded by straight lines plot the straight lines on the XY-plane as follows

$$AB \equiv x - \sqrt{3}y = -16,\quad BC \equiv \sqrt{3}x + y = -16,\quad CD \equiv x - \sqrt{3}y = 16$$

$$\&\ AD \equiv \sqrt{3}x + y = 16$$

It is obvious that pair of parallel lines AB & CD is normal to the pair of parallel lines BC & AD.

(as shown in the figure(2.17) below)

By the plot, it is clear that the plane bounded is a rectangle $(i.e.\ ABCD)$

On solving the equations, we get vertices of the rectangle ABCD

$$by\ solving,\ \ x-\sqrt{3}y=-16\ \&\sqrt{3}x+y=16$$

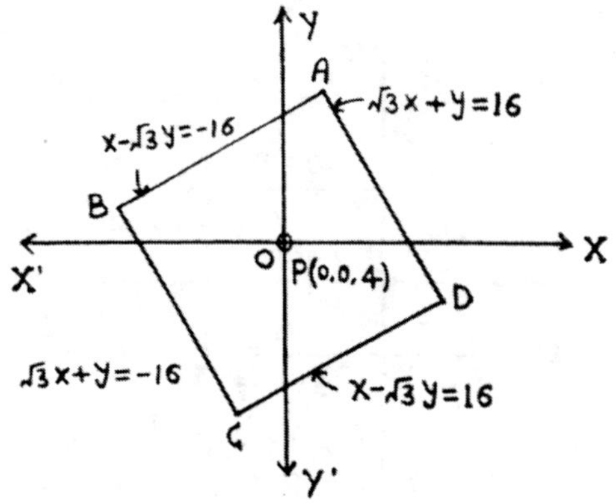

Fig (2.17)

$$\Rightarrow \sqrt{3}(\sqrt{3}y-16)+y=16\quad \Rightarrow\ 4y=16(\sqrt{3}+1)\ \ or\ y=4(\sqrt{3}+1)$$

$$x=\sqrt{3}y-16=4\sqrt{3}(\sqrt{3}+1)-16=4(\sqrt{3}-1)$$

$$\Rightarrow A\equiv\left(4(\sqrt{3}-1),4(\sqrt{3}+1)\right)$$

Similarly, we can obtain

$$B\equiv\left(-4(\sqrt{3}+1),4(\sqrt{3}-1)\right),C\equiv\left(-4(\sqrt{3}-1),-4(\sqrt{3}+1)\right)\&$$

$$D\equiv\left(4(\sqrt{3}+1),-4(\sqrt{3}-1)\right)$$

$$\Rightarrow AB=\sqrt{\{4(\sqrt{3}-1)+4(\sqrt{3}+1)\}^2+\{4(\sqrt{3}+1)-4(\sqrt{3}-1)\}^2}$$

$$=\sqrt{\{8\sqrt{3}\}^2+\{8\}^2}=16$$

$$\Rightarrow BC=\sqrt{\{-4(\sqrt{3}+1)+4(\sqrt{3}-1)\}^2+\{4(\sqrt{3}-1)+4(\sqrt{3}+1)\}^2}$$

$$=\sqrt{\{-8\}^2+\{8\sqrt{3}\}^2}=16$$

Since, adjacent sides AB & BC are equal & normal hence rectangle ABCD is a square

The point $P\equiv(0,0,4)$ is located at a normal height 4 unit on the Z-axis (normal to the plane of paper)(as indicated in the above figure)

Solid angle subtended by square plane ABCD at the point 'P'

(From the eq(26))

$$\omega_{OLAK}=4\sin^{-1}\left(\frac{a^2}{a^2+4h^2}\right)$$

$$=4\sin^{-1}\left(\frac{16^2}{(16)^2+4(4)^2}\right)$$

$$=4\sin^{-1}(0.8)=3.709\ sr$$

Hence, the flux density over the rectangular plane

$$F.D.=\frac{I\times\omega}{A}=\frac{111.4\times3.709}{(16)^2}=1.614\ lm/sq.unit\qquad Ans.$$

Example 2.43: A uniform point-source of 1kW is located at the point $P(0,-8,-3)$. Find out total radiation energy intercepted by the plane bounded by straight lines $x = 12, x = 38$ & $y = 4$ & $y = 20$.

Sol. In order to identify the geometry of plane bounded by straight lines plot the straight lines on the XY-plane as follows

$$AB \equiv y = 20, \ \ BC \equiv x = 38, \ \ CD \equiv y = 4$$

$$\& \ AD \equiv x = 12$$

It is obvious that pair of parallel lines AB & CD is normal to the pair of parallel lines BC & AD.

(as shown in the figure(2.18) below)

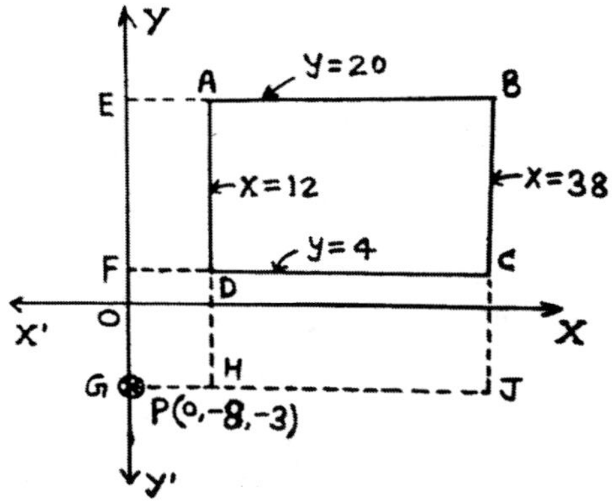

Fig (2.18)

By the plot, it is clear that the plane bounded is a rectangle($i.e.\ ABCD$). Now extend the sides AB & CD to intersect Y-axis at the points 'E' & 'F' respectively and sides AD& BCto intersect line GJ at the points 'H' & 'J' respectively. (as shown in the above figure).

$$\Rightarrow A \equiv (12,20), \ \ B \equiv (38,20\,), \ \ C \equiv (38,4) \& \ D \equiv (12,4)$$

$$\Rightarrow AB = CD = \sqrt{(12-38)^2 + (20-20)^2} = 26 \ \ \&$$

$$BC = AD = \sqrt{(38-38)^2 + (20-4)^2} = 16$$

The point $P \equiv (0,-8,-3)$ is located at a normal distance 3 unit on the vertical axis passing through the point 'G' (normally down to the plane of paper)

(as indicated in the above figure)

Now, solid angle subtended by rectangular plane GEBJ at the point 'P'

(From the eq(17))

$$\omega_{GEBJ} = \sin^{-1}\left(\frac{lb}{\sqrt{(l^2+h^2)(b^2+h^2)}}\right)$$

$$= \sin^{-1}\left(\frac{38 \times 28}{\sqrt{((38)^2+(3)^2)((28)^2+(3)^2)}}\right)$$

$$= \sin^{-1}\left(\frac{1064}{1073.419}\right) = \sin^{-1}(0.991) = 1.436\ sr$$

Similarly, solid angle subtended by rectangular plane GEAH at the same point

$$\omega_{GEAH} = \sin^{-1}\left(\frac{28 \times 12}{\sqrt{((28)^2+(3)^2)((12)^2+(3)^2)}}\right)$$

$$= \sin^{-1}\left(\frac{336}{348.323}\right) = \sin^{-1}(0.965) = 1.305\ sr$$

Similarly, solid angle subtended by rectangular plane GFCJ at the same point

$$\omega_{GFCJ} = \sin^{-1}\left(\frac{38 \times 12}{\sqrt{((38)^2 + (3)^2)((12)^2 + (3)^2)}}\right)$$

$$= \sin^{-1}\left(\frac{456}{471.496}\right) = \sin^{-1}(0.967) = 1.313\ sr$$

Similarly, solid angle subtended by rectangular plane GFDH at the same point

$$\omega_{GFDH} = \sin^{-1}\left(\frac{12 \times 12}{\sqrt{((12)^2 + (3)^2)((12)^2 + (3)^2)}}\right)$$

$$= \sin^{-1}\left(\frac{144}{153}\right) = \sin^{-1}(0.941) = 1.225\ sr$$

Now, solid angle subtended by plane GFADCJ at the same point 'P'

$$\Rightarrow \omega_{GFADCJ} = \omega_{GEAH} + \omega_{GFCJ} - \omega_{GFDH}$$

$$= 1.305 + 1.313 - 1.225 = 1.393\ sr$$

Thus, solid angle subtended by rectangular plane ABCD at the point 'P'

$$\Rightarrow \omega_{ABCD} = \omega_{GEBJ} - \omega_{GFADCJ}$$

$$= 1.436 - 1.393 = 0.046\ sr$$

Hence, the total radiation energy intercepted by bounded rectangular plane ABCD

$$\Rightarrow E_i = \left(\frac{L}{4\pi}\right) \times \omega_{ABCD} \quad (given\ that\ luminosity, L = 1kW = 1000\ W)$$

$$= \left(\frac{1000}{4\pi}\right) \times 0.046 = 3.66\ J/\sec \qquad Ans.$$

Example 2.44: A uniform point-source of 100 cd is located at the point $P(5,5,2)$. Find out flux density over the plane bounded by straight lines $x = 18, y = 5, y = -8\ \&\ y = x$. (Assume that flux is uniformly distributed over the entire plane).

Sol. In order to identify the geometry of plane bounded by straight lines plot the straight lines on the XY-plane as follows

$$AB \equiv y = 5,\ \ BC \equiv x = 18,\ \ CD \equiv y = -8$$

$$\&\ AD \equiv y = x$$

It is obvious that pair of parallel lines AB & CD is normal to the line BC.

(as shown in the figure(2.19) below)

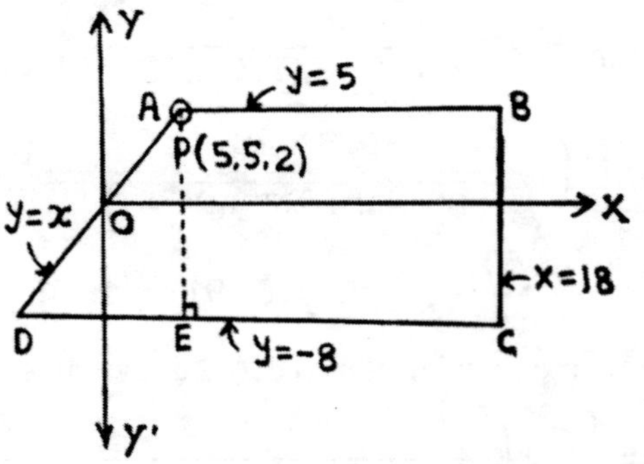

Fig (2.19)

By the plot, it is clear that the plane bounded is a trapezium($i.e.\ ABCD$). Draw perpendicular AE from the vertex 'A' to the line CD

On solving the equations of lines, we get vertices of the trapezium ABCD

$$\Rightarrow A \equiv (5,5),\ \ B \equiv (18,5\ ),\ \ C \equiv (18,-8) \&\ D \equiv (-8,-8)$$

$$\Rightarrow AB = EC = \sqrt{(5-18)^2 + (5-5)^2} = 13\ \ \&$$

$$BC = AE = \sqrt{(18-18)^2 + (5+8)^2} = 13$$

The point $P \equiv (5,5,2)$ is located at a normal height 2 unit on the vertical axis passing through the vertex 'A' (normal to the plane of paper).

(as indicated in the above figure)

Now, solid angle subtended by rectangular plane ABCE at the point 'P'

(From the eq(17))

$$\omega_{ABCE} = \sin^{-1}\left(\frac{lb}{\sqrt{(l^2+h^2)(b^2+h^2)}}\right)$$

$$= \sin^{-1}\left(\frac{13 \times 13}{\sqrt{((13)^2+(2)^2)((13)^2+(2)^2)}}\right)$$

$$= \sin^{-1}\left(\frac{169}{173}\right) = \sin^{-1}(0.977) = 1.356\ sr$$

Now, solid angle subtended by right triangular plane AED at the point 'P'

(From the eq(2))

$$\Rightarrow \omega_{\Delta AED} = \sin^{-1}\left\{\frac{b}{\sqrt{b^2+p^2}}\right\} - \sin^{-1}\left\{\left(\frac{b}{\sqrt{b^2+p^2}}\right)\left(\frac{h}{\sqrt{h^2+p^2}}\right)\right\}$$

$$= \sin^{-1}\left\{\frac{13}{\sqrt{(13)^2+(13)^2}}\right\} - \sin^{-1}\left\{\left(\frac{13}{\sqrt{(13)^2+(13)^2}}\right)\left(\frac{2}{\sqrt{(2)^2+(13)^2}}\right)\right\}$$

$$= \sin^{-1}\{0.707\} - \sin^{-1}\{0.107\}$$

$$= 0.785 - 0.107 = 0.678\ sr$$

Solid angle subtended by trapezium ABCD at the point 'P'

$$\Rightarrow \omega_{ABCD} = \omega_{ABCE} + \omega_{\Delta AED}$$

$$= 1.356 + 0.678 = 2.034\ sr$$

Area of the trapezium

$$\Rightarrow A = \frac{1}{2}(AB + CD) \times (BC) = \frac{1}{2}(13+26)(13) = 253.5\ sq.unit$$

Hence, the flux density over bounded plane ABCD

$$F.D. = \frac{I \times \omega}{A} = \frac{100 \times 2.034}{253.5} = 0.802\ lm/sq.unit \qquad Ans.$$

Example 2.45: A uniform point-source of 200 cd is located at the point $P(0,0,-5)$. Find out flux density over the plane bounded by straight lines $\sqrt{3}x - y = \pm 12\sqrt{3}$, $\sqrt{3}x + y = \pm 12\sqrt{3} \&\ y = \pm 6\sqrt{3}$. (Assume that the flux is uniformly distributed over the entire plane)

Sol. In order to identify the geometry of plane bounded by straight lines plot the straight lines on the XY-plane as follows

$$AB \equiv y = 6\sqrt{3},\ \ BC \equiv \sqrt{3}x + y = 12\sqrt{3},\ \ CD \equiv \sqrt{3}x - y = 12\sqrt{3},$$

$$DE \equiv y = -6\sqrt{3},\qquad EF \equiv \sqrt{3}x + y = -12\sqrt{3} \&\ AF \equiv \sqrt{3}x - y = -12\sqrt{3}$$

It is obvious that AB & ED, BC & EF and AF & CD are pairs of parallel lines forming a closed hexagon ABCDEF.

(as shown in the figure(2.20) below)

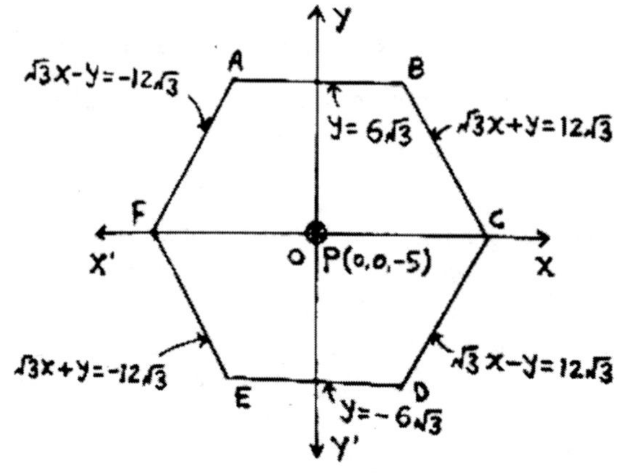

Fig (2.20)

$$\textit{on solving},\ AB \equiv y = 6\sqrt{3},\ AF \equiv \sqrt{3}x - y = -12\sqrt{3}$$

$$\Rightarrow \sqrt{3}x - 6\sqrt{3} = -12\sqrt{3}\ \ or\ x = -6 \Rightarrow A \equiv (-6, 6\sqrt{3})$$

$$\textit{on solving},\ AB \equiv y = 6\sqrt{3},\ BC \equiv \sqrt{3}x + y = 12\sqrt{3}$$

$$\Rightarrow \sqrt{3}x + 6\sqrt{3} = 12\sqrt{3}\ \ or\ x = 6 \Rightarrow B \equiv (6, 6\sqrt{3})$$

$$\textit{on solving},\ BC \equiv \sqrt{3}x + y = 12\sqrt{3}\&\ CD \equiv \sqrt{3}x - y = 12\sqrt{3},$$

$$\Rightarrow \sqrt{3}x - (12\sqrt{3} - \sqrt{3}x) = 12\sqrt{3}\ \ or\ x = \frac{24\sqrt{3}}{2\sqrt{3}} = 12$$

$$\Rightarrow y = 12\sqrt{3} - \sqrt{3}x = 12\sqrt{3} - 12\sqrt{3} \Rightarrow\ C \equiv (12, 0)$$

$$\textit{on solving},\ CD \equiv \sqrt{3}x - y = 12\sqrt{3}\ DE \equiv y = -6\sqrt{3},$$

$$\Rightarrow \sqrt{3}x - (-6\sqrt{3}) = 12\sqrt{3}\ \ or\ x = 6 \Rightarrow D \equiv (6, -6\sqrt{3})$$

$$\Rightarrow AB = \sqrt{(-6-6)^2 + \left(6\sqrt{3} - 6\sqrt{3}\right)^2} = 12$$

$$BC = \sqrt{(6-12)^2 + \left(6\sqrt{3} - 0\right)^2} = 12$$

$$CD = \sqrt{(12-6)^2 + \left(0 + 6\sqrt{3}\right)^2} = 12$$

$$OA = \sqrt{(-6-0)^2 + \left(6\sqrt{3} - 0\right)^2} = 12$$

$$OB = \sqrt{(6-0)^2 + \left(6\sqrt{3} - 0\right)^2} = 12$$

$$OD = \sqrt{(6-0)^2 + \left(-6\sqrt{3} - 0\right)^2} = 12$$

Thus, we find that plane ABCDEF is a origin centred regular hexagon with each side equal to 12 unit & the point 'P' is located at a normal height 5 unit from the origin on the Z-axis (Normally down to the plane of paper) (as indicated in the above figure)

Solid angle subtended by the bounded plane ABCDEF at the point 'P'

(From the eq(34))

$$\Rightarrow \omega_{ABCDEF} = 2\pi - 2n\sin^{-1}\left(\frac{2hsin\frac{\pi}{n}}{\sqrt{4h^2 + a^2cot^2\frac{\pi}{n}}}\right)$$

$$= 2\pi - 2\times 6\sin^{-1}\left(\frac{2(5)sin\frac{\pi}{6}}{\sqrt{4(5)^2 + (12)^2cot^2\frac{\pi}{6}}}\right)$$

$$= 2\pi - 12\sin^{-1}\left(\frac{5}{\sqrt{532}}\right) = 2\pi - 12\sin^{-1}(0.217)$$

$$= 2\pi - 2.625 = 3.658\ sr$$

Area of the regular hexagon ABCDEF,

$$\Rightarrow A = \frac{1}{4}na^2cot\frac{\pi}{n} \quad (Using\ formula\ for\ regular\ polygon)$$

$$= \frac{1}{4}\times 6(12)^2cot\frac{\pi}{6} = 374.123\ sq.unit$$

Hence, the flux density over bounded plane ABCD

$$F.D. = \frac{I\times\omega}{A} = \frac{200\times 3.658}{374.123} = 1.955\ lm/sq.unit \qquad Ans.$$

Example 2.46: A uniform point-source of 300 W is located at the point $P(14,14,2)$. Find out total radiation energy intercepted by the plane bounded by curves$x = 4, x = 24, y = 4\ \&\ y = 24\ \& x^2 + y^2 - 28x - 28y + 383 = 0$.

Sol. In order to identify the geometry of plane bounded by curves, plot the straight lines on the XY-plane as follows

$$AB \equiv y = 4,\ \ BC \equiv x = 24,\ \ CD \equiv y = 24\ \ \&$$
$$AD \equiv x = 4$$

It is obvious that pair of parallel lines AB & CD is intersecting at right angle to the pair of parallel lines BC & AB. These lines form a rectangle ABCD (as shown in the fig (2.21))

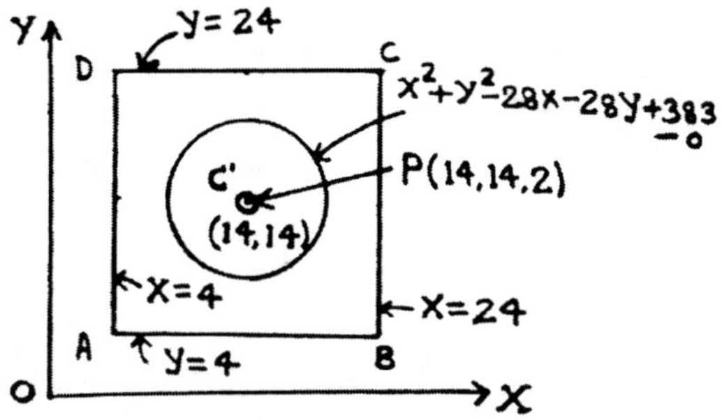

Fig (2.21)

On solving the equations of straight line we get

$$\Rightarrow A \equiv (4,4),\ \ B \equiv (24,4),\ \ C \equiv (24,24) \&\ D \equiv (4,24)$$

$$\Rightarrow AB = CD = \sqrt{(4-24)^2 + (4-4)^2} = 20\ \ \&$$

$$BC = AD = \sqrt{(24-24)^2 + (4-24)^2} = 20$$

Thus, ABCD is a square with each side 20 unit.

$$now, compare\ the\ equation,\ \ x^2 + y^2 - 28x - 28y + 383 = 0$$

With the standard form of equation of circle

$$x^2 + y^2 + 2gx + 2fy + c = 0$$

$$\Rightarrow g = -14, f = -14\ \&\ c = 383$$

$$\Rightarrow centre, C' \equiv (-g, -f) \equiv (14, 14)\&$$

$$radius, R = \sqrt{g^2 + f^2 - c} = \sqrt{(14)^2 + (14)^2 - 383} = 3$$

It is obvious that the point $P \equiv (14, 14, 2)$ is lying at a normal height 2 unit on the vertical axis passing through the centre $C' \equiv (14, 14)$ of circle

$$\Rightarrow AC' = \sqrt{(4-14)^2 + (4-14)^2} = 10\sqrt{2}\&$$

$$CC' = \sqrt{(24-14)^2 + (24-14)^2} = 10\sqrt{2}$$

It is clear that the square plane ABCD is centred at the point $C' \equiv (14, 14)$

Now, solid angle subtended by the square plane ABCD at the point 'P'

(From the eq(25))

$$\Rightarrow \omega_{ABCD} = 4\sin^{-1}\left(\frac{a^2}{a^2 + 4h^2}\right)$$

$$= 4\sin^{-1}\left(\frac{(20)^2}{(20)^2 + 4(2)^2}\right)$$

Solid angle subtended by the circular plane at the same point

(From the eq(37))

$$\Rightarrow \omega' = 2\pi\left(1 - \frac{h}{\sqrt{h^2 + R^2}}\right)$$

$$= 2\pi\left(1 - \frac{2}{\sqrt{(2)^2 + (3)^2}}\right)$$

$$= 2.798\ sr$$

Now, solid angle subtended by the plane bounded by square ABCD & circle

$$\Rightarrow \omega_o = \omega_{ABCD} - \omega'$$

$$= 5.162 - 2.798 = 2.364\ sr$$

Hence, the total radiation energy intercepted by bounded rectangular plane ABCD

$$\Rightarrow E_i = \left(\frac{L}{4\pi}\right) \times \omega_{ABCD} \quad (given\ that\ luminosity, L = 1kW = 300\ W)$$

$$= \left(\frac{300}{4\pi}\right) \times 2.364 = 56.436\ J/\sec \qquad Ans.$$

Example 2.47: A uniform point-source of 300 W is located at the point $P(0, 0, -6)$. Find out total radiation energy intercepted by the plane bounded by straight lines $x = 6, x = \infty\ \&\ y = 8\ \&\ y = 28$.

Sol. In order to identify the geometry of plane bounded by straight lines plot the straight lines on the XY-plane as follows

$$AB \equiv x = 6,\ \ BG \equiv y = 8,\ \ FG \equiv x = \infty$$

$$\&\ AF \equiv y = 28$$

It is obvious that pair of parallel lines AB & FG is normal to the pair of parallel lines AF & BG.

(as shown in the figure(2.22) below)

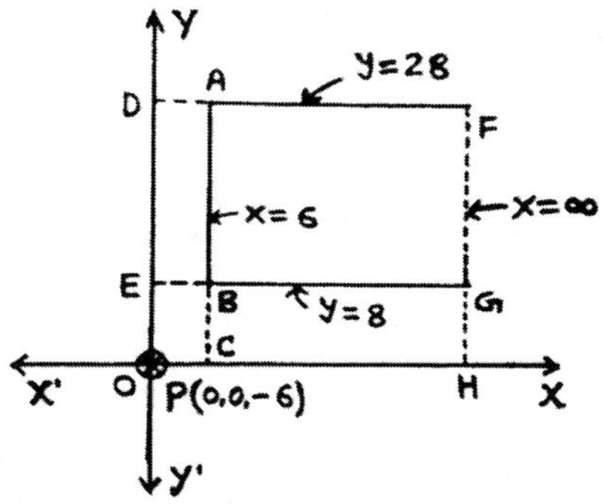

Fig (2.22)

By the plot, it is clear that the plane bounded is a rectangle ($i.e.\ ABGF$). Extend the sides AB, FG, BG & AF to intersects the axes at the points 'C', 'H', 'E' & 'D' respectively.

On solving the equations of lines, we get vertices of infinitely long rectangular plane ABGF

$$\Rightarrow A \equiv (6, 28),\ \ B \equiv (6, 8\,),\ \ G \equiv (\infty, 8) \&\ F \equiv (\infty, 28)$$

The point $P \equiv (0, 0, -6)$ is located at a normal distance 6 unit on the vertical axis passing through the origin 'O' (normally down to the plane of paper).

(as indicated in the above figure)

Now, solid angle subtended by rectangular plane ACOD at the point 'P'

(From the eq(17))

$$\omega_{ACOD} = \sin^{-1}\left(\frac{lb}{\sqrt{(l^2 + h^2)(b^2 + h^2)}}\right)$$

$$= \sin^{-1}\left(\frac{28 \times 6}{\sqrt{((28)^2 + (6)^2)((6)^2 + (6)^2)}}\right)$$

$$= \sin^{-1}\left(\frac{168}{242.981}\right) = \sin^{-1}(0.691) = 0.763\ sr$$

Similarly, solid angle subtended by rectangular plane BCOE at the same point

$$\omega_{BCOE} = \sin^{-1}\left(\frac{8 \times 6}{\sqrt{((8)^2 + (6)^2)((6)^2 + (6)^2)}}\right)$$

$$= \sin^{-1}\left(\frac{48}{84.853}\right) = \sin^{-1}(0.566) = 0.602\ sr$$

Now, solid angle subtended by infinitely long rectangular plane OEGH at the point 'P'

(From the eq(2.4))

$$\omega_{OEGH} = \sin^{-1}\left(\frac{b}{\sqrt{b^2 + h^2}}\right)(b = width\ \&h = normal\ height)$$

$$= \sin^{-1}\left(\frac{8}{\sqrt{(8)^2 + (6)^2}}\right)$$

$$= \sin^{-1}(0.8) = 0.927\ sr$$

Similarly, solid angle subtended by infinitely long rectangular plane ODFH at the point 'P'

$$\omega_{ODFH} = \sin^{-1}\left(\frac{28}{\sqrt{(28)^2 + (6)^2}}\right)$$

$$= \sin^{-1}(0.978) = 1.361\ sr$$

Solid angle subtended by the plane ODABGH at the same point 'P'

$$\Rightarrow \omega_{ODABGH} = \omega_{ACOD} + \omega_{OEGH} - \omega_{BCOE}$$
$$= 0.763 + 0.927 - 0.602 = 1.088\ sr$$

Solid angle subtended by infinitely long rectangular plane ABGF at the point 'P'

$$\Rightarrow \omega_{ABGF} = \omega_{ODFH} - \omega_{ODABGH}$$
$$= 1.361 - 1.088 = 0.273\ sr$$

Hence, the total radiation energy intercepted by bounded plane ABGF

$$\Rightarrow E_i = \left(\frac{L}{4\pi}\right) \times \omega_{ABCD} \quad (given\ that\ luminosity, L = 1kW = 300\ W)$$
$$= \left(\frac{300}{4\pi}\right) \times 0.273 = 6.517\ J/\sec \qquad Ans.$$

Summary

1. The solid angle subtended by an object at a given point is given by 'Integral Theorem' (in form of spherical co-ordinates)

$$\omega = \int_{\varphi_1}^{\varphi_2} \left| \int_{\theta_1}^{\theta_2} sin\theta\ d\theta \right| d\varphi \qquad \forall\ \theta \in [0, \pi] \&\ \varphi \in [-\pi, \pi]$$

Where, $'\theta' \&\ '\varphi'$ *are the parameters (variables) depending on the*

1. Geometrical shape of the object &

2. Orientation of shape w.r.t. to the given point (observer) in the space

2. The solid angle subtended by a plane-figure, in the first quadrant w.r.t. foot of perpendicular drawn to the plane from a given point in the space, is given by the 'Corollary of Plane-figure' (proposed by HCR)

$$\omega = \int_0^{\varphi} \frac{f(htan\varphi)d\varphi}{\sqrt{(f(htan\varphi))^2 + h^2sec^2\varphi}} \qquad \forall\ 0 \le \varphi \le \frac{\pi}{2}$$

Where, 'h' is the normal height of the point from the plane

$y = f(x) = f(htan\varphi) \neq (x = constant)$is the equation of boundary-curve of the given plane in the first quadrant w.r.t. foot of perpendicular assuming as the origin &

$'\varphi'$ is variable $\&x = htan\varphi$.

Note: Foot of perpendicular drawn from the point must lie within the boundary of plane$\&y = f(x)$ must be differential at each point of the curve in the first quadrant i.e. $y = f(x)$ is the equation of single-boundary of the given plane.

Standard formulae:

Standard formulae to be remembered are following

3. The solid angle subtended by a right triangular plane with base 'b' &perpendicular 'p' at any point lying at a height 'h' on the vertical axis passing through the acute angled vertex common to the perpendicular 'p' & hypotenuse is given as

$$\omega = \sin^{-1}\left\{\frac{b}{\sqrt{b^2 + p^2}}\right\} - \sin^{-1}\left\{\left(\frac{b}{\sqrt{b^2 + p^2}}\right)\left(\frac{h}{\sqrt{h^2 + p^2}}\right)\right\}$$

If the vertical axis passes through the vertex common to the base ‘b’ & hypotenuse, the solid angle is given as

$$\omega = \sin^{-1}\left\{\frac{p}{\sqrt{p^2+b^2}}\right\} - \sin^{-1}\left\{\left(\frac{p}{\sqrt{p^2+b^2}}\right)\left(\frac{h}{\sqrt{h^2+b^2}}\right)\right\}$$

4. The solid angle subtended by a right triangular plane with orthogonal sides ‘a’ & ‘b’ at any point lying at a height ‘h’ on the vertical axis passing through the right angled vertex is given as

$$\omega = \cos^{-1}\left\{\frac{h\left(a^2\sqrt{h^2+b^2}+b^2\sqrt{h^2+a^2}\right)}{h^2(a^2+b^2)+a^2b^2}\right\}$$

5. The solid angle subtended by a rectangular plane with sides ‘l’ & ‘b’ at any point lying at a height ‘h’ on the vertical axis passing through the centre is given as

$$\omega = 4\sin^{-1}\left(\frac{lb}{\sqrt{(l^2+4h^2)(b^2+4h^2)}}\right)$$

6. The solid angle subtended by a rectangular plane with sides ‘l’ & ‘b’ at any point lying at a height ‘h’ on the vertical axis passing through one of the vertices is given as

$$\omega = \sin^{-1}\left(\frac{lb}{\sqrt{(l^2+h^2)(b^2+h^2)}}\right)$$

7. The solid angle subtended by a square plane with each side ‘a’ at any point lying at a height ‘h’ on the vertical axis passing through the centre is given as

$$\omega = 4\sin^{-1}\left(\frac{a^2}{a^2+4h^2}\right)$$

8. The solid angle subtended by a square plane with each side ‘a’ at any point lying at a height ‘h’ on the vertical axis passing through the centre is given as

$$\omega = \sin^{-1}\left(\frac{a^2}{a^2+h^2}\right)$$

9. The solid angle subtended by a regular polygon having ‘n’ number of the sides each of length ‘a’, at any point lying at a height ‘h’ on the vertical axis passing through the centre is given as

$$\omega = 2\pi - 2n\sin^{-1}\left(\frac{2h sin\frac{\pi}{n}}{\sqrt{4h^2+a^2 cot^2\frac{\pi}{n}}}\right) \qquad \forall\ n \in N\ \&\, n \geq 3$$

10. The solid angle subtended by a circular plane with a radius ‘R’ at any point lying at a height ‘h’ on the vertical axis passing through the centre is given as

$$\omega = 2\pi\left(1-\frac{h}{\sqrt{h^2+R^2}}\right)$$

11. The solid angle subtended by an infinitely long rectangular plane having width ‘b’ at any point lying at a height ‘h’ on the vertical axis passing through the centre is given as

$$\omega = 4\sin^{-1}\left(\frac{b}{\sqrt{b^2+4h^2}}\right)$$

12. The solid angle subtended by an infinitely long rectangular plane having width ‘b’ at any point lying at a height ‘h’ on the vertical axis passing through the vertex is given as

$$\omega = \sin^{-1}\left(\frac{b}{\sqrt{b^2+h^2}}\right)$$

Exercise - 2

*** Right-triangular plane:**

Q.2.1. Find out the solid angle subtended by a right-triangular plane having perpendicular = 5cm. & base = 12 cm. at a point lying at a height 250 mm. on the vertical axis passing through the common vertex of

1. Perpendicular & hypotenuse & 2.base & hypotenuse.

Also find out the solid angle subtended by the same plane on each of the vertices.

Q.2.2. Find out the solid angle subtended by a right-triangular plane having perpendicular = 8cm. & base = 20 cm. at a point lying at a height 160 mm. on the vertical axis passing through the vertex of right angle.

Q.2.3. What is the normal height of a point lying on the vertical axis passing through the common vertex of perpendicular & hypotenuse if the solid angle subtended by a right triangular plane at the same point is $0.2\pi\ sr$.

(Take orthogonal sides of the plane as p = 3cm. &b = 4cm.)

Q.2.4. Find out the height of a given point if the solid angle subtended by the right triangular plane with orthogonal sides p = 15cm & b = 20cm at a point lying on the vertical axis passing through the common vertex of

1.) Perpendicular & hypotenuse is 0.45 sr.& 2) base & hypotenuse is 0.5 sr.

Q.2.5. What is the area of a right triangular plane subtending a solid angle 0.4sr, at a point lying at a height 5cm on the vertical axis passing through right angled vertex if the one of the orthogonal sides is 24cm.

Q.2.6. Find out the solid angle subtended by a right triangular plane having perpendicular, p = 10cm & base, b =15cm at the common vertex of

1. Perpendicular & hypotenuse, & 2.base & hypotenuse.

Q.2.7. A uniform point source having luminous intensity 350 cd is mounted at a height 3.5m on a vertical pole pitched to the right angled vertex of a right triangular plane having orthogonal sides 3m & 4m. Find out the luminous flux striking the plane & luminous flux density. If the point-source is put on the right angled vertex what is luminous flux striking the plane & luminous flux density?

(Assume uniform distribution of luminous flux over the entire plane).

Q.2.8. Find out the luminous flux striking a right triangular plane having perpendicular, p = 20cm & base, b = 50cm when a uniform point-source of 111.4 cd is put on the common vertex of

1. Perpendicular & hypotenuse, &2.base & hypotenuse.

Q.2.9. A uniform point source is mounted at a height 2.6m on a vertical pole pitched to the right angled vertex of a right triangular plane having orthogonal sides 4m & 4.5m. If luminous flux striking the plane is 151.67 lm then what is the luminous intensity of the point-source? Ans. 250 cd

Q.2.10. If the luminous flux striking a right triangular plane is 20 lm when a uniform point-source of 100 cd is placed on the common vertex of the perpendicular & hypotenuse. Find out the luminous flux density if the perpendicular, p = 22cm. (Assume uniform distribution of luminous flux over the entire plane).

Q.2.11. Find out the area of a right triangular plane if the luminous density on the plane is $4 \times 10^3\ lm/m^2$ (lux) when a uniform point-source of 100 cd is put on the common vertex of base & perpendicular = 24cm.

Q.2.12. Find out change in the luminous flux striking a right triangular plane with orthogonal sides 15cm & 20cm when a uniform point-source of 111.4 cd is placed on common vertex of smaller side & hypotenuse switched from the right angled vertex.

Q.2.13. A uniform point-source having luminous intensity 250 cd is mounted at a height 3.4m on the vertical axis passing through the right angled vertex such thattotal luminous flux falling on the right triangular plane is 300 lm. If one of the orthogonal sides of the plane is 3.5 m find out another side & luminous flux density on the plane.

(Assume uniform distribution of luminous flux over the entire plane).

Q.2.14. What will be the solid subtended by a right triangular plane at a point-source of 125 cd mounted on a vertical axis passing through the right angled vertex? The flux intercepted by the plane having perpendicular, p = 14cm & base, b = 54cm is 26 lm.

*** Isosceles-triangular plane:**

Q.2.15. Find out the solid subtended by an isosceles triangular plane at a point lying at a height 18 cm on the vertical axis passing through the common vertex of equal-sides. (Given that perpendicular, p = 7cm & base, b = 8cm.)

Q.2.16. Find out the solid subtended by an isosceles triangular plane with perpendicular, p = 12cm & base, b = 20cm at a point lying at a height 15 cm on the vertical axis passing through the common vertex of equal-sides.

Q.2.17. Find out the solid subtended by an isosceles triangular plane with perpendicular, p = 6cm & base, b = 10cm at a point lying at a height 12 cm on the vertical axis passing through the mid-point of unequal side.

Q.2.18. What is the normal height of a point lying on the vertical axis passing through the common vertex of equal sides if the solid angle subtended by an isosceles triangular plane at the same point is 0.867 sr.

(Take orthogonal sides of the plane as p = 6cm. & b = 12cm.)

Q.2.19. Find out the area of an isosceles triangular plane subtending a solid angle 1.2 sr. at a point lying at height 8cm on the vertical axis passing though the mid-point of unequal side of 14cm.

Q.2.20 What will be the solid subtended by an isosceles triangular plane at a point-source of 125 cd mounted on a vertical axis passing through the common vertex of equal sides? The flux intercepted by the plane having perpendicular, p = 12cm & base, b = 50cm is 25 lm.

Q.2.21. Find out the solid subtended by an isosceles triangular plane at a point-source emitting total flux 1400 lm mounted at a height 8cm on a vertical axis passing through the mid-point of unequal side. The flux intercepted by the plane having perpendicular, p = 18cm is 30 lm.

Q.2.22. A uniform point-source having luminous intensity 250 cd is mounted at a height 3m on the vertical axis passing through the common vertex of equal sides of an isosceles triangular plane having perpendicular, p = 100cm & base, b = 1.5m. What will be the luminous flux striking the plane & luminous flux density?

If the point-source is put on the common vertex of equal sides what will be the change in luminous flux striking the plane & luminous flux density.

(Assume uniform distribution of luminous flux over the entire plane).

Q.2.23. Find out the luminous flux striking an isosceles triangular plane having perpendicular, p = 16cm & base, b = 40cm when a uniform point-source of 111.4 cd is put on the

1. Common vertex of equal sides & 2.mid-point of the base.

Q.2.24. A uniform point source is mounted at a height 2.6m on a vertical pole pitched to the common vertex of equal sides each 1.6m of an isosceles triangular plane. If luminous flux striking the plane is 100 lm then what is the luminous intensity of the point-source? (If angle between the equal sides is 60°)

Q.2.25. If the luminous flux striking an isosceles triangular plane is 150 lm when a uniform point-source of 111.4 cd is placed on the common vertex of the equal sides. Find out the luminous flux density if the perpendicular, p = 25cm.

(Assume uniform distribution of luminous flux over the entire plane).

Q.2.26. Find out the area of an isosceles triangular plane if the luminous density on the plane is $2.5 \times 10^3\ lm/m^2$ (lux) when a uniform point-source of 110 cd is put on the common vertex of the equal sides. (Take perpendicular, p = 24cm)

Q.2.27. Find out change in the luminous flux striking an isosceles triangular plane with perpendicular = 8cm &base = 16cm when a uniform point-source of 111.4 cd is placed on common vertex of the equal sides switched from the mid-point of the base.

Q.2.28. A uniform point-source having luminous intensity 250 cd is mounted at a height 3m on the vertical pole passing through the mid-point of the base such

that the total luminous flux striking the isosceles triangular plane is 280 lm. If base, b = 4m then find out perpendicular& luminous flux density on the plane.

(Assume uniform distribution of luminous flux over the entire plane).

*** Isosceles right-triangular plane:**

Q.2.29. Find out the solid angle subtended by an isosceles right triangular plane with orthogonal sides each 20cm at a point lying at a height 12cm on the vertical axis passing through the right angled vertex.

Q.2.30. Find out the solid angle subtended by an isosceles right triangular plane with orthogonal sides each 22cm at a point lying at a height 8cm on the vertical axis passing through the acute angled vertex.

Q.2.31. Find out the solid angle subtended by an isosceles right triangular plane having orthogonal sides each 15cm at a point lying at a height 6cm on the vertical axis passing through the mid-point of the hypotenuse.

Q.2.32. What is the height of a point lying on a vertical axis passing though the right angled vertex of an isosceles right triangular plane having orthogonal sides each 12cm subtends a solid angle 1.75 sr. at the same point.

Q.2.33. Find out the height of a point lying on a vertical axis passing though the acute angled vertex of an isosceles right triangular plane having orthogonal sides each 12cm subtends a solid angle 0.78 sr. at the same point.

Q.2.34. If the solid angle subtended by an isosceles right triangular plane at a point lying on the vertical axis passing through the mid-point of the hypotenuse is 2.6 sr. What is the height of the point from the plane?

Q.2.35. The solid angle subtended by an isosceles right triangular plane at a point lying at a height 10cm on the vertical axis passing through the right angled vertex is 1.25 sr. What are the dimensions of the plane?

Q.2.36. Find out the perimeter of an isosceles right triangular plane subtending a solid angle 2.45 sr. at a point lying at a height 14cm on the vertical axis passing through the acute angle vertex.

Q.2.37. An isosceles right triangular plane subtends a solid angle 2.5 sr. at a point lying at a height 7cm on a vertical axis passing through the mid-point of the hypotenuse. What is the area of triangle?

Q.2.38. Find out the luminous flux striking the isosceles right triangular plane with orthogonal sides each 130cm if a uniform point source of 111.4 cd is mounted at a height 2.25m on the vertical axis passing through the

1. Acute angled vertex 2) right angled vertex & 3) mid-point of the hypotenuse.

Q.2.39. Find out the luminous flux density of an isosceles right triangular plane with orthogonal sides each 160cm if a uniform point source of 300 cd is mounted at a height 2.25m on the vertical axis passing through the

1.) Acute angled vertex 2) right angled vertex & 3) mid-point of the hypotenuse

Q.2.40. If an isosceles right triangular plane with orthogonal sides each 1.8 m cuts 180 lm luminous flux when a uniform point-source is mounted at a height 2.25 m on a vertical pole pitched on the right angled vertex. What is the luminous intensity of the source?

Q.2.41. An isosceles right triangular plane with orthogonal sides each 100 cm cuts 260 lm luminous flux when a uniform point-source is mounted at a height 150cm on a vertical pole pitched on the acute angled vertex. Find out the luminous intensity of the source & luminous flux density.

(Assume uniform distribution of luminous flux over the entire plane).

Q.2.42. If an isosceles right triangular plane with orthogonal sides each 2 m cuts total flux 200 lm when a uniform point-source is mounted at a height 80cm on a vertical axis passing through the mid-point of hypotenuse. What is the luminous intensity of point-source?

Q.2.43. The luminous flux density on an isosceles right triangular plane with area $256cm^2$ is $3 \times 10^3 lm/m^2$ when a uniform point source is mounted at a height 15cm on a vertical axis passing through the right angled vertex. Find out the luminous intensity of point-source.

Q.2.44. A uniform point-source of 100 cd is mounted at height 85mm on a vertical axis passing through the acute angled vertex of an isosceles right triangular plane with area $225cm^2$. What isthe luminous flux & its density on the plane?(Assume uniform distribution of luminous flux over the entire plane).

Q.2.45. Find out the area of an isosceles right triangular plane intercepting 30 lm luminous flux emitted by a uniform point-source of 125 cd mounted at a height 10 cm on a vertical axis passing through the mid-point of hypotenuse.Also find the luminous flux density.

(Assume uniform distribution of luminous flux over the entire plane).

Q.2.46. What will be the change in luminous flux if a uniform point-source of 100 cd is shifted from the acute angled vertex to the right angled vertex on an isosceles triangular plane?

Also, find out the change in the luminous flux density in lux if the point-source is shifted to the mid-point of hypotenuse & area of the plane is $289cm^2$.

(Assume uniform distribution of luminous flux over the entire plane).

Q.2.47. Find out the change in luminous flux intercepted by an isosceles right triangular plane with hypotenuse $9\sqrt{2}\ cm$ when a uniform point-source of 111.4 cd is brought down from a height 12cm to 5cm on a vertical axis passing though the right angled vertex? What will be the change in luminous flux density if flux is uniformly distributed over the entire plane?

Q.2.48. Find out the solid angle subtended by an isosceles right triangular plane at a uniform point-source emitting total flux 1400 lm & mounted at a height 20cm on a vertical axis passing through the right angled vertex. The plane intercepts total flux 50 lm.

Q.2.49. What is the solid angle subtended by an isosceles right triangular plane at a uniform point-source emitting total flux 1400 lm & mounted at a height 28cm on a vertical axis passing through the acute angled vertex if the plane intercepts total flux 20 lm?

Q.2.50. What should be the height of a uniform point-source of 100 cd on a vertical axis passing through the acute angled vertex to obtain the luminous equal to that intercepted by an isosceles right triangular plane with hypotenuse $12\sqrt{2}$cm when the same point-source is mounted at a height 12cm on a vertical axis passing through the right angled vertex?

Q.2.51. A uniform point-source of 111.4 cd is mounted at a height 13cm on a vertical axis passing through the acute angled vertex of an isosceles right triangular plane with hypotenuse 20cm. Now to obtain the same luminous flux density, what should be the height of the same point source to mount on a vertical axis passing through the right angled vertex?

(Assume uniform distribution of luminous flux over the entire plane).

Q.2.52. A uniform point-source emitting total luminous flux 1400 lm is mounted at a height 15cm on a vertical axis passing through the acute angled vertex of an isosceles right triangular plane with hypotenuse 20cm. If the same point-source is mounted at the same height on a vertical axis passing through right angled vertex what is will be the change in luminous flux intercepted by the plane?

Q.2.53. A uniform point-source of 100 cd is mounted at a height 12cm on a vertical axis passing through the right angled vertex of an isosceles right triangular plane with hypotenuse 22cm. If another point-source of 125 cd is mounted at a height 16cm on a vertical axis passing through the acute angled vertex. Find out the change in luminous flux density intercepted by the plane?

Q.2.54. Three uniform point-sources each emitting 1400 lm luminous flux are mounted at the same height 2.75m on the vertical poles pitched at the vertices of an isosceles right triangular plane with hypotenuse $3\sqrt{2}$ m. Find out the total luminous flux striking the plane. If the flux is uniformly distributed over the entire plane, what will be the luminous flux density?

Q.2.55. Two uniform point-sources each of 100 cd are mounted at the same height 2.5m on the vertical poles pitched on the acute angled vertices of an isosceles right triangular plane.

What should be the height of a uniform point-source of 200 cd mounted on a vertical axis passing through the right angled vertex such that when it is switched off luminous flux reduces to half of total flux obtained by all the sources?

Q.2.56. Two uniform point-sources each emitting total flux 1400 lm are mounted at the same height 3m on the vertical poles pitched on the acute angled vertices of an isosceles right triangular plane.

Find out the luminous intensity of a uniform point-source mounted at a height 2.25m on a vertical axis passing through the right angled vertex such that when it is switched off luminous flux reduces to half of total flux obtained by all the point-sources?

*** Equilateral-triangular plane:**

Q.2.57. Find out the solid angle subtended by an equilateral triangular plane with each side 9cm at a point lying at a height 4cm on a vertical axis passing through the one of the vertices.

Q.2.58. What is the solid angle subtended by an equilateral triangular plane with each side 12cm at a point lying at a height 8cm on a vertical axis passing through the centre.

Q.2.59. Find out the solid angle subtended by an equilateral triangular plane with each side 15cm at a point lying at a height 10cm on a vertical axis passing through the mid-point of the side.

Q.2.60. What will be the height of a point lying on a vertical axis passing through the centre of an equilateral triangular plane with each side 150mm such that the plane subtends a solid angle 1.3 sr. at the same point?

Q.2.61. A point is lying on a vertical axis passing through the vertex of an equilateral triangular plane with area 170 cm^2subtends a solid angle 2.8 sr. at the same point. Find out the height of the point from the plane.

Q.2.62. An equilateral triangular plane with each side 130mm subtends a solid angle 1.67 sr. at a point lying on a vertical axis passing through centre. What will be the height of the point from the plane?

Q.2.63. An equilateral triangular plane having area $289\sqrt{3}\ cm^2$ subtends a solid angle 2.25 sr. at a point lying on a vertical axis passing through vertex. Find out the height of the point from the plane.

Q.2.64. Solid angle subtended by an equilateral triangle at a point lying at a height 14cm on a vertical axis passing through the centre is 0.5π sr. Find out the area of the plane.

Q.2.65. Find out the circumscribed radius of an equilateral triangular plane subtending a solid angle 2 sr. at a point lying at a height 110mm on a vertical axis passing through the vertex.

Q.2.66. What is the solid angle subtended by an equilateral triangular plane with each side 16cm at a point lying at a height 15cm on a vertical axis passing through the centre if a circular hole with a radius equal to that of inscribed circle is cut from the centre of plane?

Q.2.67. A small circular hole with a radius 3cm is cut from the centre of an equilateral triangular plane with each side 200mm. What is the solid angle subtended by the plane at a point lying at a height 11cm on a vertical axis passing through the centre?

Q.2.68. An equivalent triangular hole with each side 4cm is cut from the centre of a circular plane with a radius 180mm. Find out the solid angle subtended by plane at a point lying at a height 13cm on a vertical axis passing through the centre.

Q.2.69. Find out the luminous flux striking an equilateral triangular plane with circumscribed radius $90\sqrt{3}$ cm when a uniform point-source of 100 cd is mounted at a height 1400mm on a vertical axis passing through the centre of the plane.

Q.2.70. What is the luminous flux striking an equilateral triangular plane with each side 80cm when a uniform point-source of 110 cd is mounted at a height 1.5m on a vertical axis passing through the vertex of the plane.

Q.2.71. Find out the luminous flux striking an equilateral triangular plane with circumscribed radius $100\sqrt{3}$ cm when a uniform point-source of 100 cd is mounted at a height 160cm on a vertical axis passing through the mid-point of side.

Q.2.72. Find out the luminous flux density on an equilateral triangular plane with side 20cm when a uniform point-source of 120 cd is mounted at a height 12cm on a vertical axis passing through the

1.) Vertex 2) centre & 3) mid-point

(Assume uniform distribution of luminous flux over the entire plane).

Q.2.73. A uniform point-source of 110 cd is mounted at a height 115mm on a vertical axis passing through the centre of an equilateral triangular plane intercepting 125 lm flux. Find out the circumscribed radius of the plane.

Q.2.74. An equilateral triangular plane intercepts 200 lm flux emitted from a uniform point-source of 111.4 cd mounted at a height 210mm on a vertical axis passing through the vertex. Find out the

1. Area of the plane & 2. Solid angle subtended by the plane at the point-source

Q.2.75. Find out the solid angle subtended by an equilateral triangle intercepting 50 lm flux emitted from a point-source of 150 cd mounted at a height 425mm on a vertical axis passing through the centre. If the flux is uniformly distributed over the entire plane what is the flux density on the plane?

Q.2.76. What is the flux density on an equilateral triangular plane intercepts 112 lm flux when a uniform point-source, emitting total flux 1550 lm, is mounted at a height 100mm on a vertical axis passing through the vertex?

(Assume uniform distribution of luminous flux over the entire plane).

Q.2.77. What is the total luminous flux intercepted by an equilateral triangular plane having area $4\sqrt{3}m^2$ when three bulbs of 250 cd are mounted at the same height 250cm on the vertical axes passing through the vertices of the plane?

If the flux is uniformly distributed over the entire plane, what is the flux density? (Assume bulbs as the point-sources.)

Q.2.78. Three bulbs of 150 cd are mounted at the same height 260cm on the vertical poles pitched onto the vertices of an equilateral triangular plane with each side 4.5m. Now, what will be the change in the flux density?

1.) If one of them is switched off. 2.) If two of them are switched off.

(Assume bulbs as the point-sources.)

Q.2.79. Three point-sources each emitting total flux 1400 lm are mounted at a height 254cm on the vertical poles pitched onto the vertices of an equilateral triangular plane with each side 350cm. If another point-source is mounted at a height 120cm on a vertical pole pitched onto the centre such that when it is switched off flux striking the plane reduces to half of that emitted by all the sources. What is the luminous intensity of point-source mounted at the centre?

Q.2.80. Total flux intercepted by an equilateral triangular plane with each side 230cm is 350 lm when three point-sources of same intensity 110 cd are mounted at the same height on the vertical poles pitched onto the vertices & a point-source of 150 cd is mounted at a height 200cm on a vertical pole pitched at the centre. Find out the height of the point-sources mounted above the vertices,

Q.2.81. Four bulbs of 100 cd are mounted at the same height 235cm on the vertical poles three pitched onto the vertices & one onto the centre of an equilateral triangular plane with circumscribed radius

110cm. Now, If the bulb at the centre is switched off what will be the percentage change in the flux density? (Assume bulbs as the point-sources & flux is uniformly distributed over the entire plane.)

Q.2.82. A point-source of intensity 200 cd is mounted at a height 2m on a vertical axis passing through the centre of an equilateral triangular plane with each side 3.5m. Now, if another point-source of 100cd is mounted at a height 2.2m on a vertical axis passing through the mid-point of the side what will be the change in the

1.)Flux striking the plane 2) flux density if it is uniformly distributed over the entire plane &

3) Flux striking the plane if a circular hole with a radius 30cm is made at the centre keeping point-source mounted above the mid-point off.

*** Rectangular-plane:**

Q.2.83. Find out the solid angle subtended by a rectangular plane having dimensions 10cm & 16cm at a point lying at a height 8cm on a vertical axis passing through the centre.

Q.2.84. What is the solid angle subtended by a rectangular plane having dimensions 12cm & 18cm at a point lying at a height 6cm on a vertical axis passing through the vertex?

Q.2.85. Find out the solid angle subtended by a rectangular plane of $6cm \times 15cm$ at a point lying at a height 4cm on a vertical axis passing through the mid-point of

1. Smaller side & 2.larger side

Q.2.86. If the solid angle subtended by a rectangular plane of $9cm \times 14cm$ at a point lying on a vertical axis passing through the centre is 2.2sr. What is the normal height of the point from the plane?

Q.2.87. Find out the height of a point lying on a vertical axis passing through the vertex of a rectangular plane of $12cm \times 16cm$ such that solid angle subtended by the plane at the same point is 1.2 sr.

Q.2.88. Find out the height of a point lying on a vertical axis passing through the mid-point of larger side of a rectangular plane of $12cm \times 16cm$ such that solid angle subtended by the plane at the same point is 2.5 sr.

Q.2.89. Find out the area of a rectangular plane with a side 13cm subtends a solid angle 0.9 sr. at a point lying at a height 16cm on a vertical axis passing through the centre of the plane.

Q.2.90. If a rectangular plane with a side 400mm subtends a solid angle 1.5 sr. at a height 10cm on a vertical axis passing through the vertex of the plane find out another side?

Q.2.91. Solid angle subtended by a rectangular plane having area $156\ cm^2$ at a point lying at a height 12cm on a vertical axis passing through the mid-point of larger side is 1.2 sr. Find out dimensions of the plane.

Q.2.92. Find out the change in the solid angle subtended by a rectangular plane of $14cm \times 16cm$ at point brought at a height 8cm from 15cm on a vertical axis passing through the

1. Centre & 2.vertex of the plane.

Q.2.93. Find out the solid angle subtended by a rectangular plane of $15cm \times 18cm$ having a circular hole with a radius 4cm at the centre, at a point lying at a height 11cm on a vertical axis passing through the vertex.

Q.2.94. Find out the luminous flux intercepted by a rectangular plane of $14cm \times 16cm$ emitted from a uniform point-source of 100 cd mounted at a height 25cm on vertical axis passing through the

1.) Centre 2) vertex 3) mid-point of the smaller side & 4) mid-point of the larger side.

Q.2.95. Find out the luminous flux density on a rectangular plane of $2.4m \times 3.2m$ when a uniform point-source of 110 cd mounted at a height 25cm on vertical axis passing through the

1) Centre 2) vertex 3) mid-point of the smaller side & 4) mid-point of the larger side. (Assume that luminous flux is uniformly distributed over the entire plane)

Q.2.96. Find out luminous flux intercepted by a rectangular plane of $3m \times 4m$ emitted from four identical point-sources each of 200 cd mounted at an equal height 275cm on the vertical poles pitched onto the

1) Vertices & 2) mid-points of the sides of the plane

Q.2.97. If the flux intercepted by a rectangular plane of $10m \times 14m$ emitted from a uniform point-source mounted at a height 350cm on a vertical pole pitched onto the centre is 200 lm. Find out the intensity of the source.

Q.2.98. What should be the height of a point-source 400 cd to mount on a vertical pole pitched onto the vertex of a rectangular plane of $6m \times 8m$so that the flux density is 1000 lux? (Assume that the flux is uniformly distributed over the entire plane).

Q.2.99. Find out the flux intercepted by a rectangular planar field of $6m \times 10m$ emitted by nine identical bulbs each of 400 cd mounted on the vertical poles fourpitched onto the vertices, four onto the mid-points of the sides & one onto the centre of plane.

(Assume the bulbs as uniform point-sources.)

Q.2.100. Four bulbs each of 111.4 cd are mounted at a height 180cm on the vertical poles pitched onto the vertices of a rectangular plane of $3m \times 4m$. Now another point-source of 120 cd is mounted on a vertical pole pitched onto the centre such that when it is switched off the flux intercepted by the plane reduces to three fourththe total flux obtained by all the point-sources. Find out the height of central bulb from the plane.

(Assume that the bulbs are uniform point-sources.)

Q.2.101. Find out % change in the flux intercepted by a rectangular plane of$4m \times 4.5m$when four uniform point-sources each of 100 cd are mounted at a height 200cm on the vertical poles pitched onto the vertices of the plane & another point-source of 150 cd is mounted at a height 1.5m on a vertical pole pitched onto the centre when

1.) Any of the sources mounted above the corner is switched off & 2) only central source is switched off.

Q.2.102. If the total luminous flux intercepted by a rectangular plane is 150 lm emitted by five uniform point sources four of them each of 100 cd are mounted at a height 2m on vertical poles pitched onto the vertices of a rectangular field of $16m \times 20m$& fifth is mounted at a height 3m on a verticalpole pitched onto the centre. What is the luminous intensity of the central point-source?

Q.2.103. A uniform point-source of 200 cd is mounted at a height 2.5m on a vertical pole pitched onto the centre of a rectangular plane of $8m \times 12m$& four identical point-sources are mounted at a height 2m on the vertical poles pitched on to the mid-points of the sides of the plane. If all the point-sources mounted above the mid-points of the sides are switched off the total flux intercepted by the plane reduces by 75%the total flux emitted by all the sources. Find out the

1.) Intensity of identical point-sources & 2) total flux intercepted by the plane.

3) Flux density when a) all the sources are switched on. b) four sources mounted above the mid-points are switched off. & c) only central source is switched off. (Assume uniform distribution of flux over the entire plane.)

Q.2.104. Find out the solid angle subtended by a plane bounded by the lines $x = 2, x = 10, y = -4$ & $y = 16$, at a point $(2, -4, 2)$.

Q.2.105. Calculate the solid angle subtended by a plane bounded by the lines $x = -14, x = -5, y = -1$ & $y = -16$, at a point $(-5, -1, -6)$.

Q.2.106. Find out the solid angle subtended by a plane bounded by the lines $x = -2, x = 8, y = 1$ & $y = 9$, at a point $(2, 5, -3)$.

Q.2.107. Calculate the solid angle subtended by a plane bounded by the lines $x = -5, x = 6, y = -3$ & $y = 4$, at a point $(4, -3, 2)$.

Q.2.108. Find out the solid angle subtended by a plane bounded by the lines $x = \pm\infty, y = \pm 6$, at a point $(0, 0, 13)$.

Q.2.109. Determine the solid angle subtended by a plane bounded by the lines $x = \pm\infty, y = -2$ & $y = 18$, at a point $(2, 5, -11)$.

Q.2.110. Find out the solid angle subtended by a plane bounded by the lines $x = 3, x = 16$ & $y = \pm\infty$, at a point $(3, 0, 8)$.

Q.2.111. Find out the solid angle subtended by a plane bounded by the lines $x = 4, x = \infty, y = 2$ & $y = 12$, at a point $(4, 2, -7)$.

Q.2.112. Calculate the solid angle subtended by a plane bounded by the lines $x = -2, x = \infty, y = -5$ & $y = -\infty$, at a point $(2, 4, 5)$.

Q.2.113. Find out the total flux intercepted by a plane bounded by the lines $x = -3, x = 5, y = -5$ & $y = 14$, at a point $(5, -6, -2)$.

Q.2.114. Find out the flux density over a plane bounded by the lines $x = -7, x = 7, y = -3$ & $y = 16$, at a point $(-1, -4, 3)$.

*** Square-plane:**

Q.2.115. Find out the solid angle subtended by a square plane with each side 12cm at a point lying at a height 60mm on the vertical axis passing through the 1) Centre2) vertex & 3) mid-point of the side of the plane

Q.2.116. If the solid angle subtended by a square plane of $9cm \times 9cm$ at a point lying on a vertical axis passing through the centre is 2.2sr. What is the normal height of the point from the plane?

Q.2.117. Find out the height of a point lying on a vertical axis passing through the vertex of a square plane of $12cm \times 12cm$ such that solid angle subtended by the plane at the same point is 1.4 sr.

Q.2.118. Find out the height of a point lying on a vertical axis passing through the mid-point of one of the sides of a square plane of $16cm \times 16cm$ such that solid angle subtended by the plane at the same point is 2.5 sr.

Q.2.119. Find out the area of a square plane subtends a solid angle 0.8 sr. at a point lying at a height 16cm on a vertical axis passing through the centre of the plane.

Q.2.120. If a square plane subtends a solid angle 1.6 sr. at a height 10cm on a vertical axis passing through the vertex of the plane find out the side of the plane?

Q.2.121. Solid angle subtended by a square plane at a point lying at a height 13cm on a vertical axis passing through the mid-point of the side is 1.2 sr. Find out perimeter of the plane.

Q.2.122. What is the solid angle subtended by each of the faces of a cube with each side 80cm at the centre?

Q.2.123. Find out the change in the solid angle subtended by a rectangular plane of $18cm \times 18cm$ at point brought at a height 8cm from 15cm on a vertical axis passing through the

1) Centre 2) vertex & 3) mid-point of the side of the plane

Q.2.124. Find out the solid angle subtended by a square plane of $361cm^2$ having a circular hole with a radius 5cm at the centre, at a point lying at a height 10cm on a vertical axis passing through the vertex.

Q.2.125. Find out the solid angle subtended by a square plane with each side 16cm at a point lying at a height 12cm on a vertical axis passing through a point internally dividing one of the sides in a ratio

1) 1:1 2) 1:2 & 3) 2:3.

Q.2.126. Find out the solid angle subtended by a square plane with each side 16cm at a point lying at a height 12cm on a vertical axis passing through a point externally dividing one of the sides in a ratio

1) 1:1 2) 1:2 & 3) 2:3

Q.2.127. Find out the luminous flux intercepted by a square plane of $20cm \times 20cm$ emitted from a uniform point-source of 120 cd mounted at a height 28cm on vertical axis passing through the

1) Centre 2) vertex & 3) mid-point of one of the sides

Q.2.128. Find out the luminous flux density on a square plane of $3.2m \times 3.2m$ when a uniform point-source of 110 cd mounted at a height 25cm on vertical axis passing through the

1) Centre 2) vertex & 3) mid-point of one of the sides (Assume that luminous flux is uniformly distributed over the entire plane).

Q.2.129. Find out luminous flux intercepted by a square plane of $3m \times 3m$ emitted from four identical point-sources each of 250 cd mounted at an equal height 250cm on the vertical poles pitched onto the

1) Vertices & 2) mid-points of the sides of plane

Q.2.130. If the flux intercepted by a square plane of $14m \times 14m$ emitted from a uniform point-source mounted at a height 325cm on a vertical pole pitched onto the centre is 220 lm. Whatis the intensity of the source?

Q.2.131. What should the height of a point-source 400 cd to mount on a vertical pole pitched onto the vertex of a square plane of $6m \times 6m$ so that the flux density is 650 lux. (Assume that the flux is uniformly distributed over the entire plane)

Q.2.132. Find out the flux intercepted by a square planar-field of $10m \times 10m$ emitted by nine identical bulbs each of 400 cd mounted on the vertical poles four pitched onto the vertices, four onto the mid-points of the sides & one onto the centre of the plane.

(Assume the bulbs as uniform point-sources.)

Q.2.133. Four bulbs each of 150 cd are mounted at a height 180cm on the vertical poles pitched onto the vertices of a square plane of $3m \times 3m$. Now another point-source of 120 cd is mounted on a vertical pole pitched onto the centre such that when it is switched off the flux intercepted by the plane reduces to three fourth the total flux obtained by all the point-sources. Find out the height of central bulb from the plane.

(Assume that the bulbs are uniform point-sources.)

Q.2.134. Find out % change in the flux intercepted by a rectangular plane of $4m \times 4m$ when four uniform point-sources each of 150 cd are mounted at a height 200cm on the vertical poles pitched onto the vertices of the plane & another point-source of 250 cd is mounted at a height 2m on a vertical pole pitched onto the centre when

1.) Any of the sources mounted above the corner is switched off & 2.) Only central source is switched off.

Q.2.135. If the total luminous flux intercepted by a rectangular plane is 200 lm emitted by five uniform point sources four of them each of 200 cd are mounted at a height 2m on vertical poles pitched onto the vertices of a squareplanar-field of$16m \times 16m$& fifth is mounted at a height 2.5m on a vertical pole pitched onto the centre. What is the luminous intensity of the central point-source?

Q.2.136. Find out the flux intercepted by a square plane with each side 25cm if a circular hole with a radius 45mm is made at the centre of the plane when a point-source of 200 cd is mounted at a height 128mm on a vertical axis passing through the centre of the plane.

If flux is uniformly distributed over the entire plane what would the luminous flux density?

Q.2.137. A square plane with each side 'a' intercepts one-fourth of the total flux emitted by a uniform point-source located at a height 'h' on the vertical axis passing through the centre of the plane. Prove that area of the plane is equal to

$$\frac{4h^2}{\sqrt{2}-1}.$$

Q.2.138. A square plane intercepts one-sixteenth of the total flux emitted by a uniform point-source located at a height 'h' on the vertical axis passing through one of the vertices of the plane. Prove that the ratio of side to height is given as

$$\frac{a}{h} = \frac{1}{\sqrt{\sqrt{2}-1}}$$

Q.2.139. A square plane intercepts one-third of the total flux emitted by a uniform point-source located at a height 'h' on the vertical axis passing through the centre of the plane. Prove that the ratio of side to height is given as

$$\frac{a}{h} = 2\sqrt{\frac{3}{2\sqrt{3}-3}}.$$

Q.2.140. A square plane with side 'a' intercepts one-sixth of the total flux emitted by a point-source located at a height 10 cm on the vertical axis passing through the centre of the plane. Assuming uniform distribution the flux density over the plane is 10.472 k lux find out the intensity of the point-source.

Q2.141. Determine the value of K if $a^4 = Kh^4$&the flux intercepted by a square plane with each side 'a' is $1/12$ of the total flux emitted by a uniform point-source located at a height 'h'on the vertical axis passing through one of the verticesof the plane.

Q.2.142. Find out the area of a square plane intercepting 62.832 lm flux when a uniform point-source emitting 1508 lm flux, is located at a height 4m on a vertical pole pitched onto one of the vertices of the plane.

Q.2.143. Find out the solid angle subtended by a plane bounded by the lines $x = \pm 8$ & $y = \pm 8$, at a point $(0, 0, -5)$.

Q.2.144. Find out the solid angle subtended by a plane bounded by the lines $x = 2, x = 10, y = 7$ & $y = -1$, at a point $(2, -1, -3)$.

Q.2.145. Determine the total flux intercepted by a plane bounded by the lines $y = \pm x, x + y = 5\sqrt{2}$& $x - y = 5\sqrt{2}$ when a uniform point-source of 200 cd is located at a point $\left(5/\sqrt{2}, -5/\sqrt{2}, 5\right)$.

Q.2.146. The plane bounded by the lines $x = \pm a$ & $y = \pm a$ intercepts 100 lm flux when a uniform point-source emitting 1600 lm flux is located at a point $\left(-a, a, 2 - \sqrt{2}\right)$. Find out

1) Value of constant 'a' & 2) flux density over the bounded plane (assuming uniform distribution of flux over the entire plane)

Q.2.147. Find out the flux density over a plane bounded by the lines $x + y = \pm 10, x - y = 10$ & $y = x$ when a uniform point-source of 300cd is located at a point $(10, 0, -4)$. (Assume that the flux is uniformly distributed over the entire plane.)

Q.2.148. Four point-sources each of 110 cd are mounted at an equal height 30cm on the vertical axes passing through the vertices of a square plane with each side 40cm. Now, a rectangular plane of $10cm \times 15cm$ is cut from the centre of the square plane what will the change in the

1) Flux intercepted by the plane & 2) flux density (assume that the flux is uniformly distributed over the entire plane)

Q.2.149. A square planar-ground of $100m \times 100m$ has nine point-sources each of 600 cd mounted at an equal height 2.5m on the vertical poles, four onto the vertices, four onto the mid-points of the sides& one onto the centre. If one of the sources above the vertices fails to light then in order to obtain the same luminous flux density what should be the height to mount the

1) Central source & 2) rest of three sources above the vertices

(Assume that the luminous flux is uniformly distributed over the entire plane)

Q.2.150. A uniform point-source of 200 cd is mounted at a height 275cm on a vertical pole pitched onto the centre of a square plane of $8m \times 8m$& four identical point-sources are mounted at a height 200cm on the vertical poles pitched on to the mid-points of the sides of the plane. If all the point-sources mounted above the mid-points of the sides are switched off the total flux intercepted by the plane reduces by 70% the total flux emitted by all the sources. Find out the

1) Intensity of identical point-sources & 2) total flux intercepted by the plane.

3) Flux density when a.) All the sources are switched on b) four sources mounted above the mid-points are switched off. & c) only central source is switched off. (Assume uniform distribution of flux over the entire plane.)

*** Rhombus-like plane:**

Q.2.151. Find out the solid angle subtended by a rhombus like plane with diagonals 10cm & 16cm at a point lying on the vertical axis passing through the centre.

Q.2.152. Estimate the solid angle subtended by a rhombus like plane with diagonals 16cm & 28cm at a point lying on the vertical axis passing through the centre.

Q.2.153. The solid angle subtended by a rhombus like plane with a diagonal 18cm, at the point lying at a height 13cm on the vertical axis passing through the centre, is1.3sr. Find out the area of the plane.

Q.2.154. A uniform point-source of 200 cd is located at a height 60cm on the vertical axis passing through the centre of a rhombus like plane with diagonals 13cm & 24cm. Find out

1.) Total flux intercepted by the plane & 2) flux density (assume uniform distribution of flux over the entire plane.)

Q.2.155. Find out the total flux intercepted by a plane bounded by the lines $4x + 3y = \pm 12$ & $4x - 3y = \pm 12$ if a uniform point-source of 150 cd is located at a point $(0, 0, 4)$. Also find out the flux density if the flux is uniformly distributed over the entire plane.

Q.2.156. Find out the flux density over the plane bounded by the lines $5x + 12y = \pm 30$ & $5x - 13y = \pm 30$ if a uniform point-source of 200 cd is located at a point $(0, 0, -2)$. (Assume that the flux is uniformly distributed over the entire plane.)

Q.2.157. Find out the energy density over the plane bounded by the lines $2x = y,\ y = 0, 2x - y = 26$ & $13y = 120$ if a uniform point-source of 500 W is located at a point $(0, 0, 1)$. (Assume that the energy is uniformly distributed over the entire plane.)

Q.2.158. Find out the total flux intercepted by a plane bounded by the lines $x - y = 5, y = x, y = 0$ & $y = 4.8$ if a uniform point-source of 110 cd is located at a point $(0, 0, 3)$. Also find out the flux density if the flux is uniformly distributed over the entire plane.

Q.2.159. Calculate the total energy intercepted by the plane bounded by the lines $5x + 4y = \pm 20$ & $5x - 4y = \pm 20$ when a uniform point-source of 500 W is located at a point $(0, 0, -5)$.

Q.2.160. Find out the total flux intercepted by the plane bounded by the lines $3x + 7\sqrt{2}y = \pm 21\sqrt{2}$& $3x - 7\sqrt{2}y = \pm 21\sqrt{2}$ when a uniform point-source of 100 cd is located at a point $(0, 0, 4\sqrt{2})$.

Q.2.161. Find out the flux density over the plane bounded by the lines $17x + 8y = \pm 136 + 11$ & $17x - 8y = \pm 136 + 91$ when a uniform point-source of 100 cd is located at a point $(3, -5, 2)$. (Assume that the flux is uniformly distributed over the entire plane.)

Q.2.162. Find out the energy density over the plane bounded by the lines $11x + 13y = \pm 143 + 170$ & $11x - 13y = \pm 143 - 38$ when a uniform point-source of 400 W is located at a point $(6, 8, -8)$. (Assume that the radiation energy is uniformly distributed over the entire plane.)

Q.2.163. Find out the energy density over the plane bounded by the lines $(7 + 4\sqrt{3})x + (7\sqrt{3} - 4)y = \pm 112$ &$(4\sqrt{3} - 7)x - (7\sqrt{3} + 4)y = \pm 112$ when a uniform point-source of 400 W is located at a point $(0, 0, 5)$. (Assume that the radiation energy is uniformly distributed over the entire plane.)

Q.2.164. Find out the change in the flux if a uniform point-source of 200 W located on the vertical axis passing through the centre of a rhombus like plane with diagonals 20cm & 35cm is shifted from a distance 10cm to 2cm from the centre.

Q.2.165. A uniform point-source of 300 W located on the vertical axis passing through the centre of a rhombus like plane with diagonals 20cm & 36cm, is shifted from a distance 8cm to 18 cm from the centre. What will be change in the energy density if the energy is uniformly distributed over the entire plane?

*** Right-trapezoidal plane:**

Q.2.166. Find out the solid angle subtended by the plane bounded by the lines $x = 0, y = 0, y = 8$ & $8x + 3y = 120$ at a point $(12, 8, 4)$.

Q.2.167. Find out the solid angle subtended by the plane bounded by the lines $x = 0, y = 0, y = 7$ & $7x + 4y = 126$ at the following points

1.) $(14, 7, 3)$ 2) $(18, 0, -3)$& 3) $(14, 0, -4)$.

Q.2.168. Determine the solid angle subtended by the plane bounded by the lines $x = 16, y = 0, y = 14$ & $x - y = 0$ at a point $(14, 14, -2)$.

Q.2.169. Find out the solid angle subtended by the plane bounded by the lines $x = -15, y = -2, y = 8$ & $x + y = 0$ at a point $(-8, 8, 3)$.

Q.2.170. Estimate the solid angle subtended by the plane bounded by the lines $x = -30, y = 2, y = -25$ & $x + y = 0$ at a point $(-2, 2, 4)$.

Q.2.171. Find out the solid angle subtended by the plane bounded by the lines $x = 2, y = -3, y = 10$ & $13x + 7y = 330$ at a point $(20, 10, 5)$.

Q.2.172. Find out the solid angle subtended by the plane bounded by the lines $x = -3, y = 5, y = 12$ & $7x + 4y = 83$ at a point $(5, 12, -3)$.

Q.2.173. Determine the solid angle subtended by the plane bounded by the lines $x \pm y = 0, x + y = 7\sqrt{2}$& $13x - y = 98\sqrt{2}$ at a point $(7\sqrt{2}, -7\sqrt{2}, 2)$.

Q.2.174. Find out the solid angle subtended by the plane bounded by the lines $\sqrt{3}x - y = 0, x + \sqrt{3}y = 0, x + \sqrt{3}y = 16$ &$(2\sqrt{3} + 1)x - (2 - \sqrt{3})y = 64$ at a point $(2(2 + 3\sqrt{3}), 2(2\sqrt{3} - 3), -4)$.

Q.2.175. Find out the solid angle subtended by the plane bounded by the lines $x - \sqrt{3}y = 0, \sqrt{3}x + y = 0, \sqrt{3}\,x + y = 12$ &$(3 + 2\sqrt{3})x - (3\sqrt{3} - 2)y = 84$ at a point $(7, -7\sqrt{3}, 3)$.

Q.2.176. Find out the total energy intercepted by a plane bounded by the lines $x = 0, y = 0, y = 5$ & $5x + 4y = 40$if a uniform point-source of 200 W is located at a point $(4, 5, 2\sqrt{3})$. If the energy is uniformly distributed over the entire plane find out the energy density.

Q.2.177. Find out the total flux intercepted by a plane bounded by the lines $x = 0, y = 0, y = 7$ & $7x + 6y = 126$ if a uniform point-source of 110 cd is located at a point $(18, 0, -5)$.

Q.2.178. Find out the flux density over the plane bounded by the lines $x = 2, y + 3 = 0, y - 10 = 0$ & $13x + 7y - 330 = 0$ if a uniform point-source of 200 cd is located at a point $(27, 0, 3)$. (Assume uniform distribution of flux over the entire plane)

Q.2.179. Find out the energy density over the plane bounded by the lines $x + 3 = 0, y - 5 = 0, y = 12$ & $7x + 4y - 83 = 0$ if a uniform point-source of 500 W is located at a point $(9, 0, -2)$. (Assume uniform distribution of radiation energy over the entire plane)

Q.2.180. Determine the total flux intercepted by the plane bounded by the lines $y = \pm x, x + y = 7\sqrt{2}$& $13x - y - 98\sqrt{2} = 0$when a uniform point-source of 200 cd is located at a point $\left({}^{15}/_{\sqrt{2}}, {}^{-1}/_{\sqrt{2}}, 3\right)$.

Q.2.181. Find out the energy density over the plane bounded by the lines $y = \sqrt{3}x, x + \sqrt{3}y - 16 = 0, y = {}^{-x}/_{\sqrt{3}}$&$(2\sqrt{3} + 1)x + (\sqrt{3} - 2)y = 64$ if a uniform point-source of 300 W is located at a point $(8\sqrt{3}, -8, 3)$. (Assume uniform distribution of radiation energy over the entire plane)

*** Regular-polygonal plane:**

Q.2.182. Find out the solid angle subtended by an equilateral triangular plane with each side 10cm at a point lying at a height 7cm on the vertical axis passing through the centre of the plane.

Q.2.183. Find out the solid angle subtended by a squareplane with each side 13cm at a point lying at a height 9cm on the vertical axis passing through the centre of the plane.

Q.2.184. Determine the solid angle subtended by a regular pentagonal plane with each side 14cm at a point lying at a height 11cm on the vertical axis passing through the centre of the plane.

Q.2.185. Find out the solid angle subtended by a regular hexagonal plane with each side 8cm at a point lying at a height 18cm on the vertical axis passing through the centre of the plane.

Q.2.186. Calculate the solid angle subtended by a regular polygonal plane with 15 no. of the sides each 8cm, at a point lying at a height 20cm on the vertical axis passing through the centre of the plane.

Q.2.187. Find out the solid angle subtended by the plane bounded by the lines $\pm 3x + \sqrt{3}y = 10$ & $2\sqrt{3}y + 10 = 0$ at a point $(0, 0, 4)$.

Q.2.188. Find out the solid angle subtended by the plane bounded by the lines $\pm 3x + \sqrt{3}y = 13$ & $2\sqrt{3}y = -13$ at a point $(0, 0, -3)$.

Q.2.189. Find out the solid angle subtended by the plane bounded by the lines $3x + \sqrt{3}y = 19, 3x - \sqrt{3}y + 5 = 0$ & $2\sqrt{3}y + 38 = 0$ at point $\left(-4, -5\sqrt{3}, 5\right)$.

Q.2.190. Find out the solid angle subtended by the plane bounded by the lines $\sqrt{3}x \pm 3y + 11 = 0$ & $2\sqrt{3}x = 11$ at point $(0, 0, -2)$.

Q.2.191. Find out the solid angle subtended by the plane bounded by the lines $3x + \sqrt{3}y = 1, 3x - \sqrt{3}y + 17 = 0$ & $2\sqrt{3}y + 46 = 0$ at point $\left(3, -6\sqrt{3}, -6\right)$.

Q.2.192. Determine the solid angle subtended by the plane bounded by the lines $x = \pm 5$ & $y = \pm 5$ at point $(4, 5, 2)$.

Q.2.193. Find out the solid angle subtended by the plane bounded by the lines $x = \pm 7 - 2$ & $y = \pm 7 + 5$ at points $(-2, 5, -3)$&$(5, 12, 1)$.

Q.2.194. Determine the solid angle subtended by the plane bounded by the lines $\sqrt{3}x - y = \pm 16$ & $x + \sqrt{3}y = \pm 16$ at point$\left(4\left(\sqrt{3} - 1\right), 4\left(\sqrt{3} + 1\right), -3\right)$.

Q.2.195. Find out the solid angle subtended by the plane bounded by the lines $\sqrt{3}x \pm y = \pm 5\sqrt{3}$& $y = \pm 5\sqrt{3}/2$ at a point $(0, 0, 3)$.

Q.2.196. Find out the solid angle subtended by the plane bounded by the lines $\sqrt{3}x \pm y = \pm 18$ & $y = \pm 9$ at a point $(0, 0, -7)$.

Q.2.197. Find out the solid angle subtended by the plane bounded by the lines $\sqrt{3}x + y = \pm 8\sqrt{3} - 14, \sqrt{3}x - y = \pm 8\sqrt{3} - 4$ & $y = \pm 4\sqrt{3} - 5$ at the point $\left(-3\sqrt{3}, -5, 4\right)$.

Q.2.198. Find out the solid angle subtended by the plane bounded by the lines $x = \pm 3\left(\sqrt{2} + 1\right), y = \pm 3\left(\sqrt{2} + 1\right)$& $x \pm y = \pm 3\sqrt{2}\left(\sqrt{2} + 1\right)$ at the point $(0, 0, -3)$.

Q.2.199. A uniform point-source of 100 cd is mounted at a height 2m on the vertical axis passing through the centre of anequilateral triangular plane with each side 4.5m. How much flux will be intercepted by the plane?

Q.2.200. How much radiation energy will be intercepted by a square plane with each side 4m if a uniform point-bulb of 1 kW is mounted at a height 2.6m on the vertical pole pitched onto one of the vertices.

Q.2.201. Find out the total flux intercepted by a regular pentagonal plane with each side 20cm when a uniform point-source of 200 cd is located at a height 35cm on the vertical axis passing through the centre.

Q.2.202. Find out the flux density over a regular hexagonal plane with each side 24cm when a uniform point-source of 300 cd is located at a height 11cm on the vertical axis passing through the centre. (Assume that the flux is uniformly distributed over the entire plane.)

Q.2.203. Find out the total flux intercepted by the plane bounded by the lines $\sqrt{3}x + y = 5 + 8\sqrt{3}, \sqrt{3}x - y + 5 = 0$ & $y = 5 - 2\sqrt{3}$if a uniform point-source of 150 cd is located at the point $(4, 5, 6)$.

Q.2.204. Calculate the total flux intercepted by the plane bounded by the lines $x - \sqrt{3}y = \pm 14$ &$\sqrt{3}x + y = \pm 14$ if a uniform point-source of 200 cd is located at the point $\left(-7(\sqrt{3}-1), 7(\sqrt{3}+1), 5\right)$.

Q.2.205. Find out thedensity of radiation energy intercepted by the plane bounded by the lines $\sqrt{3}x + y = \pm 12 - 11, \sqrt{3}x - y = \pm 12 - 25,\ y = 13$ & $y = 1$ if a uniform point-source of 250 W is located at the point $\left(-6\sqrt{3}, 7, -7\right)$. (Assume uniform distribution of the energy over the entire plane.)

Q.2.206. Determine the total radiation energy intercepted by the plane bounded by the lines $x = \pm 7(\sqrt{2}+1) - 4,\ y = \pm 7(\sqrt{2}+1) + 6,\ x + y = \pm 7\sqrt{2}(\sqrt{2}+1) + 2$ & $x - y = \pm 7\sqrt{2}(\sqrt{2}+1) - 10$when a uniform point-source of 500 W at the point $(-4, 6, -5)$.

*** Circular plane:**

Q.2.207. Find out the solid angle subtended by a circular plane of radius 20cm at a point lying at a height 12cm on the vertical axis passing through the centre of the plane.

Q.2.208. A circular plane subtends a solid angle 1.055sr at a point lying at a height 15cm on the vertical axis passing through the centre of the plane. What is the radius of the plane?

Ans.10cm

Q.2.209. Find out the total flux intercepted by a circular plane with a radius 12cm if a uniform point-source of 300 cd is located at a height 16cm on the vertical axis passing through the centre of the plane. If the flus is uniformly distributed over the entire plane what will be the flux density?

Q.2.210. Calculate the solid angle subtended by the plane bounded by the curve $x^2 + y^2 = 144$ at the point $(0, 0, 4)$.

Q.2.211. Calculate the solid angle subtended by the plane bounded by the curve $4x^2 + 4y^2 - 20x + 28y - 122 = 0$ at the point $\left({}^{5}/_{2}, {}^{-7}/_{2}, -3\right)$.

Q.2.212. Find out the density of light energy over a circular plane with a radius 14m if a light-bulb of 2 kW is mounted at a height 4.5m on a vertical pole pitched at the centre of the plane. (Assume that bulb is point-source emitting the light uniformly in all the directions & light is uniformly distributed over the entire plane.)

Q.2.213. Calculate the total flux intercepted by the plane bounded by the curves $x^2 + y^2 + 8\sqrt{3}x - 12y - 60 = 0,\ \sqrt{3}x + y = 6\sqrt{3} - 6, \sqrt{3}x - y + 18 + 6\sqrt{3} = 0$ & $y = 6 - 3\sqrt{3}$ if a uniform point-source of 400 cd is located at the point $\left(-4\sqrt{3}, 6, 2\right)$.

Chapter 3
Approximate Analysis of Solid Angle

This chapter mainly deals with the analysis of the approximate value of the solid angle subtended only by the symmetrical plane-figures (i.e. having at least two axes of the symmetry) at a given point lying on the axis passing through the centre* & normal as well as inclined to the plane.

So far, we have studied that the 'Fundamental Theorem', 'Integral Theorem' & 'Corollary of the Plane-figure' give solvable integrations for finding out the solid angle subtended at any point lying on the normal axis by the certain plane-figures like

1. Right triangle 2. Isosceles triangle 3. Isosceles right triangle 4. Equilateral triangle

4. Rectangle 5. Square 6. Rhombus 7. Right trapezium 8. Regular polygon (like equilateral-triangle, square, regular pentagon, regular hexagon etc.) & 9. Circle.

But, in case of the point lying on the axis inclined with the given plane (as mentioned above) at a certain angle, the integrations obtained from the 'Fundamental Theorem', 'Integral Theorem' & 'Corollary of the Plane-figure' are insolvable.

Even, in case of the given point lying on the axis passing through the centre &normal to the plane-figures (like elliptical, parallelogram etc.) insolvable integrations are obtained.

Thus, we find that the use of 'Fundamental Theorem', 'Integral Theorem' & 'Corollary of the Plane-figure'is limited for the plane-figures as well as solid-figures.

Hence, to overcome these limitations of'Fundamental Theorem', 'Integral Theorem' & 'Corollary of the Plane-figure' a new theorem named as 'Approximation-Theorem' had been proposed by H.C.R. on 7Aug, 2013.

This theorem is based on the analysis of the approximate value of the solid angle subtended by the symmetrical plane-figures (having at least two axes of symmetry) at any point lying on the axis passing through the centre & normal as well as inclined to the given plane in the space.

This theorem is useful for finding the solid angle subtended by the symmetrical planes at any point in the space i.e. it is applicable in case of all the positions of the given point & symmetrical plane-figures with minimum permissible error in the results obtained.

* In this chapter, the centre means 'the centre of the symmetry' of the given plane-figure.

Note: In this chapter, symmetrical-plane means the plane-figure having at least two axes of the symmetry Ex. Equilateral triangle, Square, Rectangle, Rhombus, Regular polygon & circle etc.

Before proceeding further, let's know the terminology of the plane-figures for ease of understanding of the Approximation-Theorem.

3.1 Terminology of plane-figures

This terminology had been proposed by H.C.R. The terminology is mainly proposed in order to make the analysis of the planes easier to understand the distinction between symmetrical & unsymmetrical plane-figures. It has the following terms relevant to the plane-figures.

3.1.1 Plane-figure (2D-figure)

"A body (or a figure) in the space having merely two geometrical dimensions is called geometrical plane-figure or simply a plane-figure."

Ex. Triangle, Rectangle, Square, Polygon, Circle, Ellipse & all other plane-laminas etc.

*A geometrical-plane-figure is massless or surrounds no space.

3.1.2 Generate-plane

"The symmetrical-plane* generated by cutting a section normally or obliquely with the axis from a solid bodyis called generate-plane"

Example

1. Circular-plane (generated by cutting a sectionnormally with the axis from a sphere or a circular cylinder or a right cone)
2. Elliptical-plane (generated by cutting a section obliquely with the axis from a circular cylinder or a right cone.) (see chapter.1. analysis of the frustum for explanation.)
3. Regular polygonal-plane (generated by cutting a section normally with the axis from a right prism or a pyramid having base as a regular polygon.)
4. Rectangular-plane (generated by cutting a section normally with the axis from a right prism or a right pyramid having rectangular base.)
5. Rhombus-like-plane (generated by cutting a section normally with the axis from a right prism or a right pyramid having rhombus-like base.)

*The plane having at least two axes of symmetry.(see art.3.1.9)

Note: All the symmetrical planes are generate-planes i.e. all the symmetrical planes can be generated by cutting the sections normally or obliquely with the axis from the solid bodies having certain geometry.

3.1.3 Degenerate-plane:

Here, we can define the same as "An unsymmetrical plane is called degenerate-plane"

Ex. Scalene triangle, Parallelogram, Trapezium etc.

Although a degenerate plane can also be generated by cutting a certain solid body, this is named so only for distinction from the symmetrical planes& their analysis.

The value of solid angle subtended by a degenerate plane may have impermissible error.

3.1.4 Real-plane:

"The plane having a certain mass (inertia) or surrounding a certain space is called real-plane"

Ex. 1. Plane-lamina 2. Foil 3. Disc 4. Sheet 5. Plate etc.

3.1.5 Centre of mass of the real-plane:

"It is a point in the space at which the entire mass of the real-plane may be assumed to be concentrated." All the planes, in actual practice, are real planes having theircertain centres of mass. (All the objects having certain mass have their certain centres of mass.)

In fact, the plane having certain mass has a certain centre of mass.

3.1.6 Centre or Centroid of the plane-figure:

"It is a point in the space at which the entire area of a given plane-figure may be assumed to be concentrated."

All the plane-figures have a certain centre or centroid depending on their geometrical shapes.

3.1.7 Symmetry of the plane:

"It is the regularity in the geometrical shape of the plane. Or simply the 'geometrical- regularity' of the plane is called symmetry of the plane."

3.1.8 Symmetrical plane:

"The plane having at least two axes of the symmetry is called symmetrical plane."

Ex. Circle, ellipse, rectangle, rhombus & regular-polygons.

3.1.9 Axis of symmetry:

"Straight line dividingthe given plane in two identical parts which are mirror-images of each other w.r.t. the same line"

(The dotted lines are the axes of the symmetry of the plane-figures as shown below)

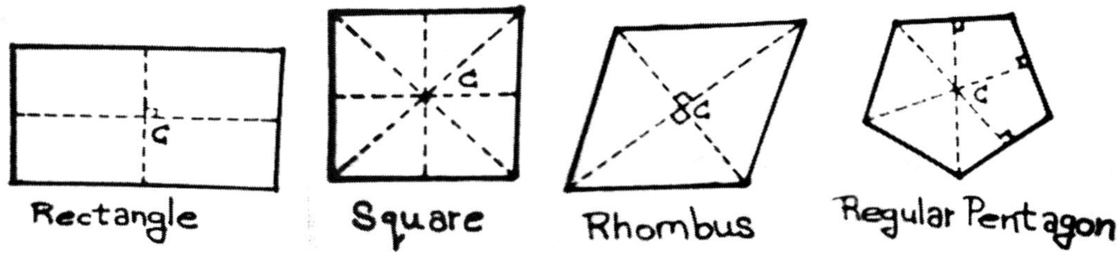

Fig (60)

3.1.10 Centre of symmetry:

"The point of intersection of the axes of symmetry (at least two) in a given plane-figure is called the centre of symmetry."

Centre of symmetry always lies within the boundary of the given symmetrical plane-figure.

(point 'C' is the 'centre of symmetry' of the plane-figures as shown above)

Note: All plane-figures have geometrical centre (centroid) but do not have the 'centre of symmetry'

Ex. Parallelogram, trapezium &irregular plane-figures

3.1.11 Point of symmetry:

"It is the point within the boundary of the given plane from which each point on the plane-boundary is at an equal & finite distance."

Or

"The point of intersection of infinite number of the axes of symmetry within the boundary of the given plane-figure is called point of symmetry."

Ex. Circle. It is the only plane-figure having point of symmetry.

* The plane-figure having a point of symmetry (i.e. infinite no. of the axes of symmetry) is called Perfect Symmetrical plane-figure (briefly P.S. plane-figure).

Ex. Circle is called P.S. plane-figure which practically doesn't exist.

(as shown in the figure(61) below)

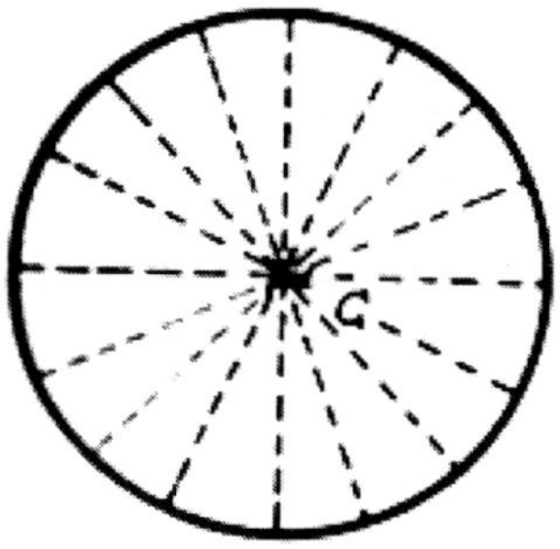

Fig (61)

3.1.12 Factor of Circularity:

"The ratio of the shortest to the longest bisecting axes of symmetry of a given plane-figure viewed from a given point in the space is called Factor of Circularity of the given plane". It is denoted by symbol "F_c".

Mathematically, it is given as follows

$$\Rightarrow \left[F_c = \frac{s}{l}\right]$$

Where 's' is the shortest axis of symmetry &

'l' is the longest axis of symmetry when the plane is vied from a given point in the space.

The shortest & the longest axes of symmetry bisect each other at a certain point lying within the boundary of the plane-figure.

Note: If all the axes of symmetry of a given plane-figure are of equal length but do not bisect each other then the Factor of Circularity of the plane figure is the ratio of smaller to the larger segments of the axis of symmetry of the given plane-figure.

Mathematically,

$$\Rightarrow \left[F_c = \frac{smaller\ segment}{larger\ segment}\right]$$

Ex. Regular polygon with odd number of sides has all the axes of symmetry of equal length but they (axes) do not bisect each other.

In any case other than above two, it is undefined/ indeterminate.

Important points:

1. It is a ratio (dimensionless quantity) such that $0 \leq F_c \leq 1$
2. Its value (F_c) depends on
   a) Geometrical shape (or dimensions) of the given plane-figure &
   b) Orientation of geometrical shape w.r.t. the given point (i.e. viewer) in the space.
3. It is undefined for those plane-figures which do not have a centre of symmetry or at least two axes of symmetry when viewed from a given point in the space.

Ex. Scalene triangle, Parallelogram, Trapezium etc.

Significance of Factor of Circularity:

1. The factor of Circularity shows that how much the given plane-figure is similar to a circle when viewed from a given point in the space. It is variable with configuration of object.
2. It shows that how much probable error may be there in the results obtained from the Approximation-formula of solid angle for a given plane-figure.
3. The plane-figure having $F_c = 1$ is called Unit Symmetrical plane-figure. In this case also, there may be certain error in the value except for a circular plane.

3.2 Approximation:

We know that

1. A circular-boundary can completely coincide with the lateral face of a right cone with a particular apex angle &apex at the given point lying on the axis normal to the given circular plane & passing through the centre &
2. An elliptical-boundary can completely coincide with the lateral face of a right cone with a particular apex angle & apex at the given point lying on the axis passing through the centre&inclined at a certain angle with the given elliptical plane.

In these two cases of complete coincidence, the solid angle subtended by circular & elliptical planes at a given point lying on the axis passing through the centre of the respective planes and normal & inclined to the respective planes, will be equal to the solid angle subtended at the apex by a right cone having a certain apex angle

i.e. when the circular & elliptical plane-boundaries are completely coincident with the lateral face of a right conewith certain apex angle then solid angle subtended by the circular & elliptical planes will be equal to that subtended by the same right cone at its apex (i.e. the given point in the space).

Mathematically, it is given from the eq(1) as follows

$$\omega = 2\pi(1 - cos\alpha) \quad \forall(0 < 2\alpha \leq 2\pi \ or \ 0 < \alpha \leq \pi) \ldots\ldots\ldots\ldots\ldots\ldots\ldots (1)$$

Where $'2\alpha'$is the apex-angle of the right cone with its lateral face completely coincident with the given circular & elliptical plane-boundaries respectively.

In all other cases except above two, the boundary of any symmetrical plane can not completely coincide with the lateral face of a right cone with any apex angle. It is because the cross-sections cut from a right cone with a certain apex angle are of two types only as follows

1. Circular-section (when cut normally with the axis of the right cone) &
2. Elliptical-section (when cut obliquely with axis of the right cone)

Thus, in other cases except above two,

Even the symmetrical plane-boundaries (except circular & elliptical plane-boundaries) are not completely coincident with the lateral face of the right cone with any apex angle, hence the above formula (1) is so far not applicable. Thus the exact value of solid angle subtended by thesymmetrical planes can not be found out from the eq(1).

But if the area of a symmetrical plane is made equivalent to the area of a circular plane then the givenplane can be replaced by the equivalent circular plane without varying the distance of the given point from the centre of the given symmetrical plane then the solid angle subtended by the equivalent circular plane at a given point will be approximately equal to that subtended by the given symmetrical plane & the approximate value can be found out from the eq(1).

Thus, the Approximation-Theorem gives the approximate value of the solid angle subtended by the symmetrical plane at a certain point by replacing the same by an equivalent circular plane without varying the distance of the given point from the centre of the given symmetrical plane.

3.3 Approximation-Theorem:

This theorem proposed by H.C.R. is applicable only for the symmetrical planes.

Statement:

"The solid angle subtended by a symmetrical plane at a given point in the space is approximately equal to that subtended at the same point by an equivalent circular plane having its centre coincident with that of the given plane, its plane normal to the line joining the given point to its centre & area equal to that of projection of the given plane in the direction of line joining its centre to the given point in the space."

Mathematically, the solid angle subtended by a symmetrical plane at a given point in the space is given by the following formula

$$\omega \cong 2\pi\left[1-\frac{1}{\sqrt{\left(1+\frac{A\cos\theta}{\pi r^2}\right)}}\right] \quad \forall\left(0 \le \theta \le \frac{\pi}{2}\right)$$

Where $'A' \rightarrow$ is the area of the given symmetrical plane

$'\theta' \rightarrow$ is the angle between the normal to the plane & the line joining the given point to the centre of the plane&

$'r' \rightarrow$ is the distance of the given point from the centre of the plane.

Note: The above formula is called Approximation-formulaof solid angle.

This formula is applicable only for the planes satisfying the following two conditions

1. The plane should have at least two axes of symmetry i.e. a centre of symmetry &
2. The factor of circularity(F_c) should be close to the unity i.e. $F_c \ge \frac{1}{2}$

If the above conditions are satisfied then the error involved in the results obtained from Approximation-formula usually would be minimum & permissible while in all other cases it may be considerable and impermissible.

Derivation:

Let there be a symmetrical plane with the centre 'C' & the area 'A' and a given point say 'P' at distance 'r' from the centre 'C' of the given plane such that

The angle between normal to the plane & line joining the given point to the centre 'C' is $'\theta'$(as shown in the figure (62) below)

It is obvious that area of projection (A_P) of the given plane having area 'A' in the direction of the line PC joining the point 'P' to the centre 'C' of the plane

$$\Rightarrow A_P = A\cos\theta$$

Where $'\theta'$ is the angle between normal to the plane & the line PC.

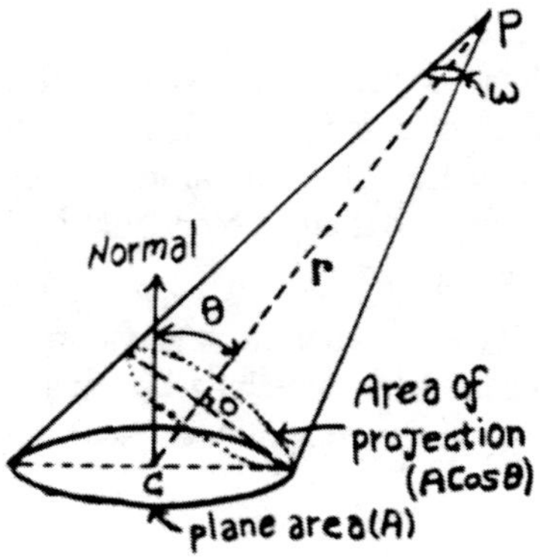

Fig (62)

Now, consider an equivalent circular plane having radius 'R', centre 'C' & its plane normal to the line PC such that

Area of the equivalent circular plane = area of projection of the given in the direction of the line PC

$$\Rightarrow \pi R^2 = A_p = A\cos\theta$$

$$\Rightarrow R^2 = \frac{A\cos\theta}{\pi} \quad \ldots\ldots\ldots\ldots\ldots (I)$$

Now, consider an imaginary right cone with the apex at the given point 'P' & the apex angle $'2\alpha'$ such that its lateral face is coincided with the boundary of the equivalent circular plane having centre 'C' & its plane normal to the line PC.

(as shown in the fig(63) below)

The imaginary right cone has been shown by the dotted lines & the boundary of the equivalent circular plane (with radius 'R') coinciding with the lateral face of the right cone

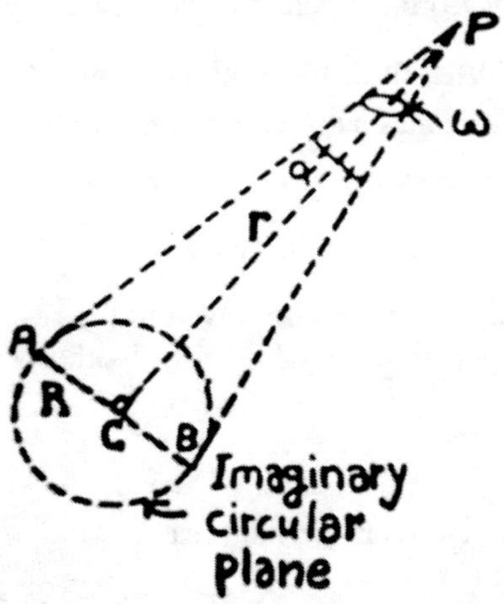

Fig (63)

in rightΔACP

$$\Rightarrow cos\alpha = \frac{CP}{AP} = \frac{CP}{\sqrt{CP^2 + AC^2}} = \frac{r}{\sqrt{r^2 + R^2}} \qquad (since, AC = BC = R\ )$$

$$\Rightarrow cos\alpha = \frac{r}{\sqrt{r^2 + \left(\frac{Acos\theta}{\pi}\right)}} \qquad (on\ setting\ the\ value\ of\ cos\alpha\ from\ eq(I))$$

$$\Rightarrow cos\alpha = \frac{1}{\sqrt{1 + \left(\frac{Acos\theta}{\pi r^2}\right)}} \quad \ldots\ldots\ldots\ldots\ldots\ldots (II)$$

In this case, the solid angle subtended by the equivalent circular plane at the point 'P'

= solid angle subtended by the right cone at the apex 'P'

We know that the solid angle subtended by the right cone at the apex is given by the eq(1) as follows

$\omega = 2\pi(1 - cos\alpha)$where $'2\alpha'$is the apex an

Now, on setting the value of $cos\alpha$from the eq(II), we get

the solid angle subtended by the given symmetrical plane at the given point 'P' in the space.

$$\Rightarrow \omega \cong 2\pi\left[1 - \frac{1}{\sqrt{1 + \left(\frac{Acos\theta}{\pi r^2}\right)}}\right] \quad \forall \left(0 \leq \theta \leq \frac{\pi}{2} \& r \geq 0\right)$$

Important deduction 1:

The solid angle subtended by a symmetrical plane at any point lying on the axis normal to the plane & passing through the centre is obtained by setting $\theta = 0$ in the above formula as follows

$$\omega \cong 2\pi\left[1 - \frac{1}{\sqrt{1 + \left(\frac{Acos0}{\pi r^2}\right)}}\right] = 2\pi\left[1 - \frac{1}{\sqrt{1 + \left(\frac{A}{\pi r^2}\right)}}\right]$$

$$\Rightarrow \omega \cong 2\pi\left[1 - \frac{1}{\sqrt{1 + \left(\frac{A}{\pi r^2}\right)}}\right] \quad \forall (r \geq 0)$$

Important deduction 2:

The solid angle subtended by any plane at any point lying on the plane completely inside the boundary is obtained by setting distance, $r = 0$ in the above formula as follows

$$\omega \cong 2\pi\left[1 - \frac{1}{\sqrt{1 + \left(\frac{A}{\pi (0)^2}\right)}}\right] = 2\pi\left[1 - \frac{1}{\sqrt{1 + (\infty)}}\right] = 2\pi[1 - 0] = 2\pi\ sr$$

Thus, $\omega = 2\pi\ sr$ is constant for all the points lying on the plane completely inside the boundary irrespective of its geometrical shape.

3.4 Limitations of Approximation-formula

It has the following limitations

1. It is applicable to find out the approximate value of the solid angle subtended by the planes satisfying the following two conditions
   a) The plane should have at least two axes of symmetry i.e. a point of symmetry &
   b) The factor of circularity(F_c) of the given plane should be close to the unity i.e. $F_c \geq 0.5$
2. The Approximation-formula gives the satisfactory results for the following plane-figures
   a) Rectangle (having the sides with small difference)
   b) Rhombus (having the diagonals with small difference)
   c) Regular polygon (like equilateral triangle, square, regular pentagon, regular hexagon etc.)
   d) Circle (eccentricity,$e = 0$)
   e) Ellipse (having small eccentricity, $e \to 0$)
3. The results obtained by this formula have certain error which increases as the factor of circularity of a given plane decreases.
4. Error involved in the results also increases as the angle $'\theta'$ between normal to the plane & the line joining the given point to the centre increases even the factor of circularity of the given plane is very close to the unity (i.e. for $F_c \cong 1$).
5. Error involved in the results also increases as the distance 'r' of the given point from the centre of the plane decreases.
6. Error involved in the results decreases in case of the more symmetrical plane (i.e. for$F_c \cong 1$), small area of the given plane, large distance(r) & smaller angle $'\theta'$
7. This formula is not applicable for finding out the precise value of the solid angle subtended by the symmetrical plane at a certain point in the space.

3.5 Applications of Approximation-formula

Hereby, we are interested to study some particular symmetrical plane-figures such as

1.Rectangle 2. Rhombus 3. Regular polygon (like equilateral triangle, square, regular pentagon, regular hexagon etc.) 4. Circle 5. Ellipse in order to find out the solid angle subtended by the planes (mentioned above) at any point lying on the axis normal to the plane & passing through the centre as well as to find out their respective factors of circularity.

Note: Hereby for ease of analysis, all the plane-figures are viewed from the point lying on the normal axis passing through the centre of symmetry & the orientation of geometrical-

shape of plane is not changed i.e. only geometrical dimensions are changed.

3.5.1. Circular plane:

Let there be a circular plane having radius 'R' &centre 'O' and a given point say P(0, 0, h) at a height 'h' lying on the axis normal to the plane & passing through the centre 'O'.

(as shown in the figure(64) below)

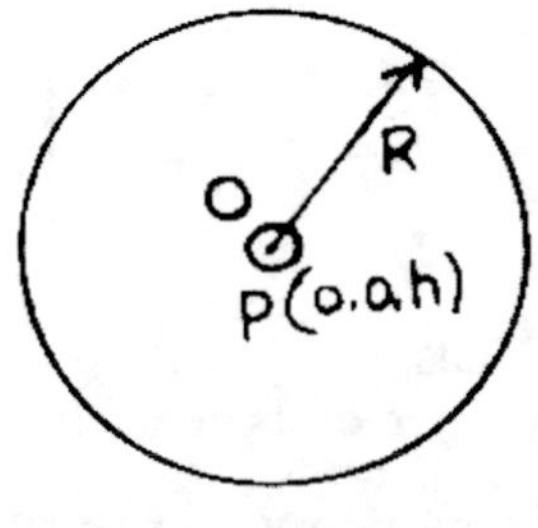

Fig (64)

Area of the circular plane $\Rightarrow A = \pi R^2$

Now, by using Approximation-formula, we get

The solid angle subtended by the circular plane at the given point 'P'

$$\Rightarrow \omega \cong 2\pi\left[1-\frac{1}{\sqrt{1+\left(\frac{A}{\pi r^2}\right)}}\right] = 2\pi\left[1-\frac{1}{\sqrt{1+\left(\frac{\pi R^2}{\pi r^2}\right)}}\right] (since, \theta = 0)$$

$$\Rightarrow \omega \cong 2\pi\left[1-\frac{1}{\sqrt{1+\left(\frac{R^2}{r^2}\right)}}\right] = 2\pi\left[1-\frac{r}{\sqrt{r^2+R^2}}\right] \quad \forall(r \geq 0) \quad or$$

$$\Rightarrow \omega = 2\pi\left[1-\frac{h}{\sqrt{h^2+R^2}}\right] \quad (equlity\ can\ be\ taken\ for\ circle\ (P.S.planefigure))$$

....................(3.1)

Where r = h = normal height of the given point from the centre of the circular plane.

The above result is same as obtained in the art.2.3.23.

Factor of circularity:

We know that the circle has infinite no. of axes of symmetry of equal length '2R', hence

The factor of circularity of the circular plane

$$\Rightarrow F_c = \frac{s}{l} = \frac{shortest\ axis\ of\ symmetry}{longest\ axis\ of\ symmetry} = \frac{2R}{2R} = 1$$

Hence, the circle is perfect symmetrical plane-figure. It is constant equal to unity iff the plane is viewed from any point lying on the normal axis passing through the centre 'O'.

3.5.2. Elliptical plane:

Let there be an elliptical plane having semi-major & semi-minor axesequal to 'a' & 'b' respectively $(a > b)$& the centre 'O' and a given point say P(0, 0, h) at a height 'h' lying on the axis normal to the plane & passing through the centre 'O'. (as shown in the figure(65) below)

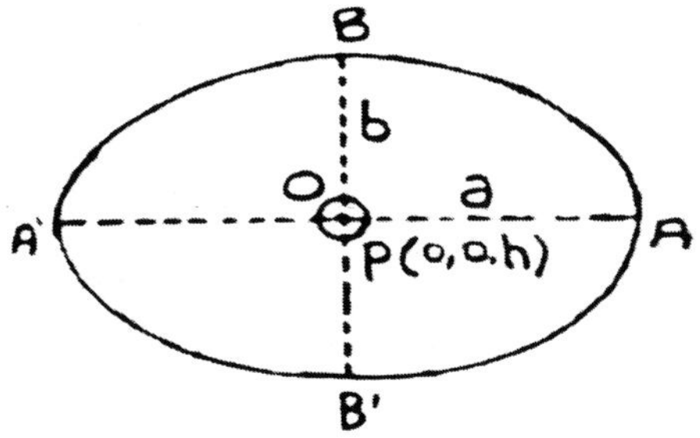

Fig (65)

We know that the area of the elliptical plane $\Rightarrow A = \pi ab$

Also, the relation among the semi-minor (b), semi-major (a) axes & the eccentricity of an ellipse is given as follows

$$e = \sqrt{1-\frac{b^2}{a^2}} \Rightarrow b^2 = a^2(1-e^2) \Rightarrow b = a\sqrt{1-e^2} \quad \forall(b < a) \ldots\ldots\ldots\ldots\ldots (I)$$

Now, by using Approximation-formula, we get

The solid angle subtended by the circular plane at the given point 'P'

$$\Rightarrow \omega \cong 2\pi\left[1-\frac{1}{\sqrt{1+\left(\frac{A}{\pi r^2}\right)}}\right] = 2\pi\left[1-\frac{1}{\sqrt{1+\left(\frac{\pi ab}{\pi r^2}\right)}}\right] (since, \theta = 0)$$

$$\Rightarrow \omega \cong 2\pi\left[1-\frac{1}{\sqrt{1+\left(\frac{ab}{r^2}\right)}}\right] \quad \forall(r\geq 0) \qquad \ldots\ldots\ldots(3.2)$$

Factor of circularity:

the dotted lines BB' & AA' are axes of symmetry of length '2b'& '2a' respectively, hence

The factor of circularity of the elliptical plane

$$F_c=\frac{s}{l}=\frac{shortest\ axis\ of\ symmetry}{longest\ axis\ of\ symmetry}=\frac{2b}{2a}=\frac{b}{a}=\frac{a\sqrt{1-e^2}}{a}=\sqrt{1-e^2}\quad (from\ eq(I))$$

$$\Rightarrow \left[F_c=\sqrt{1-e^2}\right] \qquad \ldots\ldots\ldots\ldots\ldots\ldots(3.3)$$

It is obvious that the factor of circularity of ellipse depends on the eccentricity (e) of the ellipse plane which varies as $0<F_c<1$.

Hence, for minimum permissible error the eccentricity of the ellipse should be as minimum as possible.There will be no error in the value of solid angle subtended by the elliptical planeiff e = 0 or $F_c=1$ i.e. the elliptical plane becomes a circular plane.

3.5.3 Regular polygonal plane:

Let there be a regular polygonal plane $A_1A_2A_3\ldots\ldots A_n$ having 'n' no. of the sides each equal to the length 'a' & a given point say P(0, 0, h) at a height 'h' lying on the axis normal to the plane & passing through the centre 'O'.

(as shown in the figure(66) below)

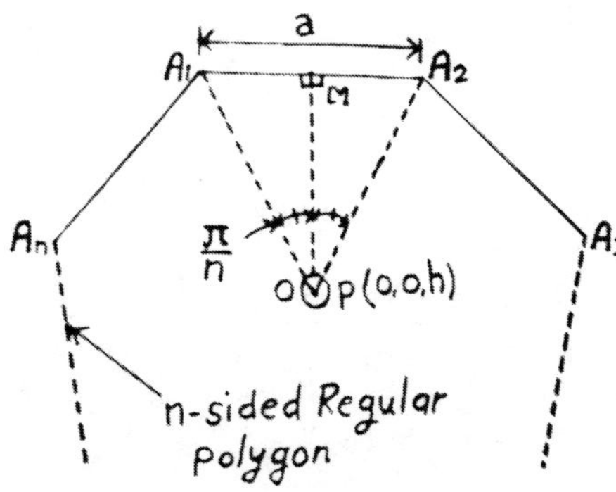

Fig (66)

Now, consider an isosceles $\Delta OA_1A_2$by joining the vertices A_1 &A_2 to the centre 'O' & drawing a perpendicular OM from the centre 'O' to the side A_1A_2

In rightΔOMA_1

$$\Rightarrow tan\left(\frac{\pi}{n}\right)=\frac{A_1M}{OM}=\frac{\left(\frac{a}{2}\right)}{OM}\Rightarrow OM=\frac{a}{2}cot\frac{\pi}{n}\ \&$$

$$\Rightarrow sin\left(\frac{\pi}{n}\right)=\frac{A_1M}{OA_1}=\frac{\left(\frac{a}{2}\right)}{OA_1}\Rightarrow OA_1=\frac{a}{2}cosec\frac{\pi}{n}$$

The area of isosceles ΔOA_1A_2

$$\Rightarrow A_1=\frac{1}{2}(base)(normal\ height)=\frac{1}{2}(A_1A_2)(OM)=\frac{1}{2}(a)\left(\frac{a}{2}cot\frac{\pi}{n}\right)=\frac{1}{4}a^2cot\frac{\pi}{n}$$

Since, the regular polygon can be divided into ‘n’ no. of the isosceles triangles which are congruent to isoscelesΔOA_1A_2.

Hence, the area (A) of the regular polygon

$$\Rightarrow A = n(A_1) = n\left(\frac{1}{4}a^2 cot\frac{\pi}{n}\right) = \frac{1}{4}na^2 cot\frac{\pi}{n}$$

Hence, the solid angle subtended by the regular polygon $A_1A_2A_3 \ldots\ldots A_n$ at the given point ‘P’ (by using Approximation-formula)

$$\Rightarrow \omega \cong 2\pi\left[1 - \frac{1}{\sqrt{1 + \left(\frac{A}{\pi r^2}\right)}}\right] = 2\pi\left[1 - \frac{1}{\sqrt{1 + \left(\frac{\frac{1}{4}na^2 cot\frac{\pi}{n}}{\pi r^2}\right)}}\right] (since, \theta = 0)$$

$$\Rightarrow \omega \cong 2\pi\left[1 - \frac{1}{\sqrt{1 + \left(\frac{na^2 cot\frac{\pi}{n}}{4\pi r^2}\right)}}\right] \forall (n \in N \,\&\, n \geq 3) \quad \ldots\ldots\ldots.(3.4)$$

Factor of Circularity:

In the above fig (66), the dotted lines OM &OA_1 are the shortest & the longest semi-axes of symmetry respectively, hence the factor of circularity of the regular polygon

$$\Rightarrow F_c = \frac{s}{l} = \frac{2OM}{2OA_1} = \frac{OM}{OA_1} = \frac{\left(\frac{a}{2}cot\frac{\pi}{n}\right)}{\left(\frac{a}{2}cosec\frac{\pi}{n}\right)} = \frac{cos\frac{\pi}{n}}{sin\frac{\pi}{n}cosec\frac{\pi}{n}} = cos\frac{\pi}{n}$$

$$\Rightarrow \left[F_c = cos\frac{\pi}{n}\right] \quad \forall (n \in N \,\&\, n \geq 3) \ \ldots\ldots\ldots\ldots (3.5)$$

(Given point (i.e. viewer) is located at normal axis passing through the centre of the plane.)

Thus, F_c is variable depending on the number (n) of sidesof the regular polygonal plane& independent of the length ‘a’ of the side.

It is obvious that the value of F_c of the regularpolygonal plane is equal to the cosine of no. of the sides (n) which varies as $0 < F_c < 1$.

Case: if$n \rightarrow \infty$then taking limit of F_c as follows

$$\Rightarrow F_c = \lim_{n\rightarrow\infty} cos\frac{\pi}{n} = cos(0) = 1$$

Thus, we find that the regular polygon having a certain circumscribed (or inscribed) radius &infinite of the sides is a Perfect Symmetrical plane-figure (P.S. plane-figure) according to the definition.

In such a case, the regular polygon becomes a circle.

Thus, the circle can be defined as “A circle is a regular-polygon witha finite radius & infinite no. of the sides.”

Thus, there will be minimum error in the approximate value of the solid angle subtended by the regular polygonal plane at a given point ‘P’ if $n \rightarrow \infty \Rightarrow F_c \rightarrow 1$

(i.e. when it becomes very likely to a circular plane)

Now, by setting the different values of 'n' we get different expressions of approximate value of solid angle subtended by the regular planes like

1.Equilateral-triangular plane, 2. Square plane, 3 Regular-pentagonal 4.Regular-hexagonal plane etc.as follows

3.5.3.1 Equilateral-triangular plane:

Let there be an equilateral-triangle planeABC having each side of length 'a' & a given point say P(0, 0, h) at a height 'h' lying on the axis normal to the plane & passing through the centre 'O'.

(as shown in the figure(67) below)

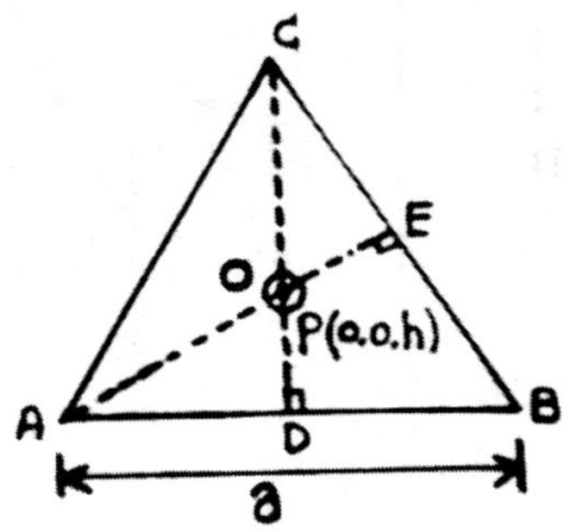

Fig (67)

Now, on setting n = 3 in the eq(3.4), we get

The solid angle subtended by the equilateral-triangular plane at the given point 'P'

$$\Rightarrow \omega \cong 2\pi\left[1-\frac{1}{\sqrt{1+\left(\frac{na^2\cot\frac{\pi}{n}}{4\pi r^2}\right)}}\right] = 2\pi\left[1-\frac{1}{\sqrt{1+\left(\frac{3a^2\cot\frac{\pi}{3}}{4\pi r^2}\right)}}\right]$$

$$\Rightarrow \boldsymbol{\omega \cong 2\pi\left[1-\frac{1}{\sqrt{1+\left(\frac{3a^2\frac{1}{\sqrt{3}}}{4\pi r^2}\right)}}\right] = 2\pi\left[1-\frac{1}{\sqrt{1+\left(\frac{\sqrt{3}a^2}{4\pi r^2}\right)}}\right]}$$

Factor of Circularity:

Now, on setting n = 3 in the eq(3.5), we get

Factor of circularity of the equilateral-triangular plane

$$\Rightarrow F_c = \cos\frac{\pi}{n} = \cos\frac{\pi}{3} = \frac{1}{2} = 0.5$$

(Given point (i.e. viewer) is located at normal axis passing through the centre of the plane.)

3.5.3.2 Square plane:

Let there be a square plane ABCD having each side of length 'a' & a given point say P(0, 0, h) at a height 'h' lying on the axis normal to the plane & passing through the centre 'O'.

(as shown in the figure(68) below)

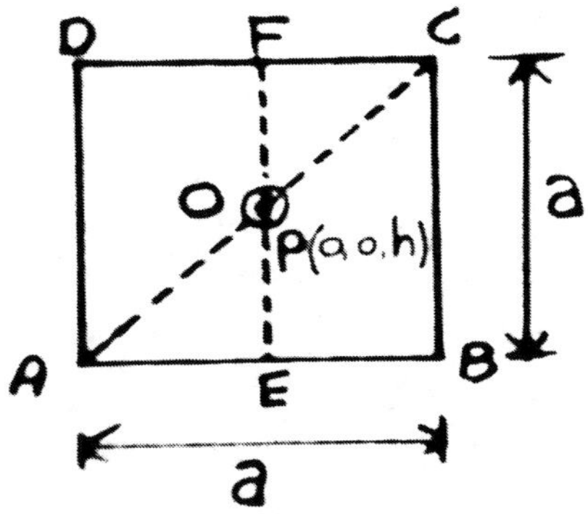

Fig (68)

Now, on setting n = 4 in the eq(3.4), we get

The solid angle subtended by the square plane at the given point 'P'

$$\Rightarrow \omega \cong 2\pi\left[1-\frac{1}{\sqrt{1+\left(\frac{na^2\cot\frac{\pi}{n}}{4\pi r^2}\right)}}\right]=2\pi\left[1-\frac{1}{\sqrt{1+\left(\frac{4a^2\cot\frac{\pi}{4}}{4\pi r^2}\right)}}\right]$$

$$\Rightarrow \boldsymbol{\omega \cong 2\pi\left[1-\frac{1}{\sqrt{1+\left(\frac{4a^2(1)}{4\pi r^2}\right)}}\right]=2\pi\left[1-\frac{1}{\sqrt{1+\left(\frac{a^2}{\pi r^2}\right)}}\right]}$$

Factor of Circularity:

Now, on setting n = 4 in the eq(3.5), we get

Factor of circularity of the square plane

$$\Rightarrow F_c = \cos\frac{\pi}{n} = \cos\frac{\pi}{4} = \frac{1}{\sqrt{2}} = 0.707$$

(Given point (i.e. viewer) is located at normal axis passing through the centre of the plane.)

3.5.3.3 Regular-pentagonal plane:

Let there be an regular-pentagonal plane ABCDE having each side of length 'a' & a given point say P(0, 0, h) at a height 'h' lying on the axis normal to the plane & passing through the centre 'O'.

(as shown in the figure(69) below)

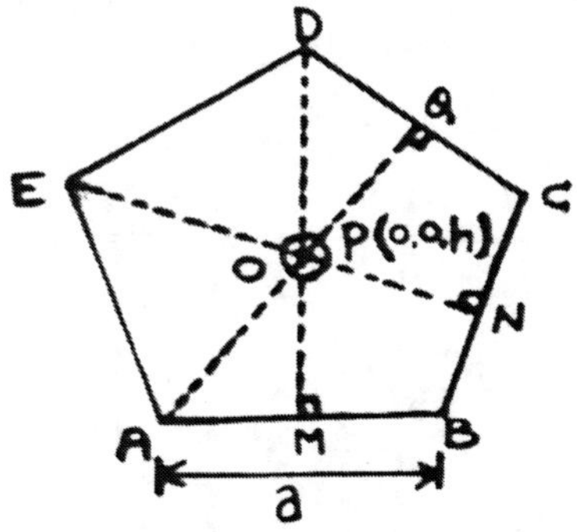

Fig (69)

Now, on setting n = 5 in the eq(3.4), we get

The solid angle subtended by the regular-pentagonal plane at the given point 'P'

$$\Rightarrow \omega \cong 2\pi\left[1-\frac{1}{\sqrt{1+\left(\frac{na^2 cot\frac{\pi}{n}}{4\pi r^2}\right)}}\right] = 2\pi\left[1-\frac{1}{\sqrt{1+\left(\frac{5a^2 cot\frac{\pi}{5}}{4\pi r^2}\right)}}\right]$$

$$\Rightarrow \boldsymbol{\omega} \cong \mathbf{2}\boldsymbol{\pi}\left[\mathbf{1}-\frac{\mathbf{1}}{\sqrt{\mathbf{1}+\left(\frac{\mathbf{5}\boldsymbol{a}^{\mathbf{2}}(\mathbf{1.376})}{\mathbf{4}\boldsymbol{\pi}\boldsymbol{r}^{\mathbf{2}}}\right)}}\right] = \mathbf{2}\boldsymbol{\pi}\left[\mathbf{1}-\frac{\mathbf{1}}{\sqrt{\mathbf{1}+\left(\frac{\mathbf{1.72}\boldsymbol{a}^{\mathbf{2}}}{\boldsymbol{\pi}\boldsymbol{r}^{\mathbf{2}}}\right)}}\right]$$

Factor of Circularity:

Now, on setting n = 5 in the eq(3.5), we get

Factor of circularity of the regular-pentagonal plane

$$\Rightarrow F_c = cos\frac{\pi}{n} = cos\frac{\pi}{5} = 0.809$$

(Given point (i.e. viewer) is located at normal axis passing through the centre of the plane.)

3.5.3.4 Regular-hexagonal plane:

Let there be an regular-hexagonal plane ABCDEF having each side of length 'a' & a given point say P(0, 0, h) at a height 'h' lying on the axis normal to the plane & passing through the centre 'O'.

(as shown in the figure(70) below)

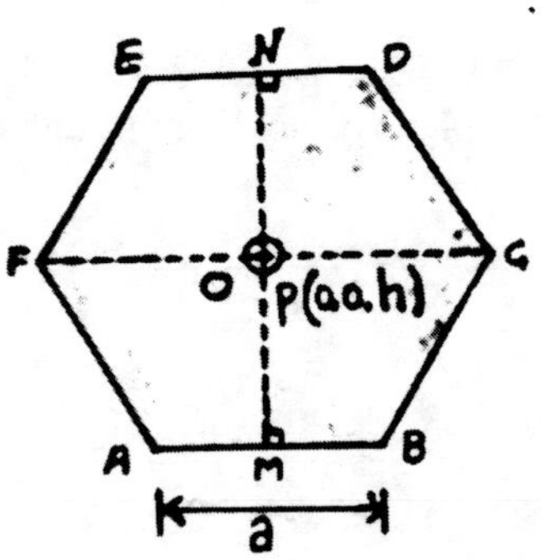

Fig (70)

Now, on setting n = 6 in the eq(3.4), we get

The solid angle subtended by the regular-hexagonal plane at the given point 'P'

$$\Rightarrow \omega \cong 2\pi\left[1-\frac{1}{\sqrt{1+\left(\frac{na^2\cot\frac{\pi}{n}}{4\pi r^2}\right)}}\right]=2\pi\left[1-\frac{1}{\sqrt{1+\left(\frac{6a^2\cot\frac{\pi}{6}}{4\pi r^2}\right)}}\right]$$

$$\Rightarrow \boldsymbol{\omega} \cong \mathbf{2\pi}\left[\mathbf{1}-\frac{\mathbf{1}}{\sqrt{\mathbf{1}+\left(\frac{\mathbf{6a^2\sqrt{3}}}{\mathbf{4\pi r^2}}\right)}}\right]=\mathbf{2\pi}\left[\mathbf{1}-\frac{\mathbf{1}}{\sqrt{\mathbf{1}+\left(\frac{\mathbf{2.598a^2}}{\mathbf{\pi r^2}}\right)}}\right]$$

Factor of Circularity:

Now, on setting n = 6 in the eq(3.5), we get

Factor of circularity of the regular-hexagonal plane

$$\Rightarrow F_c = cos\frac{\pi}{n} = cos\frac{\pi}{6} = 0.866$$

(Given point (i.e. viewer) is located at normal axis passing through the centre of the plane.)

Thus, further by setting n = 7, 8, 9, 10, ... upto n, we can obtain a no. of the expressions by following the same procedure as done above.

3.5.4. Rectangular plane:

Let there be a rectangular plane ABCD having length 'l', width 'b' ($\forall\, l > b > 0$)&the centre 'O' and a given point say P(0, 0, h) at a height 'h' lying on the normal axis passing through centre 'O' of the plane.

(as shown in the figure(71) below)

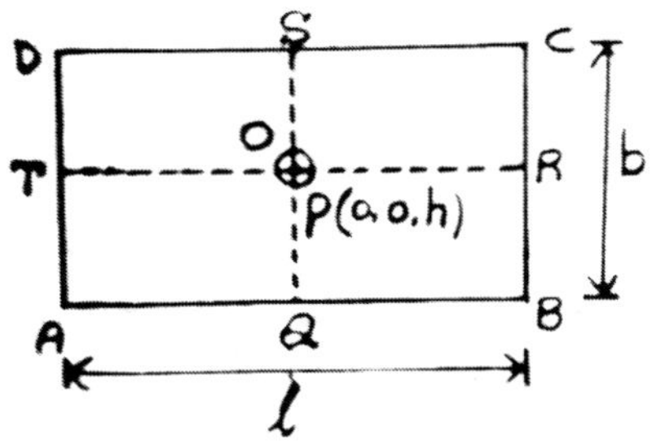

Fig (71)

Area of the rectangular plane ABCD $\Rightarrow A = lb$

Using Approximation-formula, solid angle subtended by the rectangular plane ABCD at the point 'P'

$$\Rightarrow \omega \cong 2\pi\left[1-\frac{1}{\sqrt{1+\left(\frac{A}{\pi r^2}\right)}}\right]=2\pi\left[1-\frac{1}{\sqrt{1+\left(\frac{lb}{\pi r^2}\right)}}\right](since, \theta = 0) \quad \ldots\ldots\ldots\ldots(3.7)$$

Factor of Circularity:

In the above fig (71), the dotted lines SQ & TR are the shortest & the longest axes of symmetry respectively, but the square is most symmetrical form of a rectangle i.e. factor of circularity of the rectangle $0 < F_c < 0.707$. Hence the factor of circularity of the rectangle

$$\Rightarrow F_c = \frac{s}{l} = \frac{smaller\ side}{diagonal} = \frac{AD}{AC} = \frac{b}{\sqrt{b^2 + l^2}}$$

$$\Rightarrow \left[F_c = \frac{b}{\sqrt{b^2 + l^2}}\right] \qquad \forall (l > b) \quad \ldots\ldots\ldots\ldots\ldots (3.8)$$

(Given point (i.e. viewer) is located at normal axis passing through the centre of the plane.)

Thus, F_c is variable depending on the length & the width of the rectangular plane.

It is obvious that the factor of circularity of the rectangle will be maximum

$$iff\ l = b \Rightarrow (F_c)_{max} = \frac{b}{\sqrt{b^2 + b^2}} = \frac{b}{b\sqrt{2}} = \frac{1}{\sqrt{2}} = 0.707$$

It is obvious that the value of F_c of the rectangular plane is equal to the ratio of width to diagonal which varies as $0 < F_c < 0.707$.

Thus, there will be minimum error in the approximate value of the solid angle subtended by the rectangular plane at a given point 'P' iff$F_c = 0.707$ (i.e. when it becomes a square plane (l = b))

3.5.5 Rhombus-like plane

Let there be a rhombus-like plane ABCD having diagonals of length $'2d_1'$ & $'2d_2'$

($\forall\ d_2 > d_1 > 0$)& centre 'O' and a given point say P(0, 0, h) at a height 'h' lying on the normal axis passing through centre 'O' of the plane. (as shown in the figure(72) below)

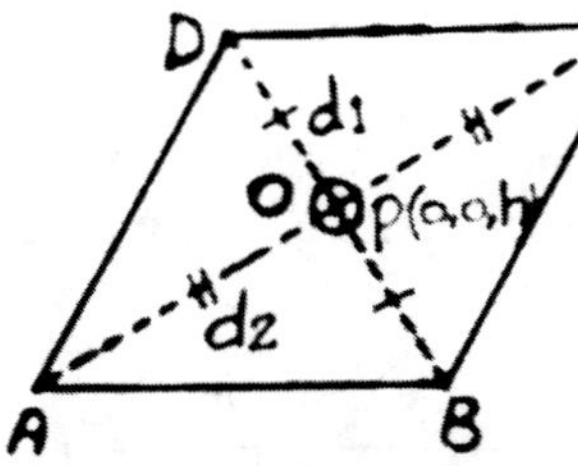

Fig (72)

Area of the rhombus-like plane ABCD

$$\Rightarrow A = 4(Area\ of\ right\ \Delta AOB) = 4\left(\frac{1}{2}d_1 d_2\right) = 2d_1 d_2$$

$$(since, \qquad \Delta AOB, \Delta AOC, \Delta COD\ \&\ \Delta AOD\ are\ congruent\ right\ triangles)$$

$$\Rightarrow in\ right\ \Delta AOB\ \Rightarrow AB = \sqrt{OB^2 + OA^2} = \sqrt{{d_1}^2 + {d_2}^2}$$

Now, using Approximation-formula, solid angle subtended by the rhombus-like plane ABCD at the given point 'P'

$$\Rightarrow \omega \cong 2\pi\left[1 - \frac{1}{\sqrt{1 + \left(\frac{A}{\pi r^2}\right)}}\right] = 2\pi\left[1 - \frac{1}{\sqrt{1 + \left(\frac{2d_1 d_2}{\pi r^2}\right)}}\right] (since, \theta = 0)$$

$$\Rightarrow \omega \cong 2\pi\left[1-\frac{1}{\sqrt{1+\left(\frac{2d_1d_2}{\pi r^2}\right)}}\right] \quad \ldots\ldots\ldots\ldots\ldots (3.9)$$

Factor of Circularity:

In the above fig (72), the dotted lines BD & AC are the shortest & the longest axes of symmetry respectively, but the square is most symmetrical form of a rhombus

i.e. factor of circularity of the rhombus $0 < F_c < 0.707$. Hence the factor of circularity of the rhombus

$$\Rightarrow F_c = \frac{s}{l} = \frac{smaller\ semidiagonal}{side} = \frac{OB}{AB} = \frac{d_1}{\sqrt{{d_1}^2+{d_2}^2}}$$

$$\Rightarrow \left[F_c = \frac{d_1}{\sqrt{{d_1}^2+{d_2}^2}}\right] \quad \forall(d_2 > d_1) \quad \ldots\ldots\ldots\ldots\ldots (3.10)$$

(Given point (i.e. viewer) is located at normal axis passing through the centre of the plane.)

Thus, F_c is variable depending on the length of the diagonals of the rhombus-like plane.

It is obvious that the factor of circularity of the rhombus will be maximum

$$iff\ d_1 = d_2 \Rightarrow (F_c)_{max} = \frac{d_2}{\sqrt{{d_2}^2+{d_2}^2}} = \frac{d_2}{d_2\sqrt{2}} = \frac{1}{\sqrt{2}} = 0.707$$

It is obvious that the value of F_c of the rhombus-like plane is equal to the ratio of smaller semi-diagonal to the side which varies as $0 < F_c < 0.707$.

Thus, there will be minimum error in the approximate value of the solid angle subtended by the rhombus-like plane at a given point 'P' iff $F_c = 0.707$ (i.e. when it becomes a square plane $(d_1 = d_2)$)

3.5.6 Solid angle subtended by a circular plane at any point in the space

Let there be a circular plane with the centre 'O' & radius 'R' and a given point say 'P' at a distance 'd' from the centre 'O' such that the angle between normal through the centre 'O' of the plane & the line OP joining the given point 'P' to the centre 'O' is $'\theta'$

(as shown in the figure (73) below)

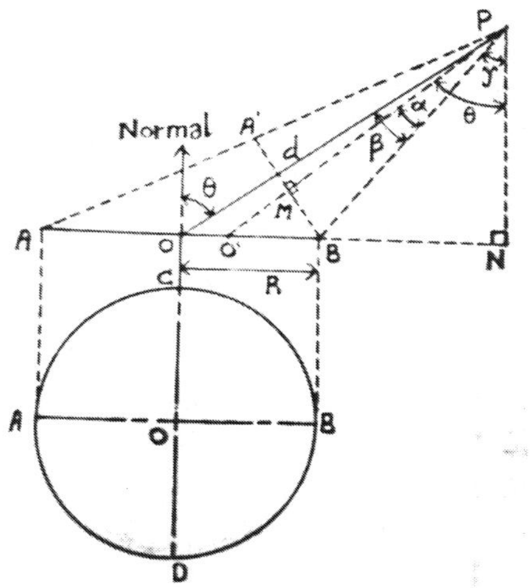

Fig (73)

$$in\ right\Delta ONP \Rightarrow sin\theta = \frac{ON}{OP} = \frac{ON}{d} \Rightarrow ON = dsin\theta\ \&\ cos\theta = \frac{PN}{OP} = \frac{PN}{d} \Rightarrow PN = dcos\theta$$

$$\Rightarrow BN = ON - OB = dsin\theta - R$$

$$in\ right\Delta PMB$$

$$\Rightarrow PB = \sqrt{PM^2 + MB^2} = \sqrt{h^2 + b^2} = k\ (let)\ where\ PM = h\ \&\ MB = A'M = b$$

$$\Rightarrow sin\ \ BPM = \frac{BM}{BP} \Rightarrow sin\alpha = \frac{b}{k}\ \&\ cos\ \ BPM = \frac{PM}{BP} \Rightarrow cos\alpha = \frac{h}{k}$$

$$in\ right\Delta PNB$$

$$\Rightarrow BN^2 + PN^2 = PB^2$$

$$\Rightarrow (dsin\theta - R)^2 + (dcos\theta)^2 = d^2sin^2\theta + R^2 - 2dRsin\theta + d^2cos^2\theta$$

$$\Rightarrow d^2 + R^2 - 2dRsin\theta = k^2 = h^2 + b^2 \ldots\ldots\ldots\ldots\ldots\ldots\ldots.(I)$$

$$\Rightarrow sin\ \ BPN = \frac{BN}{BP} \Rightarrow sin\gamma = \frac{dsin\theta - R}{k}\ \&\ cos\ \ BPM = \frac{PN}{BP} \Rightarrow cos\gamma = \frac{dcos\theta}{k}$$

In ΔAOP applying Sine-rule as follows

$$\frac{sin\ \ OPA}{AO} = \frac{sin\ \ AOP}{AP}$$

$$\Rightarrow \frac{sin\left(\alpha - (\theta - (\alpha + \gamma))\right)}{R} = \frac{sin\left(\frac{\pi}{2} + \theta\right)}{\sqrt{PN^2 + AN^2}} = \frac{cos\theta}{\sqrt{(dcos\theta)^2 + (R + dsin\theta)^2}}$$

$$\Rightarrow sin\big((2\alpha + \gamma) - \theta\big) = \frac{Rcos\theta}{\sqrt{d^2cos^2\theta + R^2 + d^2sin^2\theta + 2dRsin\theta}} = \frac{Rcos\theta}{\sqrt{d^2 + R^2 + 2dRsin\theta}}$$

$$sin\big((2\alpha + \gamma) - \theta\big) = l\ (let)\quad where\ l = \frac{Rcos\theta}{\sqrt{d^2 + R^2 + 2dRsin\theta}} \quad \ldots\ldots\ldots\ldots..(II)$$

$$\Rightarrow sin(2\alpha + \gamma)cos\theta - cos(2\alpha + \gamma)sin\theta = l$$

$$\Rightarrow (sin2\alpha cos\gamma + cos2\alpha sin\gamma)cos\theta - (cos2\alpha cos\gamma - sin2\alpha sin\gamma)sin\theta = l$$

$$\Rightarrow \big((2sin\alpha cos\alpha)cos\gamma + (2cos^2\alpha - 1)sin\gamma\big)cos\theta - \big((2cos^2\alpha - 1)cos\gamma - (2sin\alpha cos\alpha)sin\gamma\big)sin\theta = l$$

$$\Rightarrow (2sin\alpha cos\alpha cos\gamma + 2cos^2\alpha sin\gamma - sin\gamma)cos\theta - (2cos^2\alpha cos\gamma - cos\gamma - 2sin\alpha cos\alpha sin\gamma)sin\theta = l$$

Now on setting the corresponding values, we get

$$\Rightarrow \left\{2\left(\frac{b}{k}\right)\left(\frac{h}{k}\right)\left(\frac{dcos\theta}{k}\right) + 2\left(\frac{h}{k}\right)^2\left(\frac{dsin\theta - R}{k}\right) - \left(\frac{dsin\theta - R}{k}\right)\right\}cos\theta$$

$$-\left\{2\left(\frac{h}{k}\right)^2\left(\frac{dcos\theta}{k}\right) - \left(\frac{dcos\theta}{k}\right) - 2\left(\frac{b}{k}\right)\left(\frac{h}{k}\right)\left(\frac{dsin\theta - R}{k}\right)\right\}sin\theta = l$$

$$\Rightarrow \left(\frac{2(dsin\theta - R)cos\theta}{k^3} - \frac{2dsin\theta cos\theta}{k^3}\right)h^2 + b\left(\frac{2dcos^2\theta}{k^3} + \frac{2(dsin\theta - R)sin\theta}{k^3}\right)h$$

$$+\left(\frac{dsin\theta cos\theta + Rcos\theta - dsin\theta cos\theta}{k^3}\right) = l$$

$$\Rightarrow -\left(\frac{2Rcos\theta}{k^3}\right)h^2 + 2b\left(\frac{d - Rsin\theta}{k^3}\right)h + \left(\frac{Rcos\theta}{k} - l\right) = 0$$

Now, let

$$A = -2\left(\frac{R\cos\theta}{k^3}\right) \quad B = 2\left(\frac{d - R\sin\theta}{k^3}\right)$$
$$\& \ C = \left(\frac{R\cos\theta}{k} - l\right)$$

Thus, we get

$$\Rightarrow Ah^2 + Bbh + C = 0 \Rightarrow Ah^2 + C = -Bbh \ (on\ squaring\ both\ the\ sides)$$

$$\Rightarrow (Ah^2 + C)^2 = (Bbh)^2 \Rightarrow A^2h^4 + C^2 + 2ACh^2 = B^2b^2h^2$$

$$\Rightarrow A^2h^4 + C^2 + 2ACh^2 = B^2h^2(k^2 - h^2) = B^2k^2h^2 - B^2h^4 \quad (from\ eq(I))$$

$$\Rightarrow (A^2+B^2)h^4 + (2AC - B^2k^2)h^2 + C^2 = 0 \quad \ldots\ldots\ldots\ldots\ldots\ldots. (III)$$

$$where, k = \sqrt{d^2 + R^2 - 2dR\sin\theta} \ \& \ l = \frac{R\cos\theta}{\sqrt{d^2 + R^2 + 2dR\sin\theta}} \qquad from\ eq(I) \& (II)$$

The above equation is biquadratic in terms of variable 'h'. It is called Auxiliary Equation.

In this case, the projection of the circular plane in the direction of line OP at an angle $'\theta'$ with the normal through the centre 'O' of the circular plane will be an elliptical plane having centre 'M', major axis CD = diameter = 2R & minor axis A'B = 2b (let).

(as shown in the figure (74) below)

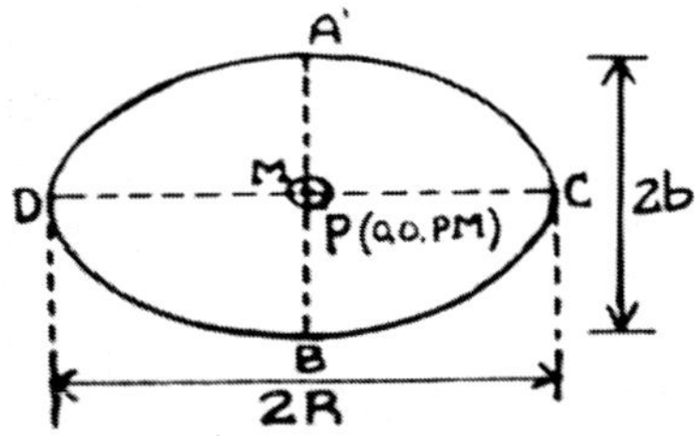

Fig (74)

In this case, the solid angle subtended by the given circular plane at the point 'P'

= solid angle subtended by the elliptical plane of projection of the circular plane at the same point 'P'

Now, by using Approximation-formula for the elliptical plane

$$\Rightarrow \omega \cong 2\pi\left[1 - \frac{1}{\sqrt{1 + \left(\frac{ab}{r^2}\right)}}\right] \quad from\ eq(3.2)$$

On setting the values of $a = DM = R, b = \sqrt{k^2 - h^2} \ \& \ r = PM = h \qquad from\ eq(I)$

Hence, we get

$$\omega \cong 2\pi\left[1 - \frac{1}{\sqrt{1 + \left(\frac{R\sqrt{k^2 - h^2}}{h^2}\right)}}\right] \qquad (\ on\ setting\ the\ value\ of\ k^2\ from\ eq(I)\ )$$

We have

$$\Rightarrow \omega \cong 2\pi\left[1-\frac{1}{\sqrt{1+\left(\frac{R\sqrt{d^2+R^2-2dRsin\theta}-h^2}{h^2}\right)}}\right] \quad \text{……………….(3.11)}$$

The value of h^2 is found out from the auxiliary equation (III) given as follows

$$\Rightarrow [(A^2+B^2)h^4+(2AC-B^2k^2)h^2+C^2=0]$$

Where

$$A=-2\left(\frac{Rcos\theta}{(d^2+R^2-2dRsin\theta)^{\frac{3}{2}}}\right), \quad B=2\left(\frac{d-Rsin\theta}{(d^2+R^2-2dRsin\theta)^{\frac{3}{2}}}\right)$$

$$C=Rcos\theta\left(\frac{1}{\sqrt{d^2+R^2-2dRsin\theta}}-\frac{1}{\sqrt{d^2+R^2+2dRsin\theta}}\right) \,\&\, k=\sqrt{d^2+R^2-2dRsin\theta}$$

Important deduction:

If the given point is lying on the normal axis passing through the centre of the circular plane then the solid angle subtended by the circular plane at the same point is obtained by setting $\theta=0$ in the above eq(3.11) as follows

$$\omega \cong 2\pi\left[1-\frac{1}{\sqrt{1+\left(\frac{R\sqrt{d^2+R^2-2dRsin0}-h^2}{h^2}\right)}}\right]=2\pi\left[1-\frac{1}{\sqrt{1+\left(\frac{R\sqrt{d^2+R^2}-h^2}{h^2}\right)}}\right] \quad \text{…………………..(IV)}$$

The value of h^2 is calculated as follows

$$A=-2\left(\frac{Rcos0}{(d^2+R^2-2dRsin0)^{\frac{3}{2}}}\right)=-2\left(\frac{R}{(d^2+R^2)^{\frac{3}{2}}}\right)$$

$$B=2\left(\frac{d-Rsin0}{(d^2+R^2-2dRsin0)^{\frac{3}{2}}}\right)=2\left(\frac{d}{(d^2+R^2)^{\frac{3}{2}}}\right)\,\&$$

$$C=Rcos(0)\left(\frac{1}{\sqrt{d^2+R^2-2dRsin0}}-\frac{1}{\sqrt{d^2+R^2+2dRsin0}}\right)$$
$$=R\left(\frac{1}{\sqrt{d^2+R^2}}-\frac{1}{\sqrt{d^2+R^2}}\right)=0\,\&$$

$$k=\sqrt{d^2+R^2-2dRsin0}=k=\sqrt{d^2+R^2}$$

Now, on setting the above values in the auxiliary equation, we have

$$(A^2+B^2)h^4+(2AC-B^2k^2)h^2+C^2=0$$

$$\Rightarrow\left(4\left(\frac{R}{(d^2+R^2)^{\frac{3}{2}}}\right)^2+4\left(\frac{d}{(d^2+R^2)^{\frac{3}{2}}}\right)^2\right)h^4$$
$$+\left(-4\left(\frac{R}{(d^2+R^2)^{\frac{3}{2}}}\right)(0)-4\left(\frac{d}{(d^2+R^2)^{\frac{3}{2}}}\right)^2\left(\sqrt{d^2+R^2}\right)^2\right)h^2+(0)^2=0$$

$$\Rightarrow \frac{4(d^2+R^2)}{(d^2+R^2)^3}h^4 - \frac{4d^2(d^2+R^2)}{(d^2+R^2)^3}h^2 = \frac{4}{(d^2+R^2)^2}(h^4 - d^2h^2) = 0$$

$$\Rightarrow (h^2 - d^2)h^2 = 0 \Rightarrow h^2 = d^2 \& h^2 = 0$$

Thus, two cases are arisen as follows

Case 1: When the given point is lying on the normal axis through the centre at a positive distance (i.e. $d > 0 \Rightarrow h^2 = d^2$)

Now, on setting $h^2 = d^2$in the above expression (IV), we get

The solid angle subtended by the circular plane at any point lying on the normal axis passing through the centre of the plane

$$\Rightarrow \omega \cong 2\pi\left[1 - \frac{1}{\sqrt{1+\left(\frac{R\sqrt{d^2+R^2-d^2}}{d^2}\right)}}\right] = 2\pi\left[1 - \frac{1}{\sqrt{1+\left(\frac{R^2}{h^2}\right)}}\right] = 2\pi\left[1 - \frac{h}{\sqrt{h^2+R^2}}\right]$$

$$\Rightarrow \omega \cong 2\pi\left[1 - \frac{h}{\sqrt{h^2+R^2}}\right] \quad or$$

$$\Rightarrow \omega = 2\pi\left[1 - \frac{h}{\sqrt{h^2+R^2}}\right] \quad (the\ equlity\ can\ be\ taken\ for\ circle\ (P.S.plane figure))$$

The above result is same as obtained in the art.2.3.23.

Case 2: When the given point is lying on the plane completely inside the boundary (i.e. $d = 0 \Rightarrow h^2 = 0$)

Now, on setting $h^2 = 0$in the above expression (IV), we get

The solid angle subtended by the circular plane at any point lying on the normal axis passing through the centre of the plane

$$\Rightarrow \omega \cong 2\pi\left[1 - \frac{1}{\sqrt{1+\left(\frac{R\sqrt{d^2+R^2-(0)^2}}{(0)^2}\right)}}\right] = 2\pi\left[1 - \frac{1}{\sqrt{1+\infty}}\right] = 2\pi[1-0]$$

$$\Rightarrow \omega \cong 2\pi \ or$$

$$\omega = 2\pi\ sr\ \ (since, the\ plane\ sees\ itself\ P.S.\ all\ around\ inside\ its\ boundary\ )$$

Thus, we find that the equality in the Approximation formula is taken in the above two cases only for a circular plane.

Projection of elliptical plane:

So far, we have studied that when a circular plane is observed from an any point (except from any point lying on the vertical axis passing through centre of the circular plane), it appears as an elliptical plane with certain eccentricity (e) depending on the radius of circular plane.

It is because when a straight line of finite length is observed from any point at a finite distance on any axis inclined with the same line & lies in a vertical plane passing through the given line generally appears shorter than it appears from the same point on the same axis normal to the same straight line.

On the contrary, an elliptical plane appears as a perfect circular plane from a point at any distance in a certain direction in a vertical plane passing through the major axis of the same plane. It is only because the projection (i.e. the length) of major axis becomes equal to that of minor axis at a certain angular position.

An elliptical plane appears as a more eccentric elliptical plane when it is observed from a point at any distance on the inclined axis passing through the centre & lies in a vertical plane passing through the minor axis of the same plane. It is only because the projection (i.e. the length) of minor axis becomes shorter than it is observed from the same point on the same axis normal to the plane.

Here, we are to study the case when a given elliptical plane appears as a perfect circular plane being observed from a certain angular position. In such a case, the exact (accurate) value of solid angle subtended by an elliptical plane at the certain point is calculated.

3.5.7 Solid angle subtended by an elliptical plane at a particular point from which it appears as a perfect circular plane

Let there be an elliptical plane with centre 'O', major axis AB = 2a & minor axis CD = 2b and a given point (i.e. observer) say 'P' lying in the vertical plane passing through the major axis at a distance d from the centre 'O' such that it appears as a perfect circular plane when the angle between the line OP & normal at the centre 'O' of the plane is $'\theta'$(as shown in the figure (75) below)

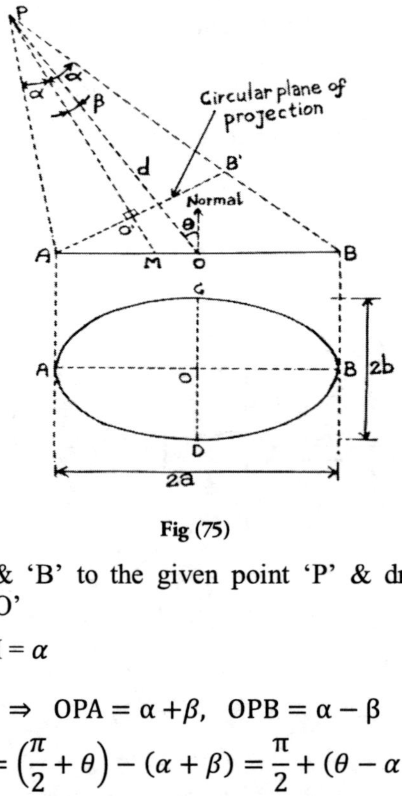

Fig (75)

Now join the end points 'A' & 'B' to the given point 'P' & draw a line AB' from the point 'A' intersecting the bisector PM at point O'

let OPM $= \beta$& APM = BPM $= \alpha$

$$\Rightarrow \text{OPA} = \alpha + \beta, \quad \text{OPB} = \alpha - \beta$$

$$\text{OAP} = \left(\frac{\pi}{2} + \theta\right) - (\alpha + \beta) = \frac{\pi}{2} + (\theta - \alpha - \beta)\&$$

$$\text{OBP} = \left(\frac{\pi}{2} - \theta\right) - (\alpha - \beta) = \frac{\pi}{2} - (\theta + \alpha - \beta)$$

$$in\, right\, \Delta AO'P$$

$$\Rightarrow sin\alpha = \frac{AO'}{AP} = \frac{b}{AP} \Rightarrow AP = bcosec\alpha \quad \left(since, AO' = O'B' = \frac{AB'}{2} = b\right)$$

Using Sine rule $in\ \Delta AOP$

$$\frac{\sin\ \text{OPA}}{AO} = \frac{\sin\ \text{OAP}}{OP} = \frac{\sin\ \text{AOP}}{AP}$$

$$\Rightarrow \frac{\sin(\alpha+\beta)}{a} = \frac{\sin\left(\frac{\pi}{2} + (\theta - \alpha - \beta)\right)}{d} = \frac{\sin\left(\frac{\pi}{2} - \theta\right)}{b\,cosec\alpha}$$

$$\Rightarrow \frac{\sin(\alpha+\beta)}{a} = \frac{\cos\left(\theta - (\alpha+\beta)\right)}{d} = \frac{\cos\theta\sin\alpha}{b}$$

$$when\ \frac{\sin(\alpha+\beta)}{a} = \frac{\cos\left(\theta - (\alpha+\beta)\right)}{d}$$

$$\Rightarrow \frac{\cos\theta\cos(\alpha+\beta) + \sin\theta\sin(\alpha+\beta)}{\sin(\alpha+\beta)} = \frac{d}{a} (since, \cos(A-B) = \cos A\cos B + \sin A\sin B)$$

$$\cos\theta\cot(\alpha+\beta) + \sin\theta = \frac{d}{a} \quad \ldots\ldots\ldots\ldots\ldots\ldots.(I)$$

$$when\ \frac{\sin(\alpha+\beta)}{a} = \frac{\cos\theta\sin\alpha}{b} \Rightarrow \sin(\alpha+\beta) = \frac{a\sin\alpha\cos\theta}{b} \quad \ldots\ldots\ldots\ldots..(II)$$

Using Sine rule *in* ΔBOP

$$\frac{\sin\ \text{OPB}}{OB} = \frac{\sin\ \text{OBP}}{OP} \Rightarrow \frac{\sin(\alpha-\beta)}{a} = \frac{\sin\left(\frac{\pi}{2} - (\theta + \alpha - \beta)\right)}{d}$$

$$\Rightarrow \frac{\sin(\alpha-\beta)}{a} = \frac{\cos\left(\theta + (\alpha-\beta)\right)}{d}$$

$$\frac{\cos\theta\cos(\alpha-\beta) - \sin\theta\sin(\alpha-\beta)}{\sin(\alpha-\beta)} = \frac{d}{a} (since, \cos(A+B) = \cos A\cos B - \sin A\sin B)$$

$$\cos\theta\cot(\alpha-\beta) - \sin\theta = \frac{d}{a} \quad \ldots\ldots\ldots\ldots\ldots\ldots.(III)$$

Now, on adding (I) & (III), we get

$$\Rightarrow \cos\theta\cot(\alpha+\beta) + \sin\theta + \cos\theta\cot(\alpha-\beta) - \sin\theta = \frac{2d}{a}$$

$$\Rightarrow \cos\theta\left(\cot(\alpha-\beta) + \cot(\alpha+\beta)\right) = \cos\theta\left(\frac{\cos(\alpha-\beta)}{\sin(\alpha-\beta)} + \frac{\cos(\alpha+\beta)}{\sin(\alpha+\beta)}\right) = \frac{2d}{a}$$

$$\Rightarrow \cos\theta\left(\frac{\sin(\alpha+\beta)\cos(\alpha-\beta) + \cos(\alpha+\beta)\sin(\alpha-\beta)}{\sin(\alpha-\beta)\sin(\alpha+\beta)}\right) = \frac{2d}{a}$$

$$\Rightarrow \cos\theta\left(\frac{\sin(\alpha+\beta+\alpha-\beta)}{\sin(\alpha-\beta)\sin(\alpha+\beta)}\right) = \frac{2d}{a} (since, \sin A\cos B + \cos A\sin B = \sin(A+B))$$

$$\Rightarrow \frac{\sin 2\alpha\cos\theta}{\sin(\alpha-\beta)\sin(\alpha+\beta)} = \frac{2\sin\alpha\cos\alpha\cos\theta}{\sin(\alpha-\beta)\sin(\alpha+\beta)} = \frac{2d}{a}$$

$$\Rightarrow \frac{2\sin\alpha\cos\alpha\cos\theta}{\sin(\alpha-\beta)\left(\frac{a\sin\alpha\cos\theta}{b}\right)} = \frac{2d}{a} \quad (on\ setting\ the\ value\ of\ \sin(\alpha+\beta)\ from (II))$$

$$\sin(\alpha-\beta) = \frac{b\cos\alpha}{d} \Rightarrow \cot(\alpha-\beta) = \frac{\cos(\alpha-\beta)}{\sin(\alpha-\beta)} = \frac{\sqrt{1-\sin^2(\alpha-\beta)}}{\sin(\alpha-\beta)}$$

$$\Rightarrow \cot(\alpha-\beta) = \frac{\sqrt{1-\left(\frac{b\cos\alpha}{d}\right)^2}}{\left(\frac{b\cos\alpha}{d}\right)} = \frac{\sqrt{d^2 - b^2\cos^2\alpha}}{b\cos\alpha}$$

Now, on setting the value of$cot(\alpha-\beta)$ in the eq(III), we get

$$\Rightarrow cos\theta\left(\frac{\sqrt{d^2-b^2cos^2\alpha}}{bcos\alpha}\right)-sin\theta=\frac{d}{a}\Rightarrow\frac{\sqrt{d^2-b^2cos^2\alpha}}{bcos\alpha}=\frac{\frac{d}{a}+sin\theta}{cos\theta}=\frac{asin\theta+d}{acos\theta}$$

On taking square of both the sides, we get

$$\frac{d^2-b^2cos^2\alpha}{b^2cos^2\alpha}=\frac{(asin\theta+d)^2}{(acos\theta)^2}=\frac{a^2sin^2\theta+d^2+2adsin\theta}{a^2cos^2\theta}$$

Adding 1 both the sides as follows

$$\frac{d^2-b^2cos^2\alpha}{b^2cos^2\alpha}+1=\frac{a^2sin^2\theta+d^2+2adsin\theta}{a^2cos^2\theta}+1$$

$$\Rightarrow\frac{d^2-b^2cos^2\alpha+b^2cos^2\alpha}{b^2cos^2\alpha}=\frac{a^2sin^2\theta+d^2+2adsin\theta+a^2cos^2\theta}{a^2cos^2\theta}$$

$$\Rightarrow\frac{d^2}{b^2cos^2\alpha}=\frac{a^2+d^2+2adsin\theta}{a^2cos^2\theta}\Rightarrow cos^2\alpha=\frac{a^2d^2cos^2\theta}{b^2(a^2+d^2+2adsin\theta)}$$

$$\Rightarrow cos\alpha=\sqrt{\frac{a^2d^2cos^2\theta}{b^2(a^2+d^2+2adsin\theta)}}=\frac{adcos\theta}{b\sqrt{a^2+d^2+2adsin\theta}}\quad\ldots\ldots\ldots\ldots\ldots(IV)$$

Now, consider a right cone with apex 'P' & apex angle $'2\alpha'$ such that the boundaries of the given elliptical plane & the circular plane of projection completely coincide with the lateral face of the right cone. (See the above fig (75) above)

In this case, the projection of major axis AB = 2a is equal to the minor axis

CD = AB' = 2b = diameter of circular plane of projection

(see the fig(76) below)

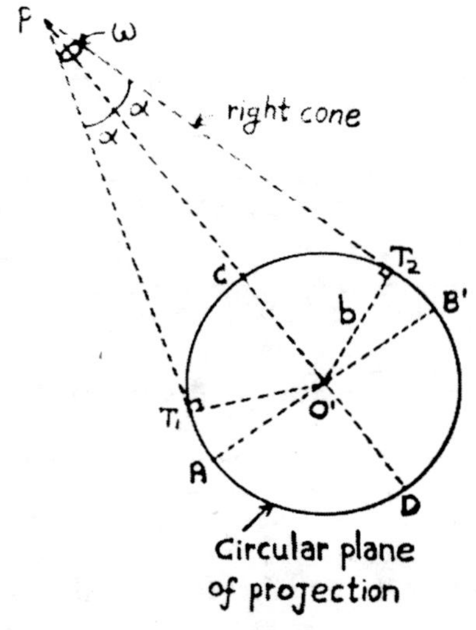

Fig (76)

Hence, the solid angle subtended by the elliptical plane at the given point 'P'

= solid angle subtended by the right cone at the apex 'P' (given point)

From the eq(1) for the right cone

$$\Rightarrow\omega=2\pi(1-cos\alpha)=2\pi\left(1-\frac{adcos\theta}{b\sqrt{a^2+d^2+2adsin\theta}}\right)$$

$$= 2\pi\left(1 - \frac{adcos\theta}{a\sqrt{1-e^2}\sqrt{a^2+d^2+2adsin\theta}}\right) = 2\pi\left(1 - \frac{dcos\theta}{\sqrt{1-e^2}\sqrt{a^2+d^2+2adsin\theta}}\right)$$

Where, eccentricity of the ellipse

$$\Rightarrow e = \sqrt{1 - \frac{b^2}{a^2}} \quad or\ b = a\sqrt{1-e^2} \quad \forall (b < a)$$

Thus,

$$\Rightarrow \left[\omega = 2\pi\left(1 - \frac{adcos\theta}{b\sqrt{a^2+d^2+2adsin\theta}}\right) = 2\pi\left(1 - \frac{dcos\theta}{\sqrt{1-e^2}\sqrt{a^2+d^2+2adsin\theta}}\right)\right]$$

$$\forall\left(0 \le \theta < \frac{\pi}{2} \,\&\, d \ge 0\right) \quad \ldots\ldots\ldots\ldots\ldots (3.12)$$

Thus, the value of ω depends on the following

1. Minor axis (2b) or Eccentricity (e) of ellipse
2. Major axis (2a)
3. Distance (d) of the given point (observer) from the centre of the plane &
4. The angle (θ) between normal to the plane & the line joining centre to the given point in the space.

Note: For any given distance (d) there is only one position (i.e. the value of θ) from which the given elliptical plane appears as a perfect circular plane.

While for a given value of θ, there are a no. of the values of distance (d) from which the given elliptical plane appears as a perfect circular plane i.e. an elliptical plane appears as a perfect circular plane only for a certain angular position $'\theta'$ w.r.t. the plane in the apace.

This arises the following case for a circular plane

Case 1: On setting $\theta = 0$, a = b = R & d = h in the eq(3.12), we get

$$\omega = 2\pi\left(1 - \frac{Rhcos(0)}{R\sqrt{R^2+h^2+2Rhsin(0)}}\right) = 2\pi\left(1 - \frac{h}{\sqrt{R^2+h^2}}\right)$$

The above result is same as obtained in the art.2.3.23 i.e. the solid angle subtended by a circular plane at any point lying on the normal axis passing through the centre.

Hence, a circular plane appears as it's only and if only it is observed from any point lying on the vertical axis passing through the centre of the same plane.

Case 2: The solid angle subtended by any plane at any point lying on the plane completely inside the boundary is obtained by setting d = 0 in the eq(3.12) as follows

$$\omega = 2\pi\left(1 - \frac{a(0)cos\theta}{b\sqrt{a^2+(0)^2+2a(0)sin\theta}}\right) = 2\pi(1-0) = 2\pi\ sr$$

The above result is correct irrespective of geometry of the plane.

3.6 Estimation of Error

As we have studied that the results obtained from the approximation formula have certain error which can be estimated for some particular symmetrical planes for certain location (generally vertical) of the given point (observer).

We have also studied that the error involved in the results obtained from the approximation formula depends on the factor of circularity (F_g) which depends on the geometry & configuration of the plane with respect to the point in the space.

Here, we are interested to estimate the error involved in the results of some common plane-figures & tabulate that for certain configuration & values of the distance of the point from the centre of plane.Before starting, let's know some important terms.

Factor of Error: "It is the difference of factor of circularity from unity"

It is denoted by $'F_e'$. Numerically it is given as follows

$$\Rightarrow [F_e = 1 - F_c]$$

If 'e' is the error involved in the value of solid angle then the error is linearly proportion to the factor of error (by inverse relation)

$$\Rightarrow e = KF_e + C = K(1 - F_c) + C$$

Where, 'K' & 'C' are constants whose values depend on actual geometrical dimensions, shape & relative distance of a given symmetrical plane. Value of 'C' is zero only for a circular plane while in all other cases, value of 'C' is always non-zero.

But practically it is difficult to find the numerical values of constants 'K' & 'C'. Hence, in order to find the exact value of error involved in the values obtained from approximation formula, we take the difference of exact & approximate values of solid angle. Thus

$$\Rightarrow \left[e = (\omega)_{exact} \sim (\omega)_{approximate}\right]$$

Where, $(\omega)_{exact}$ is calculated from the corresponding formulae obtained in the chapter 2 &

$(\omega)_{approximate}$ is calculated from approximation formula.

Note: The lines/contents highlighted by red colour must be included at their indicated places in the manuscript while rest of the material is technically error free (i.e. requiring no changes) to the best of author' knowledge.

Although, if you have any confusion or difficulty in making these minor corrections by including highlighted contents (by red colour) just leave it unchanged I will correct it when you submit me the final version of interior design in track-change mode.

Numerical values of error have been tabulated (called error-table) for some particular plane-figures that are useful for case studies of symmetrical plane-figures.

Hereby, we are to estimate the error for the following symmetrical plane-figures

1. Rectangle 2. Rhombus 3. Equilateral triangle 4. Square 5. Regular pentagon 6. Regular hexagon.

When, the given point (observer) is located at the vertical axis passing through the centre of the given plane in the space.

3.6.1 Error-table for a rectangular plane

We know that the exact (actual) value of solid angle subtended by a rectangular plane at any point lying on the axis normal to the plane & passing through the centre is given from eq(16) as follows

$$(\omega)_{actual} = 4\sin^{-1}\left\{\frac{lb}{\sqrt{(l^2 + 4h^2)(b^2 + 4h^2)}}\right\} \&$$

$$(\omega)_{approximate} = 2\pi\left[1 - \frac{1}{\sqrt{1 + \left(\frac{lb}{\pi h^2}\right)}}\right] \text{from eq(3.7) } (r = h)$$

Hence, we get the error (e) by taking the positive difference of above values for a rectangular plane as follows

$$\Rightarrow e = 4\sin^{-1}\left\{\frac{1}{\sqrt{\left(1+4\left(\frac{h}{l}\right)^2\right)\left(1+4\left(\frac{h}{b}\right)^2\right)}}\right\} \sim 2\pi\left[1-\frac{1}{\sqrt{1+\left(\frac{1}{\pi\left(\frac{h}{l}\right)\left(\frac{h}{b}\right)}\right)}}\right] \quad \forall(l \geq b)$$

Now, on setting the different values of the ratios $\frac{h}{l}$ & $\frac{h}{b}$ which are closer to each other

We can obtain error-table for a rectangular plane(having permissible error).

$$\left(for\ F_c > 0.5 \Rightarrow 0 < \left(\frac{l}{b}\right) < \sqrt{3}\right) \ where\ F_c = \frac{1}{\sqrt{\left(\frac{l}{b}\right)^2 + 1}}$$

$$\frac{1}{\sqrt{\left(\frac{l}{b}\right)^2+1}} > 0.5 \Rightarrow \sqrt{\left(\frac{l}{b}\right)^2+1} < 2 \Rightarrow \left(\frac{l}{b}\right)^2 + 1 < 4 \Rightarrow \left(\frac{l}{b}\right)^2 < 3$$

$$\Rightarrow \frac{l}{b} < \sqrt{3}$$

Note: Error table can be modified or extended for different values of dimensions.

1. $for\ {}^{l}/_{b} = 1.1\ (rectangle)$

1.) $for\ {}^{l}/_{b} = 1.1\ (rectangle)$

$\frac{h}{l}$	$\frac{h}{b} = \left(\frac{h}{l}\right) \times \left(\frac{l}{b}\right)$	$(\omega)_{act.}$ (sr.)	$(\omega)_{approx.}$ (sr.)	e (sr.)	e (%)
0.01	0.011	6.164279	6.166403	0.002124	0.034456
0.02	0.022	6.045504	6.049742	0.004238	0.070102
0.03	0.033	5.926993	5.933322	0.006329	0.106783
0.04	0.044	5.808874	5.817263	0.008389	0.144417
0.05	0.055	5.691275	5.701681	0.010406	0.182841
0.06	0.066	5.574321	5.586691	0.012370	0.221910
0.07	0.077	5.458133	5.472405	0.014272	0.261481
0.08	0.088	5.342828	5.358932	0.016104	0.301413
0.09	0.099	5.228516	5.246376	0.017860	0.341588
0.10	0.110	5.115305	5.134836	0.019531	0.381815
0.15	0.165	4.569082	4.595535	0.026453	0.578956
0.20	0.220	4.062792	4.093578	0.030786	0.757756
0.25	0.275	3.602475	3.635129	0.032654	0.906432
0.30	0.330	3.190341	3.222856	0.032515	1.019170
0.35	0.385	2.825611	2.856561	0.030950	1.095338
0.40	0.440	2.505485	2.534003	0.028518	1.138223

0.45	0.495	2.226025	2.251691	0.025666	1.152997
0.50	0.550	1.982818	2.005533	0.022715	1.145591
0.55	0.605	1.771432	1.791302	0.019870	1.121691
0.60	0.660	1.587683	1.604928	0.017245	1.086174
0.65	0.715	1.427772	1.442665	0.014893	1.043094
0.70	0.770	1.288342	1.301167	0.012825	0.995465
0.75	0.825	1.166471	1.177503	0.011032	0.945758
0.80	0.880	1.059648	1.069138	0.009490	0.895580
0.85	0.935	0.965730	0.973902	0.008172	0.846199
0.90	0.990	0.882895	0.889943	0.007048	0.798283
0.95	1.045	0.809596	0.815688	0.006092	0.752474
1.00	1.100	0.744526	0.749803	0.005277	0.708773
2.00	2.200	0.215108	0.215641	0.000533	0.247782
3.00	3.300	0.098521	0.098638	0.000117	0.118756
4.00	4.400	0.056021	0.056059	0.000038	0.067832
5.00	5.500	0.036035	0.036051	0.000016	0.044401
6.00	6.600	0.025093	0.025101	0.000008	0.031881
7.00	7.700	0.018467	0.018471	0.000004	0.021660
8.00	8.800	0.014154	0.014156	0.000002	0.014130
9.00	9.900	0.011192	0.011193	0.000001	0.008935
10.00	11.000	0.00907092	0.00907122	0.0000003	0.003307
$\frac{h}{l} > 10$	....	....	....	$\rightarrow 0$	$\rightarrow 0$

Thus, we find that for the higher values of the normal distance the error tends to zero.

3.6.2 Error (e) for a rhombus-like plane

We know that the exact (actual) value of solid angle subtended by a rhombus like plane at any point lying on the axis normal to the plane & passing through the centre is given from eq(28) as follows

$$(\omega)_{actual} = 4\cos^{-1}\left\{\frac{h\left(d_1^{\,2}\sqrt{h^2+d_2^{\,2}}+d_2^{\,2}\sqrt{h^2+d_1^{\,2}}\right)}{h^2\left(d_1^{\,2}+d_2^{\,2}\right)+d_1^{\,2}d_2^{\,2}}\right\} \&$$

$$(\omega)_{approximate} = 2\pi\left[1-\frac{1}{\sqrt{1+\left(\frac{2d_1d_2}{\pi h^2}\right)}}\right] \text{from eq(3.9) } (r = h)$$

Hence, we get the error (e) by taking the positive difference of above two values for a rhombus like plane as follows

$$e = 4\cos^{-1}\left\{\frac{h\left(d_1^{\,2}\sqrt{h^2+d_2^{\,2}}+d_2^{\,2}\sqrt{h^2+d_1^{\,2}}\right)}{h^2\left(d_1^{\,2}+d_2^{\,2}\right)+d_1^{\,2}d_2^{\,2}}\right\} \sim 2\pi\left[1-\frac{1}{\sqrt{1+\left(\frac{2d_1d_2}{\pi h^2}\right)}}\right]$$

$$e = 4\cos^{-1}\left\{\frac{h\left(\frac{d_1{}^2\sqrt{h^2+d_2{}^2}+d_2{}^2\sqrt{h^2+d_1{}^2}}{d_1{}^2d_2{}^2}\right)}{\left(\frac{h^2\left(d_1{}^2+d_2{}^2\right)+d_1{}^2d_2{}^2}{d_1{}^2d_2{}^2}\right)}\right\} \sim 2\pi\left[1-\frac{1}{\sqrt{1+\left(\frac{2}{\pi\left(\frac{h}{d_2}\right)\left(\frac{h}{d_1}\right)}\right)}}\right]$$

$$e = 4\cos^{-1}\left\{\frac{\left(\frac{h}{d_2}\right)\sqrt{\left(\frac{h}{d_2}\right)^2+1}+\left(\frac{h}{d_1}\right)\sqrt{\left(\frac{h}{d_1}\right)^2+1}}{\left(\frac{h}{d_2}\right)^2\left(\left(\frac{d_2}{d_1}\right)^2+1\right)+1}\right\} \sim 2\pi\left[1-\frac{1}{\sqrt{1+\left(\frac{2}{\pi\left(\frac{h}{d_2}\right)\left(\frac{h}{d_1}\right)}\right)}}\right]$$

$$\ldots\ldots\ldots\ldots\ldots \quad \forall(d_2 \geq d_1)$$

Now, on setting the different values of the ratios $\frac{h}{d_2}$ & $\frac{h}{d_1}$ which are closer to each other We can obtain error-table for a rhombus-like plane (having permissible errors).

$$\left(for\ F_c > 0.5 \Rightarrow 0 < \left(\frac{d_2}{d_1}\right) < \sqrt{3}\right)\ where\ F_c = \frac{d_1}{\sqrt{d_1{}^2+d_2{}^2}}$$

$$\Rightarrow F_c = \frac{1}{\sqrt{\left(\frac{d_2}{d_1}\right)^2+1}}$$

$$\frac{1}{\sqrt{\left(\frac{d_2}{d_1}\right)^2+1}} > 0.5 \Rightarrow \sqrt{\left(\frac{d_2}{d_1}\right)^2+1} < 2 \Rightarrow \left(\frac{d_2}{d_1}\right)^2+1 < 4 \Rightarrow \left(\frac{d_2}{d_1}\right)^2 < 3$$

$$\Rightarrow \frac{d_2}{d_1} < \sqrt{3}$$

Thus, by setting different values of $\frac{h}{d_2}$ for given value of $\frac{d_2}{d_1}$ we get different values of $\frac{h}{d_1}$ as follows

$$\frac{h}{d_1} = \left(\frac{h}{d_2}\right)\left(\frac{d_2}{d_1}\right)$$

Hence, we obtain different values of $'\omega'$ which can be tabulated for a rhombus-like plane.

3.6.3 Error (e) for an equilateral-triangular plane

We know that the actual value of solid angle subtended by an equilateral triangular plane at any point lying on the axis normal to the plane & passing through the centre of the plane is given by the eq(15) as follows

$$(\omega)_{actual}=2\pi-6\sin^{-1}\left\{\frac{3h}{\sqrt{12h^2+a^2}}\right\}\ \&$$

$$(\omega)_{approximate}=2\pi\left[1-\frac{1}{\sqrt{1+\left(\frac{\sqrt{3}a^2}{4\pi r^2}\right)}}\right]$$

Hence, the error (e) is equal to the difference of above two values

$$e=2\pi-6\sin^{-1}\left\{\frac{3h}{\sqrt{12h^2+a^2}}\right\}\sim 2\pi\left[1-\frac{1}{\sqrt{1+\left(\frac{\sqrt{3}a^2}{4\pi h^2}\right)}}\right](r=h)$$

$$\Rightarrow e=2\pi-6\sin^{-1}\left\{\frac{3\left(\frac{h}{a}\right)}{\sqrt{12\left(\frac{h}{a}\right)^2+1}}\right\}\sim 2\pi\left[1-\frac{\left(\frac{h}{a}\right)}{\sqrt{\left(\frac{h}{a}\right)^2+\left(\frac{\sqrt{3}}{4\pi}\right)}}\right]$$

$$\Rightarrow e=6\sin^{-1}\left\{\frac{3\left(\frac{h}{a}\right)}{\sqrt{12\left(\frac{h}{a}\right)^2+1}}\right\}\sim\frac{2\pi\left(\frac{h}{a}\right)}{\sqrt{\left(\frac{h}{a}\right)^2+\left(\frac{\sqrt{3}}{4\pi}\right)}}$$

While, the factor of circularity $F_c=0.5$ for an equilateral triangular is constant i.e. independent of geometrical dimensions of the plane

Thus, by setting different values of $\frac{h}{a}$ in the above expression, we get different values of error (e) which can be tabulated for an equilateral triangle (plane).

3.6.4 Error (e) for a square plane

We know that the actual value of solid angle subtended by a square plane at any point lying on the axis normal to the plane & passing through the centre of the plane is given by the eq(28) as follows

$$(\omega)_{actual}=4\sin^{-1}\left\{\frac{a^2}{a^2+h^2}\right\}$$

$$(\omega)_{approximate}=2\pi\left[1-\frac{1}{\sqrt{1+\left(\frac{a^2}{\pi r^2}\right)}}\right]$$

Hence, the error (e) is equal to the difference of above two values

$$e=4\sin^{-1}\left\{\frac{a^2}{a^2+h^2}\right\}\sim 2\pi\left[1-\frac{1}{\sqrt{1+\left(\frac{a^2}{\pi h^2}\right)}}\right](r=h)$$

$$\Rightarrow e = 4\sin^{-1}\left\{\frac{a^2}{\left(\frac{a^2+h^2}{a^2}\right)}\right\} \sim 2\pi\left[1 - \frac{1}{\left(\frac{a}{h}\right)\sqrt{\left(\frac{h}{a}\right)^2 + \left(\frac{1}{\pi}\right)}}\right]$$

$$\Rightarrow e = 4\sin^{-1}\left\{\frac{1}{\left(\frac{h}{a}\right)^2 + 1}\right\} \sim 2\pi\left[1 - \frac{\left(\frac{h}{a}\right)}{\sqrt{\left(\frac{h}{a}\right)^2 + \left(\frac{1}{\pi}\right)}}\right]$$

While, the factor of circularity $F_c = 0.707$ for a square plane is constant i.e. independent of geometrical dimensions of the plane

Thus, by setting different values of $\frac{h}{a}$ in the above expression, we get different values of error (e) which can be tabulated for a square (plane).

3.6.5 Error (e) for a regular-pentagonal plane

We know that the actual value of solid angle subtended by an regular pentagonal plane at any point lying on the axis normal to the plane & passing through the centre of the plane is given by the eq(35) as follows

$$(\omega)_{actual} = 2\pi - 10\sin^{-1}\left\{\frac{2h\sin\frac{\pi}{5}}{\sqrt{4h^2 + a^2\cot^2\frac{\pi}{5}}}\right\} \&$$

$$(\omega)_{approximate} = 2\pi\left[1 - \frac{1}{\sqrt{1 + \left(\frac{5a^2\cot\frac{\pi}{5}}{4\pi r^2}\right)}}\right]$$

Hence, the error (e) is equal to the difference of above two values

$$e = 2\pi - 10\sin^{-1}\left\{\frac{2h\sin\frac{\pi}{5}}{\sqrt{4h^2 + a^2\cot^2\frac{\pi}{5}}}\right\} \sim 2\pi\left[1 - \frac{1}{\sqrt{1 + \left(\frac{5a^2\cot\frac{\pi}{5}}{4\pi h^2}\right)}}\right] \quad (r = h)$$

$$\Rightarrow e = 2\pi - 10\sin^{-1}\left\{\frac{2\left(\frac{h}{a}\right)\sin\frac{\pi}{5}}{\sqrt{4\left(\frac{h}{a}\right)^2 + \cot^2\frac{\pi}{5}}}\right\} \sim 2\pi\left[1 - \frac{\left(\frac{h}{a}\right)}{\sqrt{\left(\frac{h}{a}\right)^2 + \left(\frac{5\cot\frac{\pi}{5}}{4\pi}\right)}}\right]$$

$$\Rightarrow e = 10\sin^{-1}\left\{\frac{2\left(\frac{h}{a}\right)\sin\frac{\pi}{5}}{\sqrt{4\left(\frac{h}{a}\right)^2 + \cot^2\frac{\pi}{5}}}\right\} \sim \frac{2\pi\left(\frac{h}{a}\right)}{\sqrt{\left(\frac{h}{a}\right)^2 + \left(\frac{5\cot\frac{\pi}{5}}{4\pi}\right)}}$$

While, the factor of circularity $F_c = 0.809$ for a regular pentagonal plane is constant i.e. independent of geometrical dimensions of the plane

Thus, by setting different values of $\frac{h}{a}$ in the above expression, we get different values of error (e) which can be tabulated for a regular hexagon (plane).

3.6.6Error (e) for a regular-hexagonal plane

We know that the actual value of solid angle subtended by an regular hexagonal plane at any point lying on the axis normal to the plane & passing through the centre of the plane is given by the eq(36) as follows

$$(\omega)_{actual} = 2\pi - 12\sin^{-1}\left\{\frac{h}{\sqrt{4h^2+3a^2}}\right\} \&$$

$$(\omega)_{approximate} = 2\pi\left[1 - \frac{1}{\sqrt{1+\left(\frac{6a^2\cot\frac{\pi}{6}}{4\pi r^2}\right)}}\right] = 2\pi\left[1 - \frac{1}{\sqrt{1+\left(\frac{3a^2\sqrt{3}}{2\pi r^2}\right)}}\right]$$

Hence, the error (e) is equal to the difference of above two values

$$e = 2\pi - 12\sin^{-1}\left\{\frac{h}{\sqrt{4h^2+3a^2}}\right\} \sim 2\pi\left[1 - \frac{1}{\sqrt{1+\left(\frac{3a^2\sqrt{3}}{2\pi h^2}\right)}}\right] \quad (r = h)$$

$$\Rightarrow e = 2\pi - 12\sin^{-1}\left\{\frac{\left(\frac{h}{a}\right)}{\sqrt{4\left(\frac{h}{a}\right)^2+3}}\right\} \sim 2\pi\left[1 - \frac{\left(\frac{h}{a}\right)}{\sqrt{\left(\frac{h}{a}\right)^2+\left(\frac{3\sqrt{3}}{2\pi}\right)}}\right]$$

$$\Rightarrow e = 12\sin^{-1}\left\{\frac{\left(\frac{h}{a}\right)}{\sqrt{4\left(\frac{h}{a}\right)^2+3}}\right\} \sim \frac{2\pi\left(\frac{h}{a}\right)}{\sqrt{\left(\frac{h}{a}\right)^2+\left(\frac{3\sqrt{3}}{2\pi}\right)}}$$

While, factor of circularity $F_c = 0.809$ for a regular hexagonal plane is constant i.e. it is independent of geometrical dimensions of the plane.

Thus, by setting different values of $\frac{h}{a}$ in the above expression, we get different values of error (e) which can be tabulated for a regular hexagon (plane).

Solved Examples

Example 3.1: Using Approximation find out the solid angle subtended by a rectangular plane of $12cm \times 15cm$ at a point lying at a height 18cm on the vertical axis passing through the centre &% error involved in the result. Also calculate the factor of circularity& discuss the results.

Sol. given that

$$length, l = 15cm, width, b = 12cm \;\&\; normal\ height, h = 18cm$$

$$\therefore area\ of\ the\ rectangular\ plane, A = l \times b = 15 \times 12 = 180\ cm^2$$

We know that the approximate value of solid angle subtended by a rectangular plane with sides 'l' & 'b' at any point lying at a normal distance 'r' from the centre is given from the eq(3.7) as

$$\omega_{approx.} = 2\pi\left[1 - \frac{1}{\sqrt{\left(1 + \frac{lb}{\pi r^2}\right)}}\right]$$

On setting the values in the above expression, we get approximate value of solid angle

$$\omega_{approx.} = 2\pi\left[1 - \frac{1}{\sqrt{\left(1 + \frac{180}{\pi(18)^2}\right)}}\right] = 2\pi\left[1 - \frac{1}{\sqrt{1.177}}\right]$$

$$= 2\pi[1 - 0.922] = 0.491\ sr \qquad \textit{Ans.}$$

But the actual value of solid angle, subtended by a rectangular plane with sides 'l' & 'b' at any point lying at a normal height 'h' from the centre, is given from the eq(16) as follows

$$\omega_{act.} = 4\sin^{-1}\left(\frac{lb}{\sqrt{(l^2 + 4h^2)(b^2 + 4h^2)}}\right)$$

On setting the values, we get

$$\omega_{act.} = 4\sin^{-1}\left(\frac{15 \times 12}{\sqrt{((15)^2 + 4(18)^2)((12)^2 + 4(18)^2)}}\right)$$

$$= 4\sin^{-1}\left(\frac{180}{\sqrt{2190240}}\right) = 4\sin^{-1}(0.122) = 0.489\ sr$$

$$\therefore \%\ error\ (e) = \left(\frac{\omega_{approx.} - \omega_{act.}}{\omega_{act.}}\right) \times 100$$

$$= \left(\frac{0.491 - 0.489}{0.489}\right) \times 100 = 0.41\ \% \qquad \textit{Ans.}$$

Now, factor of circularity of a rectangular plane is given as follows

$$F_c = \frac{b}{\sqrt{l^2 + b^2}}$$

$$\Rightarrow F_c = \frac{12}{\sqrt{(15)^2 + (12)^2}} = 0.625 \qquad \textit{Ans.}$$

We find that error in the value of solid angle is 0.41 % which is permissible because the factor of circularity is closer to unity height (distance) of the point from the plane is greater as compared to the dimensions of the plane.

Example 3.2: Using Approximation find out the solid angle subtended by a rectangular plane of $8cm \times 25cm$ at a point lying at a height 10cm on the vertical axis passing through the centre &%error involved in the result. Also calculate the factor of circularity& discuss the results.

Sol. given that

$$length, l = 25cm, width, b = 8cm\ \&\ normal\ height, h = 10cm$$

$$\therefore area\ of\ the\ rectangular\ plane, A = l \times b = 25 \times 8 = 200\ cm^2$$

We know that the approximate value of solid angle subtended by a rectangular plane with sides 'l' & 'b' at any point lying at a normal distance 'r' from the centre is given from the eq(3.7) as

$$\omega_{approx.} = 2\pi\left[1 - \frac{1}{\sqrt{\left(1 + \frac{lb}{\pi r^2}\right)}}\right]$$

On setting the values in the above expression, we get approximate value of solid angle

$$\omega_{approx.} = 2\pi\left[1 - \frac{1}{\sqrt{\left(1 + \frac{200}{\pi(10)^2}\right)}}\right] = 2\pi\left[1 - \frac{1}{\sqrt{1.636}}\right]$$

$$= 2\pi[1 - 0.781] = 1.372\ sr \qquad Ans.$$

But the actual value of solid angle, subtended by a rectangular plane with sides 'l' & 'b' at any point lying at a normal height 'h' from the centre, is given from the eq(16) as follows

$$\omega_{act.} = 4\sin^{-1}\left(\frac{lb}{\sqrt{(l^2 + 4h^2)(b^2 + 4h^2)}}\right)$$

On setting the values, we get

$$\omega_{act.} = 4\sin^{-1}\left(\frac{25 \times 8}{\sqrt{((25)^2 + 4(10)^2)((8)^2 + 4(10)^2)}}\right)$$

$$= 4\sin^{-1}\left(\frac{200}{\sqrt{475600}}\right) = 4\sin^{-1}(0.29) = 1.177\ sr$$

$$\therefore\ \%\ error\ (\%\ e) = \left(\frac{\omega_{approx.} - \omega_{act.}}{\omega_{act.}}\right) \times 100$$

$$= \left(\frac{1.372 - 1.177}{1.177}\right) \times 100 = 16.57\% \qquad Ans.$$

Now, the factor of circularity of a rectangular plane is given as follows

$$F_c = \frac{b}{\sqrt{l^2 + b^2}}$$

$$\Rightarrow F_c = \frac{8}{\sqrt{(25)^2 + (8)^2}} = 0.305 \qquad Ans.$$

We find that the error involved in the value of solid angle is 16.57% which is not permissible. It is because the

1. Distance of the point is smaller as compared to the dimensions of rectangle
2. Factor of circularity (F_c) is much smaller than unity ($even\ F_c < 0.5$)

Example 3.3: Using Approximation find out the solid angle subtended by a rectangular plane of $5cm \times 20cm$ at a point lying at a height 50cm on the vertical axis passing through the centre &% error involved in the result. Also calculate the factor of circularity & discuss the results.

Sol. given that

$$length, l = 20cm, width, b = 5cm\ \&\ normal\ height, h = 50cm$$

$$\therefore\ area\ of\ the\ rectangular\ plane, A = l \times b = 20 \times 5 = 100\ cm^2$$

We know that the approximate value of solid angle subtended by a rectangular plane with sides 'l' & 'b' at any point lying at a normal distance 'r' from the centre is given from the eq(3.7) as

$$\omega_{approx.} = 2\pi\left[1 - \frac{1}{\sqrt{\left(1 + \frac{lb}{\pi r^2}\right)}}\right]$$

On setting the values in the above expression, we get approximate value of solid angle

$$\omega_{approx.} = 2\pi\left[1 - \frac{1}{\sqrt{\left(1 + \frac{100}{\pi(50)^2}\right)}}\right] = 2\pi\left[1 - \frac{1}{\sqrt{1.0127}}\right]$$

$$= 2\pi[1 - 0.994] = 0.0396\ sr \qquad Ans.$$

But the actual value of solid angle, subtended by a rectangular plane with sides 'l' & 'b' at any point lying at a normal height 'h' from the centre, is given from the eq(16) as follows

$$\omega_{act.} = 4\sin^{-1}\left(\frac{lb}{\sqrt{(l^2 + 4h^2)(b^2 + 4h^2)}}\right)$$

On setting the values, we get

$$\omega_{act.} = 4\sin^{-1}\left(\frac{20 \times 5}{\sqrt{((20)^2 + 4(50)^2)((5)^2 + 4(50)^2)}}\right)$$

$$= 4\sin^{-1}\left(\frac{100}{\sqrt{104260000}}\right) = 4\sin^{-1}(0.00979) = 0.0392\ sr$$

$$\therefore\ \%\ error\ (\%\ e) = \left(\frac{\omega_{approx.} - \omega_{act.}}{\omega_{act.}}\right) \times 100$$

$$= \left(\frac{0.0396 - 0.0392}{0.0392}\right) \times 100 = 1.02\% \qquad Ans.$$

Now, the factor of circularity of a rectangular plane is given as follows

$$F_c = \frac{b}{\sqrt{l^2 + b^2}}$$

$$\Rightarrow F_c = \frac{5}{\sqrt{(20)^2 + (5)^2}} = 0.242 \quad Ans.$$

We find that the error involved in the value of solid angle is 1.02% which is permissible. It is because the distance of the point from the plane is greater as compared to the dimensions of the plane although the factor of circularity if much smaller than unity ($even\ F_c < 0.5$)

Thus, factor of the distance is dominant. So error is permissible (less).

Example 3.4: Using Approximation find out the solid angle subtended by a rhombus-like plane having diagonals 24cm & 32cm at a point lying at a height 13cm on the vertical axis passing through the centre & error involved in the result.

Sol. given that

$$diagonal\ of\ rhombus, 2d_1 = 24cm\ , 2d_2 = 32cm\ \&$$

$$height\ of\ the\ point, h = 13cm$$

We know that the approximate value of solid angle subtended by a rectangular plane with sides '$2d_1$' & '$2d_2$' at any point lying at a normal distance 'r' from the centre is given from the eq(3.7) as

$$\omega_{approx.} = 2\pi\left[1 - \frac{1}{\sqrt{\left(1 + \frac{2d_1d_2}{\pi r^2}\right)}}\right]$$

On setting the values in the above expression, we get approximate value of solid angle

$$\omega_{approx.} = 2\pi\left[1 - \frac{1}{\sqrt{\left(1 + \frac{2 \times 12 \times 16}{\pi(13)^2}\right)}}\right] = 2\pi\left[1 - \frac{1}{\sqrt{1.723}}\right]$$

$$= 2\pi[1 - 0.762] = 1.497\ sr \qquad Ans.$$

We know that the solid angle subtended by a rhombus like plane with diagonals $2d_1\&2d_2$ at any point lying at a normal height h is given by the eq(28) as follows

$$\omega_{act.} = 4\cos^{-1}\left(\frac{h\left({d_1}^2\sqrt{h^2+{d_2}^2}+{d_1}^2\sqrt{h^2+{d_2}^2}\right)}{h^2({d_1}^2+{d_2}^2)+{d_1}^2{d_2}^2}\right)$$

On setting the values, we get

$$\omega_{act.} = 4\cos^{-1}\left(\frac{13\left((12)^2\sqrt{(13)^2+(16)^2}+(16)^2\sqrt{(13)^2+(12)^2}\right)}{(13)^2((12)^2+(16)^2)+(12)^2(16)^2}\right)$$

$$= 4\cos^{-1}\left(\frac{97470.599}{104464}\right) = 4\cos^{-1}(0.933) = 1.472\ sr$$

$$\therefore\ \%\ error\ (\%\ e) = \left(\frac{\omega_{approx.} - \omega_{act.}}{\omega_{act.}}\right) \times 100$$

$$= \left(\frac{1.497 - 1.472}{1.472}\right) \times 100 = 1.7\% \qquad Ans.$$

Now, the factor of circularity of a rectangular plane is given as follows

$$F_c = \frac{d_1}{\sqrt{{d_1}^2+{d_2}^2}}$$

$$\Rightarrow F_c = \frac{12}{\sqrt{(12)^2+(16)^2}} = 0.6 \qquad Ans.$$

We find that the error involved in the value of solid angle is 1.7% which is permissible. It is because the distance of the point from the plane is smaller as compared to the dimensions of the plane the factor of circularity if much close to unity ($F_c > 0.5$)

Thus, factor of distance is dominant. So error is permissible (less).

Example 3.5: Using Approximation find out the solid angle subtended by a rhombus-like plane having diagonals 8cm & 28cm at a point lying at a height 11cm on the vertical axis passing through the centre & error involved in the result.

Sol. given that

$$diagonal\ of\ rhombus, 2d_1 = 8cm, 2d_2 = 28cm\ \&$$
$$height\ of\ the\ point, h = 11cm$$

We know that the approximate value of solid angle subtended by a rectangular plane with sides '$2d_1$' & '$2d_2$' at any point lying at a normal distance 'r' from the centre is given from the eq(3.7) as

$$\omega_{approx.} = 2\pi\left[1 - \frac{1}{\sqrt{\left(1+\frac{2d_1d_2}{\pi r^2}\right)}}\right]$$

On setting the values in the above expression, we get approximate value of solid angle

$$\omega_{approx.} = 2\pi\left[1 - \frac{1}{\sqrt{\left(1+\frac{2\times4\times14}{\pi(11)^2}\right)}}\right] = 2\pi\left[1-\frac{1}{\sqrt{1.295}}\right]$$

$$= 2\pi[1 - 0.879] = 0.761\ sr \qquad Ans.$$

We know that the solid angle subtended by a rhombus like plane with diagonals $2d_1\&2d_2$ at any point lying at a normal height h is given by the eq(28) as follows

$$\omega_{act.} = 4\cos^{-1}\left(\frac{h\left({d_1}^2\sqrt{h^2 + {d_2}^2} + {d_1}^2\sqrt{h^2 + {d_2}^2}\right)}{h^2\left({d_1}^2 + {d_2}^2\right) + {d_1}^2{d_2}^2}\right)$$

On setting the values, we get

$$\omega_{act.} = 4\cos^{-1}\left(\frac{11\left((4)^2\sqrt{(11)^2 + (14)^2} + (14)^2\sqrt{(11)^2 + (4)^2}\right)}{(11)^2((4)^2 + (14)^2) + (4)^2(14)^2}\right)$$

$$= 4\cos^{-1}\left(\frac{28368.924}{28788}\right) = 4\cos^{-1}(0.985) = 0.683\ sr$$

$$\therefore\ \%\ error\ (\%\ e) = \left(\frac{\omega_{approx.} - \omega_{act.}}{\omega_{act.}}\right) \times 100$$

$$= \left(\frac{0.761 - 0.683}{0.683}\right) \times 100 = 11.4\% \qquad Ans.$$

Now, the factor of circularity of a rectangular plane is given as follows

$$F_c = \frac{d_1}{\sqrt{{d_1}^2 + {d_2}^2}}$$

$$\Rightarrow F_c = \frac{4}{\sqrt{(14)^2 + (4)^2}} = 0.275 \qquad Ans.$$

We find that the error involved in the value of solid angle is 11.4% which is not permissible. It is because the distance of the point from the plane is smaller as compared to the dimensions of the plane & more over factor of circularity is much smaller than unity $(even\ F_c < 0.5)$

Example 3.6: Using Approximation find out the solid angle subtended by a rhombus-like plane having diagonals 6cm & 20cm at a point lying at a height 40cm on the vertical axis passing through the centre & error involved in the result.

Sol. given that

$$diagonal\ of\ rhombus, 2d_1 = 6cm\ , 2d_2 = 20cm\ \&$$
$$height\ of\ the\ point, h = 40cm$$

We know that the approximate value of solid angle subtended by a rectangular plane with sides '$2d_1$' & '$2d_2$' at any point lying at a normal distance 'r' from the centre is given from the eq(3.7) as

$$\omega_{approx.} = 2\pi\left[1 - \frac{1}{\sqrt{\left(1 + \frac{2d_1d_2}{\pi r^2}\right)}}\right]$$

On setting the values in the above expression, we get approximate value of solid angle

$$\omega_{approx.} = 2\pi\left[1 - \frac{1}{\sqrt{\left(1 + \frac{2\times 3\times 10}{\pi(40)^2}\right)}}\right] = 2\pi\left[1 - \frac{1}{\sqrt{1.0119}}\right]$$

$$= 2\pi[1 - 0.994] = 0.0372\ sr \qquad Ans.$$

We know that the solid angle subtended by a rhombus like plane with diagonals $2d_1\&2d_2$ at any point lying at a normal height h is given by the eq(28) as follows

$$\omega_{act.} = 4\cos^{-1}\left(\frac{h\left(d_1^{\,2}\sqrt{h^2+d_2^{\,2}}+d_1^{\,2}\sqrt{h^2+d_2^{\,2}}\right)}{h^2\left(d_1^{\,2}+d_2^{\,2}\right)+d_1^{\,2}d_2^{\,2}}\right)$$

On setting the values, we get

$$\omega_{act.} = 4\cos^{-1}\left(\frac{40\left((3)^2\sqrt{(40)^2+(10)^2}+(10)^2\sqrt{(40)^2+(3)^2}\right)}{(40)^2((3)^2+(10)^2)+(3)^2(10)^2}\right)$$

$$= 4\cos^{-1}\left(\frac{175292.549}{175300}\right) = 4\cos^{-1}(0.999) = 0.0369\ sr$$

$$\therefore\ \%\ error\ (\%\ e) = \left(\frac{\omega_{approx.}-\omega_{act.}}{\omega_{act.}}\right)\times 100$$

$$= \left(\frac{0.0372-0.0369}{0.0369}\right)\times 100 = 0.81\% \qquad Ans.$$

Now, the factor of circularity of a rectangular plane is given as follows

$$F_c = \frac{d_1}{\sqrt{d_1^{\,2}+d_2^{\,2}}}$$

$$\Rightarrow F_c = \frac{3}{\sqrt{(10)^2+(3)^2}} = 0.287 \qquad Ans.$$

We find that the error involved in the value of solid angle is 0.81%which is permissible. It is because the distance of the point from the plane is greater as compared to the dimensions of the plane although factor of circularity is much smaller than unity $(even\ F_c < 0.5)$

Thus the factor of distance is dominant so error is permissible.

Example 3.7: Find out the solid angle subtended by a circular plane with a radius 14cm at any point lying at a normal height 18cm from the centre of plane.

Sol. given that

$$radius\ of\ circular\ plane, R = 14cm\ \&\ normal\ height, h = 18cm$$

$$\omega = 2\pi\left[1-\frac{1}{\sqrt{1+\frac{R^2}{h^2}}}\right]$$

On setting the values, we get

$$\omega = 2\pi\left[1-\frac{1}{\sqrt{1+\frac{(14)^2}{(18)^2}}}\right] = 2\pi\left[1-\frac{1}{\sqrt{1.605}}\right]$$

$$= 2\pi[1-0.789] = 1.323\ sr \qquad Ans.$$

Since, the circular plane is perfect symmetrical hence there is no error in the above value of solid angle

Example 3.8: Find out the solid angle subtended by an elliptical plane with minor & major axes 10cm & 16 at any point lying at a normal height 8cm from the centre of plane.Also find the factor of circularity.

Sol. given that

$$major\ axis\ of\ plane, 2a = 16cm, minor\ axis, 2b = 12cm\ \&\ normal\ height, h = 8cm$$

$$\omega = 2\pi\left[1 - \frac{1}{\sqrt{1 + \frac{ab}{h^2}}}\right]$$

On setting the values, we get

$$\omega = 2\pi\left[1 - \frac{1}{\sqrt{1 + \frac{5 \times 8}{(8)^2}}}\right] = 2\pi\left[1 - \frac{1}{\sqrt{1.625}}\right]$$

$$= 2\pi[1 - 0.784] = 1.354\ sr \qquad Ans.$$

Now, the factor of circularity of an elliptical plane is given as follows

$$F_c = \frac{b}{\sqrt{a^2 + b^2}}$$

$$\Rightarrow F_c = \frac{6}{\sqrt{(8)^2 + (6)^2}} = 0.6 \qquad Ans.$$

Example 3.9: Using Approximation find out the solid angle subtended by an equilateral triangular plane having each side 20cm at a point lying at a height 8cm on the vertical axis passing through the centre & error involved in the result.

Sol. given that

$$number\ of\ the\ sides, n = 3,\ \ length\ of\ each\ side, a = 20cm\ \&$$

$$\boldsymbol{normal\ height, h = 13cm}$$

We know that the approximate value of solid angle subtended by a regular polygonal plane with 'n' number of the sides each of length 'a' at any point lying at a normal height 'h' from the centre of the plane is given by the eq(3.4) as follows

$$\omega \cong 2\pi\left[1 - \frac{1}{\sqrt{1 + \left(\frac{na^2 \cot\frac{\pi}{n}}{4\pi r^2}\right)}}\right]$$

On setting the values in the above expression, we get

$$\omega_{approx.} = 2\pi\left[1 - \frac{1}{\sqrt{1 + \left(\frac{3(20)^2 \cot\frac{\pi}{3}}{4\pi(13)^2}\right)}}\right] = 2\pi\left[1 - \frac{1}{\sqrt{1.326}}\right]$$

$$= 2\pi[1 - 0.868] = 0.827\ sr$$

We know that the exact value of solid angle subtended by a regular polygonal plane with 'n' number of the sides each of length 'a' at any point lying at a normal height 'h' from the centre of the plane is given by the eq(3.4) as follows

$$\omega_{act.} = 2\pi - 2n\sin^{-1}\left(\frac{2hsin\frac{\pi}{n}}{\sqrt{4h^2 + a^2cot^2\frac{\pi}{n}}}\right)$$

Now, on setting the values in the above equation we get

$$\omega_{act.} = 2\pi - 2\times 3\sin^{-1}\left(\frac{2\times 13sin\frac{\pi}{3}}{\sqrt{4(13)^2 + (20)^2cot^2\frac{\pi}{3}}}\right)$$

$$= 2\pi - 6\sin^{-1}\left(\frac{22.517}{\sqrt{809.333}}\right) = 2\pi - 6\sin^{-1}(0.791) = 2\pi - 5.475 = 0.808\ sr$$

$$\therefore\ \%\ error\ (\%\ e) = \left(\frac{\omega_{approx.} - \omega_{act.}}{\omega_{act.}}\right)\times 100$$

$$= \left(\frac{0.827 - 0.808}{0.808}\right)\times 100 = 2.35\ \% \qquad Ans.$$

Thus, we find that the error involved is permissible.

Example 3.10: Using Approximation find out the solid angle subtended by a square plane having each side 16cm at a point lying at a height 6cm on the vertical axis passing through the centre & error involved in the result.

Sol. given that

$$number\ of\ the\ sides, n = 4,\ \ length\ of\ each\ side, a = 16cm\ \&$$
$$normal\ height, h = 6cm$$

We know that the approximate value of solid angle subtended by a regular polygonal plane with 'n' number of the sides each of length 'a' at any point lying at a normal height 'h' from the centre of the plane is given by the eq(3.4) as follows

$$\omega \cong 2\pi\left[1 - \frac{1}{\sqrt{1 + \left(\frac{na^2cot\frac{\pi}{n}}{4\pi r^2}\right)}}\right]$$

On setting the values in the above expression, we get

$$\omega_{approx.} = 2\pi\left[1 - \frac{1}{\sqrt{1 + \left(\frac{4(16)^2cot\frac{\pi}{4}}{4\pi(6)^2}\right)}}\right] = 2\pi\left[1 - \frac{1}{\sqrt{3.263}}\right]$$

$$= 2\pi[1 - 0.553] = 2.805\ sr$$

We know that the exact value of solid angle subtended by a regular polygonal plane with 'n' number of the sides each of length 'a' at any point lying at a normal height 'h' from the centre of the plane is given by the eq(3.4) as follows

$$\omega_{act.} = 2\pi - 2n\sin^{-1}\left(\frac{2hsin\frac{\pi}{n}}{\sqrt{4h^2 + a^2cot^2\frac{\pi}{n}}}\right)$$

Now, on setting the values in the above equation we get

$$\omega_{act.} = 2\pi - 2 \times 4 \sin^{-1}\left(\frac{2 \times 6 sin\frac{\pi}{4}}{\sqrt{4(6)^2 + (16)^2 cot^2\frac{\pi}{4}}}\right)$$

$$= 2\pi - 8\sin^{-1}\left(\frac{8.485}{\sqrt{400}}\right) = 2\pi - 8\sin^{-1}(0.424) = 2\pi - 3.503 = 2.78\ sr$$

$$\therefore\ \%\ error\ (\%\ e) = \left(\frac{\omega_{approx.} - \omega_{act.}}{\omega_{act.}}\right) \times 100$$

$$= \left(\frac{2.805 - 2.78}{2.78}\right) \times 100 = 0.9\ \% \qquad Ans.$$

Thus, we find that the error involved is permissible.

Example 3.11: Using Approximation find out the solid angle subtended by a regular pentagonal plane having each side 12cm at a point lying at a height 10cm on the vertical axis passing through the centre & error involved in the result.

Sol. given that

$$number\ of\ the\ sides, n = 5,\ \ length\ of\ each\ side, a = 12cm\ \&$$
$$normal\ height, h = 10cm$$

We know that the approximate value of solid angle subtended by a regular polygonal plane with 'n' number of the sides each of length 'a' at any point lying at a normal height 'h' from the centre of the plane is given by the eq(3.4) as follows

$$\omega \cong 2\pi\left[1 - \frac{1}{\sqrt{1 + \left(\frac{na^2 cot\frac{\pi}{n}}{4\pi r^2}\right)}}\right]$$

On setting the values in the above expression, we get

$$\omega_{approx.} = 2\pi\left[1 - \frac{1}{\sqrt{1 + \left(\frac{5(12)^2 cot\frac{\pi}{5}}{4\pi(10)^2}\right)}}\right] = 2\pi\left[1 - \frac{1}{\sqrt{1.789}}\right]$$

$$= 2\pi[1 - 0.748] = 1.585\ sr$$

We know that the exact value of solid angle subtended by a regular polygonal plane with 'n' number of the sides each of length 'a' at any point lying at a normal height 'h' from the centre of the plane is given by the eq(3.4) as follows

$$\omega_{act.} = 2\pi - 2n\sin^{-1}\left(\frac{2hsin\frac{\pi}{n}}{\sqrt{4h^2 + a^2cot^2\frac{\pi}{n}}}\right)$$

Now, on setting the values in the above equation we get

$$\omega_{act.} = 2\pi - 2 \times 5 \sin^{-1}\left(\frac{2 \times 10 sin\frac{\pi}{5}}{\sqrt{4(10)^2 + (12)^2 cot^2\frac{\pi}{5}}}\right)$$

$$= 2\pi - 10\sin^{-1}\left(\frac{11.756}{\sqrt{672.797}}\right) = 2\pi - 10\sin^{-1}(0.453)$$

$$= 2\pi - 4.701 = 1.582\ sr$$

$$\therefore\ \%\ error\ (\%\ e) = \left(\frac{\omega_{approx.} - \omega_{act.}}{\omega_{act.}}\right) \times 100$$

$$= \left(\frac{1.585 - 1.582}{1.582}\right) \times 100 = 0.19\ \% \qquad Ans.$$

Thus, we find that the error involved is negligible (permissible).

Where, factor of circularity is given as

$$\Rightarrow F_C = cos\frac{\pi}{n} = cos\frac{\pi}{5} = 0.809$$

The value of F_C of a regular pentagonal plane is very close to unity.

Example 3.12: Using Approximation find out the solid angle subtended by a regular hexagonal plane having each side 15cm at a point lying at a height 8cm on the vertical axis passing through the centre & error involved in the result.

Sol. given that

$$number\ of\ the\ sides, n = 6,\ \ length\ of\ each\ side, a = 15cm\ \&$$
$$normal\ height, h = 8cm$$

We know that the approximate value of solid angle subtended by a regular polygonal plane with 'n' number of the sides each of length 'a' at any point lying at a normal height 'h' from the centre of the plane is given by the eq(3.4) as follows

$$\omega \cong 2\pi\left[1 - \frac{1}{\sqrt{1 + \left(\frac{na^2 cot\frac{\pi}{n}}{4\pi r^2}\right)}}\right]$$

On setting the values in the above expression, we get

$$\omega_{approx.} = 2\pi\left[1 - \frac{1}{\sqrt{1 + \left(\frac{6(15)^2 cot\frac{\pi}{6}}{4\pi(8)^2}\right)}}\right] = 2\pi\left[1 - \frac{1}{\sqrt{3.907}}\right]$$

$$= 2\pi[1 - 0.506] = 3.104\ sr$$

We know that the exact value of solid angle subtended by a regular polygonal plane with 'n' number of the sides each of length 'a' at any point lying at a normal height 'h' from the centre of the plane is given by the eq(3.4) as follows

$$\omega_{act.} = 2\pi - 2n\sin^{-1}\left(\frac{2hsin\frac{\pi}{n}}{\sqrt{4h^2 + a^2 cot^2\frac{\pi}{n}}}\right)$$

Now, on setting the values in the above equation we get

$$\omega_{act.} = 2\pi - 2\times 6\sin^{-1}\left(\frac{2\times 8sin\frac{\pi}{6}}{\sqrt{4(8)^2 + (15)^2 cot^2\frac{\pi}{6}}}\right)$$

$$= 2\pi - 12\sin^{-1}\left(\frac{8}{\sqrt{931}}\right) = 2\pi - 12\sin^{-1}(0.262)$$

$$= 2\pi - 3.181 = 3.102\ sr$$

$$\therefore\ \%\ error\ (\%\ e) = \left(\frac{\omega_{approx.} - \omega_{act.}}{\omega_{act.}}\right) \times 100$$

$$= \left(\frac{3.104 - 3.102}{3.102}\right) \times 100 = 0.06\ \% \qquad Ans.$$

Thus, we find that the error involved is negligible (permissible).

Where, factor of circularity is given as

$$\Rightarrow F_C = cos\frac{\pi}{n} = cos\frac{\pi}{6} = 0.866$$

The value of $F_C$of a regular hexagonal plane is very close to unity.

Example 3.13: Using approximation, find out the solid angle subtended by a plane bounded by the curve $4x^2 + 9y^2 + 16x + 18y - 11 = 0$ at the point$(-2, -1, 4)$.

Sol. given that

The equation of the curve,

$$4x^2 + 9y^2 + 16x + 18y - 11 = 0$$

Compare the above equation with the general form as follows

$$ax^2 + by^2 + 2hxy + 2gx + 2fy + c = 0$$

$$\Rightarrow a = 4, b = 9, h = 0, g = 8, f = 9\ \&\ c = -11$$

Now, let's check whether the equation represents a pair of straight lines or a conic

$$\Delta = abc + 2fgh - af^2 - bg^2 - ch^2$$
$$= (4)(9)(-11) + 2(9)(8)(0) - (4)(9)^2 - (9)(8)^2 - (-11)(0)^2$$

$$= -396 - 324 - 576 = -12960 \ \Rightarrow\ \Delta \neq 0$$

Thus, the given equation represents a conic, we have

$$h^2 - ab = (0)^2 - (4)(9) = -36 \Rightarrow h^2 - ab < 0\ \&\ a \neq b$$

Hence, the equation represents an ellipse, thus we have

$$4x^2 + 9y^2 + 16x + 18y - 11 = 0 \Rightarrow 4(x^2 + 4x + 4) + 9(x^2 + 2x + 1) = 36$$

$$\Rightarrow \frac{4(x+2)^2}{36} + \frac{9(y+1)^2}{36} = 1\ \ or\ \ \frac{(x+2)^2}{3^2} + \frac{(y+1)^2}{2^2} = 1$$

On comparing the above equation with standard form of equation of ellipse as given below

$$\frac{X^2}{A^2} + \frac{Y^2}{B^2} = 1$$

We get,

$$A = 3, B = 2\ \&\ centre\ (X, Y) = 0 \Rightarrow (x + 2 = 0, y + 1 = 0)\ or\ (-2, -1)$$

Since, the given point $(-2, -1, 4)$ lies on the axis at a normal distance, h = 4 unit from the centre $(-2, -1)$ of the ellipse

Hence, the solid angle subtended by the elliptical plane at the given point

$$\omega = 2\pi\left[1 - \frac{1}{\sqrt{1 + \frac{ab}{h^2}}}\right]$$

On setting the values, we get

$$\omega_{approx.} = 2\pi\left[1-\frac{1}{\sqrt{1+\frac{3\times 2}{(4)^2}}}\right] = 2\pi\left[1-\frac{1}{\sqrt{1.375}}\right]$$

$$= 2\pi[1-0.853] = 0.925\ sr \qquad Ans.$$

Example 3.14: Using approximation, find out the solid angle subtended by a plane bounded by the curves $x^2+y^2=256\ \&9x^2+25y^2=225$ at the point$(0,0,-5)$.

Sol. given that

Equations of the curves are as follows

$$x^2+y^2=256 \Rightarrow x^2+y^2=16^2$$

The above equation represents a circle with a radius R = 16 unit & centre at the origin$(0,0,0)$. Since the given point$(0,0,-5)$ lies on the Z-axis (towards $-ve$ direction) at a normal height h = 5 unit from the centre (i.e. origin) of the circle,

Hence, the solid angle subtended by circle at the given point $(0,0,-5)$

$$\omega = 2\pi\left[1-\frac{1}{\sqrt{1+\frac{R^2}{h^2}}}\right]$$

On setting the values, we get

$$\omega_1 = 2\pi\left[1-\frac{1}{\sqrt{1+\frac{(16)^2}{(5)^2}}}\right] = 2\pi\left[1-\frac{1}{\sqrt{11.24}}\right]$$

$$= 2\pi[1-0.298] = 4.409\ sr \qquad \&$$

$$9x^2+25y^2=225 \Rightarrow \frac{9x^2}{225}+\frac{25y^2}{225}=1\ or\ \frac{x^2}{5^2}+\frac{y^2}{3^2}=1$$

The above equation shows an ellipse with semi major & minor axes as a = 5 & b = 3 and centre at the origin$(0,0,0)$. Since the given point$(0,0,-5)$ lies on the Z-axis (towards $-ve$ direction) at a normal height h = 5 unit from the centre (i.e. origin) of the ellipse,

Hence, the solid angle subtended by ellipse at the given point $(0,0,-5)$

$$\omega = 2\pi\left[1-\frac{1}{\sqrt{1+\frac{ab}{h^2}}}\right]$$

On setting the values, we get

$$\omega_2 = 2\pi\left[1-\frac{1}{\sqrt{1+\frac{5\times 3}{(5)^2}}}\right] = 2\pi\left[1-\frac{1}{\sqrt{1.6}}\right]$$

$$= 2\pi[1-0.79] = 1.316\ sr$$

Since, the circle and ellipse both are centred at the origin and the ellipse is completely enclosed by the circle (since, radius, $R > a, b$) hence, the region bounded by curves will be a common region between circle and ellipse

Hence, the solid angle subtended by bounded region at the given point $(0,0,-5)$

$$\Rightarrow \omega_{approx.} = \omega_1 - \omega_2 = 4.409 - 1.316 = 3.093\ sr \qquad Ans.$$

Summary

1. Approximate value of the solid angle subtended by a symmetrical plane at any point in the space is given by 'Approximation-theorem' as follows

$$\omega \cong 2\pi\left[1 - \frac{1}{\sqrt{\left(1 + \frac{A\cos\theta}{\pi r^2}\right)}}\right] \quad \forall \left(0 \le \theta \le \frac{\pi}{2}\right)$$

Where $'A' \rightarrow$ is the area of the given symmetrical plane

$'\theta' \rightarrow$ is the angle between the normal to the plane & the line joining the given point to the centre of the plane &

$'r' \rightarrow$ is the distance of the given point from the centre of the plane

Limitation:

This formula is applicable only for the planes satisfying the following two conditions

i. The plane should have at least two axes of symmetry i.e. a point of symmetry &

ii. The factor of circularity(F_c) should be close to the unity i.e. $F_c \ge 0.5$

If the above conditions are satisfied then the error involved in the results obtained from Approximation-formula will be minimum & permissible.

2. The solid angle subtended by an elliptical plane with major & minor axes 'a' & 'b' at any point lying at a normal distance 'r' from the centreis given as

$$\omega \cong 2\pi\left[1 - \frac{1}{\sqrt{\left(1 + \frac{ab}{r^2}\right)}}\right]$$

Factor of circularity of elliptical plane,

$$F_c = \sqrt{1 - e^2}\left[\forall\ (0 < F_c < 1) \&\ e = \sqrt{1 - \frac{b^2}{a^2}}\right]$$

3. The solid angle subtended by a circular plane with a radius 'R' at any point lying at a normal distance 'r' from the centre is given as

$$\omega = 2\pi\left[1 - \frac{1}{\sqrt{\left(1 + \frac{R^2}{r^2}\right)}}\right]$$

Factor of circularity of circular plane,

$$F_c = 1$$

4. The solid angle subtended by a regular polygonal plane with 'n' number of the sides each of length 'a' at any point lying at a normal distance 'r' from the centre is given as

$$\omega \cong 2\pi\left[1-\frac{1}{\sqrt{1+\left(\frac{na^2cot\frac{\pi}{n}}{4\pi r^2}\right)}}\right] \qquad \forall(n \in N \;\&\; n \geq 3)$$

Factor of circularity of regular polygonal plane,

$$F_c = cos\frac{\pi}{n}\left[\forall\; n \in N \;\&\; n \geq 3 \;\Rightarrow\left(\frac{1}{2} \leq F_c < 1\right)\right]$$

Higher is the number of sides higher will be the factor of circularity of a regular polygon.

5. The solid angle subtended by a rectangular plane with sides 'l' & 'b' at any point lying at a normal distance 'r' from the centre is given as

$$\omega \cong 2\pi\left[1-\frac{1}{\sqrt{\left(1+\frac{lb}{\pi r^2}\right)}}\right]$$

Factor of circularity of rectangular plane,

$$F_c = \frac{b}{\sqrt{b^2+l^2}}\left[\forall\; b < l \;\Rightarrow(0 < F_c < 0.707)\right]$$

6. The solid angle subtended by a rhombus-like plane with diagonals $'d_1'$ & $'d_2'$ at any point lying at a normal distance 'r' from the centre is given as

$$\omega \cong 2\pi\left[1-\frac{1}{\sqrt{\left(1+\frac{d_1 d_2}{2\pi r^2}\right)}}\right]$$

Factor of circularity of rhombus-like plane,

$$F_c = \frac{d_1}{\sqrt{{d_1}^2+{d_2}^2}}\left[\forall\; d_1 < d_2 \Rightarrow(0 < F_c < 0.707)\right]$$

7. The solid angle subtended by a circular plane at any point in the space is given as

$$\omega \cong 2\pi\left[1-\left\{1+\left(\frac{R\sqrt{d^2+R^2-2dRsin\theta}-h^2}{h^2}\right)\right\}^{-\frac{1}{2}}\right]$$

The value of h^2 is found out from the auxiliary equation (III) given as follows

$$\Rightarrow[(A^2+B^2)h^4+(2AC-B^2k^2)h^2+C^2=0]$$

Where

$$A = -2\left(\frac{Rcos\theta}{(d^2+R^2-2dRsin\theta)^{\frac{3}{2}}}\right), \qquad B = 2\left(\frac{d-Rsin\theta}{(d^2+R^2-2dRsin\theta)^{\frac{3}{2}}}\right)$$

$$C = Rcos\theta\left(\frac{1}{\sqrt{d^2+R^2-2dRsin\theta}}-\frac{1}{\sqrt{d^2+R^2+2dRsin\theta}}\right) \;\&\; k = \sqrt{d^2+R^2-2dRsin\theta}$$

8. The solid angle subtended by an elliptical plane with major axis 'a', minor axis 'b' & eccentricity 'e', at a distance 'd' from the centre to a particular point from which it appears perfect circular plane is given as

$$\Rightarrow \left[\omega = 2\pi\left(1 - \frac{adcos\theta}{b\sqrt{a^2 + d^2 + 2adsin\theta}}\right) = 2\pi\left(1 - \frac{dcos\theta}{\sqrt{1-e^2}\sqrt{a^2 + d^2 + 2adsin\theta}}\right)\right]$$

$$\forall\left(0 \le \theta < \frac{\pi}{2} \& \, d \ge 0\right)$$

Where, $'\theta'$ is the angle between normal to the plane & the line joining the particular point to the centre of the plane.

Exercise - 3

Q.3.1. Using Approximationfind out the solid angle subtended by a rectangular plane of $30cm \times 40cm$ at a point lying at a height 20cm on the vertical axis passing through the centre & error involved in the result.

Q.3.2. Find out the solid angle subtended by a rectangular plane $45cm \times 60cm$ at a point lying at a height 50cm on the vertical axis passing through the vertex. Use 'Approximation'&also find out the error involved in the result.

Q.3.3. Using Approximation find out the solid angle subtended by a rhombus-like plane having diagonals 20cm & 30cm at a point lying at a height 13cm on the vertical axis passing through the centre & error involved in the result.

Q.3.4. Using Approximation find out the solid angle subtended by a rhombus-like plane with diagonals 22cm & 28cm at a point lying at a height 35cm on the vertical axis passing through one of the vertices & factor of circularity.

Q.3.5. Find out the solid angle subtended by a equilateral triangular plane with each side 20cm at a point lying at a height 25cm on the vertical axis passing through the centre of the plane. Use 'Approximation' & also find out the error involved in the result.

Q.3.6. Find out the solid angle subtended by a equilateral triangular plane with circumscribed radius $12\sqrt{3}$ cm at a point lying at a height 25cm on the vertical axis passing through the vertex of the plane. Use 'Approximation' & also find out the error involved in the result.

Q.3.7. Using Approximation find out the solid angle subtended by a equilateral triangular plane with inscribed radius $9\sqrt{3}$ cm at a point lying at a height 40cm on the vertical axis passing through a point internally dividing one of the sides in a ratio 1) 1: 1 2) 1: 2 & 3) 2: 3

Q.3.8. Using Approximation find out the solid angle subtended by a square plane with each side 20cm at a point lying at a height 12cm on a vertical axis passing through the 1. Centre & 2.vertex 3.mid-point of the side.

Q.3.9. Using Approximation find out the solid angle subtended by a square plane with having diagonal 50cm at a point lying at a height 10cm on a vertical axis passing through a point internally dividing the side into a ratio 2:3. Also find out the error involved in the result.

Q.3.10. Find out the solid angle subtended by a regular pentagonal plane with each side 13cm at a point lying at a height 12cm on a vertical axis passing through the centre of the plane. Use 'Approximation' & also find the error involved in the result.

Q.3.11. Using Approximation find out the solid angle subtended by a regular pentagonal plane with each side 16cm at a point lying at a height 9cm on a vertical axis passing through

1. One of the vertices & 2.mid-point of one of the sides of the plane.

Q.3.12. Find out the approximate value of solid angle subtended by a by a regular hexagonal plane with each side 10cm at a point lying at a height 60cm on a vertical axis passing through the centre of the plane. Use 'Approximation' & also find the error involved in the result.

Q.3.13. Using Approximation find out the solid angle subtended by a regular hexagonal plane with each side 14cm at a point lying at a height 10cm on a vertical axis passing through

1. One of the vertices & 2.mid-point of one of the sides of the plane.

Q.3.14. Use Approximation&find out the solid angle subtended by hexagonal plane having each side 12cm at a point lying at a height 150mm on a vertical axis passing through the 1) Centre 2) vertex & 3) a point dividing one of the sides in a ratio 2:5

Q.3.15. Find out the solid angle subtended by a polygonal plane having twenty sides each 6cm at a point lying at a height 110mm on a vertical axis passing through the centre. Use 'Approximation' & also find the error involved in the result.

Q.3.16. Using Approximation find out the solid angle subtended by an elliptical plane with minor & major axes 20cm & 30cm respectively at a point lying at a height 16cm on a vertical axis passing through the centre& also find the factor of circularity.

Q.3.17. Find out the solid angle subtended by a plane bounded by the lines $x = \pm 12$ & $y = \pm 5$ at a point $(0, 0, 6)$. Use 'Approximation formula'.

Q.3.18. Calculate the solid angle subtended by a plane bounded by the lines $x = \pm 13$ & $y = \pm 13$ at a point $(0, 0, -4)$. Use 'Approximation formula'.

Q.3.19. Find out the solid angle subtended by a plane bounded by the lines $7x + 4y = \pm 28$ & $7x - 4y = \pm 28$ at a point $(0, 0, 7)$. Use 'Approximation formula'.

Q.3.20. Find out the solid angle subtended by a plane bounded by the lines $y = \pm 7\sqrt{3}, \sqrt{3}x + y = \pm 14\sqrt{3}$&$\sqrt{3}x - y = \pm 14\sqrt{3}$ at a uniform point-source of 100 cd located at the point$(0, 0, 5)$. If the flux is uniformly distributed over the entire plane what is the luminous flux density using approximation?

Q.3.21. Find out approximate value of the solid angle subtended by a plane bounded by the circle $x^2 + y^2 = 169$ at the point $(0, 0, -11)$. Use 'Approximation formula'.

Q.3.22. Using approximation, find out the solid angle subtended by a plane bounded by the curve $x^2 + y^2 - 6x - 14y - 86 = 0$ at the point $(3, 7, 9)$.

Q.3.23. Using approximation, find out the solid angle subtended by a plane bounded by the curve $4x^2 + 9y^2 = 36$ at the point$(0, 0, 5)$.

Q.3.24. Using approximation, find out the solid angle subtended by a plane bounded by the curve $9x^2 + 16y^2 - 18x - 64y - 71 = 0$ at the point $(1, 2, 6)$.

Q.3.25. Using approximation, find out the solid angle subtended by a plane bounded by the curves $x^2 + y^2 = 4$ &$9x^2 + 25y^2 = 225$ at the point $(0, 0, 3)$.

Q.3.26. Find out approximate value of the solid angle subtended by a plane bounded by the curves $x^2 + y^2 = 144$ &$25x^2 + 36y^2 = 900$ at the point $(0, 0, -7)$.

Q.3.27. Find out approximate value of the solid angle subtended by a plane bounded by the curves $x^2 + y^2 - 10x + 14y - 47 = 0$ &$4x^2 + 9y^2 - 40x + 126y + 505 = 0$ at the point $(5, -7, -12)$.

Q.3.28. Find out approximate value of the solid angle subtended by a plane bounded by the curves $15x^2 + 16y^2 + 210x - 96y + 639 = 0$ &$4x^2 + 9y^2 + 56x - 54y + 241 = 0$ at the point $(7, 3, -4)$.

Chapter 4
Analysis of 3D-Figures

In this chapter, some particular & regular three dimensional figures (both solid and hollow) are analysed in order to find the solid angle subtended by the same at some particular point (i.e. location of the observer) in the space.

The 3D-figures which are analysed are as follows

1. Sphere
   a) solid sphere& b) hemispherical shell (cap).
2. Circular cylinder
   a) solid cylinder& b) cylindrical shell (hollow)
3. Right cone
   a) solid right cone & b) conical shell (hollow)
4. Right pyramid (with regular polygonal base)
   a) solid right pyramid & b) right pyramidal shell (hollow)
5. Right prism (with regular polygonal cross-section)
   a) solid right prism b) prismatic shell (hollow)

Whenever, we observe an objecthaving two or three finite dimensions it seen that the same object can be visualised as a two dimensional figure (i.e. a plane-figure). It is because every object with finite two or three dimensions in the space can be completely eclipsed by a finite plane with certain geometrical shape& dimensions. In such a way, the solid angle subtended by a given object at a certain point in the space would be equal to that subtended by a plane completely eclipsing the object at the same point.

Thus, the basic formulae of the plane-figures are used in order to find the solid angle subtended by three dimensional figures.

4.1 SOLID ANGLE SUBTENDED BY A CLOSED SURFACE AT ANY POINT COMPLETELY INSIDE ITS BOUNDARY

Method 1: Using Fundamental theorem

Let there be a closed surface & a given point say 'P' lying completely inside the surface boundary.

(as shown in the figure (77) below)

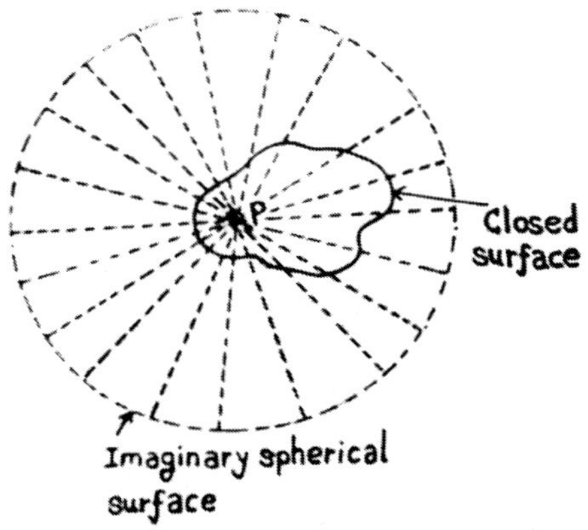

Fig (77)

Now, consider an imaginary spherical surface with centre at the given point 'P' & a radius say 'R' such that it completely encloses the given surface.

Then in such case, the projection ofclosed surface w.r.t. the point 'P' on the spherical surface will be the internal spherical surface as a whole.

$$\Rightarrow A = area\ of\ projection\ of\ closed\ surface = surface\ area\ of\ sphere = 4\pi R^2$$

Hence, the solid angle subtended by the closed surface at the point 'P'

= solid angle subtended by the spherical surface at the centre (i.e. point 'P')

Using fundamental theorem

$$\Rightarrow \omega = \int_S \frac{dA}{r^2} = \int_S \frac{dA}{R^2} = \frac{1}{R^2}\int_S dA \quad (since,\ 'R' is\ constant\ for\ all\ the\ points)$$

$$\Rightarrow \omega = \frac{A}{R^2} = \frac{4\pi R^2}{R^2} = 4\pi\ sr$$

Thus, the above result is independent of the geometrical shape & the dimensions of closed surface.

Method 2: Using Integral theorem

We know the solid angle subtended by any object at a given point in the space is given as

$$\Rightarrow \left[\omega = \int_{\varphi_1}^{\varphi_2}\left|\int_{\theta_1}^{\theta_2} sin\theta d\theta\right| d\varphi\right] \qquad \forall(0 \le \theta \le \pi\ \&-\pi \le \varphi \le \pi)$$

Whereθ & φ are parameters (variables).

Similarly, consider an imaginary spherical surface with centre at the given point 'P' & a radius say 'R' such that it completely encloses the given surface. (see fig(76) above)

the solid angle subtended by the closed surface at the point 'P'

= solid angle subtended by the spherical surface at the centre (i.e. point 'P')

Now, applying the limits of the parameters $\theta\ from\ \theta_1 = 0\ to\ \theta_2 = \pi$ &

$\varphi\ from\ \varphi_1 = -\pi\ to\ \varphi_2 = \pi$ for the whole spherical surface

(see the fig(78) below)

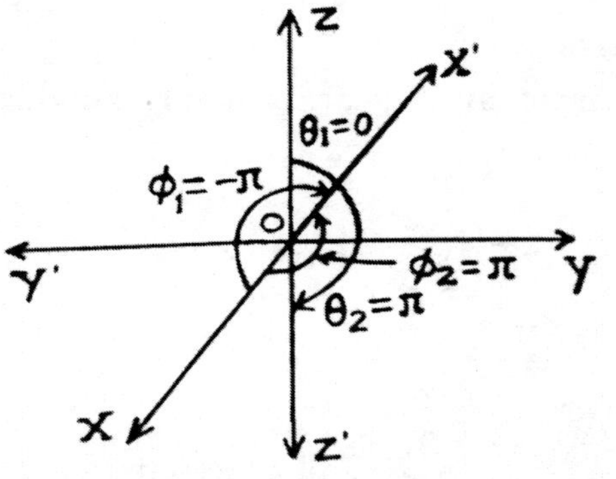

Fig (78)

$$\Rightarrow \omega = \int_{\varphi_1=-\pi}^{\varphi_2=\pi}\left|\int_{\theta_1=0}^{\theta_2=\pi} sin\theta d\theta\right| d\varphi = \int_{\varphi_1=-\pi}^{\varphi_2=\pi} |[-cos\theta]_0^\pi| d\varphi$$

$$\int_{\varphi_1=-\pi}^{\varphi_2=\pi} |-cos\pi - (-cos0)| d\varphi = \int_{\varphi_1=-\pi}^{\varphi_2=\pi} |-(-1) - (-1)| d\varphi = \int_{\varphi_1=-\pi}^{\varphi_2=\pi} |2| d\varphi$$

$$= 2\int_{\varphi_1=-\pi}^{\varphi_2=\pi} d\varphi = 2[\varphi]_{-\pi}^{\pi} = 2(\pi - (-\pi)) = 4\pi\ sr$$

Thus, the above result is independent of the geometrical shape & the dimensions of closed surface.

4.2 SOLID ANGLE SUBTENDED AT ANY POINT LYING ON A PLANE FULLY COVERED BY A CAP

Method 1: Using Fundamental theorem

Let there be a cap (irrespective of geometry) fully covering a plane &given point say 'P' lying on the plane completely inside the cap.

(as shown in the figure (79) below)

Now, consider an imaginary spherical surface with centre at the given point 'P' & a radius say 'R' such that it completely encloses the cap.

Then in such case, the projection of the cap w.r.t. the point 'P' on the spherical surface will be half of internal spherical surface.

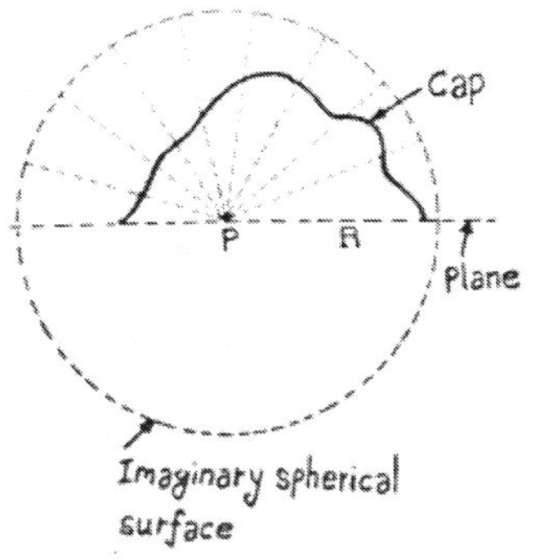

Fig (79)

$$\Rightarrow A = area\ of\ projection\ of\ the\ cap = surface\ area\ of\ hemisphere$$
$$= 2\pi R^2$$

Hence, the solid angle subtended by the cap at the point 'P'

= solid angle subtended by the hemispherical surface at the centre (i.e. point 'P')

Using fundamental theorem

$$\Rightarrow \omega = \int_S \frac{dA}{r^2} = \int_S \frac{dA}{R^2} = \frac{1}{R^2}\int_S dA\ \ (since,\ 'R' is\ constant\ for\ all\ the\ points)$$

$$\Rightarrow \omega = \frac{A}{R^2} = \frac{2\pi R^2}{R^2} = 2\pi\ sr$$

Thus, the above result is independent of the geometrical shape & the dimensions of the cap.

Method 2: Using Integral theorem

We know the solid angle subtended by any object at a given point in the space is given as

$$\Rightarrow \left[\omega = \int_{\varphi_1}^{\varphi_2}\left|\int_{\theta_1}^{\theta_2} sin\theta d\theta\right| d\varphi\right] \qquad \forall(0 \le \theta \le \pi\ \&-\pi \le \varphi \le \pi)$$

Whereθ & φ are parameters (variables).

Similarly, consider an imaginary spherical surface with centre at the given point 'P' & a radius say 'R' such that it completely encloses the given cap. (see fig(79) above)

the solid angle subtended by the cap at the point 'P'

= solid angle subtended by the hemispherical surface at the centre (i.e. point 'P')

Now, applying the limits of the parameters $\theta\ from\ \theta_1 = 0\ to\ \theta_2 = \frac{\pi}{2}$&

$\varphi\ from\ \varphi_1 = -\pi\ to\ \varphi_2 = \pi$ for the hemispherical surface (see the fig(80) below)

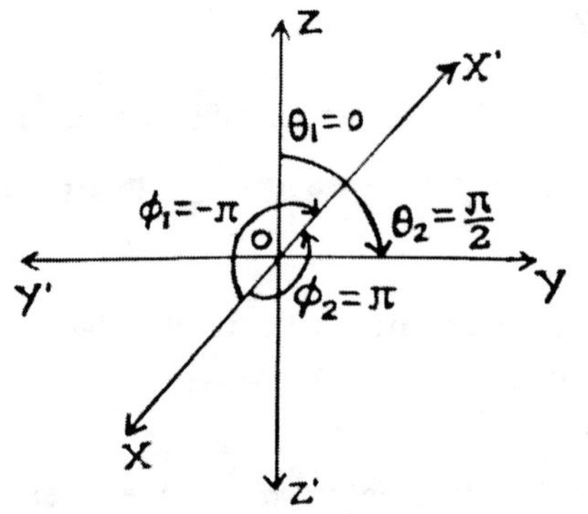

Fig (80)

$$\Rightarrow \omega = \int_{\varphi_1=-\pi}^{\varphi_2=\pi}\left|\int_{\theta_1=0}^{\theta_2=\frac{\pi}{2}} \sin\theta d\theta\right| d\varphi = \int_{\varphi_1=-\pi}^{\varphi_2=\pi}\left|[-\cos\theta]_0^{\frac{\pi}{2}}\right| d\varphi$$

$$\int_{\varphi_1=-\pi}^{\varphi_2=\pi}\left|-\cos\frac{\pi}{2} - (-\cos 0)\right| d\varphi = \int_{\varphi_1=-\pi}^{\varphi_2=\pi} |-(0) - (-1)| d\varphi = \int_{\varphi_1=-\pi}^{\varphi_2=\pi} |1| d\varphi$$

$$= \int_{\varphi_1=-\pi}^{\varphi_2=\pi} d\varphi = [\varphi]_{-\pi}^{\pi} = (\pi - (-\pi)) = 2\pi\ sr$$

Thus, the above result is independent of the geometrical shape & the dimensions of the cap.

4.3 SPHERE

In this section, we are to study & make solutions of the solid or the hollow sphere, spherical shell (cap) & minor frustum of the sphere.

4.3.1 Solid angle subtended by a sphere at a given point in the space

Let there a sphere having centre 'O' & radius 'R' and a given point say 'P' at a distance d from the centre 'O' of the sphere. (as shown in the figure(81) below)

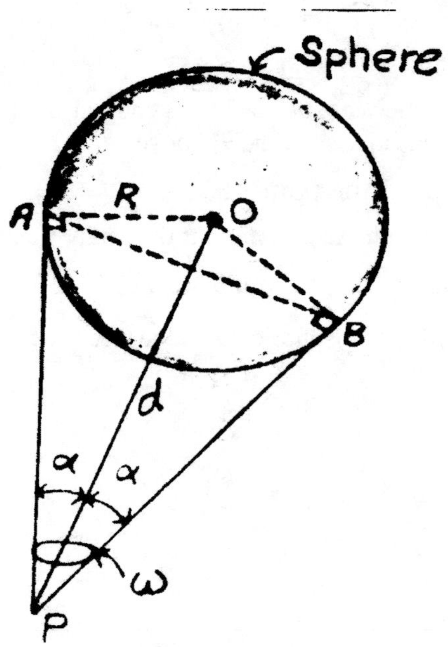

Fig (81)

Now, draw two tangents PA & PB from the point 'P' to the surface of the sphere such that the maximum angle between them is2α. Join the points 'A' & 'B' to the centre 'O'.

$$\Rightarrow OA = OB = R \;\&\; OP = d$$

In right ΔOAP

$$\Rightarrow PA = \sqrt{OP^2 - OA^2} = \sqrt{d^2 - R^2}\;\&\; cos\alpha = \frac{PA}{OP} = \frac{\sqrt{d^2 - R^2}}{d}$$

Now, consider an imaginary right cone with apex at the point 'P' & minimum apex angle 2α such that its lateral face encloses the sphere intercepting at a circle with a diameter AB (see the above fig(81)).

Hence, the solid angle subtended by the sphere at the point 'P'

= solid angle subtended by the imaginary right cone at the apex 'P'

From the eq(1), we have

$$\Rightarrow \omega = 2\pi(1 - cos\alpha)$$

$$\Rightarrow \omega = 2\pi\left[1 - \frac{\sqrt{d^2 - R^2}}{d}\right] \quad \forall(d \geq R) \qquad \dots\dots\dots\dots\dots\dots (39)$$

Important deductions:

1. The solid angle subtended by the sphere at any point lying on its surface is obtained by putting d = R in the above eq(39) as follows

$$\omega = 2\pi\left[1 - \frac{\sqrt{R^2 - R^2}}{d}\right] = 2\pi(1 - 0) = 2\pi\ sr$$

2. The solid angle subtended by the sphere at any point lying at the infinity is obtained by taking the limit of ω at $x \to \infty$ from the above eq(39) as follows

$$\Rightarrow \omega = \lim_{x\to\infty} 2\pi\left[1 - \frac{\sqrt{d^2 - R^2}}{d}\right] = 2\pi \lim_{x\to\infty}\left[1 - \sqrt{1 - \left(\frac{R}{d}\right)^2}\right] = 2\pi\left[1 - \sqrt{1 - \left(\frac{R}{\infty}\right)^2}\right]$$

$$= 2\pi\left[1 - \sqrt{1 - 0}\right] = 2\pi[1 - 1] = 0$$

Thus the sphere behaves as a point for the observer lying at the infinity.

4.3.2 Solid angle subtended by a hemispherical shell (cap) at any point lying on the geometrical axis

Let there be a hemispherical shell with centre 'O' & radius R and a given point say 'P' lying on the geometrical axis at a distance 'x' towards inside the shell (as shown in the figure (82) below)

Now, draw a reference line AB diametrically passing through the centre 'O' of the circular opening (mouth of the cap)& join the point 'P' with the points 'A & 'B'

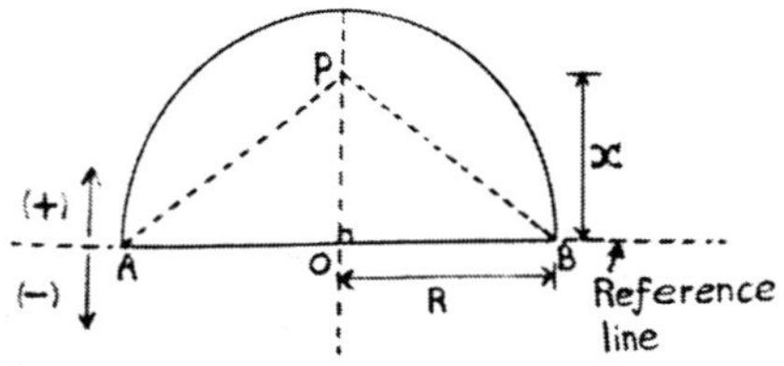

Fig (82)

The solid angle subtended by the circular opening of the shell at the point 'P' is given from the eq(37) as follows

$$\omega = 2\pi\left[1 - \frac{h}{\sqrt{h^2 + R^2}}\right] = 2\pi\left[1 - \frac{x}{\sqrt{x^2 + R^2}}\right] (since, height\ (h) = x\ )$$

Assume that the circular opening of the shell is closed then the solid angle subtended by the closed shell at the point 'P' (completely inside its surface-boundary) = 4π

Hence, the solid angle subtended by the hemispherical shell (cap) at the given point 'P'

= (solid angle subtended by the closed shell at the point 'P') – (solid angle subtended by the circular opening at the point 'P')

$$\Rightarrow \omega = 4\pi - 2\pi\left[1 - \frac{x}{\sqrt{x^2+R^2}}\right] = 4\pi - 2\pi + \frac{2\pi x}{\sqrt{x^2+R^2}} = 2\pi + \frac{2\pi x}{\sqrt{x^2+R^2}}$$

$$\Rightarrow \omega = 2\pi\left[1 + \frac{x}{\sqrt{x^2+R^2}}\right] \quad \forall\left(x < R \text{ or } x \in (-\infty, R)\right) \quad \dots\dots\dots\dots\dots\dots\dots.(40)$$

Sign convention: Distance (x) (measured on the geometrical axis of the spherical shell)

1. is positive above the reference line & towards inside the shell and
2. is negative below the reference line & away from the circular opening.

Case 1: when the given point lies on the geometrical axis below the reference line

The given point 'P' is lying on the geometrical axis of the spherical shell at a distance x from the centre 'O' below the reference line AB & away from the circular opening. (as shown in the figure (83) below)

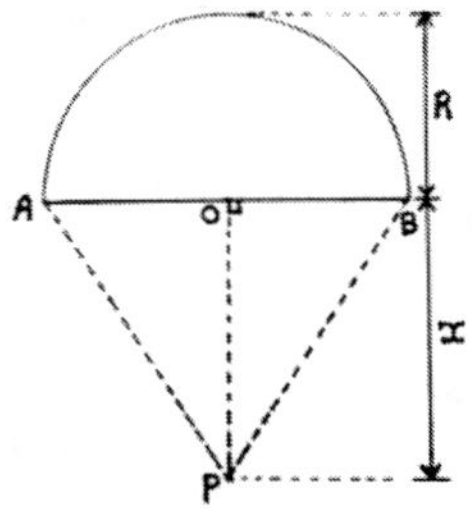

Fig (82)

In this case, the solid angle subtended by the hemispherical shell at the given point 'P'

= solid angle subtended by the circular opening of the shell at the same point 'P'

Since, the point is lying below the reference line hence by setting $x = -x$ in the above eq(40) as follows

$$\Rightarrow \omega = 2\pi\left[1 + \frac{(-x)}{\sqrt{(-x)^2+R^2}}\right]$$

$$= 2\pi\left[1 - \frac{x}{\sqrt{x^2+R^2}}\right] \quad \forall \quad (x \geq 0) \dots\dots\dots\dots\dots\dots\dots.(41)$$

Case 2: when the given point lies in the plane of the open base inside the shell

The given point lies on the centre 'O' of the shell (as shown in the figure (84) below)

In this case, the solid angle subtended by the hemispherical shell at the centre 'O' is obtained by setting x = 0 in the eq(40) or (41) as follows

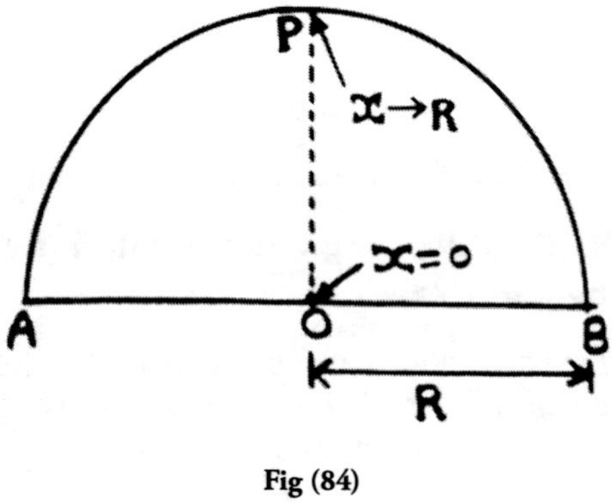

Fig (84)

$$\Rightarrow \omega = 2\pi\left[1 + \frac{(0)}{\sqrt{(0)^2+R^2}}\right] = 2\pi[1+0]$$

$$= 2\pi \ sr$$

or

$$\Rightarrow \omega = 2\pi\left[1 - \frac{(0)}{\sqrt{(0)^2+R^2}}\right] = 2\pi[1-0]$$

$$= 2\pi \ sr$$

Note: The solid angle subtended by the hemispherical shell (cap) at any point lying in the plane circular opening inside the shell (cap) is always $2\pi \ sr$

Case 3: when the given point lies on the geometrical axis & touching the surface inside the shell($x \to R^-$) (see the above fig(84))

In this case, solid angle subtended by the hemispherical shell at the same point can be found out by taking the limit of ω at $x = R$ *when* $x \to R^-$ from the eq(40) as follows

$$\Rightarrow \omega = \lim_{x \to R^-} 2\pi\left[1 + \frac{x}{\sqrt{x^2 + R^2}}\right] = 2\pi \lim_{k \to 0}\left[1 + \frac{(R-k)}{\sqrt{(R-k)^2 + R^2}}\right] \Rightarrow x = R - k \;\&\; k \to 0$$

$$= 2\pi\left[1 + \frac{(R-0)}{\sqrt{(R-0)^2 + R^2}}\right] (since, k \neq 0)$$

$$= 2\pi\left[1 + \frac{R}{\sqrt{R^2 + R^2}}\right] = 2\pi\left[1 + \frac{R}{R\sqrt{2}}\right] = 2\pi\frac{(\sqrt{2}+1)}{\sqrt{2}} = \pi(2 + \sqrt{2}) = 3.414\pi\ sr$$

4.3.3Solid angle subtended by a hemispherical shell (cap) at any point lying on the geometrical axis towards the surface ($x \geq R$)

Let there be a hemispherical shell with centre 'O' & radius 'R' and a given point say 'P' lying on the geometrical axis of the shell at a distance 'x' from the centre 'O'. (as shown in the figure (85) below)

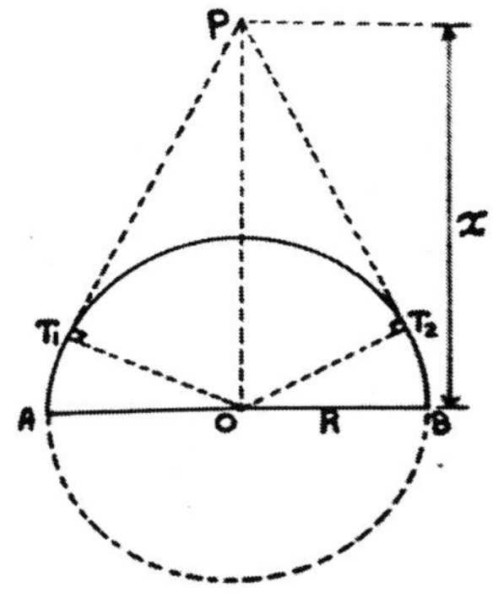

Fig (85)

In this case, the hemispherical shell will behave as a solid perfect sphere

Hence, solid angle subtended by the hemispherical shell at the point 'P'

= solid angle subtended by the sphere at the same point 'P

$$\Rightarrow \omega = 2\pi\left[1 - \frac{\sqrt{d^2 - R^2}}{d}\right]$$

On setting d = distance from the centre 'O' = x, we get

$$\Rightarrow \omega = 2\pi\left[1 - \frac{\sqrt{x^2 - R^2}}{x}\right] \forall (x \geq R) \qquad \dots\dots\dots\dots\dots\dots\dots (42)$$

4.3.4 Graphical presentation of variation of solid angle subtended by a hemispherical shell along the geometrical axis

We know that the solid angle subtended by the hemispherical shell at any point lying on the geometrical axis is given as from eq(40) & (42) as follows

$$\omega = 2\pi\left[1 + \frac{x}{\sqrt{x^2 + R^2}}\right] \quad \forall (x < R)$$

$$= 2\pi\left[1 - \frac{\sqrt{x^2 - R^2}}{x}\right] \quad \forall (x \geq R)$$

The above expressions show that the variation of ω is non-linear with distance x(variable) from the centre 'O'.

The curve is discontinuous at x = R because of infinitesimal thin boundary of the shell.

Now, let's find the following values

1. The value of ω at x = 0

$$\Rightarrow \omega = 2\pi\left[1+\frac{(0)}{\sqrt{(0)^2+R^2}}\right] = 2\pi[1+0] = 2\pi\ sr$$

2. The value of ω at $x \to R^-$

$$\Rightarrow \omega = \lim_{x\to R^-} 2\pi\left[1+\frac{x}{\sqrt{x^2+R^2}}\right] = 2\pi \lim_{k\to 0}\left[1+\frac{(R-k)}{\sqrt{(R-k)^2+R^2}}\right] \Rightarrow x = R-k \;\&\; k\to 0$$

$$= 2\pi\left[1+\frac{(R-0)}{\sqrt{(R-0)^2+R^2}}\right] (since, k \neq 0)$$

$$= 2\pi\left[1+\frac{R}{\sqrt{R^2+R^2}}\right] = 2\pi\left[1+\frac{R}{R\sqrt{2}}\right] = 2\pi\frac{(\sqrt{2}+1)}{\sqrt{2}} = \pi(2+\sqrt{2}) = 3.414\pi\ sr$$

3. The value of ω at $x = R$

$$\Rightarrow \omega = 2\pi\left[1-\frac{\sqrt{R^2-R^2}}{R}\right] = 2\pi[1-0] = 2\pi\ sr$$

If solid angle (ω) is taken on the Y-axis & distance 'x' is taken on X-axis

The non-linear variation of ω has been shown by the graph below.

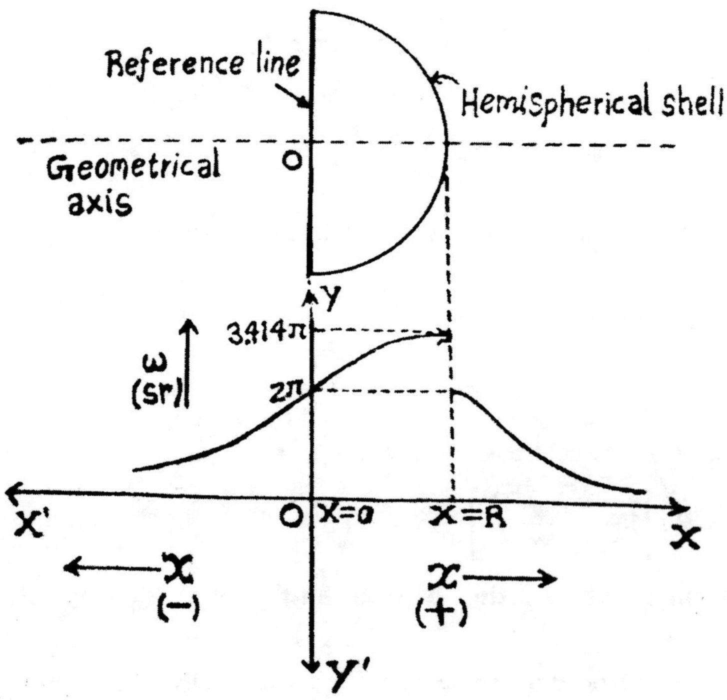

Fig (86)

*A typical Solid angle versus Distance (ω-x) graph for a hemispherical shell

*** Frustum of the sphere**

Here, we are to analyse the minor frustum of sphere having one circular face & another spherical surface & an angle 2α at the centre of parent sphere such that $0 < 2\alpha < \pi$.

4.3.5 Solid angle subtended by a minor frustum of the sphere at any point lying on the geometrical axis towards circular face

Let there be a minor frustum of the sphere with radius 'R' subtends an angle '2α'

$\forall(0 < 2\alpha < \pi)$ at the centre 'C' of the sphere& a given point say 'P' at a height 'h' lying on the geometrical axis of sphere towards the circular face of the frustum. (as shown in the figure (87) below)

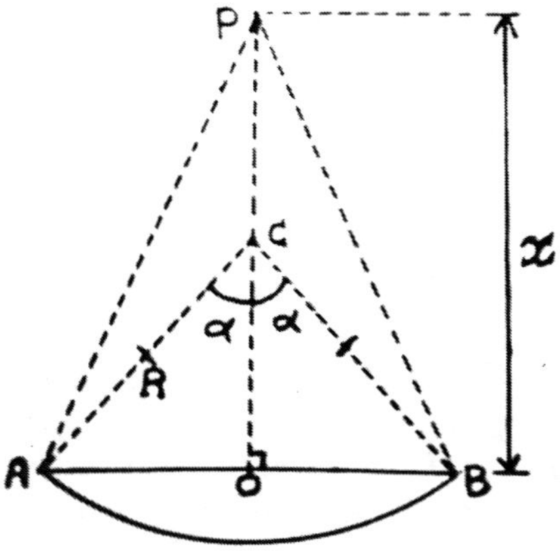

Fig (87)

In right ΔAOC

$$\Rightarrow sin\alpha = \frac{OA}{AC} = \frac{OA}{R} \Rightarrow OA = Rsin\alpha \quad (since, AC = BC = radius\ of\ sphere = R)$$

In this case, the frustum (with circular face) will appear like a circular plane with a radius $OA = Rsin\alpha$ to an observer lying at the given point 'P'

Hence, the solid angle subtended by the minor frustum of the sphere at the given point 'P'

= solid angle subtended by the circular face of the frustum at the same point 'P'

Using formula from eq(37) as follows

$$\omega = 2\pi\left[1 - \frac{h}{\sqrt{h^2 + R^2}}\right]$$

Now, on setting $R = Rsin\alpha$ & $h = height = x$ in the above expression as follows

$$\Rightarrow \omega = 2\pi\left[1 - \frac{x}{\sqrt{x^2 + (Rsin\alpha)^2}}\right]$$

$$\Rightarrow \omega = 2\pi\left[1 - \frac{x}{\sqrt{x^2 + R^2sin^2\alpha}}\right] \quad \forall\left(x \geq 0\ and\ 0 < \alpha < \frac{\pi}{2}\right) \ \ldots\ldots\ldots (43)$$

Important deduction: If the given point 'P' lies at the centre 'C' of the sphere then the solid angle subtended byany frustumat the centre 'C' of the sphere is obtained by setting $x = OC = Rcos\alpha$ in the above equation as follows

$$\omega = 2\pi\left[1 - \frac{(Rcos\alpha)}{\sqrt{(Rcos\alpha)^2 + R^2sin^2\alpha}}\right] = 2\pi\left[1 - \frac{Rcos\alpha}{\sqrt{R^2cos^2\alpha + R^2sin^2\alpha}}\right] = 2\pi\left[1 - \frac{Rcos\alpha}{R}\right]$$

$$\Rightarrow \omega = 2\pi[1 - cos\alpha] \quad \forall(0 < \alpha \leq \pi) \ \ldots\ldots\ldots\ldots\ldots (44)$$

Note: The above eq(44) is applicable for any frustum (minor or major) as well as for a perfect sphere in order to find solid angle subtended at their respective centres only.

Hence, the following cases can be proved as follows

1. Solid angle subtended by the hemisphere (also a frustum) at its centre is obtained by setting $\alpha = \frac{\pi}{2}$ in the eq(44) as follows

$$\omega = 2\pi\left[1 - cos\frac{\pi}{2}\right] = 2\pi[1 - 0] = 2\pi\ sr$$

2. Solid angle subtended by the sphere (not a frustum) at its centre is can be obtained by setting $\alpha = \pi$ in the eq(44) as follows

$$\omega = 2\pi[1 - cos\pi] = 2\pi[1 - (-1)] = 4\pi \ sr$$

4.3.6 Solid angle subtended by a minor frustum of the sphere at any point lying on the geometrical axis towards spherical surface

Let there be a minor frustum of the sphere with radius 'R' subtends an angle $'2\alpha'$ $\forall(0 < 2\alpha < \pi)$at the centre 'C' of the sphere & a given point say 'P' at a height 'h' lying on the geometrical axis of sphere towards the spherical surface of the frustum. (as shown in the figure (88) below)

In this case, it is to be noted that as the given point (i.e. observer) moves along the geometrical axis from the initial point 'P' on the surface up to the point 'P_1' the frustum will appear as a perfect sphere i.e. for each location of the given point (i.e. observer) between the points 'P' &'P_1'on the geometrical axis, the frustum appears as perfect sphere while for location of the observer a little away from the point 'P_1' on the geometrical axis, the frustum appears (can be visualised) as a perfect circular plane.

Thus, we are to find out these ranges of distance (x) measured from the centre 'O' of circular face of the frustum on its geometrical axis.

Now, in rightΔAOC

$$\Rightarrow cos\alpha = \frac{OC}{AC} = \frac{OC}{R} \Rightarrow OC = Rcos\alpha \quad (since, AC = PC = BC = radius\ of\ sphere = R)$$

$$\Rightarrow sin\alpha = \frac{OA}{AC} = \frac{OA}{R} \Rightarrow OA = Rsin\alpha$$

In rightΔP_1AC

$$\Rightarrow cos\alpha = \frac{AC}{P_1C} = \frac{R}{P_1C} \Rightarrow P_1C = \frac{R}{cos\alpha} = Rsec\alpha \ \&$$

$$\Rightarrow OP_1 = P_1C - OC = Rsec\alpha - Rcos\alpha = R\left(\frac{1}{cos\alpha} - cos\alpha\right) = R\left(\frac{1 - cos^2\alpha}{cos\alpha}\right) = \frac{Rsin^2\alpha}{cos\alpha}$$

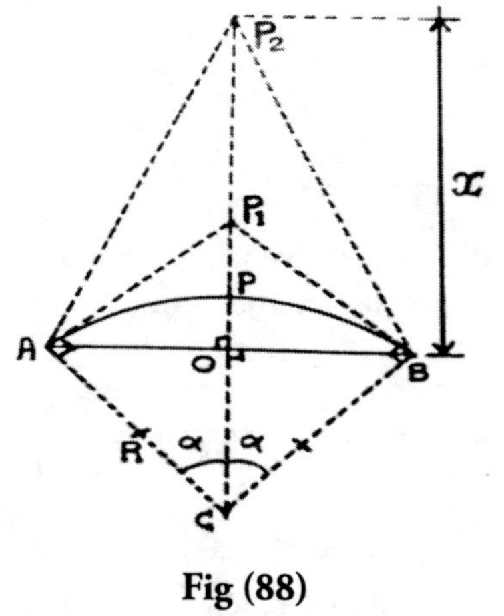

Fig (88)

$$\Rightarrow OP_1 = Rsin\alpha\left(\frac{sin\alpha}{cos\alpha}\right) = Rsin\alpha tan\alpha \ \&$$

$$\Rightarrow OP = PC - OC = R - Rcos\alpha = R(1 - cos\alpha)$$

Thus, we have

1. For the given point (observer) lying on the geometrical axis between the points 'P' & 'P_1' i.e. for $R(1 - cos\alpha) \leq x \leq Rsin\alpha tan\alpha$ (where 'x' is the distance measured from centre 'O')

The solid angle subtended by the minor frustum at the given point will be equal to that subtended by the sphere at the same point.

Hence, solid angle subtended by the minor frustum of sphere at any point lying on the geometrical axis between the points 'P' & 'P_1' is given from the eq(39) as follows

$$\Rightarrow \omega = 2\pi\left[1 - \frac{\sqrt{d^2 - R^2}}{d}\right]$$

Now, on setting d = $x + Rcos\alpha$ in the above expression, we get

$$\omega = 2\pi\left[1 - \frac{\sqrt{(x + Rcos\alpha)^2 - R^2}}{x + Rcos\alpha}\right]$$

$$\forall\left(R(1 - cos\alpha) \le x \le Rsin\alpha tan\alpha \; \& \; 0 < \alpha < \frac{\pi}{2}\right) \quad \dots\dots\dots\dots\dots (45)$$

2. For the given point (observer) lying on the geometrical axis between the point 'P_1' & infinity i.e. for $(x > Rsin\alpha tan\alpha)$ (where 'x' is the distance measured from centre 'O')

   The solid angle subtended by the minor frustum at the given point will be equal to that subtended by the circular plane with radius $OA = Rsin\alpha$ & centre 'O' at the same point.

   Hence, solid angle subtended by the minor frustum of sphere at any point lying on the geometrical axis between the point 'P_1' & infinity is given from the eq(37) as follows

$$\Rightarrow \omega = 2\pi\left[1 - \frac{h}{\sqrt{h^2 + R^2}}\right]$$

Now, on setting h = x $\&R = radius\ of\ circular\ plane = OA = Rsin\alpha$ in the above expression, as follows

$$\Rightarrow \omega = 2\pi\left[1 - \frac{x}{\sqrt{x^2 + (Rsin\alpha)^2}}\right] = 2\pi\left[1 - \frac{x}{\sqrt{x^2 + R^2sin^2\alpha}}\right]$$

$$\Rightarrow \omega = 2\pi\left[1 - \frac{x}{\sqrt{x^2 + R^2sin^2\alpha}}\right] \quad \forall\left(x > Rsin\alpha tan\alpha \; \& 0 < \alpha < \frac{\pi}{2}\right) \quad \dots\dots\dots\dots\dots (46)$$

4.4 CIRCULAR CYLINDER

In this section, we are to study & make solutions of the following

1. Solid or hollow circular cylinder (with finite & infinite length)
2. Cylindrical shell (with both the ends open & with one end open) &
3. Symmetrical & longitudinal frustum of the cylindrical shell.

4.4.1 Solid angle subtended by a finite circular cylinder (solid or hollow) at any point lying on the transverse axis (neglecting the effect of curvature ofcircular ends w.r.t. the observer)

Let there a finite circular cylinder having length 'L' & radius 'R' and a given point say 'P' lying on the transverse axis at a normal distance 'x' from the axis of cylinder such that the distance between the point 'O' and the centre 'C_1' of circular end of the cylinder is 'y'.

(as shown in the figure(89) below with upper one as front-view & lower one as top-view of circular section with centre 'O' & its plane passing through the given point 'P')

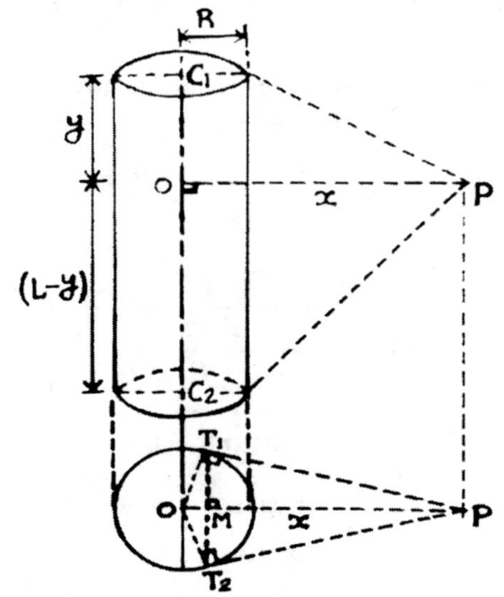

Fig (89)

Now, draw the smallest tangents PT_1 & PT_2 from the point 'P' to the surface of cylinder. (see the lower one as top-view in the above fig(89))

In rightΔOT_1P

$$\Rightarrow PT_1 = \sqrt{OP^2 - OT_1{}^2} = \sqrt{x^2 - R^2}\&$$

We know that in a right triangle

$$perpendicular\ to\ the\ hypotenuse = \frac{(height) \times (base)}{\sqrt{(height)^2 + (base)^2}} (using\ formula\ *)$$

$$\therefore\ T_1M = \frac{(OT_1) \times (PT_1)}{\sqrt{OT_1{}^2 + PT_1{}^2}} = \frac{R\sqrt{x^2 - R^2}}{\sqrt{R^2 + \left(\sqrt{x^2 - R^2}\right)^2}} = \frac{R\sqrt{x^2 - R^2}}{\sqrt{R^2 + x^2 - R^2}} \quad (OT_1 = radius = R)$$

$$\Rightarrow T_1M = T_2M = \frac{R\sqrt{x^2 - R^2}}{x} \Rightarrow T_1T_2 = 2T_1M = \frac{2R\sqrt{x^2 - R^2}}{x}$$

Now, in rightΔOMT_1

$$\Rightarrow OM = \sqrt{OT_1{}^2 - T_1M^2} = \sqrt{R^2 - \left(\frac{R\sqrt{x^2 - R^2}}{x}\right)^2} = \sqrt{\frac{R^2x^2 - R^2x^2 + R^4}{x^2}} = \frac{R^2}{x}$$

$$\therefore PM = OP - OM = x - \frac{R^2}{x} = \frac{x^2 - R^2}{x}$$

Now, draw the tangents from the given point 'P' at all the points of surface of the cylinder thus we get a rectangular plane ABCD with length 'L' & width T_1T_2 due to interception of the tangents with surface of the cylinder. (as shown in the figure (90) below)

In this case, the solid angle subtended by the circular cylinder at the given point 'P' lying on the transverse axis at a normal distance 'x' from the axis of the cylinder

= solid angle subtended by the rectangular plane ABCD at the same point 'P' lying on the transverse (vertical) axis passing through the point 'M' at a normal distance PM $= \frac{x^2 - R^2}{x}$

(Point 'M' is the foot of perpendicular drawn from the given point 'P' to the plane ABCD)

* The derivation of this formula is not necessary here.

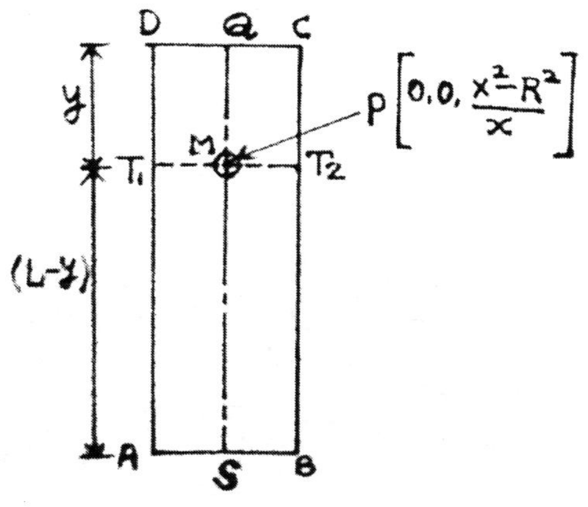

Fig (90)

Now, divide the rectangular plane ABCD into four elementary rectangular planes MT_1DQ, $ASMT_1$, MT_2CQ & SBT_2M. Thus, we get

$$\Rightarrow \omega = [\omega]_{ABCD} = [\omega]_{MT_1DQ} + [\omega]_{ASMT_1} + [\omega]_{MT_2CQ} + [\omega]_{SBT_2M}$$

$$\Rightarrow \omega = 2\left([\omega]_{MT_1DQ} + [\omega]_{ASMT_1}\right) \qquad (using\ symmetry\ of\ plane\ about\ line\ QS)$$

Now, we know that the solid angle subtended by the rectangular plane at any point lying on the vertical axis passing through the vertex is given by the eq(17) as follows

$$\omega = \sin^{-1}\left\{\frac{lb}{\sqrt{(l^2 + h^2)(b^2 + h^2)}}\right\}$$

On setting the following values

$$l = length = y, b = width = T_1M = \frac{R\sqrt{x^2 - R^2}}{x} \;\&\; h = height = PM = \frac{x^2 - R^2}{x}$$

in the above expression, we have

$$[\omega]_{MT_1DQ} = \sin^{-1}\left\{\frac{y\left(\frac{R\sqrt{x^2 - R^2}}{x}\right)}{\sqrt{\left(y^2 + \left(\frac{x^2 - R^2}{x}\right)^2\right)\left(\left(\frac{R\sqrt{x^2 - R^2}}{x}\right)^2 + \left(\frac{x^2 - R^2}{x}\right)^2\right)}}\right\}$$

$$= \sin^{-1}\left\{\frac{y\left(\frac{R}{x}\right)\sqrt{x^2 - R^2}}{\sqrt{\left(y^2 + \frac{(x^2 - R^2)^2}{x^2}\right)\left(\frac{R^2}{x^2}(x^2 - R^2) + \frac{(x^2 - R^2)^2}{x^2}\right)}}\right\}$$

$$[\omega]_{MT_1DQ} = \sin^{-1}\left\{\frac{Ry\sqrt{x^2 - R^2}}{\sqrt{(x^2y^2 + (x^2 - R^2)^2)(x^2 - R^2)}}\right\} = \sin^{-1}\left\{\frac{Ry}{\sqrt{x^2y^2 + (x^2 - R^2)^2}}\right\}$$

Similarly, On setting the following values

$$l = length = (L - y), b = width = T_2M = \frac{R\sqrt{x^2 - R^2}}{x} \;\&\; h = height = PM = \frac{x^2 - R^2}{x}$$

in the above expression, we have

$$[\omega]_{ASMT_1} = \sin^{-1}\left\{\frac{(L-y)\left(\frac{R\sqrt{x^2-R^2}}{x}\right)}{\sqrt{\left((L-y)^2+\left(\frac{x^2-R^2}{x}\right)^2\right)\left(\left(\frac{R\sqrt{x^2-R^2}}{x}\right)^2+\left(\frac{x^2-R^2}{x}\right)^2\right)}}\right\}$$

$$= \sin^{-1}\left\{\frac{(L-y)\left(\frac{R}{x}\right)\sqrt{x^2-R^2}}{\sqrt{\left((L-y)^2+\frac{(x^2-R^2)^2}{x^2}\right)\left(\frac{R^2}{x^2}(x^2-R^2)+\frac{(x^2-R^2)^2}{x^2}\right)}}\right\}$$

$$= \sin^{-1}\left\{\frac{R(L-y)\sqrt{x^2-R^2}}{\sqrt{(x^2(L-y)^2+(x^2-R^2)^2)(x^2-R^2)}}\right\} = \sin^{-1}\left\{\frac{R(L-y)}{\sqrt{x^2(L-y)^2+(x^2-R^2)^2}}\right\}$$

Hence by adding above values, we get

$$\Rightarrow \omega = 2\left[\sin^{-1}\left\{\frac{Ry}{\sqrt{x^2y^2+(x^2-R^2)^2}}\right\} + \sin^{-1}\left\{\frac{R(L-y)}{\sqrt{x^2(L-y)^2+(x^2-R^2)^2}}\right\}\right] \quad \dots\dots\dots (47)$$

$$\forall (x \geq R \;\&\; 0 \leq y \leq L)$$

Now, the following cases are arisen for a circular cylinder.

4.4.2 Solid angle subtended by a finite circular cylinder (solid or hollow) at any point lying on the transverse axis passing through the mid-point of longitudinal axis

Let there a finite circular cylinder having length 'L' & radius 'R' and a given point say 'P' lying on the transverse axis passing through the mid-point of longitudinal axis at a normal distance 'x' from the longitudinal axis of cylinder. (as shown in the figure(91) below)

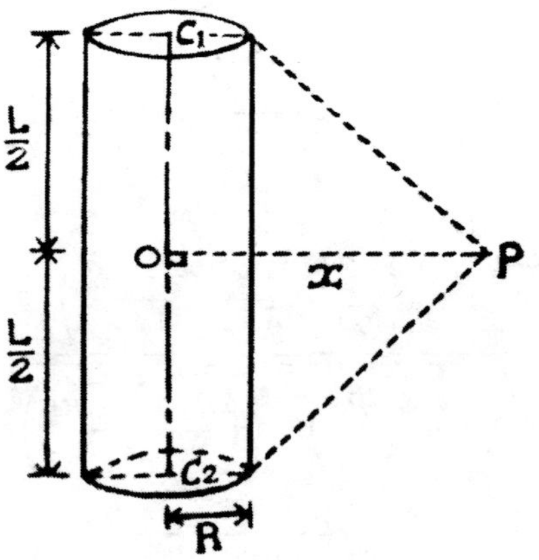

Fig (91)

In this case, the solid angle subtended by the circular cylinder at the point 'P' is obtained by setting $y = \frac{L}{2}$ in the eq(47) as follows

$$\Rightarrow \omega = 2\left[\sin^{-1}\left\{\frac{R\left(\frac{L}{2}\right)}{\sqrt{x^2\left(\frac{L}{2}\right)^2+(x^2-R^2)^2}}\right\} + \sin^{-1}\left\{\frac{R\left(L-\frac{L}{2}\right)}{\sqrt{x^2\left(L-\frac{L}{2}\right)^2+(x^2-R^2)^2}}\right\}\right]$$

$$= 2\left[\sin^{-1}\left\{\frac{RL}{\sqrt{x^2L^2+4(x^2-R^2)^2}}\right\} + \sin^{-1}\left\{\frac{RL}{\sqrt{x^2L^2+4(x^2-R^2)^2}}\right\}\right]$$

$$\Rightarrow \omega = 4\sin^{-1}\left\{\frac{RL}{\sqrt{x^2L^2 + 4(x^2 - R^2)^2}}\right\} \quad \forall(x \geq R) \quad \dots\dots\dots\dots\dots\dots\dots\dots (48)$$

4.4.3 Solid angle subtended by a circular cylinder at any point lying on the surface between itscircular ends.

Let there be circular cylinder with radius 'R' & length 'L' and a given point say 'P' lying on the surface of the cylinder between the circular ends. (as shown in the figure (92) below)

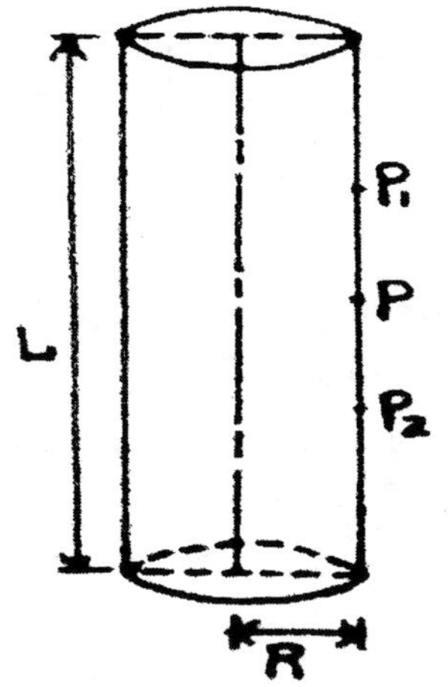

Fig (92)

In this case, the solid angle is obtained by setting x = R in the above eq(48) as follows

$$\omega = 4\sin^{-1}\left\{\frac{RL}{\sqrt{R^2L^2 + 4(R^2 - R^2)^2}}\right\} = 4\sin^{-1}\left\{\frac{RL}{RL}\right\} = 4\sin^{-1}\{1\} = 4\frac{\pi}{2} = 2\pi \ sr$$

4.4.4 Solid angle subtended by a finite circular cylinder (solid or hollow) at any point lying on the transverse axis passing through the centre of circular end

Let there a finite circular cylinder having length 'L' & radius 'R' and a given point say 'P_1' or 'P_2' lying on the transverse axis passing through the centre 'C_1' or 'C_2' of circular ends at a normal distance 'x' from the longitudinal axis of cylinder. (as shown in the figure(93) below)

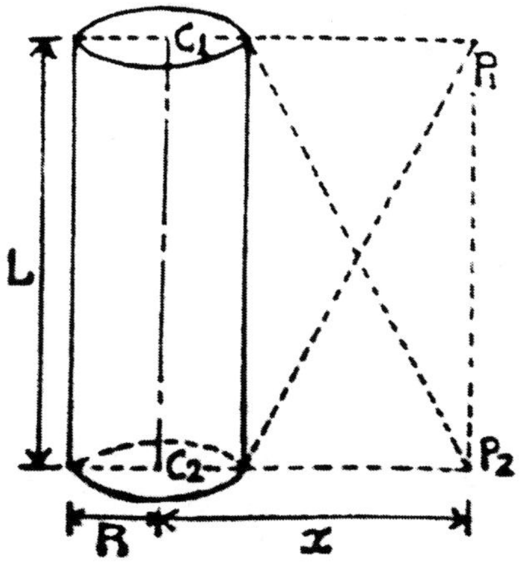

Fig (93)

In this case, the solid angle subtended by the circular cylinder at the point 'P_1' or 'P_2' is obtained by setting y =0in the eq(47) as follows

$$\Rightarrow \omega = 2\left[\sin^{-1}\left\{\frac{R(0)}{\sqrt{x^2(0)^2 + (x^2 - R^2)^2}}\right\} + \sin^{-1}\left\{\frac{R(L-0)}{\sqrt{x^2(L-0)^2 + (x^2 - R^2)^2}}\right\}\right]$$

$$= 2\left[\sin^{-1}\{0\} + \sin^{-1}\left\{\frac{RL}{\sqrt{x^2L^2 + (x^2 - R^2)^2}}\right\}\right]$$

$$\Rightarrow \omega = 2\sin^{-1}\left\{\frac{RL}{\sqrt{x^2L^2 + (x^2 - R^2)^2}}\right\} \quad \forall (x \geq R) \quad \ldots\ldots\ldots\ldots\ldots\ldots\ldots\ldots (49)$$

Important deduction: The solid angle subtended by circular cylinder at any point lying on the surface between the circular ends is obtained by setting x = R in the above eq(49) as follows

$$\omega = 2\sin^{-1}\left\{\frac{RL}{\sqrt{R^2L^2 + (R^2 - R^2)^2}}\right\} = 2\sin^{-1}\left\{\frac{RL}{RL}\right\} = 2\sin^{-1}\{1\} = 2\frac{\pi}{2} = \pi\ sr$$

4.4.5 Solid angle subtended by an infinitely long circular cylinder (solid or hollow) at any point lying on the transverse axis passing through the centre of circular end ($L \to \infty$)

Let there an infinitely long circular cylinder having radius 'R' and a given point say 'P' lying on the transverse axis passing through the centre 'C_1' of circular end at a normal distance 'x' from the longitudinal axis of cylinder. (as shown in the figure(94) below)

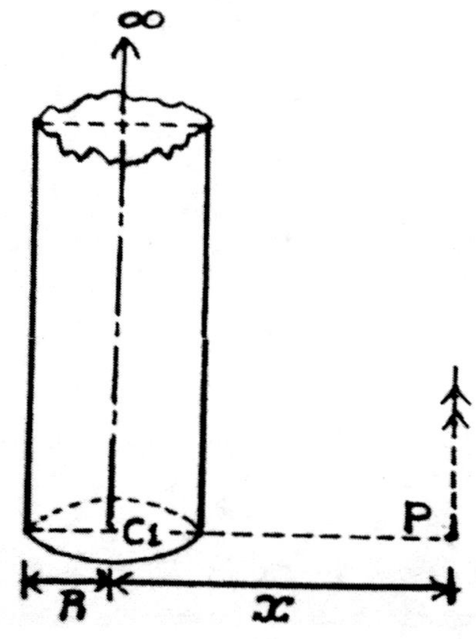

Fig (94)

In this case, the solid angle subtended by the infinitely long circular cylinder at the point 'P' is obtained by taking the limit of ω at $L \to \infty$ from the eq(49)as follows

$$\Rightarrow \omega = \lim_{L\to\infty}\left[2\sin^{-1}\left\{\frac{RL}{\sqrt{x^2L^2 + (x^2 - R^2)^2}}\right\}\right] = 2\lim_{K\to 0}\left[\sin^{-1}\left\{\frac{R}{\sqrt{x^2 + \frac{(x^2 - R^2)^2}{L^2}}}\right\}\right]$$

$$= 2\lim_{K\to 0}\left[\sin^{-1}\left\{\frac{R}{\sqrt{x^2 + K^2(x^2 - R^2)^2}}\right\}\right] \qquad where, K = \frac{1}{L} \Rightarrow K \to 0$$

$$= 2\left[\sin^{-1}\left\{\frac{R}{\sqrt{x^2 + (0)^2(x^2 - R^2)^2}}\right\}\right] = 2\sin^{-1}\left\{\frac{R}{\sqrt{x^2 + 0}}\right\} = 2\sin^{-1}\left\{\frac{R}{x}\right\}$$

$$\Rightarrow \boxed{\omega = 2\sin^{-1}\left(\frac{R}{x}\right)} \quad \forall (x \geq R) \quad \ldots\ldots\ldots\ldots\ldots\ldots\ldots\ldots (50)$$

4.4.6 Solid angle subtended by an infinitely long circular cylinder (solid or hollow) at any point lying on the transverse axis passing through the centre of longitudinal axis ($L \to \infty$)

Let there an infinitely long circular cylinder having radius 'R' and a given point say 'P' lying on the transverse axis passing through the centre longitudinal axis at a normal distance 'x' from the same axis of cylinder. (as shown in the figure(95) below)

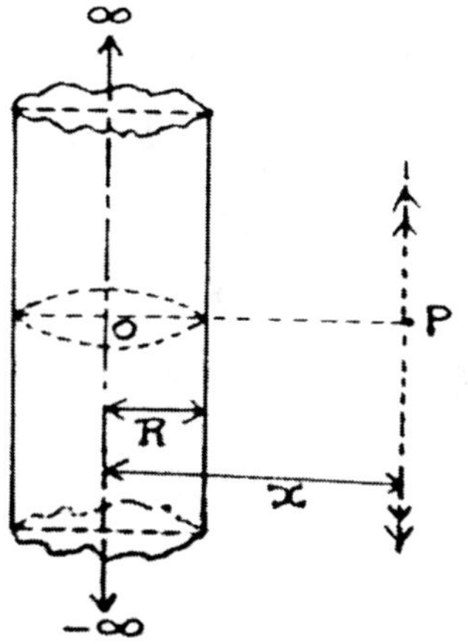

Fig (95)

In this case, the solid angle subtended by the infinitely long circular cylinder at the point 'P' is obtained by taking the limit of ω at $L \to \infty$ from the eq(48) as follows

$$\Rightarrow \omega = \lim_{L\to\infty}\left[4\sin^{-1}\left\{\frac{RL}{\sqrt{x^2L^2 + 4(x^2 - R^2)^2}}\right\}\right] = 4\lim_{L\to\infty}\left[\sin^{-1}\left\{\frac{R}{\sqrt{x^2 + \frac{4(x^2-R^2)^2}{L^2}}}\right\}\right]$$

$$= 4\left[\sin^{-1}\left\{\frac{R}{\sqrt{x^2 + 4K^2(x^2 - R^2)^2}}\right\}\right] = 4\sin^{-1}\left\{\frac{R}{\sqrt{x^2+0}}\right\} = 4\sin^{-1}\left\{\frac{R}{x}\right\} \quad K = \frac{1}{L} \Rightarrow K \to 0$$

$$\Rightarrow \left[\omega = 4\sin^{-1}\left(\frac{R}{x}\right)\right] \quad \forall (x \geq R) \quad \ldots\ldots\ldots\ldots\ldots\ldots\ldots\ldots (51)$$

The above result shows that the value of solid angle is double the value as obtained by the eq(50) because in this case the cylinder is equivalent to two infinitely long cylinders connected co-axially at the point 'O' of the longitudinal axis.

Important deduction: The solid angle subtended by the infinitely long cylinder at any point lying on its surface between the ends at infinity is obtained by setting x = R in the above expression as follows

$$\omega = 4\sin^{-1}\left(\frac{R}{R}\right) = 4\sin^{-1}(1) = 4\frac{\pi}{2} = 2\pi\ sr$$

4.4.7Solid angle subtended by a cylinder with front end closed at any point lying on the longitudinal axis

Let there be a solid cylinder with radius 'R' & length 'L' and a given point say 'P' lying on the longitudinal axis at a distance 'x' from the centre 'O' of closed front circular-end (as shown in the figure (96) below)

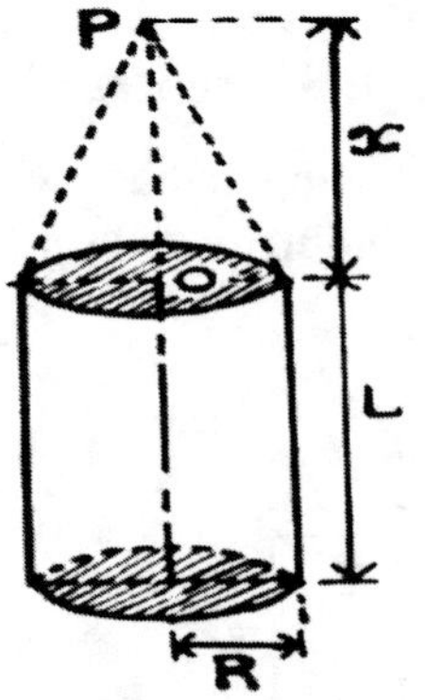

Fig (96)

In this case, the whole cylinder appears as a circular plane with centre 'C' & radius 'R' (i.e. identical to the front circular end of the cylinder) irrespective of the length 'L' of the cylinder.

Hence, the solid angle subtended by the solid cylinder at the given point 'P'

= solid angle subtended by the front circular end of the solid cylinder at the same point 'P'

Now, on setting h = height = x in the eq(37), we get

$$\Rightarrow \omega = 2\pi\left[1 - \frac{x}{\sqrt{x^2 + R^2}}\right] \quad \forall (x \geq 0) \quad \ldots\ldots\ldots\ldots\ldots\ldots\ldots\ldots\ldots\ldots\ldots\ldots\ldots\ldots (52)$$

Thus, the value of solid angle is dependent on the radius 'R' & the distance 'x' measured from the centre of the front face and independent of length of the solid cylinder.

Important deduction: Solid angle subtended by the solid cylinder at any point lying on the circular face completely inside the boundary is obtained by setting x = 0 in the above expression as follows

$$\boldsymbol{\omega = 2\pi\left[1 - \frac{(0)}{\sqrt{(0)^2 + R^2}}\right] = 2\pi[1 - 0] = 2\pi\, sr}$$

4.4.8 Solid angle subtended by a hollow cylindrical shell at any point lying on the longitudinal axis inside it

Let there be a hollow cylindrical shell with radius 'R' & length 'L' and a given point say 'P' lying on the longitudinal axis at a distance 'x' from the centre 'C_1' of circular-end inside the shell. (as shown in the figure (97) below)

The solid angle subtended by the open circular-end at the point 'P' lying on the vertical axis C_2C_1 at a distance 'x' from the centre 'C_1'.

$$\Rightarrow \omega_1 = 2\pi\left[1 - \frac{x}{\sqrt{x^2 + R^2}}\right] \qquad \textit{from eq(37) for a circular plane}$$

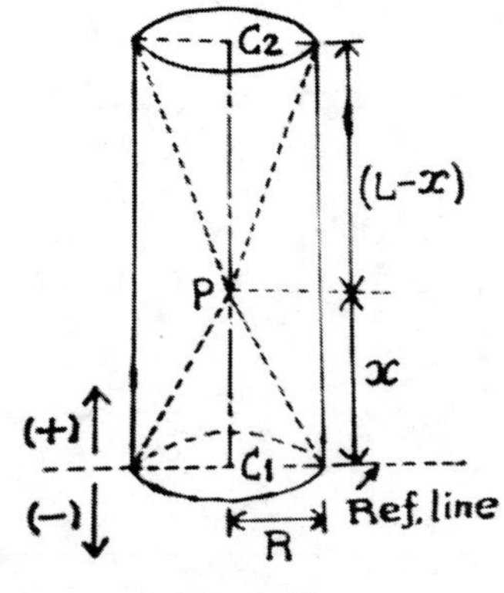

Fig (97)

Similarly, solid angle subtended by the open circular-end at the point 'P' lying on the vertical axis C_2C_1 at a distance '$(L-x)$' from the centre 'C_2'.

$$\Rightarrow \omega_2 = 2\pi\left[1-\frac{(L-x)}{\sqrt{(L-x)^2+R^2}}\right] \qquad from\ eq(37)$$

Assume that both the ends of shell are closed then the solid angle subtended by the closed cylindrical shell at the given point 'P' completely inside it

$$\Rightarrow \omega' = 4\pi$$

Hence, the solid angle subtended by the hollow cylindrical shell (i.e. with open ends) at the point 'P'

$$\Rightarrow \omega = \omega' - \omega_1 - \omega_2 = 4\pi - 2\pi\left[1-\frac{x}{\sqrt{x^2+R^2}}\right] - 2\pi\left[1-\frac{(L-x)}{\sqrt{(L-x)^2+R^2}}\right]$$

$$\Rightarrow \omega = 4\pi - 2\pi + \frac{2\pi x}{\sqrt{x^2+R^2}} - 2\pi + \frac{2\pi(L-x)}{\sqrt{(L-x)^2+R^2}}$$

$$\Rightarrow \omega = 2\pi\left[\frac{x}{\sqrt{x^2+R^2}} + \frac{(L-x)}{\sqrt{(L-x)^2+R^2}}\right] \quad \forall x \in (-\infty, +\infty) \ \dots\dots\dots\dots\dots\dots\dots (53)$$

Sign convention: Distance 'x' measured from the centre of the nearest endof hollow cylindrical shell (i.e. from the reference line) is taken

1. Positive when the point lies on the longitudinal axis inside the shell &
2. Negative when the point lies on the longitudinal axis outside the shell

* Maximum value of solid angle subtended by a hollow cylindrical shell at the point of maxima on longitudinal axis:

Since, the value of solid angle (ω) depends on the variable distance 'x'. Hence on differentiating the eq(53) both the sides w.r.t. variable 'x' as follows

$$\Rightarrow \frac{d\omega}{dx} = 2\pi\frac{d}{dx}\left[\frac{x}{\sqrt{x^2+R^2}} + \frac{(L-x)}{\sqrt{(L-x)^2+R^2}}\right]$$

$$\Rightarrow \frac{d\omega}{dx} = 2\pi\frac{d}{dx}\left(\frac{x}{\sqrt{x^2+R^2}}\right) + 2\pi\frac{d}{dx}\left(\frac{(L-x)}{\sqrt{(L-x)^2+R^2}}\right)$$

$$= 2\pi\left[\frac{\sqrt{x^2+R^2}\frac{d(x)}{dx} - x\frac{d}{dx}\left(\sqrt{x^2+R^2}\right)}{\left(\sqrt{x^2+R^2}\right)^2}\right]$$

$$+2\pi\left[\frac{\sqrt{(L-x)^2+R^2}\frac{d(L-x)}{dx} - (L-x)\frac{d}{dx}\left(\sqrt{(L-x)^2+R^2}\right)}{\left(\sqrt{(L-x)^2+R^2}\right)^2}\right]$$

$$= 2\pi\left[\frac{\sqrt{x^2+R^2}.1 - \frac{x.(2x)}{2\sqrt{x^2+R^2}}}{x^2+R^2}\right] + 2\pi\left[\frac{\sqrt{(L-x)^2+R^2}.(-1) - \frac{(L-x).(-2)(L-x)}{2\sqrt{(L-x)^2+R^2}}}{(L-x)^2+R^2}\right]$$

$$= 2\pi\left[\frac{R^2+x^2-x^2}{(x^2+R^2)^{3/2}}\right] + 2\pi\left[\frac{-R^2-(L-x)^2+(L-x)^2}{((L-x)^2+R^2)^{3/2}}\right]$$

$$\Rightarrow \frac{d\omega}{dx} = 2\pi R^2\left[\frac{1}{(x^2+R^2)^{3/2}} - \frac{1}{((L-x)^2+R^2)^{3/2}}\right] \quad \dots\dots\dots\dots\dots\dots (I)$$

Again differentiating the above eq(I) w.r.t. 'x' as follows

$$\Rightarrow \frac{d^2\omega}{dx^2} = 2\pi R^2 \left[\frac{d}{dx}\left((x^2+R^2)^{-3/2}\right) - \frac{d}{dx}\left(((L-x)^2+R^2)^{-3/2}\right)\right]$$

$$= 2\pi R^2 \left[\frac{-3}{2}(x^2+R^2)^{-5/2}(2x) - \left(\frac{-3}{2}\right)((L-x)^2+R^2)^{-5/2}(-2(L-x))\right]$$

$$\Rightarrow \frac{d^2\omega}{dx^2} = -6\pi R^2 \left[x(x^2+R^2)^{-5/2} + (L-x)((L-x)^2+R^2)^{-5/2}\right] \quad \ldots\ldots\ldots\ldots\ldots\ldots (II)$$

Now, for maxima or minima $\frac{d\omega}{dx} = 0$

$$\therefore 2\pi R^2 \left[\frac{1}{(x^2+R^2)^{3/2}} - \frac{1}{((L-x)^2+R^2)^{3/2}}\right] = 0 \Rightarrow (x^2+R^2)^{3/2} = ((L-x)^2+R^2)^{3/2}$$

$$\Rightarrow x^2+R^2 = (L-x)^2+R^2 \Rightarrow x^2 = L^2+x^2-2xL \quad \Rightarrow 2xL = L^2$$

$$\Rightarrow x = \frac{L}{2}$$

Now, on setting this value in the eq(II), we get

$$\left(\frac{d^2\omega}{dx^2}\right)_{x=\frac{L}{2}} = -6\pi R^2 \left[\left(\frac{L}{2}\right)\left(\left(\frac{L}{2}\right)^2 + R^2\right)^{-5/2} + \left(L-\frac{L}{2}\right)\left(\left(L-\frac{L}{2}\right)^2 + R^2\right)^{-5/2}\right]$$

$$\Rightarrow \left(\frac{d^2\omega}{dx^2}\right)_{x=\frac{L}{2}} = -6\pi R^2 L\left(\left(\frac{L}{2}\right)^2 + R^2\right)^{-5/2} = -ve\ (value)$$

Hence, the value of solid angle is maximum at $x = \frac{L}{2}$ i.e. at the mid-point of the cylinder shell

Now, on setting $x = \frac{L}{2}$ in the eq(53) we have

$$(\omega)_{max} = 2\pi \left[\frac{\left(\frac{L}{2}\right)}{\sqrt{\left(\frac{L}{2}\right)^2 + R^2}} + \frac{\left(L-\frac{L}{2}\right)}{\sqrt{\left(L-\frac{L}{2}\right)^2 + R^2}}\right] = 2\pi\left[\frac{L}{\sqrt{L^2+4R^2}} + \frac{L}{\sqrt{L^2+4R^2}}\right]$$

$$\Rightarrow (\omega)_{max} = 4\pi\left[\frac{L}{\sqrt{L^2+4R^2}}\right] \quad \ldots\ldots\ldots\ldots\ldots\ldots (III)$$

Thus, we find that $(\omega)_{max}$ depends on the length (L) & radius (R) of hollow cylindrical shell.

4.4.9 Solid angle subtended by a hollow cylindrical shell at any point lying on the longitudinal axis outside it

Method 1:

Let there be a hollow cylindrical shell with radius 'R' & length 'L' and a given point say 'P' lying on the longitudinal axis at a distance 'x' from the centre 'C_1' of circular-end outside the shell. (as shown in the figure (98) below)

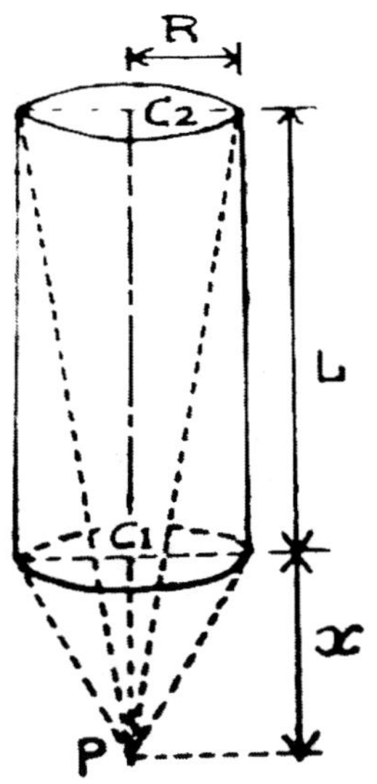

Fig (98)

The solid angle subtended by the front open circular-end with centre C_1 at the point 'P' lying on the vertical axis C_2C_1 at a distance 'x'.

$$\Rightarrow \omega_1 = 2\pi\left[1 - \frac{x}{\sqrt{x^2 + R^2}}\right] \quad \textit{from eq(37) for a circular plane}$$

Similarly, solid angle subtended by the open circular-end with centre C_2 at the point 'P' lying on the vertical axis C_2C_1 at a distance $(L + x)$

$$\Rightarrow \omega_2 = 2\pi\left[1 - \frac{(L+x)}{\sqrt{(L+x)^2 + R^2}}\right] \quad \textit{from eq(37)}$$

Assume that the front circular-end of shell with centre 'C_1' is closed then the solid angle subtended by the cylindrical shell at the given point 'P' outside it

= solid angle subtended by the front circular end with centre 'C_1' at the point 'P'

$$\Rightarrow \omega = \omega_1 - \omega_2 \quad \text{(Since, the end with centre '}C_2\text{' is open)}$$

Hence, the solid angle subtended by the hollow cylindrical shell (i.e. with open ends) at the point 'P' outside the shell

$$\Rightarrow \omega = \omega_1 - \omega_2 = 2\pi\left[1 - \frac{x}{\sqrt{x^2 + R^2}}\right] - 2\pi\left[1 - \frac{(L+x)}{\sqrt{(L+x)^2 + R^2}}\right]$$

$$\Rightarrow \omega = 2\pi - \frac{2\pi x}{\sqrt{x^2 + R^2}} - 2\pi + \frac{2\pi(L+x)}{\sqrt{(L+x)^2 + R^2}}$$

$$\Rightarrow \omega = 2\pi\left[\frac{(L+x)}{\sqrt{(L+x)^2 + R^2}} - \frac{x}{\sqrt{x^2 + R^2}}\right] \quad \forall\ (x \geq 0)\ \dots\dots\dots\dots\dots\dots\dots\dots (54)$$

Method 2: According to the sign convention taking negative sign of distance 'x' for the point lying on the longitudinal axis outside the hollow cylindrical shell i.e. by setting $x = -x$ in the eq(53) as follows

$$\omega = 2\pi\left[\frac{(-x)}{\sqrt{(-x)^2 + R^2}} + \frac{(L-(-x))}{\sqrt{(L-(-x))^2 + R^2}}\right] = 2\pi\left[\frac{-x}{\sqrt{x^2 + R^2}} + \frac{(L+x)}{\sqrt{(L+x)^2 + R^2}}\right]$$

$$\Rightarrow \omega = 2\pi\left[\frac{(L+x)}{\sqrt{(L+x)^2 + R^2}} - \frac{x}{\sqrt{x^2 + R^2}}\right] \quad \forall\ (x \geq 0)\ \dots\dots\dots\dots\dots\dots\dots\dots (54)$$

4.4.10 Solid angle subtended by a hollow cylindrical shell at the mid-point of longitudinal axis

Let there be a hollow cylindrical shell with radius 'R' & length 'L' and 'P' is the mid-point of longitudinal axis of the shell (as shown in the figure (99) below)

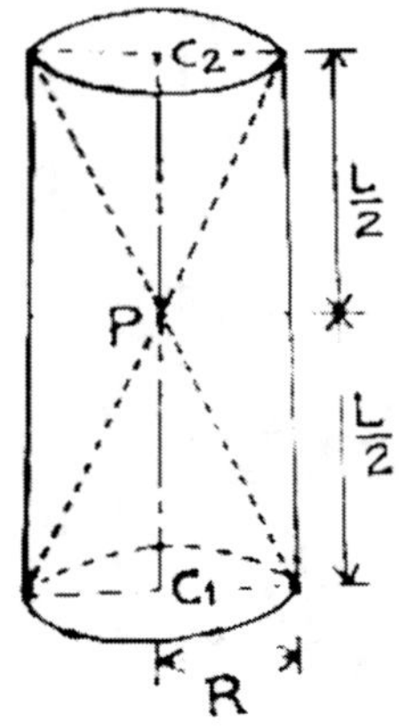

Fig (99)

In this case, the solid angle subtended by the given shell at the point 'P' is obtained by setting x $= \frac{L}{2}$ in the eq(53) as follows

$$\omega = 2\pi\left[\frac{\left(\frac{L}{2}\right)}{\sqrt{\left(\frac{L}{2}\right)^2 + R^2}} + \frac{\left(L - \frac{L}{2}\right)}{\sqrt{\left(L - \frac{L}{2}\right)^2 + R^2}}\right] = 2\pi\left[\frac{L}{\sqrt{L^2 + 4R^2}} + \frac{L}{\sqrt{L^2 + 4R^2}}\right]$$

$$\Rightarrow \omega = 4\pi\left[\frac{L}{\sqrt{L^2 + 4R^2}}\right] \quad \ldots\ldots\ldots\ldots\ldots\ldots (55)$$

4.4.11 Solid angle subtended by a hollow cylindrical shell at the centre of any one of the open ends

Let there be a hollow cylindrical shell with radius 'R', length 'L' and centres of circular ends as C_1&C_2 . (as shown in the figure (100) below)

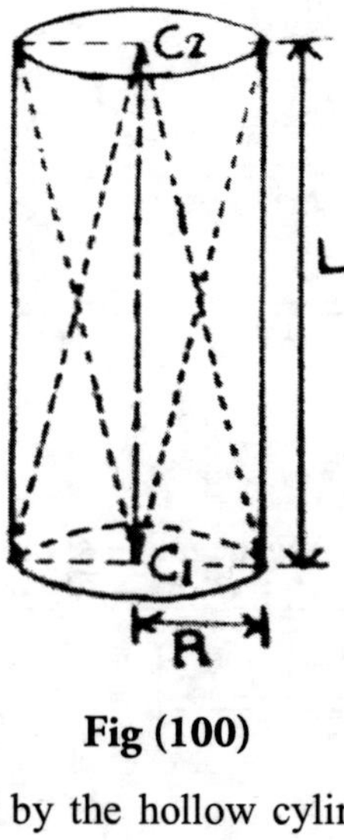

Fig (100)

In this case, the solid angle subtended by the hollow cylindrical shell at the point 'P' is obtained by setting x = 0 in the eq(53) as follows

$$\omega = 2\pi\left[\frac{(0)}{\sqrt{(0)^2 + R^2}} + \frac{(L-0)}{\sqrt{(L-0)^2 + R^2}}\right] = 2\pi\left[0 + \frac{L}{\sqrt{L^2 + R^2}}\right]$$
$$= 2\pi\left[\frac{L}{\sqrt{L^2 + R^2}}\right]$$

Or

By setting x = 0 in the eq(54) as follows

$$\omega = 2\pi\left[\frac{(L+0)}{\sqrt{(L+0)^2 + R^2}} - \frac{(0)}{\sqrt{(0)^2 + R^2}}\right] = 2\pi\left[\frac{L}{\sqrt{L^2 + R^2}} - 0\right]$$
$$\Rightarrow \omega = 2\pi\left[\frac{L}{\sqrt{L^2 + R^2}}\right] \quad \dots\dots\dots\dots\dots\dots.(56)$$

The above value is also the maximum value of solid angle subtended by the hollow cylindrical shell at the mid-point of the longitudinal axis.

4.4.12 Solid angle subtended by an infinitely long hollow cylindrical shell at any point inside the boundary of open end

Let there be an infinitely long hollow cylindrical shell with radius 'R' and an arbitrary point say 'P' inside the boundary of open end of the shell. (as shown in the figure (101) below)

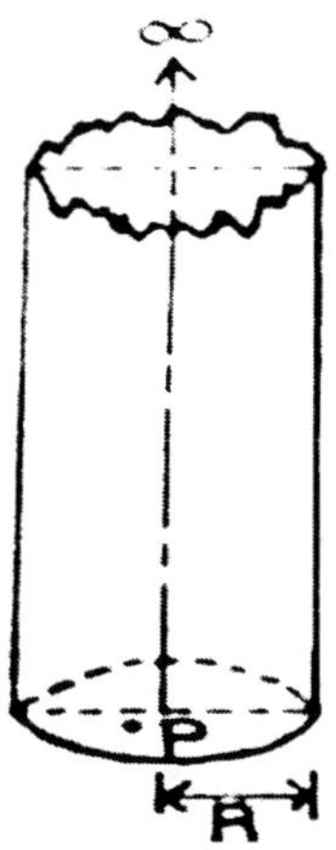

Fig (101)

In this case, solid angle subtended by the infinitely long hollow cylindrical shell at the point 'P' is obtained by taking limit of $\omega \; at \; L \to \infty$ from the eq(56) as follows

$$\omega = \lim_{L\to\infty} 2\pi\left[\frac{L}{\sqrt{L^2 + R^2}}\right] = 2\pi \lim_{L\to\infty}\left[\frac{1}{\sqrt{1 + \left(\frac{R}{L}\right)^2}}\right] = 2\pi\left[\frac{1}{\sqrt{1 + (0)^2}}\right] = 2\pi\left[\frac{1}{\sqrt{1+0}}\right] = 2\pi \; sr$$

Thus, an infinitely long hollow cylindrical shell behaves as a cap fully covering all the points in the plane of open end of the shell irrespective of the radius.

4.4.13 Solid angle subtended by an infinitely long hollow cylindrical shell at any point ofcentral section inside the shell

Let there be an infinitely long hollow cylindrical shell with radius 'R' and an arbitrary point say 'P'of central section inside the shell. (as shown in the figure (102) below)

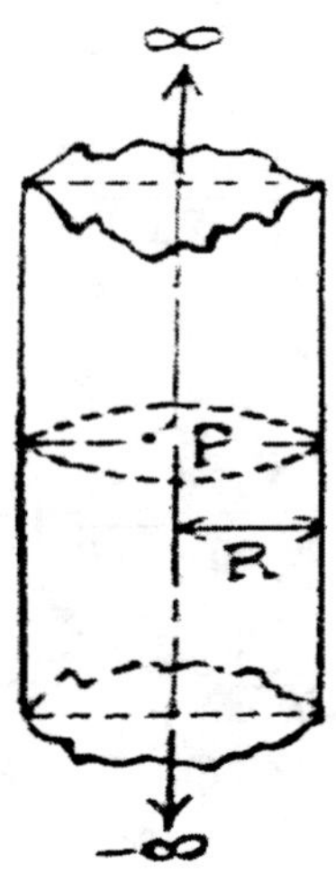

Fig (102)

In this case, solid angle subtended by the infinitely long hollow cylindrical shell at the point 'P' is obtained by taking limit of $\omega\ at\ L \to \infty$ from the eq(55) as follows

$$\omega = \lim_{L\to\infty} 4\pi \left[\frac{L}{\sqrt{L^2+4R^2}}\right] = 4\pi \lim_{L\to\infty}\left[\frac{1}{\sqrt{1+4\left(\frac{R}{L}\right)^2}}\right] = 4\pi\left[\frac{1}{\sqrt{1+4(0)^2}}\right] = 4\pi\left[\frac{1}{\sqrt{1+0}}\right]$$

$$\Rightarrow \omega = 4\pi\ sr$$

It is obvious that the infinitely long cylindrical shell behaves as a completely closed surface for each point of the central section inside the shell irrespective of the radius.

4.4.14 Graphical presentation of variation ofsolid angle subtended by a hollow cylindrical shell along longitudinal axis

Since, we know that the solid angle subtended by a hollow cylindrical shell at any point on its longitudinal axis is given by the eq(53) as follows

$$\omega = 2\pi\left[\frac{x}{\sqrt{x^2+R^2}} + \frac{(L-x)}{\sqrt{(L-x)^2+R^2}}\right] \quad \forall x \in (-\infty, +\infty)$$

Where 'x' is the distance measured from the centre of the nearest open end of the shell & the value of ω varies non-linearly with variable 'x'. let us find the following values

1. at x = 0

$$\Rightarrow \omega = 2\pi\left[\frac{(0)}{\sqrt{(0)^2+R^2}} + \frac{(L-0)}{\sqrt{(L-0)^2+R^2}}\right] = 2\pi\left[\frac{L}{\sqrt{L^2+R^2}}\right]$$

2. at $x = \frac{L}{2}$

$$\Rightarrow \omega = 2\pi\left[\frac{\left(\frac{L}{2}\right)}{\sqrt{\left(\frac{L}{2}\right)^2+R^2}} + \frac{\left(L-\frac{L}{2}\right)}{\sqrt{\left(L-\frac{L}{2}\right)^2+R^2}}\right] = 4\pi\left[\frac{L}{\sqrt{L^2+4R^2}}\right] \&$$

3. at x = L

$$\Rightarrow \omega = 2\pi\left[\frac{L}{\sqrt{L^2+R^2}}+\frac{(L-L)}{\sqrt{(L-L)^2+R^2}}\right] = 2\pi\left[\frac{L}{\sqrt{L^2+R^2}}\right]$$

Now, if the distance (x) is taken as abscissa & solid angle (ω) is taken as ordinate then the ω vs $x$graph is plotted.

A typical ω-x graph for a hollow cylindrical shell is shown in the fig(103) below

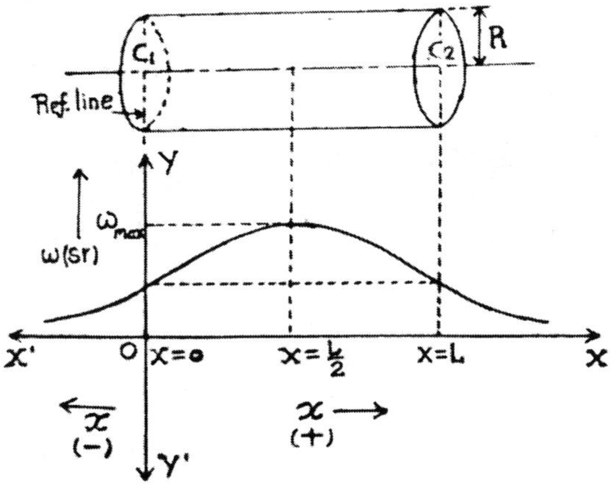

Fig (103)

* A typical ω-x graph for a hollow cylindrical shell.

*** Comparison of ω-x curves**

It is to be noted that the value of solid angle increases with decreasing radius keeping the length (L) of the shell constant. Thus if we take three hollow cylindrical shells of equal length (L) with radii $R_1, R_2 \& R_3$ respectively such that $R_1 > R_2 > R_3$ then

1. for the shell of radius R_1

$$\Rightarrow \omega_1 = 2\pi\left[\frac{(L+x)}{\sqrt{(L+x)^2+{R_1}^2}}-\frac{x}{\sqrt{x^2+{R_1}^2}}\right]$$

2. for the shell of radius R_2

$$\Rightarrow \omega_2 = 2\pi\left[\frac{(L+x)}{\sqrt{(L+x)^2+{R_2}^2}}-\frac{x}{\sqrt{x^2+{R_2}^2}}\right]$$

3. for the shell of radius R_3

$$\Rightarrow \omega_2 = 2\pi\left[\frac{(L+x)}{\sqrt{(L+x)^2+{R_3}^2}}-\frac{x}{\sqrt{x^2+{R_3}^2}}\right]$$

From the above three expressions, it is obvious if the length 'L' of all the shells is kept constant then for each value of distance (x),the values of solid angle subtended by the respective shells are related as

$$\Rightarrow \omega_1 < \omega_2 < \omega_3 \qquad \forall(R_1 > R_2 > R_3)$$

the relative variation of solid angle for the shells is shown by a typical ω-x graph in the fig(104) below

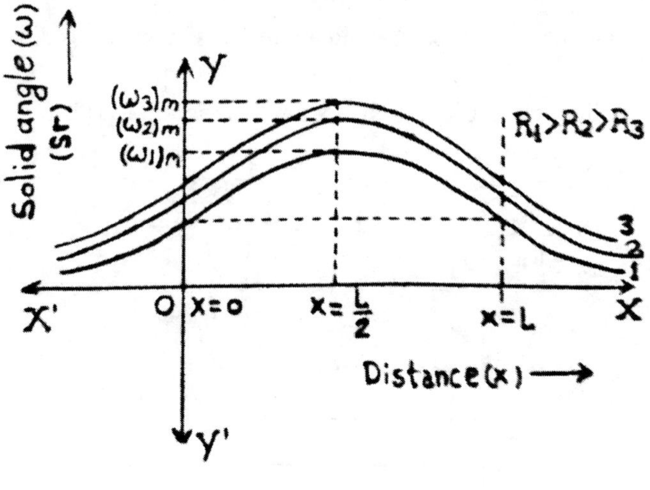

Fig (104)

4.4.15 Solid angle subtended by acylindrical shell with one end open at any point lying on the longitudinal axis inside it

Let there be a cylindrical shell with radius 'R', length 'L' & centres 'C_1'of open end & 'C_2' of closed end and a given point say 'P' lying on the longitudinal axis at a distance 'x' from the centre 'C_1' of open end inside the shell. (as shown in the figure (105) below)

The solid angle subtended by the open circular-end at the point 'P' lying on the vertical axis C_2C_1 at a distance 'x' from the centre 'C_1'.

Using formula from theeq(37) for a circular plane as follows

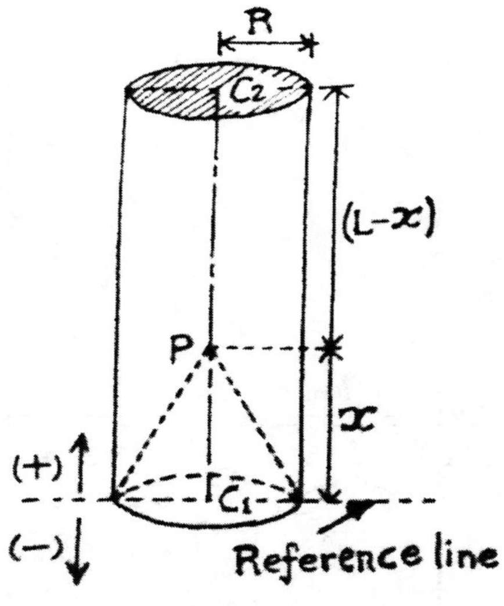

Fig (105)

$$\Rightarrow \omega_1 = 2\pi\left[1 - \frac{x}{\sqrt{x^2 + R^2}}\right]$$

Assume that the end with centre 'C_1' is closed then the solid angle subtended by the closed shell at the same point 'P'

$$\omega_2 = 4\pi$$

But, one end of the shell is open hence the solid angle subtended by the cylindrical shell at the given point 'P'

$$\Rightarrow \omega = \omega_2 - \omega_1 = 4\pi - 2\pi\left[1 - \frac{x}{\sqrt{x^2+R^2}}\right] = 4\pi - 2\pi + \frac{2\pi x}{\sqrt{x^2+R^2}} = 2\pi + \frac{2\pi x}{\sqrt{x^2+R^2}}$$

$$\Rightarrow \omega = 2\pi\left[1 + \frac{x}{\sqrt{x^2+R^2}}\right] \quad \forall(x < L) \quad \dots\dots\dots\dots\dots\dots\dots\dots (57)$$

Sign convention: The distance (x) measured from the centre of open end of the cylindrical shell is taken as

1. Positive when the point lies on the longitudinal axis inside the shell &
2. Negative when the point lies on the longitudinal axis outside the shell

It is obvious the value of solid angle is independent of the length of the shell within the range of distance (x)

Let's consider the point lying on the longitudinal axis outside the shell

4.4.16 Solid angle subtended by a cylindrical shell with one end open at any point lying on the longitudinal axis outside it

Let there be a cylindrical shell with radius 'R', length 'L' & centres 'C_1'of open end & 'C_2' of closed end and a given point say 'P_1' or 'P_2' lying on the longitudinal axis at a distance 'x' from the centre 'C_1' or 'C_2' of nearest end outside the shell. (as shown in the figure (106) below)

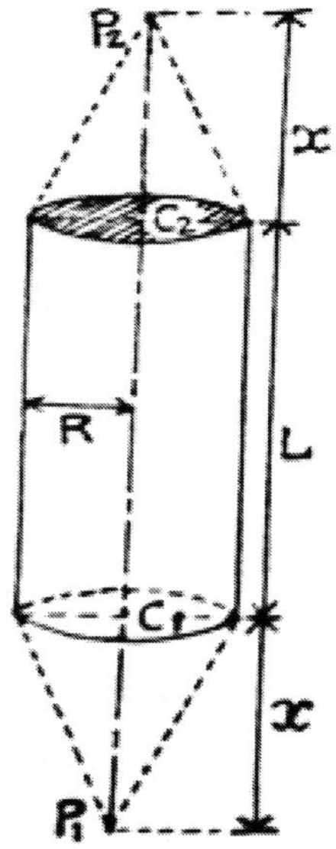

Fig (106)

In this case, the cylindrical shell will appear as circular plane with radius 'R' & centre 'C_1' when viewed from the point 'P_1' or with centre 'C_2' when viewed from the point 'P_2' lying on longitudinal axis.

Thus, the solid angle subtended by the cylindrical shell at any point lying on the longitudinal axis outside & on either side of the shell will be equal.

Now, using formula from the eq(37) for a circular plane

The solid angle subtended by the cylindrical shell at any point 'P_1' or 'P_2'

$$\Rightarrow \omega = 2\pi\left[1 - \frac{x}{\sqrt{x^2+R^2}}\right]$$

Or

by setting $x = -x$ in the eq(57) for the point outside the shell as follows

$$\Rightarrow \omega = 2\pi\left[1 + \frac{(-x)}{\sqrt{(-x)^2+R^2}}\right] = 2\pi\left[1 - \frac{x}{\sqrt{x^2+R^2}}\right] \quad \forall(x \geq 0) \quad \dots\dots\dots\dots\dots\dots\dots\dots (58)$$

4.4.17 Solid angle subtended by a cylindrical shell with one end open at the centre of closed end inside the shell

Let there be a cylindrical shell with radius 'R', length 'L' & centres 'C_1'of open end & 'C_2' of closed end. (as shown in the figure (107) below)

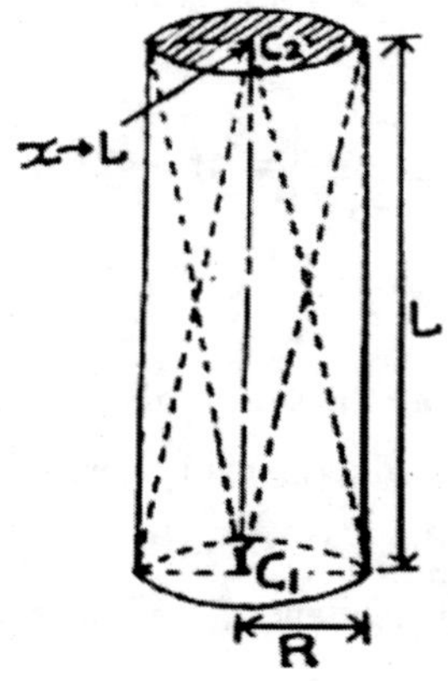

Fig (107)

In this case, the solid angle subtended by the cylindrical shell at the centre 'C_2' of closed end inside the shell is obtained by taking the limit of $\omega\ at\ x \to L^-$ from the eq(57) as follows

$i.e. x = L - k\ \&\ k \to 0$. 'k' is the infinitesimal small thickness of the boundary of the shell.

$$\Rightarrow \omega = \lim_{x \to L^-} 2\pi\left[1 + \frac{x}{\sqrt{x^2 + R^2}}\right] = 2\pi \lim_{k \to 0}\left[1 + \frac{(L-k)}{\sqrt{(L-k)^2 + R^2}}\right] = 2\pi\left[1 + \frac{(L-k)}{\sqrt{(L-k)^2 + R^2}}\right]$$

$$= 2\pi\left[1 + \frac{(L-0)}{\sqrt{(L-0)^2 + R^2}}\right] = 2\pi\left[1 + \frac{L}{\sqrt{L^2 + R^2}}\right] (since, k \neq 0)$$

$$\Rightarrow \omega = 2\pi\left[1 + \frac{L}{\sqrt{L^2 + R^2}}\right] \quad \ldots\ldots\ldots\ldots\ldots\ldots\ldots\ldots\ldots\ldots (59)$$

Note: if the shell is thick enough then the length 'L' can be displaced by the actual depth of the shell.

4.4.18 Solid angle subtended by a cylindrical shell with one end open at any point in the plane of open end inside its boundary

Let there be a cylindrical shell with radius 'R', length 'L' & centres 'C_1'of open end & 'C_2' of closed end& an arbitrary point say 'P' in the plane of open end inside the boundary. (as shown in the figure (108) below)

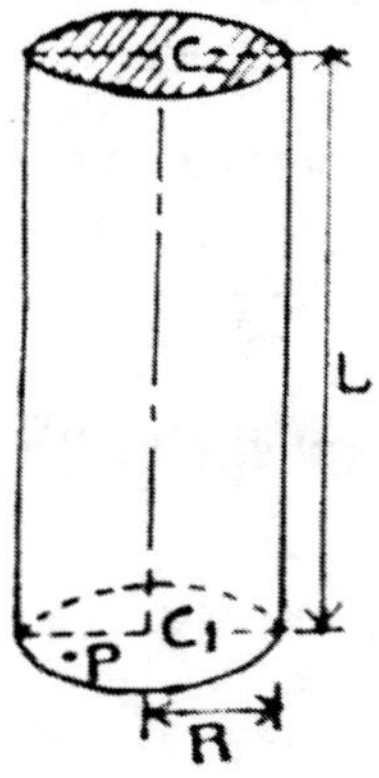

Fig (108)

In this case, the solid angle subtended by the cylindrical shell at any point 'P' in the plane of open end inside the boundary is obtained by setting x = 0 in the eq(57) as follows

$$\Rightarrow \omega = 2\pi\left[1+\frac{(0)}{\sqrt{(0)^2+R^2}}\right] = 2\pi[1+0] = 2\pi\ sr$$

Or by setting x = 0 in the eq(58) as follows

$$\Rightarrow \omega = 2\pi\left[1-\frac{(0)}{\sqrt{(0)^2+R^2}}\right] = 2\pi[1-0] = 2\pi\ sr$$

Thus, the cylindrical shell with one end open behaves as cap fully covering all the points of plane of open end inside its boundary.

4.4.19 Solid angle subtended by an infinitely long cylindrical shell with one end closed & another open to infinity at any point inside it

Let there be an infinitely long cylindrical shell with one end closed & another open to infinity & an arbitrary point say 'P' inside the shell (as shown in the figure (109) below)

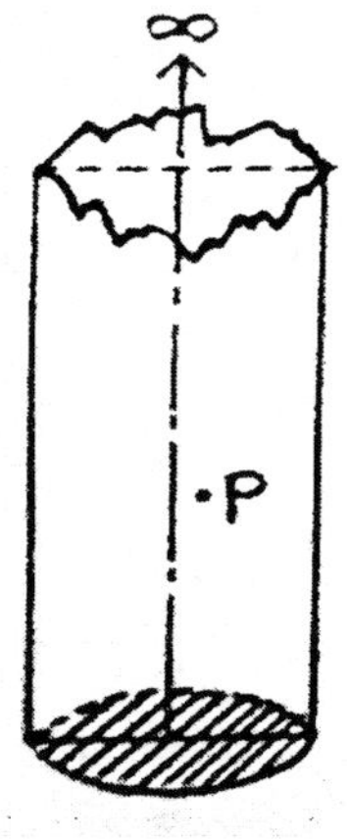

Fig (109)

In this case, the solid angle subtended by the infinitely long cylindrical shell at the point 'P' inside the shell is obtained by taking the limit of $\omega\ at\ L \to \infty$ from the eq(59) as follows

$$\Rightarrow \omega = \lim_{L\to\infty} 2\pi\left[1+\frac{L}{\sqrt{L^2+R^2}}\right] = 2\pi \lim_{L\to\infty}\left[1+\frac{1}{\sqrt{1+\left(\frac{R}{L}\right)^2}}\right] = 2\pi\left[1+\frac{1}{\sqrt{1+(0)^2}}\right]$$

$$\Rightarrow \omega = 2\pi\left[1+\frac{1}{\sqrt{1+0}}\right] = 2\pi[1+1] = 4\pi\ sr$$

Thus, an infinitely long cylindrical shell with one end closed & another open to infinity behaves as a closed surface for each point inside it.

4.4.20 Graphical presentation of variation of solid angle subtended by a cylindrical shell with one end open & another open along its longitudinal axis

We know that the solid angle subtended by a cylindrical shell with one end open is given by the eq(57)& (58) as follows

$$\Rightarrow \omega = 2\pi\left[1+\frac{x}{\sqrt{x^2+R^2}}\right] \quad \forall(x < L)$$

$$= 2\pi\left[1+\frac{(x-L)}{\sqrt{(x-L)^2+R^2}}\right] \quad \forall(x \geq L)$$

Where the distance 'x' is measured from the centre 'C_1' of the open end (taken as reference)

The curve is discontinuous at x = L because of infinitesimal thin boundary of the shell.

Let's find the following values of 'ω'.

1. at x = 0

$$\Rightarrow \omega = 2\pi\left[1 + \frac{(0)}{\sqrt{(0)^2 + R^2}}\right] = 2\pi[1 + 0] = 2\pi\ sr$$

2. at $x = \frac{L}{2}$

$$\Rightarrow \omega = 2\pi\left[1 + \frac{\left(\frac{L}{2}\right)}{\sqrt{\left(\frac{L}{2}\right)^2 + R^2}}\right] = 2\pi\left[1 + \frac{L}{\sqrt{L^2 + 4R^2}}\right]$$

3. at $x \to L^-$

$$\Rightarrow \omega = \lim_{x\to L^-} 2\pi\left[1 + \frac{x}{\sqrt{x^2 + R^2}}\right] = 2\pi\lim_{k\to 0}\left[1 + \frac{(L-k)}{\sqrt{(L-k)^2 + R^2}}\right] \quad where, k\ thickness\ of\ shell$$

$$= 2\pi\left[1 + \frac{(L-0)}{\sqrt{(L-0)^2 + R^2}}\right] = 2\pi\left[1 + \frac{L}{\sqrt{L^2 + R^2}}\right](since, k \neq 0)$$

4. at x = L

$$\Rightarrow \omega = 2\pi\left[1 + \frac{(L-L)}{\sqrt{(L-L)^2 + R^2}}\right] = 2\pi[1 + 0] = 2\pi\ sr$$

It is obvious that the value of solid angle (ω) varies non-linearly with the distance (x).

Now, if solid angle (ω) is taken as abscissa and the distance (x) as ordinate then we get a ω-x graph. A typical ω-x graph is shown in the fig(110) below.

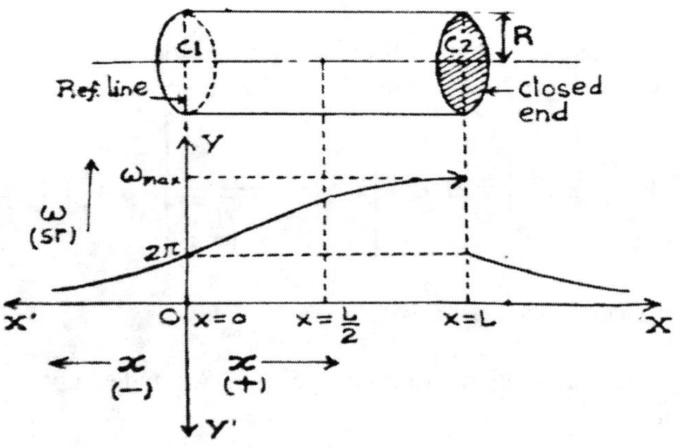

Fig (110)

* A typical ω-x graph for a cylindrical shell with one end closed & another open.

* Comparison of ω-x curves

It is to be noted that the value of solid angle increases with decreasing radius keeping the length (L) of the shell constant. Thus if we take three cylindrical shells with one end closed & another open of equal length (L) with radii $R_1, R_2 \& R_3$ respectively such that $R_1 > R_2 > R_3$ then

1. for the shell of radius R_1

$$\Rightarrow \omega_1 = 2\pi\left[1+\frac{x}{\sqrt{x^2+{R_1}^2}}\right] \qquad \forall(x<L)$$

$$= 2\pi\left[1+\frac{(x-L)}{\sqrt{(x-L)^2+{R_1}^2}}\right] \qquad \forall(x\geq L)$$

2. for the shell of radius R_2

$$\Rightarrow \omega_2 = 2\pi\left[1+\frac{x}{\sqrt{x^2+{R_2}^2}}\right] \qquad \forall(x<L)$$

$$= 2\pi\left[1+\frac{(x-L)}{\sqrt{(x-L)^2+{R_2}^2}}\right] \qquad \forall(x\geq L)$$

3. for the shell of radius R_3

$$\Rightarrow \omega_3 = 2\pi\left[1+\frac{x}{\sqrt{x^2+{R_3}^2}}\right] \qquad \forall(x<L)$$

$$= 2\pi\left[1+\frac{(x-L)}{\sqrt{(x-L)^2+{R_3}^2}}\right] \qquad \forall(x\geq L)$$

From the above three expressions, it is obvious if the length 'L' of all the shells is kept constant then for each value of distance (x) of the given point from the centre of the open end,the values of solid angle subtended by the respective shells are related as

1. $\omega_1 > \omega_2 > \omega_3 \qquad \forall(x < 0\,\&\, R_1 > R_2 > R_3)$
2. $\omega_1 = \omega_2 = \omega_3 = 2\pi \qquad \forall(x = 0\,\&\, R_1 > R_2 > R_3)$
3. $\omega_1 < \omega_2 < \omega_3 \qquad \forall(0 < x < L\,\&\, R_1 > R_2 > R_3)$
4. $\omega_1 = \omega_2 = \omega_3 = 2\pi \qquad \forall(x = L\,\&\, R_1 > R_2 > R_3)$
5. $\omega_1 > \omega_2 > \omega_3 \qquad \forall(x > L\,\&\, R_1 > R_2 > R_3)$

The relative variation of solid angle for the shells is shown by a typical ω-x graph in the fig(111) below

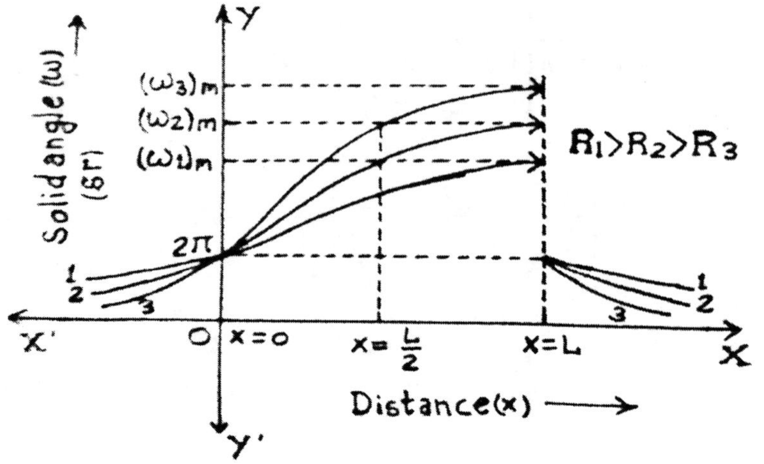

Fig (111)

4.4.21 Solid angle subtended by a longitudinal & symmetrical frustum of a cylinder at any point lying on the transverse axis

Let there be a longitudinal & symmetrical frustum of the cylinder with radius 'R', length 'L' & angle of curvature subtended by the frustum at the longitudinal axis is '2α' ($\forall\ 0 < \alpha < \frac{\pi}{2}$) and a given point say P' lying on the transverse axis at a normal distance 'x' from the longitudinal axis C_1C_2& 'y' the distance of the point 'O' from the centre 'C_1'. (as shown in the figure(112) below)

In this case, we find that as the given point (observer) moves on transverse axis from the point 'N' then the frustum appears as a complete cylinder up to the point 'P' which is the point of intersection of the tangents at the end points T_1&T_2& transverse axis while on moving further a little away from the point 'P' it appears as a rectangular plane. Thus there are two ranges of the distance 'x'. Let's find the ranges of 'x' & corresponding values of solid angle.

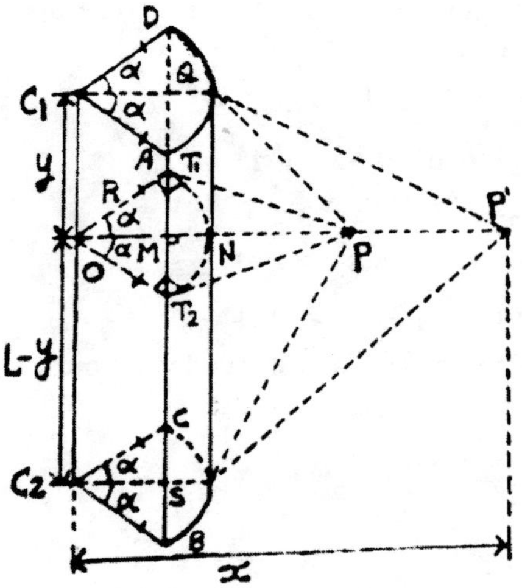

Fig (112)

In rightΔOMT_1

$$\Rightarrow \cos\alpha = \frac{OM}{OT_1} = \frac{OM}{R} \Rightarrow OM = R\cos\alpha \ \ \& \ \ \sin\alpha = \frac{MT_1}{OT_1} = \frac{MT_1}{R} \Rightarrow MT_1 = R\sin\alpha = MT_2$$

In right ΔOT_1P

$$\Rightarrow cos\alpha = \frac{OT_1}{OP} = \frac{R}{OP} \Rightarrow OP = Rsec\alpha$$

$$\Rightarrow T_1T_2 = 2MT_1 = 2Rsin\alpha \quad \& P'M = OP' - OM = x - Rcos\alpha$$

Now, let us consider the following cases

1.) For $R \leq x \leq Rsec\alpha \ \forall\left(0 < \alpha < \frac{\pi}{2}\right) \& \forall(0 \leq y \leq L)$

The solid angle subtended by the frustum at any point lying on the transverse axis between the points 'N' & 'P' is given from the eq(53) as follows (Neglecting effect of end-curvature)

$$\Rightarrow \omega = 2\left[\sin^{-1}\left\{\frac{Ry}{\sqrt{x^2y^2 + (x^2 - R^2)^2}}\right\} + \sin^{-1}\left\{\frac{R(L-y)}{\sqrt{x^2(L-y)^2 + (x^2 - R^2)^2}}\right\}\right]$$

2.) For $x > Rsec\alpha \ \forall(0 < \alpha < \frac{\pi}{2})$

The frustum will appear as rectangular plane ABCD

(as shown in the figure (113) below)

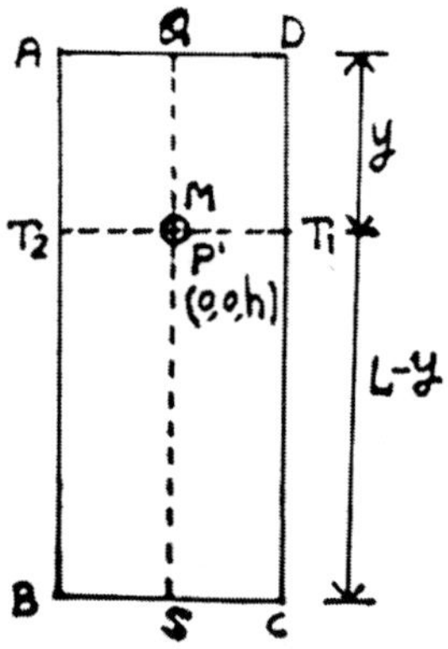

Fig (113)

The given pointP'may be assumed to be lying at a height 'h' on the vertical axis passing through the point 'M'

Now, divide the rectangle ABCD into four elementary rectangles MT_2AQ, MT_2BS, $MSCT_1\&MT_1DQ$ then the solid angle subtended by the frustum of the cylinder at the point 'P'

$$\Rightarrow \omega = \omega_{ABCD} = \omega_{MT_2AQ} + \omega_{MT_2BS} + \omega_{MSCT_1} + \omega_{MT_1DQ} = 2\left(\omega_{MT_2AQ} + \omega_{MT_2BS}\right)$$

(Since, the plane ABCD is symmetrical about the line QS)

Now, we know that the solid angle subtended by the rectangular plane at any point lying on the vertical axis passing through the vertex is given by the eq(17) as follows

$$\omega = \sin^{-1}\left\{\frac{lb}{\sqrt{(l^2 + h^2)(b^2 + h^2)}}\right\}$$

On setting the following values

$$l = length = y, b = width = MT_2 = Rsin\alpha \,\&\, h = height = P'M = x - Rcos\alpha$$

in the above expression, we have

$$[\omega]_{MT_2AQ} = \sin^{-1}\left\{\frac{y(Rsin\alpha)}{\sqrt{(y^2 + (x - Rcos\alpha)^2)((Rsin\alpha)^2 + (x - Rcos\alpha)^2)}}\right\}$$

$$\Rightarrow [\omega]_{MT_2AQ} = \sin^{-1}\left\{\frac{Rysin\alpha}{\sqrt{(y^2 + (x - Rcos\alpha)^2)(R^2sin^2\alpha + (x - Rcos\alpha)^2)}}\right\}$$

Similarly, On setting the following values

$$l = length = (L - y), b = width = MT_2 = Rsin\alpha \;\&\; h = height = P'M = x - Rcos\alpha$$

in the above expression, we have

$$[\omega]_{MT_2BS} = \sin^{-1}\left\{\frac{(L-y)(Rsin\alpha)}{\sqrt{((L-y)^2 + (x - Rcos\alpha)^2)((Rsin\alpha)^2 + (x - Rcos\alpha)^2)}}\right\}$$

$$= \sin^{-1}\left\{\frac{R(L-y)sin\alpha}{\sqrt{((L-y)^2 + (x - Rcos\alpha)^2)(R^2sin^2\alpha + (x - Rcos\alpha)^2)}}\right\}$$

Hence by adding above values, we get

$$\Rightarrow \omega = 2\left[\sin^{-1}\left\{\frac{Rysin\alpha}{\sqrt{(y^2 + (x - Rcos\alpha)^2)(R^2sin^2\alpha + (x - Rcos\alpha)^2)}}\right\}\right]$$

$$+2\left[\sin^{-1}\left\{\frac{Rysin\alpha}{\sqrt{(y^2 + (x - Rcos\alpha)^2)(R^2sin^2\alpha + (x - Rcos\alpha)^2)}}\right\}\right]$$

$$\forall (x \geq R \;\&\; 0 \leq y \leq L)$$

Important deduction: If $\alpha \to \frac{\pi}{2}$ then we get the following expressions for a longitudinally half cylinder

$$\Rightarrow R \leq x \leq Rsec\frac{\pi}{2} \;\; or \; R \leq x < \infty \;\&\; 0 \leq y \leq L$$

$$\Rightarrow \omega = 2\left[\sin^{-1}\left\{\frac{Ry}{\sqrt{x^2y^2 + (x^2 - R^2)^2}}\right\} + \sin^{-1}\left\{\frac{R(L-y)}{\sqrt{x^2(L-y)^2 + (x^2 - R^2)^2}}\right\}\right]$$

It is obvious that longitudinally half cylinder appears as a perfect cylinder at any distance on the transverse axis (passing through the centre of longitudinal axis).

4.4.22 Solid angle subtended by a longitudinal & symmetrical frustum of the hollow cylindrical shell at any point lying on the longitudinal axis

Let there be a longitudinal & symmetrical frustum of the hollow cylindrical shell with radius 'R', length 'L' & angle of curvature subtended by the frustum at the longitudinal axis is 'θ' and a given point say 'P' lying on the longitudinal axis at a distance 'x' from the centre 'C_1' of the end. (as shown in the figure(114) below)

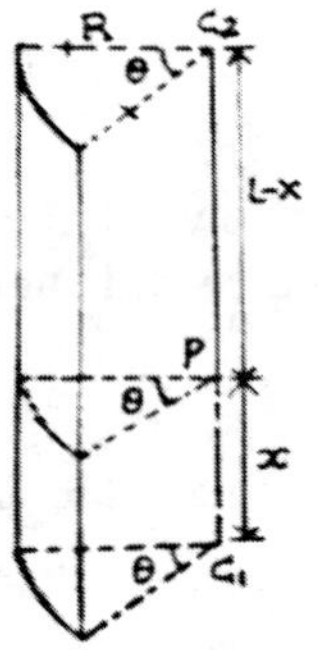

Fig (114)

We know that the solid angle subtended by the a hollow cylindrical shell is given from the eq(53) as follows

$$\omega_{\theta=2\pi} = 2\pi\left[\frac{x}{\sqrt{x^2 + R^2}} + \frac{(L-x)}{\sqrt{(L-x)^2 + R^2}}\right]$$

Since, the shell is symmetrical about the longitudinal axis C_1C_2 thus if we take a frustum symmetrically about the longitudinal axis with an angle of curvature 'θ' then by using symmetry, the solid angle subtended by the frustum at the point 'p'

$$\Rightarrow \omega = \frac{\theta}{2\pi}(\omega_{\theta=2\pi}) = \frac{\theta}{2\pi}2\pi\left[\frac{x}{\sqrt{x^2 + R^2}} + \frac{(L-x)}{\sqrt{(L-x)^2 + R^2}}\right]$$

$$\Rightarrow \omega = \theta\left[\frac{x}{\sqrt{x^2 + R^2}} + \frac{(L-x)}{\sqrt{(L-x)^2 + R^2}}\right] \quad \forall (x \in (-\infty, \infty) \& \; 0 < \theta \leq 2\pi) \quad \ldots\ldots\ldots\ldots(60)$$

* The above expression is in generalised form for a hollow cylindrical shell.

4.4.23 Solid angle subtended by a longitudinal & symmetrical frustum of the hollow cylindrical shell at mid-point of the longitudinal axis

Let there be a longitudinal & symmetrical frustum of the hollow cylindrical shell with radius 'R', length 'L' & angle of curvature subtended by the frustum at the longitudinal axis is 'θ' and 'P' is the mid-point of the longitudinal axis C_2C_1 (as shown in the figure(115) below)

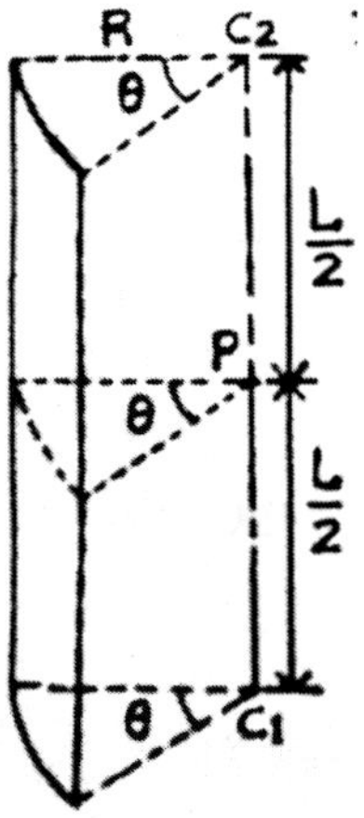

Fig (115)

In this case, the solid angle subtended by the frustum at the mid-point 'p' is obtained by setting $x = \frac{L}{2}$in the eq(60) as follows

$$\Rightarrow \omega = \theta\left[\frac{\left(\frac{L}{2}\right)}{\sqrt{\left(\frac{L}{2}\right)^2 + R^2}} + \frac{(L - \frac{L}{2})}{\sqrt{(L - \frac{L}{2})^2 + R^2}}\right] = \theta\left[\frac{L}{\sqrt{L^2 + 4R^2}} + \frac{L}{\sqrt{L^2 + 4R^2}}\right]$$

$$\Rightarrow \omega = 2\theta\left[\frac{L}{\sqrt{L^2 + 4R^2}}\right] \quad \forall\, (0 < \theta \leq 2\pi) \qquad \dots\dots\dots\dots\dots\dots\dots\dots (61)$$

*** Solid angle subtended by the frustum of a hollow cylindrical shell at the centre: is obtained by setting x = 0 in the eq(60) as follows**

$$\Rightarrow \omega = \theta\left[\frac{(0)}{\sqrt{(0)^2 + R^2}} + \frac{(L - 0)}{\sqrt{(L - 0)^2 + R^2}}\right] = \theta\left[0 + \frac{L}{\sqrt{L^2 + R^2}}\right]$$

$$\Rightarrow \omega = \theta\left[\frac{L}{\sqrt{L^2 + R^2}}\right] \qquad \forall\, (0 < \theta \leq 2\pi) \qquad \dots\dots\dots\dots\dots\dots\dots\dots (62)$$

4.5 RIGHT CONE

In this section, we are to study & make solutions of the following

1. Solid or hollow right cone (with finite & infinite height)
2. Hollow conical shell (with the open circular base) &
3. Symmetrical & longitudinal frustum of the hollow conical shell.

4.5.1 Solid angle subtended by a right cone at the apex point

Let there be a right cone with radius 'R', normal height 'H' & the apex point say ''P (as shown in the figure (116) below)

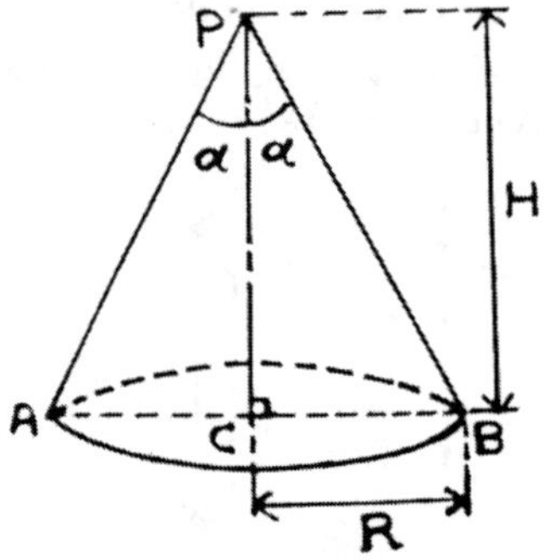

Fig (116)

In right ΔBCP

$$\Rightarrow cos\alpha = \frac{PC}{BP} = \frac{PC}{\sqrt{PC^2 + BC^2}} = \frac{H}{\sqrt{H^2 + R^2}}$$

Now, we know that the solid angle subtended by the right cone at the apex is given from the eq(1) as follows

$\omega = 2\pi(1 - cos\alpha)$ where '2α' is the apex angle

Hence, on setting the value of $cos\alpha$ in the above expression, we get

$$\Rightarrow \omega = 2\pi\left[1 - \frac{H}{\sqrt{H^2 + R^2}}\right] \quad \ldots\ldots\ldots\ldots\ldots\ldots (63)$$

4.5.2 Solid angle subtended a right cone (hollow or solid) at any external point lying in the plane of the circular base

Let there be a right cone with radius 'R' & normal height 'H' & a given external point say ''P lying in the plane of the circular base at a distance 'x' from the centre 'C' of the base.

(as shown in the figure (117) below with upper one as front-view & lower one as top-view of circular base with centre 'C' & its plane passing through the given point 'P')

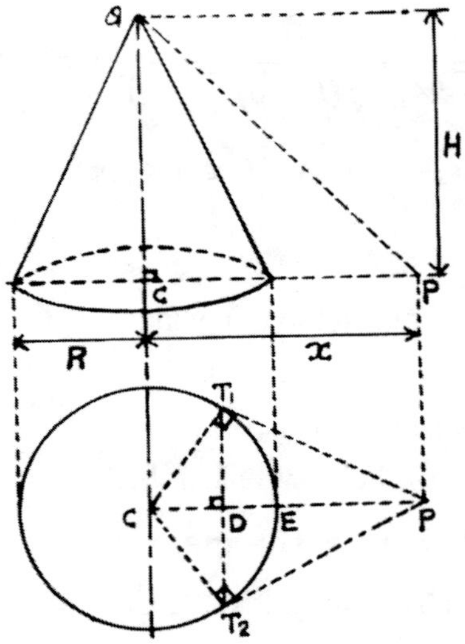

Fig (117)

Now, draw the smallest tangents on the lateral surface of the cone as PT_1 & PT_2 from the point 'P'

(see the lower one as top-view in the above fig.)

In right ΔCT_1P

$$\Rightarrow PT_1 = \sqrt{CP^2 - {CT_1}^2} = \sqrt{x^2 - R^2} = PT_2 \&$$

We know that in a right triangle

$$perpendicular\ to\ the\ hypotenuse = \frac{(height) \times (base)}{\sqrt{(height)^2 + (base)^2}} (using\ formula *)$$

$$\therefore\ T_1M = \frac{(CT_1) \times (PT_1)}{\sqrt{{CT_1}^2 + {PT_1}^2}} = \frac{R\sqrt{x^2 - R^2}}{\sqrt{R^2 + \left(\sqrt{x^2 - R^2}\right)^2}} = \frac{R\sqrt{x^2 - R^2}}{\sqrt{R^2 + x^2 - R^2}} \quad (CT_1 = radius = R)$$

$$\Rightarrow T_1D = T_2D = \frac{R\sqrt{x^2 - R^2}}{x} \Rightarrow T_1T_2 = 2T_1D = \frac{2R\sqrt{x^2 - R^2}}{x}$$

Now, in rightΔCDT_1

$$\Rightarrow CD = \sqrt{{CT_1}^2 - T_1D^2} = \sqrt{R^2 - \left(\frac{R\sqrt{x^2 - R^2}}{x}\right)^2} = \sqrt{\frac{R^2x^2 - R^2x^2 + R^4}{x^2}} = \frac{R^2}{x}$$

$$\therefore\ PD = CP - CD = x - \frac{R^2}{x} = \frac{x^2 - R^2}{x}$$

Now, draw the tangents from the given point 'P' at all the points of surface of the cone thus we get an isosceles triangular plane $QT_1T_2$with base T_1T_2due to interception of the tangents with surface of the cone. (as shown in the figure (118) below)

It is obvious that the right cone would appear like an isosceles triangular plane QT_1T_2 when viewed from the given point 'P' & the point 'P' may be assumed to be lying on the vertical axis passing through the mid-point 'D' of the base. Thus, right cone may be replaced by the triangular plane.

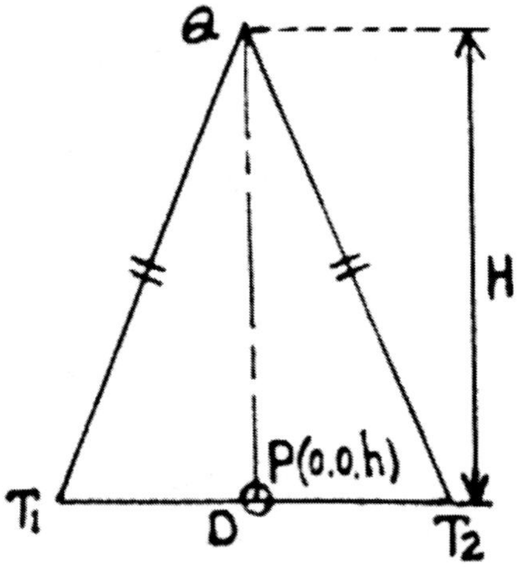

Fig (118)

In this case, the solid angle subtended by the right cone at the given point 'P' lying in the plane of the circular base at a normal distance 'x' from the centre 'C' of the base.

= solid angle subtended by the isosceles plane QT_1T_2 at the same point 'P' lying on the vertical axis passing through the mid- point 'D' at a normal distance PD = $\frac{x^2-R^2}{x}$

(Point 'D' is the foot of perpendicular drawn from the given point 'P' to the plane QT_1T_2)

We know that the solid angle subtended by an isosceles triangular plane at any point lying on the vertical axis passing through the mid-point of the base (unequal side) is given by the eq(9) as follows

$$\omega = 2\cos^{-1}\left\{\frac{h\left(2p^2\sqrt{4h^2+b^2} + b^2\sqrt{h^2+p^2}\right)}{h^2(4p^2+b^2)+p^2b^2}\right\}$$

Now, on setting the value of b = base = $T_1T_2 = \frac{2R\sqrt{x^2-R^2}}{x}$ & p = perpendicular = QD = H & h = height = PD = $\frac{x^2-R^2}{x}$ as follows, the value of solid angle

(* The derivation of this formula is not necessary here.)

$$= 2\cos^{-1}\left\{\frac{\left(\frac{x^2-R^2}{x}\right)\left(2H^2\sqrt{4\left(\frac{x^2-R^2}{x}\right)^2+\left(\frac{2R\sqrt{x^2-R^2}}{x}\right)^2} + \left(\frac{2R\sqrt{x^2-R^2}}{x}\right)^2\sqrt{\left(\frac{x^2-R^2}{x}\right)^2+H^2}\right)}{\left(\frac{x^2-R^2}{x}\right)^2\left(4H^2+\left(\frac{2R\sqrt{x^2-R^2}}{x}\right)^2\right)+H^2\left(\frac{2R\sqrt{x^2-R^2}}{x}\right)^2}\right\}$$

$$= 2\cos^{-1}\left\{\frac{\left(\frac{x^2-R^2}{x}\right)\left(2H^2\left(\frac{2\sqrt{x^2-R^2}}{x}\right)\sqrt{x^2-R^2+R^2} + \left(\frac{4R^2(x^2-R^2)}{x^2}\right)\left(\frac{1}{x}\right)\sqrt{(x^2-R^2)^2+H^2x^2}\right)}{\frac{(x^2-R^2)^2}{x^2}\left(\frac{4}{x^2}\right)\left(H^2x^2+R^2(x^2-R^2)\right)+\frac{4H^2R^2(x^2-R^2)}{x^2}}\right\}$$

$$= 2\cos^{-1}\left\{\frac{\left(\frac{x^2-R^2}{x}\right)\left(4H^2x\left(\frac{\sqrt{x^2-R^2}}{x}\right) + \left(\frac{4R^2(x^2-R^2)}{x^3}\right)\sqrt{(x^2-R^2)^2+H^2x^2}\right)}{\frac{4(x^2-R^2)^2}{x^4}\left(R^2(x^2-R^2)+H^2x^2\right)+\frac{4H^2R^2(x^2-R^2)}{x^2}}\right\}$$

$$\Rightarrow \omega = 2\cos^{-1}\left\{\frac{\left(H^2x^3\sqrt{x^2-R^2} + R^2(x^2-R^2)\sqrt{(x^2-R^2)^2+H^2x^2}\right)}{(x^2-R^2)\left(R^2(x^2-R^2)+H^2x^2\right)+H^2R^2x^2}\right\} \quad \forall (x \geq R)$$

.... (64)

Important deduction: If the given point lies on the surface of a right cone in the plane of the base (i.e. for which the distance from the centre 'C' is x = R) is obtained by setting x = R in the above eq(64) as follows

$$\omega = 2\cos^{-1}\left\{\frac{\left(H^2R^3\sqrt{R^2-R^2} + R^2(R^2-R^2)\sqrt{(R^2-R^2)^2+H^2R^2}\right)}{(R^2-R^2)\left(R^2(R^2-R^2)+H^2R^2\right)+H^2R^2R^2}\right\}$$

$$\Rightarrow \omega = 2\cos^{-1}\left\{\frac{\left(H^2R^3(0) + R^2(0)\sqrt{0+H^2R^2}\right)}{(0)\left(R^2(0)+H^2R^2\right)+H^2R^2R^2}\right\} = 2\cos^{-1}\{0\} = 2\frac{\pi}{2} = \pi\ sr$$

4.5.3 Solid angle subtended by an infinitely long right cone at any external point lying in the plane of the circular base

Let there be a right cone with radius 'R' &normal height 'H' & a given external point say ''P lying in the plane of the circular base at a distance 'x' from the centre 'C' of the base. (as shown in the figure (119) below)

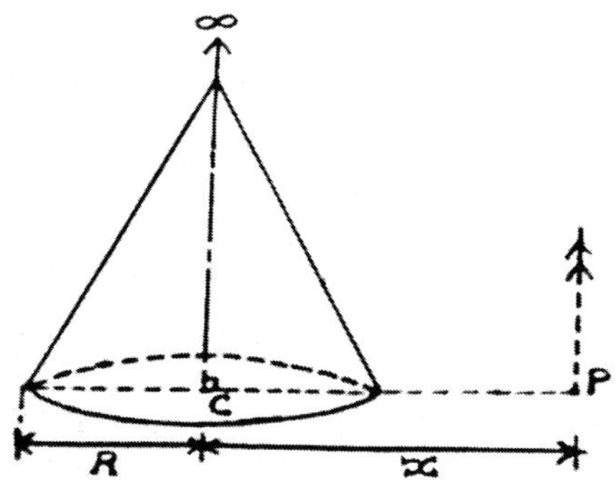

Fig (119)

In this case, the solid angle subtended by the infinitely long right cone at the point 'P' is obtained by taking limit of ω at $H \to \infty$ from the above eq(64) as follows

$$\Rightarrow \omega = \lim_{H\to\infty} 2\cos^{-1}\left\{\frac{\left(H^2x^3\sqrt{x^2-R^2} + R^2(x^2-R^2)\sqrt{(x^2-R^2)^2+H^2x^2}\right)}{(x^2-R^2)(R^2(x^2-R^2)+H^2x^2)+H^2R^2x^2}\right\}$$

$$= 2\lim_{H\to\infty}\cos^{-1}\left\{\frac{\left(x^3\sqrt{x^2-R^2} + R^2\left(\frac{x^2-R^2}{H}\right)\sqrt{\left(\frac{x^2-R^2}{H}\right)^2+x^2}\right)}{(x^2-R^2)\left(R^2\left(\frac{x^2-R^2}{H^2}\right)+x^2\right)+R^2x^2}\right\}$$

$$= 2\cos^{-1}\left\{\frac{\left(x^3\sqrt{x^2-R^2} + R^2\left(\frac{x^2-R^2}{\infty}\right)\sqrt{\left(\frac{x^2-R^2}{\infty}\right)^2+x^2}\right)}{(x^2-R^2)\left(R^2\left(\frac{x^2-R^2}{(\infty)^2}\right)+x^2\right)+R^2x^2}\right\}$$

$$\Rightarrow \omega = 2\cos^{-1}\left\{\frac{\left(x^3\sqrt{x^2-R^2} + R^2(0)\sqrt{0+x^2}\right)}{(x^2-R^2)(R^2(0)+x^2)+R^2x^2}\right\} = 2\cos^{-1}\left\{\frac{x^3\sqrt{x^2-R^2}}{(x^2-R^2)x^2+R^2x^2}\right\}$$

$$= 2\cos^{-1}\left\{\frac{x^3\sqrt{x^2-R^2}}{x^4-R^2x^2+R^2x^2}\right\} = 2\cos^{-1}\left\{\frac{x^3\sqrt{x^2-R^2}}{x^4}\right\} = 2\cos^{-1}\left\{\frac{\sqrt{x^2-R^2}}{x}\right\}$$

$$= 2\cos^{-1}\left\{\sqrt{1-\left(\frac{R}{x}\right)^2}\right\} = 2\sin^{-1}\left(\frac{R}{x}\right) \quad since, \cos^{-1}\sqrt{1-x^2} = \sin^{-1}x \ \forall\, x \in (-1,1)$$

$$\Rightarrow \omega = 2\sin^{-1}\left(\frac{R}{x}\right) \quad \forall (x \geq R) \qquad \ldots\ldots\ldots\ldots\ldots\ldots (65)$$

The above result is same to that of eq(50) it implies that an infinitely long right cone is similar to an infinitely long cylinder.

Important deduction: If the given point lies on the surface of an infinitely long right cone in the plane of the base (i.e. for which the distance from the centre 'C' is x = R) is obtained by setting x = R in the above eq(65) as follows

$$\omega = 2\cos^{-1}\left\{\sqrt{1-\left(\frac{R}{R}\right)^2}\right\} = 2\cos^{-1}\{\sqrt{1-1}\} = 2\cos^{-1}\{0\} = 2\frac{\pi}{2} = \pi\ sr$$

Thus, we find that above value of solid angle is equal to that obtained for a right cone with finite height. It is because when the point lies on the surface then the solid angle subtended at the same point is not affected by the geometrical shape & its dimensions.

4.5.4 Solid angle subtended by a hollow right conical shell at any internal point lying on the longitudinal axis

Let there be a hollow right conical shell with radius 'R' & normal height 'H' & a given internal point say 'P' lying on the longitudinal axis at a distance 'x' from the centre 'C' of the base. (as shown in the figure (120) below)

Now, assume that the circular base of the shell is closed then the solid angle subtended the closed shell at the point 'P' is

$$\Rightarrow \omega_1 = 4\pi$$

And the solid angle subtended by the circular base with radius 'R' & centre 'C' at the point 'P' lying on the vertical axis at a height 'x' from the centre 'C'. Using formula from the eq(37) as follows

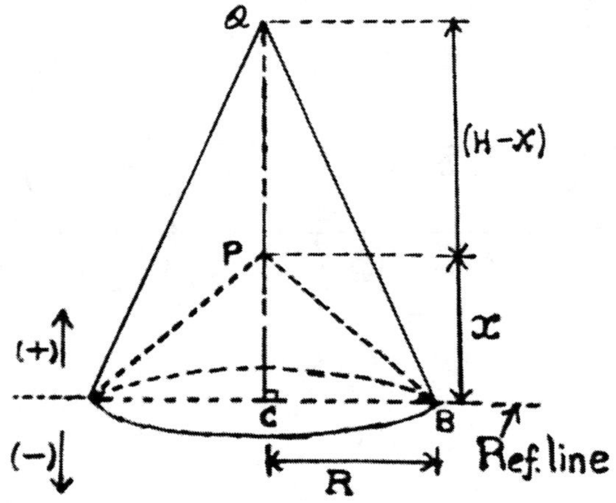

Fig (120)

$$\Rightarrow \omega_2 = 2\pi\left[1 - \frac{x}{\sqrt{x^2 + R^2}}\right]$$

But the base is open hence the solid angle subtended by the circular base is subtracted from that subtended by the closed shell

Hence, the solid angle subtended by the hollow right conical shell at the point 'P' is obtained as follows

$$\Rightarrow \omega = \omega_1 - \omega_2 = 4\pi - 2\pi\left[1 - \frac{x}{\sqrt{x^2 + R^2}}\right] = 4\pi - 2\pi + \frac{2\pi x}{\sqrt{x^2 + R^2}} = 2\pi + \frac{2\pi x}{\sqrt{x^2 + R^2}}$$

$$\Rightarrow \omega = 2\pi\left[1 + \frac{x}{\sqrt{x^2 + R^2}}\right] \quad \forall (x < H) \quad \ldots\ldots\ldots\ldots\ldots\ldots\ldots\ldots\ldots\ldots (66)$$

The above value is dependent on the radius (R) & the distance 'x' from the centre of the circular base but independent of height of the conical shell.

Hence, hence if the

Sign convention: The distance (x) measured from the centre of circular base of the conical shell is taken as

1. Positive when the point lies on the longitudinal axis inside the shell &
2. Negative when the point lies on the longitudinal axis outside the shell

4.5.5 Solid angle subtended by a hollow right conical shell at any external point lying on the longitudinal axis

Let there be a hollow right conical shell with radius 'R' & normal height 'H' & a given external point say 'P_1' towards the base & 'P_2' towards the apex (peak) lying on the longitudinal axis at a distance 'x' from the centre 'C' of the base. (as shown in the figure (121) below)

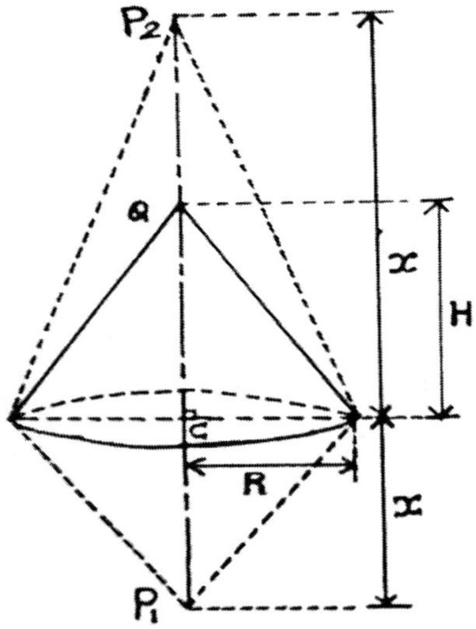

Fig (121)

In both the cases, when the hollow conical shell is viewed from the point 'P_1' & 'P_2' it appears as a circular plane at a normal distance 'x' from the respective points lying on the longitudinal axis outside the shell.

Hence, from the eq(37) used for a circular plane, the solid angle subtended by the hollow conical shell at the point 'P_1' or 'P_2'

$$\Rightarrow \omega = 2\pi\left[1 - \frac{x}{\sqrt{x^2 + R^2}}\right]$$

Or by using sign convention i.e. on setting x = −x in the eq(66) as follows

$$\Rightarrow \omega = 2\pi\left[1 + \frac{(-x)}{\sqrt{(-x)^2 + R^2}}\right] = 2\pi\left[1 - \frac{x}{\sqrt{x^2 + R^2}}\right]$$

Hence, we have two cases for the range of distance 'x' as follows

1. For the point lying towards the base of the conical shell

$$\Rightarrow \omega = 2\pi\left[1 - \frac{x}{\sqrt{x^2 + R^2}}\right] \quad \forall (x \geq 0) \qquad \dots\dots\dots\dots\dots\dots\dots (67.1)$$

2. For the point lying towards the apex of the conical shell

$$\Rightarrow \omega = 2\pi\left[1 - \frac{x}{\sqrt{x^2 + R^2}}\right] \quad \forall (x \geq H) \qquad \dots\dots\dots\dots\dots\dots\dots (67.2)$$

4.5.6 Solid angle subtended by a hollow right conical shell at the apex inside the shell

Let there be a hollow right conical shell with radius 'R', normal height 'H' &apex point 'Q'

(as shown in the figure (122) below)

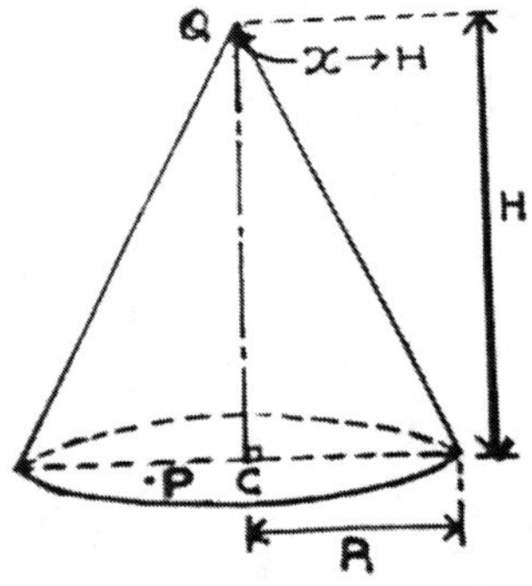

Fig (122)

In this case, the solid angle subtended by the conical shell at the apex 'Q' inside the shell is obtained by taking the limit of ω at $x \to H^-$ from the eq(66) as follows

$i.e. at\ x = H - k\ \&\ k \to 0$. 'k' is the infinitesimal small thickness of the boundary of the shell.

$$\Rightarrow \omega = \lim_{x \to H^-} 2\pi\left[1 + \frac{x}{\sqrt{x^2 + R^2}}\right] = 2\pi \lim_{k \to 0}\left[1 + \frac{(H-k)}{\sqrt{(H-k)^2 + R^2}}\right] = 2\pi\left[1 + \frac{(H-k)}{\sqrt{(H-k)^2 + R^2}}\right] \quad where\ x$$

$$= (H-k) \Rightarrow k \to 0$$

$$= 2\pi\left[1 + \frac{(H-0)}{\sqrt{(H-0)^2 + R^2}}\right] = 2\pi\left[1 + \frac{H}{\sqrt{H^2 + R^2}}\right] (since, k \neq 0)$$

$$\Rightarrow \omega = 2\pi\left[1 + \frac{H}{\sqrt{H^2 + R^2}}\right] \quad \ldots\ldots\ldots\ldots (68)$$

4.5.7 Solid angle subtended by a hollow right conical shell at any point of the circular base inside the shell

Let there be a hollow right conical shell with radius 'R', normal height 'H' & an arbitrary point say 'P' in the plane of the base inside the shell. (as shown in the figure (122) above)

In this case, the solid angle subtended by the hollow conical shell at the point 'P' is obtained by setting x = 0 in the eq(66) as follows

$$\omega = 2\pi\left[1 + \frac{(0)}{\sqrt{(0)^2 + R^2}}\right] = 2\pi[1 + 0] = 2\pi\ sr$$

It is obvious that the hollow conical shell behaves as a cap for each point of the base inside the shell.

4.5.8 Graphical presentation of variation of solid angle subtended by a hollow right conical shell along the longitudinal axis

We know that solid angle subtended by the hollow conical shell at any point lying on the longitudinal axis is given by the eq(66) and (67.2) as follows

$$\Rightarrow \omega = 2\pi\left[1 + \frac{x}{\sqrt{x^2 + R^2}}\right] \quad \forall (x < H)$$

$$= 2\pi\left[1 - \frac{x}{\sqrt{x^2 + R^2}}\right] \quad \forall (x \geq H)$$

Where, 'x' is the distance of the given point lying on the longitudinal axis from the centre of base of the shell.

The curve is discontinuous at x = H because of infinitesimal thin boundary of the shell.

Let's find the following values of 'ω'.

1. at x = 0

$$\Rightarrow \omega = 2\pi\left[1+\frac{(0)}{\sqrt{(0)^2+R^2}}\right] = 2\pi[1+0] = 2\pi\ sr$$

2. at $x = \frac{H}{2}$

$$\Rightarrow \omega = 2\pi\left[1+\frac{\left(\frac{H}{2}\right)}{\sqrt{\left(\frac{H}{2}\right)^2+R^2}}\right] = 2\pi\left[1+\frac{H}{\sqrt{H^2+4R^2}}\right]$$

3. at $x \to H^-$

$$\Rightarrow \omega = \lim_{x\to L^-} 2\pi\left[1+\frac{x}{\sqrt{x^2+R^2}}\right] = 2\pi\lim_{k\to 0}\left[1+\frac{(H-k)}{\sqrt{(H-k)^2+R^2}}\right] \quad where, k\ thickness\ of\ shell$$

$$= 2\pi\left[1+\frac{(H-0)}{\sqrt{(H-0)^2+R^2}}\right] = 2\pi\left[1+\frac{H}{\sqrt{H^2+R^2}}\right] (since, k \neq 0)$$

4. at x = H

$$\Rightarrow \omega = 2\pi\left[1-\frac{H}{\sqrt{H^2+R^2}}\right]$$

It is obvious that the value of solid angle (ω) varies non-linearly with the distance (x).

Now, if solid angle (ω) is taken as abscissa and the distance (x) as ordinate then we get a ω-x graph. A typical ω-x graph is shown in the fig(123) below.

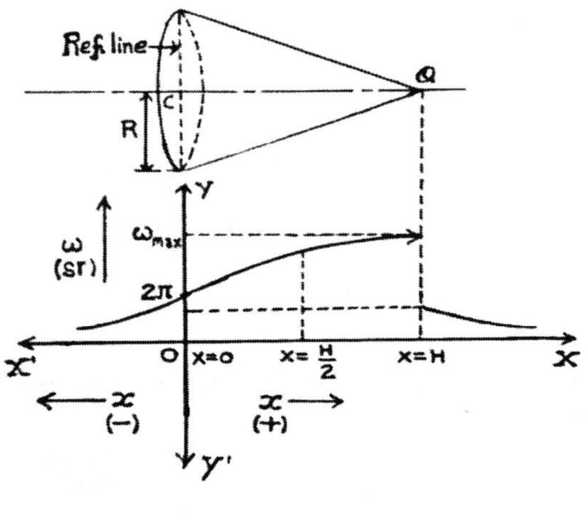

Fig (123)

* A typical ω-x graph for a hollow right conical shell.

* Comparison of ω-x curves

It is to be noted that the value of solid angle increases with decreasing radius (R) keeping the normal height (H) of the shell constant. Thus if we take three hollow conical shells of equal height (H) with radii $R_1, R_2 \& R_3$ respectively such that $R_1 > R_2 > R_3$ then

1. for the shell of radius R_1

$$\Rightarrow \omega_1 = 2\pi\left[1+\frac{x}{\sqrt{x^2+{R_1}^2}}\right] \quad \forall(x < H)$$

$$= 2\pi\left[1 - \frac{x}{\sqrt{x^2 + {R_1}^2}}\right] \qquad \forall(x \geq H)$$

2. for the shell of radius R_2

$$\Rightarrow \omega_2 = 2\pi\left[1 + \frac{x}{\sqrt{x^2 + {R_2}^2}}\right] \qquad \forall(x < H)$$

$$= 2\pi\left[1 - \frac{x}{\sqrt{x^2 + {R_2}^2}}\right] \qquad \forall(x \geq H)$$

3. for the shell of radius R_3

$$\Rightarrow \omega_3 = 2\pi\left[1 + \frac{x}{\sqrt{x^2 + {R_3}^2}}\right] \qquad \forall(x < H)$$

$$= 2\pi\left[1 - \frac{x}{\sqrt{x^2 + {R_3}^2}}\right] \qquad \forall(x \geq H)$$

From the above three expressions, it is obvious if the normal height 'H' of all the shells is kept constant then for each value of distance (x) of the given point from the centre of the base,the values of solid angle subtended by the respective shells are related as

1. $\omega_1 > \omega_2 > \omega_3$ $\qquad \forall(x < 0 \& R_1 > R_2 > R_3)$
2. $\omega_1 = \omega_2 = \omega_3 = 2\pi$ $\qquad \forall(x = 0 \ \& \ R_1 > R_2 > R_3)$
3. $\omega_1 < \omega_2 < \omega_3$ $\qquad \forall(0 < x < H \ \& \ R_1 > R_2 > R_3)$
4. $\omega_1 > \omega_2 > \omega_3$ $\qquad \forall(x = H \ \& \ R_1 > R_2 > R_3)$
5. $\omega_1 > \omega_2 > \omega_3$ $\qquad \forall(x > H \ \& \ R_1 > R_2 > R_3)$

The relative variation of solid angle for the shells is shown by a typical ω-x graph in the fig(124) below

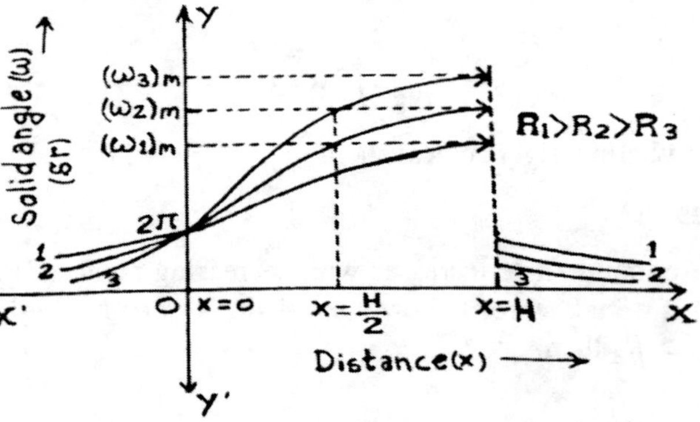

Fig (124)

4.5.9 Solid angle subtended by an infinitely long hollow right conical shell at any point lying on the longitudinal axis

Let there be an infinitely long hollow conical shell with radius 'R' & a given point say 'P' lying on the longitudinal axis inside the shell (as shown in the figure (125) below)

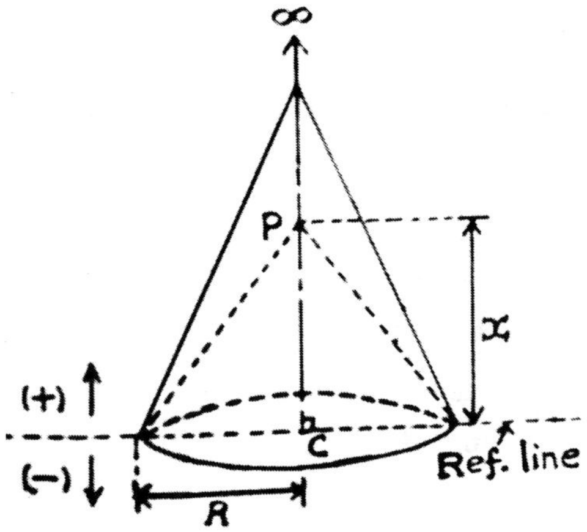

Fig (125)

We know that the solid angle subtended by the hollow conical shell at any point lying on the longitudinal axis is independent on the normal height of the shell. Hence, this case is similar to the previous article 4.5.4. Only range of the distance (x) will change accordingly. Hence the solid is given by the eq(66) as follows

$$\Rightarrow \omega = 2\pi\left[1 + \frac{x}{\sqrt{x^2 + R^2}}\right] \quad \forall\, x \in (-\infty, \infty) \quad \ldots\ldots\ldots\ldots\ldots\ldots\ldots\ldots\ldots\ldots (69)$$

Sign convention: The distance (x) measured from the centre of circular base of the conical shell is taken as

1. Positive when the point lies on the longitudinal axis inside the shell &
2. Negative when the point lies on the longitudinal axis outside the shell

4.5.10 Solid angle subtended by a longitudinal & symmetrical frustum of a hollow right conical shell at any point lying on the longitudinal axis

Let there be a longitudinal & symmetrical frustum of the hollow conical shell with radius 'R', normal height 'H'& angle of curvature 'θ' of the lateral face and a given point say 'P' lying on the longitudinal axis at a distance 'x' from the centre 'C'. (as shown in the figure (126) below)

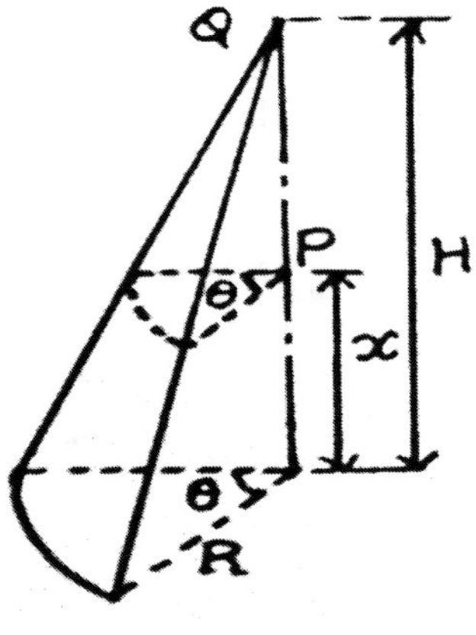

Fig (126)

Since, the frustum of the conical shell is symmetrical about the longitudinal hence using symmetry we get the solid angle subtended by the frustum of the hollow conical shell at the point 'P' by the eq(66) as follows

$$\Rightarrow \omega = 2\pi\left[1+\frac{x}{\sqrt{x^2+R^2}}\right] \quad \forall(x<H) \qquad \dots\dots\dots\dots\dots\dots\dots\dots\dots\dots\dots (66)$$

$$\Rightarrow \omega = 2\pi\left[1-\frac{x}{\sqrt{x^2+R^2}}\right] \quad \forall(x\geq H) \qquad \dots\dots\dots\dots\dots\dots\dots\dots\dots\dots\dots (67.2)$$

Note: The symmetry can be used for those 3D-figures having an axis of symmetry such as

1. Sphere 2.Cylinder & 3.Cone.

But, the symmetry can't be used for right pyramid & prism as they don't have any axis of symmetry.

Hence, the analysis of the frustum of these figures is not necessary here.

4.6 RIGHT PYRAMID WITH REGULAR POLYGONAL BASE

In this section, we are to study & make solutions of the following

1. Solid or hollow right pyramid (with finite & infinite length)
2. Hollow right pyramidal shell (with regular polygonal base base) &

In this section, the equation used for the regular polygonal plane is very significant for the analysis of the right pyramid.

4.6.1 Solid angle subtended by a polygonal right pyramid (solid or hollow) at the apex

Let there be a right pyramid with normal height 'H', apex point 'P', angle between the consecutive lateral edges 'α' & base as a regular polygon with 'n' no. of the sides each of length 'a' (as shown in the figure (127) below the upper one is showing the front view & lower one is showing the top view)

Now, join centre 'O' of the base to all the vertices $A_1, A_2, A_3 \dots\dots\dots A_n$ of the polygonal base we obtain a regular polygon $A_1A_2A_3 \dots\dots A_n$

We know that the angle subtended at the centre 'O' by each side of regular polygonal base$A_1A_2A_3 \dots\dots A_n$

$$\Rightarrow \ A_1OA_2 = \frac{2\pi}{n} \Rightarrow \ A_1OM = \frac{1}{2}(\ A_1OA_2) = \frac{2\pi}{2n} = \frac{\pi}{n}$$

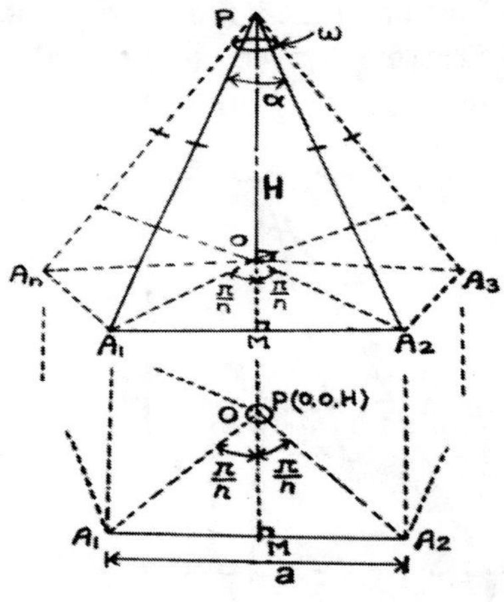

Fig (127)

In regular polygonal base$A_1A_2A_3 \dots\dots A_n$, we have

$$A_1A_2 = A_2A_3 = \ \dots = A_{n-1}A_n = a \ \& \ OA_1 = OA_2 = \ \dots\dots = OA_n = cicumscribed\ radius$$

Thus, $\Delta OA_1A_2, \Delta OA_2A_3, \dots\dots\dots\dots\dots, \Delta OA_{n-1}A_n$ are congruent isosceles triangles which the regular polygon consists of.

Thus we obtain 'n' no. of the congruent isosceles triangles as $\Delta OA_1A_2, \Delta OA_2A_3, \Delta OA_3A_4 \dots\dots\dots\dots \Delta OA_nA_1$. For ease of calculation consider an isosceles triangle ΔOA_1A_2& draw a perpendicular PM to the side A_1A_2(see the top-view in fig(127))

In right ΔOMA_1

$$\Rightarrow tan\left(\frac{\pi}{n}\right) = \frac{A_1M}{OM} = \frac{\left(\frac{a}{2}\right)}{OM} \Rightarrow OM = \frac{a}{2}cot\frac{\pi}{n}$$

Since, the polygonal base consists of 'n' no. of the congruent isosceles triangles

Hence, the solid angle subtended by the polygonal right pyramid at the apex point 'P'

= solid angle subtended by the regular polygonal base at the apex point 'P'

$$\Rightarrow \omega = \omega_{\Delta OA_1A_2} + \omega_{\Delta OA_2A_3} + \omega_{\Delta OA_3A_4} + \dots\dots\dots. + \omega_{\Delta OA_nA_1} = n\left(\omega_{\Delta OA_1A_2}\right)$$

We know that the solid angle subtended by the isosceles triangular plane at any point lying on the axis normal to the plane & passing through the common vertex of the equal sides is given by the eq(7) as follows

$$\omega = 2\left[sin^{-1}\left\{\frac{b}{\sqrt{b^2+4p^2}}\right\} - sin^{-1}\left\{\left(\frac{b}{\sqrt{b^2+4p^2}}\right)\left(\frac{h}{\sqrt{h^2+p^2}}\right)\right\}\right]$$

Now, on setting the values of

$b = base = A_1A_2 = a, p = perpendicular = OM = \frac{a}{2}cot\frac{\pi}{n}$ & h = height of pyramid = H in the above expression, we get

$$\omega_{\Delta OA_1A_2} = 2\left[sin^{-1}\left\{\frac{a}{\sqrt{a^2+4\left(\frac{a}{2}cot\frac{\pi}{n}\right)^2}}\right\} - sin^{-1}\left\{\left(\frac{a}{\sqrt{a^2+4\left(\frac{a}{2}cot\frac{\pi}{n}\right)^2}}\right)\left(\frac{H}{\sqrt{H^2+\left(\frac{a}{2}cot\frac{\pi}{n}\right)^2}}\right)\right\}\right]$$

$$\Rightarrow \omega_{\Delta OA_1A_2} = 2\left[\sin^{-1}\left\{\frac{a}{a\sqrt{1+cot^2\frac{\pi}{n}}}\right\} - \sin^{-1}\left\{\left(\frac{a}{a\sqrt{1+cot^2\frac{\pi}{n}}}\right)\left(\frac{2H}{\sqrt{4H^2+a^2cot^2\frac{\pi}{n}}}\right)\right\}\right]$$

$$= 2\left[\sin^{-1}\left\{\frac{1}{cosec\frac{\pi}{n}}\right\} - \sin^{-1}\left\{\left(\frac{1}{cosec\frac{\pi}{n}}\right)\left(\frac{2H}{\sqrt{4H^2+a^2cot^2\frac{\pi}{n}}}\right)\right\}\right]$$

$$= 2\left[\sin^{-1}\left\{sin\frac{\pi}{n}\right\} - \sin^{-1}\left\{\left(sin\frac{\pi}{n}\right)\left(\frac{2H}{\sqrt{4H^2+a^2cot^2\frac{\pi}{n}}}\right)\right\}\right]$$

$$= 2\left[\frac{\pi}{n} - \sin^{-1}\left\{\frac{2Hsin\frac{\pi}{n}}{\sqrt{4H^2+a^2cot^2\frac{\pi}{n}}}\right\}\right]$$

Hence, we get

$$\Rightarrow \omega = n\left(\omega_{\Delta OA_1A_2}\right) = 2n\left[\frac{\pi}{n} - \sin^{-1}\left\{\frac{2Hsin\frac{\pi}{n}}{\sqrt{4H^2+a^2cot^2\frac{\pi}{n}}}\right\}\right]$$

$$\Rightarrow \omega = 2\pi - 2n\sin^{-1}\left\{\frac{2Hsin\frac{\pi}{n}}{\sqrt{4H^2+a^2cot^2\frac{\pi}{n}}}\right\} \forall(n \in N \,\&\, n \geq 3) \quad \ldots\ldots\ldots..(I)$$

Now, join the apex point 'P' to the point 'M'

In right ΔA_1MP

$$\Rightarrow tan\ A_1PM = \frac{A_1M}{PM} \Rightarrow tan\frac{\alpha}{2} = \frac{\left(\frac{a}{2}\right)}{PM} \Rightarrow PM = \frac{a}{2}cot\frac{\alpha}{2}$$

By joining the point 'P' to the centre 'O' and the point 'M' we obtain a right ΔPOM

(as shown in the figure (128) below)

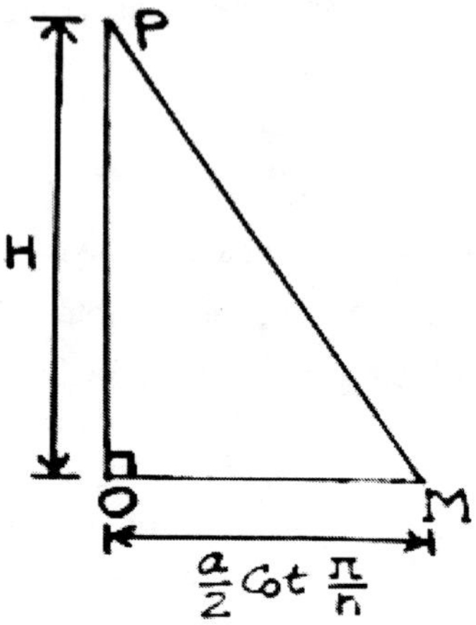

Fig (128)

In right ΔPOM

$$\Rightarrow PM^2 = OP^2 + OM^2 \Rightarrow \left(\frac{a}{2}cot\frac{\alpha}{2}\right)^2 = H^2 + \left(\frac{a}{2}cot\frac{\pi}{n}\right)^2$$

$$\Rightarrow H^2 = \frac{a^2}{4}\left(cot^2\frac{\alpha}{2} - cot^2\frac{\pi}{n}\right)$$

$$\Rightarrow H = \frac{a}{2}\sqrt{cot^2\frac{\alpha}{2} - cot^2\frac{\pi}{n}} \quad \ldots\ldots\ldots..(II)$$

Now, on setting the value of H in the above eq(70), we get

$$\omega = 2\pi - 2n\sin^{-1}\left\{\frac{2\left(\frac{a}{2}\sqrt{cot^2\frac{\alpha}{2} - cot^2\frac{\pi}{n}}\right)sin\frac{\pi}{n}}{\sqrt{4\left(\frac{a}{2}\sqrt{cot^2\frac{\alpha}{2} - cot^2\frac{\pi}{n}}\right)^2 + a^2cot^2\frac{\pi}{n}}}\right\}$$

$$= 2\pi - 2n\sin^{-1}\left\{\frac{asin\frac{\pi}{n}\sqrt{cot^2\frac{\alpha}{2} - cot^2\frac{\pi}{n}}}{\sqrt{a^2cot^2\frac{\alpha}{2} - a^2cot^2\frac{\pi}{n} + a^2cot^2\frac{\pi}{n}}}\right\}$$

$$= 2\pi - 2n\sin^{-1}\left\{\frac{asin\frac{\pi}{n}\sqrt{\frac{1}{tan^2\frac{\alpha}{2}} - \frac{1}{tan^2\frac{\pi}{n}}}}{acot\frac{\alpha}{2}}\right\} = 2\pi - 2n\sin^{-1}\left\{\frac{asin\frac{\pi}{n}\sqrt{tan^2\frac{\pi}{n} - tan^2\frac{\alpha}{2}}}{acot\frac{\alpha}{2}tan\frac{\alpha}{2}tan\frac{\pi}{n}}\right\}$$

$$= 2\pi - 2n\sin^{-1}\left\{\frac{sin\frac{\pi}{n}\sqrt{tan^2\frac{\pi}{n} - tan^2\frac{\alpha}{2}}}{\left(\frac{sin\frac{\pi}{n}}{cos\frac{\pi}{n}}\right)}\right\} = 2\pi - 2n\sin^{-1}\left\{cos\frac{\pi}{n}\sqrt{tan^2\frac{\pi}{n} - tan^2\frac{\alpha}{2}}\right\}$$

Thus, we have

$$\Rightarrow \omega = 2\pi - 2n\sin^{-1}\left\{\frac{2Hsin\frac{\pi}{n}}{\sqrt{4H^2 + a^2cot^2\frac{\pi}{n}}}\right\} = 2\pi - 2n\sin^{-1}\left\{cos\frac{\pi}{n}\sqrt{tan^2\frac{\pi}{n} - tan^2\frac{\alpha}{2}}\right\}$$

$$\forall\left(n \in N \,\&\, n \geq 3 \,\&\, 0 \leq \alpha \leq \frac{2\pi}{n}\right) \quad \ldots\ldots\ldots\ldots\ldots\ldots\ldots\ldots (70)$$

And the relation between the angle 'α' & normal height 'H' of the right pyramid is given by the eq(II) as follows

$$\Rightarrow H = \frac{a}{2}\sqrt{cot^2\frac{\alpha}{2} - cot^2\frac{\pi}{n}} \quad \ldots\ldots\ldots\ldots\ldots\ldots\ldots\ldots\ldots\ldots (71)$$

The above eq(70) is very important for the analysis of different 3d-figures. Both the expressions require at least two parameters whether 'H' & 'a' or 'n' & 'α'

The different expressions can be obtained by setting different values of the 'n' (the no. of the sides of base or no. of the lateral edges) as follows

4.6.1.1 Solid angle subtended by a right pyramid (solid or hollow) with equilateral triangularbase at the apex

Let there be a right pyramid with normal height 'H', angle between the consecutive lateral edges 'α' & 'a' be the length of each side of equilateral triangular base (as shown in the figure (129) below)

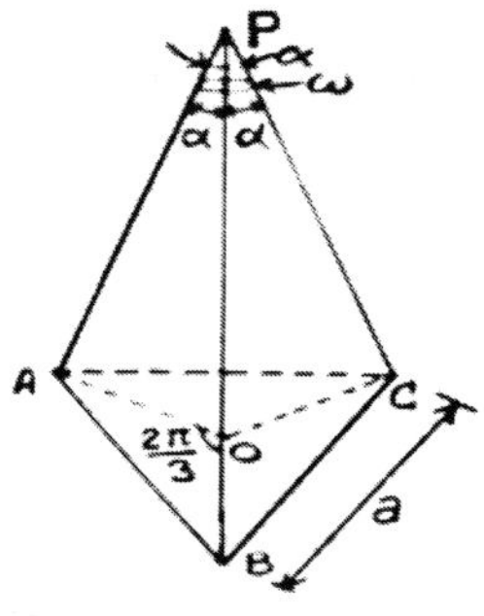

Fig (129)

The solid angle subtended by the right pyramid at the apex 'P' is obtained by setting n = 3 in the eq(70) as follows

$$\Rightarrow \omega = 2\pi - 2 \times 3\sin^{-1}\left\{\frac{2Hsin\frac{\pi}{3}}{\sqrt{4H^2 + a^2cot^2\frac{\pi}{3}}}\right\} = 2\pi - 2 \times 3\sin^{-1}\left\{cos\frac{\pi}{3}\sqrt{tan^2\frac{\pi}{3} - tan^2\frac{\alpha}{2}}\right\}$$

$$= 2\pi - 6\sin^{-1}\left\{\frac{2H\left(\frac{\sqrt{3}}{2}\right)}{\sqrt{4H^2 + a^2\left(\frac{1}{\sqrt{3}}\right)^2}}\right\} = 2\pi - 6\sin^{-1}\left\{\left(\frac{1}{2}\right)\sqrt{\left(\sqrt{3}\right)^2 - tan^2\frac{\alpha}{2}}\right\}$$

$$\Rightarrow \omega = 2\pi - 6\sin^{-1}\left\{\frac{3H}{\sqrt{12H^2 + a^2}}\right\} = 2\pi - 6\sin^{-1}\left\{\frac{1}{2}\sqrt{3 - tan^2\frac{\alpha}{2}}\right\} \qquad \forall\, 0 \leq \alpha \leq \frac{2\pi}{3}$$

Thus we find the value of the solid angle depends on the normal height (H) & the length (a) of the side of the equilateral triangular base

Or depends on the no. of the sides (n) & the angle (α) of the lateral face (i.e. the angle between the consecutive lateral edges) of the pyramid. i.e. two parameters whether 'H' & 'a' or 'n' & 'α' should be known at all.

4.6.1.2 Solid angle subtended by a right pyramid (solid or hollow) with square base at the apex

Let there be a right pyramid with normal height 'H', angle between the consecutive lateral edges 'α' & 'a' be the length of each side of the square base (as shown in the figure (130) below)

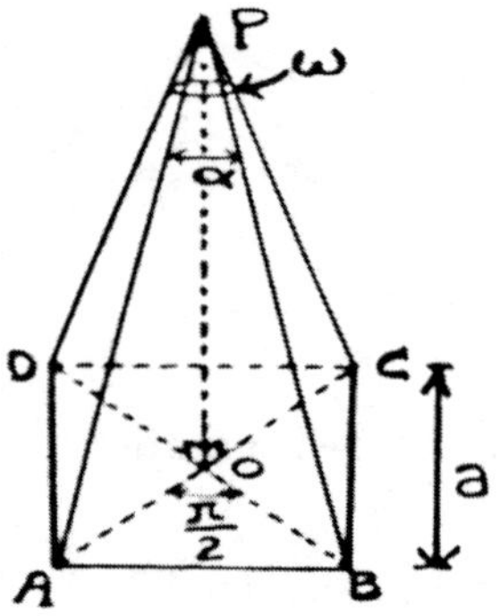

Fig (130)

The solid angle subtended by the right pyramid at the apex 'P' is obtained by setting n = 4 in the eq(70) as follows

$$\Rightarrow \omega = 2\pi - 2\times 4\sin^{-1}\left\{\frac{2H\sin\frac{\pi}{4}}{\sqrt{4H^2 + a^2\cot^2\frac{\pi}{4}}}\right\} = 2\pi - 2\times 4\sin^{-1}\left\{\cos\frac{\pi}{4}\sqrt{\tan^2\frac{\pi}{4} - \tan^2\frac{\alpha}{2}}\right\}$$

$$= 2\pi - 8\sin^{-1}\left\{\frac{2H\left(\frac{1}{\sqrt{2}}\right)}{\sqrt{4H^2 + a^2(1)^2}}\right\} = 2\pi - 8\sin^{-1}\left\{\left(\frac{1}{\sqrt{2}}\right)\sqrt{(1)^2 - \tan^2\frac{\alpha}{2}}\right\}$$

$$\Rightarrow \omega = 2\pi - 8\sin^{-1}\left\{\frac{\sqrt{2}H}{\sqrt{4H^2 + a^2}}\right\} = 2\pi - 8\sin^{-1}\left\{\frac{1}{\sqrt{2}}\sqrt{1 - \tan^2\frac{\alpha}{2}}\right\} \qquad \forall\, 0 \le \alpha \le \frac{\pi}{2}$$

Thus we find the value of the solid angle depends on the normal height (H) & the length (a) of the side of the square base.

Or depends on the no. of the sides (n) & the angle (α) of the lateral face (i.e. the angle between the consecutive lateral edges) of the pyramid. i.e. two parameters whether 'H' & 'a' or 'n' & 'α' should be known at all.

4.6.1.3 Solid angle subtended by a right pyramid (solid or hollow) with regular pentagonal base at the apex

Let there be a right pyramid with normal height 'H', angle between the consecutive lateral edges 'α' & 'a' be the length of each side of the regular pentagonal base (as shown in the figure (131) below)

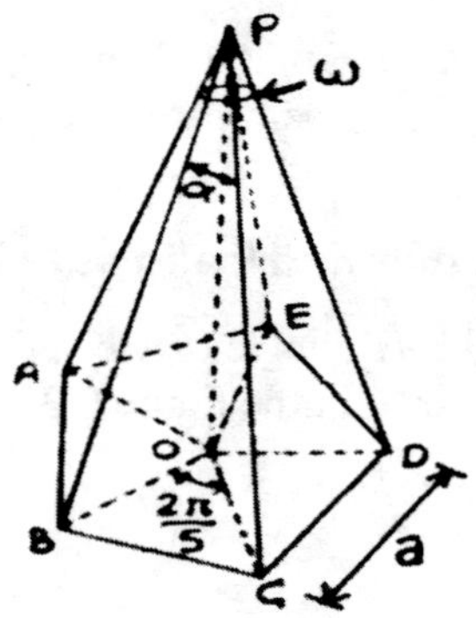

Fig (131)

The solid angle subtended by the right pyramid at the apex 'P' is obtained by setting n = 5 in the eq(70) as follows

$$\Rightarrow \omega = 2\pi - 2\times 5\sin^{-1}\left\{\frac{2H\sin\frac{\pi}{5}}{\sqrt{4H^2 + a^2\cot^2\frac{\pi}{5}}}\right\} = 2\pi - 2\times 5\sin^{-1}\left\{\cos\frac{\pi}{5}\sqrt{\tan^2\frac{\pi}{5} - \tan^2\frac{\alpha}{2}}\right\}$$

$$= 2\pi - 10\sin^{-1}\left\{\frac{2H(0.588)}{\sqrt{4H^2 + a^2(1.376)^2}}\right\} = 2\pi - 10\sin^{-1}\left\{(0.809)\sqrt{(0.726)^2 - \tan^2\frac{\alpha}{2}}\right\}$$

$$\Rightarrow \omega = 2\pi - 10\sin^{-1}\left\{\frac{1.176H}{\sqrt{4H^2 + 1.893a^2}}\right\}$$

$$= 2\pi - 10\sin^{-1}\left\{(0.809)\sqrt{0.527 - \tan^2\frac{\alpha}{2}}\right\} \qquad \forall\, 0 \le \alpha \le \frac{2\pi}{5}$$

Thus we find the value of the solid angle depends on the normal height (H) & the length (a) of the side of the regular pentagonal base.

Or depends on the no. of the sides (n) & the angle (α) of the lateral face (i.e. the angle between the consecutive lateral edges) of the pyramid. i.e. two parameters whether 'H' & 'a' or 'n' & 'α' should be known at all.

4.6.1.4 Solid angle subtended by a right pyramid (solid or hollow) with regular hexagonal base at the apex

Let there be a right pyramid with normal height 'H', angle between the consecutive lateral edges 'α' & 'a' be the length of each side of regular hexagonal base (as shown in the figure (132) below)

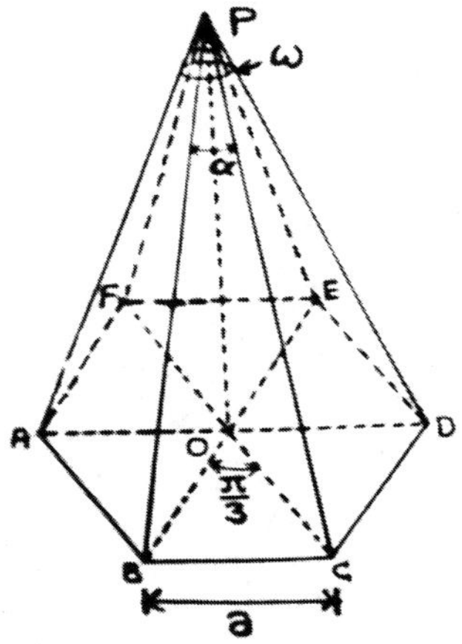

Fig (132)

The solid angle subtended by the right pyramid at the apex 'P' is obtained by setting n = 6 in the eq(70) as follows

$$\Rightarrow \omega = 2\pi - 2\times 6\sin^{-1}\left\{\frac{2H\sin\frac{\pi}{6}}{\sqrt{4H^2 + a^2\cot^2\frac{\pi}{6}}}\right\} = 2\pi - 2\times 6\sin^{-1}\left\{\cos\frac{\pi}{6}\sqrt{\tan^2\frac{\pi}{6} - \tan^2\frac{\alpha}{2}}\right\}$$

$$= 2\pi - 12\sin^{-1}\left\{\frac{2H\left(\frac{1}{2}\right)}{\sqrt{4H^2 + a^2\left(\sqrt{3}\right)^2}}\right\} = 2\pi - 12\sin^{-1}\left\{\left(\frac{\sqrt{3}}{2}\right)\sqrt{\left(\frac{1}{\sqrt{3}}\right)^2 - \tan^2\frac{\alpha}{2}}\right\}$$

$$\Rightarrow \omega = 2\pi - 12\sin^{-1}\left\{\frac{H}{\sqrt{4H^2 + 3a^2}}\right\} = 2\pi - 12\sin^{-1}\left\{\frac{1}{2}\sqrt{1 - 3\tan^2\frac{\alpha}{2}}\right\} \qquad \forall\, 0 \le \alpha \le \frac{\pi}{3}$$

Thus we find the value of the solid angle depends on the normal height (H) & the length (a) of the side of the regular hexagonal base

Or depends on the no. of the sides (n) & the angle (α) of the lateral face (i.e. the angle between the consecutive lateral edges) of the pyramid i.e. two parameters whether 'H' & 'a' or 'n' & 'α' should be known at all.

4.6.2 Solid angle subtended by a hollow right pyramidal shell with regular polygonal base at any internal point lying on the longitudinal axis

Let there be a hollow right pyramidal shell with normal height 'H', angle between the consecutive lateral edges 'α', 'a' be the length of each side of 'n' no. of sides of the regular polygonal& a given point 'P' lying on the longitudinal axis at a distance 'x' from the centre 'O' inside the shell. (as shown in the figure (133) below)

Now, assume that the pyramid shell is closed with the base then the solid angle subtended by the closed shell at the point 'P'

$$\Rightarrow \omega_1 = 4\pi$$

Hence, the solid angle subtended by the polygonal pyramidal shell at the given point 'P' = solid angle subtended by the regular polygonal base at the point 'P'

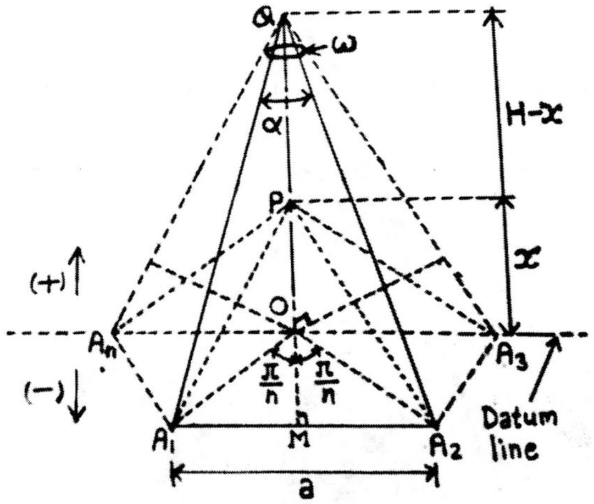

Fig (133)

We know that the solid angle subtended by the regular polygonal plane at any point lying on the vertical axis passing through the centre is given by the eq(34) as follows

$$\omega = 2\pi - 2n\sin^{-1}\left\{\frac{2h\sin\frac{\pi}{n}}{\sqrt{4h^2 + a^2\cot^2\frac{\pi}{n}}}\right\}$$

On setting h = x = height of the point 'P' from the centre of the polygonal base we get

$$\omega_2 = 2\pi - 2n\sin^{-1}\left\{\frac{2x\sin\frac{\pi}{n}}{\sqrt{4x^2 + a^2\cot^2\frac{\pi}{n}}}\right\}$$

But the base is not closed hence the solid angle by the hollow pyramidal shell at the point will equal to the difference of above two values as follows

$$\Rightarrow \omega = \omega_1 - \omega_2 = 4\pi - \left[2\pi - 2n\sin^{-1}\left\{\frac{2x\sin\frac{\pi}{n}}{\sqrt{4x^2 + a^2\cot^2\frac{\pi}{n}}}\right\}\right]$$

$$= 4\pi - 2\pi + 2n\sin^{-1}\left\{\frac{2x\sin\frac{\pi}{n}}{\sqrt{4x^2 + a^2\cot^2\frac{\pi}{n}}}\right\} = 2\pi + 2n\sin^{-1}\left\{\frac{2x\sin\frac{\pi}{n}}{\sqrt{4x^2 + a^2\cot^2\frac{\pi}{n}}}\right\}$$

$$\Rightarrow \omega = 2\pi + 2n\sin^{-1}\left\{\frac{2x\sin\frac{\pi}{n}}{\sqrt{4x^2 + a^2\cot^2\frac{\pi}{n}}}\right\} \quad \forall (x < H, n \in N \;\&\; n \geq 3) \quad \ldots\ldots\ldots\ldots (72)$$

Where the relation between the height 'H', length of side 'a', no. of the sides 'n' & the angle 'α' of the lateral face is given by the eq(71) as follows

$$\Rightarrow H = \frac{a}{2}\sqrt{\cot^2\frac{\alpha}{2} - \cot^2\frac{\pi}{n}}$$

With the help of above relation if the any three parameters are known then the fourth one can be found.

Sign convention: The distance (x) measured from the centre of open end of the pyramidal shell is taken as

1. Positive when the point lies on the longitudinal axis inside the shell &
2. Negative when the point lies on the longitudinal axis outside the shell

It is obvious the value of solid angle is independent of the length of the shell within the range of distance (x)

Now, let's consider the point lying on the longitudinal axis outside the shell

4.6.3 Solid angle subtended by a hollow right pyramidal shell with regular polygonal base at any external point lying on the longitudinal axis

Let there be a hollow right pyramidal shell with normal height 'H', angle between the consecutive lateral edges 'α', 'a' be the length of each side of 'n' no. of sides of the regular polygonal & a given point 'P_1' towards base or 'P_2' towards apex point 'Q' lying on the longitudinal axis at a distance 'x' from the centre 'O' outside the shell. (as shown in the figure (134) below)

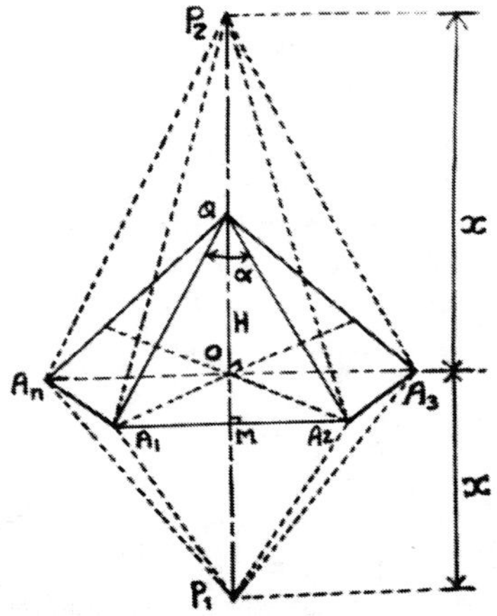

Fig (134)

From both the positions of the given point 'P_1' or 'P_2' the pyramid appears as a regular polygonal plane

Hence, the solid angle subtended by the polygonal pyramidal shell at the given point 'P_1' or 'P_2' = solid angle subtended by the regular polygonal base (plane) at the point 'P_1' or 'P_2'

We know that the solid angle subtended by the regular polygonal plane at any point lying on the vertical axis passing through the centre is given by the eq(34) as follows

$$\omega = 2\pi - 2n\sin^{-1}\left\{\frac{2h\sin\frac{\pi}{n}}{\sqrt{4h^2 + a^2\cot^2\frac{\pi}{n}}}\right\}$$

On setting h = x = height of the point 'P_1' from the centre of the polygonal base thus we get the solid angle subtended by the shell at the external point 'P_1'

$$\omega_1 = 2\pi - 2n\sin^{-1}\left\{\frac{2x\sin\frac{\pi}{n}}{\sqrt{4x^2 + a^2\cot^2\frac{\pi}{n}}}\right\} \forall (x \geq 0, n \in N \,\&\, n \geq 3) \quad \ldots\ldots\ldots\ldots\ldots\ldots (73.1)$$

Similarly, the solid angle subtended by the shell at the external point 'P_2'

$$\omega_2 = 2\pi - 2n\sin^{-1}\left\{\frac{2x\sin\frac{\pi}{n}}{\sqrt{4x^2 + a^2\cot^2\frac{\pi}{n}}}\right\} \forall (x \geq H, n \in N \,\&\, n \geq 3) \quad \ldots\ldots\ldots\ldots\ldots\ldots (73.2)$$

Where the relation between the height 'H', length of side 'a', no. of the sides 'n' & the angle 'α' of the lateral face is given by the eq(71) as follows

$$\Rightarrow H = \frac{a}{2}\sqrt{\cot^2\frac{\alpha}{2} - \cot^2\frac{\pi}{n}}$$

With the help of above relation if the any three parameters are known then the fourth one can be found. The eq(73.2) is also applicable for a solid right pyramid.

4.6.4 Solid angle subtended by a hollow right pyramidal shell with regular-polygonal base at the apex point inside the shell

Let there be a hollow right pyramidal shell with normal height 'H', angle between the consecutive lateral edges 'α', 'a' be the length of each side of 'n' no. of sides of the regular-polygonal & the apex point 'Q' (as shown in the fig(135) below)

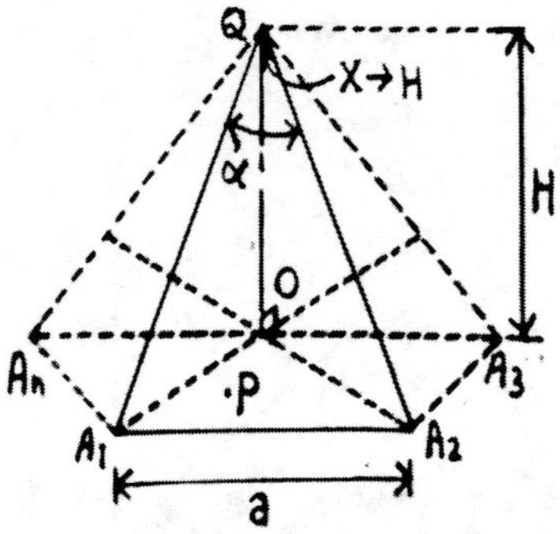

Fig (135)

In this case, the solid angle subtended by the pyramidal shell at the internal apex point is obtained by taking the limit of ω from the eq(72) at $x \to H^-$or at $x = (H - k)$, 'k' is the infinitesimal small thickness of the shell such as $k \to 0 \;\&\, k \neq 0$

Hence, we have

$$\Rightarrow \omega = \lim_{x\to H^-}\left[2\pi + 2n\sin^{-1}\left\{\frac{2x\sin\frac{\pi}{n}}{\sqrt{4x^2 + a^2\cot^2\frac{\pi}{n}}}\right\}\right]$$

$$= \lim_{k\to 0}\left[2\pi + 2n\sin^{-1}\left\{\frac{2(H-k)\sin\frac{\pi}{n}}{\sqrt{4(H-k)^2 + a^2\cot^2\frac{\pi}{n}}}\right\}\right] \quad where\ x = (H-k) \Rightarrow k \to 0$$

$$= 2\pi + 2n\sin^{-1}\left\{\frac{2(H-0)\sin\frac{\pi}{n}}{\sqrt{4(H-0)^2 + a^2\cot^2\frac{\pi}{n}}}\right\} (since, k \neq 0)$$

$$\Rightarrow \omega = 2\pi + 2n\sin^{-1}\left\{\frac{2H\sin\frac{\pi}{n}}{\sqrt{4H^2 + a^2\cot^2\frac{\pi}{n}}}\right\} \quad \forall\ (n \in N\ \&\ n \geq 3) \qquad \ldots\ldots\ldots\ldots\ldots (74)$$

It is obvious from the above eq(74) that

The value of solid angle subtended by the hollow right pyramidal shell at the internal apex point exceeds that subtended at the external apex point

Mathematically,

$$\Rightarrow (\omega)_{x\to H^-} = 4\pi - (\omega)_{x=H}$$

$$or \quad (\omega)_{x=H} + (\omega)_{x\to H^-} = 4\pi$$

$$or \quad (\omega)_{at\ external\ apex} + (\omega)_{at\ internal\ apex} = 4\pi$$

Note: The above relation also holds true for the following

1. Hollow right conical shell, intersecting planes
2. Planes (at least three or more) intersecting one another at a single point in the space

   Example: lateral faces of a tetrahedron intersecting one another at a vertex, orthogonal planes intersecting one another at the origin etc.
3. Straight lines (at least three or more) intersecting one another at a single point in the space

   Example: orthogonal axes intersecting one another at the origin.

Where the relation between the height 'H', length of side 'a', no. of the sides 'n' & the angle 'α' of the lateral face is given by the eq(71) as follows

$$\Rightarrow H = \frac{a}{2}\sqrt{\cot^2\frac{\alpha}{2} - \cot^2\frac{\pi}{n}}$$

4.6.5 Solid angle subtended by a hollow right pyramidal shell with regular polygonal base at any internal point lying in the plane of the base

Let there be a hollow right pyramidal shell with normal height 'H', angle between the consecutive lateral edges 'α', 'a' be the length of each side of 'n' no. of sides of the regular polygonal & an arbitrary point say 'P' (as shown in the fig(135) above)

In this case, the solid angle subtended by the pyramidal shell at the internal apex point is obtained by setting x = 0 the eq(72) as follows

$$\omega = 2\pi + 2n\sin^{-1}\left\{\frac{2(0)\sin\frac{\pi}{n}}{\sqrt{4(0)^2 + a^2\cot^2\frac{\pi}{n}}}\right\} = 2\pi + 2n\sin^{-1}\left\{\frac{(0)}{a\cot\frac{\pi}{n}}\right\} = 2\pi + 2n\sin^{-1}\{0\} = 2\pi\ sr$$

Or by setting x = 0 in the eq(73.1) as follows

$$\omega = 2\pi - 2n\sin^{-1}\left\{\frac{2(0)sin\frac{\pi}{n}}{\sqrt{4(0)^2 + a^2cot^2\frac{\pi}{n}}}\right\} = 2\pi - 2n\sin^{-1}\left\{\frac{(0)}{acot\frac{\pi}{n}}\right\} = 2\pi - 2n\sin^{-1}\{0\} = 2\pi\ sr$$

Thus, a hollow right pyramidal shell behaves as a cap for each point lying in the plane of the base of the shell.

The equation (70) is very useful for the analysis of regular tetrahedron, the intersecting planes & the equally inclined straight lines in the space.

* **Regular tetrahedron**: "Regular tetrahedron is a three dimensional figure surrounded by four finite equilateral-triangular planes." or

"It is a right pyramid having six finite & equal edges& angle between the consecutive edges is60^o."

Thus, for a regular tetrahedron the angle between the consecutive lateral edges $\alpha = \frac{\pi}{3} = 60^o$

4.6.6 Solid angle subtended by a regular tetrahedron at anyvertex externally& internally

Let there be a regular tetrahedron with the edge length 'a' & vertices 'A', 'B', 'C' & 'O' (as shown in the figure (136) below)

The solid angle subtended by the regular tetrahedron at any of the vertices externally is obtained by setting n = 3 &$\alpha = \frac{\pi}{3}$ or $\frac{\alpha}{2} = \frac{\pi}{6}$ in the eq(70) as follows

$$\Rightarrow \omega_o = 2\pi - 2 \times 3\sin^{-1}\left\{cos\frac{\pi}{3}\sqrt{tan^2\frac{\pi}{3} - tan^2\frac{\pi}{6}}\right\} = 2\pi - 6\sin^{-1}\left\{\frac{1}{2}\sqrt{\left(\sqrt{3}\right)^2 - \left(\frac{1}{\sqrt{3}}\right)^2}\right\}$$

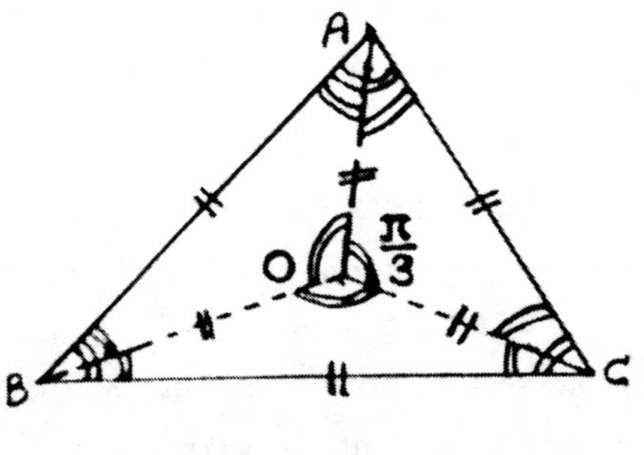

Fig (136)

$$\Rightarrow \omega_o = 2\pi - 6\sin^{-1}\left\{\frac{1}{2}\sqrt{\frac{9-1}{3}}\right\} = 2\pi - 6\sin^{-1}\left\{\frac{1}{2}\frac{2\sqrt{2}}{\sqrt{3}}\right\} = 2\pi - 6\sin^{-1}\left\{\sqrt{\frac{2}{3}}\right\}$$

$$\Rightarrow \omega_o = 2\pi - 6 \times 0.955316 = 6.283185 - 5.731896 = 0.551289 \approx 0.55\ sr\ \&$$

We know that for the hollow right pyramidal shell

The value of solid angle subtended by the hollow right pyramidal shell at the internal apex point (vertex) exceeds that subtended at the external apex point (vertex)

Mathematically,

$$\Rightarrow (\omega)_{x\to H^-} = 4\pi - (\omega)_{x=H}$$

Hence, solid angle subtended by the hollow right pyramidal shell at the internal apex

$$\Rightarrow \omega_i = 4\pi - \omega_o = 12.566371 - 0.551289 = 12.016371 \approx 12.02\ sr$$

4.6.7 Solid angle subtended by an infinitely long right pyramid at the apex point

Let there be an infinitely long right pyramid having 'n' of the edges, the angle between the consecutive lateral edges is $'\alpha'$& apex point 'P' (as shown in the figure(137) below)

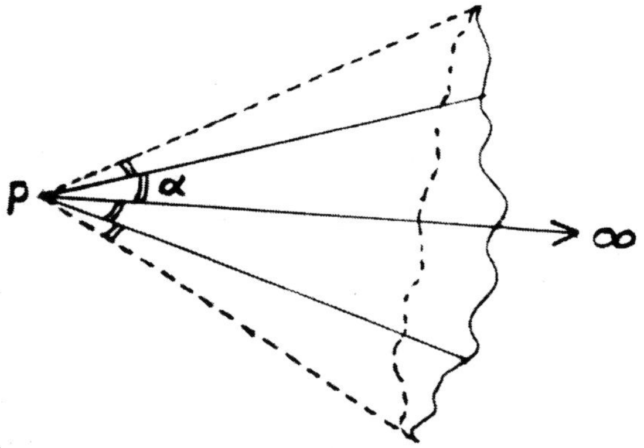

Fig (137)

In this case the value of the solid angle subtended by the infinitely long right pyramid at the apex 'P'should be independent of the height & length of side of the base of the pyramid

In this case, the solid angle subtended by the infinitely long right pyramid at the apex is given by the eq(70) as follows

$$\Rightarrow \omega= 2\pi - 2n\sin^{-1}\left\{cos\frac{\pi}{n}\sqrt{tan^2\frac{\pi}{n} - tan^2\frac{\alpha}{2}}\right\} \ \forall \left(0 \leq \alpha \leq \frac{2\pi}{n}\right) \quad \dots\dots\dots\dots\dots\dots \ (70.1)$$

Thus, only two parameters 'n' & 'α' of the right pyramid should be known which are independent of the height & length of side of the base.

4.6.8 Solid angle subtended by 'n' no. of the intersecting & equally inclined planes at the point of intersection in the space

Let there be 'n' no. of the planes intersecting one another at a single point in the space such that the angle between any two consecutive lines of intersection is $'\alpha'$(as shown in the figure (138) below)

In this case, the intersecting planes appear like an infinitely long right pyramid when viewed from the point 'P' of intersection of 'n' no. of the planes intersecting one another at a single point in the space.

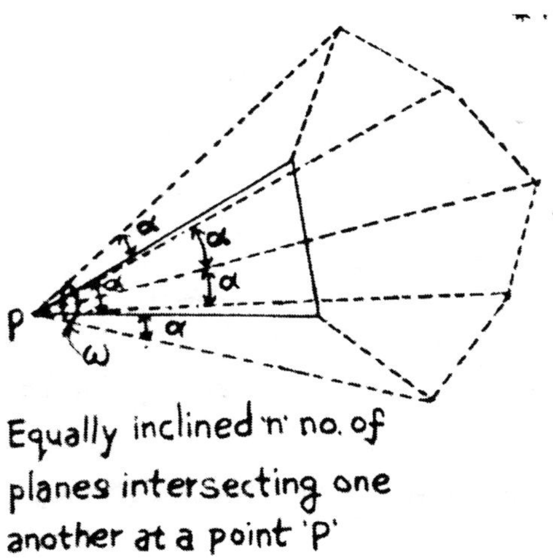

Fig (138)

Hence, the solid angle subtended by 'n' no. of the intersecting planes at the point of the intersection 'P'

$$\Rightarrow \omega = 2\pi - 2n\sin^{-1}\left\{\cos\frac{\pi}{n}\sqrt{\tan^2\frac{\pi}{n} - \tan^2\frac{\alpha}{2}}\right\} \ \forall \left(0 \le \alpha \le \frac{2\pi}{n}\right) \quad \dots\dots\dots\dots\dots\dots \ (70.1)$$

Important deductions:

1. The solid angle subtended by the orthogonal planes XY, YZ & ZX planes at the origin point 'O'

   The angle between any two orthogonal planes XY, YZ & ZX is $\alpha = \frac{\pi}{2}$ (as shown in the figure (139) below)

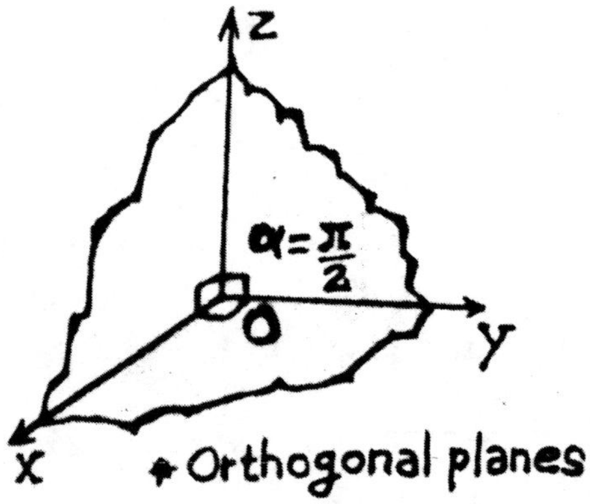

Fig (139)

In this the solid angle subtended by the orthogonal planes at the origin is obtained by setting n = 3 &$\alpha = \frac{\pi}{2}$or $\frac{\alpha}{2} = \frac{\pi}{4}$ as follows

$$\Rightarrow \omega = 2\pi - 2 \times 3\sin^{-1}\left\{\cos\frac{\pi}{3}\sqrt{\tan^2\frac{\pi}{3} - \tan^2\frac{\pi}{4}}\right\} = 2\pi - 6\sin^{-1}\left\{\left(\frac{1}{2}\right)\sqrt{\left(\sqrt{3}\right)^2 - (1)^2}\right\}$$

$$= 2\pi - 6\sin^{-1}\left\{\left(\frac{1}{2}\right)\sqrt{3-1}\right\} = 2\pi - 6\sin^{-1}\left\{\frac{1}{\sqrt{2}}\right\} = 2\pi - 6\frac{\pi}{4} = 2\pi - \frac{3\pi}{2} = \frac{\pi}{2}\ sr$$

Also, the solid angle subtended by each out of eight quadrants of X, Y & Z axes at the origin is obtained by dividing total solid angle of 4π as follows

$$\omega = \frac{4\pi}{8} = \frac{\pi}{2}\ sr$$

2. The solid angel subtended by the planes when 'n' no. of the intersecting planes at the point of intersection when $\frac{\alpha}{2} = \frac{\pi}{n}$

$$\omega = 2\pi - 2n\sin^{-1}\left\{\cos\frac{\pi}{n}\sqrt{\tan^2\frac{\pi}{n} - \tan^2\frac{\pi}{n}}\right\} = 2\pi - 6\sin^{-1}\left\{\cos\frac{\pi}{n}(0)\right\} = 2\pi - 0 = 2\pi\ sr$$

In this case, all the planes intersect one another in the same plane i.e. they coincide with a single plane in the space.

4.6.9 Solid angle subtended by 'n' no. of the intersecting & equally inclined straight lines at the point of intersection in the space

Let there be 'n' no. of the straight lines intersecting one another at a single point in the space such that the angle between any two consecutive straight lines is $'\alpha'$(as shown in the figure (140) below)

In this case, the intersecting planes appear like an infinitely long right pyramid when viewed from the point 'P' of intersection of 'n' no. of the planes intersecting one another at a single point in the space.

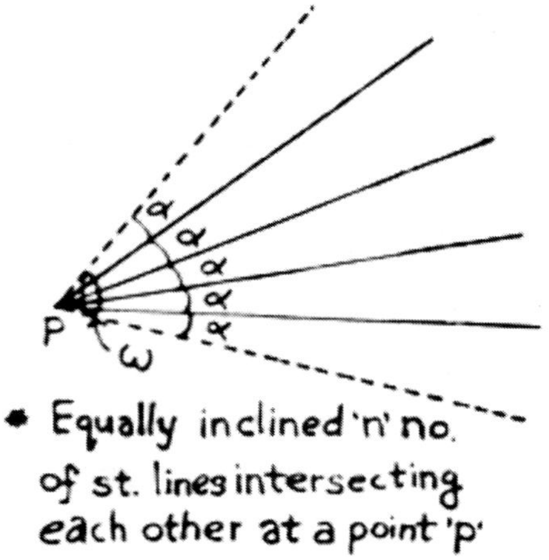

Fig (140)

Hence, the solid angle subtended by 'n' no. of the intersecting straight lines at the point of the intersection 'P'

$$\Rightarrow \omega = 2\pi - 2n\sin^{-1}\left\{\cos\frac{\pi}{n}\sqrt{\tan^2\frac{\pi}{n} - \tan^2\frac{\alpha}{2}}\right\} \ \forall \left(0 \le \alpha \le \frac{2\pi}{n}\right) \quad \ldots\ldots\ldots\ldots\ldots\ldots \quad (70.1)$$

Important deductions:

1. The solid angle subtended by the orthogonal axes X, Y & Z at the origin point 'O'

   The angle between any two orthogonal axes X, Y & Z is $\alpha = \frac{\pi}{2}$ (as shown in the figure (141) below)

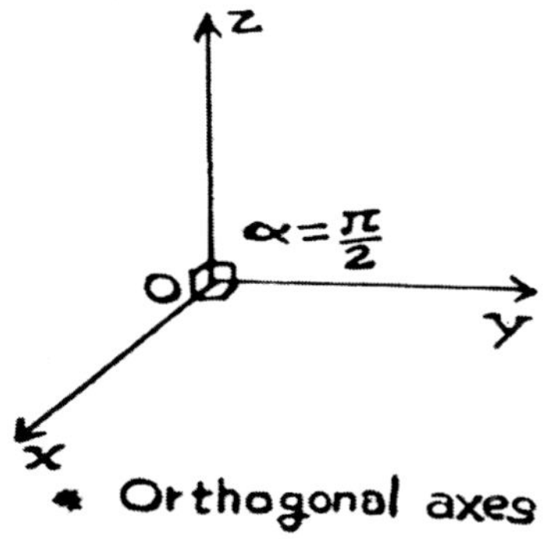

Fig (141)

In this the solid angle subtended by the orthogonal axes at the origin is obtained by setting n = 3 & $\alpha = \frac{\pi}{2}$ or $\frac{\alpha}{2} = \frac{\pi}{4}$ as follows

$$\Rightarrow \omega = 2\pi - 2 \times 3\sin^{-1}\left\{\cos\frac{\pi}{3}\sqrt{\tan^2\frac{\pi}{3} - \tan^2\frac{\pi}{4}}\right\} = 2\pi - 6\sin^{-1}\left\{\left(\frac{1}{2}\right)\sqrt{(\sqrt{3})^2 - (1)^2}\right\}$$

$$= 2\pi - 6\sin^{-1}\left\{\left(\frac{1}{2}\right)\sqrt{3-1}\right\} = 2\pi - 6\sin^{-1}\left\{\frac{1}{\sqrt{2}}\right\} = 2\pi - 6\frac{\pi}{4} = 2\pi - \frac{3\pi}{2} = \frac{\pi}{2}\ sr$$

Also, the solid angle subtended by each out of eight quadrants of X, Y & Z axes at the origin is obtained by equally dividing total solid angle of 4π as follows

$$\omega = \frac{4\pi}{8} = \frac{\pi}{2}\ sr$$

2. The solid angel subtended by 'n' no. of the intersecting straight lines at the point of intersection when $\frac{\alpha}{2} = \frac{\pi}{n}$

$$\omega = 2\pi - 2n\sin^{-1}\left\{cos\frac{\pi}{n}\sqrt{tan^2\frac{\pi}{n} - tan^2\frac{\pi}{n}}\right\} = 2\pi - 6\sin^{-1}\left\{cos\frac{\pi}{n}(0)\right\} = 2\pi - 0 = 2\pi\ sr$$

In this case, all the lines intersect one another in the same plane i.e. they coincide with a single plane in the space.

4.7 RIGHT PRISM WITH REGULAR POLYGONAL BASE

In this section, we are to study & make solutions of the following

1. Solid or hollow right prism (with regular polygonal base base)
2. Hollow right prismatic shell (with regular polygonal base base) &

4.7.1 Solid angle subtended by a solid right prism with regular polygonal cross-section at any point lying on the longitudinal axis

Let there be a right prism with length 'L'& cross-section as a regular polygon with 'n' no. of the sides each of length 'a' and a given point say 'P' lying on the longitudinal axis at a distance 'x' from the centre 'C' of the front end. (as shown in the figure (142) below)

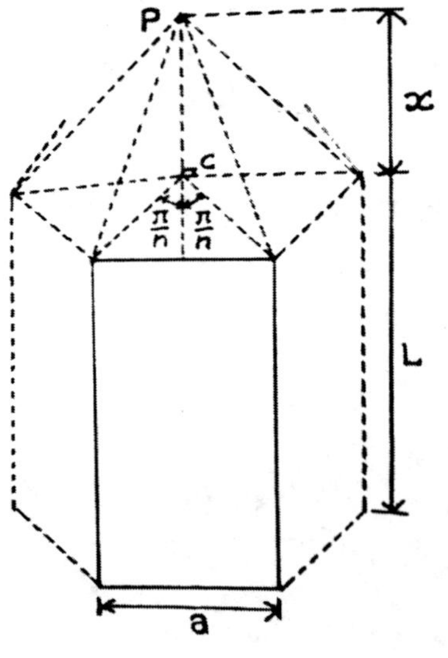

Fig (142)

In this case, when the prism is viewed from the point 'P' it appears as a regular polygonal plane

Hence the solid angle subtended by the right prism at the point 'P' is given from the eq(34) as follows

$$\omega = 2\pi - 2n\sin^{-1}\left\{\frac{2hsin\frac{\pi}{n}}{\sqrt{4h^2 + a^2cot^2\frac{\pi}{n}}}\right\}$$

Now, on setting h = normal height = x, we get

$$\Rightarrow \left[\omega = 2\pi - 2n\sin^{-1}\left\{\frac{2xsin\frac{\pi}{n}}{\sqrt{4x^2 + a^2cot^2\frac{\pi}{n}}}\right\}\right] \quad \forall(x \geq 0, n \in N\ \&\ n \geq 3) \ldots\ldots\ldots(75)$$

Important deduction1: The formulae for the right prisms with different cross-sections can be derived by setting the different values of 'n' in the eq(75) as follows

Case 1: Solid angle subtended by aright prism with equilateral triangularcross-section at any point lying on the longitudinal axis

Is obtained by setting n = 3 in the eq(34), we get

$$\omega = 2\pi - 2(3)\sin^{-1}\left\{\frac{2x sin\frac{\pi}{3}}{\sqrt{4x^2 + a^2 cot^2\frac{\pi}{3}}}\right\} = 2\pi - 6\sin^{-1}\left\{\frac{2x\frac{\sqrt{3}}{2}}{\sqrt{4x^2 + a^2\left(\frac{1}{\sqrt{3}}\right)^2}}\right\}$$

$$\Rightarrow \omega = 2\pi - 6\sin^{-1}\left\{\frac{3x}{\sqrt{12x^2 + a^2}}\right\}$$

Case 2: Solid angle subtended by a right prism with square cross-section at any point lying on the longitudinal axis

Is obtained by setting n = 4 in the eq(34), we get

$$\omega = 2\pi - 2(4)\sin^{-1}\left\{\frac{2x sin\frac{\pi}{4}}{\sqrt{4x^2 + a^2 cot^2\frac{\pi}{4}}}\right\} = 2\pi - 8\sin^{-1}\left\{\frac{2x\frac{1}{\sqrt{2}}}{\sqrt{4x^2 + a^2(1)^2}}\right\}$$

$$\Rightarrow \omega = 2\pi - 8\sin^{-1}\left\{\frac{x\sqrt{2}}{\sqrt{4x^2 + a^2}}\right\} = 4\left[\frac{\pi}{2} - 2\sin^{-1}\left\{\frac{x\sqrt{2}}{\sqrt{4x^2 + a^2}}\right\}\right]$$

$$= 4\left[\frac{\pi}{2} - \sin^{-1}\left\{2\left(\frac{x\sqrt{2}}{\sqrt{4x^2 + a^2}}\right)\left(\frac{\sqrt{2x^2 + a^2}}{\sqrt{4x^2 + a^2}}\right)\right\}\right]$$

$$\left(since,\ \ 2\sin^{-1}x = \sin^{-1}\left(2x\sqrt{1 - x^2}\right) \qquad \forall(-1 \le x \le 1)\right)$$

$$= 4\left[\frac{\pi}{2} - \sin^{-1}\left\{\frac{2x\sqrt{2}\sqrt{2x^2 + a^2}}{4x^2 + a^2}\right\}\right] = 4\cos^{-1}\left\{\frac{2x\sqrt{2}\sqrt{2x^2 + a^2}}{4x^2 + a^2}\right\}$$

$$\left(since,\ \ \frac{\pi}{2} - \sin^{-1}x = \cos^{-1}x \qquad \forall(-1 \le x \le 1)\right)$$

$$= 4\sin^{-1}\left\{\frac{a^2}{4x^2 + a^2}\right\}$$

$$\left(since,\ \ \cos^{-1}x = \sin^{-1}\sqrt{1 - x^2} \qquad \forall(-1 \le x \le 1)\right)$$

Case 3: Solid angle subtended by a right prism with regular pentagonal cross-section at any point lying on the longitudinal axis

Is obtained by setting n = 5 in the eq(34), we get

$$\omega = 2\pi - 2(5)\sin^{-1}\left\{\frac{2x sin\frac{\pi}{5}}{\sqrt{4x^2 + a^2 cot^2\frac{\pi}{5}}}\right\} = 2\pi - 10\sin^{-1}\left\{\frac{2x(0.587)}{\sqrt{4x^2 + a^2(1.376)^2}}\right\}$$

$$\Rightarrow \omega = 2\pi - 10\sin^{-1}\left\{\frac{1.175x}{\sqrt{4x^2 + 1.893a^2}}\right\} \ \forall(x \ge 0) \ \ldots\ldots\ldots\ldots\ldots\ldots\ldots (75.1)$$

Case 4: Solid angle subtended by a right prism with regular hexagonal cross-section at any point lying on the longitudinal axis

Is obtained by setting n = 6 in the eq(34), we get

$$\omega = 2\pi - 2(6)\sin^{-1}\left\{\frac{2x\sin\frac{\pi}{6}}{\sqrt{4x^2 + a^2\cot^2\frac{\pi}{6}}}\right\} = 2\pi - 12\sin^{-1}\left\{\frac{2x\frac{1}{2}}{\sqrt{4x^2 + a^2(\sqrt{3})^2}}\right\}$$

$$\Rightarrow \omega = 2\pi - 12\sin^{-1}\left\{\frac{x}{\sqrt{4x^2 + 3a^2}}\right\} \quad \forall (x \geq 0) \ldots\ldots\ldots\ldots\ldots\ldots\ldots\ldots\ldots\ldots (75.2)$$

Thus, further by setting n = 7, 8, 9…….upto n, different formulae can be derived.

4.7.2 Solid angle subtended by a hollow right prismatic shell with regular polygonal cross-section at any internal point lying on the longitudinal axis

Let there be a hollow right prismatic shell with length 'L' & regular polygonal cross-section with 'n' no. of the sides each of length 'a' & a given point say 'P' lying on the longitudinal axis inside the shell. (as shown in the figure (143) below)

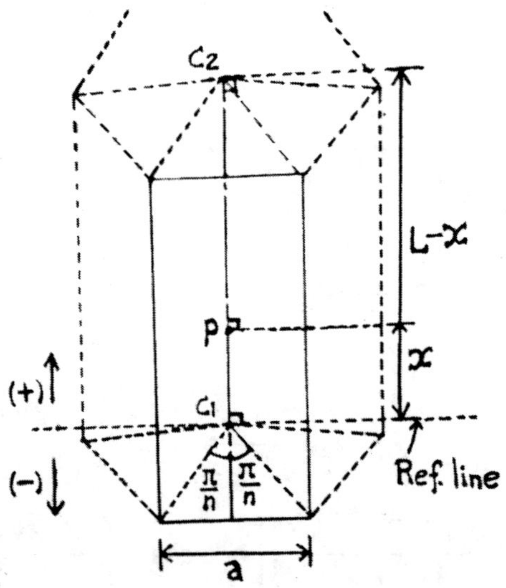

Fig (143)

Assume that both the ends of the shell are closed then the solid angle subtended by the closed shell at the point 'P'

$$\Rightarrow \omega_o = 4\pi$$

The solid angle subtended by the regular polygonal face with centre 'C_1' at the point 'P' is given by the eq(34) as follows

$$\omega = 2\pi - 2n\sin^{-1}\left\{\frac{2h\sin\frac{\pi}{n}}{\sqrt{4h^2 + a^2\cot^2\frac{\pi}{n}}}\right\}$$

Now, on setting h = height = x in the above expression, we get

$$\omega_1 = 2\pi - 2n\sin^{-1}\left\{\frac{2x\sin\frac{\pi}{n}}{\sqrt{4x^2 + a^2\cot^2\frac{\pi}{n}}}\right\}$$

Similarly, the solid angle subtended by the polygonal face with centre 'C_2' at the point 'P' is obtained by setting h = (L− x) in the eq(34), we get

$$\omega_2 = 2\pi - 2n\sin^{-1}\left\{\frac{2(L-x)sin\frac{\pi}{n}}{\sqrt{4(L-x)^2 + a^2cot^2\frac{\pi}{n}}}\right\}$$

But the ends are not closed hence by subtracting ω_1&ω_2 from ω_o, we get

The solid angle subtended by the hollow prismatic shell at the point 'P'

$$\Rightarrow \omega = \omega_o - \omega_1 - \omega_2$$

$$\Rightarrow \omega = 4\pi - \left[2\pi - 2n\sin^{-1}\left\{\frac{2xsin\frac{\pi}{n}}{\sqrt{4x^2 + a^2cot^2\frac{\pi}{n}}}\right\} + 2\pi - 2n\sin^{-1}\left\{\frac{2(L-x)sin\frac{\pi}{n}}{\sqrt{4(L-x)^2 + a^2cot^2\frac{\pi}{n}}}\right\}\right]$$

$$= 4\pi - 2\pi + 2n\sin^{-1}\left\{\frac{2xsin\frac{\pi}{n}}{\sqrt{4x^2 + a^2cot^2\frac{\pi}{n}}}\right\} - 2\pi + 2n\sin^{-1}\left\{\frac{2(L-x)sin\frac{\pi}{n}}{\sqrt{4(L-x)^2 + a^2cot^2\frac{\pi}{n}}}\right\}$$

$$\Rightarrow \omega = 2n\left[\sin^{-1}\left\{\frac{2xsin\frac{\pi}{n}}{\sqrt{4x^2 + a^2cot^2\frac{\pi}{n}}}\right\} + \sin^{-1}\left\{\frac{2(L-x)sin\frac{\pi}{n}}{\sqrt{4(L-x)^2 + a^2cot^2\frac{\pi}{n}}}\right\}\right]$$

$$\forall(x \in R, n \in N \,\&\, n \geq 3) \quad \dots\dots\dots\dots\dots\dots..(76)$$

Sign convention: The distance (x) measured from the centre of open end of the right prismatic shell is taken as

1. Positive when the point lies on the longitudinal axis inside the shell &
2. Negative when the point lies on the longitudinal axis outside the shell

It is obvious the value of solid angle is dependent of the length of the shell with the range of distance (x)

Now, let's consider the point lying on the longitudinal axis outside the shell

4.7.3 Solid angle subtended by a hollow right prismatic shell with regular polygonal cross-section at any external point lying on the longitudinal axis

Let there be a hollow right prismatic shell with length 'L' & regular polygonal cross-section with 'n' no. of the sides each of length 'a' & a given point say 'P' lying on the longitudinal axis outside the shell. (as shown in the figure (144) below)

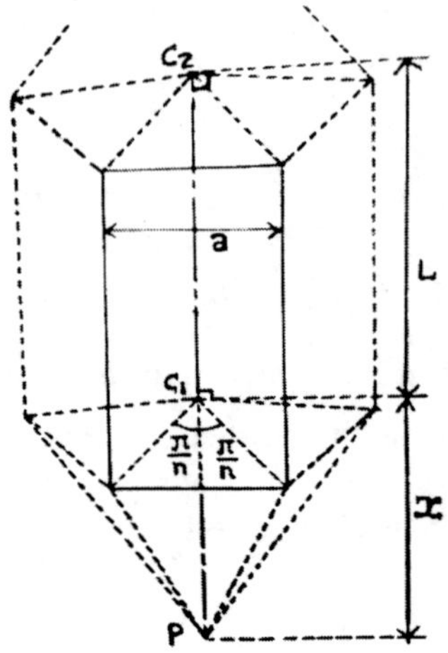

Fig (144)

Assume that the front face of the shell is closed the shell will appear as a polygonal plane

Hence, the solid angle subtended by the regular polygonal face with centre 'C_1' at the point 'P' is given by the eq(34) as follows

$$\omega = 2\pi - 2n\sin^{-1}\left\{\frac{2h\sin\frac{\pi}{n}}{\sqrt{4h^2 + a^2\cot^2\frac{\pi}{n}}}\right\}$$

Now, on setting h = height = x in the above expression, we get

$$\omega_1 = 2\pi - 2n\sin^{-1}\left\{\frac{2x\sin\frac{\pi}{n}}{\sqrt{4x^2 + a^2\cot^2\frac{\pi}{n}}}\right\}$$

Similarly, the solid angle subtended by the polygonal face with centre 'C_2' at the point 'P' is obtained by setting h = (L + x) in the eq(34), we get

$$\omega_2 = 2\pi - 2n\sin^{-1}\left\{\frac{2(L+x)\sin\frac{\pi}{n}}{\sqrt{4(L+x)^2 + a^2\cot^2\frac{\pi}{n}}}\right\}$$

But the front face is not closed then in this case, the rear face with centre 'C_2' will appear open end.(void) then by subtracting ω_2 fromω_1, we get

The solid angle subtended by the hollow prismatic shell at the point 'P'

$$\Rightarrow \omega = \omega_1 - \omega_2$$

$$\Rightarrow \omega = 2\pi - 2n\sin^{-1}\left\{\frac{2x\sin\frac{\pi}{n}}{\sqrt{4x^2 + a^2\cot^2\frac{\pi}{n}}}\right\} - 2\pi + 2n\sin^{-1}\left\{\frac{2(L+x)\sin\frac{\pi}{n}}{\sqrt{4(L+x)^2 + a^2\cot^2\frac{\pi}{n}}}\right\}$$

$$= \quad -2n\sin^{-1}\left\{\frac{2x\sin\frac{\pi}{n}}{\sqrt{4x^2 + a^2\cot^2\frac{\pi}{n}}}\right\} + 2n\sin^{-1}\left\{\frac{2(L+x)\sin\frac{\pi}{n}}{\sqrt{4(L+x)^2 + a^2\cot^2\frac{\pi}{n}}}\right\}$$

$$\Rightarrow \omega = 2n\left[\sin^{-1}\left\{\frac{2(L+x)sin\frac{\pi}{n}}{\sqrt{4(L+x)^2+a^2cot^2\frac{\pi}{n}}}\right\} - \sin^{-1}\left\{\frac{2xsin\frac{\pi}{n}}{\sqrt{4x^2+a^2cot^2\frac{\pi}{n}}}\right\}\right]$$

The above result can also be obtained by setting $x = -x$ in the eq(76) for the external point according to the sign convention as follows

$$\omega = 2n\left[\sin^{-1}\left\{\frac{2(-x)sin\frac{\pi}{n}}{\sqrt{4(-x)^2+a^2cot^2\frac{\pi}{n}}}\right\} + \sin^{-1}\left\{\frac{2(L-(-x))sin\frac{\pi}{n}}{\sqrt{4(L-(-x))^2+a^2cot^2\frac{\pi}{n}}}\right\}\right]$$

$$= 2n\left[-\sin^{-1}\left\{-\frac{2xsin\frac{\pi}{n}}{\sqrt{4x^2+a^2cot^2\frac{\pi}{n}}}\right\} + \sin^{-1}\left\{\frac{2(L+x)sin\frac{\pi}{n}}{\sqrt{4(L+x)^2+a^2cot^2\frac{\pi}{n}}}\right\}\right]$$

$$\Rightarrow \omega = 2n\left[\sin^{-1}\left\{\frac{2(L+x)sin\frac{\pi}{n}}{\sqrt{4(L+x)^2+a^2cot^2\frac{\pi}{n}}}\right\} - \sin^{-1}\left\{\frac{2xsin\frac{\pi}{n}}{\sqrt{4x^2+a^2cot^2\frac{\pi}{n}}}\right\}\right]$$

$$\forall (x \geq 0, n \in N \,\&\, n \geq 3) \quad \dots\dots\dots\dots\dots (77)$$

4.7.4 Solid angle subtended by a hollow right prismatic shell with regular polygonal cross-section at the mid-point of the longitudinal axis

Let there be a hollow right prismatic shell with length 'L' & regular polygonal cross-section with 'n' no. of the sides each of length 'a' & 'P' is the mid-point of the longitudinal axis (as shown in the figure (145) below)

In this case, the solid angle subtended by the hollow right prismatic shell at the mid-point 'P' is obtained by setting $x = \frac{L}{2}$ in the eq(76) as follows

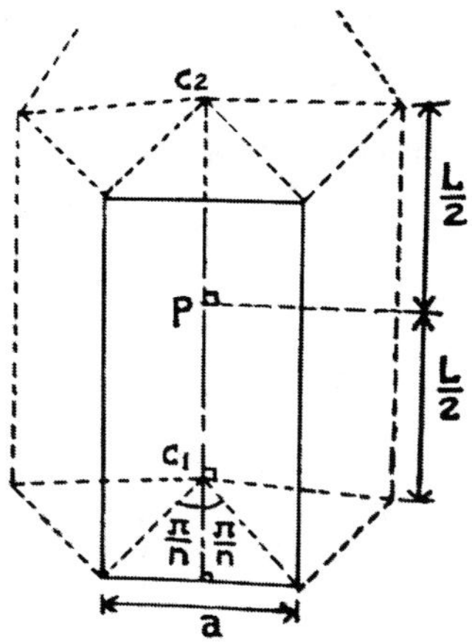

Fig (145)

$$\omega = 2n\left[\sin^{-1}\left\{\frac{2\left(\frac{L}{2}\right)sin\frac{\pi}{n}}{\sqrt{4\left(\frac{L}{2}\right)^2+a^2cot^2\frac{\pi}{n}}}\right\} + \sin^{-1}\left\{\frac{2(L-\frac{L}{2})sin\frac{\pi}{n}}{\sqrt{4(L-\frac{L}{2})^2+a^2cot^2\frac{\pi}{n}}}\right\}\right]$$

$$= 2n\left[\sin^{-1}\left\{\frac{Lsin\frac{\pi}{n}}{\sqrt{L^2 + a^2cot^2\frac{\pi}{n}}}\right\} + \sin^{-1}\left\{\frac{Lsin\frac{\pi}{n}}{\sqrt{L^2 + a^2cot^2\frac{\pi}{n}}}\right\}\right]$$

$$\Rightarrow \omega = 4n\sin^{-1}\left\{\frac{Lsin\frac{\pi}{n}}{\sqrt{L^2 + a^2cot^2\frac{\pi}{n}}}\right\} \qquad \forall(n \in N \,\&\, n \geq 3) \ldots\ldots\ldots\ldots\ldots (78)$$

The above value is maximum value of solid angle at $x = \frac{L}{2}$ i.e. at the mid-point of the longitudinal axis.

4.7.5 Solid angle subtended by a hollow right prismatic shell with regular polygonal cross-section at the centre of the end face

Let there be a hollow right prismatic shell with length 'L' & regular polygonal cross-section with 'n' no. of the sides each of length 'a'. (as shown in the figure (145) above)

In this case, the solid angle subtended by the hollow right prismatic shell at the centre of the end face 'C_1' or 'C_2' is obtained by setting x = 0 in the eq(76) as follows

$$\omega = 2n\left[\sin^{-1}\left\{\frac{2(0)sin\frac{\pi}{n}}{\sqrt{4(0)^2 + a^2cot^2\frac{\pi}{n}}}\right\} + \sin^{-1}\left\{\frac{2(L-0)sin\frac{\pi}{n}}{\sqrt{4(L-0)^2 + a^2cot^2\frac{\pi}{n}}}\right\}\right]$$

$$= 2n\left[0 + \sin^{-1}\left\{\frac{2Lsin\frac{\pi}{n}}{\sqrt{4L^2 + a^2cot^2\frac{\pi}{n}}}\right\}\right]$$

$$\Rightarrow \omega = 2n\sin^{-1}\left\{\frac{2Lsin\frac{\pi}{n}}{\sqrt{4L^2 + a^2cot^2\frac{\pi}{n}}}\right\} \qquad \forall(n \in N \,\&\, n \geq 3) \ldots\ldots\ldots\ldots\ldots (79)$$

4.7.6 Solid angle subtended by an infinitely long hollow right prismatic shell with regular polygonal cross-section at the mid-point of the longitudinal axis

In this case, the solid angle subtended by an infinitely long hollow right prismatic shell at the mid-point of longitudinal axis is obtained by taking the limit of ω at $L \to \infty$ from the eq(78) as follows

$$\Rightarrow \omega = \lim_{L\to\infty} 4n\sin^{-1}\left\{\frac{Lsin\frac{\pi}{n}}{\sqrt{L^2 + a^2cot^2\frac{\pi}{n}}}\right\} = 4n\lim_{L\to\infty}\sin^{-1}\left\{\frac{sin\frac{\pi}{n}}{\sqrt{1 + \left(\frac{a}{L}\right)^2 cot^2\frac{\pi}{n}}}\right\}$$

$$= 4n\sin^{-1}\left\{\frac{sin\frac{\pi}{n}}{\sqrt{1 + \left(\frac{a}{\infty}\right)^2 cot^2\frac{\pi}{n}}}\right\} = 4n\sin^{-1}\left\{\frac{sin\frac{\pi}{n}}{\sqrt{1+0}}\right\} = 4n\sin^{-1}\left\{sin\frac{\pi}{n}\right\}$$

$$\Rightarrow \omega = 4n\frac{\pi}{n} = 4\pi\ sr$$

Thus, an infinitely long hollow right prismatic shell behaves as a closed surface,

4.7.7 Solid angle subtended by an infinitely long hollow right prismatic shell with regular polygonal cross-section at any point of the end face inside the shell

In this case, the solid angle subtended by an infinitely long hollow right prismatic shell at any point of the end face is obtained by taking the limit of $\omega\ at\ L \to \infty$ from the eq(79) as follows

$$\Rightarrow \omega = \lim_{L\to\infty} 2n\sin^{-1}\left\{\frac{2Lsin\frac{\pi}{n}}{\sqrt{4L^2 + a^2cot^2\frac{\pi}{n}}}\right\} = 2n\lim_{L\to\infty}\sin^{-1}\left\{\frac{2sin\frac{\pi}{n}}{\sqrt{4 + \left(\frac{a}{L}\right)^2 cot^2\frac{\pi}{n}}}\right\}$$

$$= 2n\sin^{-1}\left\{\frac{2sin\frac{\pi}{n}}{\sqrt{4 + \left(\frac{a}{\infty}\right)^2 cot^2\frac{\pi}{n}}}\right\} = 2n\sin^{-1}\left\{\frac{2sin\frac{\pi}{n}}{\sqrt{4+0}}\right\} = 2n\sin^{-1}\left\{sin\frac{\pi}{n}\right\}$$

$$\Rightarrow \omega = 2n\frac{\pi}{n} = 2\pi\ sr$$

Thus, an infinitely long hollow right prismatic shell behaves as a cap for all the points of the end face.

4.7.8 Solid angle subtended by a right prismatic shell (with one end open & another closed) at any internal point lying on the longitudinal axis

Let there be a right prismatic shell with length 'L', cross-section as a regular polygon with 'n' no. of the sides each of length 'a' & one end closed & another open and a given internal point say 'P' lying on the longitudinal axis at a distance 'x' from the centre 'C_1' of the open end (reference plane) (as shown in the figure(146) below)

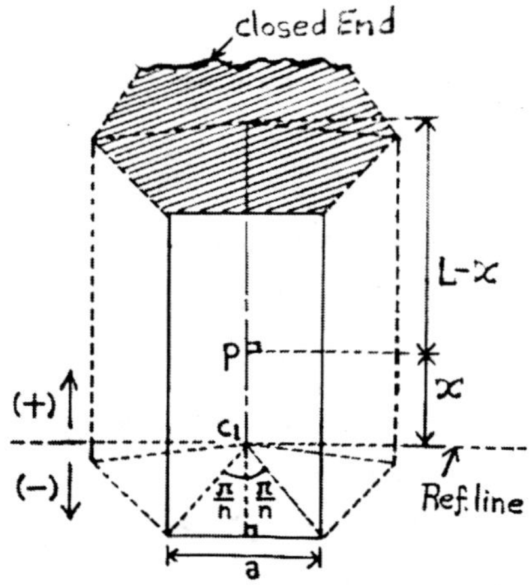

Fig (146)

Now, assume that the end with centre 'C_1' is closed then the solid angle subtended by the closed shell at the point 'P'

$$\Rightarrow \omega_o = 4\pi$$

The solid angle subtended by the regular polygonal end-face with centre 'C_1' at the point 'P' is given by the eq(34) as follows

$$\omega = 2\pi - 2n\sin^{-1}\left\{\frac{2hsin\frac{\pi}{n}}{\sqrt{4h^2 + a^2cot^2\frac{\pi}{n}}}\right\}$$

Now, on setting h = height = x in the above expression, we get

$$\omega_1 = 2\pi - 2n\sin^{-1}\left\{\frac{2x sin\frac{\pi}{n}}{\sqrt{4x^2 + a^2 cot^2\frac{\pi}{n}}}\right\}$$

But the end with centre 'C_1' is not closed hence by subtracting ω_1 from ω_o, we get

The solid angle subtended by the hollow prismatic shell at the point 'P'

$$\Rightarrow \omega = \omega_o - \omega_1$$

$$\Rightarrow \omega = 4\pi - \left[2\pi - 2n\sin^{-1}\left\{\frac{2x sin\frac{\pi}{n}}{\sqrt{4x^2 + a^2 cot^2\frac{\pi}{n}}}\right\}\right]$$

$$= 4\pi - 2\pi + 2n\sin^{-1}\left\{\frac{2x sin\frac{\pi}{n}}{\sqrt{4x^2 + a^2 cot^2\frac{\pi}{n}}}\right\} = 2\pi + 2n\sin^{-1}\left\{\frac{2x sin\frac{\pi}{n}}{\sqrt{4x^2 + a^2 cot^2\frac{\pi}{n}}}\right\}$$

$$\Rightarrow \omega = 2\pi + 2n\sin^{-1}\left\{\frac{2x sin\frac{\pi}{n}}{\sqrt{4x^2 + a^2 cot^2\frac{\pi}{n}}}\right\} \forall (x < L, n \in N \ \& \ n \geq 3) \quad \dots\dots\dots\dots\dots (80)$$

Sign convention: The distance (x) measured from the centre of open end of the right prismatic shell is taken as

1. Positive when the point lies on the longitudinal axis inside the shell &
2. Negative when the point lies on the longitudinal axis outside the shell

It is obvious the value of solid angle is dependent of the length of the shell with the range of distance (x)

Now, let's consider the point lying on the longitudinal axis outside the shell

4.7.9 Solid angle subtended by a right prismatic shell (with one end open & another closed) at any external point lying on the longitudinal axis

Let there be a right prismatic shell with length 'L', cross-section as a regular polygon with 'n' no. of the sides each of length 'a' & one end closed & another open and a given external point say 'P_1' (towards open end) or 'P_2' (towards closed end) lying on the longitudinal axis at a distance 'x' from the centre 'C_1' of the open end (reference plane) (as shown in the figure(147) below)

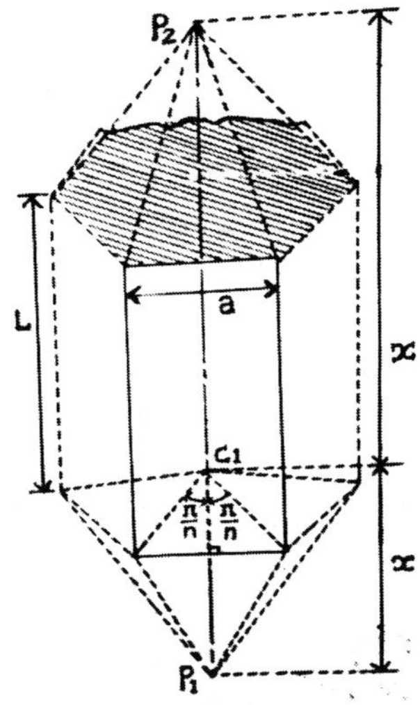

Fig (147)

In this case, when the shell is viewed from the point 'P_1' or point 'P_1' lying on the longitudinal axis it appears as a regular polygonal plane hence the solid angle subtended by the shell will be equal to that subtended by the regular polygonal plane at the same point 'P_1' or 'P_2' &

We know that the solid angle subtended by the regular polygonal plane at any point lying on the vertical axis passing through the centre of the plane is given from the eq(34) as follows

$$\omega = 2\pi - 2n\sin^{-1}\left\{\frac{2h\sin\frac{\pi}{n}}{\sqrt{4h^2 + a^2\cot^2\frac{\pi}{n}}}\right\}$$

Now, on setting h = height = x in the above expression, we get the solid angle subtended by the shell at the point 'P_1'

$$\Rightarrow \omega = 2\pi - 2n\sin^{-1}\left\{\frac{2x\sin\frac{\pi}{n}}{\sqrt{4x^2 + a^2\cot^2\frac{\pi}{n}}}\right\}$$

Similarly, on setting h = height = $(L - x)$ in the above expression, we get the solid angle subtended by the shell at the point 'P_1'

$$\Rightarrow \omega = 2\pi - 2n\sin^{-1}\left\{\frac{2(L-x)\sin\frac{\pi}{n}}{\sqrt{4(L-x)^2 + a^2\cot^2\frac{\pi}{n}}}\right\}$$

The above results can also be obtained by setting $x = -x$ in the eq(80) according to the sign convention for the external point, we get

$$\omega = 2\pi + 2n\sin^{-1}\left\{\frac{2(-x)\sin\frac{\pi}{n}}{\sqrt{4(-x)^2 + a^2\cot^2\frac{\pi}{n}}}\right\} = 2\pi + 2n\sin^{-1}\left\{\frac{-2x\sin\frac{\pi}{n}}{\sqrt{4x^2 + a^2\cot^2\frac{\pi}{n}}}\right\}$$

$$= 2\pi - 2n\sin^{-1}\left\{\frac{2x\sin\frac{\pi}{n}}{\sqrt{4x^2 + a^2\cot^2\frac{\pi}{n}}}\right\} \quad (since\ \sin^{-1}(-x) = -\sin^{-1}x\ \ \forall -1 \le x \le 1)$$

Now, applying the limits of range of the distance 'x' measured from the centre 'C_1' of the open end of the prismatic shell

Hence,we get the solid angle subtended by the shell at the point 'P_1'

$$\Rightarrow \omega = 2\pi - 2n\sin^{-1}\left\{\frac{2x\sin\frac{\pi}{n}}{\sqrt{4x^2 + a^2\cot^2\frac{\pi}{n}}}\right\} \quad \forall (x \geq 0, n \in N \;\&\; n \geq 3) \dots\dots\dots\dots\dots (80.1)$$

Similarly, on setting x = $(L - x)$, solid angle subtended by the shell at the point 'P_2'

$$\Rightarrow \omega = 2\pi - 2n\sin^{-1}\left\{\frac{2(L - x)\sin\frac{\pi}{n}}{\sqrt{4(L - x)^2 + a^2\cot^2\frac{\pi}{n}}}\right\} \quad \forall (x \geq L, n \in N \;\&\; n \geq 3) \dots\dots\dots (80.2)$$

4.7.10 Solid angle subtended by a right prismatic shell (with one end open & another closed) at any internal point lying in the plane of the open end

In this case, the solid angle subtended by the shell is obtained by setting x = 0 in the eq(80) as follows

$$\omega = 2\pi + 2n\sin^{-1}\left\{\frac{2(0)\sin\frac{\pi}{n}}{\sqrt{4(0)^2 + a^2\cot^2\frac{\pi}{n}}}\right\} = 2\pi + 2n\sin^{-1}\{0\} = 2\pi \; sr$$

Or by setting x = 0 in the eq(80.1) as follows

$$\omega = 2\pi - 2n\sin^{-1}\left\{\frac{2(0)\sin\frac{\pi}{n}}{\sqrt{4(0)^2 + a^2\cot^2\frac{\pi}{n}}}\right\} = 2\pi - 2n\sin^{-1}\{0\} = 2\pi \; sr$$

Hence, it's obvious that the shell behaves as cap for each internal point lying in the plane of the open end.

4.7.11 Solid angle subtended by a right prismatic shell (with one end open & another closed) at the mid-point of the longitudinal axis

In this case, the solid angle subtended by the shell at the mid-point of the longitudinal axis is obtained by setting x = $\frac{L}{2}$ in the eq(80) as follows

$$\omega = 2\pi + 2n\sin^{-1}\left\{\frac{2\left(\frac{L}{2}\right)\sin\frac{\pi}{n}}{\sqrt{4\left(\frac{L}{2}\right)^2 + a^2\cot^2\frac{\pi}{n}}}\right\} = 2\pi + 2n\sin^{-1}\left\{\frac{L\sin\frac{\pi}{n}}{\sqrt{L^2 + a^2\cot^2\frac{\pi}{n}}}\right\}$$

$$\Rightarrow \omega = 2\pi + 2n\sin^{-1}\left\{\frac{L\sin\frac{\pi}{n}}{\sqrt{L^2 + a^2\cot^2\frac{\pi}{n}}}\right\} \quad \forall (n \in N \;\&\; n \geq 3) \dots\dots\dots\dots\dots\dots (81)$$

4.7.12 Solid angle subtended by a right prismatic shell (with one end open & another closed) at the centre of the closed end inside the shell

In this case, the solid angle subtended by the prismatic shell at the internal apex point is obtained by taking the limit of ω from the eq(72) at $x \to L^-$ or at $x = (L - k)$, 'k' is the infinitesimal small thickness of the shell such as $k \to 0 \;\&\; k \neq 0$

Hence, we have

$$\Rightarrow \omega = \lim_{x\to L^-}\left[2\pi + 2n\sin^{-1}\left\{\frac{2x\sin\frac{\pi}{n}}{\sqrt{4x^2 + a^2\cot^2\frac{\pi}{n}}}\right\}\right]$$

$$= \lim_{k\to 0}\left[2\pi + 2n\sin^{-1}\left\{\frac{2(L-k)\sin\frac{\pi}{n}}{\sqrt{4(L-k)^2 + a^2\cot^2\frac{\pi}{n}}}\right\}\right] \quad where\ x = (L-k) \Rightarrow k \to 0$$

$$= 2\pi + 2n\sin^{-1}\left\{\frac{2(L-0)\sin\frac{\pi}{n}}{\sqrt{4(L-0)^2 + a^2\cot^2\frac{\pi}{n}}}\right\} (since, k \neq 0)$$

$$\Rightarrow \omega = 2\pi + 2n\sin^{-1}\left\{\frac{2L\sin\frac{\pi}{n}}{\sqrt{4L^2 + a^2\cot^2\frac{\pi}{n}}}\right\} \quad \forall\ (n \in N\ \&\ n \geq 3) \quad \ldots\ldots\ldots\ldots (82)$$

4.7.13 Solid angle subtended by a symmetrical & longitudinal frustum of the hollow right prismatic shell at any point of the longitudinal axis

Let there be a symmetrical & longitudinal frustum of the hollow right prismatic shell having 'm' no. of the lateral faces& given point say 'P' lying on the longitudinal axis at a distance 'x' from the centre of the end-face. (as shown in the figure(148) below)

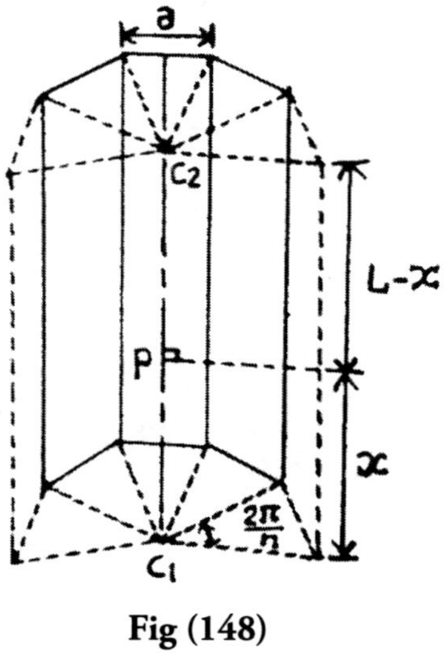

Fig (148)

We know that the solid angle subtended by a hollow complete shell (with 'n' no. of the lateral faces) at any point lying on the longitudinal axis is given by the eq(76) as follows

$$\omega = 2n\left[\sin^{-1}\left\{\frac{2x\sin\frac{\pi}{n}}{\sqrt{4x^2 + a^2\cot^2\frac{\pi}{n}}}\right\} + \sin^{-1}\left\{\frac{2(L-x)\sin\frac{\pi}{n}}{\sqrt{4(L-x)^2 + a^2\cot^2\frac{\pi}{n}}}\right\}\right]$$

Hence, the solid angle subtended by the single lateral face at the same point

$$\Rightarrow \omega = \frac{1}{n}2n\left[\sin^{-1}\left\{\frac{2x\sin\frac{\pi}{n}}{\sqrt{4x^2 + a^2\cot^2\frac{\pi}{n}}}\right\} + \sin^{-1}\left\{\frac{2(L-x)\sin\frac{\pi}{n}}{\sqrt{4(L-x)^2 + a^2\cot^2\frac{\pi}{n}}}\right\}\right]$$

$$= 2\left[\sin^{-1}\left\{\frac{2x\sin\frac{\pi}{n}}{\sqrt{4x^2 + a^2\cot^2\frac{\pi}{n}}}\right\} + \sin^{-1}\left\{\frac{2(L-x)\sin\frac{\pi}{n}}{\sqrt{4(L-x)^2 + a^2\cot^2\frac{\pi}{n}}}\right\}\right]$$

Hence, the solid angle subtended by the frustum with 'm' no. of the lateral faces

$$\Rightarrow \omega = 2m\left[\sin^{-1}\left\{\frac{2x\sin\frac{\pi}{n}}{\sqrt{4x^2 + a^2\cot^2\frac{\pi}{n}}}\right\} + \sin^{-1}\left\{\frac{2(L-x)\sin\frac{\pi}{n}}{\sqrt{4(L-x)^2 + a^2\cot^2\frac{\pi}{n}}}\right\}\right] \quad \dots\dots\dots (83)$$

$$\forall\, (m, n \in N \;\&\; m \le n)$$

4.8 TORUS

It is a circular tube with a certain diameter. Let's analyse a torus having inner radius 'r' & outer radius 'R'.

4.8.1 Solid angle subtended by a torus at any point lying on its axis

Let there be a torus with inner & outer radii 'r' & 'R' respectively & a given point say 'P' lying on the axis at a height 'h' from the centre 'O'

(As shown in the fig (148.1) below)

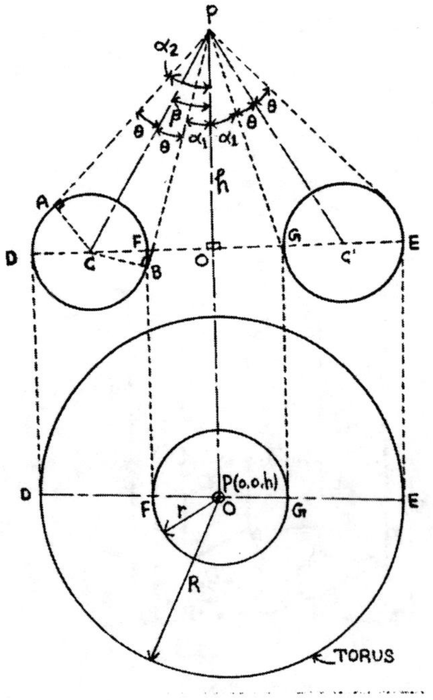

Fig (148.1)

Now, draw the tangents PA & PB from the point 'P' at the surface of torus (upper part of figure showing sectional view of torus)

$$\Rightarrow FD = OD - OB = R - r$$

$$\Rightarrow FC = CD = CA = CB = \frac{FD}{2} = \frac{R-r}{2} \;\&$$

$$OC = OF + FC = r + \frac{R-r}{2} = \frac{R+r}{2}$$

In right ΔCOP

$$\Rightarrow CP = \sqrt{OC^2 + OP^2} = \sqrt{\left(\frac{R+r}{2}\right)^2 + (h)^2} = \frac{1}{2}\sqrt{(R+r)^2 + 4h^2} \&$$

$$\Rightarrow sin\ \ OPC = \frac{OC}{CP} \Rightarrow sin\beta = \frac{\left(\frac{R+r}{2}\right)}{\left(\frac{1}{2}\sqrt{(R+r)^2 + 4h^2}\right)} = \frac{R+r}{\sqrt{(R+r)^2 + 4h^2}}$$

In right ΔPAC

$$\Rightarrow sin\ \ APC = \frac{CA}{CP} \Rightarrow sin\theta = \frac{\left(\frac{R-r}{2}\right)}{\left(\frac{1}{2}\sqrt{(R+r)^2 + 4h^2}\right)} = \frac{R-r}{\sqrt{(R+r)^2 + 4h^2}}$$

Now, let's considertwo right cones with common apex point 'P', equal normal height OP = h andmaximum apex angle 2 $APO = 2\alpha_2$& minimum apex angle 2 $BPO = 2\alpha_1$respectively such that torus is completely covered by both the cones

In such case, solid angle subtended by the torus at the given point 'P'

$$= \omega_{(right\ cone\ (with\ apex\ angle\ 2\alpha_2)} - \omega_{(right\ cone\ (with\ apex\ angle\ 2\alpha_1)}$$

$$\Rightarrow \omega = 2\pi(1 - cos\alpha_2) - 2\pi(1 - cos\alpha_1) \qquad from\ eq(1)$$

$$= 2\pi\left(1 - cos(\beta + \theta)\right) - 2\pi\left(1 - cos(\beta - \theta)\right)$$

$$= 2\pi\left(cos(\beta - \theta) - cos(\beta + \theta)\right)$$

$$= 2\pi(cos\beta cos\theta + sin\beta sin\theta - cos\beta cos\theta + sin\beta sin\theta)$$

$$= 4\pi sin\beta sin\theta$$

On setting the values of $sin\beta$ & $sin\theta$ in the above expression, we get

$$\Rightarrow \omega = 4\pi \times \frac{R+r}{\sqrt{(R+r)^2 + 4h^2}} \times \frac{R-r}{\sqrt{(R+r)^2 + 4h^2}}$$

$$= 4\pi \times \frac{(R+r)(R-r)}{(R+r)^2 + 4h^2} = 4\pi \frac{R^2 - r^2}{(R+r)^2 + 4h^2}$$

$$\Rightarrow \omega = 4\pi\left(\frac{R^2 - r^2}{(R+r)^2 + 4h^2}\right) \qquad \forall\, R \geq r \,\&\, R, r, h \in [0, \infty) \quad \ldots\ldots\ldots\ldots\ldots\ldots (83.1)$$

4.8.2 Solid angle subtended by a torus at its centre

As shown in the figure (148.2) below

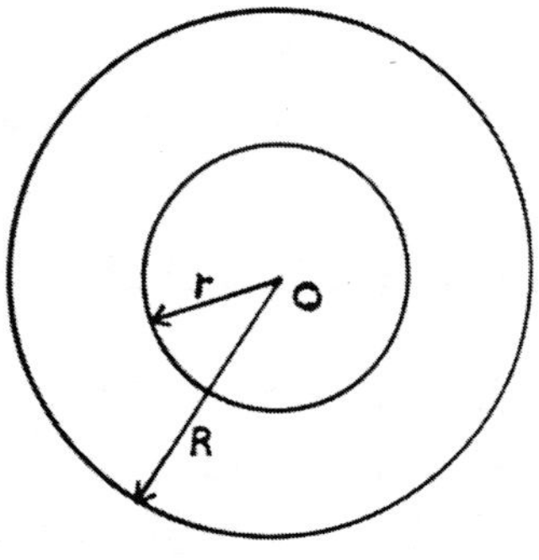

Fig (148.2)

In this case, solid angle subtended by the torus is obtained by setting as follows

$$\omega = 4\pi\left(\frac{R^2 - r^2}{(R+r)^2 + 4(0)^2}\right)$$

$$= 4\pi\left(\frac{R^2 - r^2}{(R+r)^2}\right) = 4\pi\frac{(R+r)(R-r)}{(R+r)^2} = 4\pi\frac{R-r}{R+r}$$

$$\Rightarrow \omega = 4\pi\left(\frac{R-r}{R+r}\right) \quad \ldots\ldots\ldots\ldots\ldots\ldots\ldots (83.2)$$

Important deductions:

1. On setting R = r in the above eq(83.1) or (83.2), we get

$$\omega = 4\pi\left(\frac{r^2 - r^2}{(r+r)^2 + 4h^2}\right) = 0 \ or$$

$$\omega = 4\pi\left(\frac{r-r}{r+r}\right) = 0$$

   It is clear from the above result that solid angle subtended a torus with diameter $R - r = 0$ i.e. circle at any point on its axis is always zero.

2. It is obvious from the eq(83.1) that solid angle subtended by a torus with finite radii at the infinity tends to zero i.e.$\Rightarrow \omega \to 0 \ as\ h \to \infty$.

Solved examples

*** Sphere:**

Example 4.1: A point-source of 200 cd is completely enclosed by a surface. What is the total flux striking the surface?

Sol. given that $intensity\ of\ point\ source, I = 200\ cd$

We know that solid angle subtended a closed surface at any point completely inside it is $4\pi\ sr$ hence the total flux striking the surface

$$\Rightarrow F_t = 4\pi \times I = 4\pi \times 200 = 2513.274\ lm \qquad Ans.$$

Example 4.2: Find out the solid angle subtended by a sphere with a radius 22cm at a point lying at a distance 1.25m from the centre.

Sol. given that

$$radius\ of\ the\ sphere, R = 20cm\ \&\ distance, d = 1.25m = 125cm$$

We know that solid angle subtended by a sphere with a radius 'R' at any point lying at a distance d from the centre is given as

$$\omega = 2\pi\left[1 - \frac{\sqrt{d^2 - R^2}}{d}\right]$$

Hence, on setting the values we get the solid angle subtended by the sphere at the given point

$$\omega = 2\pi\left[1 - \frac{\sqrt{(125)^2 - (22)^2}}{125}\right] = 2\pi[1 - 0.984]$$

$$= 0.1\ sr \qquad Ans.$$

Example 4.3: If the solid angle subtended by a sphere with a diameter 108cm at a point in the space is 0.68 sr. what is the distance of given point from the centre of sphere?

Sol. given that

$$radius\ of\ the\ sphere, R = \frac{108}{2} = 54\ cm\ \&\ solid\ angle, \omega = 0.68\ sr$$

Let the distance of given point from the centre of sphere d then

Solid angle subtended by a sphere with a radius 'R' at any point lying at a distance d from the centre is given as

$$\omega = 2\pi\left[1 - \frac{\sqrt{d^2 - R^2}}{d}\right]$$

Hence, on setting the values we get

$$\omega = 0.68 = 2\pi\left[1 - \frac{\sqrt{d^2 - (54)^2}}{d}\right]$$

$$\frac{\sqrt{d^2 - 2916}}{d} = 1 - \frac{0.68}{2\pi} = 0.892$$

$$d^2 - 2916 = 0.796d^2\ \ or\ d^2 = \frac{2916}{0.204} = 14294.118\ \ or\ d = 119.558cm$$

$$\therefore\ distance, d = 119.558\ cm\ or\ 1.195m \qquad Ans.$$

Example 4.4: Find out the distance of a point from the centre of a sphere with a radius 40cm subtending a solid angle $\pi\ sr$ at the same point.

Sol. given that

$$radius\ of\ the\ sphere, R = 40\ cm\ \&\ solid\ angle, \omega = \pi\ sr$$

Let the distance of given point from the centre of sphere d then we have

Solid angle subtended by a sphere with a radius 'R' at any point lying at a distance d from the centre is given as

$$\omega = 2\pi\left[1 - \frac{\sqrt{d^2 - R^2}}{d}\right]$$

Hence, on setting the values we get

$$\omega = \pi = 2\pi\left[1 - \frac{\sqrt{d^2 - (40)^2}}{d}\right]$$

$$\frac{\sqrt{d^2 - 1600}}{d} = 1 - \frac{1}{2} = 0.5$$

$$d^2 - 1600 = 0.25d^2\ \ or\ d^2 = \frac{1600}{0.75} = 2133.333\ \ or\ d = 46.188\ cm$$

$$\therefore\ distance, d = 46.188\ cm \qquad Ans.$$

Example 4.5: Find out the radius of a sphere subtending a solid angle 0.35 sr. at a point lying at a distance 76cm from the centre.

Sol. given that

$$distance, d = 76\ cm\ \&\ solid\ angle, \omega = 0.35\ sr$$

Let the radius of the sphere be R then we have

Solid angle subtended by a sphere with a radius 'R' at any point lying at a distance d from the centre is given as

$$\omega = 2\pi\left[1 - \frac{\sqrt{d^2 - R^2}}{d}\right]$$

Hence, on setting the values we get

$$\omega = 0.35 = 2\pi\left[1 - \frac{\sqrt{(76)^2 - R^2}}{76}\right]$$

$$\frac{\sqrt{5776 - R^2}}{76} = 1 - \frac{0.35}{2\pi} = 0.944$$

$$5776 - R^2 = (71.744)^2 = 5147.201$$

$$\Rightarrow \quad R = \sqrt{5776 - 5147.201} = 25.076\ cm \qquad Ans.$$

Example 4.6: Find out the solid angle subtended by a hemispherical shell with a radius 13cm at a point lying at a distance 18cm from the centre of circular-section, on the axis towards the

1) Section & 2) surface

Sol. given that

$$radius\ of\ the\ hemisphere, R = 13cm\ \&\ distance, d = 18\ cm$$

1. When the point lies on the axis towards the surface

We know that solid angle subtended by a hemi-sphere with a radius 'R' at any point lying, at a distance d from the centre of circular sectionis given as

$$\omega = 2\pi\left[1 - \frac{\sqrt{d^2 - R^2}}{d}\right]$$

Hence, on setting the values we get the solid angle subtended by the hemisphere at the given point

$$\omega = 2\pi\left[1 - \frac{\sqrt{(18)^2 - (13)^2}}{18}\right] = 2\pi[1 - 0.692]$$

$$= 1.935\ sr \qquad Ans.$$

2. When the point lies on the axis towards the circular section

We know that solid angle subtended by a hemi-sphere with a radius 'R' at any point lying, at a distance d from the centre of circular section is given as

$$\omega = 2\pi\left[1 - \frac{d}{\sqrt{d^2 + R^2}}\right]$$

Hence, on setting the values we get the solid angle subtended by the hemisphere at the given point

$$\omega = 2\pi\left[1 - \frac{18}{\sqrt{(18)^2 + (13)^2}}\right] = 2\pi[1 - 0.811]$$

$$= 1.187\ sr \qquad Ans.$$

Example 4.7: Find out the solid angle subtended by a hemispherical shell with a radius 120mm at a point lying at a distance 4cm from the centre of circular-section on the axis inside the shell.

Sol. given that

$$radius\ of\ the\ hemisphere, R = 120mm = 12cm\ \&\ distance, d = 4\ cm$$

We know that solid angle subtended by a hemi-spherical shell with a radius 'R' at any point lying, at a distance d from the centre of circular section, on the axis inside the shell is given as

$$\omega = 2\pi\left[1 + \frac{d}{\sqrt{d^2 + R^2}}\right]$$

Hence, on setting the values we get the solid angle subtended by the hemispherical shell at the given point

$$\omega = 2\pi\left[1 + \frac{4}{\sqrt{(4)^2 + (12)^2}}\right] = 2\pi[1 + 0.316]$$

$$= 8.269\ sr \qquad Ans.$$

Example 4.8: Find out the solid angle subtended by a hemispherical shell with a radius 14cm ata point lying, at a distance 150mm from the centre of circular-section, on the axis towards the surface of shell.

Sol. given that

$$radius\ of\ the\ hemisphere, R = 14cm\ \&\ distance, d = 150\ mm = 15cm$$

We know that solid angle subtended by a hemi-sphere with a radius 'R' at any point lying, at a distance d from the centre of circular section, on the axis towards the surface is given as

$$\omega = 2\pi\left[1 - \frac{\sqrt{d^2 - R^2}}{d}\right]$$

Hence, on setting the values we get the solid angle subtended by the hemisphere at the given point

$$\omega = 2\pi\left[1 - \frac{\sqrt{(15)^2 - (14)^2}}{15}\right] = 2\pi[1 - 0.359]$$

$$= 4.027\ sr \qquad Ans.$$

Example 4.9: What is the solid angle subtended by a hemispherical shell with a radius 28cm at the centre of mass.

Sol. given that

$$radius\ of\ the\ hemisphere, R = 28cm$$

The distance of centre of mass of a hemispherical shell with a radius 'R' from the centre of circular section is given as

$$d = \frac{R}{2} = \frac{28}{2} = 14cm$$

$$\omega = 2\pi\left[1 + \frac{d}{\sqrt{d^2 + R^2}}\right]$$

Hence, the solid angle subtended by the hemispherical shell at the given point

$$\omega = 2\pi\left[1 + \frac{d}{\sqrt{d^2 + R^2}}\right] = 2\pi\left[1 + \frac{4}{\sqrt{(4)^2 + (12)^2}}\right]$$

$$= 2\pi[1 + 0.316] = 8.269\ sr \qquad Ans.$$

Example 4.10: A frustum of a spherical shell with a radius 18 cm subtends an angle 90^0 at the centre of the shell. What will be solid angle subtended by the frustum at a point lying at a distance 110mm from the centre of circular section, on the axis towards the section?

Sol. given that

$$radius\ of\ the\ hemisphere, R = 18cm\ \&\ distance, d = 110mm = 11cm$$

Now, the radius of circular section,

$$r = 18 sin\frac{90^o}{2} = 18 sin 45^o = 12.728\ cm$$

Hence, the solid angle subtended by the frustum of hemispherical shell at the given point

$$\omega = 2\pi\left[1 - \frac{d}{\sqrt{d^2 + r^2}}\right] = 2\pi\left[1 - \frac{11}{\sqrt{(11)^2 + (12.728)^2}}\right]$$

$$= 2\pi[1 - 0.654] = 2.174\ sr \qquad Ans.$$

Example 4.11: A frustum of a spherical shell with a radius 12 cm subtends an angle 120^0 at the centre of the shell. Find out the solid angle subtended by the frustum at a point lying, at a distance 126mm from the centre of its circular section, on the axis towards the surface.

Sol. given that

$$radius\ of\ shell, R = 12cm, angle, 2\alpha = 120^o \&\ distance, x = 126mm = 12.6cm$$

First let us find the location of point,

$$\Rightarrow R(1 - cos\alpha) = 12(1 - cos60^o) = 6cm\ \&$$

$$Rsin\alpha tan\alpha = 12sin60^o tan60^o = 18cm$$

Thus we find that the value of distance, x = 12.6 cm lies within range as follows

$$R(1 - cos\alpha) \leq x \leq Rsin\alpha tan\alpha$$

$$or\ \ 6 \leq 12.6 \leq 18$$

Hence, the solid angle subtended by the frustum at the given point is given by the eq(45) as follows

$$\omega = 2\pi\left[1 - \frac{\sqrt{(x + Rcos\alpha)^2 - R^2}}{x + Rcos\alpha}\right]$$

Now, on setting the values we get

$$= 2\pi\left[1 - \frac{\sqrt{(12.6 + 12cos60^o)^2 - (12)^2}}{12.6 + 12cos60^o}\right] = 2\pi\left[1 - \frac{14.211}{18.6}\right]$$

$$= 2\pi[1 - 0.764] = 1.483\ sr \qquad Ans.$$

Example 4.12: A frustum of a spherical shell with a radius 10 cm subtends an angle 110^0 at the centre of the shell. Find out the solid angle subtended by the frustum at a point lying, at a distance 144mm from the centre of its circular section, on the axis towards the surface.

Sol. given that

$$radius\ of\ spherical\ shell, R = 10cm, angle, 2\alpha = 110^o \&$$

$$distance, x = 144mm = 14.4cm$$

First let us find the location of point,

$$\Rightarrow R(1 - cos\alpha) = 10(1 - cos55^o) = 4.264cm\ \&$$

$$Rsin\alpha tan\alpha = 10sin55^o tan55^o = 11.699cm$$

Thus we find that the value of distance, x = 14.4 cm lies within range as follows

$$x > Rsin\alpha tan\alpha$$

$$or\ \ 14.4 > 11.699$$

Hence, the solid angle subtended by the frustum at the given point is given by the eq(46) as follows

$$\omega = 2\pi\left[1 - \frac{x}{\sqrt{x^2 + R^2 sin^2\alpha}}\right]$$

Now, on setting the values we get

$$= 2\pi\left[1 - \frac{14.4}{\sqrt{(14.4)^2 + (10)^2 sin^2 55^o}}\right] = 2\pi\left[1 - \frac{14.4}{16.57}\right]$$

$$= 2\pi[1 - 0.869] = 0.823\ sr \qquad Ans.$$

Example 4.13: Determine the solid angle subtended by a frustum of a spherical shell with a radius 25 cm subtends an angle 100^0 at the centre, at its centre of mass.

Sol. given that

$$Radius\ of\ the\ shell, R = 25cm\ \&$$
$$angle\ subtended\ at\ the\ centre, 2\alpha = 100^o$$

We know that the distance, of the centre of mass of the frustum, of a spherical shell with a radius 'R', subtending an angle 2α at the centre of parent shell, is given as

$$\bar{x} = \frac{R}{4}\left(\frac{1 - cos2\alpha}{1 - cos\alpha}\right)$$

On setting the values we get

$$\bar{x} = \frac{25}{4}\left(\frac{1 - cos100^o}{1 - cos50^o}\right) = 20.535\ cm$$

The distance of the centre of mass of the frustum from the centre of its circular section

$$\Rightarrow x = \bar{x} - Rcos\alpha = 20.535 - 25cos50^o = 4.465\ cm$$

Now, the radius of circular section of the frustum

$$r = Rsin\alpha = 25sin50^o = 19.151\ \ cm$$

Hence, the solid angle subtended by the frustum of spherical shell at the given point

$$\omega = 2\pi\left[1 + \frac{x}{\sqrt{x^2 + r^2}}\right] = 2\pi\left[1 + \frac{4.465}{\sqrt{(4.465)^2 + (19.151)^2}}\right]$$
$$= 2\pi[1 + 0.227] = 7.709\ sr \qquad Ans.$$

Example 4.14: A frustum of a spherical shell with a radius 16 cm subtends a solid angle 0.987 sr. at a point lying, at a distance 130mm from the centre of circular section, on the axis towards the section. Determine the plane-angle subtended by the frustum at the centre of its parent sphere.

Sol. given that

$$solid\ angle, \omega = 0.987\ sr$$

Let the radius of the spherical shell be 'R'&plane-angle subtended by the frustum at the centre of parent shell be $'2\alpha'$ then

The distance of the given point from the centre of circular section

$$d = Rcos\alpha\ \&$$

Radius of the circular section of the frustum,

$$r = Rsin\alpha$$

Hence the solid angle subtended by the frustum at the given point

$$\omega = 2\pi\left[1 - \frac{d}{\sqrt{d^2 + r^2}}\right]$$

$$= 2\pi\left[1 - \frac{Rcos\alpha}{\sqrt{(Rcos\alpha)^2 + (Rsin\alpha)^2}}\right] = 2\pi\left[1 - \frac{Rcos\alpha}{R\sqrt{sin^2\alpha + cos^2\alpha}}\right]$$

$$\omega = 2\pi[1 - cos\alpha] \Rightarrow\ 0.987 = 2\pi[1 - cos\alpha]$$

$$\Rightarrow cos\alpha = 1 - \frac{0.987}{2\pi} = 1 - 0.157 = 0.843$$

$$\Rightarrow \alpha = \cos^{-1}(0.843) = 32.54^o$$

$$\therefore\ 2\alpha = 2 \times 32.54^o = 64.08^o \qquad Ans.$$

Example 4.15: A uniform point-source of 150cd is mounted on tip of a 60cm long straight rod vertically pitched at the hemispherical summit of a temple. Determine luminous flux intercepted by the summit if outer radius of summit is 2.5m. (Neglect thickness of rod)

Sol. given that

$$intensity, I = 150\ cd, radius\ of\ hemispherical\ summit, R = 2.5m = 250cm\ \&$$
$$length\ of\ rod, l = 60cm$$

$$distance\ (d)\ of\ the\ point\ source\ from\ the\ centre\ of\ hemisphere = R + l$$
$$d = 60 + 250 = 310cm$$

Solid angle subtended by the hemisphere at the point-source

$$\omega = 2\pi\left[1 - \frac{\sqrt{d^2 - R^2}}{d}\right]$$

Hence, on setting the values we get the solid angle subtended by the hemisphere at the given point

$$\omega = 2\pi\left[1 - \frac{\sqrt{(310)^2 - (250)^2}}{310}\right] = 2\pi[1 - 0.591]$$
$$= 2.57\ sr$$

$$hence, total\ flux\ intercepted\ by\ the\ summit,$$

$$\Rightarrow \quad F = I \times \omega = 150 \times 2.57 = 385.5\ lm \qquad Ans.$$

Example 4.16: Find out the luminous flux density over a frustum, of a spherical shell with a radius 28cm, subtending an angle 90° at the centre of shell when a uniform point-source of 111cd is mounted at a point lying, on the axis at a distance 40cm from the centre of the circular section of frustum,towards the surface. (Assume uniform distribution of flux over the entire surface of frustum)

Sol. given that

$$radius\ of\ spherical\ shell, R = 28cm, angle, 2\alpha = 90^o,$$
$$distance, x = 40cm\ \&\ intensity, I = 111\ cd$$

Surface area of the frustum of a spherical shell is given as

$$A = 2\pi R^2(1 - cos\alpha)$$
$$= 2\pi(28)^2(1 - cos45^o) = 1442.797\ cm^2 = 1442.797 \times 10^{-4}m^2$$

1. When the point lies on the axis towards the surface

Let us find the location of point,

$$\Rightarrow R(1 - cos\alpha) = 28(1 - cos45^o) = 8.201cm\ \&$$
$$Rsin\alpha tan\alpha = 28sin45^o tan45^o = 19.799cm$$

Thus we find that the value of distance, x = 40 cm lies within range as follows

$$x > Rsin\alpha tan\alpha$$
$$or\ \ 40 > 19.799$$

Hence, the solid angle subtended by the frustum at the given point is given by the eq(46) as follows

$$\omega = 2\pi\left[1 - \frac{x}{\sqrt{x^2 + R^2 sin^2\alpha}}\right]$$

Now, on setting the values we get

$$= 2\pi\left[1 - \frac{40}{\sqrt{(40)^2 + (28)^2 sin^2 45^o}}\right] = 2\pi\left[1 - \frac{40}{44.632}\right]$$
$$= 2\pi[1 - 0.896] = 0.653\ sr$$

Total flux intercepted by the frustum,

$$F = I \times \omega = 111 \times 0.653 = 72.483\ lm$$

Hence, flux density over the surface of the frustum

$$\Rightarrow \quad F.D. = \frac{F}{A} = \frac{72.483}{1442.797 \times 10^{-4}} = 502.378\ lux \qquad Ans.$$

Example 4.17: A uniform point-source of 250cd is mounted at the centre of mass of hemi-spherical shell with a radius 20cm. Determine the luminous flux intercepted by the shell & flux density.

(Assume that the flux is uniformly distributed over the entire surface of the shell.)

Sol. given that

$$radius\ of\ the\ hemispherical\ shell, R = 20cm\ \&\ intensity, I = 250\ cd$$

We know that the distance (x) of centre of mass of a hemispherical shell from the centre of parent shell is given as

$$x = \frac{R}{2} = \frac{20}{2} = 10cm$$

Hence, the solid angle subtended by the hemispherical shell at the centre of mass

$$\omega = 2\pi \left[1 + \frac{x}{\sqrt{x^2 + R^2}}\right]$$

Hence, on setting the values we get the solid angle subtended by the hemisphere at the given point

$$\omega = 2\pi \left[1 + \frac{10}{\sqrt{(10)^2 + (20)^2}}\right] = 2\pi[1 + 0.447]$$

$$= 9.092\ sr$$

$$hence, total\ flux\ intercepted\ by\ the\ summit,$$

$$\Rightarrow \quad F = I \times \omega = 250 \times 9.092 = 2273\ lm \qquad Ans.$$

Hence, flux density over the surface of the frustum

$$\Rightarrow \quad F.D. = \frac{F}{A} = \frac{2273}{2\pi(0.2)^2} = 9044\ lux \qquad Ans.$$

Example 4.18: There are two spheres A & B with radii $R_A \& R_B$ respectively, radiating energy uniformly in all the directions. If $x_A \& x_B$ are the fractions of total energy intercepted by bodies A & B respectively then prove that

$$\frac{R_A}{R_B} = \sqrt{\frac{x_A - {x_A}^2}{x_B - {x_B}^2}} \quad \forall \left(x_A, x_B \leq \frac{1}{2}\right).$$

Sol. given that

$$radii\ of\ spheres\ A\ \&\ B\ are\ R_A \& R_B \&$$

$$fractions\ of\ total\ radiation\ are\ \ x_A \& x_B$$

Let the luminosities (total radiations) of the spheres A & B be $L_A \& L_B$ respectively & distance between their centres be d

Solid angle subtended by the sphere-A at the centre of sphere-B

$$\omega_A = 2\pi \left[1 - \frac{\sqrt{d^2 - {R_A}^2}}{d}\right]$$

Since, each of the spheres emits energy radially so these can be treated as the uniform point-sources hence

We know that radiation energy, emitted by a uniform point-source with luminosity 'L', intercepted by a sphere with a radius 'R' at a distance d from the centre to the point-source is given as

$$E_i = \left(\frac{L}{4\pi}\right) \times \omega = \left(\frac{L}{4\pi}\right) \times 2\pi\left[1 - \frac{\sqrt{d^2 - R^2}}{d}\right] \qquad (see\ art. 5.3.7)$$

Hence, the total radiation energy, emitted by sphere-B, intercepted by sphere-A

$$E_A = \left(\frac{L_B}{4\pi}\right) \times \omega_A = \left(\frac{L_B}{4\pi}\right) \times 2\pi\left[1 - \frac{\sqrt{d^2 - {R_A}^2}}{d}\right]$$

But, total radiation energy, emitted by sphere-B, intercepted by sphere-A

$$E_A = x_A \times (total\ radiation\ emitted\ by\ sphere\ B) = x_A \times (L_B)$$

Equating both the results, we get

$$\left(\frac{L_B}{4\pi}\right) \times 2\pi\left[1 - \frac{\sqrt{d^2 - {R_A}^2}}{d}\right] = x_A \times (L_B)$$

$$\Rightarrow \left[1 - \frac{\sqrt{d^2 - {R_A}^2}}{d}\right] = 2x_A$$

$$\Rightarrow \frac{\sqrt{d^2 - {R_A}^2}}{d} = (1 - 2x_A) \ \ or \ \ d^2 - {R_A}^2 = d^2(1 - 2x_A)^2$$

$$\Rightarrow {R_A}^2 = d^2\{1 - (1 - 2x_A)^2\}$$

$$= d^2\{1 - 1 - 4{x_A}^2 + 4x_A\} = 4d^2(x_A - {x_A}^2)$$

$$\Rightarrow R_A = 2d\sqrt{x_A - {x_A}^2} \qquad \dots\dots\dots\dots\dots\dots (I)$$

Similarly, we can obtain the radius of sphere-B as follows

$$R_B = 2d\sqrt{x_B - {x_B}^2} \qquad \dots\dots\dots\dots\dots\dots (II)$$

Hence, by diving the eq(I) by (II) we get

$$\frac{R_A}{R_B} = \frac{2d\sqrt{x_A - {x_A}^2}}{2d\sqrt{x_B - {x_B}^2}} = \frac{\sqrt{x_A - {x_A}^2}}{\sqrt{x_B - {x_B}^2}} = \sqrt{\frac{x_A - {x_A}^2}{x_B - {x_B}^2}}$$

We know that, the solid angle subtended by a sphere can't exceed $2\pi\ sr$ hence, the maximum radiation energy intercepted by a sphere

$$(E_i)_{max} = \left(\frac{L}{4\pi}\right) \times \omega_{max} = \left(\frac{L}{4\pi}\right) \times 2\pi = \frac{L}{2}$$

If the sphere intercepts x fraction of total radiation energy (equal Luminosity, L) then we get

$$E_i = x \times (luminosity\ of\ point\ source) = xL$$

$$\Rightarrow E_i \leq (E_i)_{max} \quad or\ xL \leq \frac{L}{2}$$

$$\Rightarrow x \leq \frac{1}{2}$$

Similarly, we have

$$x_A,\ x_B \le \frac{1}{2}$$

Hence, we get

$$\frac{R_A}{R_B} = \sqrt{\frac{x_A - {x_A}^2}{x_B - {x_B}^2}} \quad \forall \left(x_A,\ x_B \le \frac{1}{2}\right) \qquad \textit{proved.}$$

Example 4.19: Two red hot spheres A & B with radii 3cm & 7cm respectively, radiating energy uniformly in all the directions, are kept touching one another. Find out the factions of total radiation-energy intercepted by them.

Sol. given that

$$radii\ of\ the\ red\ hot\ spheres, R_A = 3cm\ \&\&R_B = 7cm$$

Let $x_A \& x_B$ be the fractions of radiation energy intercepted by the spheres A & B with luminosities (total radiations) $L_A \& L_B$ respectively

Distance (d) between the centres of the spheres

$$d = R_A + R_A = 3 + 7 = 10cm \qquad (since, spheres\ are\ touching\ each\ other)$$

Solid angle subtended by the sphere-A at the centre of sphere-B

$$\omega_A = 2\pi\left[1 - \frac{\sqrt{d^2 - {R_A}^2}}{d}\right]$$

On setting the values, we get

$$\omega_A = 2\pi\left[1 - \frac{\sqrt{(10)^2 - (3)^2}}{10}\right] = 2\pi[1 - 0.954] = 0.289\ sr$$

Similarly, solid angle subtended by the sphere-A at the centre of sphere-B

$$\omega_B = 2\pi\left[1 - \frac{\sqrt{d^2 - {R_B}^2}}{d}\right]$$

On setting the values, we get

$$\omega_B = 2\pi\left[1 - \frac{\sqrt{(10)^2 - (7)^2}}{10}\right] = 2\pi[1 - 0.714] = 1.797\ sr$$

Thus, the total radiation energy, emitted by sphere-B, intercepted by sphere-A

$$E_A = \left(\frac{L_B}{4\pi}\right) \times \omega_A = x_A \times (total\ radiation\ emitted\ by\ sphere\ B) = x_A \times (L_B)$$

$$x_A = \frac{\omega_A}{4\pi} = \frac{0.289}{4\pi} = 0.023$$

Similarly, the value of fraction x_B can be obtained as follows

$$x_B = \frac{\omega_B}{4\pi} = \frac{1.797}{4\pi} = 0.143$$

Hence, the fractions $(x_A \& x_B)$ of the radiation energy intercepted by the spheres A& B respectively

$$\Rightarrow x_A = 0.023\ \& x_B = 0.143 \qquad Ans.$$

*** Circular cylinder:**

Example 4.20: Find out the solid angle subtended by a solid cylinder with radius 11cm at a point lying on the longitudinal axis at a distance 160mm from the centre of front-end. If the length of cylinder isincreased what will be change in the value of solid angle?

Sol. given that

$$radius\ of\ solid\ cylinder, R = 11cm\ \&\ distance, x = 160mm = 16cm$$

Solid angle, subtended by a cylinder with a radius 'R' at any point lying on the longitudinal axis at a distance x from the centre of front-end, is given as

$$\omega = 2\pi\left[1 - \frac{x}{\sqrt{x^2 + R^2}}\right]$$

On setting the values, we get

$$\omega = 2\pi\left[1 - \frac{16}{\sqrt{(16)^2 + (11)^2}}\right] = 2\pi[1 - 0.824] = 1.106\ sr \qquad Ans.$$

Since, the value of solid angle does not depend on the length of the cylinder hence there will be no change in the value of solid angle.

Example 4.21:A cylinder subtends a plane-angle 80^0 at a point lying on the longitudinal axis. Find out the solid angle subtended by the cylinder at the same point.

Sol. given that

$$angle, 2\alpha = 80^o\ or\ \alpha = 40^o$$

We know that the solid angle, subtended by a cylinder with a radius 'R' at any point lying on the longitudinal axis at a distance x from the centre of front-end, is given as

$$\omega = 2\pi\left[1 - \frac{x}{\sqrt{x^2 + R^2}}\right]$$

If 2α is the plane-angle subtended by the cylinder at the same point then

$$cos\alpha = \frac{x}{\sqrt{x^2 + R^2}} \Rightarrow \omega = 2\pi[1 - cos\alpha]$$

Hence, the solid angle subtended by the cylinder at the given point

$$\omega = 2\pi[1 - cos40^o] = 1.47\ sr \qquad Ans.$$

Example 4.22: Find out the solid angle subtended by a solid cylinder with a radius 10cm & length 20cm at a point lying on the transverse axis at a distance 30cm from the centre of cylinder. (Neglect the effect of curvature of the circular-ends)

Sol. given that

$$radius\ of\ cylinder, R = 10cm,\ \ Length, L = 20cm\ \&$$
$$distance\ of\ the\ given\ point, x = 30cm$$

We know that the solid angle, subtended by a cylinder with a radius 'R' & length 'L'at any point lying on the transverse axis at a distance x from the centre of longitudinal axis, is given by the eq(48) as follows

$$\omega = 4\sin^{-1}\left(\frac{RL}{\sqrt{x^2L^2 + 4(x^2 - R^2)^2}}\right)$$

On setting the values, we have

$$\omega = 4\sin^{-1}\left(\frac{10 \times 20}{\sqrt{(30)^2(20)^2 + 4((30)^2 - (10)^2)^2}}\right)$$

$$= 4\sin^{-1}\left(\frac{200}{1708.801}\right) = 4\sin^{-1}(0.117) = 0.469\ sr \qquad Ans.$$

Example 4.23: Find out the solid angle subtended by an infinite long cylinder with a radius 18cm at a point lying on the transverse axis at a distance 3m from the centre of longitudinal axis.

Sol. given that

$$radius\ of\ cylinder, R = 18cm\ \&\ distance, x = 3m = 300cm$$

We know that the solid angle subtended by an infinitely long cylinder at any point lying on the transverse axis at a distance 'x' from the centre of longitudinal axis is given from axis is given from the eq() as follows

$$\omega = 4\sin^{-1}\left(\frac{R}{x}\right)$$

On setting the values we have,

$$\omega = 4\sin^{-1}\left(\frac{18}{300}\right) = 4 \times 0.06 = 0.24\ sr \qquad Ans.$$

Example 4.24: An infinite long cylinder with a radius 33cm subtends a solid angle 1.25 sr. at a point lying on the transverse axis. What will the distance of the point from the centre of longitudinal axis?

Sol. given that

$$radius\ of\ cylinder, R = 33cm\ \&\ solid\ angle, \omega = 1.25sr$$

Let the distance of the given point from the centre be x

Hence, the solid angle subtended by the cylinder at the same point

$$\omega = 4\sin^{-1}\left(\frac{R}{x}\right)$$

On setting the values we have,

$$\omega = 1.25 = 4\sin^{-1}\left(\frac{33}{x}\right)$$

$$\sin^{-1}\left(\frac{33}{x}\right) = \frac{1.25}{4} = 0.3125\ or\ \frac{33}{x} = \sin(0.3125) = 0.307$$

$$\Rightarrow x = \frac{33}{0.307} = 107.492\ cm \qquad Ans.$$

Example 4.25: If the solid angle subtended by an infinite long cylinder at a point lying on the transverse axis at a distance 4m from the centre of the longitudinal axis is 0.113 sr. then determine the diameter of cylinder.

Sol. given that

$$distance, x = 4m\ \&\ solid\ angle, \omega = 0.113sr$$

Let the radius of cylinder be R

Then, the solid angle subtended by the cylinder at the same point

$$\omega = 4\sin^{-1}\left(\frac{R}{x}\right)$$

On setting the values we have,

$$\omega = 0.113 = 4\sin^{-1}\left(\frac{R}{4}\right)$$

$$\sin^{-1}\left(\frac{R}{4}\right) = \frac{0.113}{4} = 0.0282\ or\ \frac{R}{4} = \sin(0.0282) = 0.0282$$

$$\Rightarrow R = 4 \times 0.0282 = 0.113\ m\ or\ 11.3cm$$

$$hence,\ the\ diameter\ of\ cylinder = 2R = 2 \times 11.3 = 22.6\ cm \qquad Ans.$$

Example 4.26:A long cylindrical pole with a diameter 900mm & height 25m standing on a plane surface. What will be the solid angle subtended by the pole at a point lying in the plane of base at a distance 2m from the centre of base?

(Neglect the effect of curvature of the circular ends)

Sol. given that

$$diameter\ of\ cylinder, D = 900mm = 0.9\ m,\quad Length, L = 25\ m\ \&$$
$$distance\ of\ the\ given\ point, x = 2m$$

We know that the solid angle, subtended by a cylinder with a radius 'R' & length 'L' at any point lying in the plane of base at a distance x from the centre of base, is given by the eq(49) as follows

$$\omega = 2\sin^{-1}\left(\frac{RL}{\sqrt{x^2L^2 + (x^2 - R^2)^2}}\right)$$

On setting the values, we have

$$\omega = 2\sin^{-1}\left(\frac{0.45 \times 25}{\sqrt{(2)^2(25)^2 + ((2)^2 - (0.45)^2)^2}}\right)$$
$$= 2\sin^{-1}\left(\frac{11.25}{50.155}\right) = 2\sin^{-1}(0.224) = 0.452\ sr \qquad Ans.$$

Example 4.27: Find out the solid angle subtended by an infinite long cylindrical pole with a radius 34cm at a point lying at a distance 45cm from the longitudinal axis in the plane of the base.

Sol. given that

$$radius\ of\ cylinder, R = 34cm\ \&\ distance, x = 45cm$$

We know that the solid angle subtended by an infinitely long cylinder at any point lying in the plane of base at a distance 'x' from the longitudinal axis is given from axis is given from the eq() as follows

$$\omega = 2\sin^{-1}\left(\frac{R}{x}\right)$$

On setting the values we have,

$$\omega = 2\sin^{-1}\left(\frac{34}{45}\right) = 2 \times 0.856 = 1.713\ sr \qquad Ans.$$

Example 4.28: Determine the solid angle subtended by a hollow cylindrical shell with a radius 8cm & length 18cm at an internal point lying on the longitudinal axis at a distance 6cm from the centre of one of the ends.

Sol. given that

$$radius\ of\ hollow\ cylindrical\ shell, R = 8cm, length, L = 18cm\ \&$$
$$distance, x = 6cm$$

We know that solid angle, subtended by a hollow cylindrical shell with a radius 'R' & length 'L' at any internal point lying on the longitudinal axis at a distance x from the centre of one of the ends, is given from the eq(53) as follows

$$\omega = 2\pi\left[\frac{x}{\sqrt{x^2 + R^2}} + \frac{(L - x)}{\sqrt{(L - x)^2 + R^2}}\right]$$

On setting the values, we get

$$\omega = 2\pi\left[\frac{6}{\sqrt{6^2 + 8^2}} + \frac{(18 - 6)}{\sqrt{(18 - 6)^2 + 8^2}}\right]$$
$$= 2\pi[0.6 + 0.832] = 8.997\ sr \qquad Ans.$$

Example 4.29: Determine the solid angle subtended by a hollow cylindrical shell with a radius 5cm & length 20cm at an external point lying on the longitudinal axis at a distance 11cm from the centre of front-end.

Sol. given that

$$radius\ of\ hollow\ cylindrical\ shell, R = 5cm, length, L = 20cm\ \&$$
$$distance, x = 11cm$$

We know that solid angle, subtended by a hollow cylindrical shell with a radius 'R' & length 'L' at any external point lying on the longitudinal axis at a distance x from the centre of front-end, is given from the eq(54) as follows

$$\omega = 2\pi\left[\frac{(L+x)}{\sqrt{(L+x)^2+R^2}} - \frac{x}{\sqrt{x^2+R^2}}\right]$$

On setting the values, we get

$$\omega = 2\pi\left[\frac{(20+11)}{\sqrt{(20+11)^2+(5)^2}} - \frac{11}{\sqrt{(11)^2+(5)^2}}\right]$$
$$= 2\pi[0.987 - 0.910] = 0.484\ sr \qquad Ans.$$

Example 4.30: Determine the solid angle subtended by a hollow cylindrical shell with a radius 75cm & length 6m at the centre of

1) Longitudinal axis & 2) one of the ends

Sol. given that

$$radius, R = 75cm = 0.75m\ \&\ length, L = 5.4cm$$

1. Solid angle subtended by the shell at the mid-point of longitudinal axis

$$\omega = 4\pi\left(\frac{L}{\sqrt{L^2+4R^2}}\right)$$

On setting the values, we get

$$\omega = 4\pi\left(\frac{5.4}{\sqrt{(5.4)^2+4(0.75)^2}}\right) = 4\pi(0.963) = 12.101\ sr \qquad Ans.$$

2. Solid angle subtended by the shell at the centre of one of the ends

$$\omega = 2\pi\left(\frac{L}{\sqrt{L^2+R^2}}\right)$$

On setting the values, we get

$$\omega = 2\pi\left(\frac{5.4}{\sqrt{(5.4)^2+(0.75)^2}}\right) = 2\pi(0.99) = 6.22\ sr \qquad Ans.$$

Example 4.31: Determine the length of a hollow cylindrical shell with a radius 10 cm subtending a solid angle 8.512 sr. at the centre of shell.

Sol. given that

$$radius, R = 10\ cm\ \&\ solid\ angle, \omega = 8.512\ sr$$

Let the length of shell be 'L'. Now we have

Solid angle subtended by the shell at the mid-point of longitudinal axis

$$\omega = 4\pi\left(\frac{L}{\sqrt{L^2+4R^2}}\right)$$

On setting the values, we get

$$\omega = 8.512 = 4\pi\left(\frac{L}{\sqrt{L^2+4(0.1)^2}}\right)$$

$$\frac{L}{\sqrt{L^2+0.04}}=\frac{8.512}{4\pi}=0.677$$

$$L^2+0.04=2.182L^2 \Rightarrow 1.182L^2=0.04$$

$$or\ L=0.184m\ or\ 18.4cm \qquad Ans.$$

Example: 4.32: Find out the total flux intercepted by a cylindrical shell with one end closed, another open, a radius 14cm & length 32cm when a uniform point-source emitting 1400 lm flux at the mid-point of the axis.

Sol. given that

$$radius, R=14cm, length, L=32cm\ \&\ total\ flux, F_t=1400\ lm$$

Luminous intensity (I) of the source,

$$\Rightarrow I=\frac{F_t}{4\pi}=\frac{1400}{4\pi}=111.408\ cd$$

Solid angle, subtended by a cylindrical shell with one end closed, another open, radius 'R' & length 'L' at any point lying on the longitudinal axis, is given from the equation (57) as follows

$$\omega=2\pi\left[1+\frac{x}{\sqrt{x^2+R^2}}\right]$$

On setting the values in the above expression we have

$$\omega=2\pi\left[1+\frac{16}{\sqrt{(16)^2+(14)^2}}\right]$$

$$=2\pi[1+0.752]=11.008\ sr$$

Hence the total flux intercepted by the shell

$$\Rightarrow F=I\times\omega=111.408\times 11.008=1226.379\ lm \qquad Ans.$$

Example 4.33: A hollow cylindrical shell with radius 'R' & length 'L' intercepts x fraction of the total radiation energy emitted by a uniform point-source located at the centre of one of the ends of the shell. Prove that the area (A) of lateral surface is given as

$$A=\frac{4\pi x R^2}{\sqrt{1-4x^2}} \quad \forall\left(x<\frac{1}{2}\right).$$

Sol.

Let W be the total radiation energy emitted by the point-source & 'L' be the length of shell

Then the solid angle subtended by the shell at the point-source

$$\omega=2\pi\left(\frac{L}{\sqrt{L^2+R^2}}\right)$$

The total energy intercepted by the shell

$$\Rightarrow E_i=\left(\frac{W}{4\pi}\right)\times\omega=\left(\frac{W}{4\pi}\right)\times 2\pi\left(\frac{L}{\sqrt{L^2+R^2}}\right) \qquad \dots\dots\dots\dots\dots\dots(I)$$

But it is given that

$$E_i=x\times(total\ energy\ emiited\ by\ the\ sourec)=x\times(W) \qquad \dots\dots\dots\dots\dots(II)$$

Equating the eq(I) & (II), we get

$$\left(\frac{W}{4\pi}\right)\times 2\pi\left(\frac{L}{\sqrt{L^2+R^2}}\right)=x\times(W)$$

$$\Rightarrow \frac{L}{\sqrt{L^2+R^2}}=2x\ \ or\ \ \frac{L^2}{L^2+R^2}=4x^2$$

$$\Rightarrow 4x^2L^2+4x^2R^2=L^2\ \ or\ \ L^2(1-4x^2)=4x^2R^2$$

$$\Rightarrow L = \sqrt{\frac{4x^2R^2}{(1-4x^2)}} = \frac{2xR}{\sqrt{1-4x^2}}$$

Hence, the area of lateral surface (A)

$$A = 2\pi RL = 2\pi R \times \frac{2xR}{\sqrt{1-4x^2}} = \frac{4\pi xR^2}{\sqrt{1-4x^2}}$$

Since, the point-source is located at the centre of one of the ends of the shell hence the half of the total energy will escape from one end and a small amount of total energy will escape from another end

Thus, radiation energy intercepted by the shell will be less than half

$$(E_i)_{max} = \left(\frac{W}{4\pi}\right) \times \omega_{max} = \left(\frac{W}{4\pi}\right) \times 2\pi = \frac{W}{2}$$

$$\Rightarrow E_i < (E_i)_{max} \ or \ Wx < \frac{W}{2} \Rightarrow x < \frac{1}{2}$$

Hence, we get

$$A = \frac{4\pi xR^2}{\sqrt{1-4x^2}} \quad \forall \left(x < \frac{1}{2}\right) \qquad \textit{proved.}$$

*** Right cone:**

Example 4.34: Find out the solid angle subtended at the apex-point by a 13cm long right cone with a radius 6cm.

Sol. given that

$$normal\ height, H = 13cm\ \&\ radiu, R = 6cm$$

Hence, the solid angle subtended by the right cone at the apex-point

$$\omega = 2\pi\left[1 - \frac{H}{\sqrt{H^2+R^2}}\right]$$

On setting the values, we get

$$\omega = 2\pi\left[1 - \frac{13}{\sqrt{(13)^2+(6)^2}}\right]$$

$$= 2\pi[1 - 0.908] = 0.578\ sr \qquad \textit{Ans.}$$

Example 4.35: A 25cm long right cone subtends a solid angle 0.612sr at the apex-point. Find out the radius of base of the cone.

Sol. given that

$$normal\ height, H = 25cm\ \&\ solid\ angle, \omega = 0.612\ sr$$

Let the R be the radius of the right cone.

Hence, the solid angle subtended by the right cone at the apex-point

$$\omega = 2\pi\left[1 - \frac{H}{\sqrt{H^2+R^2}}\right]$$

On setting the values, we get

$$\omega = 0.612 = 2\pi\left[1 - \frac{25}{\sqrt{(25)^2+R^2}}\right] = 2\pi\left[1 - \frac{25}{\sqrt{625+R^2}}\right]$$

$$\frac{25}{\sqrt{625+R^2}} = 1 - \frac{0.612}{2\pi} = 0.902$$

$$625 + R^2 = 768.187 \ \Rightarrow R = \sqrt{143.187} = 11.966\ cm$$

$$\therefore \quad radius\ of\ the\ shell, R = 11.966\ cm \qquad Ans.$$

Example 4.36: Find out the solid angle subtended by a 28cm long right cone with a radius 6cm at a point lying in the plane of base at a normal distance 16cm from the axis of cone.

Sol. given that

$$height\ of\ right\ cone, H = 28cm, radius, R = 6cm\ \&$$

$$normal\ distance, x = 16cm$$

We know that the solid angle subtended by a right cone with a radius R & normal height H at any point lying in the plane of base is given from the eq(64) as follows

$$\omega = 2\cos^{-1}\left\{\frac{\left(H^2x^3\sqrt{x^2-R^2} + R^2(x^2-R^2)\sqrt{(x^2-R^2)^2+H^2x^2}\right)}{(x^2-R^2)(R^2(x^2-R^2)+H^2x^2)+H^2R^2x^2}\right\}$$

On setting the values in the above expression, we get

$$\omega = 2\cos^{-1}\left\{\frac{\left(28^2 16^3\sqrt{16^2-6^2} + 6^2(16^2-6^2)\sqrt{(16^2-6^2)^2+28^2 16^2}\right)}{(16^2-6^2)(6^2(16^2-6^2)+28^2 16^2)+28^2 6^2 16^2}\right\}$$

$$= 2\cos^{-1}\left\{\frac{47630742.44 + 3952897.31}{45897280 + 7225344}\right\} = 2\cos^{-1}\left(\frac{51583639.75}{53122624}\right)$$

$$= 2\cos^{-1}(0.971) = 0.483\ sr \qquad Ans.$$

Example 4.37: A uniform point-source of 150 cd is located at a normal distance 1.5m from the axis of the cone, in the plane of the base of an infinitely long right cone with a radius 21cm. Calculate the total flux intercepted by the right cone.

Sol. given that

$$intensity, I = 150\ cd,\ \ distance, x = 1.5m = 150cm\ \&$$

$$radius, R = 21cm$$

Solid angle subtended by an infinitely long right cone with a radius R at any point lying in the plane of base at a distance x from the centre of base, is given by eq(65) as follows

$$\omega = 2\sin^{-1}\left(\frac{R}{x}\right)$$

On setting the values in the expression, we get

$$\omega = 2\sin^{-1}\left(\frac{21}{150}\right) = 2 \times 0.14 = 0.28\ sr$$

Hence, the flux intercepted by the right cone

$$\Rightarrow F = I \times \omega = 150 \times 0.28 = 42\ lm \qquad Ans.$$

Example 4.38: A uniform point-source is located at a normal distance 56cm from the axis of an infinitely long right cone with a radius 11cm, in the plane of the base. Calculate the intensity of point-source if the total flux intercepted by the right cone is 79.2 lm.

Sol. given that

$$radius, R = 11cm\ \ distance, x = 56cm\ \&$$

$$flux\ intercepted\ by\ the\ right\ cone, F = 79.2\ lm$$

Solid angle subtended by an infinitely long right cone with a radius R at any point lying in the plane of base at a distance x from the centre of base, is given by eq(65) as follows

$$\omega = 2\sin^{-1}\left(\frac{R}{x}\right)$$

On setting the values in the expression, we get

$$\omega = 2\sin^{-1}\left(\frac{11}{56}\right) = 2 \times 0.0.198 = 0.396\ sr$$

Let 'I' be the intensity of the point-source then we have

The total flux intercepted by the right cone

$$F = 79.2 = I \times \omega = I \times 0.396$$

$$\Rightarrow Intensity, I = \frac{79.2}{0.396} = 200\ cd \qquad Ans.$$

Example 4.39: Determine the solid angle subtended by a 50cm long hollow conical shell with a radius 10cm, at an internal point lying on the axis at a distance 12cm from the peak-point.

Sol. given that

$$radius\ of\ hollow\ conical\ shell, R = 10cm,\ \ normal\ height, H = 50cm\ \&$$

$$distance\ from\ the\ peak\ point = 12cm$$

$$\therefore\ distance\ from\ the\ centre\ of\ base, x = H - 12 = 50 - 12 = 38cm$$

We know that solid angle subtended by a hollow conical shell with a radius 'R' at any internal point lying on the axis at a distance 'x' from the centre of base, is given from the eq(66) as follows

$$\omega = 2\pi\left[1 + \frac{x}{\sqrt{x^2 + R^2}}\right]$$

On setting the values in the above expression we have

$$\omega = 2\pi\left[1 + \frac{38}{\sqrt{38^2 + 12^2}}\right] = 2\pi[1 + 0.953] = 12.271\ sr \qquad Ans.$$

Example 4.40: Determine the solid angle subtended by a hollow conical shell with a radius 10cm, at an external point lying on the axis at a distance 30cm from the centre of base towards peak.

Sol. given that

$$radius\ of\ hollow\ conical\ shell, R = 10cm\ \&$$

$$distance\ from\ the\ centre\ of\ base, x = 30cm$$

We know that solid angle subtended by a hollow conical shell with a radius 'R' at any external point lying on the axis at a distance 'x' from the centre of base, is given from the eq(67.2) as follows

$$\omega = 2\pi\left[1 - \frac{x}{\sqrt{x^2 + R^2}}\right]\ (\forall\ x \geq H)$$

On setting the values in the above expression we have

$$\omega = 2\pi\left[1 - \frac{30}{\sqrt{30^2 + 10^2}}\right] = 2\pi[1 - 0.949] = 0.32\ sr \qquad Ans.$$

Example 4.41:A uniform point-source of 200 cd is located at the mid-point of axis of a 36cm long hollow conical shell with a radius 13cm. Find out the flux intercepted by the shell.

Sol. given that

$$Intensity, I = 200\ cd, radius\ of\ hollow\ conical\ shell, R = 13cm\ \&$$

$$normal\ height\ of\ shell, H = 36cm$$

$$distance\ from\ the\ centre\ of\ base, x = \frac{36}{2} = 18cm$$

We know that solid angle subtended by a hollow conical shell with a radius 'R' at any internal point lying on the axis at a distance 'x' from the centre of base, is given from the eq(66) as follows

$$\omega = 2\pi\left[1 + \frac{x}{\sqrt{x^2 + R^2}}\right]$$

On setting the values in the above expression we have

$$\omega = 2\pi\left[1 + \frac{18}{\sqrt{18^2 + 13^2}}\right] = 2\pi[1 + 0.811] = 11.379\ sr$$

Hence, the flux intercepted by the shell

$$\Rightarrow F = I \times \omega = 200 \times 11.379 = 2275.8\ lm \qquad \textit{Ans.}$$

Example 4.42:A uniform point-source of 200 W is located insidea 22cm long hollow conical shell with a radius 8cm, on the axis at a distance 5cm from the apex. Find out the total radiation energy intercepted by the shell.

Sol. given that

$$luminosity, L = 200\ W,\ \ radius\ of\ hollow\ conical\ shell, R = 8cm,$$

$$normal\ height, H = 22cm\ \&\ distance\ from\ the\ peak\ point = 5cm$$

$$\therefore\ distance\ from\ the\ centre\ of\ base, x = H - 5 = 22 - 5 = 17cm$$

We know that solid angle subtended by a hollow conical shell with a radius 'R' at any internal point lying on the axis at a distance 'x' from the centre of base, is given from the eq(66) as follows

$$\omega = 2\pi\left[1 + \frac{x}{\sqrt{x^2 + R^2}}\right]$$

On setting the values in the above expression we have

$$\omega = 2\pi\left[1 + \frac{17}{\sqrt{17^2 + 8^2}}\right] = 2\pi[1 + 0.905] = 11.969\ sr$$

Hence, the total radiation energy intercepted by the shell

$$\Rightarrow E_i = \left(\frac{L}{4\pi}\right) \times \omega = \left(\frac{200}{4\pi}\right) \times 11.969 = 190.49\ j/sec \qquad \textit{Ans.}$$

Example 4.43: Find out the total radiation energy intercepted by a hollow conical shell with a radius 14cm if a uniform point-source of 300 W is located outside on the axis at a distance 30cm from the centre of base.

Sol. given that

$$luminosity, L = 300\ W,\ \ radius\ of\ hollow\ conical\ shell, R = 14cm\ \&$$

$$distance\ from\ the\ centre\ of\ base, x = 30cm$$

We know that solid angle subtended by a hollow conical shell with a radius 'R' at any external point lying on the axis at a distance 'x' from the centre of base, is given from the eq(67.1) or (67.2) as follows

$$\omega = 2\pi\left[1 - \frac{x}{\sqrt{x^2 + R^2}}\right]$$

On setting the values in the above expression we have

$$\omega = 2\pi\left[1 - \frac{30}{\sqrt{30^2 + 14^2}}\right] = 2\pi[1 - 0.906] = 0.591\ sr$$

Hence, the total radiation energy intercepted by the shell

$$\Rightarrow E_i = \left(\frac{L}{4\pi}\right) \times \omega = \left(\frac{300}{4\pi}\right) \times 0.591 = 14.109\ j/sec \qquad \textit{Ans.}$$

Example 4.44:A hollow conical shell with a radius 'R' intercepts x fraction of the total radiation emitted by a uniform point-source located at the apex-point inside the shell. Prove that volume (V) of the shell is given as

$$V = \frac{1}{6}\pi R^3 \left(\frac{2x - 1}{\sqrt{x - x^2}}\right) \quad \forall\ x \in \left(\frac{1}{2}, 1\right).$$

Sol. given that

$$radius\ of\ the\ conical\ shell = R\ \&\ fraction\ of\ radiation\ energy = x$$

Let the normal height of the shell be 'H' & 'L' be the luminosity of the point-source then we have

Solid angle subtended by the conical shell at the apex point inside the shell

$$\omega = 2\pi\left[1 + \frac{H}{\sqrt{H^2+R^2}}\right]$$

Now, radiation energy intercepted by the shell

$$E_i = \left(\frac{L}{4\pi}\right)\times\omega = \left(\frac{L}{4\pi}\right)\times 2\pi\left[1 + \frac{H}{\sqrt{H^2+R^2}}\right]$$

But it is given that intercepted energy (E_i)

$$E_i = x \times (total\ energy\ emitted\ by\ the\ source) = xL$$

Equating both the results, we get

$$\left(\frac{L}{4\pi}\right)\times 2\pi\left[1 + \frac{H}{\sqrt{H^2+R^2}}\right] = xL$$

$$\left[1 + \frac{H}{\sqrt{H^2+R^2}}\right] = 2x \quad or \quad \frac{H^2}{H^2+R^2} = (2x-1)^2$$

$$\frac{H^2+R^2}{H^2} = \frac{1}{(2x-1)^2}$$

On subtracting 1 from both the sides, we get

$$\frac{H^2+R^2}{H^2} - 1 = \frac{1}{(2x-1)^2} - 1 \ or \ \frac{H^2+R^2-H^2}{H^2} = \frac{1-(2x-1)^2}{(2x-1)^2}$$

$$\Rightarrow \frac{R^2}{H^2} = \frac{1-4x^2-1+4x}{(2x-1)^2} = \frac{4(x-x^2)}{(2x-1)^2}$$

$$\Rightarrow H^2 = \frac{R^2(2x-1)^2}{4(x-x^2)} \quad or \quad H = \frac{R(2x-1)}{2\sqrt{x-x^2}}$$

Hence the volume (V) of the conical shell

$$\Rightarrow V = \frac{1}{3}\pi R^2 H = \frac{1}{3}\pi R^2\left(\frac{R(2x-1)}{2\sqrt{x-x^2}}\right)$$

$$= \frac{1}{6}\pi R^3\left(\frac{2x-1}{\sqrt{x-x^2}}\right)$$

We know that the solid angle subtended by the conical shell at internal apex

$$\omega = 2\pi\left[1 + \frac{H}{\sqrt{H^2+R^2}}\right]$$

It is obvious that the value of solid angle subtended by the shell varies as follows

$$\Rightarrow \ 2\pi < \omega < 4\pi$$

Thus, we find that energy of interception varies as follows

$$\left(\frac{L}{4\pi}\right)\times 2\pi < E_i < \left(\frac{L}{4\pi}\right)\times 4\pi$$

$$\Rightarrow \left(\frac{L}{4\pi}\right)\times 2\pi < xL < \left(\frac{L}{4\pi}\right)\times 4\pi \qquad (since, E_i = xL)$$

$$\Rightarrow \frac{1}{2} < x < 1 \ or\ x \in \left(\frac{1}{2}, 1\right)$$

Hence, we get

$$V = \frac{1}{6}\pi R^3 \left(\frac{2x-1}{\sqrt{x-x^2}}\right) \quad \forall\ x \in \left(\frac{1}{2}, 1\right) \qquad \textit{proved.}$$

Example 4.45: A hollow conical shell with a radius 'R' intercepts x fraction of the total radiation emitted by a uniform point-source located at the centre of mass. Prove that

$$V = \frac{1}{2}\pi R^3 \left(\frac{2x-1}{\sqrt{x-x^2}}\right) \quad \forall\ x \in \left(\frac{1}{2}, 1\right).$$

Sol. given that

$$radius\ of\ the\ conical\ shell = R\ \&\ fraction\ of\ radiation\ energy = x$$

Let the normal height of the shell be 'H' & 'L' be the luminosity of the point-source then we have

We know that the solid angle subtended by the conical shell with a radius R at any internal point lying on the axis at a distance x from the centre of base

$$\omega = 2\pi\left[1 + \frac{x}{\sqrt{x^2+R^2}}\right]$$

But the distance of centre of mass of a conical shell, with a normal height H, lies at a height $H/3$ from the centre of base

Hence, on setting $x = H/3$ in the above equation, we get value of solid angle

$$\omega = 2\pi\left[1 + \frac{\left(\frac{H}{3}\right)}{\sqrt{\left(\frac{H}{3}\right)^2 + R^2}}\right] = 2\pi\left[1 + \frac{H}{\sqrt{H^2+9R^2}}\right]$$

Now, radiation energy intercepted by the shell

$$E_i = \left(\frac{L}{4\pi}\right) \times \omega = \left(\frac{L}{4\pi}\right) \times 2\pi\left[1 + \frac{H}{\sqrt{H^2+9R^2}}\right]$$

But it is given that intercepted energy (E_i)

$$E_i = x \times (total\ energy\ emitted\ by\ the\ source) = xL$$

Equating both the results, we get

$$\left(\frac{L}{4\pi}\right) \times 2\pi\left[1 + \frac{H}{\sqrt{H^2+9R^2}}\right] = xL$$

$$\left[1 + \frac{H}{\sqrt{H^2+9R^2}}\right] = 2x \quad or \quad \frac{H^2}{H^2+9R^2} = (2x-1)^2$$

$$\frac{H^2+9R^2}{H^2} = \frac{1}{(2x-1)^2}$$

On subtracting 1 from both the sides, we get

$$\frac{H^2+9R^2}{H^2} - 1 = \frac{1}{(2x-1)^2} - 1 \ or \ \frac{H^2+9R^2-H^2}{H^2} = \frac{1-(2x-1)^2}{(2x-1)^2}$$

$$\Rightarrow \frac{9R^2}{H^2} = \frac{1-4x^2-1+4x}{(2x-1)^2} = \frac{4(x-x^2)}{(2x-1)^2}$$

$$\Rightarrow H^2 = \frac{9R^2(2x-1)^2}{4(x-x^2)} \ or \ H = \frac{3R(2x-1)}{2\sqrt{x-x^2}}$$

Hence the volume (V) of the conical shell

$$\Rightarrow V = \frac{1}{3}\pi R^2 H = \frac{1}{3}\pi R^2 \left(\frac{3R(2x-1)}{2\sqrt{x-x^2}}\right)$$

$$= \frac{1}{2}\pi R^3 \left(\frac{2x-1}{\sqrt{x-x^2}}\right)$$

We know that the solid angle subtended by the conical shell at the centre of mass

$$\omega = 2\pi\left[1 + \frac{H}{\sqrt{H^2+9R^2}}\right]$$

It is obvious that the value of solid angle subtended by the shell varies as follows

$$\Rightarrow \ 2\pi < \omega < 4\pi$$

Thus, we find that energy of interception varies as follows

$$\left(\frac{L}{4\pi}\right) \times 2\pi < E_i < \left(\frac{L}{4\pi}\right) \times 4\pi$$

$$\Rightarrow \left(\frac{L}{4\pi}\right) \times 2\pi < xL < \left(\frac{L}{4\pi}\right) \times 4\pi \qquad (since, E_i = xL)$$

$$\Rightarrow \frac{1}{2} < x < 1 \ \ or \ x \in \left(\frac{1}{2}, 1\right)$$

Hence, we get

$$V = \frac{1}{2}\pi R^3 \left(\frac{2x-1}{\sqrt{x-x^2}}\right) \quad \forall\ x \in \left(\frac{1}{2}, 1\right) \qquad proved.$$

*** Right pyramid:**

Example 4.46: Find out the solid angle subtended by a right pyramid having base as a square with each side 9cm & height 28cm, at the apex-point.

Sol. given that

$$number\ of\ the\ sides\ of\ base, n = 4,\ \ side\ of\ base, a = 9cm\ \&$$
$$normal\ height\ of\ right\ pyramid, H = 28cm$$

We know that solid angle subtended by a right pyramid, with base as a regular polygon with 'n' number of the sides each of length 'a' & normal height 'H', at the apex point is given from the eq(70) as follows

$$\omega = 2\pi - 2n\sin^{-1}\left(\frac{2H sin\frac{\pi}{n}}{\sqrt{4H^2 + a^2 cot^2\frac{\pi}{n}}}\right)$$

On setting the values in the above expression, we get

$$\omega = 2\pi - 2\times 4\sin^{-1}\left(\frac{2(28)sin\frac{\pi}{4}}{\sqrt{4(28)^2 + (9)^2 cot^2\frac{\pi}{4}}}\right)$$

$$= 2\pi - 8\sin^{-1}\left(\frac{39.598}{\sqrt{3217}}\right) = 2\pi - 8\sin^{-1}(0.698)$$

$$= 2\pi - 6.181 = 0.102\ sr \qquad Ans.$$

Example 4.47: Determine the solid angle subtended by a right pyramid with base as a regular hexagon & an angle between consecutive edges 30° at the apex-point.

Sol. given that

$$number\ of\ the\ sides\ of\ base, n = 6\ \&\ angle\ between\ cosecutive\ edges, \alpha = 30^o$$

We know that solid angle subtended by a right pyramid, with base as a regular polygon with 'n' number of the sides & angle between the consecutive edges 'α', at the apex point is given from the eq(70) as follows

$$\omega = 2\pi - 2n\sin^{-1}\left(cos\frac{\pi}{n}\sqrt{tan^2\frac{\pi}{n} - tan^2\frac{\alpha}{2}}\right)$$

On setting the values, we get

$$\omega = 2\pi - 2\times 6\sin^{-1}\left(cos\frac{\pi}{6}\sqrt{tan^2\frac{\pi}{6} - tan^2\frac{30^o}{2}}\right)$$

$$= 2\pi - 12\sin^{-1}\left(0.866\sqrt{0.333 - 0.072}\right)$$

$$= 2\pi - 12\sin^{-1}(0.442) = 2\pi - 5.494 = 0.789\ sr \qquad Ans.$$

Example 4.48: Calculate the solid angle subtended by a right pyramid with base as a regular decagon & an angle between consecutive edges 20° at the apex-point.

Sol. given that

$$number\ of\ the\ sides\ of\ base, n = 10\ \&\ angle\ between\ cosecutive\ edges, \alpha = 20^o$$

We know that solid angle subtended by a right pyramid, with base as a regular polygon with 'n' number of the sides & angle between the consecutive edges 'α', at the apex point is given from the eq(70) as follows

$$\omega = 2\pi - 2n\sin^{-1}\left(cos\frac{\pi}{n}\sqrt{tan^2\frac{\pi}{n} - tan^2\frac{\alpha}{2}}\right)$$

On setting the values, we get

$$\omega = 2\pi - 2\times 10\sin^{-1}\left(cos\frac{\pi}{10}\sqrt{tan^2\frac{\pi}{10} - tan^2\frac{20^o}{2}}\right)$$

$$= 2\pi - 20\sin^{-1}\left(0.951\sqrt{0.105 - 0.031}\right)$$

$$= 2\pi - 20\sin^{-1}(0.259) = 2\pi - 5.24 = 1.043\ sr \qquad Ans.$$

Example 4.49:A uniform point-source emitting the total flux 1600 lm is located at the apex-point of a right pyramid with base as a regular pentagon with each side 6cm & height 24cm. Find out the total flux intercepted by the pyramid.

Sol. given that

$$toatl\ flux, F_t = 1600\ lm,\ \ number\ of\ the\ sides\ of\ base, n = 5,$$

$$side\ of\ base, a = 6cm\ \ \&\ normal\ height\ of\ right\ pyramid, H = 24cm$$

Intensity of the point-source,

$$I = \frac{F_t}{4\pi} = \frac{1600}{4\pi} = 127.324\ cd$$

We know that solid angle subtended by a right pyramid, with base as a regular polygon with 'n' number of the sides each of length 'a' & normal height 'H', at the apex point is given from the eq(70) as follows

$$\omega = 2\pi - 2n\sin^{-1}\left(\frac{2Hsin\frac{\pi}{n}}{\sqrt{4H^2 + a^2cot^2\frac{\pi}{n}}}\right)$$

On setting the values in the above expression, we get

$$\omega = 2\pi - 2\times 5\sin^{-1}\left(\frac{2(24)sin\frac{\pi}{5}}{\sqrt{4(24)^2 + (6)^2cot^2\frac{\pi}{5}}}\right)$$

$$= 2\pi - 10\sin^{-1}\left(\frac{28.214}{\sqrt{2372.199}}\right) = 2\pi - 10\sin^{-1}(0.579)$$

$$= 2\pi - 6.175 = 0.108\ sr$$

Hence, the total flux intercepted by the right pyramid

$$\Rightarrow F = I \times \omega = 127.324 \times 0.108 = 13.751\ lm \qquad Ans.$$

Example 4.50: Calculate the solid angle subtended by a hollow right pyramidal shell with base as a regular decagon, angle between the consecutive lateral edges 18° & height 20cm at an internal point lying on the axis at a distance 15cm from the centre of base.

Sol. given that

$$number\ of\ the\ sides, n = 10, angle\ between\ the\ consecutive\ edges, \alpha = 18^o$$

$$normal\ height\ of\ right\ pyramid, H = 20cm\ \&\ distance, x = 15cm$$

Let 'a' be the length of each side of the base then we have

$$H = \frac{a}{2}\sqrt{cot^2\frac{\alpha}{2} - cot^2\frac{\pi}{n}}$$

$$\Rightarrow a = \frac{2H}{\sqrt{cot^2\frac{\alpha}{2} - cot^2\frac{\pi}{n}}} = \frac{2\times 20}{\sqrt{cot^2\frac{18^o}{2} - cot^2\frac{\pi}{10}}} = \frac{40}{5.513} = 7.255\ cm$$

We know that solid angle subtended by a right pyramid, with base as a regular polygon with 'n' number of the sides each of length 'a' at any internal point lying on the axis at a distance 'x' from the centre of base is given from the eq(72) as follows

$$\omega = 2\pi + 2n\sin^{-1}\left(\frac{2x sin\frac{\pi}{n}}{\sqrt{4x^2 + a^2 cot^2\frac{\pi}{n}}}\right)$$

On setting the values, we get

$$\omega = 2\pi + 2\times 10\sin^{-1}\left(\frac{2\times 15 sin\frac{\pi}{10}}{\sqrt{4(15)^2 + (7.255)^2 cot^2\frac{\pi}{10}}}\right)$$

$$= 2\pi + 20\sin^{-1}\left(\frac{9.27}{\sqrt{1398.566}}\right) = 2\pi + 20\sin^{-1}(0.248)$$

$$= 2\pi + 5.012 = 11.295\ sr \qquad Ans.$$

Example 4.51: A uniform point-source of 100 W is located on the axis inside a hollow right pyramidal shell at a distance 5cm from the centre of regular heptagonal base with each side 10cm. Find out the total radiation energy intercepted by the shell.

Sol. given that

$$number\ of\ the\ sides, n = 7, \qquad length\ of\ each\ side\ of\ base, a = 10cm,$$

$$luminosity, L = 100\ W\ \&\ distance\ from\ the\ centre\ of\ base, x = 5cm$$

We know that solid angle subtended by a right pyramid, with base as a regular polygon with 'n' number of the sides each of length 'a' at any internal point lying on the axis at a distance 'x' from the centre of base is given from the eq(72) as follows

$$\omega = 2\pi + 2n\sin^{-1}\left(\frac{2x\sin\frac{\pi}{n}}{\sqrt{4x^2 + a^2\cot^2\frac{\pi}{n}}}\right)$$

On setting the values, we get

$$\omega = 2\pi + 2\times 7\sin^{-1}\left(\frac{2\times 5\sin\frac{\pi}{7}}{\sqrt{4(5)^2 + (10)^2\cot^2\frac{\pi}{7}}}\right)$$

$$= 2\pi + 14\sin^{-1}\left(\frac{4.339}{\sqrt{531.194}}\right) = 2\pi + 14\sin^{-1}(0.188)$$

$$= 2\pi + 2.648 = 8.931\ sr$$

Hence, the total radiation energy intercepted by the right pyramidal shell

$$E_i = \left(\frac{L}{4\pi}\right)\times\omega = \left(\frac{100}{4\pi}\right)\times 8.931 = 71.071\ j/\sec \qquad \textit{Ans.}$$

Example 4.52: Find out the total radiation energy intercepted by a hollow right pyramidal shell with base as a regular polygon with 12 sides, angle between the consecutive lateral edges 22.5^o & height 20cm when a uniform point-source of 300 W is located at the mid-point of the axis.

Sol. given that

$$number\ of\ the\ sides, n = 12, angle\ between\ the\ consecutive\ edges, \alpha = 22.5^o$$

$$normal\ height\ of\ right\ pyramid, H = 20cm, distance, x = \frac{20}{2} = 10cm\ \&$$

$$luminosity, L = 300\ W$$

Let 'a' be the length of each side of the base then we have

$$H = \frac{a}{2}\sqrt{\cot^2\frac{\alpha}{2} - \cot^2\frac{\pi}{n}}$$

$$\Rightarrow a = \frac{2H}{\sqrt{\cot^2\frac{\alpha}{2} - \cot^2\frac{\pi}{n}}} = \frac{2\times 20}{\sqrt{\cot^2\frac{22.5^o}{2} - \cot^2\frac{\pi}{12}}} = \frac{40}{3.368} = 11.876\ cm$$

We know that solid angle subtended by a right pyramid, with base as a regular polygon with 'n' number of the sides each of length 'a' at any internal point lying on the axis at a distance 'x' from the centre of base is given from the eq(72) as follows

$$\omega = 2\pi + 2n\sin^{-1}\left(\frac{2x\sin\frac{\pi}{n}}{\sqrt{4x^2 + a^2\cot^2\frac{\pi}{n}}}\right)$$

On setting the values, we get

$$\omega = 2\pi + 2\times 12\sin^{-1}\left(\frac{2\times 10\sin\frac{\pi}{12}}{\sqrt{4(10)^2 + (11.876)^2\cot^2\frac{\pi}{12}}}\right)$$

$$= 2\pi + 24\sin^{-1}\left(\frac{5.176}{\sqrt{2364.425}}\right) = 2\pi + 20\sin^{-1}(0.106)$$

$$= 2\pi + 2.124 = 8.407\ sr$$

Hence, the total radiation energy intercepted by the right pyramidal shell

$$E_i = \left(\frac{L}{4\pi}\right) \times \omega = \left(\frac{300}{4\pi}\right) \times 8.407 = 200.702\ j/\sec \qquad Ans.$$

Example 4.53: Find out the total flux intercepted by a hollow right pyramidal shell with base as a regular polygon with 15 sides, angle between the consecutive lateral edges 18°& height 24cm when a uniform point-source of 200 cd is located outside the shell on the axis at a distance 8cm from the centre of base. If the flux is uniformly distributed over the intercepted surface then find out the flux density.

Sol. given that

$$number\ of\ the\ sides, n = 15, angle\ between\ the\ consecutive\ edges, \alpha = 18^o$$
$$normal\ height\ of\ right\ pyramid, H = 24cm, distance, x = 8cm\ \&$$
$$intensity, I = 200\ cd$$

Let 'a' be the length of each side of the base then we have

$$H = \frac{a}{2}\sqrt{cot^2\frac{\alpha}{2} - cot^2\frac{\pi}{n}}$$

$$\Rightarrow a = \frac{2H}{\sqrt{cot^2\frac{\alpha}{2} - cot^2\frac{\pi}{n}}} = \frac{2 \times 24}{\sqrt{cot^2\frac{18^o}{2} - cot^2\frac{\pi}{15}}} = \frac{48}{4.211} = 11.399\ cm$$

We know that solid angle subtended by a right pyramid, with base as a regular polygon with 'n' number of the sides each of length 'a' at any external point lying on the axis at a distance 'x' from the centre of base is given from the eq(72) as follows

$$\omega = 2\pi - 2n\sin^{-1}\left(\frac{2xsin\frac{\pi}{n}}{\sqrt{4x^2 + a^2cot^2\frac{\pi}{n}}}\right)$$

On setting the values, we get

$$\omega = 2\pi - 2 \times 15\sin^{-1}\left(\frac{2 \times 8sin\frac{\pi}{15}}{\sqrt{4(8)^2 + (11.399)^2cot^2\frac{\pi}{15}}}\right)$$

$$= 2\pi - 30\sin^{-1}\left(\frac{3.326}{\sqrt{3131.971}}\right) = 2\pi - 30\sin^{-1}(0.059)$$

$$= 2\pi - 1.771 = 4.512\ sr$$

Hence, the total flux intercepted by the right pyramid

$$\Rightarrow F = I \times \omega = 200 \times 4.512 = 902.4\ lm \qquad Ans.$$

Let's find area of one of the lateral triangular faces

Normal height (h) from the apex to the mid-point of one of the sides of the base

$$\Rightarrow tan\frac{\alpha}{2} = \frac{a}{2h}\ or\ h = \frac{a}{2}cot\frac{\alpha}{2}$$

On setting the values in the above expression, we get

$$h = \frac{11.399}{2}cot\frac{18^o}{2} = 35.985\ cm$$

Total area of lateral surface of the right pyramid

$$\Rightarrow A = 15 \times (area\ of\ one\ lateral\ face) = 15 \times \left(\frac{1}{2} \times base\ (a) \times height(h)\right)$$

$$= 15 \times \left(\frac{1}{2} \times 11.399 \times 35.985\right) = 3076.448\ cm^2$$

$$\therefore flux\ density\ over\ the\ lateral\ surface,$$

$$\Rightarrow F.D. = \frac{F}{A} = \frac{902.4}{3076.448 \times 10^{-4}} = 2933.253\ lux\ or\ 2.933\ klux \qquad Ans.$$

Note: Area of lateral surface of a right pyramid, with base as a regular polygon with 'n' number of the sides each of length 'a', angle between consecutive edges α& normal height 'H', is given as

$$A_L = \frac{1}{4}na\sqrt{4H^2 + a^2cot^2\frac{\pi}{n}} = \frac{1}{4}na^2cot\frac{\alpha}{2}$$

Where,

$$H = \frac{a}{2}\sqrt{cot^2\frac{\alpha}{2} - cot^2\frac{\pi}{n}}$$

Example 4.54:A hollow right pyramidal shell, with base as a regular polygon with 'n' no. of the sides each of length 'a', intercepts x fraction of the total radiation emitted by a uniform point-source located at the apex-point inside the shell. Prove that area of lateral surface (A_L) of the shell is given as

$$A_L = \frac{na^2cos\frac{\pi}{n}}{4\sqrt{sin^2\frac{\pi}{n} - sin^2\left\{\frac{\pi}{n}(2x-1)\right\}}} \qquad \forall\ x \in \left(\frac{1}{2}, 1\right)$$

Sol. given that

$$number\ of\ the\ sides = n, length\ of\ each\ side = a\ \&$$

$$frction\ of\ the\ radiation\ energy\ intercepted\ by\ the\ shell = x$$

Let 'H' be the normal height of the pyramidal shell& luminosity of the point-source be 'L' then we have

Solid angle subtended by the right pyramidal shell at the apex point

$$\omega = 2\pi + 2n\sin^{-1}\left(\frac{2Hsin\frac{\pi}{n}}{\sqrt{4H^2 + a^2cot^2\frac{\pi}{n}}}\right)$$

Hence, the radiation intercepted by the shell is given as follows

$$E_i = \left(\frac{L}{4\pi}\right) \times \omega$$

$$= \left(\frac{L}{4\pi}\right) \times \left\{2\pi + 2n\sin^{-1}\left(\frac{2Hsin\frac{\pi}{n}}{\sqrt{4H^2 + a^2cot^2\frac{\pi}{n}}}\right)\right\}$$

But it is given that the fraction of total radiation energy (emitted by the point-source (L)) intercepted by the shell is 'x' hence we have

$$E_i = x \times (total\ radiation\ energy\ (L)) = xL$$

Equating both the results, we get

$$\left(\frac{L}{4\pi}\right) \times \left\{2\pi + 2n\sin^{-1}\left(\frac{2Hsin\frac{\pi}{n}}{\sqrt{4H^2 + a^2cot^2\frac{\pi}{n}}}\right)\right\} = xL$$

$$\Rightarrow 2n\sin^{-1}\left(\frac{2Hsin\frac{\pi}{n}}{\sqrt{4H^2 + a^2cot^2\frac{\pi}{n}}}\right) = 4\pi x - 2\pi$$

$$\Rightarrow \frac{2Hsin\frac{\pi}{n}}{\sqrt{4H^2+a^2cot^2\frac{\pi}{n}}} = sin\left\{\frac{2\pi(2x-1)}{2n}\right\} = sin\left\{\frac{\pi}{n}(2x-1)\right\}$$

On squaring both sides, we get

$$\frac{4H^2sin^2\frac{\pi}{n}}{4H^2+a^2cot^2\frac{\pi}{n}} = sin^2\left\{\frac{\pi}{n}(2x-1)\right\}$$

$$\Rightarrow 4H^2sin^2\frac{\pi}{n} = 4H^2sin^2\left\{\frac{\pi}{n}(2x-1)\right\} + a^2cot^2\frac{\pi}{n}sin^2\left\{\frac{\pi}{n}(2x-1)\right\}$$

$$\Rightarrow 4H^2\left(sin^2\frac{\pi}{n} - sin^2\left\{\frac{\pi}{n}(2x-1)\right\}\right) = a^2cot^2\frac{\pi}{n}sin^2\left\{\frac{\pi}{n}(2x-1)\right\}$$

$$\Rightarrow \frac{4H^2}{a^2} = \frac{cot^2\frac{\pi}{n}sin^2\left\{\frac{\pi}{n}(2x-1)\right\}}{sin^2\frac{\pi}{n} - sin^2\left\{\frac{\pi}{n}(2x-1)\right\}} \quad \ldots\ldots\ldots\ldots\ldots\ldots (I)$$

Let α be the angle between the consecutive edges.

Normal height (h) from the apex to the mid-point of one of the sides of the base

$$\Rightarrow tan\frac{\alpha}{2} = \frac{a}{2h} \quad or\ h = \frac{a}{2}cot\frac{\alpha}{2}$$

The area of one of the lateral face (an isosceles triangle)

$$= \frac{1}{2} \times base \times height = \frac{1}{2} \times a \times \frac{a}{2}cot\frac{\alpha}{2} = \frac{1}{4}a^2cot\frac{\alpha}{2}$$

Since, there are 'n' number of the lateral faces which are congruent to each other hence area of lateral surface (A_L) of the pyramidal shell

$$\Rightarrow A_L = n \times \frac{1}{4}a^2cot\frac{\alpha}{2} = \frac{1}{4}na^2cot\frac{\alpha}{2} \quad \ldots\ldots\ldots\ldots\ldots\ldots (II)$$

But we know the relation between α & H as follows

$$H = \frac{a}{2}\sqrt{cot^2\frac{\alpha}{2} - cot^2\frac{\pi}{n}}$$

$$\Rightarrow cot^2\frac{\alpha}{2} = \frac{4H^2}{a^2} + cot^2\frac{\pi}{n} = \frac{4H^2+a^2cot^2\frac{\pi}{n}}{a^2}$$

$$or\ cot\frac{\alpha}{2} = \sqrt{\frac{4H^2+a^2cot^2\frac{\pi}{n}}{a^2}} = \sqrt{\frac{4H^2}{a^2} + cot^2\frac{\pi}{n}}$$

On setting this value in the above eq(II), we get area of lateral surface

$$A_L = \frac{1}{4}na^2\sqrt{\frac{4H^2}{a^2} + cot^2\frac{\pi}{n}}$$

On setting the value of $\frac{4H^2}{a^2}$ from the eq(I), we get

$$\Rightarrow A_L = \frac{1}{4}na^2\sqrt{\frac{cot^2\frac{\pi}{n}sin^2\left\{\frac{\pi}{n}(2x-1)\right\}}{sin^2\frac{\pi}{n} - sin^2\left\{\frac{\pi}{n}(2x-1)\right\}} + cot^2\frac{\pi}{n}}$$

$$=\frac{1}{4}na^2\sqrt{\frac{cot^2\frac{\pi}{n}sin^2\left\{\frac{\pi}{n}(2x-1)\right\}+cot^2\frac{\pi}{n}sin^2\frac{\pi}{n}-cot^2\frac{\pi}{n}sin^2\left\{\frac{\pi}{n}(2x-1)\right\}}{sin^2\frac{\pi}{n}-sin^2\left\{\frac{\pi}{n}(2x-1)\right\}}}$$

$$=\frac{1}{4}na^2\sqrt{\frac{cot^2\frac{\pi}{n}sin^2\frac{\pi}{n}}{sin^2\frac{\pi}{n}-sin^2\left\{\frac{\pi}{n}(2x-1)\right\}}}=\frac{1}{4}na^2\frac{cos\frac{\pi}{n}}{\sqrt{sin^2\frac{\pi}{n}-sin^2\left\{\frac{\pi}{n}(2x-1)\right\}}}$$

$$\Rightarrow A_L=\frac{na^2cos\frac{\pi}{n}}{4\sqrt{sin^2\frac{\pi}{n}-sin^2\left\{\frac{\pi}{n}(2x-1)\right\}}}$$

We know that the solid angle subtended by the conical shell at the internal apex-point

$$\omega=2\pi+2n\sin^{-1}\left(\frac{2Hsin\frac{\pi}{n}}{\sqrt{4H^2+a^2cot^2\frac{\pi}{n}}}\right)$$

It is obvious that the value of solid angle subtended by the shell varies as follows

$$\Rightarrow\ 2\pi<\omega<4\pi$$

Thus, we find that energy of interception varies as follows

$$\left(\frac{L}{4\pi}\right)\times 2\pi<E_i<\left(\frac{L}{4\pi}\right)\times 4\pi$$

$$\Rightarrow\left(\frac{L}{4\pi}\right)\times 2\pi<xL<\left(\frac{L}{4\pi}\right)\times 4\pi\qquad(since, E_i=xL)$$

$$\Rightarrow\frac{1}{2}<x<1\ or\ x\in\left(\frac{1}{2},1\right)$$

Hence, we get

$$A_L=\frac{na^2cos\frac{\pi}{n}}{4\sqrt{sin^2\frac{\pi}{n}-sin^2\left\{\frac{\pi}{n}(2x-1)\right\}}}\qquad\forall\ x\in\left(\frac{1}{2},1\right)\qquad proved.$$

Example 4.55: Find out the solid angle subtended at the apex by an infinitely long right pyramid having 18 lateral edges such that the angle between consecutive edges is 10°.

Sol. given that

$$number\ of\ the\ lateral\ edges, n=18\ \&$$

$$angle\ between\ consecutive\ lateral\ edges, \alpha=10^o$$

We know that solid angle subtended by an infinitely long pyramid at the apex is given from the eq(70.1) as follows

$$\omega=2\pi-2n\sin^{-1}\left(cos\frac{\pi}{n}\sqrt{tan^2\frac{\pi}{n}-tan^2\frac{\alpha}{2}}\right)$$

On setting the values in the above expression, we get

$$=2\pi-2\times 18\sin^{-1}\left(cos\frac{\pi}{18}\sqrt{tan^2\frac{\pi}{18}-tan^2\frac{10^o}{2}}\right)$$

$$=2\pi-36\sin^{-1}(0.151)$$

$$=2\pi-5.457=0.826\ sr\qquad Ans.$$

Example 4.56: Six planes are intersecting each other at a single point in the space such that the angle between any two consecutive lines of intersection is 30°. Find out the solid angle subtended by the planes at the point of intersection.

Sol. given that

$$number\ of\ the\ planes, n = 6\ \&$$

$$angle\ between\ consecutive\ lines\ of\ intersection, \alpha = 30^o$$

We know that solid angle, subtended at the point of intersection by 'n' number of the planes intersecting each other at a single point in the space such that angle between consecutive lines of intersection is α, is given from the eq(70.1) as follows

$$\omega = 2\pi - 2n\sin^{-1}\left(cos\frac{\pi}{n}\sqrt{tan^2\frac{\pi}{n} - tan^2\frac{\alpha}{2}}\right)$$

On setting the values in the above expression, we get

$$= 2\pi - 2\times 6\sin^{-1}\left(cos\frac{\pi}{6}\sqrt{tan^2\frac{\pi}{6} - tan^2\frac{30^o}{2}}\right)$$

$$= 2\pi - 12\sin^{-1}(0.443)$$

$$= 2\pi - 5.507 = 0.776\ sr \qquad Ans.$$

*** Right prism**

Example 4.57: Find out the solid angle subtended by a solid right prism with cross-section as an equilateral triangle with each side 12cm at a point lying on the longitudinal axis at a distance 8cm from the centre of front-end.

Sol. given that

$$number\ of\ the\ sides, n = 3, length\ of\ each\ side, a = 12cm\ \&$$

$$distance\ of\ the\ point\ from\ the\ centre\ of\ fron\ end, x = 8cm$$

We know that the solid angle, subtended by a solid right prism with cross section as a regular polygon with 'n' number of the sides each of length 'a' at any point lying on the longitudinal axis at a distance 'x' from the centre of the front end, is given by the eq(75) as follows

$$\omega = 2\pi - 2n\sin^{-1}\left(\frac{2xsin\frac{\pi}{n}}{\sqrt{4x^2 + a^2cot^2\frac{\pi}{n}}}\right)$$

On setting the values, we get

$$\omega = 2\pi - 2\times 3\sin^{-1}\left(\frac{2\times 8sin\frac{\pi}{3}}{\sqrt{4(8)^2 + (12)^2cot^2\frac{\pi}{3}}}\right)$$

$$= 2\pi - 6\sin^{-1}\left(\frac{13.856}{\sqrt{305}}\right)$$

$$= 2\pi - 6\sin^{-1}(0.793) = 2\pi - 5.494 = 0.789\ sr \qquad Ans.$$

Example 4.58: Find out the solid angle subtended by a solid right prism with cross-section as a square with each side 17cm at a point lying on the longitudinal axis at a distance 9cm from the centre of front-end.

Sol. given that

$$number\ of\ the\ sides, n = 4, length\ ofeach\ side, a = 17cm\ \&$$

$$distance\ of\ the\ point\ from\ the\ centre\ of\ fron\ end, x = 9cm$$

We know that the solid angle, subtended by a solid right prism with cross section as a regular polygon with 'n' number of the sides each of length 'a' at any point lying on the longitudinal axis at a distance 'x' from the centre of the front end, is given by the eq(75) as follows

$$\omega = 2\pi - 2n\sin^{-1}\left(\frac{2xsin\frac{\pi}{n}}{\sqrt{4x^2 + a^2cot^2\frac{\pi}{n}}}\right)$$

On setting the values, we get

$$\omega = 2\pi - 2\times 4\sin^{-1}\left(\frac{2\times 9sin\frac{\pi}{4}}{\sqrt{4(9)^2 + (17)^2cot^2\frac{\pi}{4}}}\right)$$

$$= 2\pi - 8\sin^{-1}\left(\frac{12.728}{\sqrt{613}}\right)$$

$$= 2\pi - 8\sin^{-1}(0.514) = 2\pi - 4.319 = 1.964\ sr \qquad Ans.$$

Example 4.59: The solid angle subtended by a solid right prism with cross-section as a regular hexagon with each side 20cm, at a point lying on the longitudinal axis, is 1.56sr. Find out the distance from the centre of front-end.

Sol. given that

$$number\ of\ the\ sides, n = 6,\ \ length\ of\ each\ side, a = 21cm\ \&$$

$$solid\ angle, \omega = 1.56\ sr$$

$$let\ the\ distance\ of\ the\ point\ from\ the\ centre\ of\ fron\ end\ be\ x$$

We know that the solid angle, subtended by a solid right prism with cross section as a regular polygon with 'n' number of the sides each of length 'a' at any point lying on the longitudinal axis at a distance 'x' from the centre of the front end, is given by the eq(75) as follows

$$\omega = 2\pi - 2n\sin^{-1}\left(\frac{2xsin\frac{\pi}{n}}{\sqrt{4x^2 + a^2cot^2\frac{\pi}{n}}}\right)$$

On setting the values, we get

$$\omega = 1.56 = 2\pi - 2\times 6\sin^{-1}\left(\frac{2xsin\frac{\pi}{6}}{\sqrt{4x^2 + (21)^2cot^2\frac{\pi}{6}}}\right)$$

$$\Rightarrow \sin^{-1}\left(\frac{x}{\sqrt{4x^2 + 1323}}\right) = \frac{2\pi - 1.56}{12} = 0.393$$

$$\frac{x^2}{4x^2 + 1323} = sin^2(0.393) = 0.147$$

$$\Rightarrow x^2 = 0.588x^2 + 194.481 \quad or\ x = \sqrt{\frac{194.481}{(1 - 0.588)}} = 21.726\ cm$$

$$hence,\ \ the\ distance,\ \ x = 21.726\ cm \qquad Ans.$$

Example 4.60: Find out the total radiation energy intercepted by a solid right prism with cross-section as a regular polygon with 36 sides each 12cm when a uniform point source emitting total radiation 500 W is located on the longitudinal axis at a distance 4cm from the centre of front-end.

Sol. given that

$$number\ of\ the\ sides, n = 36, length\ of each\ side, a = 12cm$$
$$distance\ of\ point\ source\ from\ the\ centre\ of\ fron\ end, x = 4cm\ \&$$
$$luminosity\ of\ the\ point\ source, L = 500\ W$$

We know that the solid angle, subtended by a solid right prism with cross section as a regular polygon with 'n' number of the sides each of length 'a' at any point lying on the longitudinal axis at a distance 'x' from the centre of the front end, is given by the eq(75) as follows

$$\omega = 2\pi - 2n\sin^{-1}\left(\frac{2xsin\frac{\pi}{n}}{\sqrt{4x^2 + a^2cot^2\frac{\pi}{n}}}\right)$$

On setting the values, we get

$$\omega = 2\pi - 2\times 36\sin^{-1}\left(\frac{2\times 4sin\frac{\pi}{36}}{\sqrt{4(4)^2 + (12)^2cot^2\frac{\pi}{36}}}\right)$$
$$= 2\pi - 8\sin^{-1}\left(\frac{0.697}{\sqrt{18877.038}}\right)$$
$$= 2\pi - 8\sin^{-1}(5.073\times 10^{-3}) = 2\pi - 0.0406 = 6.242\ sr$$

Hence, the radiation energy intercepted by the right prism

$$E_i = \left(\frac{L}{4\pi}\right)\times\omega$$
$$= \left(\frac{500}{4\pi}\right)\times 6.242 = 248.361\ j/sec \qquad Ans.$$

Example 4.61: A solid right prism with cross-section as a regular polygon with 'n' no. of the sides each of length 'a', intercepts x fraction of the total radiation emitted by a uniform point-source located on the longitudinal axis at a distance 'd' from the centre of front-end. Prove that

$$\frac{d}{a} = \frac{cot\frac{\pi}{n}sin\left\{\frac{\pi}{n}(1-2x)\right\}}{2\sqrt{sin^2\frac{\pi}{n} - sin^2\left\{\frac{\pi}{n}(1-2x)\right\}}}\quad \forall\ \left(x < \frac{1}{2}\right).$$

Sol. given that

$$number\ of\ the\ sides = n,\ \ length\ of each\ side = a,$$
$$fraction\ of\ radiation\ intercepted\ by\ the\ prism = x\ \&$$
$$distance\ of\ point\ source\ from\ the\ centre\ of\ fron\ end = d$$
$$let\ luminosity\ of\ the\ point\ source\ be\ 'L'$$

We know that the solid angle, subtended by a solid right prism with cross section as a regular polygon with 'n' number of the sides each of length 'a' at any point lying on the longitudinal axis at a distance 'd' from the centre of the front end, is given by the eq(75) as follows

$$\omega = 2\pi - 2n\sin^{-1}\left(\frac{2dsin\frac{\pi}{n}}{\sqrt{4d^2 + a^2cot^2\frac{\pi}{n}}}\right)$$

Now, the radiation energy intercepted by the right prism

$$E_i = \left(\frac{L}{4\pi}\right)\times\omega$$

$$= \left(\frac{L}{4\pi}\right) \times \left(2\pi - 2n\sin^{-1}\left(\frac{2d sin\frac{\pi}{n}}{\sqrt{4d^2 + a^2 cot^2\frac{\pi}{n}}}\right)\right)$$

But, it is given that the fraction of radiation energy intercepted by the prism is x hence, we get

$$E_i = x \times (toatl\ radiation\ energy\ emitted\ by\ the\ source\ (L)) = xL$$

Now, equating both the results above, we get

$$\left(\frac{L}{4\pi}\right) \times \left(2\pi - 2n\sin^{-1}\left(\frac{2d sin\frac{\pi}{n}}{\sqrt{4d^2 + a^2 cot^2\frac{\pi}{n}}}\right)\right) = xL$$

$$\Rightarrow \sin^{-1}\left(\frac{2d sin\frac{\pi}{n}}{\sqrt{4d^2 + a^2 cot^2\frac{\pi}{n}}}\right) = \frac{2\pi - 4\pi x}{2n} = \frac{\pi}{n}(1-2x)$$

$$or\quad \frac{2d sin\frac{\pi}{n}}{\sqrt{4d^2 + a^2 cot^2\frac{\pi}{n}}} = sin\left\{\frac{\pi}{n}(1-2x)\right\}$$

On squaring both sides, we get

$$\frac{4d^2 sin^2\frac{\pi}{n}}{4d^2 + a^2 cot^2\frac{\pi}{n}} = sin^2\left\{\frac{\pi}{n}(1-2x)\right\}$$

$$\Rightarrow 4d^2 sin^2\frac{\pi}{n} = 4d^2 sin^2\left\{\frac{\pi}{n}(1-2x)\right\} + a^2 cot^2\frac{\pi}{n} sin^2\left\{\frac{\pi}{n}(1-2x)\right\}$$

$$\Rightarrow 4d^2\left(sin^2\frac{\pi}{n} - sin^2\left\{\frac{\pi}{n}(1-2x)\right\}\right) = a^2 cot^2\frac{\pi}{n} sin^2\left\{\frac{\pi}{n}(1-2x)\right\}$$

$$\Rightarrow \frac{d^2}{a^2} = \frac{cot^2\frac{\pi}{n} sin^2\left\{\frac{\pi}{n}(1-2x)\right\}}{4\left(sin^2\frac{\pi}{n} - sin^2\left\{\frac{\pi}{n}(1-2x)\right\}\right)}$$

$$or\quad \frac{d}{a} = \frac{cot\frac{\pi}{n} sin\left\{\frac{\pi}{n}(1-2x)\right\}}{2\sqrt{sin^2\frac{\pi}{n} - sin^2\left\{\frac{\pi}{n}(1-2x)\right\}}}$$

It is obvious that the solid angle subtended by the right prism varies as follows

$$\omega = 2\pi - 2n\sin^{-1}\left(\frac{2d sin\frac{\pi}{n}}{\sqrt{4d^2 + a^2 cot^2\frac{\pi}{n}}}\right) \Rightarrow 0 < \omega < 2\pi$$

Hence, the radiation energy (E_i) intercepted by the right prism varies as follows

$$0 < E_i < \left(\frac{L}{4\pi}\right) \times 2\pi \Rightarrow 0 < xL < \left(\frac{L}{4\pi}\right) \times 2\pi$$

$$\Rightarrow 0 < x < \frac{1}{2} \quad or\ x < \frac{1}{2} (since, fraction\ (x)\ is\ always\ positive\ )$$

Hence, we get

$$\frac{d}{a}=\frac{cot\frac{\pi}{n}sin\left\{\frac{\pi}{n}(1-2x)\right\}}{2\sqrt{sin^2\frac{\pi}{n}-sin^2\left\{\frac{\pi}{n}(1-2x)\right\}}}\quad\forall\ \left(x<\frac{1}{2}\right)\qquad proved.$$

Example 4.62: Find out the total radiation energy intercepted by a hollow right prismatic shell with cross-section as a square each 8cm & length 40cm when a uniform point source emitting total radiation 300 W is located on the longitudinal axis at a distance 16cm from the centre of one of the ends inside the shell.

Sol. given that

$$number\ of\ the\ sides, n=4, length\ of each\ side, a=8cm$$

$$distance\ of\ point\ source\ from\ the\ centre\ of\ one\ of\ the\ ends, x=16cm$$

$$length\ of\ the\ shell, L=40cm\ \&\ luminosity\ of\ point\ source, L'=300\ W$$

We know that the solid angle, subtended by a hollow right prismatic shell with cross section as a regular polygon with 'n' number of the sides each of length 'a' & length 'L' at any point lying on the longitudinal axis at a distance 'x' from the centre of one of the ends inside the shell, is given by the eq(76) as follows

$$\omega=2n\left[\sin^{-1}\left(\frac{2xsin\frac{\pi}{n}}{\sqrt{4x^2+a^2cot^2\frac{\pi}{n}}}\right)+\sin^{-1}\left(\frac{2(L-x)sin\frac{\pi}{n}}{\sqrt{4(L-x)^2+a^2cot^2\frac{\pi}{n}}}\right)\right]$$

On setting the values, we get

$$\omega=2\times4\left[\sin^{-1}\left(\frac{2\times16sin\frac{\pi}{4}}{\sqrt{4(16)^2+(8)^2cot^2\frac{\pi}{4}}}\right)+\sin^{-1}\left(\frac{2(40-16)sin\frac{\pi}{4}}{\sqrt{4(40-16)^2+(8)^2cot^2\frac{\pi}{4}}}\right)\right]$$

$$=8\left[\sin^{-1}\left(\frac{22.627}{\sqrt{1088}}\right)+\sin^{-1}\left(\frac{33.941}{\sqrt{2368}}\right)\right]=8[\sin^{-1}(0.686)+\sin^{-1}(0.697)]$$

$$=8[1.527]=12.216\ sr$$

Hence, the radiation energy intercepted by the right prism

$$E_i=\left(\frac{L'}{4\pi}\right)\times\omega$$

$$=\left(\frac{300}{4\pi}\right)\times12.216=291.635\ j/sec\qquad Ans.$$

Example 4.63: Find out the total flux intercepted by a hollow right prismatic shell with cross-section as a regular pentagon with each side 11cm, & length 36cm when a uniform point source emitting 1600 lm flux is located at the mid-point of longitudinal axis

Sol. given that

$$number\ of\ the\ sides, n=5, length\ of each\ side, a=11cm$$

$$distance\ of\ point\ source\ from\ the\ centre\ of\ one\ of\ the\ ends, x=\frac{36}{2}=18cm$$

$$length\ of\ the\ shell, L=36cm\ \&\ total\ flux\ of\ point\ source, F_t=1600\ lm$$

Intensity of the point-source,

$$I=\frac{F_t}{4\pi}=\frac{1600}{4\pi}=127.324\ cd$$

We know that the solid angle, subtended by a hollow right prismatic shell with cross section as a regular polygon with 'n' number of the sides each of length 'a' & length 'L' at any point lying on the longitudinal axis at a distance 'x' from the centre of one of the ends inside the shell, is given by the eq(76) as follows

$$\omega = 2n\left[\sin^{-1}\left(\frac{2x sin\frac{\pi}{n}}{\sqrt{4x^2 + a^2 cot^2\frac{\pi}{n}}}\right) + \sin^{-1}\left(\frac{2(L-x)sin\frac{\pi}{n}}{\sqrt{4(L-x)^2 + a^2 cot^2\frac{\pi}{n}}}\right)\right]$$

On setting the values, we get

$$\omega = 2\times 5\left[\sin^{-1}\left(\frac{2\times 18 sin\frac{\pi}{5}}{\sqrt{4(18)^2 + (11)^2 cot^2\frac{\pi}{5}}}\right) + \sin^{-1}\left(\frac{2(36-18)sin\frac{\pi}{5}}{\sqrt{4(36-18)^2 + (11)^2 cot^2\frac{\pi}{5}}}\right)\right]$$

$$= 10\left[\sin^{-1}\left(\frac{21.16}{\sqrt{1525.226}}\right) + \sin^{-1}\left(\frac{21.16}{\sqrt{1525.226}}\right)\right] = 20[\sin^{-1}(0.542)]$$

$$= 20[0.573] = 11.46\ sr$$

Hence, the radiation energy intercepted by the right prism

$$F = I\times\omega$$

$$= 127.324\times 11.46 = 1459.133\ lm \qquad \textit{Ans.}$$

Example 4.64:.Determine the total flux intercepted by a hollow right prismatic shell with cross-section as an equilateral triangle with each side 20cm, & length 80cm if a uniform point-source emitting 1585 lm flux is located at a point lying on the longitudinal axis at a distance 12cm from the centre of front-end outside the shell.

Sol. given that

$$number\ of\ the\ sides, n = 3, length\ of each\ side, a = 20cm$$

$$distance\ of\ point\ source\ from\ the\ centre\ of\ one\ of\ the\ ends, x = 12cm$$

$$length\ of\ the\ shell, L = 80cm\ \&\ total\ flux\ of\ point\ source, F_t = 1585\ lm$$

Intensity of the point-source,

$$I = \frac{F_t}{4\pi} = \frac{1585}{4\pi} = 150\ cd$$

We know that the solid angle subtended by a hollow right prismatic shell with cross-section as a regular polygon with 'n' number of the sides each of length 'a' & length 'L', at any external point lying on the longitudinal axis at a distance 'x' from the centre of front end is given as

$$\omega = 2n\left[\sin^{-1}\left(\frac{2(L+x)sin\frac{\pi}{n}}{\sqrt{4(L+x)^2 + a^2 cot^2\frac{\pi}{n}}}\right) - \sin^{-1}\left(\frac{2x sin\frac{\pi}{n}}{\sqrt{4x^2 + a^2 cot^2\frac{\pi}{n}}}\right)\right]$$

On setting the values, we get

$$\omega = 2\times 3\left[\sin^{-1}\left(\frac{2(80+12)sin\frac{\pi}{3}}{\sqrt{4(80+12)^2 + (20)^2 cot^2\frac{\pi}{3}}}\right) - \sin^{-1}\left(\frac{2\times 12 sin\frac{\pi}{3}}{\sqrt{4(12)^2 + (20)^2 cot^2\frac{\pi}{3}}}\right)\right]$$

$$= 6\left[\sin^{-1}\left(\frac{159.349}{\sqrt{33989.333}}\right) - \sin^{-1}\left(\frac{20.785}{\sqrt{709.333}}\right)\right]$$

$$= 6[\sin^{-1}(0.864) - \sin^{-1}(0.78)]$$

$$= 6[1.043 - 0.895] = 0.888\ sr$$

Hence, the radiation energy intercepted by the right prism

$$F = I\times\omega$$

$$= 150\times 0.888 = 133.2\ lm \qquad \textit{Ans.}$$

Example 4.65: Calculate the total flux density intercepted by a hollow right prismatic shell with cross-section as a regular octagon with each side 14cm, & length 150cm if a uniform point source emitting 3142 lm flux is located at the centre of the shell. (Assume that flux is uniformly distributed over the entire surface)

Sol. given that

$$number\ of\ the\ sides, n = 8, length\ of each\ side, a = 14cm$$

$$distance\ of\ point\ source\ from\ the\ centre\ of\ one\ of\ the\ ends, x = \frac{150}{2} = 75cm$$

$$length\ of\ the\ shell, L = 150cm\ \&\ total\ flux\ of\ point\ source, F_t = 3142\ lm$$

Intensity of the point-source,

$$I = \frac{F_t}{4\pi} = \frac{3142}{4\pi} = 250\ cd$$

We know that the solid angle, subtended by a hollow right prismatic shell with cross section as a regular polygon with 'n' number of the sides each of length 'a' & length 'L' at any point lying on the longitudinal axis at a distance 'x' from the centre of one of the ends inside the shell, is given by the eq(76) as follows

$$\omega = 2n\left[\sin^{-1}\left(\frac{2x sin\frac{\pi}{n}}{\sqrt{4x^2 + a^2cot^2\frac{\pi}{n}}}\right) + \sin^{-1}\left(\frac{2(L-x)sin\frac{\pi}{n}}{\sqrt{4(L-x)^2 + a^2cot^2\frac{\pi}{n}}}\right)\right]$$

On setting the values, we get

$$\omega = 2 \times 8\left[\sin^{-1}\left(\frac{2 \times 75sin\frac{\pi}{8}}{\sqrt{4(75)^2 + (14)^2cot^2\frac{\pi}{8}}}\right) + \sin^{-1}\left(\frac{2(150-75)sin\frac{\pi}{8}}{\sqrt{4(150-75)^2 + (14)^2cot^2\frac{\pi}{8}}}\right)\right]$$

$$= 16\left[\sin^{-1}\left(\frac{57.402}{\sqrt{23642.372}}\right) + \sin^{-1}\left(\frac{57.402}{\sqrt{23642.372}}\right)\right] = 32[\sin^{-1}(0.373)]$$

$$= 32[0.382\,] = 12.224\ sr$$

Hence, the radiation energy intercepted by the right prism

$$F = I \times \omega$$

$$= 250 \times 12.224 = 3056\ lm$$

Area (A) of the lateral surface of the shell

$$\Rightarrow A = no. of\ the\ lateral\ faces \times area\ of\ one\ lateral\ face$$

$$= n \times a \times L = 8 \times 14 \times 150 = 16800\ cm^2\ \ or\ 1.68\ m^2$$

Hence, the flux density,

$$F.D. = \frac{F}{A} = \frac{3056}{1.68} = 1819.048\ lux \qquad Ans.$$

Example 4.66: Determine the total flux intercepted by a hollow right prismatic shell with cross-section as a regular polygon with 24sides each 11cm & length 60cm if a uniform point-source emitting 1400 lm fluxis located at the centre of one of the ends.

Sol. given that

$$number\ of\ the\ sides, n = 24, length\ of each\ side, a = 11cm$$

$$distance\ of\ point\ source\ from\ the\ centre\ of\ one\ of\ the\ ends, x = 0$$

$$length\ of\ the\ shell, L = 60cm\ \&\ total\ flux\ of\ point\ source, F_t = 1400\ lm$$

Intensity of the point-source,

$$I = \frac{F_t}{4\pi} = \frac{1400}{4\pi} = 111.408\ cd$$

We know that the solid angle, subtended by a hollow right prismatic shell with cross section as a regular polygon with 'n' number of the sides each of length 'a' & length 'L' at any point lying on the longitudinal axis at a distance 'x' from the centre of one of the ends inside the shell, is given by the eq(76) as follows

$$\omega = 2n\left[\sin^{-1}\left(\frac{2x\sin\frac{\pi}{n}}{\sqrt{4x^2 + a^2\cot^2\frac{\pi}{n}}}\right) + \sin^{-1}\left(\frac{2(L-x)\sin\frac{\pi}{n}}{\sqrt{4(L-x)^2 + a^2\cot^2\frac{\pi}{n}}}\right)\right]$$

On setting the values, we get

$$\omega = 2\times 24\left[\sin^{-1}\left(\frac{2\times(0)\sin\frac{\pi}{24}}{\sqrt{4(0)^2 + (11)^2\cot^2\frac{\pi}{24}}}\right) + \sin^{-1}\left(\frac{2(60-0)\sin\frac{\pi}{24}}{\sqrt{4(60-0)^2 + (11)^2\cot^2\frac{\pi}{24}}}\right)\right]$$

$$= 48\left[\sin^{-1}(0) + \sin^{-1}\left(\frac{15.663}{\sqrt{21381.153}}\right)\right] = 48[\sin^{-1}(0.107)]$$

$$= 48[0.107] = 5.146\ sr$$

Hence, the total flux intercepted by the right prism

$$F = I\times\omega$$

$$= 111.408\times 5.146 = 573.305\ lm \qquad \textit{Ans.}$$

Example 4.67: Find out the total radiation energy intercepted by a right prismatic shell with one end closed & another open, cross-section as a squarewith each side 16cm if a uniform point-source of 1kW is located on the longitudinal axis at a distance 18cm from the centre of open end inside the shell.

Sol. given that

$$number\ of\ the\ sides, n = 4, length\ of each\ side, a = 16cm$$

$$distance\ of\ point\ source\ from\ the\ centre\ of\ open\ end, x = 18cm\ \&$$

$$luminosity\ of\ point\ source, L = 1kW = 1000W$$

We know that the solid angle subtended by a right pyramidal shell with one open & another closed, cross-section as a regular polygon with 'n' number of the sides each of length 'a' & length 'L' at any internal point lying on the longitudinal axis at a distance 'x' from the centre of open end is given from the eq(80) as follows

$$\omega = 2\pi + 2n\sin^{-1}\left(\frac{2x\sin\frac{\pi}{n}}{\sqrt{4x^2 + a^2\cot^2\frac{\pi}{n}}}\right)$$

On setting the values in the above expression, we get

$$\omega = 2\pi + 2\times 4\sin^{-1}\left(\frac{2\times 18\sin\frac{\pi}{4}}{\sqrt{4(18)^2 + (16)^2\cot^2\frac{\pi}{4}}}\right)$$

$$= 2\pi + 8\sin^{-1}\left(\frac{25.456}{\sqrt{1552}}\right)$$

$$= 2\pi + 8\sin^{-1}(0.646) = 2\pi + 5.619 = 11.902\ sr$$

Hence, total radiation energy intercepted by the right prism

$$E_i = \left(\frac{L}{4\pi}\right)\times\omega$$

$$= \left(\frac{1000}{4\pi}\right) \times 11.902 = 947.131\ j/sec \qquad Ans.$$

Example 4.68: Find out the total radiation energy intercepted by a right prismatic shell with one end closed & another open & cross-section as a regular pentagon with each side 8cm if a uniform point-source of 200W is located on the longitudinal axis at a distance 10cm from the centre of open end outside the shell.

Sol. given that

$$number\ of\ the\ sides, n = 5, length\ of each\ side, a = 8cm$$

$$distance\ of\ point\ source\ from\ the\ centre\ of\ open\ end, x = 10cm\ \&$$

$$luminosity\ of\ point\ source, L = 200W$$

We know that the solid angle subtended by a right pyramidal shell with one open & another closed, cross-section as a regular polygon with 'n' number of the sides each of length 'a' & length 'L' at any external point lying on the longitudinal axis at a distance 'x' from the centre of open end is given from the eq(80) as follows

$$\omega = 2\pi - 2n\sin^{-1}\left(\frac{2xsin\frac{\pi}{n}}{\sqrt{4x^2 + a^2cot^2\frac{\pi}{n}}}\right)$$

On setting the values in the above expression, we get

$$\omega = 2\pi - 2 \times 5\sin^{-1}\left(\frac{2 \times 10sin\frac{\pi}{5}}{\sqrt{4(10)^2 + (8)^2cot^2\frac{\pi}{5}}}\right)$$

$$= 2\pi - 10\sin^{-1}\left(\frac{11.756}{\sqrt{521.243}}\right)$$

$$= 2\pi - 10\sin^{-1}(0.515) = 2\pi - 5.41 = 0.873\ sr$$

Hence, total radiation energy intercepted by the right prism

$$E_i = \left(\frac{L}{4\pi}\right) \times \omega$$

$$= \left(\frac{200}{4\pi}\right) \times 0.873 = 13.894\ j/sec \qquad Ans.$$

Example 4.69: A hollow right prismatic shell with cross-section as a regular polygon with 'n' no. of the sides each of length 'a' intercepts x fraction of the total radiation emitted by a uniform point-source located at the mid-point of longitudinal axis. Prove that surface area (A) of the shell is given as

$$A = \frac{na^2cot\frac{\pi}{n}sin\frac{\pi x}{n}}{\sqrt{sin^2\frac{\pi}{n} - sin^2\frac{\pi x}{n}}} \quad \forall\ (x < 1)\ .$$

Sol. given that

$$number\ of\ the\ sides = n,\ \ length\ of each\ side\ of\ cross\ section = a,$$

$$let\ the\ luminosity\ of\ point\ source = L' \&\ length\ of\ shell = L\ then$$

$$distance\ of\ point\ source\ from\ the\ centre\ of\ open\ end, x = \frac{L}{2}$$

We know that the solid angle, subtended by a hollow right prismatic shell with cross section as a regular polygon with 'n' number of the sides each of length 'a' & length 'L' at any point lying on the longitudinal axis at a distance 'x' from the centre of one of the ends inside the shell, is given by the eq(76) as follows

$$\omega = 2n\left[\sin^{-1}\left(\frac{2x\sin\frac{\pi}{n}}{\sqrt{4x^2 + a^2\cot^2\frac{\pi}{n}}}\right) + \sin^{-1}\left(\frac{2(L-x)\sin\frac{\pi}{n}}{\sqrt{4(L-x)^2 + a^2\cot^2\frac{\pi}{n}}}\right)\right]$$

On setting $x = {}^{L}/_{2}$ in the above expression, we get

$$\omega = 2n\left[\sin^{-1}\left(\frac{2\left(\frac{L}{2}\right)\sin\frac{\pi}{n}}{\sqrt{4\left(\frac{L}{2}\right)^2 + a^2\cot^2\frac{\pi}{n}}}\right) + \sin^{-1}\left(\frac{2\left(L-\frac{L}{2}\right)\sin\frac{\pi}{n}}{\sqrt{4\left(L-\frac{L}{2}\right)^2 + a^2\cot^2\frac{\pi}{n}}}\right)\right]$$

$$= 2n\left[\sin^{-1}\left(\frac{L\sin\frac{\pi}{n}}{\sqrt{L^2 + a^2\cot^2\frac{\pi}{n}}}\right) + \sin^{-1}\left(\frac{L\sin\frac{\pi}{n}}{\sqrt{L^2 + a^2\cot^2\frac{\pi}{n}}}\right)\right]$$

$$\Rightarrow \omega = 4n\sin^{-1}\left(\frac{L\sin\frac{\pi}{n}}{\sqrt{L^2 + a^2\cot^2\frac{\pi}{n}}}\right)$$

Hence, the radiation energy intercepted by the right prism

$$E_i = \left(\frac{L'}{4\pi}\right) \times \omega$$

$$= \left(\frac{L'}{4\pi}\right) \times 4n\sin^{-1}\left(\frac{L\sin\frac{\pi}{n}}{\sqrt{L^2 + a^2\cot^2\frac{\pi}{n}}}\right)$$

But it is given that the fraction of the radiation energy intercepted by the shell is 'x' hence we get

$$E_i = x \times (total\ radiation\ energy\ emitted\ by\ the\ point\ source\ (L')) = xL'$$

Now, equating both the results, we get

$$\left(\frac{L'}{4\pi}\right) \times 4n\sin^{-1}\left(\frac{L\sin\frac{\pi}{n}}{\sqrt{L^2 + a^2\cot^2\frac{\pi}{n}}}\right) = xL'$$

$$\Rightarrow \sin^{-1}\left(\frac{L\sin\frac{\pi}{n}}{\sqrt{L^2 + a^2\cot^2\frac{\pi}{n}}}\right) = \frac{\pi x}{n} \quad or \quad \frac{L\sin\frac{\pi}{n}}{\sqrt{L^2 + a^2\cot^2\frac{\pi}{n}}} = \sin\frac{\pi x}{n}$$

On squaring both the sides, we get

$$\frac{L^2\sin^2\frac{\pi}{n}}{L^2 + a^2\cot^2\frac{\pi}{n}} = \sin^2\frac{\pi x}{n}$$

$$\Rightarrow L^2\sin^2\frac{\pi}{n} = L^2\sin^2\frac{\pi x}{n} + a^2\cot^2\frac{\pi}{n}\sin^2\frac{\pi x}{n}$$

$$\Rightarrow L^2\left(\sin^2\frac{\pi}{n} - \sin^2\frac{\pi x}{n}\right) = a^2\cot^2\frac{\pi}{n}\sin^2\frac{\pi x}{n}$$

$$\Rightarrow L^2 = \frac{a^2\cot^2\frac{\pi}{n}\sin^2\frac{\pi x}{n}}{\sin^2\frac{\pi}{n} - \sin^2\frac{\pi x}{n}}$$

$$or\ L = \frac{acot\frac{\pi}{n}sin\frac{\pi x}{n}}{\sqrt{sin^2\frac{\pi}{n} - sin^2\frac{\pi x}{n}}} \quad \dots\dots\dots\dots\dots\dots\dots (I)$$

Now, the area (A) of the lateral surface of the shell,

$$\Rightarrow A = n \times (area\ of\ one\ lateral\ face) = n \times (a \times L) = naL$$

On setting the value of L from the eq(I), we get surface area of the shell

$$A = na \times \frac{acot\frac{\pi}{n}sin\frac{\pi x}{n}}{\sqrt{sin^2\frac{\pi}{n} - sin^2\frac{\pi x}{n}}} = \frac{na^2cot\frac{\pi}{n}sin\frac{\pi x}{n}}{\sqrt{sin^2\frac{\pi}{n} - sin^2\frac{\pi x}{n}}}$$

Since, the solid angle subtended by the shell at the centre varies as follows

$$\omega = 4n \sin^{-1}\left(\frac{Lsin\frac{\pi}{n}}{\sqrt{L^2 + a^2cot^2\frac{\pi}{n}}}\right)$$

$$\Rightarrow 0 < \omega < 4\pi \quad \left(since,\ \sin^{-1}\left(\frac{Lsin\frac{\pi}{n}}{\sqrt{L^2 + a^2cot^2\frac{\pi}{n}}}\right) < \frac{\pi}{n}\{\forall\ (a, L) > 0\}\right)$$

Hence, the energy of interception varies as follows

$$0 < E_i < \left(\frac{L'}{4\pi}\right) \times 4\pi$$

$$\Rightarrow 0 < xL' < \left(\frac{L'}{4\pi}\right) \times 4\pi \qquad (since, \quad E_i = xL\ )$$

$$\Rightarrow 0 < x < 1\ or\ x < 1 \qquad (since, the\ fraction\ (x) is\ always\ positive)$$

Hence we get

$$A = \frac{na^2cot\frac{\pi}{n}sin\frac{\pi x}{n}}{\sqrt{sin^2\frac{\pi}{n} - sin^2\frac{\pi x}{n}}} \qquad \forall (\ x < 1)$$

Example 4.70: A hollow right prismatic shell with cross-section as a regular polygon with 'n' no. of the sides each of length 'a' & length 'L' intercepts x fraction of the total radiation emitted by a uniform point-source located at the centre of one of the ends. Prove that surface area (A) of the shell is given as

$$A = \frac{2nL^2\sqrt{sin^2\frac{\pi}{n} - sin^2\frac{2\pi x}{n}}}{cot\frac{\pi}{n}sin\frac{2\pi x}{n}} \qquad \forall \left(x < \frac{1}{2}\right)$$

Sol. given that

$$number\ of\ the\ sides = n, \quad length\ of\ the\ shell = L,$$

$$let\ the\ luminosity\ of\ point\ source = L' \&\ length\ of\ each\ side\ of\ cross\ section = a$$

Since, the point lies at the centre of one of the ends hence

$$distance\ of\ point\ source\ from\ the\ centre\ of\ open\ end, x = 0$$

We know that the solid angle, subtended by a hollow right prismatic shell with cross section as a regular polygon with 'n' number of the sides each of length 'a' & length 'L' at any point lying on the longitudinal axis at a distance 'x' from the centre of one of the ends inside the shell, is given by the eq(76) as follows

$$\omega = 2n\left[\sin^{-1}\left(\frac{2x\sin\frac{\pi}{n}}{\sqrt{4x^2 + a^2\cot^2\frac{\pi}{n}}}\right) + \sin^{-1}\left(\frac{2(L-x)\sin\frac{\pi}{n}}{\sqrt{4(L-x)^2 + a^2\cot^2\frac{\pi}{n}}}\right)\right]$$

On setting $x = 0$ in the above expression, we get

$$\omega = 2n\left[\sin^{-1}\left(\frac{2(0)\sin\frac{\pi}{n}}{\sqrt{4(0)^2 + a^2\cot^2\frac{\pi}{n}}}\right) + \sin^{-1}\left(\frac{2(L-0)\sin\frac{\pi}{n}}{\sqrt{4(L-0)^2 + a^2\cot^2\frac{\pi}{n}}}\right)\right]$$

$$= 2n\left[\sin^{-1}(0) + \sin^{-1}\left(\frac{2L\sin\frac{\pi}{n}}{\sqrt{4L^2 + a^2\cot^2\frac{\pi}{n}}}\right)\right]$$

$$\Rightarrow \omega = 2n\sin^{-1}\left(\frac{2L\sin\frac{\pi}{n}}{\sqrt{4L^2 + a^2\cot^2\frac{\pi}{n}}}\right)$$

Hence, the radiation energy intercepted by the right prism

$$E_i = \left(\frac{L'}{4\pi}\right) \times \omega$$

$$= \left(\frac{L'}{4\pi}\right) \times 2n\sin^{-1}\left(\frac{2L\sin\frac{\pi}{n}}{\sqrt{4L^2 + a^2\cot^2\frac{\pi}{n}}}\right)$$

But it is given that the fraction of the radiation energy intercepted by the shell is 'x' hence we get

$$E_i = x \times (total\ radiation\ energy\ emitted\ by\ the\ point\ source\ (L')) = xL'$$

Now, equating both the results, we get

$$\left(\frac{L'}{4\pi}\right) \times 2n\sin^{-1}\left(\frac{2L\sin\frac{\pi}{n}}{\sqrt{4L^2 + a^2\cot^2\frac{\pi}{n}}}\right) = xL'$$

$$\Rightarrow \sin^{-1}\left(\frac{2L\sin\frac{\pi}{n}}{\sqrt{4L^2 + a^2\cot^2\frac{\pi}{n}}}\right) = \frac{2\pi x}{n} \quad or \quad \frac{2L\sin\frac{\pi}{n}}{\sqrt{4L^2 + a^2\cot^2\frac{\pi}{n}}} = \sin\frac{2\pi x}{n}$$

On squaring both the sides, we get

$$\frac{4L^2\sin^2\frac{\pi}{n}}{4L^2 + a^2\cot^2\frac{\pi}{n}} = \sin^2\frac{2\pi x}{n}$$

$$\Rightarrow 4L^2\sin^2\frac{\pi}{n} = 4L^2\sin^2\frac{2\pi x}{n} + a^2\cot^2\frac{\pi}{n}\sin^2\frac{2\pi x}{n}$$

$$\Rightarrow 4L^2\left(\sin^2\frac{\pi}{n} - \sin^2\frac{2\pi x}{n}\right) = a^2\cot^2\frac{\pi}{n}\sin^2\frac{2\pi x}{n}$$

$$\Rightarrow a^2 = \frac{4L^2\left(\sin^2\frac{\pi}{n} - \sin^2\frac{2\pi x}{n}\right)}{\cot^2\frac{\pi}{n}\sin^2\frac{2\pi x}{n}}$$

$$or \quad a=\frac{2L\sqrt{sin^2\frac{\pi}{n}-sin^2\frac{2\pi x}{n}}}{cot\frac{\pi}{n}sin\frac{2\pi x}{n}} \qquad \dots\dots\dots\dots\dots\dots (I)$$

Now, the area (A) of the lateral surface of the shell,

$$\Rightarrow A=n\times(area\ of\ one\ lateral\ face)=n\times(a\times L)=naL$$

On setting the value of L from the eq(I), we get surface area of the shell

$$A=nL\times\frac{2L\sqrt{sin^2\frac{\pi}{n}-sin^2\frac{2\pi x}{n}}}{cot\frac{\pi}{n}sin\frac{2\pi x}{n}}=\frac{2nL^2\sqrt{sin^2\frac{\pi}{n}-sin^2\frac{2\pi x}{n}}}{cot\frac{\pi}{n}sin\frac{2\pi x}{n}}$$

Since, the solid angle subtended by the shell at the centre varies as follows

$$\omega=2n\sin^{-1}\left(\frac{2Lsin\frac{\pi}{n}}{\sqrt{4L^2+a^2cot^2\frac{\pi}{n}}}\right)$$

$$or\ \omega=2n\sin^{-1}\left(\frac{Lsin\frac{\pi}{n}}{\sqrt{L^2+\left(\frac{a^2cot^2\frac{\pi}{n}}{4}\right)}}\right)$$

$$\Rightarrow 0<\omega<2\pi \qquad \left(since,\ \ \sin^{-1}\left(\frac{Lsin\frac{\pi}{n}}{\sqrt{L^2+\left(\frac{a^2cot^2\frac{\pi}{n}}{4}\right)}}\right)<\frac{\pi}{n}\{\forall\ (a,L)>0\}\right)$$

Hence, the energy of interception varies as follows

$$0<E_i<\left(\frac{L'}{4\pi}\right)\times 2\pi$$

$$\Rightarrow\ 0<xL'<\left(\frac{L'}{4\pi}\right)\times 2\pi \qquad (since, \qquad E_i=xL\ )$$

$$\Rightarrow 0<x<\frac{1}{2}\ or\ x<\frac{1}{2}(since, the\ fraction\ (x) is\ always\ positive)$$

Hence we get

$$A=\frac{2nL^2\sqrt{sin^2\frac{\pi}{n}-sin^2\frac{2\pi x}{n}}}{cot\frac{\pi}{n}sin\frac{2\pi x}{n}} \qquad \forall\left(x<\frac{1}{2}\right) \qquad proved.$$

Example 4.71: Find out the solid angle subtended by a torus with inner & outer radii 12cm & 25cm respectively at a point lying at a normal distance 20cm from the centre.

Sol.given that

$$inner\ radius, r=12cm, outer\ radius, R=25cm\ \&\ normal\ distance, h=20cm$$

On setting the values in the eq(83.1)

We get solid angle subtended by the torus at the point 'P'

$$\Rightarrow \omega = 4\pi\left(\frac{R^2 - r^2}{(R + r)^2 + 4h^2}\right)$$

$$= 4\pi\left(\frac{(25)^2 - (12)^2}{(25 + 12)^2 + 4(20)^2}\right) = 4\pi\left(\frac{481}{2969}\right) = 2.036\ sr \qquad Ans.$$

Example 4.72: A uniform point-source of 1400 lm is located at the centre of a torus with inner & outer radii 16cm & 40cm respectively. What will be the total flux intercepted by the torus?

Sol.given that

$$inner\ radius, r = 12cm, outer\ radius, R = 25cm\ \&\ total\ flux, F_t = 1400\ lm$$

On setting the values in the eq(83.2)

We get solid angle subtended by the torus at its centre

$$\Rightarrow \omega = 4\pi\left(\frac{R - r}{R + r}\right)$$

$$= 4\pi\left(\frac{40 - 16}{40 + 16}\right) = 5.385\ sr$$

Hence, the total flux intercepted by the torus

$$\Rightarrow F = I \times \omega = \left(\frac{F_t}{4\pi}\right) \times \omega$$

$$= \left(\frac{1400}{4\pi}\right) \times 5.385 = 599.934\ lm \qquad Ans.$$

Summary

* Solid angle subtended by the Universe at any point in the space is always $4\pi\ sr$.

Sphere:

1. The solid angle subtended by a closed surface at any point completely inside its boundary is always $4\pi\ sr$ $\quad(\forall\ 0 \leq \omega \leq 4\pi\ )$.
2. The solid angle subtended at any point lying on the plane by a cap fully covering the plane is always $2\pi\ sr$
3. The solid angle subtended by a sphere with a radius 'R' at any point lying at a distance d from the centre is given as

$$\omega = 2\pi\left[1 - \frac{\sqrt{d^2 - R^2}}{d}\right] \quad \forall\ d \geq R$$

4. The solid angle subtended by a hemispherical shell with a radius 'R' at any internal point lying on the axis at a distance 'x' from the centre is given as

$$\omega = 2\pi\left[1 + \frac{x}{\sqrt{x^2 + R^2}}\right] \quad \forall\ 0 \leq x < R$$

5. The solid angle subtended by a hemispherical shell with a radius 'R' at any external point lying on the axis at a distance 'x' from the centre of circular-opening is given as

$$\omega = 2\pi\left[1 - \frac{x}{\sqrt{x^2 + R^2}}\right]$$

6. The solid angle subtended by a hemispherical shell at any point lying inside the shell on the plane of circular opening, is always $2\pi\ sr$
7. The solid angle subtended by a hemispherical shell at thepeak-point of the surface lying inside the shell, is always $3.414\pi\ sr$ irrespective of the radius.
8. The solid angle subtended by a hemispherical shell at the centre of mass is always $2.894\pi\ sr$irrespective of the radius.

Circular cylinder:

9. The solid angle subtended by the solid cylinder at any point lying on the longitudinal axis at a distance 'x' from the centre of front-end is given as

$$\omega = 2\pi\left[1 - \frac{x}{\sqrt{x^2 + R^2}}\right]$$

10. The solid angle subtended by a cylinder with radius 'R' & length 'L' at any point lying on the transverse axis at a distance 'x' from the centre (neglecting the effect of curvature of circular ends) is given as

$$\omega = 4\sin^{-1}\left[\frac{RL}{\sqrt{x^2L^2 + 4(x^2 - R^2)^2}}\right] \quad \forall\ [\ x \geq R\ ]$$

11. The solid angle subtended by an infinitely long cylinder with a radius 'R' at any point lying on the transverse axis at a distance 'x' from the centre is given as

$$\omega = 4\sin^{-1}\left(\frac{R}{x}\right) \quad \forall\ [x \geq R]$$

12. The solid angle subtended by a hollow cylindrical shell with radius 'R' & length 'L' at any internal point lying on the longitudinal axis at a distance 'x' from the centre of one of the ends is given as

$$\omega = 2\pi\left[\frac{x}{\sqrt{x^2 + R^2}} + \frac{(L - x)}{\sqrt{(L - x)^2 + R^2}}\right] \quad \forall\, x \in [0, L]$$

13. The solid angle subtended by a hollow cylindrical shell with radius 'R' & length 'L' at any external point lying on the longitudinal axis at a distance 'x' from the centre of front end is given as

$$\omega = 2\pi\left[\frac{(L + x)}{\sqrt{(L + x)^2 + R^2}} - \frac{x}{\sqrt{x^2 + R^2}}\right] \quad \forall\, [x \geq 0\,]$$

14. The solid angle subtended by a hollow cylindrical shell with radius 'R' & length 'L' at the centre is given as

$$\omega = 4\pi\left[\frac{L}{\sqrt{L^2 + 4R^2}}\right]$$

Right cone:

15. The solid angle subtended by a right cone with normal height 'H' & radius 'R' at the apex point is given as

$$\omega = 2\pi\left[1 - \frac{H}{\sqrt{H^2 + R^2}}\right]$$

16. The solid angle subtended by a hollow right conical shell with normal height 'H' & radius 'R' at any internal point lying on the axis at a distance 'x' from the centre of base is given as

$$\omega = 2\pi\left[1 + \frac{x}{\sqrt{x^2 + R^2}}\right] \quad \forall\ 0 \leq x < H$$

17. The solid angle subtended by a hollow right conical shell with normal height 'H' & radius 'R' at any external point lying on the axis at a distance 'x' from the centre of base is given as

$$\omega = 2\pi\left[1 - \frac{x}{\sqrt{x^2 + R^2}}\right]$$

18. The solid angle subtended by a hollow right conical shell with normal height 'H' & radius 'R' at the apex point inside the shell is given as

$$\omega = 2\pi\left[1+\frac{H}{\sqrt{H^2+R^2}}\right]$$

19. The solid angle subtended by a hollow right conical shell with normal height 'H' & radius 'R' at the centre of mass is given as

$$\omega = 2\pi\left[1+\frac{H}{\sqrt{H^2+9R^2}}\right]$$

Right pyramid:

20. The solid angle subtended by a right pyramid with base as a regular polygon with 'n' number of the sides each of length 'a', normal height 'H' & angle between consecutive lateral edges $'\alpha'$, at any internal point lying on the axis at a distance 'x' from the centre of baseis given as

$$\omega = 2\pi - 2n\sin^{-1}\left(\frac{2Hsin\frac{\pi}{n}}{\sqrt{4H^2+a^2cot^2\frac{\pi}{n}}}\right) = 2\pi - 2n\sin^{-1}\left(cos\frac{\pi}{n}\sqrt{tan^2\frac{\pi}{n}-tan^2\frac{\alpha}{2}}\right)$$

Where, the relation between $a,\ H,\ \alpha\ \&\ n$ is given as

$$H = \frac{a}{2}\sqrt{cot^2\frac{\alpha}{2}-cot^2\frac{\pi}{n}} \quad \forall\left(n\in N, n\geq 3\ \&\ 0<\alpha\leq\frac{2\pi}{n}\right)$$

21. The solid angle subtended by a right pyramidal shell with base as a regular polygon with 'n' number of the sides each of length 'a' & normal height 'H', at any internal point lying on the axis at a distance 'x' from the centre of base is given as

$$\omega = 2\pi + 2n\sin^{-1}\left(\frac{2xsin\frac{\pi}{n}}{\sqrt{4x^2+a^2cot^2\frac{\pi}{n}}}\right) \quad \forall\ 0\leq x<H$$

22. The solid angle subtended by a right pyramidal shell with base as a regular polygon with 'n' number of the sides each of length 'a' & normal height 'H', at any external point lying on the axis at a distance 'x' from the centre of base is given as

$$\omega = 2\pi - 2n\sin^{-1}\left(\frac{2xsin\frac{\pi}{n}}{\sqrt{4x^2+a^2cot^2\frac{\pi}{n}}}\right)$$

23. The solid angle subtended by a right pyramid with base as a regular polygon with 'n' number of the sides each of length 'a', normal height 'H' & angle between consecutive lateral edges $'\alpha'$ at apex point inside the shell is given as

$$\omega = 2\pi + 2n\sin^{-1}\left(\frac{2Hsin\frac{\pi}{n}}{\sqrt{4H^2+a^2cot^2\frac{\pi}{n}}}\right) = 2\pi + 2n\sin^{-1}\left(cos\frac{\pi}{n}\sqrt{tan^2\frac{\pi}{n}-tan^2\frac{\alpha}{2}}\right)$$

24. The solid angle subtended by a regular tetrahedron at each of the vertices (externally), is 0.55 sr.
25. The solid angle subtended by a regular tetrahedron at each of the vertices (internally), is 6.83 sr.
26. The solid angle subtended by each of the octants of orthogonal planes at the origin point is 0.5π sr.

Right prism:

27. The solid angle subtended by a solid right prism with cross-section as a regular polygon with 'n' number of the sides each of length 'a' & length 'L', at any point lying on the longitudinal axis at a distance 'x' from the centre of front end is given as

$$\omega = 2\pi - 2n\sin^{-1}\left(\frac{2x\sin\frac{\pi}{n}}{\sqrt{4x^2 + a^2\cot^2\frac{\pi}{n}}}\right)$$

In this case, value of solid angle is independent of the length of the right prism.

28. The solid angle subtended by a hollow right prismatic shell with cross-section as a regular polygon with 'n' number of the sides each of length 'a' & length 'L', at any internal point lying on the longitudinal axis at a distance 'x' from the centre of one of the ends is given as

$$\omega = 2n\left[\sin^{-1}\left(\frac{2x\sin\frac{\pi}{n}}{\sqrt{4x^2 + a^2\cot^2\frac{\pi}{n}}}\right) + \sin^{-1}\left(\frac{2(L-x)\sin\frac{\pi}{n}}{\sqrt{4(L-x)^2 + a^2\cot^2\frac{\pi}{n}}}\right)\right]$$

$$\forall\ x \in [0, L],\ \ n \in N\ \&\ n \geq 3$$

29. The solid angle subtended by a hollow right prismatic shell with cross-section as a regular polygon with 'n' number of the sides each of length 'a' & length 'L', at any external point lying on the longitudinal axis at a distance 'x' from the centre of front end is given as

$$\omega = 2n\left[\sin^{-1}\left(\frac{2(L+x)\sin\frac{\pi}{n}}{\sqrt{4(L+x)^2 + a^2\cot^2\frac{\pi}{n}}}\right) - \sin^{-1}\left(\frac{2x\sin\frac{\pi}{n}}{\sqrt{4x^2 + a^2\cot^2\frac{\pi}{n}}}\right)\right] \quad \forall\, x \geq 0$$

30. The solid angle subtended by a hollow right prismatic shell with cross-section as a regular polygon with 'n' number of the sides each of length 'a' & length 'L' at the centre is given as

$$\omega = 4n\sin^{-1}\left(\frac{L\sin\frac{\pi}{n}}{\sqrt{L^2 + a^2\cot^2\frac{\pi}{n}}}\right)$$

31. The solid angle subtended by a hollow right prismatic shell with cross-section as a regular polygon with 'n' number of the sides each of length 'a' & length 'L' at the centre of one of the ends is given as

$$\omega = 2n\sin^{-1}\left(\frac{2L\sin\frac{\pi}{n}}{\sqrt{4L^2 + a^2\cot^2\frac{\pi}{n}}}\right)$$

32. The solid angle subtended by a right pyramidal shell with one open & another closed, cross-section as a regular polygon with 'n' number of the sides each of length 'a' & length 'L', at any internal point lying on the longitudinal axis at a distance 'x' from the centre of open end is given as

$$\omega = 2\pi + 2n\sin^{-1}\left(\frac{2x\sin\frac{\pi}{n}}{\sqrt{4x^2 + a^2\cot^2\frac{\pi}{n}}}\right)\ \left[\ \forall\ 0 \leq x < L,\ \ n \in N\ \&\ n \geq 3\ \right]$$

33. The solid angle subtended by a right pyramidal shell with one open & another closed, cross-section as a regular polygon with 'n' number of the sides each of length 'a' & length 'L' at any internal point lying on the longitudinal axis at a distance 'x' from the centre of front end (open or closed) is given as

$$\omega = 2\pi - 2n\sin^{-1}\left(\frac{2x\sin\frac{\pi}{n}}{\sqrt{4x^2 + a^2\cot^2\frac{\pi}{n}}}\right) [\forall\ x \geq 0]$$

Torus:

34. Solid angle subtended by a torus with inner & outer radii 'r' & 'R' respectively at any point lying at a height 'h' from the centre is given as

$$\omega = 4\pi\left(\frac{R^2 - r^2}{(R + r)^2 + 4h^2}\right) \qquad \forall\ R \geq r\ \&\ R, r, h \in [0, \infty)$$

35. Solid angle subtended by a torus with inner & outer radii 'r' & 'R' respectively at the centre is given as

$$\omega = 4\pi\left(\frac{R - r}{R + r}\right)$$

Exercise - 4

*** Sphere:**

Q.4.1. Find out the solid angle subtended by a sphere with a radius 20cm at a point lying at a distance 1.2m from the centre.

Q.4.2. If the solid angle subtended by a sphere with a diameter 100cm at a point in the space is 0.88 sr. what is the distance of the given point from the centre of the sphere?

Q.4.3. Find out the distance of a point from the centre of a sphere with a radius 40cm subtending a solid angle $\pi\ sr$ at the same point.

Q.4.4. Find out the radius of a sphere subtending a solid angle 0.5 sr. at a point lying at a distance 90cm from the centre.

Q.4.5. Find out the solid angle subtended by a hemispherical shell with a radius 12cm at a point lying at a distance 15cm from the centre of circular-section, on the axis towards the 1)Section& 2) surface

Q.4.6. Find out the solid angle subtended by a hemispherical shell with a radius 150mm at a point lying at a distance 5cm from the centre of circular-section on the axis inside the shell.

Q.4.7. Find out the solid angle subtended by a hemispherical shell with a radius 130mm at a point lying at a distance 165mm from the centre of circular-section on the axis towards the surface of the shell.

Q.4.8. What is the solid angle subtended by a hemispherical shell with a radius 24cm at the centre of mass.

Q.4.9. A frustum of a spherical shell with a radius 16 cm subtends an angle 90^0 at the centre of the shell. What will be solid angle subtended by the frustum at a point lying, at a distance 140mm from the centre of circular section, on the axis towards the section?

Q.4.10. A frustum of a spherical shell with a radius 16 cm subtends an angle 150^0 at the centre of the shell. Find out the solid angle subtended by the frustum at a point lying at a distance 120mm from the centre of its circular section, on the axis towards the surface.

Q.4.11. Determine the solid angle subtended by a frustum of a spherical shell with a radius 20 cm subtends an angle 120^0 at the centre at its centre of mass.

Q.4.12. A frustum of a spherical shell subtends a solid angle 1.4 sr. at a point lying on the axis towards the section. Determine the plane-angle subtended by the frustum at the centre of its parent sphere.

Q.4.13. Find out the plane-angle subtended by a frustum of the sphere at the centre of its parent sphere with a radius 'R' if the solid angle subtended by it at a point lying at a distance $\frac{2R}{\sqrt{3}}$ from the centre of circular section, on the axis towards the surface is $\pi\ sr$.

Q.4.14. A hemisphere with a radius 16 cm subtends a solid angle 1.8 sr. at a point lying on the axis towards the surface. Find out the distance of the point from the centre of the circular section.

Q.4.15. A uniform point-source of 100cd is mounted on tip of a 60cm long thin rod vertically pitched at hemispherical summit of a temple. Determine luminous flux intercepted by the summit if the outer radius of summit is 2.5m.

Q.4.16. Find out the luminous flux intercepted by a hemispherical shell with a radius 20cm when a uniform point-source of 111cd is mounted at a point lying on the axis at a distance 40cm from the centre of the circular section 1. towards the section & 2.towards the surface.

Q.4.17. A uniform point-source of 110cd is mounted at the centre of mass of hemi-spherical shell with a radius 25cm. Determine the luminous flux intercepted by the shell & flux density.

(Assume that the flux is uniformly distributed over the entire surface of the shell.)

Q.4.18. A uniform point-source emitting total flux 1400 lm is mounted at the centre of mass of a hemi-spherical shell such that the total flux intercepted by the shell is 900 lm. Determine the radius of the shell.

Q.4.19. There are two spheres A & B with surface areas A_A&A_B respectively,radiating energy uniformly in all the directions. If x_A&x_B are the fractions of total energy intercepted by bodies A &B respectively then prove that

$$\frac{A_A}{A_B} = \frac{x_A - {x_A}^2}{x_B - {x_B}^2} \quad \forall \left(x_A, x_B \leq \frac{1}{2}\right).$$

*** Circular-cylinder:**

Q.4.20. Find out the solid angle subtended by a solid cylinder with radius 12cm at a point lying on the longitudinal axis at a distance 180mm from the centre of the front-face.

Q.4.21. Find out the solid angle subtended by a solid cylinder with radius 20cm at a point lying on the longitudinal axis at a distance 30 cm from the centre of the front-end.

Q.4.22. A cylinder subtends a plane-angle 60^0 at a point lying on the longitudinal axis. Find out the solid angle subtended by the cylinder at the same point.

Q.4.23. Find out the solid angle subtended by a solid cylinder with a radius 10cm & length 25cm at a point lying on the transverse axis at a distance 1m from the centre of longitudinal axis. (Neglect the effect of curvature of the circular ends)

Q.4.24. Find out the solid angle subtended by an infinite long cylinder with a radius 16cm at a point lying on the transverse axis at a distance 3m from the centre of longitudinal axis.

Q.4.25. An infinite long cylinder with a radius 20cm subtends a solid angle 1.5 sr. at a point lying on the transverse axis. What will the distance of the point from the centre of longitudinal axis?

Q.4.26. If the solid angle subtended by an infinite long cylinder at a point lying on the transverse axis at a distance 5m from the centre of the longitudinal axis is 0.102 sr. Determine the radius of cylinder.

Q.4.27. A long cylindrical pole with a diameter 920mm & height 30m standing on a plane surface. What will be the solid angle subtended by the pole at a point lying in the plane of base at a distance 2.5m from the centre of base?

(Neglect the effect of curvature of the circular ends)

Q.4.28. Find out the solid angle subtended by an infinite long cylindrical pole with a radius 30cm at a point lying at a distance 50cm from the longitudinal axis in the plane of base.

Q.4.29. An infinite long cylinder pole with a radius 36cm subtends a solid angle 1.8 sr. at a point lying in the plane of the base. What will the distance of the point from the centre of base?

Q.4.30. If the solid angle subtended by an infinite long cylindrical pole at a point lying at a distance 2m from the longitudinal axis in the plane of the base is 0.118 sr. Determine the radius of pole.

Q.4.31. Determine the solid angle subtended by a hollow cylindrical shell with a radius 5cm & length 15cm at an internal point lying on the longitudinal axis at a distance 8cm from the centre of one of the ends.

Q.4.32. Determine the solid angle subtended by a hollow cylindrical shell with a radius 1.5m & length 6m at the centre of 1. Longitudinal axis & 2.one of the ends.

Q.4.33. Determine the length of a hollow cylindrical shell with a radius 30 cm subtending a solid angle 2.5 sr. at the centre of axis.

Q.4.34. Find out the radius of a hollow cylindrical shell 30 cm long, subtending a solid angle 1.2 sr. at the centre of axis.

Q.4.35. Determine the length of a hollow cylindrical shell with a radius 25 cm subtending a solid angle 1.5 sr. at the centre of one of the ends.

Q.4.36. What is the radius of a hollow cylindrical shell 40 cm long subtending a solid angle 2.2 sr. at the centre of one of the ends.

Q.4.37. If the solid angle subtended by a hollow cylinder with 2m internal diameter at an internal point lying on the axis at a distance 1.4m from the centre of one of the ends, is 0.5 sr. What is the length of the cylinder?

Q.4.38. Find out the solid angle subtended by a hollow cylindrical shell with a radius 10cm & length 1m at an external point lying on the longitudinal axis at a distance 20cm from the centre of the front end.

Q.4.39. A hollow cylinder with an internal diameter 1.6m subtends a solid angle 0.45 sr. at an external point lying on the axis at a distance 30 cm from the centre of the front-end. Determine the length of the cylinder.

Q.4.40. Find out the solid angle subtended by acylindrical shell with one end close & another open, having a radius 5cm & length 30cm at an internal point lying on the axis at a distance 1) 8cm from the centre of the open end& 2)10cm from the closed end

Q.4.41. Determine the solid angle subtended by a cylindrical shell with one end close & another open, radius 5cm & length 30cm at internal centre of the closed end.

Q.4.42. Find out the luminous flux directly escaping a hollow cylindrical shell with radius 10cm & length 30cm when a uniform point-source of 200 cd is located at the mid-point of the longitudinal axis.

Q.4.43. A hollow cylindrical shell with a radius 5cm & length 20cm intercepts 2248 lm flux emitted by a uniform point-source located at the mid-point of longitudinal axis. Find out the luminous intensity of source.

Q.4.44. A uniform point-source is located at the mid-point of the longitudinal axis of a hollow cylindrical shell. If the shell intercepts half of the total flux emitted from the source find out the ratio of length to radius of the shell.

Q.4.45. A hollow cylindrical shell with radius $5\sqrt{2}\ cm$ intercepts one-third of the total flux emitted by a uniform point-source located at the mid-point of the longitudinal axis. Find out the length of the shell.

Q.4.46. A hollow cylindrical shell with radius 'R' & length 'L' intercepts one seventh of the total flux emitted by a uniform point-source located at the mid-point of the longitudinal axis. Prove that

$$L = \frac{R}{2\sqrt{3}}.$$

Q.4.47. A hollow cylindrical shell with radius 'R' & length 'L' intercepts x fraction of the total radiation energy emitted by a uniform point-source located at the mid-point of the longitudinal axis. Prove that the area (A) of the lateral face is given as

$$A = \frac{4\pi x R^2}{\sqrt{1-x^2}} \quad or \quad \frac{L}{R} = \frac{2x}{\sqrt{1-x^2}} \quad \forall\ (x < 1)\ .$$

Q.4.48. A uniform point-source of 200 cd is located on the longitudinal axis at a distance 10cm from the centre of end of a hollow cylindrical shell with radius 6cm & length 15cm. Find out the total flux intercepted by the shell.

Q.4.49. One of the two uniform point-sources each of 200 W is located at the centre longitudinal axis & another at the centre of one of the ends of a hollow cylindrical shell with a radius 8cm & length 25cm. Determine total energy intercepted by the shell.

Q.4.50. A hollow cylindrical shell with a radius 'R' & length 'L' intercepts F_1 flux emitted by a uniform point-source with intensity I_1 located at the centre of the longitudinal axis while shell intercepts F_2 flux emitted by a uniform point-source with intensity I_2 located at the centre of one of the ends of the shell. Prove that

$$\frac{F_1}{F_2} = \frac{2I_1}{I_2}\sqrt{\frac{L^2 + R^2}{L^2 + 4R^2}}.$$

Q.4.51. A hollow cylindrical shell with radius 'R' & length 'L' intercepts x fraction of the total radiation energy emitted by a uniform point-source located at the centre of one of the ends of the shell. Prove that the area (A) of lateral surface is given as

$$A = \frac{4\pi x R^2}{\sqrt{1 - 4x^2}} \quad or \quad \frac{L}{R} = \frac{2x}{\sqrt{1 - 4x^2}} \ \forall \left(x < \frac{1}{2}\right).$$

Q.4.52. A uniform point-source of 500 W is located at the centre of longitudinal axis of a hollow cylindrical shell with a radius 12cm and length 40cm. What will be the change in the energy intercepted by the shell if the point source is shifted at the centre of one of the ends of shell?

Q.4.53. Find out the change in the flux intercepted by a hollow cylindrical shell with a radius 30cm & length 2m if a uniform point-source of 110 cd inside the shell is shifted from a distance 5cm to 40cm from the centre of one of the ends of the shell.

Q.4.54. A uniform point-source is located at the centre of longitudinal axis of a hollow cylindrical shell with a radius 10cm & length 50cm such that the flux density over the internal surface is found to be 3713.906 lux. Determine the intensity of the source if the flux is uniformly distributed over the internal surface.

Q.4.55. A uniform point-source emitting 1400 lm flux is located at the centre of a 70cm long closed hollow cylindrical shell. Now a small hole of 4cm radius is made at the centre of one of the ends. Calculate the total flux directly escaping through the hole.

Q.4.56. A uniform point-source emitting 1500 lm flux is located on the transverse axis of an infinitely long cylinder with radius 20cm. If the total flux intercepted by the cylinder is 500 lm find out the normal distance of point-source from the centre of the cylinder.

Q.4.57. The radiation energy intercepted by an infinite long cylinder is one-sixth of the total energy emitted by a uniform point-source located on the transverse axis at a normal distance 20cm from the centre of cylinder. What is the radius of cylinder?

Q.4.58. An infinitely long cylinder with radius 'R'intercepts x fraction of the total radiation emitted by a uniform point-source located on the transverse axis at a normal distance 'd' from the centre. Prove that

$$R = d\sin\pi x \ \forall \left(x \le \frac{1}{2}\right).$$

Q.4.59. A uniform point-source of 5 kW is located at a normal distance 4m from the longitudinal axis in the plane of base of an infinitely long vertical pole with a diameter 40cm. Find out the total radiation energy intercepted by the pole.

Q.4.60. An infinitely long cylinder with radius 'R' intercepts x fraction of the total radiation emitted by a uniform point-source located on the transverse axis at a normal distance 'd' from the centre of one of the ends. Prove that

$$R = d\sin 2\pi x \ \forall \left(x \leq \frac{1}{4}\right).$$

Q.4.61. An infinitely long cylinder with radius 6cm intercepts 377 lm flux emitted by a uniform point-source of 150 cd located on the transverse axis passing through the centre of one of the ends of cylinder. Find out the normal distance of the point-source from the longitudinal axis.

*** Right cone:**

Q.4.62. Find out the solid angle subtended at the apex-point by a 10cm long right cone with a radius 2.8cm.

Q.4.63. A 20cm long right cone subtends a solid angle 0.54sr at the apex-point. Find out the radius of base of the cone.

Q.4.64. Find out the length of a right cone with a radius 20cm subtending a solid angle 1.4sr at the apex point.

Q.4.65. Find out the solid angle subtended by a 25cm long right cone with a radius 4cm at a point lying in the plane of base at a normal distance 16cm from the axis of cone.

Q.4.66. Find out the solid angle subtended by an infinitely long right cone with a radius 28cm at a point lying in the plane of the base at a normal distance 1m from the axis of the cone.

Q.4.67. An infinitely long right cone with a radius 15cm subtends a solid angle 1.65sr at a point lying in the plane of the base. Find out the normal distance from the axis of the cone.

Q.4.68. An infinitely long right cone subtends a solid angle 1.36sr at a point lying in the plane of the base at a normal distance 16cm from the axis of the cone. Find out the radius of right cone.

Q.4.69. A uniform point-source of 150 cd is located at a normal distance 1.5m from the axis of the cone, in the plane of the base of an infinitely long right cone with a radius 27cm. Calculate the total flux intercepted by the right cone.

Q.4.70. A uniform point-source is located at a normal distance 56cm from the axis ofan infinitely long right cone with a radius 12cm, in the plane of base. Calculate the intensity of point-source if the total flux intercepted by the right cone is 64.79 lm.

Q.4.71. A uniform point-source of 110 cd is located at a normal distance 1.4m from the axis of an infinitely long right cone in the plane of base. Calculate the radius of cone if the total flux intercepted by the right cone is 34.67 lm.

Ans. 21.971cm

Q.4.72. A uniform point-source of 100 cd is located in the plane of the base of an infinitely long right cone with a radius 18cm. Calculate the normal distance of point-source from the axis of cone if the total flux intercepted by the cone is 70.78 lm.

Ans. 51.939cm

Q.4.73. An infinitely long right cone with radius 'R' intercepts x fraction of the total radiation emitted by a uniform point-source located in the plane of base at a normal distance 'd' from the axis of cone. Prove that

$$R = d\sin 2\pi x \ \forall \left(x \leq \frac{1}{4}\right).$$

Q.4.74. Calculate the total flux intercepted by a 40cm long right cone with a radius 10cm if a uniform point-source of 110 cd is located at the apex-point.

Ans. 20.636 lm

Q.4.75. Determine the total flux intercepted by a right cone with apex angle 80^oif a uniform point-source emitting 1400 lm flux is located at the apex-point.

Ans. 578.446 lm

Q.4.76. Find out the solid angle subtended by a 35cm long hollow conical shell with a radius 7cm, at an internal point lying on the axis at a distance 13cm from the centre of base.

Q.4.77. Find out the solid angle subtended by a 20cm long hollow conical shell with a radius 8cm, at the centre of the mass.

Q.4.78. Find out the solid angle subtended by a 45cm long hollow conical shell with a radius 12cm at an internal point lying on the axis at a distance 40cm from the centre of base.

Q.4.79. Determine the solid angle subtended by a hollow conical shell with a radius 9cm at an external point lying on the axis at a distance 14cm from the centre of base towards base.

Q.4.80. Determine the solid angle subtended by a hollow conical shell with a radius 6cm, at an external point lying on the axis at a distance 36cm from the centre of base towards peak.

Q.4.81. What will be the change in the value of solid angle subtended by a 24cm long hollow conical shellwith a radius 5cmat an internal point lying on the axis when shifts from a distance 4cm to 18cm from the apex?

Q.4.82. A uniform point-source of 200 cd is located at the mid-point of axis of a 44cm long hollow conical shellwith a radius 11cm.Find out the flux intercepted by the shell.

Q.4.83. A uniform point-source of 200 W is located insidea hollow conical shell on the axis at a distance 10cm from the centre of the base of with a radius 8cm. Find out the total radiation energy intercepted by the shell.

Q.4.84. Find out the flux intercepted by a hollow conical shell with a radius 14cm if a uniform point-source of 300 cd is located outside on the axis at a distance 11cm from the centre of the base.

Q.4.85. A uniform point-bulb of 500 W is suspended vertically from the peak-point at a distance 25cm from it in a conical tent with apex angle 120°& radius of base 1.6m. Find out the total energy striking the lateral surface of tent.

Q.4.86. A hollow conical shell with height 'H' & a radius 'R' intercepts x fraction of the total radiation emitted by a uniform point-source located at the apex-point inside the shell. Prove that

$$\frac{R}{H}=\frac{2\sqrt{x-x^2}}{2x-1}\quad \forall\, x \in \left(\frac{1}{2},1\right).$$

Q.4.87. A hollow conical shell with height 'H' & a radius 'R' intercepts x fraction of the total radiation emitted by a uniform point-source located at the centre of mass. Prove that

$$\frac{R}{H}=\frac{2\sqrt{x-x^2}}{3(2x-1)}\quad \forall\, x \in \left(\frac{1}{2},1\right).$$

Q.4.88. A hollow conical shell with height 'H' & a radius 'R' intercepts x fraction of the total radiation emitted by a uniform point-source located at the mid-point of the axis. Prove that

$$\frac{R}{H}=\frac{\sqrt{x-x^2}}{2x-1}\quad \forall\, x \in \left(\frac{1}{2},1\right).$$

Q.4.89. A hollow conical shell with a radius 13cm, intercepts 1050 lm flux emitted from a uniform point-source of 200 cd is located at the centre of mass of the shell. Find out height of the shell.

Q.4.90. A hollow conical shell with height 35cm, intercepts 900 lm flux emitted from a uniform point-source of 120 cd is located at the centre of axis of the shell. Find out radius of the shell.

Q.4.91. A hollow conical shell with height 40cm, intercepts 960 lm flux emitted from a uniform point-source of 120 cd is located at the peak-point inside the shell. Find out radius of the shell.

Q.4.92. A hollow conical shell with height 'H' & a radius 'R' intercepts x fraction of the total radiation emitted by a uniform point-source located at a point dividing the axis into an internal ratio of m: n from the centre of base. Prove that

$$\frac{R}{H}=2\left(\frac{m}{m+n}\right)\frac{\sqrt{x-x^2}}{2x-1}\quad \forall\, x \in \left(\frac{1}{2},1\right).$$

Q.4.93. A hollow conical shell with height 'H' & a radius 'R' intercepts x fraction of the total radiation emitted by a uniform point-source located towards the base at a point externally dividing the axis into a ratio of m: n. Prove that

$$\frac{R}{H}=2\left(\frac{m}{n-m}\right)\frac{\sqrt{x-x^2}}{1-2x}\ \forall\ \left(x<\frac{1}{2}\&\,m<n\right).$$

*** Right pyramid:**

Q.4.94. Find out the solid angle subtended by a right pyramid having base as a square with each side 10cm & height 32cm at the apex-point.

Q.4.95. Find out the solid angle subtended by a right pyramid having base as a regular hexagon with each side 8cm & height 20cm at the apex-point.

Q.4.96. Determine the solid angle subtended by a right pyramid with base as a regular heptagon & an angle between consecutive edges 30° at the apex-point

Q.4.97. Calculate the solid angle subtended by a right pyramid with base as a regular octagon & an angle between consecutive edges 22.5^o at the apex-point.

Q.4.98. A uniform point-source emitting the total flux 1500 lm is located at the apex-point of a right pyramid with base as a regular hexagon with each side 5cm & height 24cm. Find out the total flux intercepted by the pyramid.

Q.4.99. A uniform point-source of 200W is located at the apex-point of a right pyramid with base as a regular pentagon with each side 12cm & height 20cm.Find out total radiation energy intercepted by the pyramid.

Q.4.100. A uniform point-source of 500 W is located at the apex-point of a right pyramid with base as an equilateral triangle & an angle between lateral edges 60°.Find out total radiation energy intercepted by the pyramid.

Q.4.101. Find out the solid angle subtended by a hollow right pyramidal shell with base as a regular hexagon with each side 11cm& height 30cmat an internal point lying on the axis at a distance 8cm from the centre of the base.

Q.4.102. Find out the solid angle subtended by a hollow right pyramidal shell with base as a regular heptagon with each side 6cm & height 28cm at an internal point lying on the axis at a distance 25cm from the centre of the base.

Also find out the angle between the consecutive lateral edges of pyramid.

Q.4.103. Calculate the solid angle subtended by a hollow right pyramidal shell with base as a regular nonagon, angle between the consecutive lateral edges 30°& height 32cm at an internal point lying on the axis at a distance 18cm from the centre of the base.

Q.4.104. Calculate the solid angle subtended by a hollow right pyramidal shell with base as a regular pentagon, angle between the consecutive lateral edges 60°& height 40cm at the mid-point of axis.

Q.4.105. Find out the solid angle subtended by a hollow right pyramidal shell with base as a regular polygon with 12 sides, angle betweenthe consecutive lateral edges 20°& height 52cm at internal apex-point of the shell.

Q.4.106. Find out the solid angle subtended by a hollow right pyramidal shell with base as a regular pentagon with 15 sides, angle between the consecutive lateral edges 15°& height 26cm, at internal apex-point of the shell.

Q.4.107. Find out the solid angle subtended by a hollow right pyramidal shell with base as a regular heptagon with each side 8cm & height 44cm, at an external point lying on the axis at a distance 10cm from the centre of the base.

Also find out the angle between the consecutive lateral edges of pyramid.

Q.4.108. Calculate the solid angle subtended by a hollow right pyramidal shell with base as a regular decagon, angle between the consecutive lateral edges 20°& height 22cm at an internal point lying on the axis at a distance 15cm from the centre of the base.

Q.4.109. A uniform point-source of 150 W is located on the axis inside a hollow right pyramidal shell at a distance 5cm from the centre of regular heptagonal base with each side 9cm. Find out the total radiation energy intercepted by the shell.

Q.4.110. Find out the total flux intercepted by a hollow right pyramidal shell with base as a regular polygon with 12 sides, angle between theconsecutive lateral edges 22.5^o& height 24cm when a uniform point-source of 300 cd is located at the mid-point of the axis.

Q.4.111. Find out the total flux intercepted by a hollow right pyramidal shell with base as a regular polygon with 15 sides, angle between theconsecutive lateral edges 12°& height 20cm when a uniform point-source of 200 cd is located outside the shell on the axis at a distance 10cm from the centre of base. If the flux is uniformly distributed over the intercepted surface then find out the flux density.

Q.4.112. A hollow right pyramidal shell, with base as a regular polygon with 'n' no. of the sides each of length 'a', angle between theconsecutive edges 'α' & normal height 'H', intercepts x fraction of the total radiation emitted by a uniform point-source located at the apex-point inside the shell. Prove that

$$\frac{H}{a}=\frac{cot\frac{\pi}{n}sin\left\{\frac{\pi}{n}(2x-1)\right\}}{2\sqrt{sin^2\frac{\pi}{n}-sin^2\left\{\frac{\pi}{n}(2x-1)\right\}}}\ \&$$

$$\alpha=2\tan^{-1}\left(sec\frac{\pi}{n}\sqrt{sin^2\frac{\pi}{n}-sin^2\left\{\frac{\pi}{n}(2x-1)\right\}}\right)\ \forall\ x\in\left(\frac{1}{2},1\right)$$

Q.4.113. A hollow right pyramidal shell with a baseas regular polygon with 10 sides each of 6cm, intercepts 1507.96 lm flux emitted by a uniform point-source of 200 cd located at the apex-point inside the shell. Find out Height of the shell.

Q.4.114. A hollow right pyramidal shell with base as a regular polygon with 15 sides & height 36cm, intercepts 879.64 lm flux emitted by a uniform point-source of 100 cd located at the apex-point inside the shell. Find out 1) Length of side of base & 2) angle between two consecutive lateral edges

Q.4.115. A hollow right pyramidal shell, with base as a regular polygon with 'n' no. of the sides each of length 'a', angle between consecutive edges 'α' & normal height 'H', intercepts x fraction of the total radiation emitted by a uniform point-source located at the mid-point of axis. Prove that

$$\frac{H}{a}=\frac{cot\frac{\pi}{n}sin\left\{\frac{\pi}{n}(2x-1)\right\}}{\sqrt{sin^2\frac{\pi}{n}-sin^2\left\{\frac{\pi}{n}(2x-1)\right\}}}\ \&$$

$$\alpha=2\tan^{-1}\left(tan\frac{\pi}{n}\sqrt{\frac{sin^2\frac{\pi}{n}-sin^2\left\{\frac{\pi}{n}(2x-1)\right\}}{sin^2\frac{\pi}{n}+3sin^2\left\{\frac{\pi}{n}(2x-1)\right\}}}\right)\ \forall\ x\in\left(\frac{1}{2},1\right)$$

Q.4.116. A hollow right pyramidal shell with base as a pentagon &height 46cm, intercepts 898.49 lm flux emitted by a uniform point-source of 110 cd located at the mid-point of axis. Find out the length of side of the base.

Q.4.117. A hollow right pyramidal shell with a base as regular heptagon with each side 8cm, intercepts 2261.95 lm flux emitted by a uniform point-source of 300 cd located at the mid-point of axis. Find out 1) Height of the shell& 2) angle between two consecutive lateral edges

Q.4.118. A uniform point-source of 100 cd is located at the apex-point inside a hollow right pyramid shell with square base & angle between the consecutive lateral edges 60°. Find out the total flux intercepted by the shell.

Q.4.119. A uniform point-source of 150 cd is located at the apex-point inside a hollow right pyramid shell with base as an equilateral triangle with each side 15cm & angle between the consecutive lateral edges 45°. Find out 1) Total flux intercepted by the shell & 2) height of the shell.

Q.4.120. A uniform point-source of 200 cd is located at the mid-point of axis of a hollow right pyramid shell with square base & angle between the consecutive lateral edges 30°. Find out the total flux intercepted by the shell.

Q.4.121. A uniform point-source of 111 cd is located at the mid-point of axis of a hollow right pyramid shell with base as an equilateral triangle, height 28cm & angle between the consecutive lateral edges 75°.Find out 1) Total flux intercepted by the shell & 2) length of side of the base

Q.4.122. A hollow right pyramidal shell, with base as a regular polygon with 'n' no. of the sides each of length 'a', angle between the consecutive edges 'α' & normal height 'H', intercepts x fraction of the total radiation emitted by a uniform point-source located at a point internally dividing the axis into a ratio of $m_1: m_2$ from the centre of base. Prove that

$$\frac{H}{a}=\frac{(m_1+m_2)cot\frac{\pi}{n}sin\left\{\frac{\pi}{n}(2x-1)\right\}}{2m_1\sqrt{sin^2\frac{\pi}{n}-sin^2\left\{\frac{\pi}{n}(2x-1)\right\}}}\&$$

$$\alpha=2\tan^{-1}\left(m_1tan\frac{\pi}{n}\sqrt{\frac{sin^2\frac{\pi}{n}-sin^2\left\{\frac{\pi}{n}(2x-1)\right\}}{{m_1}^2sin^2\frac{\pi}{n}+({m_2}^2+2m_1m_2)sin^2\left\{\frac{\pi}{n}(2x-1)\right\}}}\right)$$

$$\forall\ x\in\left(\frac{1}{2},1\right)$$

Q.4.123. Find out the height of a hollow right pyramidal shell with base as a square with each side 7.5cm, intercepting 210 W radiation energyemitted by a uniform point-source of 300 W located at a point internally dividing the axis into a ratio 2: 3.

Q.4.124. Determine the angle between the consecutive lateral edges of a hollow right pyramidal shell with base as an equilateral triangle, intercepting 325 W radiation energy emitted by a uniform point-source of 500 W located at a point internally dividing the axis into a ratio 1: 2.

Q.4.125. A hollow right pyramidal shell, with base as a regular polygon with 'n' no. of the sides each of length 'a', angle between the consecutive edges 'α' & normal height 'H', intercepts x fraction of the total radiation emitted by a uniform point-source located towards the base at a point externally dividing the axis into a ratio of $m_1: m_2$. Prove that

$$\frac{H}{a}=\frac{(m_2-m_1)cot\frac{\pi}{n}sin\left\{\frac{\pi}{n}(1-2x)\right\}}{2m_1\sqrt{sin^2\frac{\pi}{n}-sin^2\left\{\frac{\pi}{n}(1-2x)\right\}}}\&$$

$$\alpha=2\tan^{-1}\left(m_1tan\frac{\pi}{n}\sqrt{\frac{sin^2\frac{\pi}{n}-sin^2\left\{\frac{\pi}{n}(1-2x)\right\}}{{m_1}^2sin^2\frac{\pi}{n}+({m_2}^2-2m_1m_2)sin^2\left\{\frac{\pi}{n}(1-2x)\right\}}}\right)$$

$$\forall\ x<\frac{1}{2}\&m_2>m_1$$

Q.4.126. Determine the angle between the consecutive lateral edges of a hollow right pyramidal shell with base as an equilateral triangle, intercepting 3769.91lm flux emitted by a uniform point-source of 500 cd located towards the base at a point externally dividing the axis into a ratio 3:4 .

Q.4.127. Find out the height of a hollow right pyramidal shell with base as a square with each side 9cm, intercepting 350 W radiation energy emitted by a uniform point-source of 1kW located towards the baseat a point externally dividing the axis into a ratio 1: 3. Also, find out the angle between the consecutive lateral edges of the shell.

* Intersection/concurrency ofthe straight-lines& the planes:

Q.4.128. Find out the solid angle subtended by four concurrent straight lines equally inclined with one another at an angle 60° at the point of intersection.

Q.4.129. Find out the solid angle subtended by three concurrent straight lines equally inclined with one another at an angle 100° at the point of intersection.

Q.4.130. Calculate the solid angle subtended by six concurrent straight lines equally inclined with one another at an angle 45° at the point of intersection.

Q.4.131. Find out the angle between four concurrent straight lines subtending a solid angle 1.2sr at the point of concurrency.

Q.4.132. Determine the solid angle subtended by the orthogonal axes: X, Y & Z at the origin.

Ans.$^{\pi}/_{2}$ $sr.$

Q.4.133. Find out the solid angle subtended at the point of concurrency by five concurrent planessuch that the angle between the consecutive line of intersection is60^o.

Q.4.134. Determine the solid angle subtended at the point of concurrency by three concurrent planes such that the angle between the consecutive line of intersection is 75°.

Q.4.135. Find out the solid angle subtended at the point of concurrency by nine concurrent planes such that the angle between the consecutive lines of intersection is 35°.

Ans. 5.724sr

Q.4.136. If the solid angle subtended at the point of concurrency by eight equally inclined concurrent planes is 3.344sr find out the angle between the consecutive lines of intersection.

Ans. 40°

Q.4.137. If the solid angle subtended at the point of concurrency by 15 equally inclined concurrent planes is 1.923sr find out the angle between the consecutive lines of intersection.

Ans. 20°

* Right prism:

Q.4.138. Find out the solid angle subtended by a solid right prism with cross-section as a equilateral triangle with each side 15cm at a point lying on the longitudinal axis at adistance 10cm from the centre of front-end. Ans. 0.772sr

Q.4.139. Find out the solid angle subtended by a solid right prism with cross-section as a square with each side 20cm at a point lying on the longitudinal axis at a distance 13cm from the centre of front-end.

Q.4.140. Calculate the solid angle subtended by a solid right prism with cross-section as a regular pentagon with each side 11cm, at a point lying on the longitudinal axis at a distance 8cm from the centre of front-end.

Q.4.141. Find out the solid angle subtended by a solid right prism with cross-section as a regular hexagon with each side 14cm, at a point lying on the longitudinal axis at a distance 16cm from the centre of front-end.

Q.4.142. The solid angle subtended by a solid right prism with cross-section as a square with each side 20cm, at a point lying on the longitudinal axis, is 1.532sr. Find out the distance from the centre of front-end.

Q.4.143. Find out the length of side of equilateral triangular base of a solid right prism subtending a solid angle 2.212sr at a point lying on the longitudinal axis at a distance 7cm from the centre of front-end.

Q.4.144. The solid angle subtended by a solid right prism with cross-section as a regular hexagon with each side 8cm, at a point lying on the longitudinal axis, is 0.846sr. Find out the distance from the centre of front-end.

Q.4.145. Find out the length of side of regular-pentagonal cross-section of a solid right prism subtending a solid angle 1.12sr at a point lying on the longitudinal axis at a distance 12.5cm from the centre of front-end.

Q.4.146. Find out the total flux intercepted by a solid right prism with cross-section as equilateral triangle with each side 12.8cm when a uniform point source emitting total flux 1400 lm is located on the longitudinal axis at a distance 6cm from the centre of front-end.

Q.4.147. Determine the total radiation energy intercepted by a solid right prism with cross-section as a square with each side 14cm when a uniform point source emitting total flux 1500 lm is located on the longitudinal axis at a distance 26cm from the centre of front-end.

Q.4.148. Find out the total flux intercepted by a solid right prism with cross-section as regular pentagon with each side 12.8cm when a uniform point source of 200 cd is located on the longitudinal axis at a distance 7.6cm from the centre of front-end.

Q.4.149. Determine the total radiation energy intercepted by a solid right prism with cross-section as regular hexagon with each side 16cm when a uniform point source of 250 cd is located on the longitudinal axis at a distance 12.6cm from the centre of front-end.

Q.4.150. A solid right prism , with cross-section as a regular polygon with 'n' no. of the sides each of length 'a', intercepts x fraction of the total radiation emitted by a uniform point-source located on the longitudinal axis at a distance 'd' from the centre of front-end. Prove that

$$\frac{d}{a}=\frac{\cot\frac{\pi}{n}\sin\left\{\frac{\pi}{n}(1-2x)\right\}}{2\sqrt{\sin^2\frac{\pi}{n}-\sin^2\left\{\frac{\pi}{n}(1-2x)\right\}}}\quad\forall\ \left(x<\frac{1}{2}\right).$$

Q.4.151. A solid right prism with cross-section as equilateral triangle with each side 18cm intercepts 200 W energy emitted by a uniform point-source of 1 kW located on the longitudinal axis. Find out the distance of point-source from the centre of front-end.

Q.4.152. The total flux intercepted by a solid right prism with cross-section as square is 300 lm emitted by a uniform point-source emitting 1500 lm flux located on the longitudinal axis at a distance 118mm from the centre of front-end. Determine the length of side of square cross-section.

Q.4.153. A solid right prism with cross-section as regular hexagon with each side 12cm intercepts 225 W energy emitted by a uniform point-source of 1.8 kW located on the longitudinal axis. Calculate the distance of point-source from the centre of front-end.

Q.4.154. The total flux intercepted by a solid right prism with cross-section as regular octagon is 200 lm emitted by a uniform point-source of 100 cd located on the longitudinal axis at a distance 118mm from the centre of front-end. Determine the length of side of square cross-section.

Q.4.155. Find out the solid angle subtended by a hollow right prismatic shell, with cross-section as an equilateral triangle with each side 12cm& length 45cm, at an internal point lying on the longitudinal axis at a distance 10cm from the centre of one of the ends.

Q.4.156. Calculate the solid angle subtended by a hollow right prismatic shell with cross-section as a square with each side 14cm, & length 36cm at an internal point lying on the longitudinal axis at a distance 11cm from the centre of one of the ends.

Q.4.157. Find out the solid angle subtended by a hollow right prismatic shell with cross-section as a regular pentagon with each side 13cm, & length 39cm at the mid-point of longitudinal axis.

Q.4.158. Calculate the solid angle subtended by a hollow right prismatic shell with cross-section as a regular octagon with each side 17cm, & length 51cm at the centre of one of the ends.

Q.4.159. Determine the solid angle subtended by a hollow right prismatic shell with cross-section as a square with each side 25cm, & length 80cm at an external point lying on the longitudinal axis at a distance 7cm from the centre of front-end.

Q.4.160. Estimate the solid angle subtended by a hollow right prismatic shell, with cross-section as a regular heptagon with each side 25cm, & length 80cm, at an external point lying on the longitudinal axis at a distance 7cm from the centre of front-end.

Q.4.161. Calculate the solid angle subtended by a hollow right prismatic shell, with cross-section as a regular octagon with each side 16cm& length 64cm, at an external point lying on the longitudinal axis at a distance 18cm from the centre of front-end.

Q.4.162. Find out the solid angle subtended by a hollow right prismatic shell with cross-section as a regular octagon with each side 16cm, & length 64cm at an external point lying on the longitudinal axis at a distance 18cm from the centre of front-end.

Q.4.163. Find out the solid angle subtended by a right prismatic shell, with one end closed & another open, cross-section as a regular hexagon with each side 6cm, & length 50cm, at an internal point lying on the longitudinal axis at a distance 14cm from the centre of open-end.

Q.4.164. Determine the solid angle subtended by a right prismatic shell, with one end closed & another open, cross-section as a regular hexagon with each side 6cm& length 50cm, at the centre of closed end inside the shell.

Q.4.165. Find out the total flux intercepted by a hollow right prismatic shell with cross-section as an equilateral triangle with each side 11cm& length 44cm if a uniform point-source of 200 cd is located on the longitudinal axis at a distance 8cm from the centre of one of the ends.

Q.4.166. Find out the total radiation energy intercepted by a hollow right prismatic shell with cross-sectionas a square with each side 12cm& length 51cm if a uniform point-source of 500 W is located on the longitudinal axis at a distance 14cm from the centre of one of the ends.

Q.4.167. Find out the total flux intercepted by a hollow right prismatic shell with cross-section as a regular nonagon with each side 9cm& length 52cm if a uniform point-source of 100 cd located on the longitudinal axis at a distance 17cm from the centre of one of the ends.

Q.4.168. Find out the total radiation energy intercepted by a hollow right prismatic shell with cross-sectionas a square with each side 10cm& length 33cm if a uniform point-source of 300 W located at the centre of longitudinal axis

Q.4.169. Determine the total flux intercepted by a hollow right prismatic shell with cross-section as an equilateral triangle with each side 11cm& length 44cm if a uniform point-source of 110 cd located at the centre of one of the ends.

Q.4.170. Find out the total radiation energy intercepted by a right prismatic shell with one end closed & another open, cross-section as a square with each side 16cm if a uniform point-source of 1kW is located on the longitudinal axis at a distance 13cm from the centre of open end outside the shell.

Q.4.171. Calculate the total radiation energy intercepted by a right prismatic shell with one end closed & another open, cross-section as a regular hexagon with each side 10cm& length 63cm when a uniform point-source of 400 W located at the centre of longitudinal axis.

Q.4.172. Determine the total flux intercepted by a right prismatic shell withone end closed & another open, cross-section as a regular pentagon with each side 4cm& length 66cm if a uniform point-source of 250 cd located at the centre of closed-end inside the shell.

Q.4.173. A hollow right prismatic shell with cross-section as a regular polygon with 'n' no. of the sides each of length 'a'& length 'L' intercepts x fraction of the total radiation emitted by a uniform point-source located at the mid-point (centre) of the longitudinal axis. Prove that

$$\frac{L}{a} = \frac{cot\frac{\pi}{n} sin\frac{\pi x}{n}}{\sqrt{sin^2\frac{\pi}{n} - sin^2\frac{\pi x}{n}}} \quad \forall\ (x < 1)\ .$$

Q.4.174. A hollow right prismatic shell with cross-section as a regular polygon with 'n' no. of the sides each of length 'a'& length 'L' intercepts x fraction of the total radiation emitted by a uniform point-source located at the centre of one of the ends. Prove that

$$\frac{L}{a}=\frac{cot\frac{\pi}{n}sin\frac{2\pi x}{n}}{2\sqrt{sin^2\frac{\pi}{n}-sin^2\frac{2\pi x}{n}}}\quad\forall\left(x<\frac{1}{2}\right).$$

Q.4.175. A uniform point-source of 200 cd is located at the centre of a hollow right prismatic shell with cross-section as a square with each side 20cm such that it intercepts total flux 2010.62 lm. Find out the length of the shell.

Q.4.176. A uniform point-source of 1 kW is located at the centre of a hollow right prismatic shell with cross-section as an equilateral triangle, & length 60cm such that it intercepts total energy 760 W. Find out the length of side of cross-section.

Q.4.177. A uniform point-source emitting total flux 1600 lm is located at the centre of one of the ends of a hollow right prismatic shell with cross-section as a regular pentagon& length 43cm such that it intercepts total energy 720 lm. Find out the length of side of cross-section.

Q.4.178. A uniform point-source of 1.5 kW is located at the centre of one of the ends of a hollow right prismatic shell with cross-section as a square with each side 20cm such that it intercepts total flux 375 W. Find out the length of the shell.

Q.4.179. What will be the change in the flux intercepted by a hollow right prismatic shell with cross-section as a square with each side 16cm, & length 48cm if a uniform point-source of 210 cd is shifted on the longitudinal axis from a distance 5cm to 30cm from the centre of one of the ends inside the shell.

Q.4.180. A right prismatic shell, with one end closed & another open, cross-section as a regular polygon with 'n' no. of the sides each of length 'a', & length 'L', intercepts x fraction of the total radiation emitted by a uniform point-source located at the centre of the shell. Prove that

$$\frac{L}{a}=\frac{cot\frac{\pi}{n}sin\left\{\frac{\pi}{n}(2x-1)\right\}}{\sqrt{sin^2\frac{\pi}{n}-sin^2\left\{\frac{\pi}{n}(2x-1)\right\}}}\quad\forall\, x\in\left(\frac{1}{2},1\right).$$

Q.4.181. A right prismatic shell with one end closed & another open, cross-section as a square with each side 14cm, intercepts the total flux 1344 lm emitted by a uniform point-source of1600 lm located at the centre ofthe shell. Find out length of the shell.

Q.4.182. A right prismatic shell, with one end closed & another open, cross-section as a regular polygon with 'n' no. of the sides each of length 'a', & length 'L', intercepts x fraction of the total radiation emitted by a uniform point-source located at a point internally dividing the longitudinal axis into a ratio of $m_1:m_2$ from the centre of open-end. Prove that

$$\frac{L}{a}=\frac{(m_1+m_2)cot\frac{\pi}{n}sin\left\{\frac{\pi}{n}(2x-1)\right\}}{2m_1\sqrt{sin^2\frac{\pi}{n}-sin^2\left\{\frac{\pi}{n}(2x-1)\right\}}}\quad\forall\, x\in\left(\frac{1}{2},1\right).$$

Q.4.183. A right prismatic shell, with one end closed & another open, cross-section as a regular polygon with 'n' no. of the sides each of length 'a', & length 'L', intercepts x fraction of the total radiation emitted by a uniform point-source located towards the open-end at a point externally dividing the longitudinal axis into a ratio of $m_1:m_2$. Prove that

$$\frac{L}{a}=\frac{(m_2-m_1)cot\frac{\pi}{n}sin\left\{\frac{\pi}{n}(1-2x)\right\}}{2m_1\sqrt{sin^2\frac{\pi}{n}-sin^2\left\{\frac{\pi}{n}(1-2x)\right\}}}\quad\forall\, x<\frac{1}{2}\,\&m_2>m_1.$$

Chapter 5
Significance & Applications of Solid Angle

In this chapter, we are to study & analyse the solid angle subtended by different smaller & huge objects related to the following

1. Solid state (Crystal structure)
   a) Analysis of arrangement of the spheresin a single layer. b.) Analysis of tetrahedral void & octahedral void in the unit cell.
2. Photometry
   a) Relation between luminous intensity & luminous flux density &
   b) Derivation of Lambert' cosine formula
3. Space science (Analysis of the terrestrial bodies)
   a) Sun. b.) Earth. c.) Moon.

5.1 SOLID STATE (ARRANGEMENT OF THE SPHERES)

Here, we are interested to analyse the different structures of identical spheres arranged in a certain fashion.

5.1.1 Solid angle subtended by 'n' number of the identical spheres arranged,touching one another, in acomplete circular-fashion at any point lying on the verticalaxis passing through the centre of reference plane

Let there be 'n' of the identical spheres with a radius 'R' & centres 'C_1', 'C_2', 'C_3', 'C_4', ……… 'C_{n-1}' & 'C_n'arranged, touching one another, in a complete circular-fashion in the plane of the paper and a given point say 'P' at a height 'h' lying on the vertical axis passing through the centre 'O' of the reference plane*. (as shown in the figure (149) below)

When this arrangement of the identical spheres is observed from the point 'P' (normal to the plane of paper as indicated in the figure.) each sphere appears as whole clearly without eclipsing one another but merely touching one another at single point.

(as indicated 'M' is the point of contact of the spheres with centres 'C_1' & 'C_2')

* Reference plane: it is the regular-polygonal plane obtained by joining the centres of all the spheres in a circular-fashion. (The circular fashion must be complete.)

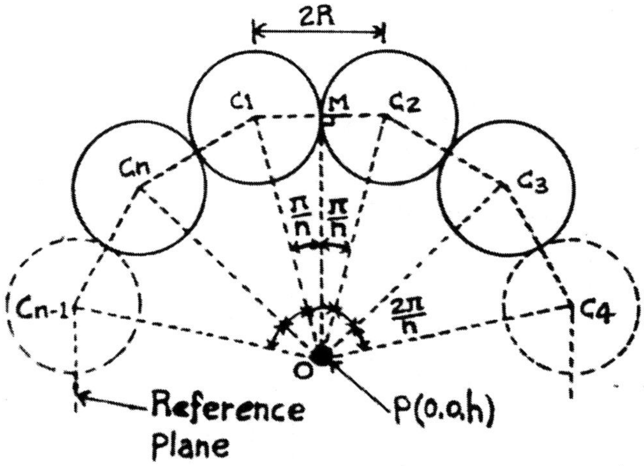

Fig (149)

Join the centres of all the spheres to obtain a regular polygonal plane $C_1C_2C_3C_4 \dots \dots \dots \dots . C_{n-1}C_n$ with centre 'O' say the reference plane.

Now, consider any two spheres with centres say 'C_1' & 'C_2' touching each other at the point 'M' & join the points $C_1, C_2 \& M$ to the centre 'O'.

Thus, we have

$$\Rightarrow \ C_1OC_2 = \frac{2\pi}{n} \ \ or \ \ C_1OM = \frac{\pi}{n} \& C_1C_2 = 2R \ or \ C_1M = R$$

In rightΔC_1MO

$$\Rightarrow sin \ \ C_1OM = \frac{C_1M}{OC_1} \Rightarrow sin\frac{\pi}{n} = \frac{R}{OC_1} \ or \ OC_1 = Rcosec\frac{\pi}{n} \qquad \dots \dots \dots \dots \dots \dots (I)$$

Now, consider a single sphere with centre $'C_1'$

(as shown in the figure(150) below)

In rightΔC_1OP

$$\Rightarrow C_1P = \sqrt{PO^2 + OC_1{}^2} = \sqrt{h^2 + \left(Rcosec\frac{\pi}{n}\right)^2} = \sqrt{h^2 + R^2cosec^2\frac{\pi}{n}}$$

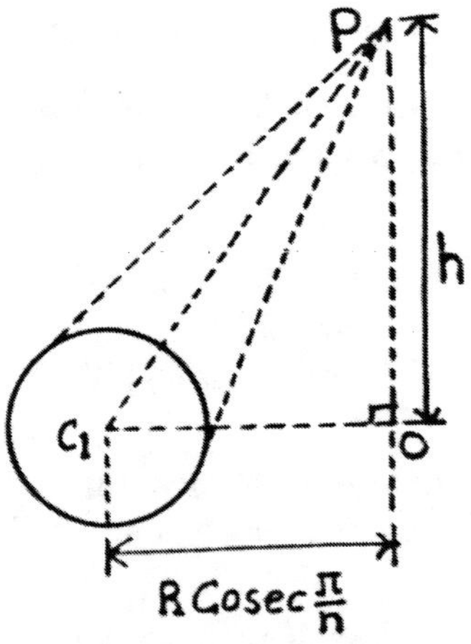

Fig (150)

We know that the solid angle subtended by the sphere at any point in the space is given by the eq(39) as follows

$$\omega = 2\pi\left[1 - \frac{\sqrt{d^2 - R^2}}{d}\right]$$

Hence, by setting d = $C_1P = \sqrt{h^2 + R^2cosec^2\frac{\pi}{n}}$

Solid angle subtended by the sphere with centre 'C_1' ay the point 'P'

$$\Rightarrow \omega_1 = 2\pi\left[1 - \frac{\sqrt{\left(\sqrt{h^2 + R^2cosec^2\frac{\pi}{n}}\right)^2 - R^2}}{\sqrt{h^2 + R^2cosec^2\frac{\pi}{n}}}\right] = 2\pi\left[1 - \frac{\sqrt{h^2 + R^2cosec^2\frac{\pi}{n} - R^2}}{\sqrt{h^2 + R^2cosec^2\frac{\pi}{n}}}\right]$$

$$= 2\pi\left[1 - \frac{\sqrt{h^2 + R^2\left(cosec^2\frac{\pi}{n} - 1\right)}}{\sqrt{h^2 + R^2 cosec^2\frac{\pi}{n}}}\right] = 2\pi\left[1 - \frac{\sqrt{h^2 + R^2 cot^2\frac{\pi}{n}}}{\sqrt{h^2 + R^2 cosec^2\frac{\pi}{n}}}\right]$$

Hence, the sum of solid angles subtended by the 'n' no. of the identical spheres at the given point 'P'

$$\Rightarrow \omega = n \times (solid\ angle\ subtended\ by\ the\ single\ sphere\ at\ the\ point\ 'P') = n \times \omega_1$$

$$\Rightarrow \omega = n \times 2\pi\left[1 - \frac{\sqrt{h^2 + R^2 cot^2\frac{\pi}{n}}}{\sqrt{h^2 + R^2 cosec^2\frac{\pi}{n}}}\right]$$

$$\Rightarrow \omega = 2n\pi\left[1 - \sqrt{\frac{h^2 + R^2 cot^2\frac{\pi}{n}}{h^2 + R^2 cosec^2\frac{\pi}{n}}}\right] \quad \forall (h \geq 0, n \in N \ \&\ n \geq 3) \quad \ldots\ldots\ldots\ldots\ldots (84)$$

5.1.2 Solid angle subtended by 'n' no. of the identical spheres arranged,touching one another, in acomplete circular-fashion at the centre of the reference plane

In this case, the point 'P' coincides with the centre 'O' of the reference plane (i.e. the regular polygonal plane) (as shown in the figure (151) below)

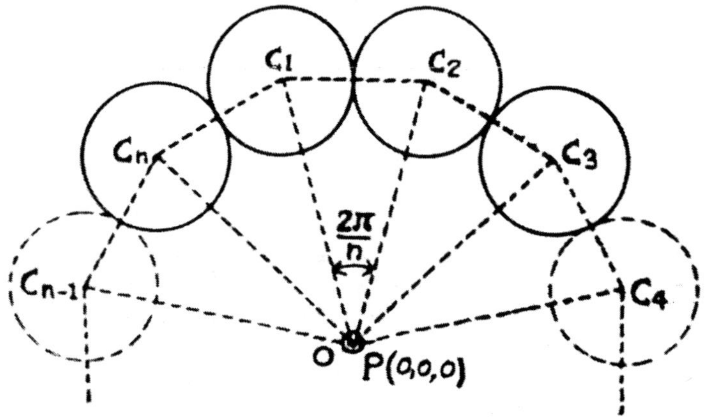

Fig (151)

In this case, the solid angle subtended by the 'n' no. of the identical spheres each of radius 'R' arranged,touching one another, in a complete circular-fashion at the centre 'O' is obtained by setting h = 0 in the above eq(84) as follows

$$\Rightarrow \omega_o = 2n\pi\left[1 - \sqrt{\frac{(0)^2 + R^2 cot^2\frac{\pi}{n}}{(0)^2 + R^2 cosec^2\frac{\pi}{n}}}\right] = 2n\pi\left[1 - \sqrt{\frac{\left(\frac{cos^2\frac{\pi}{n}}{sin^2\frac{\pi}{n}}\right)}{\left(\frac{1}{sin^2\frac{\pi}{n}}\right)}}\right] = 2n\pi\left[1 - \sqrt{cos^2\frac{\pi}{n}}\right]$$

$$= 2n\pi\left[1 - cos\frac{\pi}{n}\right] = 2n\pi\left[1 - \left(1 - 2sin^2\frac{\pi}{2n}\right)\right] = 2n\pi\left[1 - 1 + 2sin^2\frac{\pi}{2n}\right] = 4n\pi sin^2\frac{\pi}{2n}$$

$$\Rightarrow \omega_o = 4n\pi sin^2\left(\frac{\pi}{2n}\right) \quad \forall (n \in N \ \&\ n \geq 3) \quad \ldots\ldots\ldots\ldots\ldots\ldots\ldots\ldots\ldots (85)$$

It is obvious from the above eq(85) that the solid angle subtended by the 'n' no. of the identical spheres at the centre of the reference plane is independent of the radius 'R' of the spheres but merely depends on the no. of the spheres 'n'.

Also, when the 'n' no. of the spheres increases the value of the solid angle 'ω_o' decreases as the spheres move farther from the centre of the.

Thus if we take different samples of equal no. of the identical spheres &arrange them touching one another in a complete circular-fashion about a given point in the space then the solid angle subtended by them remains constant whether the spheres have atomic size or planetary size. Let's call such a point as Universal point.

* **Universal-Point**: "It is the centre of the regular-polygon formed by joining the centres of all the identical spheres arranged, touching one another, in a complete circular-fashion."

The solid angle subtended by the 'n' no. of the spheres at the Universal-Point is given by the eq(85) as follows

$$\omega_o = 4n\pi sin^2\left(\frac{\pi}{2n}\right) \quad \forall(n \in N \;\&\; n \geq 3)$$

It is obvious at least three spheres (here identical) are required for circular arrangement of the spheres. The distance 'D_o' of the Universal-Point from the centre of each of the sphere is given as

$$\Rightarrow D_o = OC_1 = OC_2 = OC_3 = \dots\dots\dots = OC_{n-1} = OC_n = Rcosec\frac{\pi}{n} \qquad \forall(n \in N \;\&\; n \geq 3)$$

Thus, as the no. of the identical spheres in the arrangement increases the distance of their centre from the Universal-Point also increases keeping their radius constant.

Important deduction 1: The different expressions can be derived for different no. of the identical spheres.

1. Solid angle subtended by three identical spheres arranged,touching one another, in a complete circular-fashion at the Universal-Point (i.e. centre of the reference plane)

In this case the solid angle subtended by the entire arrangement of three identical spheres each with radius 'R' is obtained by setting n = 3 in the eq(85) as follows

$$\Rightarrow \omega_o = 4(3)\pi sin^2\left(\frac{\pi}{2(3)}\right) = 12\pi sin^2\left(\frac{\pi}{6}\right) = 12\pi\left(\frac{1}{2}\right)^2 = 12\pi\frac{1}{4} = 3\pi \; sr \;\; \&$$

The distance (D_o) from the centre of each sphere to the Universal-Point

$$\Rightarrow D_o = Rcosec\frac{\pi}{3} = R\left(\frac{2}{\sqrt{3}}\right) = \frac{2R}{\sqrt{3}}$$

2. Solid angle subtended by four identical spheres arranged, touching one another, in a complete circular-fashion at the Universal-Point (i.e. centre of the reference plane)

In this case the solid angle subtended by the entire arrangement of four identical spheres each with radius 'R' is obtained by setting n = 4 in the eq(85) as follows

$$\Rightarrow \omega_o = 4(4)\pi sin^2\left(\frac{\pi}{2(4)}\right) = 16\pi sin^2\left(\frac{\pi}{8}\right) = 16\pi(0.382683)^2 = 16\pi(0.146446)$$

$$= 7.361193 \approx 7.36 \; sr \;\&$$

The distance (D_o) from the centre of each sphere to the Universal-Point

$$\Rightarrow D_o = Rcosec\frac{\pi}{4} = R(\sqrt{2}) = R\sqrt{2}$$

3. Solid angle subtended by five identical spheres arranged, touching one another, in a complete circular-fashion at the Universal-Point (i.e. centre of the reference plane)

In this case the solid angle subtended by the entire arrangement of five identical spheres each with radius 'R' is obtained by setting n = 5 in the eq(85) as follows

$$\Rightarrow \omega_o = 4(5)\pi sin^2\left(\frac{\pi}{2(5)}\right) = 20\pi sin^2\left(\frac{\pi}{10}\right) = 20\pi(0.309017)^2 = 20\pi(0.095491)$$

$$= 5.999908 \approx 6\ sr$$

& the distance (D_o) from the centre of each sphere to the Universal-Point

$$\Rightarrow D_o = Rcosec\frac{\pi}{5} = R(1.701301) \approx 1.7R$$

4. Solid angle subtended by six identical spheres arranged, touching one another, in a complete circular-fashion at the Universal-Point (i.e. centre of the reference plane)

In this case the solid angle subtended by the entire arrangement of six identical spheres each with radius 'R' is obtained by setting n = 6 in the eq(85) as follows

$$\Rightarrow \omega_o = 4(6)\pi sin^2\left(\frac{\pi}{2(6)}\right) = 24\pi sin^2\left(\frac{\pi}{12}\right) = 24\pi(0.258819)^2 = 24\pi(0.066987)$$

$$= 5.050723 \approx 5\ sr$$

& the distance (D_o) from the centre of each sphere to the Universal-Point

$$\Rightarrow D_o = Rcosec\frac{\pi}{6} = R(2) = 2R$$

Thus, following the same procedure & further setting n = 7, 8, 9, 10 …..upto 'n' we can derive a number of the expressions for different numbers of the spheres.

Important deduction 2: Solid angle subtended by infinite number of the identical spheres arranged, touching one another, in a complete circular-fashion at the Universal-Point (i.e. centre of the reference plane)

In this case, the solid angle subtended by infinite no. of the spheres at the Universal-Point is obtained by taking the limit of 'ω_o' at $n \to \infty$ from the eq(85) as follows

$$\Rightarrow \omega_o = \lim_{n\to\infty} 4n\pi sin^2\left(\frac{\pi}{2n}\right) = 4\pi \lim_{k\to 0}\left(\frac{1}{k}\right) sin^2\left(\frac{k\pi}{2}\right) \qquad where\ k = \frac{1}{n}\ \&\ k \to 0\ as\ n \to \infty$$

$$= \pi^3 \lim_{k\to 0} k\left(\frac{sin\left(\frac{k\pi}{2}\right)}{\left(\frac{k\pi}{2}\right)}\right)^2 = \pi^3 \lim_{k\to 0} k \times \left[\lim_{k\to 0}\left(\frac{sin\left(\frac{k\pi}{2}\right)}{\left(\frac{k\pi}{2}\right)}\right)\right]^2 \quad (\,by\ the\ proprty\ of\ limit)$$

$$= \pi^3(0) \times (1)^2 = 0 \qquad since, k \neq 0\ \&\ \lim_{x\to 0}\frac{sinx}{x} = 1$$

& the distance of the Universal-Point from the centre of each of the spheres

$$\Rightarrow D_o = \lim_{n\to\infty} Rcosec\frac{\pi}{n} = R\lim_{k\to 0} cosec(k\pi) \qquad where\ k = \frac{1}{n}\ \&\ k \to 0\ as\ n \to \infty$$

$$= R\big(cosec(0)\big) = \infty$$

Thus, the solid angle subtended by infinite no. of the spheres at the Universal-Point is zero instead of being infinite because there distance becomes infinite as their no. becomes infinite.

Important deduction 3: Solid angle subtended by the circular arrangement of 'n' no. of the identical spheres at the point lying at a height h = R on the vertical axis passing through the centre of the reference plane

In this case, the solid angle subtended by the arrangement at the given point is obtained by setting h = R in the eq(84) as follows

$$\omega = 2n\pi\left[1 - \sqrt{\frac{R^2 + R^2cot^2\frac{\pi}{n}}{R^2 + R^2cosec^2\frac{\pi}{n}}}\right] = 2n\pi\left[1 - \sqrt{\frac{1 + cot^2\frac{\pi}{n}}{1 + cosec^2\frac{\pi}{n}}}\right]$$

$$= 2n\pi\left[1 - \sqrt{\frac{cosec^2\frac{\pi}{n}}{1 + cosec^2\frac{\pi}{n}}}\right] = 2n\pi\left[1 - \sqrt{\frac{cosec^2\frac{\pi}{n}sin^2\frac{\pi}{n}}{sin^2\frac{\pi}{n} + 1}}\right] = 2n\pi\left[1 - \sqrt{\frac{1}{1 + sin^2\frac{\pi}{n}}}\right]$$

$$\Rightarrow \omega = 2n\pi\left[1 - \frac{1}{\sqrt{1 + sin^2\frac{\pi}{n}}}\right] \quad \forall(n \in N \,\&\, n \geq 3) \quad \dots\dots\dots\dots\dots\dots\dots (86)$$

5.1.3 Solid angle subtended by the spheres forming a simple cubic cell (i.e. a simple crystalline unit cell)at the centre

Let there be a simple cubic cell consisting of eight identical spheres each with radius 'R' placed at the corners of the simple cubic cell with the centre 'C' (as shown in the figure (152) below, a simple network structure of the cell)

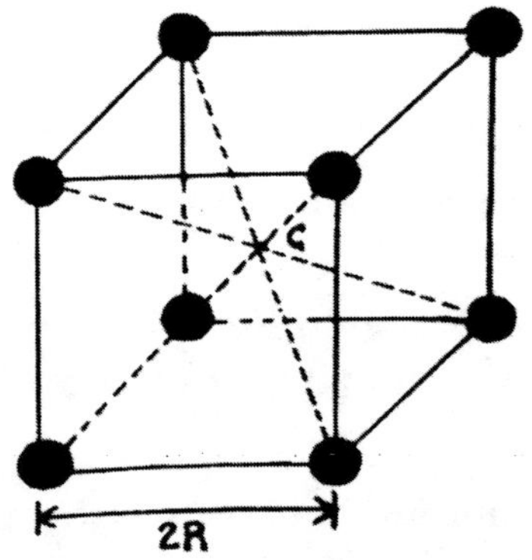

Fig (152)

We can divide the unit cell into two identical layers of the four spheres arranged touching one another in a complete circular-fashion i.e. one layer above another.

Now, consider a single layer by thoroughly separating both the layers by a reference plane with centre 'O'(i.e. by the square plane$C_1C_2C_3C_4$ obtained by joining the centres 'C_1', 'C_2', 'C_3' & 'C_4' of the spheres) (as shown in the figure (153) below)

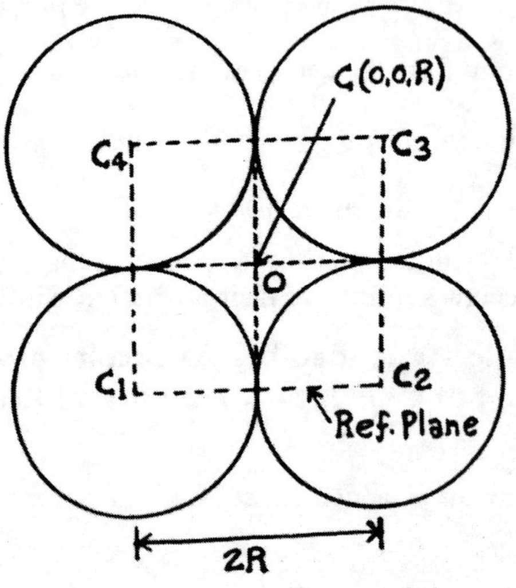

Fig (153)

Since, centre ‘C’ of the cell is lying on the vertical axis at a height ‘R’ from the centre ‘O’of the reference plane.(see the above fig.)

We know that the solid angle subtended by the circular arrangement of ‘n’ no. of the identical spheres at the point lying at a height h = R on the vertical axis passing through the centre of the reference plane is given by the eq(86) as follows

$$\omega = 2n\pi\left[1 - \frac{1}{\sqrt{1 + sin^2\frac{\pi}{n}}}\right]$$

Hence, the solid angle subtended by the single layer of four identical spheres at the centre ‘C’ of the shell is obtained by setting n = 4 as follows

$$\omega_1 = 2(4)\pi\left[1 - \frac{1}{\sqrt{1 + sin^2\frac{\pi}{n}}}\right] = 8\pi\left[1 - \frac{1}{\sqrt{1 + \left(\frac{1}{\sqrt{2}}\right)^2}}\right] = 8\pi\left[1 - \sqrt{\frac{2}{3}}\right]$$

But, there are two layers each of four spheres, hence the solid angle subtended by all the eight spheres (i.e. two identical layers) at the centre ‘C’ of the simple cubic cell

$$\Rightarrow \omega = 2 \times \omega_1 = 16\pi\left[1 - \sqrt{\frac{2}{3}}\right] \quad or \Rightarrow \omega = 16\pi[1 - 0.816496] = 16\pi[0.183504] = 9.223917$$

$$\approx 9.224\ sr$$

5.1.4 Solid angle subtended by the spheres forming a tetrahedral void at the centre

Let there a tetrahedral void with centre ‘O’ consisting of the four identical spheres each with radius ‘R’ (as shown in the figure (154) below)

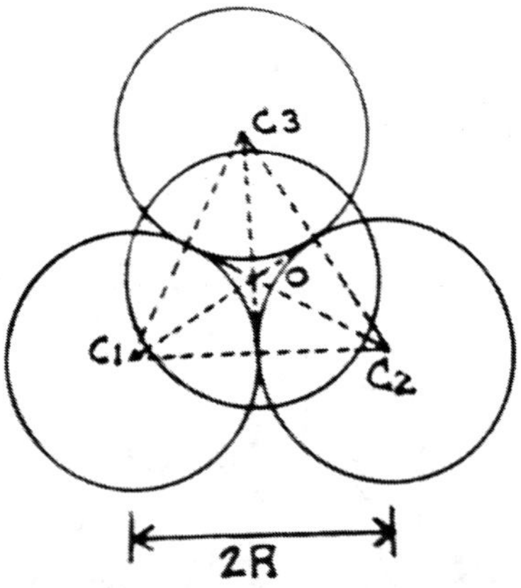

Fig (154)

Now, join the centres ‘C_1’, ‘C_2’ & ‘C_3’ thus we have, $C_1C_2 = C_2C_3 = C_3C_1 = 2R$ & $C_1OC_2 = 2\cos^{-1}\left(\frac{1}{\sqrt{3}}\right) = 109.47^o$

In equilateral ΔC_1OC_2

$$\Rightarrow sin\left(\frac{C_1OC_2}{2}\right) = \frac{\left(\frac{C_1C_2}{2}\right)}{OC_1} = \frac{R}{OC_1} \Rightarrow OC_1 = \frac{R}{sin\left(\frac{C_1OC_2}{2}\right)}$$

$$OC_1 = \frac{R}{\sqrt{1 - cos^2\left(\frac{C_1OC_2}{2}\right)}} = \frac{R}{\sqrt{1 - \left(\frac{1}{\sqrt{3}}\right)^2}} = R\sqrt{\frac{3}{2}}$$

We know that the solid angle subtended by the sphere at any point in the space is given by the eq(39) as follows

$$\omega = 2\pi\left[1 - \frac{\sqrt{d^2 - R^2}}{d}\right]$$

Hence, by setting d = $OC_1 = R\sqrt{\frac{3}{2}}$

Solid angle subtended by the sphere with centre 'C_1' ay the centre 'O' of the tetrahedral void

$$\Rightarrow \omega_R = 2\pi\left[1 - \frac{\sqrt{\left(R\sqrt{\frac{3}{2}}\right)^2 - R^2}}{\left(R\sqrt{\frac{3}{2}}\right)}\right] = 2\pi\left[1 - \frac{\sqrt{\frac{3}{2}R^2 - R^2}}{\left(R\sqrt{\frac{3}{2}}\right)}\right] = 2\pi\left[1 - \frac{R}{R\sqrt{3}}\right]$$

$$= 2\pi\left[1 - \frac{1}{\sqrt{3}}\right] = 2\pi(0.423) = 0.846\pi\ sr\ or\ 2.6556\ sr$$

Since, the tetrahedral void consists of four identical spheres hence the solid angle subtended by all the four spheres at the centre 'O' of tetrahedral void

$$\Rightarrow \omega = 4\omega_R = 4 \times 2.6556 = 10.622\ sr$$

It implies that the centre of the tetrahedral void is not completely enclosed by all four spheres and each sphere subtends a solid angle 2.655 sr at the centre of tetrahedral void. Hence about $(4\pi - 10.622)/_{4\pi} = 0.1547\ or\ 15.47\%$ of tetrahedral void is exposed to the outside space.

$$\Rightarrow [\omega_R = 2.655\ sr]$$

The radius 'r' of the smaller sphere which can be completely trapped (inserted) in the tetrahedral void is given as

$$\Rightarrow r = OC_1 - R = R\sqrt{\frac{3}{2}} - R = R\left(\sqrt{\frac{3}{2}} - 1\right) = R(1.2247 - 1) \approx 0.225R$$

$$\Rightarrow \left[r = R\left(\sqrt{\frac{3}{2}} - 1\right) \approx 0.225R\right]$$

5.1.5 Solid angle subtended by the largest sphere fully trapped in a tetrahedral void at the centre of each sphere forming the void

We know that the solid angle subtended by the sphere at any point in the space is given by the eq(39) as follows

$$\omega = 2\pi\left[1 - \frac{\sqrt{d^2 - R^2}}{d}\right]$$

Now, by setting d = $OC_1 = R\sqrt{\frac{3}{2}}$ & R = r = $R\left(\sqrt{\frac{3}{2}} - 1\right)$

$$\Rightarrow \omega_r = 2\pi\left[1-\frac{\sqrt{\left(R\sqrt{\frac{3}{2}}\right)^2-\left(R\left(\sqrt{\frac{3}{2}}-1\right)\right)^2}}{\left(R\sqrt{\frac{3}{2}}\right)}\right] = 2\pi\left[1-\frac{\sqrt{\frac{2}{3}R^2-\frac{2}{3}R^2-R^2+2R^2\sqrt{\frac{3}{2}}}}{\left(R\sqrt{\frac{3}{2}}\right)}\right]$$

$$= 2\pi\left[1-\sqrt{2\times\frac{2}{3}\sqrt{\frac{3}{2}}-\frac{2}{3}}\right] = 2\pi[1-0.983] \approx 0.106814 \approx 0.107\ sr$$

$$\Rightarrow \boxed{\omega_r = 2\pi\left[1-\sqrt{\left(\sqrt{\frac{8}{3}}-\frac{2}{3}\right)}\right] \approx 0.107\ sr}$$

5.1.6 Solid angle subtended by the spheres forming an octahedral voidat the centre

Let there an octahedral void with centre 'O' consisting of the six identical spheres each with radius 'R' (as shown in the figure (155) below)

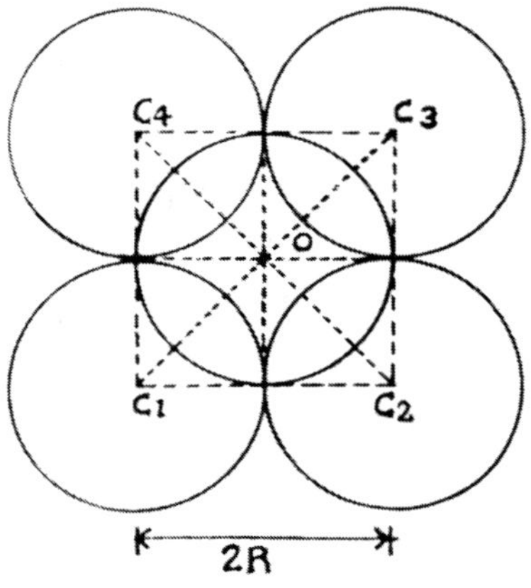

Fig (155)

Now, join the centres 'C_1', 'C_2', 'C_3' & 'C_4' & we have, $C_1C_2 = C_2C_3 = C_3C_4 = C_4C_1 = 2R$

In equilateral $\Delta C_1C_2C_3$

$$\Rightarrow C_1C_3 = \sqrt{C_1C_2^{\ 2}+C_2C_3^{\ 2}} = \sqrt{(2R)^2+(2R)^2} = 2R\sqrt{2} \Rightarrow OC_1 = \frac{C_1C_3}{2} = R\sqrt{2}$$

Now, consider a single sphere with centre 'C_1' & then setting d = $OC_1 = R\sqrt{2}$ in the eq(39), we get the solid angle subtended by the sphere with centre 'C_1' at the centre 'O' of the octahedral void

$$\Rightarrow \omega_R = 2\pi\left[1-\frac{\sqrt{\left(R\sqrt{2}\right)^2-R^2}}{\left(R\sqrt{2}\right)}\right] = 2\pi\left[1-\frac{\sqrt{2R^2-R^2}}{\left(R\sqrt{2}\right)}\right] = 2\pi\left[1-\frac{1}{\sqrt{2}}\right]$$

Since, the octahedral void consists of six identical spheres hence the solid angle subtended by all the six spheres at the centre 'O' of octahedral void

$$\Rightarrow \omega = 6\omega_R = 6\times 2\pi\left[1-\frac{1}{\sqrt{2}}\right] = 12\pi\left[1-\frac{1}{\sqrt{2}}\right] \approx 11.041814 \approx 11.042\ sr$$

It implies that the centre of the tetrahedral void is not completely enclosed by all six spheres i.e. about $(4\pi - 11.042)/_{4\pi} = 12.13\%$ of the octahedral void is exposed to the surrounding space through eight minute openings of the void and each of six spheres forming void subtends an approximate solid angle 1.84 sr at the centre of octahedral void.

Hence the solid angle subtended by the single sphere at the centre of the octahedral void

$$\Rightarrow \left[\omega_R = 2\pi\left[1 - \frac{1}{\sqrt{2}}\right] \approx 1.840302 \approx 1.84\ sr\right]$$

The radius 'r' of the smaller sphere which can be completely trapped (inserted) in the octahedral void is given as

$$\Rightarrow r = OC_1 - R = R\sqrt{2} - R = R(\sqrt{2} - 1) = R(1.414 - 1) \approx 0.414R$$

$$\Rightarrow [r = R(\sqrt{2} - 1) \approx 0.414R]$$

5.1.7 Solid angle subtended by the largest sphere fully trapped in an octahedral void at the centre of the spheres forming the void

We know that the solid angle subtended by the sphere at any point in the space is given by the eq(39) as follows

$$\omega = 2\pi\left[1 - \frac{\sqrt{d^2 - R^2}}{d}\right]$$

Now, by setting d = $OC_1 = R\sqrt{2}$& R = r = $R(\sqrt{2} - 1)$

$$\Rightarrow \omega_r = 2\pi\left[1 - \frac{\sqrt{(R\sqrt{2})^2 - (R(\sqrt{2} - 1))^2}}{(R\sqrt{2})}\right] = 2\pi\left[1 - \frac{\sqrt{2R^2 - 2R^2 - R^2 + 2\sqrt{2}R^2}}{(R\sqrt{2})}\right]$$

$$= 2\pi\left[1 - \frac{\sqrt{2\sqrt{2} - 1}}{\sqrt{2}}\right] = 2\pi\left[1 - \frac{\sqrt{4\sqrt{2} - 2}}{2}\right] \approx 0.275548 \approx 0.275\ sr$$

$$\Rightarrow \left[\omega_r = 2\pi\left[1 - \frac{\sqrt{4\sqrt{2} - 2}}{2}\right] \approx 0.275\ sr\right]$$

5.2 PHOTOMETRY

Luminosity of curved-surface: When a given surface is illuminated by a given luminous-source of constant intensity then the flux intercepted by the surface is distributed over the entire surface. The distribution of the flux may be uniform or non-uniform depending on the certain factors such as geometry, relative configuration of the source & surface.

In-fact, the distribution of the luminous flux emitted by a given luminous-source over a given surface depends on the following factors

1. Geometry of the luminous-source (i.e. pointed, linear or extended) & the surface (i.e. plane or curved neglecting the microscopic properties such as roughness, irregularities etc.) &
2. Relative configuration of the source & the surface in the space.

   Thus, if we divide a given surface into a large number of very small finite-elementary parts (even like square dots on a computer-screen with higher resolution (dpi)) then these elementary parts (dots) may be assumed to be planar then we find that their configurations w.r.t. a given luminous-source may be same or different from each other (although this difference is negligible for the elements(dots) lying in the closest neighbourhood.) hence the luminous flux density on each of the elements (dots) also may be equal or different from each other i.e. the luminous flux density may be constant throughout the given surface or may be varying from point to point over the surface. It totally depends on the two factors as mentioned above.

For equal distribution of the flux over a given, the surface should be symmetrical about the given luminous-source in the space.

It is to be noted that the luminous flux emitted from a given luminous-source intercepted by each finite-elementary part (dot) of a given surface will be constant if and only if

"All the finite-elementary parts (dots) of a finite surface are at an equal distance from the luminous-source."

In fact, the above condition is satisfied in the following two cases

1. Uniform point-source located at the centre of a spherical surface &
2. Uniform linear-source coincident with the axis of a cylindrical surface.

In above two cases, each of the finite elements (dots)is at an equal distance from the source hence the flux intercepted by each element is also constant so is the flux density.

It is the reason that for uniform illumination of the fluorescent screen (like conventional computer & television screens, screen for the projector in the cinema-hall) it is made curved i.e. certain curvature is given to the surface being illuminated by a uniform point-source in order to have uniform luminosity throughout the surface.

Luminosity of plane-surface: If we consider a plane-surface & a uniform point-source illuminating the plane & divide plane-surface into a large number of finite-elementary parte (even like square dots on a computer-screen with high resolution (dpi)). Since the plane-surface is no more symmetrical about the point-source i.e. the distance of each element (dot) from the point-source is not constant but it varies constantly from point to point all over the plane this results in the variation of the flux intercepted by each elementary part of the plane hence the flux density of the plane varies from point to point.

It is to be noted that the luminosity of the element (dot) located at the foot of perpendicular (say maxima) drawn from the point-source is maximum since the distance from the source is minimum. Thus the luminosity of dots on a plane-surface away from the maxima successively decreases as the distance from the illuminating point-source increases.

Thus the luminous flux emitted by a uniform point-source is no more uniformly distributed over the plane-surface for any configuration of the plane with respect to the point-source in the space i.e. luminous flux density on a plane-surface always varies from point to point (i.e. on each elementary area of the plane-surface) irrespective of the luminous source (like intensity, geometry etc.) & its configuration w.r.t. the plane.

We find that the curved surfaces are more likely to have the uniform flux density as compared

5.2.1 Relation between the luminous intensity (I) & luminous flux density/luminosity (E)

Let there a point source of luminous intensity 'I' uniformly emitting the light in the space at a distance 'r' from the centre 'O' of a given finite plane-surface having area 'A' such that the angle between normal to the plane & line joining the centre 'O' to the point source 'S' is 'θ'. (as shown in the figure (156) below)

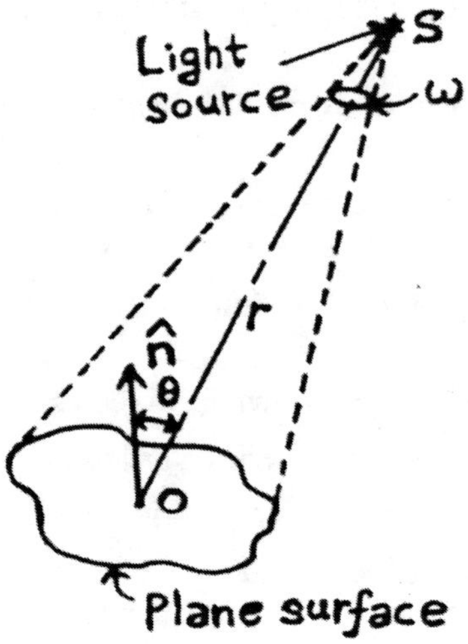

Fig (156)

If the solid angle subtended by the plane-surface at the point source is 'ω' then the total luminous flux (F) striking the plane

$$\Rightarrow F = I\omega$$

Hence, the luminous flux density (luminosity)

$$\Rightarrow E = \frac{total\ luminous\ flux\ striking\ the\ plane}{area\ of\ the\ plane} = \frac{F}{A} = \frac{I\omega}{A} \quad \dots\dots\dots\dots\dots\dots (I)$$

Where, the value of solid angle subtended by the plane is given by the fundamental theorem as follows

$$\Rightarrow \omega = \int_S \frac{dA}{r^2}$$

$$\therefore \left[E = \frac{I\omega}{A} = \frac{I}{A}\int_S \frac{dA}{r^2}\right]$$

Where, 'r' is the distance of any arbitrary point on the given surface from the point source which is constant for all the points lying on the elementary area 'dA'

Above formula is applicable for all the plane-surfaces but it is difficult to find out the solid angle subtended by the irregular plane-surface at the given point-source. But the approximate value of the solid angle subtended by the symmetrical planes can be found out using 'Approximation formula'

5.2.2 Derivation of Lambert's cosine formula

From the above figure (156), the area of projection of the plane-surface in the direction of the line joining the centre 'O' to the point source 'S'

$$\Rightarrow A' = A\cos\theta$$

Now, if area of the plane-surface is sufficiently small or the distance 'r' is sufficiently large then the area of projection A' may be assumed to be a part of the spherical surface with radius 'r' & centre at the point source 'S' i.e. the distance 'r' may be assumed to be constant for all the points of the plane-surface. (as shown in the figure (157) below)

Hence, the solid angle subtended by the plane-surface at the point source

$$\Rightarrow \omega = \int_S \frac{dA'}{r^2} = \frac{1}{r^2}\int_S dA' = \frac{A'}{r^2} = \frac{A\cos\theta}{r^2}$$

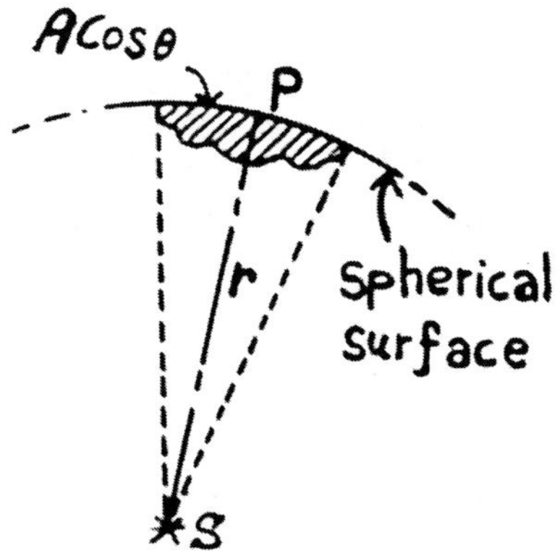

Fig (157)

Hence, the luminous flux density

$$\Rightarrow E = \frac{I\omega}{A} = \frac{I}{A}\frac{A\cos\theta}{r^2} = \frac{I\cos\theta}{r^2}$$

The above formula is called 'Lambert's cosine formula' which is applicable only for the small plane-surface and long distance because the error involved in the results increase for the extended surface & shorter distance. Thus the cosine formula has limitations used in the photometry.

Note: The luminous flux emitted by a uniform point-source is no more uniformly distributed over the plane-surface for any configuration of the plane with respect to the point-source in the space i.e. luminous flux density on a plane-surface always varies from point to point (i.e. on each elementary area of the plane-surface) irrespective of the luminous source (like intensity, geometry etc.) & its configuration w.r.t. the plane.

5.3 SPACE SCIENCE (CELESTIAL BODIES)

In this section, we are interested to analyse the solid angle subtended by some common terrestrial bodies like Sun, Earth & Moon etc.

Assumption: Here, we are to make an assumption that the concerned celestial body has spherical shape. This assumption makes the analysis easier to make calculations.

5.3.1 Solid angle subtended by the terrestrial bodies at their respective centres

Let there be two spherical terrestrial bodies with centres 'C_1' & 'C_1' and radii 'R_1' & 'R_2' respectively at a distance d between their centres. (as shown in the figure (158) below)

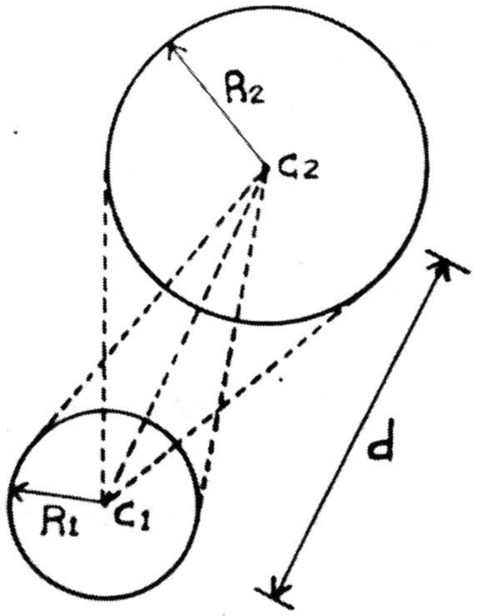

Fig (158)

We know that the solid angle subtended by the sphere at any point in the space is given by the eq(39) as follows

$$\omega = 2\pi\left[1 - \frac{\sqrt{d^2 - R^2}}{d}\right]$$

Hence, the solid angle subtended by the body-1 with centre 'C_1' at the centre 'C_2' of the body-2

$$\Rightarrow \omega_{12} = 2\pi\left[1 - \frac{\sqrt{d^2 - {R_1}^2}}{d}\right]$$

Hence, the solid angle subtended by the body-2 with centre 'C_2' at the centre 'C_1' of the body-1

$$\Rightarrow \omega_{21} = 2\pi\left[1 - \frac{\sqrt{d^2 - {R_2}^2}}{d}\right]$$

It is obvious that the solid angle subtended by the larger body will be higher than that subtended by the smaller one at their respective centres.

Here, we are to analyse the solid angle subtended by the common bodies assuming their shape to be spherical.

5.3.2 Solid angle subtended by the Earth & the Sun at their respective centres

We have the following data based on the experiments

$$Mean\ radius\ of\ the\ Earth, R_e = 6371\ Km = 6371000\ m$$

$$Mean\ radius\ of\ the\ Sun, R_s = 695500\ Km = 695500000\ m\quad \&$$

$$Average\ distance\ between\ the\ Earth\ \&\ the\ Sun, D_{es} = 149.59787\ millionKm\ = 149597870000\ m$$

Now, on setting the corresponding values, the solid angle subtended by the Sun at the centre of the Earth

$$\Rightarrow \omega_{se} = 2\pi\left[1 - \frac{\sqrt{{D_{es}}^2 - {R_s}^2}}{D_{es}}\right] = 2\pi\left[1 - \frac{\sqrt{(149597870000)^2 - (695500000)^2}}{(149597870000)}\right]$$

$$= 2\pi \times 1.0807266 \times 10^{-5} \approx 6.79 \times 10^{-5}\ sr\ or\ 67.90\ \mu sr$$

Similarly, solid angle subtended by the Earth at the centre of the Sun

$$\Rightarrow \omega_{es} = 2\pi\left[1 - \frac{\sqrt{{D_{es}}^2 - {R_e}^2}}{D_{es}}\right] = 2\pi\left[1 - \frac{\sqrt{(149597870000)^2 - (6371000)^2}}{(149597870000)}\right]$$

$$= 2\pi \times 9.09 \times 10^{-10} \approx 5.71 \times 10^{-9}\ sr\ or\ 5.71\ nsr$$

Now, we have

$$\frac{\omega_{se}}{\omega_{es}} = \frac{2\pi \times 1.0807266 \times 10^{-5}}{2\pi \times 9.09 \times 10^{-10}} = 0.118891815 \times 10^5 \approx 11891.42$$

Thus, the solid angle subtended by the Sun at the centre of the Earth is about 11891.42 times that subtended by the Earth at the centre of the Sun.

5.3.3 Solid angle subtended by the Earth & the Moon at their centres

We have the following data based on the experiments

$$Mean\ radius\ of\ the\ Earth, R_e = 6371\ Km = 6371000\ m$$

$$Mean\ radius\ of\ the\ Moon, R_m = 1737.10\ Km = 1737100\ m \quad \&$$

$$Average\ distance\ between\ the\ Earth\ \&\ the\ Moon, D_{em} = 384400Km \qquad = 384400000\ m$$

Now, on setting the corresponding values, the solid angle subtended by the Moon at the centre of the Earth

$$\Rightarrow \omega_{me} = 2\pi\left[1 - \frac{\sqrt{D_{em}{}^2 - R_m{}^2}}{D_{em}}\right] = 2\pi\left[1 - \frac{\sqrt{(384400000)^2 - (1737100)^2}}{(384400000)}\right]$$

$$= 2\pi \times 1.0210689 \times 10^{-5} \approx 6.415 \times 10^{-5}\ sr\ or\ 64.15\ \mu sr$$

Similarly, solid angle subtended by the Earth at the centre of the Moon

$$\Rightarrow \omega_{em} = 2\pi\left[1 - \frac{\sqrt{D_{em}{}^2 - R_e{}^2}}{D_{em}}\right] = 2\pi\left[1 - \frac{\sqrt{(384400000)^2 - (6371000)^2}}{(384400000)}\right]$$

$$= 2\pi \times 1.37356204 \times 10^{-4} \approx 8.6303 \times 10^{-4}\ sr\ or\ 863.03\ \mu sr$$

Now, we have

$$\frac{\omega_{em}}{\omega_{me}} = \frac{2\pi \times 1.37356204 \times 10^{-4}}{2\pi \times 1.0210689 \times 10^{-5}} = 13.4521974 \approx 13.45$$

Thus, the solid angle subtended by the Earth at the centre of the Moon is about 13.45 times that subtended by the Moon at the centre of the Earth.

5.3.4 Luminosity of star: It can be defined as the 'Total radiation energy emitted by a star per unit time'. It is simply denoted by 'L'. Its unit is j/sec or W.

Let there be a star with a (effective or mean) radius 'R_e' (assuming the star as a spherical body) has effective temperature 'T_e' (absolute).

The total surface area of the star

$$S = 4\pi R_e{}^2$$

We know the total radiation energy per unit area emitted by a perfect black body ($emissivity, \epsilon = 1$) is given from the Stefan's equation as follows

$$E = \sigma T^4$$

Where, σ is called Stefan's constant & $\sigma = 5.67 \times 10^{-8}\ Wm^{-2}K^{-4}$.

Now, the totalradiation energy emitted by the star

$$E_t = surface\ area \times energy\ emitted\ per\ unit\ surface\ area$$

$$= 4\pi R_e{}^2 \times \sigma T_e{}^4 = 4\pi\sigma R_e{}^2 T_e{}^4$$

Hence, the luminosity of the star

$$\Rightarrow L = 4\pi\sigma R_e{}^2 T_e{}^4 \qquad \ldots\ldots\ldots\ldots\ldots\ldots (87)$$

5.3.5 Brightness of star: It can be defined as the 'Radiation energy striking per unit time per unit area of an imaginary spherical surface with centre coincident with that of star & radius equal to the distance between the centre & the given point in the space'.

It is can simply be denoted by the symbol 'B'. Its unit is W/m^2.

Let there be a star with a radius say 'R_e' (assuming the star as a spherical body) has effective temperature 'T_e' (absolute) & a given point at a distance 'x' from the centre of star.

We know that the total energy emitted by a star is given as

$$E_t = 4\pi\sigma R_e{}^2 T_e{}^4$$

Now, the surface area of the imaginary spherical surface with centre coincident with that of the star & radius equal to 'x'

$$S = 4\pi x^2$$

Hence, radiation energy striking per unit time per unit area of the imaginary spherical surface

$$= \frac{E_t}{S} = \frac{4\pi\sigma {R_e}^2 {T_e}^4}{4\pi x^2} = \sigma {T_e}^4 \left(\frac{R_e}{x}\right)^2$$

$$\Rightarrow B = \sigma {T_e}^4 \left(\frac{R_e}{x}\right)^2 \qquad \ldots\ldots\ldots\ldots\ldots\ldots (88)$$

While for a uniform point-source with luminosity 'L', the brightness at a distance 'x' is given as

$$\Rightarrow B = \frac{total\ radiation\ energy}{area\ of\ imaginary\ spherical\ surface} = \frac{E_t}{S} = \frac{L}{4\pi x^2}$$

Note: The star is assumed as a perfect black body i.e. emissivity (ϵ) of star is taken unity.

In other cases, emissivity (ϵ)of source can be considered.

5.3.6 Relation between the Luminosity & Brightness of a star

On dividing the equation (88) by (87) as follows

$$\Rightarrow \frac{B}{L} = \frac{\sigma {T_e}^4 \left(\frac{R_e}{x}\right)^2}{4\pi\sigma {R_e}^2 {T_e}^4} = \frac{1}{4\pi x^2}$$

$$\Rightarrow B = \frac{L}{4\pi x^2} \qquad \ldots\ldots\ldots\ldots\ldots\ldots (89)$$

5.3.7 Energy of interception

Let there be a spherical (or point) source (assuming perfect black body, e = 1) with luminosity 'L' emitting the radiation energy equally in all the directions and agiven object subtending a solid angle $'\omega'$ at the centre of the source.

Now, the radiation energy emitted by the source per unit time per unit solid angle

$$= \frac{total\ energy}{4\pi} = \frac{L}{4\pi}$$

Hence, the total radiation energy intercepted by the object

$$E_i = \omega \times \frac{L}{4\pi} = \frac{\omega L}{4\pi}$$

Where the solid angle subtended by the spherical body

$$\Rightarrow \omega = 2\pi\left(1 - \frac{\sqrt{d^2 - R^2}}{d}\right) \&\ luminosity, L = 4\pi\sigma {R_e}^2 {T_e}^4$$

$$\Rightarrow E_i = \frac{4\pi\sigma {R_e}^2 {T_e}^4}{4\pi} \times 2\pi\left(1 - \frac{\sqrt{d^2 - R^2}}{d}\right) = 2\pi\sigma {R_e}^2 {T_e}^4 \left(1 - \frac{\sqrt{d^2 - R^2}}{d}\right)$$

$$\therefore E_i = \frac{\omega L}{4\pi} = 2\pi\sigma {R_e}^2 {T_e}^4 \left(1 - \frac{\sqrt{d^2 - R^2}}{d}\right) \qquad \ldots\ldots\ldots\ldots\ldots\ldots (90)$$

Where,R_e is the effective or mean radius &T_e is the effective temperature of a spherical source.

5.3.8 Area of interception of a spherical body:

It is the area of visible surface of a spherical body when viewed from a certain point in the space or area of surface of the spherical body intercepted by a given point in the space.

If 'R' is the radius of a sphere & a given point is lying at a distance d from the centre of the body then area of interception is given as

$$\Rightarrow A_i = 2\pi R^2(1 - cos\theta)$$

The value of θ can be found out as follows

$$cos\theta = \frac{R}{d} \quad \forall R \le d$$

$$\Rightarrow A_i = 2\pi R^2 \left(1 - \frac{R}{d}\right)$$

In actual practice, the energy of interception is non-uniformly distributed over the area of interception of a spherical body.

Solved Examples

Example 5.1: Find out the solid angle subtended by three identical spheres each with a radius 4cm arranged, touching one another, in a complete circular fashion at a point lying at a height 12cm on the axis passing through the centre of the regular-polygonal plane (called reference plane) obtained by joining the centres of all the spheres.

Sol. given that

$$number\ of\ the\ spheres, n = 3, radius\ of\ spheres, R = 4cm\ \&$$
$$height\ of\ the\ point\ from\ the\ centre\ of\ reference\ plane, h = 12cm$$

Hence, solid angle subtended by the spheres at the given point (using formula)

$$\omega = 2n\pi\left[1 - \sqrt{\frac{h^2 + R^2cot^2\frac{\pi}{n}}{h^2 + R^2cosec^2\frac{\pi}{n}}}\right]$$

On setting corresponding values in the above expression, we get

$$\omega = 2 \times 3\pi\left[1 - \sqrt{\frac{12^2 + 4^2cot^2\frac{\pi}{3}}{12^2 + 4^2cosec^2\frac{\pi}{3}}}\right] = 6\pi\left[1 - \sqrt{\frac{149.333}{165.333}}\right]$$

$$= 6\pi[1 - 0.950] = 0.942\ sr \qquad Ans.$$

Example 5.2: Determine the solid angle subtended by fifteen identical spheres each with a radius 11cm arranged, touching one another, in a complete circular fashion at a point lying at a height 21cm on the axis passing through the centre of the regular-polygonal plane obtained by joining the centres of all the spheres.

Sol. given that

$$number\ of\ the\ spheres, n = 15,\ radius\ of\ spheres, R = 11cm\ \&$$
$$height\ of\ the\ point\ from\ the\ centre\ of\ reference\ plane, h = 21cm$$

Hence, solid angle subtended by the spheres at the given point (using formula)

$$\omega = 2n\pi\left[1 - \sqrt{\frac{h^2 + R^2cot^2\frac{\pi}{n}}{h^2 + R^2cosec^2\frac{\pi}{n}}}\right]$$

On setting corresponding values in the above expression, we get

$$\omega = 2 \times 15\pi\left[1 - \sqrt{\frac{21^2 + 11^2 cot^2\frac{\pi}{15}}{21^2 + 11^2 cosec^2\frac{\pi}{15}}}\right] = 30\pi\left[1 - \sqrt{\frac{3119.159}{3240.159}}\right]$$

$$= 30\pi[1 - 0.981] = 1.791\ sr \qquad Ans.$$

Example 5.3: Thirty identical spheres are arranged, touching one another, in a complete circular fashion. What will be the solid angle subtended by the spheres at the centre of the regular polygonal plane obtained by joining the centres of all the spheres?

Sol. given that

$$number\ of\ the\ identical\ spheres, n = 30$$

We know that solid angle subtended by 'n' number of the identical spheres arranged, touching one another, in a complete circular fashion at the centre of reference plane is given as (irrespective radius of the spheres)

$$\omega = 4n\pi sin^2\left(\frac{\pi}{2n}\right)$$

On setting n = 30 in the above expression, we get

$$= 4 \times 30\pi sin^2\left(\frac{\pi}{2 \times 30}\right)$$

$$= 120\pi sin^2 3^o = 1.032\ sr \qquad Ans.$$

Example 5.4: Twenty identical spheres are arranged, touching one another, in a complete circular fashion. Determine the solid angle subtended by the spheres at the centre of the regular polygonal plane obtained by joining the centres of all the spheres. If radius of all the spheres is changed what will be the change in the solid angle?

Sol. given that

$$number\ of\ the\ identical\ spheres, n = 20$$

Hence, solid angle subtended by the spheres at the centre of reference plane is given as follows

$$\omega = 4n\pi sin^2\left(\frac{\pi}{2n}\right)$$

On setting n = 20 in the above expression, we get

$$= 4 \times 20\pi sin^2\left(\frac{\pi}{2 \times 20}\right)$$

$$= 80\pi sin^2 4.5^o = 1.547\ sr \qquad Ans.$$

Since, the solid angle subtended by the spheres does not depend on the radius of the spheres hence there will be no change in the value of solid angle.

Example 5.5: sixty identical spheres are arranged, touching one another, in a complete circular fashion. Determine the solid angle subtended by the spheres at the centre of

1) Regular polygonal plane obtained by joining the centres of all the spheres & 2) plane congruent & parallel to the first one & touching all the spheres.

Sol. given that

$$number\ of\ the\ identical\ spheres, n = 60$$

1. Solid angle subtended by the spheres at the centre of reference plane is given as follows

$$\omega = 4n\pi sin^2\left(\frac{\pi}{2n}\right)$$

On setting n = 20 in the above expression, we get

$$= 4 \times 60\pi sin^2\left(\frac{\pi}{2 \times 60}\right)$$

$$= 240\pi sin^2 1.5^o = 0.517\ sr \qquad Ans.$$

2. We know that the solid angle subtended by the circular arrangement of 'n' no. of the identical spheres each of radius 'R' at the point lying at a height h on the vertical axis passing through the centre of the reference plane is given by the eq(86) as follows

$$\omega = 2n\pi\left[1 - \sqrt{\frac{h^2 + R^2 cot^2\frac{\pi}{n}}{h^2 + R^2 cosec^2\frac{\pi}{n}}}\right]$$

On setting h = R in the above equation, we get

$$\omega = 2n\pi\left[1 - \sqrt{\frac{R^2 + R^2 cot^2\frac{\pi}{n}}{R^2 + R^2 cosec^2\frac{\pi}{n}}}\right] = 2n\pi\left[1 - \sqrt{\frac{1 + cot^2\frac{\pi}{n}}{1 + cosec^2\frac{\pi}{n}}}\right]$$

$$= 2n\pi\left[1 - \sqrt{\frac{cosec^2\frac{\pi}{n}}{1 + \frac{1}{sin^2\frac{\pi}{n}}}}\right] = 2n\pi\left[1 - \frac{1}{\sqrt{1 + sin^2\frac{\pi}{n}}}\right]$$

Now, on setting n = 60, we get

$$\omega = 2 \times 60\pi\left[1 - \frac{1}{\sqrt{1 + sin^2\frac{\pi}{60}}}\right] = 120\pi\left[1 - \frac{1}{1.0013}\right]$$

$$= 120\pi[1 - 0.998] = 0.754\ sr \qquad Ans.$$

Example 5.6: If the solid angle subtended by each of the spheres arranged, touching one another, in a complete circular fashion is 95.455 msr at the centre of the regular polygonal plane obtained by joining the centres of all the spheres. Find out the number of spheres.

Sol. given that

$$solid\ angle\ subtended\ by\ each\ of\ the\ spheres, \omega = 95.455\ msr$$

Let the number of the identical spheres be 'n' then solid angle subtended by each of the spheres at the centre of reference plane is given as follows

$$\omega = \frac{1}{n}\left(4n\pi sin^2\left(\frac{\pi}{2n}\right)\right) = 4\pi sin^2\left(\frac{\pi}{2n}\right)$$

On setting the value of ω in the above expression, we get

$$0.095455 = 4\pi sin^2\left(\frac{\pi}{2n}\right)$$

$$sin\left(\frac{\pi}{2n}\right) = \sqrt{\frac{0.095455}{4\pi}} = 0.0871\ or\ \frac{180^o}{2n} = \sin^{-1}(0.0871) = 5^o$$

$$n = \frac{180^o}{10^o} = 18 \qquad Ans.$$

Example 5.7: If the solid angle subtended by each of the spheres arranged, touching one another, in a complete circular fashion is 0.379 sr at the centre of the regular polygonal plane obtained by joining the centres of all the spheres. Find out the number of spheres.

Sol. given that

$$solid\ angle\ subtended\ by\ each\ of\ the\ spheres, \omega = 0.379\ sr$$

Let the number of the identical spheres be 'n' then solid angle subtended by each of the spheres at the centre of reference plane is given as follows

$$\omega = \frac{1}{n}\left(4n\pi sin^2\left(\frac{\pi}{2n}\right)\right) = 4\pi sin^2\left(\frac{\pi}{2n}\right)$$

On setting the value of ω in the above expression, we get

$$0.379 = 4\pi sin^2\left(\frac{\pi}{2n}\right)$$

$$sin\left(\frac{\pi}{2n}\right) = \sqrt{\frac{0.379}{4\pi}} = 0.174 \ or \ \frac{180^o}{2n} = \sin^{-1}(0.174) = 10^o$$

$$n = \frac{180^o}{20^o} = 9 \qquad Ans.$$

Example 5.8: If the solid angle subtended by each of the spheres arranged, touching one another, in a complete circular fashion is 3.473 sr at the centre of a plane which is parallel & congruent to the regular polygonal plane obtained by joining the centres of all the spheres, & touching all the spheres at peak points. Find out the number of spheres.

Sol. given that

$$solid\ angle\ subtended\ by\ each\ of\ the\ spheres, \omega = 3.473\ sr$$

Let the number of the identical spheres be 'n' then solid angle subtended by each of the spheres at the centre of a plane, parallel & congruent to the reference plane, & touching all the spheres at peak points is given as follows

$$\Rightarrow \omega = \frac{1}{n} \times 2n\pi\left[1 - \frac{1}{\sqrt{1 + \frac{1}{sin^2\frac{\pi}{n}}}}\right] = 2\pi\left[1 - \frac{1}{\sqrt{1 + \frac{1}{sin^2\frac{\pi}{n}}}}\right]$$

Now, on setting the value of ω in the above expression, we get

$$3.473 = 2\pi\left[1 - \frac{1}{\sqrt{1 + \frac{1}{sin^2\frac{\pi}{n}}}}\right] \Rightarrow \frac{1}{\sqrt{1 + \frac{1}{sin^2\frac{\pi}{n}}}} = 1 - \frac{3.473}{2\pi} = 0.447$$

$$1 + \frac{1}{sin^2\frac{\pi}{n}} = \left(\frac{1}{0.447}\right)^2 \ or \ \frac{1}{sin^2\frac{\pi}{n}} = 4.005$$

$$or\ sin\frac{\pi}{n} = \sqrt{\frac{1}{4.005}} = 0.4997 \ or \ \frac{\pi}{n} = \sin^{-1}(0.4997)$$

$$or \ \frac{180^o}{n} = 30^o \Rightarrow \quad n = 6 \qquad Ans.$$

Example 5.9: Find out the luminous flux intercepted by eight identical spheres each with a radius 11cm arranged, touching one another, in a complete circular fashion if a uniform point-source of 110 cd is located at a point lying on the vertical axis at height 17cm from the centre of regular polygon obtained by joining the centres of all the spheres.

Sol. given that

$$number\ of\ spheres, n = 8, radius\ of\ spheres, R = 11cm, intensity, I = 110\ cd\ \&$$

$$height\ of\ the\ point, h = 17cm$$

Now, solid angle subtended by the spheres at the given point (using formula)

$$\omega = 2n\pi\left[1 - \sqrt{\frac{h^2 + R^2cot^2\frac{\pi}{n}}{h^2 + R^2cosec^2\frac{\pi}{n}}}\right]$$

On setting corresponding values in the above expression, we get

$$\omega = 2 \times 8\pi\left[1 - \sqrt{\frac{17^2 + 11^2cot^2\frac{\pi}{8}}{17^2 + 11^2cosec^2\frac{\pi}{8}}}\right] = 16\pi\left[1 - \sqrt{\frac{994.24}{1115.24}}\right]$$

$$= 16\pi[1 - 0.944] = 2.815\ sr$$

Hence, the total flux intercepted by the spheres

$$\Rightarrow\ \ F = I \times \omega = 110 \times 2.815 = 309.65\ lm \qquad Ans.$$

Example 5.10: Determine the luminous flux intercepted by twelve no. of the identical spheres arranged, touching one another, in a complete circular fashion when a uniform point source of 200 cd is located at the centre of regular polygonal plane obtained by joining the centres of all the spheres. Also find the flux directly exposed to the surroundings.

Sol. given that

$$number\ of\ spheres, n = 12\ \ \&\ intensity, I = 200\ cd$$

Now, solid angle subtended by the spheres at the centre of reference plane is given as follows

$$\omega = 4n\pi sin^2\left(\frac{\pi}{2n}\right)$$

On setting n = 12 in the above expression, we get

$$= 4 \times 12\pi sin^2\left(\frac{\pi}{2 \times 12}\right)$$

$$= 48\pi sin^2 7.5^o = 2.569\ sr$$

Hence, the total flux intercepted by the spheres

$$\Rightarrow\ \ F = I \times \omega = 200 \times 2.569 = 513.8\ lm \qquad Ans.$$

We know that total flux emitted by uniform point-source

$$\Rightarrow F_t = 4\pi \times I = 4\pi \times 200 = 2513.274\ lm$$

Hence, the flux directly exposed to the surroundings

$$= F_t - F = 2513.274 - 513.8 = 1999.474\ lm \qquad Ans.$$

Example 5.11: What will the luminous flux intercepted by twenty number of the identical spheres arranged, touching one another, in a complete circular fashion when a uniform point source of 110 cd is located at the centre of a plane congruent & parallel, to the regular polygonal plane obtained by joining the centres of all the spheres, & touching all the spheres at peak points.

Sol. given that

$$number\ of\ spheres, n = 20\ \ \&\ intensity, I = 110\ cd$$

Let the number of the identical spheres be 'n' then solid angle subtended by each of the spheres at the centre of a plane, parallel & congruent to the reference plane, & touching all the spheres at peak points is given as follows

$$\omega = 2n\pi\left[1 - \frac{1}{\sqrt{1 + sin^2\frac{\pi}{n}}}\right]$$

Now, on setting n = 20, we get

$$\omega = 2 \times 20\pi\left[1 - \frac{1}{\sqrt{1 + sin^2\frac{\pi}{20}}}\right] = 40\pi\left[1 - \frac{1}{1.0122}\right]$$

$$= 40\pi[1 - 0.988] = 1.508\ sr$$

Hence, the total flux intercepted by the spheres

$$\Rightarrow \quad F = I \times \omega = 110 \times 1.508 = 165.88\ lm \qquad Ans.$$

Example 5.12: Determine the total flux directly escaping the tetrahedral void when a uniform point-source of 250 cd is placed at the centre of the void.

Sol. given that

$$Intensity\ of\ the\ uniform\ point\ source, I = 250\ cd$$

We know that the solid angle subtended by all the spheres forming a tetrahedral void at the centre is 10.622 sr.

Hence, the flux intercepted by the spheres or trapped in the void

$$\Rightarrow \quad F = I \times \omega = 250 \times 10.622 = 2655.5\ lm$$

We know that total flux emitted by uniform point-source

$$\Rightarrow F_t = 4\pi \times I = 4\pi \times 250 = 3141.593\ lm$$

Hence, the flux directly escaping the void (i.e. exposed to the surroundings)

$$= F_t - F = 3141.593 - 2655.5 = 486.093\ lm \qquad Ans.$$

Example 5.13: Determine the total flux directly escaping the octahedral void when a uniform point-source emitting 1400 lm flux is placed at the centre of the void.

Sol. given that

$$total\ flux\ emitted\ by\ the\ uniform\ point\ source, F_t = 1400\ lm$$

We know that total flux emitted by the source is given as

$$\Rightarrow F_t = 4\pi \times I \quad \Rightarrow I = \frac{1400}{4\pi} = 111.408\ cd$$

Since, the solid angle subtended by all the spheres forming an octahedral void at the centre is 11.042 sr.

Hence, the flux intercepted by the spheres or trapped in the void

$$\Rightarrow \quad F = I \times \omega = 111.408 \times 11.042 = 1230.167\ lm$$

Hence, the flux directly escaping the void (i.e. exposed to the surroundings)

$$= F_t - F = 1400 - 1230.167 = 169.833\ lm \qquad Ans.$$

Example 5.14: A uniform point-source of 110 cd is located at a distance 1m from the centre of a sphere with a radius 22cm. If the flux if uniformly distributed over the intercepted surface what will be the flux density?

Sol. given that

$$intensity\ of\ point\ source, I = 110\ cd, radius, R = 22cm = 0.22m\ \&$$

$$distance\ of\ the\ point\ from\ the\ centre\ of\ the\ sphere, d = 1m$$

We know that solid angle subtended by a sphere at the given point in the space

$$\omega = 2\pi\left[1 - \frac{\sqrt{d^2 - R^2}}{d}\right]$$

$$= 2\pi\left[1 - \frac{\sqrt{(1)^2 - (0.22)^2}}{(1)}\right] = 2\pi[1 - 0.975] = 2\pi \times 0.025 = 0.157\ sr.$$

Hence, the flux intercepted by the sphere

$$\Rightarrow \quad F = I \times \omega = 110 \times 0.157 = 17.27\ lm$$

Intercepted area of the sphere w.r.t. to the given point

$$\Rightarrow A_i = 2\pi R^2(1 - cos\alpha) = 2\pi R^2\left(1 - \frac{R}{d}\right)$$

$$= 2\pi(0.22)^2\left(1 - \frac{(0.22)}{1}\right) = 0.237\ m^2$$

Hence, the flux density over the intercepted area

$$F.D. = \frac{F}{A_i} = \frac{17.27}{0.237} = 72.869\ lux \qquad Ans.$$

Example 5.15: A uniform point-source emitting total flux 1600 lm is located at the centre of a simple cubic cell. Determine the luminous flux directly escaping the cell.

Sol. given that

$$total\ flux\ emitted\ by\ the\ uniform\ point\ source, F_t = 1600\ lm$$

We know that total flux emitted by the source is given as

$$\Rightarrow F_t = 4\pi \times I \quad \Rightarrow I = \frac{1600}{4\pi} = 127.324\ cd$$

Since, the solid angle subtended by all the spheres forming a simple cubic cell at the centre is 9.224 *sr*.

Hence, the flux intercepted by the spheres or trapped in the void

$$\Rightarrow \quad F = I \times \omega = 127.324 \times 9.224 = 1174.436\ lm$$

Hence, the flux directly escaping the void (i.e. exposed to the surroundings)

$$= F_t - F = 1600 - 1174.436 = 425.564\ lm \qquad Ans.$$

Example 5.16: 'n' number of the identical spheres each of radius 'R' arranged, touching each other, in a complete circular fashion such that all the spheres intercept x fraction of the total flux emitted by a uniform point-source located at a normal height 'h' from the centre of a regular polygon obtained by joining the centres of all the spheres. Prove that

$$\frac{h}{R} = \frac{1}{2}\sqrt{\frac{(n-2x)^2 cosec^2\frac{\pi}{n} - n^2 cot^2\frac{\pi}{n}}{x(n-x)}} \quad (\forall\ x < 1).$$

Sol. Let the intensity of the point-source be 'I'.

The solid angle subtended by the spheres at the point source

$$\omega = 2n\pi\left[1 - \sqrt{\frac{h^2 + R^2 cot^2\frac{\pi}{n}}{h^2 + R^2 cosec^2\frac{\pi}{n}}}\right]$$

Hence the total flux intercepted by the spheres

$$F = I \times \omega = I \times 2n\pi\left[1 - \sqrt{\frac{h^2 + R^2cot^2\frac{\pi}{n}}{h^2 + R^2cosec^2\frac{\pi}{n}}}\right] = x(total\ flux)$$

$$I \times 2n\pi\left[1 - \sqrt{\frac{h^2 + R^2cot^2\frac{\pi}{n}}{h^2 + R^2cosec^2\frac{\pi}{n}}}\right] = x(4\pi \times I)$$

$$or\ 1 - \sqrt{\frac{h^2 + R^2cot^2\frac{\pi}{n}}{h^2 + R^2cosec^2\frac{\pi}{n}}} = \frac{2x}{n} \Rightarrow \sqrt{\frac{h^2 + R^2cot^2\frac{\pi}{n}}{h^2 + R^2cosec^2\frac{\pi}{n}}} = 1 - \frac{2x}{n} = \frac{(n-2x)}{n}$$

$$or\ \frac{h^2 + R^2cot^2\frac{\pi}{n}}{h^2 + R^2cosec^2\frac{\pi}{n}} = \frac{(n-2x)^2}{n^2}$$

$$or\ n^2h^2 + n^2R^2cot^2\frac{\pi}{n} = (n-2x)^2h^2 + (n-2x)^2R^2cosec^2\frac{\pi}{n}$$

$$\Rightarrow h^2\{n^2 - (n-2x)^2\} = R^2\left\{(n-2x)^2cosec^2\frac{\pi}{n} - n^2cot^2\frac{\pi}{n}\right\}$$

$$\Rightarrow h^2\{(n - n + 2x)(n + n - 2x)\} = R^2\left\{(n-2x)^2cosec^2\frac{\pi}{n} - n^2cot^2\frac{\pi}{n}\right\}$$

$$h^2\{4x(n-x)\} = R^2\left\{(n-2x)^2cosec^2\frac{\pi}{n} - n^2cot^2\frac{\pi}{n}\right\}$$

$$\Rightarrow \frac{h^2}{R^2} = \frac{(n-2x)^2cosec^2\frac{\pi}{n} - n^2cot^2\frac{\pi}{n}}{4x(n-x)}$$

$$or\ \frac{h}{R} = \sqrt{\frac{(n-2x)^2cosec^2\frac{\pi}{n} - n^2cot^2\frac{\pi}{n}}{4x(n-x)}}$$

$$\Rightarrow \frac{h}{R} = \frac{1}{2}\sqrt{\frac{(n-2x)^2cosec^2\frac{\pi}{n} - n^2cot^2\frac{\pi}{n}}{x(n-x)}}$$

Since, the spheres can't intercept the total flux emitted by the point-source, hence the value of fraction x is always less than unity.

Thus we have,

$$\frac{h}{R} = \frac{1}{2}\sqrt{\frac{(n-2x)^2cosec^2\frac{\pi}{n} - n^2cot^2\frac{\pi}{n}}{x(n-x)}}\ (\forall\, x < 1) \qquad proved.$$

Example 5.17: Nine identical spheres arranged, touching each other, in a complete circular fashion such that all the spheres intercept 124 lm flux emitted by a uniform point-source of 200 cd located at a normal height 19cm from the centre of a regular polygon obtained by joining the centres of all the spheres. Find out the radius of the spheres.

Sol. given that

$$number\ of\ the\ spheres, n = 9, intercepted\ flux, F = 124\ lm,$$
$$intensity, I = 200\ cd\ \&\ height\ of\ point\ source, h = 19\ cm$$

The solid angle subtended by the spheres at the point source

$$\omega = 2n\pi\left[1 - \sqrt{\frac{h^2 + R^2 cot^2\frac{\pi}{n}}{h^2 + R^2 cosec^2\frac{\pi}{n}}}\right]$$

Hence the total flux intercepted by the spheres

$$F = I \times \omega = I \times 2n\pi\left[1 - \sqrt{\frac{h^2 + R^2 cot^2\frac{\pi}{n}}{h^2 + R^2 cosec^2\frac{\pi}{n}}}\right]$$

Now, on setting the values, we have

$$\Rightarrow\ 124 = 200 \times 2 \times 9\pi\left[1 - \sqrt{\frac{(19)^2 + R^2 cot^2\frac{\pi}{9}}{(19)^2 + R^2 cosec^2\frac{\pi}{9}}}\right]$$

$$\sqrt{\frac{361 + 7.549R^2}{361 + 8.549R^2}} = 1 - \frac{124}{3600\pi} = 0.989$$

$$\frac{361 + 7.549R^2}{361 + 8.549R^2} = (0.989)^2 = 0.978$$

$$361 + 7.549R^2 = 353.058 + 8.361R^2 \Rightarrow R^2 = \frac{7.942}{0.812} = 9.781$$

$$radius\ of\ the\ spheres, R = 3.127\ cm \qquad Ans.$$

Example 5.18: Find out the flux density on a plane surface having area 2.345 m^2&subtending a solid angle 2.45 sr. at a uniform point-source emitting a total flux 1257 lm.

Sol. given that

$$Area, A = 2.345\ m^2, solid\ angle, \omega = 2.45\ sr\ \&$$
$$total\ flux\ emitted\ by\ source, F_t = 1257\ lm$$

We know that the total flux emitted by a point-source is given as

$$F_t = 4\pi \times I\ \Rightarrow I = \frac{F_t}{4\pi} = \frac{1257}{4\pi} = 100\ cd$$

Total flux intercepted by the plane

$$F = I \times \omega = 100 \times 2.45 = 245\ lm$$

Hence, flux density over the plane

$$F.D. = \frac{F}{A} = \frac{245}{2.45} = 100\ lux \qquad Ans.$$

Example 5.19: Determine the flux ona square plane with each side 18cm if a uniform point-source of 150 cd is mounted at a normal height 14cm from the centre. (Assume that the flux is uniformly distributed over the entire plane)

Sol. given that

$$side\ of\ square\ plane, a = 18cm,\ \ intensity, I = 150\ cd\ \&$$
$$normal\ height, h = 14cm$$

We know that the solid angle subtended by a square plane at any point lying on the vertical axis passing through the centre of plane is given as

$$\omega = 4\sin^{-1}\left(\frac{a^2}{a^2 + 4h^2}\right)$$

$$= 4\sin^{-1}\left(\frac{(18)^2}{(18)^2 + 4(14)^2}\right) = 4\sin^{-1}(0.292) = 1.185\ sr$$

Hence, total flux intercepted by the plane

$$F = I \times \omega = 150 \times 1.185 = 177.75\ lm$$

Hence, flux density over the plane

$$F.D. = \frac{F}{A} = \frac{177.75}{(0.18)^2} = 5486.111\ lux \qquad Ans.$$

Example 5.20: Find out the luminous intensity of a uniform point-source mounted at a height 15cm on a vertical axis passing through the centre of a rectangular plane of $12cm \times 22cm$if the total flux intercepted by the plane is 177.4 lm.

Sol. given that

$$length, l = 22cm, \quad width, b = 12cm, \quad flux\ intercepted, F = 177.4\ lm\ \&$$
$$normal\ height, h = 15cm$$

We know that the solid angle subtended by a rectangular plane at any point lying on the vertical axis passing through the centre of plane is given as

$$\omega = 4\sin^{-1}\left(\frac{lb}{\sqrt{(l^2 + 4h^2)(b^2 + 4h^2)}}\right)$$

$$= 4\sin^{-1}\left(\frac{(22)(12)}{\sqrt{((22)^2 + 4(15)^2)((12)^2 + 4(15)^2)}}\right) = 4\sin^{-1}(0.22) = 0.887\ sr$$

If the intensity of source be 'I' the total flux intercepted by the plane

$$F = I \times \omega \Rightarrow 177.4 = I \times .887\ or\ I = \frac{177.4}{0.887}$$

$$intensity, \quad I = 200\ cd \qquad Ans.$$

Example 5.21: Find out the luminous flux density of a square plane with each side 32cm if a uniform point emitting total flux 1257 lm is located at a height 24cm on a vertical axis passing through one of the vertices of plane.

(Assume that the flux is uniformly distributed over the entire plane.)

Sol. given that

$$side\ of\ square\ plane, a = 32cm, \quad total\ flux, F_t = 1257\ lm\ \&$$
$$normal\ height, h = 24cm$$

We know that the total flux emitted by a point-source is given as

$$F_t = 4\pi \times I \Rightarrow I = \frac{F_t}{4\pi} = \frac{1257}{4\pi} = 100\ cd$$

We know that solid angle subtended by a square plane at any point lying on the vertical axis passing through the centre of plane is given as

$$\omega = \sin^{-1}\left(\frac{a^2}{a^2 + h^2}\right)$$

$$= \sin^{-1}\left(\frac{(32)^2}{(32)^2 + (24)^2}\right) = \sin^{-1}(0.64) = 0.694\ sr$$

Hence, total flux intercepted by the plane

$$F = I \times \omega = 100 \times 0.694 = 69.4\ lm$$

Hence, flux density over the plane

$$F.D. = \frac{F}{A} = \frac{69.4}{(0.32)^2} = 677.734\ lux \qquad Ans.$$

Example 5.22: Find out the flux density of a circular plane with a radius 28cm when a uniform point-source emitting 1885 lm flux is located at a height 120cm on a vertical axis passing through the centre of the plane.(Assume that the flux is uniformly distributed over the entire plane.)

Sol. given that

$$radius\ of\ circular\ plane, R = 28cm,\ \ toatl\ flux\ , F_t = 1885\ lm\ \&$$
$$normal\ height, h = 120cm$$

We know that the total flux emitted by a point-source is given as

$$F_t = 4\pi \times I \ \Rightarrow I = \frac{F_t}{4\pi} = \frac{1885}{4\pi} = 150\ cd$$

We know that the solid angle subtended by a circular plane at any point lying on the vertical axis passing through the centre of plane is given as

$$\omega = 2\pi\left[1 - \frac{h}{\sqrt{h^2 + R^2}}\right]$$

$$= 2\pi\left[1 - \frac{120}{\sqrt{(120)^2 + (28)^2}}\right] = 2\pi[1 - 0.974] = 0.163\ sr$$

Hence, total flux intercepted by the plane

$$F = I \times \omega = 150 \times 0.163 = 24.45\ lm$$

Hence, flux density over the plane

$$F.D. = \frac{F}{A} = \frac{24.45}{\pi(0.28)^2} = 99.269\ lux \qquad Ans.$$

Summary

Solid state (arrangement of identical spheres):

1. The solid angle subtended by 'n' number of the identical spheres each with a radius 'R' arranged, touching one another, in a complete circular fashion at a point lying on the vertical axis at a height 'h' from the centre of reference plane is given as

$$\omega = 2n\pi\left[1 - \sqrt{\frac{h^2 + R^2cot^2\frac{\pi}{n}}{h^2 + R^2cosec^2\frac{\pi}{n}}}\right][\ \forall\ n \in N\ \&\ n \geq 3\ ]$$

2. The solid angle subtended by 'n' number of the identical spheres each with a radius 'R' arranged, touching one another, in a complete circular fashion at the centre of reference plane is given as

$$\omega = 4n\pi sin^2\left(\frac{\pi}{2n}\right)[\ \forall n \in N\ \&\ n \geq 3\ ]$$

3. The solid angle subtended by the spheres forming a simple cubic cell at the centre is9.224 sr. Each of the spheres subtends 1.153 sr.

4. The solid angle subtended by the spheres forming a tetrahedral void at the centre is 10.622 sr. Each of the spheres subtends 2.655 sr.

5. The ratio of radius (r) of the largest sphere fully trapped in a tetrahedral void to the radius (R) of spheres forming the void, is $^r/_R = 0.225$
6. The solid angle subtended by largest sphere fully trapped in a tetrahedral void at the centre of each of the spheres forming the void is $0.107\ sr$.
7. The solid angle subtended by the spheres forming a octahedral void at the centre is $11.042\ sr$. Each of the spheres subtends $1.84\ sr$.
8. The ratio of radius (r) of the largest sphere fully trapped in a octahedral void to the radius (R) of spheres forming the void, is $^r/_R = 0.414$
9. The solid angle subtended by largest sphere fully trapped in an octahedral void at the centre of each of the spheres forming the void is $0.275\ sr$.

* About 15.47% of tetrahedral void &12.13% of octahedral void is exposed to their outside (surrounding) space. Thus an octahedral void is more packed than a tetrahedral void. (HCR)

Photometry/Radiometry:

10. The radiation (visible light or other E.M. radiation) energy emitted by uniform point-source located at the centre of a spherical surface is uniformly distributed over the entire the surface.
11. The radiation (visible light or other E.M. radiation) energy emitted by uniform linear-source coincident with the axis of a cylindrical surface is uniformly distributed over the entire the surface.
12. The distribution of radiation energy, emitted by a uniform point-source, over a plane surface is no more constant. Energy density at the foot of perpendicular drawn from the point-source to the plane is maximum & decreases successively at the points lying away from point of maxima.
13. The total luminous flux, emitted by a uniform point-source with luminous intensity I, intercepted by an object with visible surface 'S', is given as

$$F = I \times \omega = I \times \int_S \frac{dA}{r^2}$$

Where, $'\omega'$ is the solid angle subtended by the same object at the given point-source.

14. The total radiation energy, emitted by a uniform point-source with luminosity L, intercepted by an object with visible surface 'S', is given as

$$E = \left(\frac{L}{4\pi}\right) \times \omega = \left(\frac{L}{4\pi}\right) \int_S \frac{dA}{r^2}$$

Where, $'\omega'$ is the solid angle subtended by the same object at the given point-source.

15. Density of luminous flux (or luminosity) emitted by a uniform point-source with intensity I, over a plane-surface (S) with a total area 'A' (assuming uniform distribution of intercepted flux over the entire plane), is given as

$$F.D. = \frac{Total\ flux\ (F)\ intercepted\ by\ the\ plane}{Area\ (A)} = \frac{I \times \omega}{A} = \left(\frac{I}{A}\right) \int_S \frac{dA}{r^2}$$

16. Lambert's cosine formula:

Density of luminous flux (or luminosity (E)) emitted by a uniform point-source with intensity 'I' over a plane-surface with area 'A' (assuming uniform distribution of intercepted flux over the entire plane), is given

$$E = \frac{I cos\theta}{r^2}$$

Where, 'r' is the distance of point-source from the centre of the plane &

$'\theta'$ is the angle between normal to the plane & line joining point-source to the centre.

Limitation: The above formula is applicable for smaller plane surfaces and longer distances for minimum error involved in the results obtained.

Space-science (Celestial bodies):

17. The solid angle subtended by a celestial body with a mean radius at any point in the space (assuming that the body has a perfect spherical shape) is given as

$$\omega = 2\pi\left[1 - \frac{\sqrt{d^2 - R^2}}{d}\right] (\forall\ d \geq R)$$

18. The solid angle subtended by the Sun at the centre of the Earth is about $6.79 \times 10^{-5}\ sr$.
19. The solid angle subtended by the Earth at the centre of the Sun is about $5.71 \times 10^{-9}\ sr$.
20. Solid angle subtended by the Sun at the centre of the Earth is about 11891.42 times that subtended by the Earth at the centre of the Sun.
21. The total radiation energy, emitted by the Sun, incident on the Earth is about $1.75 \times 10^{17}\ {}^{j}/_{sec}$& energy density over the intercepted surface of the Earth is about $686.43\ {}^{W}/_{m^2}$ (Assuming no part of the radiation, emitted by the Sun, travelling to the Earth is absorbed or reflected).
22. The solid angle subtended by the Moon at the centre of the Earth is about $6.41 \times 10^{-5}\ sr$.
23. The solid angle subtended by the Earth at the centre of the Sun is about $8.63 \times 10^{-4}\ sr$.
24. Solid angle subtended by the Earth at the centre of the Moon is about 13.45 times that subtended by the Moon at the centre of the Earth.
25. The luminosity (L) of a star with a mean radius 'R_o', effective temperature 'T_e' (absolute) & emissivity 'e' is given as

$$L = 4\pi e\sigma {R_o}^2 {T_e}^4$$

Where, $\sigma = 5.67 \times 10^{-8}\ Wm^{-2}K^{-4}$

If the star is assumed as a perfect black body i.e. e = 1then we have

$$L = 4\pi\sigma {R_o}^2 {T_e}^4$$

26. Brightness (B) of a star with a mean radius 'R_o', effective temperature 'T_e' (absolute) & emissivity 'e' at a point lyingat a distance 'x' from the centre is given as

$$B = e\sigma {T_e}^4 \left(\frac{R_o}{x}\right)^2$$

If the star is assumed as a perfect black body i.e. e = 1 then we have

$$B = \sigma {T_e}^4 \left(\frac{R_o}{x}\right)^2$$

27. The relation between luminosity (L) & brightness (B) of a star at a particular point lying at a distance 'x' from the centre, is given as

$$B = \frac{L}{4\pi x^2}$$

28. Energy emitted by a uniform spherical-source with mean radius R_o, luminosity 'L', effective temperature 'T_e' & emissivity 'e', intercepted by a spherical body with a radius 'R' lying at a distance 'd' between their centres, is given as

$$E_i = \left(\frac{L}{2}\right)\left(1 - \frac{\sqrt{d^2 - R^2}}{d}\right) = 2\pi e\sigma {R_o}^2 {T_e}^4 \left(1 - \frac{\sqrt{d^2 - R^2}}{d}\right)$$

If the spherical-source is assumed as a perfect black body i.e. emissivity, e = 1 then we have

$$E_i = \left(\frac{L}{2}\right)\left(1 - \frac{\sqrt{d^2 - R^2}}{d}\right) = 2\pi\sigma R_o{}^2 T_e{}^4\left(1 - \frac{\sqrt{d^2 - R^2}}{d}\right)$$

29. Area of interception of a spherical body with a radius 'R' w.r.t. a given point lying at a distance d from the centre is given as

$$A_i = 2\pi R^2\left(1 - \frac{R}{d}\right)[\forall\, d \geq R\,]$$

30. If the radiation emitted, by a uniform spherical-source with mean radius R_o, effective temperature 'T_e', emissivity 'e' &luminosity 'L', is uniformly distributed over the area of interception of a spherical body with a radius 'R' & lying at a distance d from the point-source to its centre, is given as

$$E.D. = \left(\frac{L}{4\pi R^2}\right)\left(\frac{d - \sqrt{d^2 - R^2}}{d - R}\right) = e\sigma T_e{}^4\left(\frac{R_o}{R}\right)^2\left(\frac{d - \sqrt{d^2 - R^2}}{d - R}\right)[\forall\, d > (R_o + R)]$$

The above expression is applicable for uniform spherical-source like a star. If the source is assumed as perfect black body, take e =1 in the above expression.

Exercise - 5

* Solid state (Arrangement of spherical bodies):

Q.5.1. Find out the solid angle subtended by four identical spheres each with a radius 5cm arranged,touching one another, in a complete circular fashion at a point lying at a height 20cm on the axis passing through the centre of the regular-polygonal plane obtained by joining the centres of all the spheres.

Q.5.2. Determine the solid angle subtended by twelve identical spheres each with a radius 10cm arranged, touching one another, in a complete circular fashion at a point lying at a height 25cm on the axis passing through the centre of the regular-polygonal plane obtained by joining the centres of all the spheres.

Q.5.3. Three identical spheres are arranged, touching one another, in a complete circular fashion. What will be the solid angle subtended by the spheres at the centre of the regular polygonal plane obtained by joining the centres of all the spheres?

Q.5.4. Twenty identical spheres are arranged, touching one another, in a complete circular fashion. Determine the solid angle subtended by the spheres at the centre of the regular polygonal plane obtained by joining the centres of all the spheres. If radius of all the spheres is changed what will be change in the solid angle?

Q.5.5. Hundred identical spheres are arranged, touching one another, in a complete circular fashion. Determine the solid angle subtended by the spheres at the centre of the

1) Regular polygonal plane obtained by joining the centres of all the spheres & 2) plane congruent & parallel to the first one & touching all the spheres.

Q.5.6. If the solid angle subtended by each of the spheres arranged, touching one another, in a complete circular fashion is 49.54 msr at the centre of the regular polygonal plane obtained by joining the centres of all the spheres. Find out the number of spheres.

Q.5.7. If the solid angle subtended by each of the spheres arranged, touching one another, in a complete circular fashion is 0.478 sr at the centre of the regular polygonal plane obtained by joining the centres of all the spheres. Find out the number of spheres.

Q.5.8. Find out the number of spheres arranged, touching one another, in a complete circular fashion if each of the spheres subtends a solid angle 1.84 sr. at the centre of the regular polygonal plane obtained by joining the centres of all the spheres.

Q.5.9. Determine the number of spheres arranged, touching one another, in a complete circular fashion if each of the spheres subtends a solid angle 0.663 sr. at the centre of a plane touching all the spheres, congruent & parallel to a regular polygonal plane obtained by joining the centres of all the spheres.

Q.5.10. Find out the luminous flux intercepted by six identical spheres each with a radius 12cm arranged, touching one another, in a complete circular fashion if a uniform point-source of 100 cd is located at a point lying on the vertical axis at height 20cm from the centre of a regular polygon obtained by joining the centres of all the spheres.

Q.5.11. Determine the luminous flux intercepted by eight no. of the identical spheres arranged, touching one another, in a complete circular fashion when a uniform point source of 120 cd is located at the centre of the regular polygonal plane obtained by joining the centres of all the spheres.

Q.5.12. What will the luminous flux intercepted by five number of the identical spheres arranged, touching one another, in a complete circular fashion when a uniform point source of 100 cd is located at the centre of a plane congruent & parallel, to the regular polygonal plane obtained by joining the centres of all the spheres, & touching all the spheres at peak points.

Q.5.13. Determine the total flux directly escaping the tetrahedral void when a uniform point-source of 120 cd is placed at the centre of the void.

Q.5.14. Determine the total flux directly escaping the octahedral void when a uniform point-source of 100 cd is placed at the centre of the void.

Q.5.15. A uniform point-source of 111 cd is located at a distance 1.5m from the centre of a sphere with a radius 20cm. If the flux if uniformly distributed over the intercepted surface what will be the flux density?

Q.5.16. A uniform point-source emitting total flux 1400 lm is located at the centre of a simple cubic cell. Determine the luminous flux directly escaping the cell.

*** Photometry:**

Q.5.17. Find out the luminous flux intercepted by a plane surface subtending a solid angle 2.4 sr. at a uniform point-source emitting a total flux 1257 lm.

Q.5.18. Determine the flux intercepted by a square plane with each side 16cm if a uniform point-source of 110 cd is mounted at a height 18cm on a vertical axis passing through the centre.

Q.5.19. Find out the luminous intensity of a uniform point-source mounted at a height 20cm on a vertical axis passing through the centre of a rectangular plane of $10cm \times 20cm$if the total flux intercepted by the plane is 60 lm.

Q.5.20. Find out the luminous flux density of a square plane with each side 36cm if a uniform point emitting total flux 1257 lm is located at a height 20cm on a vertical axis passing through the vertex of the plane.

(Assume that the flux is uniformly distributed over the entire plane.)

Q.5.21. Find out the flux density of a circular plane with a radius 15cm when a uniform point-source of 150 cd is located at a height 1m on a vertical axis passing through the centre of the plane.(Assume that the flux is uniformly distributed over the entire plane.)

Q.5.22. Find out the flux intercepted by a plane bounded by the lines x = 20, x = 0, y = 12 & y = 0 on the X-Y plane when a uniform point-source of 150 cd is located at a point (0, 0, 15).

Q.5.23. A uniform point-source of 120 cd is located at a point (10, 0, 15). Find out the flux intercepted by a plane bounded by the curves x = 16, x = 0, y = 12 & y = 0 on the X-Y plane.

Q.5.24. If a uniform point-source of 100 cd is located at a point (16, 0, 15). Find out the flux intercepted by a plane bounded by the straight lines x = 9, x = 0, y = 11 & y = 0 on the X-Y plane.

Q.5.25. Find out the flux intercepted by a plane bounded by the straight lines x = 18, x = 0, y =13 & y = 0 on the X-Y plane if a uniform point-source of 150 cd is located at a point (0, 4, 10).

Q.5.26. A uniform point-source of 200 cd is located at a point (0, 20, 10). Find out the flux intercepted by a plane bounded by the straight lines x = 11, x = 0, y =8& y = 0 on the X-Y plane.

Q.5.27. If a uniform point-source of 100 cd is located at a point (3, 5, 8). Find out the flux intercepted by a plane bounded by the straight lines x = 12, x = 0, y = 8 & y = 0 on the X-Y plane.

Q.5.28. A uniform point-source of 200 cd is located at a point (15, 18, 8). Find out the flux intercepted by a plane bounded by the straight lines x = 12, x = 0, y = 8 & y = 0 on the X-Y plane.

Q.5.29. Determine the luminous flux density on a plane bounded by the straight lines y = 7, y = 0, z = 13 & z = 0 on Y-Z plane if a uniform point-source of 120 cd is located at a point (25, 4, 7). (Take all the dimensions in cm. & assume that the flux is uniformly distributed over the entire plane.)

Q.5.30. What will be the luminous flux density on a plane bounded by the straight lines y = 7, y = 0, z = 13 & z = 0 on Y-Z plane if a uniform point-source of 120 cd is located at a point (6, 12, 27). (Take all the dimensions in cm. & assume that the flux is uniformly distributed over the entire plane.)

Q.5.31. Find out the luminous flux density on a plane bounded by the straight lines x = 7, x = 0, z = 13& z = 0 on Z-X plane if a uniform point-source of 100 cd is located at a point (26, 3, 29). (Take all the dimensions in cm. & assume that the flux is uniformly distributed over the entire plane.)

Q.5.32. Determine the flux density on a cylindrical surface with a radius 6cm enveloping a uniform linear-source coincident with the longitudinal axis of the cylinder, emitting 9333 lm/m flux.

Q.5.33. A light bulb of 100 W is located at a point (12, 15, 17). Find out the total light energy striking the plane bounded by the lines x = 4, x = 0, y = 12 & y = 0 on the X-Y plane. (Assume the bulb as a point-source emitting the light energy uniformly in all the directions).

Q.5.34. A light bulb of 500 W is mounted at a height 1.5m on the axis passing through the vertex of a square plane with each side 4m. Find out the total light energy striking the plane.(Assume the bulb as a point-source emitting the light energy uniformly in all the directions).

Q.5.35. A uniform point-source with luminosity 200 W is located at a height 20cm on a vertical axis passing through the centre of a circular plane with a radius 50cm. What will the energy density on the plane if the energy is uniformly distributed over the entire plane?

Q.5.36. A uniform point-source with luminosity 2 kW is located at a height 5m on a vertical axis passing through the right angled vertex of a right triangular plane with orthogonal sides 10m & 12m. What will the energy density on the plane if the energy is uniformly distributed over the entire plane?

Q.5.37. Total light energy intercepted by a rectangular plane having sides 15cm & 13cm, is 50 j/sec when a uniform point-source is located at a height 8cm on a vertical axis passing through the centre. What is the luminosity of the source? Also find out brightness of the source at a distance 3m.

Q.5.38. The total light energy intercepted by a circular plane having radius 40cm, is 90.54 j/sec when a uniform point-source is located at a height 112mm on a vertical axis passing through the centre. Determine the brightness of the source at a distance 2.5m.

Q.5.39. A uniform point-source having brightness 0.398 W/m^2 at a distance 10m, is mounted at a height 4m on a vertical pole pitched onto the vertex of a square plane which intercepts a total light energy 20j/sec. Find out the energy density if the intercepted energy is uniformly distributed over the entire plane.

Q.5.40. Determine the total light energy intercepted by a plane bounded by the lines x = 2, x = 13, y = 5 & y = 25 on the X-Y plane when a uniform point-source of 100 W is located at a point (0, 0, 8).

Q.5.41. Determine the total light energy intercepted by a plane bounded by the lines $x = -1, x = 3, y = -2$ & $y = 7$ on the X-Y plane when a uniform point-source of 250 W is located at a point$(1, 3, 7)$.

Q.5.42. Calculate the total light energy intercepted by a plane bounded by the lines $x = -3, x = 5, y = 1$ & $y = 8$ on the X-Y plane if a uniform point-source of 220 W is located at a point $(-1, 13, 10)$.

Q.5.43. Determine the total flux intercepted by a plane bounded by the lines $x = -2, x = 10, y = -5$ & $y = 12$ on the X-Y plane when a uniform point-source of 200 cd is located at a point $(-2, -8, 4)$.

Q.5.44. Find out the total light energy intercepted by a plane bounded by the lines $x = 5, x = 24, y = -3$ & $y = 14$ on the X-Y plane when a uniform point-source of 250 W is located at a point $(3, -9, 7)$.

Q.5.45. Determine the total light energy intercepted by a plane bounded by the lines $x = -10, x = -3, y = 2$ & $y = 17$ on the X-Y plane when a uniform point-source of 500 W is located at a point $(2, -4, 4)$.

Q.5.46. Determine the total flux intercepted by a plane bounded by the curves $x = 0, y = 0$ &$x^2 + y^2 = 16$ on the X-Y plane when a uniform point-source of 100 cd is located at a point $(0, 0, 8)$.

Q.5.47. Calculate the total light energy intercepted by a plane bounded by the lines $x = -11, x = 13, y = -1$ & $y = 27$ on the X-Y plane when a uniform point-source of 300 W is located at a point $(-3, -6, 7)$.

Q.5.48. Find out the total flux intercepted by a plane bounded by the lines $x = 12, y = 0, y = 3x$ & $y = 17$ on the X-Y plane when a uniform point-source of 100 cd is located at a point $(0, 0, 13)$.

Q.5.49. Find out the total light energy intercepted by a plane bounded by the lines $x = 16, y = 0, y = 2x$ & $y = 14$ on the X-Y plane if a uniform point-source of 200 W is located at a point $(7, 14, 27)$.

Q.5.50. Determine the total flux intercepted by a plane bounded by the lines $x = -22, y = 0, y = -3x$ & $y = 18$ on the X-Y plane when a uniform point-source of 200 cd is located at a point $(-6, 0, 15)$.

Q.5.51. Find out the total flux intercepted by a plane bounded by the lines $x = 4, y = 2, y = 14$ & $y = -2x + 36$ on the X-Y plane when a uniform point-source of 100 cd is located at a point

1) $(11, 14, 5)$ 2) $(11, 2, 12)$& 3) $(17, 2, 3)$.

Q.5.52. A uniform point-source of 300 W is located at a point $(0, 0, 15)$. Find out the total energy intercepted by the plane bounded by the lines$x = 0, y = 6,\ y = 2x$.

Q.5.53. A uniform point-source of 200 W is located at a point $(6, -6, 7)$. Find out the total energy intercepted by the plane bounded by the lines $7x + y = 0, x - y = 0,\ x + y = 12$.

Q.5.54. A uniform point-source of 100 cd is located at a point $(-4, 24, 5)$. Calculate the total flux intercepted by the plane bounded by the lines $y = 0.5x + 3, y = -2x + 16,\ y = -6x$.

Q.5.55. Calculate the total light energy intercepted by a plane bounded by the lines $x = 0, y = 0, y = 12$ & $y = -2x + 18$ on the X-Y plane when a uniform point-source of 500 W is located at a point $(9, 0, 7)$.

Q.5.56. Find out the total flux intercepted by a plane bounded by the lines $8x - 3y + 24 = 0, y = -15$ & $8x + 3y = 24$ on the X-Y plane when a uniform point-source of 200 cd is located at a point $(0, 8, 4)$.

Q.5.57. Calculate the flux density over a plane bounded by the lines $7x + 9y = 0, x = 9$ & $9y = 7x$ on the X-Y plane when a uniform point-source of 300 cd is located at a point $(9, 0, 7)$. (Assume that the flux is uniformly distributed over the entire plane.)

Q.5.58. Determine the energy density over a plane bounded by the lines $17x - 4y = 0, y = -17$ & $17x + 4y = 0$ on the X-Y plane when a uniform point-source of 500 W is located at a point $(0, -17, 2)$.(Assume that the energy is uniformly distributed over the entire plane.)

Q.5.59. Calculate the total energy intercepted by a plane bounded by the lines $x = -3, x = 27, y = -8, y = 12$ & $x^2 + y^2 = 9$ on the X-Y plane when a uniform point-source of 200 W is located at a point $(0, 0, 5)$.

Q.5.60. Calculate the energy density over a plane bounded by the curves $x = -3, x = 7, y = -1, y = 5$ & $x^2 + y^2 = 169$ on the X-Y plane when a uniform point-source of 300 W is located at a point $(0, 0, 6)$.(Assume that the energy is uniformly distributed over the entire plane.)

Q.5.61. A uniform point-source emitting total flux 1400 lm is located at the point $(0, 0, 11)$. Determine the total flux intercepted by a plane bounded by the curves $x^2 + y^2 = 49$ & $x^2 + y^2 = 289$.

Q.5.62. Calculate the total energy intercepted by a plane bounded by the curves $x = 2, x = 27, y = 0, y = 12$ & $x^2 + y^2 = 144$ on the X-Y plane when a uniform point-source of 100 W is located at a point $(0, 0, 7)$.

Q.5.63. Determine the energy density over a plane bounded by the curves $x = -3, x = -24, y = -1, y = -15$ & $x^2 + y^2 = 49$ on the X-Y plane when a uniform point-source of 200 W is located at a point $(0, 0, -4)$.(Assume that the energy is uniformly distributed over the entire plane.)

Q.5.64. Find out the solid angle subtended by a plane bounded by the lines $y = \pm 3\sqrt{3}, \sqrt{3}x + y = \pm 6\sqrt{3}$ & $\sqrt{3}x - y = \pm 6\sqrt{3}$ at a uniform point-source of 200 cd located at the point $(0, 0, 7)$. If the flux is uniformly distributed over the entire plane what is the luminous flux density?

Q.5.65. A uniform point-source of 200 W is located at a point $(2, -3, -6)$. Determine the total energy intercepted by the plane bounded by the lines $x = -8, y = -3, y = 7$ & $\frac{x}{3.4} + \frac{y}{17} = 1$

Q.5.66. A uniform point-source emitting total flux 1400 lm, is located at a point $(7, -1, 5)$. Determine the value of constant $'a'$ ($\forall\, a > 0$) if 78.49 lm is the total flux intercepted by a plane bounded by the lines $x = -2, x = 7, y = -1$ & $y = a$. If the flux is uniformly distributed over the entire plane what is the luminous fluxdensity? *Ans:* $a = 9$, $0.872\ lm\ per\ sq.unit$.

Q.5.67. Calculate the total energy intercepted by a plane bounded by the curves $x = -3, x = 9, y = -2, y = 12$ & $x^2 + y^2 = 1$ on the X-Y plane when a uniform point-source of 100 W is located at a point $(0, 0, -4)$.

Q.5.68. Find out the total flux intercepted by a plane bounded by the curves $2x + 3y - 13 = 0$ & $x^2 + y^2 - 4x - 6y + 12 = 0$ on the X-Y plane when a uniform point-source of 200 cd is located at a point $(2, 3, -3)$.

Q.5.69. Determine the total energy intercepted by a plane bounded by the curves $x + y - 9 = 0$ & $4x^2 + 4y^2 - 36x - 36y + 158 = 0$ when a uniform point-source of 500 W is located at a point $(4.5, 4.5, -5)$.

Q.5.70. Calculate the total flux intercepted by a plane bounded by the curves $x - y = 0, y = 0, x = -15$ & $x^2 + y^2 + 10x + 14y + 58 = 0$ when a uniform point-source of 100 cd is located at a point $(-5, -7, 3)$.

Q.5.71. A uniform point-source of 250 W is located at a point $(-10, 0, 6)$. Determine the total energy intercepted by the plane bounded by the curves $y = -3$ & $x^2 + y^2 + 20x + 51 = 0$. If the energy is uniformly distributed over the entire plane what is the energy density?

Q.5.72. Calculate the total flux intercepted by a plane bounded by the curves $x = 4, x = 16, y = -2, y = -12$ & $x^2 + y^2 - 8x + 24y + 135 = 0$ when a uniform point-source of 300 cd is located at a point $(4, -12, -12)$.

Q.5.73. Determine the total flux intercepted by a plane bounded by the curves $x = 4, x = 16, y = -2, y = -12$ & $x^2 + y^2 - 8x + 24y + 12 = 0$ if a uniform point-source of 150 cd is located at a point $(4, -12, 2)$.

Q.5.74. A uniform point-source of 2kW is located at a point $(4, 0, 11)$. Determine the total energy intercepted by the plane bounded by the curves $x = -5, x = 17, y = -3, y = 13$ & $x^2 + y^2 - 8x + 7 = 0$. If the energy is uniformly distributed over the entire plane what is the energy density?

Q.5.75. A point-sourceemitting 2800 lm flux uniformly in all the directions, is located at the point $(-5,-7,2)$. What will be the total flux intercepted by a plane bounded by the curves $x=-5$, $y=-7, 7x-3y+35=0$ &$x^2+y^2+10x+14y+70=0$.

Q.5.76. A uniform point-source of 500 W is located at a point $(-6,-9,8)$. Determine the total energy intercepted by the plane bounded by the curves $x=0, y=0, 2y=3x$ &$x^2+y^2+12x+18y-244=0$.If the energy is uniformly distributed over the entire plane what is the energy density?

Q.5.77. Calculate the total flux intercepted by a plane bounded by the curves $x=-9, x=-4$ &$x^2+y^2+10x+24y-27=0$if a uniform point-source of 222 cd is located at a point $(-5,-12,3)$.

Infinite plane:

Q.5.78. A uniform point-source of 500 W is located at a point $(0,0,5)$. What will be the total energy intercepted by a plane bounded by the curves $x=\pm\infty,\ y=2$ & $y=10$.

Q.5.79. A uniform point-source of 200 cd is located at a point $(0,0,8)$.What will be the total flux intercepted by a plane bounded by the curves $x=\pm\infty, y=\pm 15$.

Q.5.80. A uniform point-source of 100 W is located at a point $(0,0,-13)$. Find out the total energy intercepted by a plane bounded by the curves $x=\pm\infty$ & $y=\pm\infty$.

Q.5.81. Find out the total flux intercepted by a plane bounded by the curves $x=8, x=\infty, y=2$ & $y=16$ if a uniform point-source of 500 cd is located at the point $(2,9,-9)$.

Q.5.82. Find out the total energy intercepted by a plane bounded by the curves $x=\pm\infty, y=25, y=-18$ &$x^2+y^2+10x+24y-27=0$ if a uniform point-source of 250 W is located at the point $(-5,0,15)$.

Q.5.83. Determine the total flux intercepted by a plane bounded by the curves $x=0, y=-3, y=10, x=\infty$ &$x^2+y^2=16$ if a uniform point-source of 200 cd is located at the point $(0,0,5)$.

Q.5.84. Calculate the total energy intercepted by a plane bounded by the lines$x=0, y=-2, y=16$ & $x=\infty$ if a uniform point-source of 500 W is located at the point $(-6,-4,8)$.

Q.5.85. Find out the total flux intercepted by a plane bounded by the lines$x=-8, x=\infty, y=-4$ & $y=18$ if a uniform point-source of 200 cd is located at the point $(2,3,4)$.

Q.5.86. If a uniform point-source of 150 cd is located at the point $(0,0,8)$ find out the total flux intercepted by a plane bounded by the lines$x=3, x=\infty, y=-4$ & $y=-\infty$.

Q.5.87. A uniform point-source of 500 W is located at the point $(2,-3,13)$. Find out the total energy intercepted by a plane bounded by the lines$x=-4, x=\infty, y=5$ & $y=-\infty$.

Q.5.88. If a uniform point-source of 300 cd is located at the point $(2,6,8)$ find out the total flux intercepted by a plane bounded by the lines$x=\pm\infty, y=2$ & $y=16$.

Q.5.89. Find out the total energy intercepted by a plane bounded by the lines$x=0, x=\infty, y=16, y=-15$ &$x^2+y^2-10x+12y-45=0$ when a uniform point-source of 300 W is located at the point $(2,3,6)$.

Q.5.90. A uniform point-source of 200 W is located at the point $(2,6,8)$ find out the total energy intercepted by a plane bounded by the lines$x=0, x=2, x=16, x=\infty, y=0, y=9, y=20$ & $y=\infty$.

Q.5.91. A uniform point-source of 110 cd is located at the point $(8,6,7)$ find out the total energy intercepted by a plane bounded by the lines$x=0, x=8, x=20, x=\infty, y=2, y=6, y=10$ & $y=15$.

Q.5.92. Find out the total energy intercepted by a plane bounded by the lines$x=3, x=6, x=18, y=-3, y=7$ & $y=20$ when a uniform point-source of 150 W is located at the point $(6,0,8)$.

Q5.93. Determine the total flux intercepted by a plane bounded by the lines $y=4, y=16, x-y=0$ & $x=\infty$ when a uniform point-source of 100 cd is located at the point $(4,4,-7)$.

Q.5.94. A planebounded by the lines $x = \pm\frac{a}{2}$ & $y = \pm\frac{a}{2}$intercepts one-sixth of the total flux emitted by a uniform point-source located at the point $(0, 0, 6)$. If the flux density over the plane is 1.6 lm per sq. unit what is the intensity of the point-source? (Assume uniform distribution of the flux over the entire plane.) Ans.110 cd

Q.5.95. A square plane with each side 'a' intercepts one-fourth of the total flux emitted by a uniform point-source located at a height 'h' on the vertical axis passing through the centre of the plane. Prove that the ratio of side to height is given as

$$\frac{a}{h} = \frac{2}{\sqrt{\sqrt{2} - 1}}.$$

Q.5.96. A square plane intercepts one-sixteenth of the total flux emitted by a uniform point-source located at a height 'h' on the vertical axis passing through one of the vertices of the plane. Prove that the area of plane is equal to $\frac{h^2}{\sqrt{2}-1}$.

Q.5.97. A square plane intercepts one-third of the total flux emitted by a uniform point-source located at a height 'h' on the vertical axis passing through the centre of the plane. Prove that the area of plane is equal to $\frac{4\sqrt{3}h^2}{2-\sqrt{3}}$.

Q.5.98. A square plane with each side 'a' intercepts $1/12$ of the total flux emitted by a uniform point-source located at a height 'h' on the vertical axis passing through one of the vertices of the plane. Prove that the ratio of side to height is given as

$$\frac{a}{h} = \sqrt{\frac{3}{2\sqrt{3} - 3}}.$$

Q.5.99. Prove that the area of a square plane, intercepting $1/24$ of the total flux emitted by a uniform point-source located at a height 'h' on the vertical axis passing through one of the vertices of the plane, is h^2.

Q.5.100. A circular plane intercepts one-fourth of the total flux emitted by a uniform point-source located at a height 'h' on the vertical axis passing through the centre of plane. Prove that the area of plane is equal to $3\pi h^2$.

Q.5.101. A circular plane with radius 'R' intercepts x fraction of the total flux emitted by a uniform point-source located at a height 'h' on the vertical axis passing through centre of the plane. Prove that the ratio of radius to height is given as

$$\frac{R}{h} = \frac{2\sqrt{x - x^2}}{1 - 2x} \quad \forall\, x < \frac{1}{2}$$

Q.5.102. If a circular plane intercepts one-sixth of the total flux emitted by a uniform point-source located at a height 'h' on the vertical axis passing through the centre of plane. Prove that the area of plane is equal to $5\pi h^2/4$.

Q.5.103. Prove that a circular plane having area equal to $7\pi h^2/9$ always intercepts one-eighth of the total flux emitted by a uniform point-source located at a height 'h' on the vertical axis passing through the centre of plane.

Q.5.104. A circular plane intercepts one-third of the total flux emitted by a uniform point-source located at a height 'h' on the vertical axis passing through the centre of plane. Prove that the radius of plane is equal to $2h\sqrt{2}$.

Q.5.105. A square plane intercepts one-fourth of the total flux emitted by a uniform point-source located at a height 'h' on the vertical axis passing through one of the vertices of plane. Prove that the side of plane is equal to $h/\sqrt{\sqrt{2} - 1}$.

Q.5.106. A square plane with each side 'a' intercepts x fraction of the total radiation emitted by a uniform point-source located at a height 'h' on the vertical axis passing through centre of the plane . Prove that the ratio of side to height is given as

$$\frac{R}{h} = 2\sqrt{\frac{sin\pi x}{1 - sin\pi x}} \quad \forall\, x < \frac{1}{2}.$$

Q.5.107. If a square plane intercepts x fraction of the total radiation emitted by a uniform point-source located at a height 'h' on the vertical axis passing through one of the vertices of the plane then prove that

$$area\ of\ plane = \frac{h^2 sin4\pi x}{1 - sin4\pi x} \quad \forall\, x < \frac{1}{8}.$$

Q.5.108. A square plane intercepts x fraction of the total radiation emitted by a uniform point-source located at a height 'h' on the vertical axis passing through mid-point of one of the sides of the plane. Prove that

$$area\ of\ plane = h^2\left(\frac{5 + \sqrt{25 + 4cot^2 2\pi x}}{2cot^2 2\pi x}\right) \quad \forall\, x < \frac{1}{4}.$$

Q.5.109. An isosceles right-triangular plane intercepts $1/_{24}$ of the total flux emitted by a uniform point-source located at a height 'h' on the vertical axis passing through right angled vertex of the plane. Prove that the area of plane is h^2.

Q.5.110. An isosceles right-triangular plane intercepts one-sixteenth of the total radiation emitted by a uniform point-source located at a height 'h' on the vertical axis passing through right angled vertex of the plane. Prove that the area of plane is $h^2(\sqrt{2} + 1)$.

Q.5.111. An isosceles right-triangular plane intercepts x fraction of the total radiation emitted by a uniform point-source located at a height 'h' on the vertical axis passing through acute angled vertex of the plane. Prove that

$$area\ of\ plane = \frac{h^2}{4}\sqrt{cot^2\left\{\frac{\pi}{4}(1 - 16x)\right\} - 1} \quad \forall\, x < \frac{1}{16}.$$

Q.5.112. Prove that an isosceles right-triangular plane with each orthogonal side 'h' intercepts $1/_{48}$ of the total radiation emitted by a uniform point-source located at a height 'h' on the vertical axis passing through the acute angled vertex of the plane.

Q.5.113. An isosceles right-triangular plane intercepts $1/_{12}$ of the total flux emitted by a uniform point-source located at a height 'h' on the vertical axis passing through right angled vertex of the plane. Prove that the hypotenuse of plane is equal to $2h\sqrt{3 + 2\sqrt{3}}$.

Q.5.114. An isosceles right-triangular plane intercepts x fraction of the total radiation emitted by a uniform point-source located at a height 'h' on the vertical axis passing through right angled vertex of the plane. Prove that

$$area\ of\ plane = h^2 tan4\pi x(sec4\pi x + tan4\pi x) \quad \forall\, x < \frac{1}{8}.$$

Q.5.115. A rectangular plane with sides 'a' & 'b' intercepts x fraction of the total radiation emitted by a uniform point-source located at a height 'h' on the vertical axis passing through one of the vertices of the plane. Prove that

$$h = \sqrt{\frac{\sqrt{(a^2 + b^2)^2 + 4a^2b^2cot^2 4\pi x} - (a^2 + b^2)}{2}} \quad \forall\, x < \frac{1}{8}.$$

*** Space science: (Smaller spherical-bodies)**

Q.5.116. Find out the total radiation energy intercepted by a spherical body with a radius 25cm if a point-source of 1 kW is located at a distance 1m from the centre of the sphere.

Q.5.117. Find out the density of radiation energy intercepted by a sphere with a radius 30cm when a uniform point-source of 5 kW is placed at a distance 2m from the centre of the sphere. (Assume that the intercepted energy is uniformly distributed over intercepted surface of the sphere.)

Q.5.118. A uniform point-source of light is located at a distance 5m from the centre of a sphere with a radius 50cm. what is the area of illuminated surface?

If the luminosity of source is 10kW find out the energy intercepted by the sphere.

Q. 5.119. A spherical body with radius 15cm intercepts one third part of total radiation emitted by a uniform point-source. What is the distance of point-source from the centre of the sphere?

Q.5.120. A spherical body intercepts one-eighth part of the total radiation emitted from a uniform point-source located at a distance 1.5m from the centre. Determine the radius of the sphere.

Q.5.121. A sphere with a radius 15cm intercepts 6.254 W radiation energy emittedby a uniform point-source located at a distance 3m from the centre of the sphere. Find out the luminosity of the point-source.

Q.5.122. A sphere with a radius 20cm intercepts 8 W radiation energy emitted by a uniform point-source located at a distance 5m from the centre of the sphere. Find out the brightness of the point-source at a distance 5m.

Q.5.123. Determine the radiation energy intercepted by a sphere with a radius 25cm if a uniform point-source with brightness 0.159 $W/_{m^2}$at a distance 10m is located at a distance 2m from the centre of the sphere.

Q.5.124. A point-bulb of light of 100 W is located at a height 1m vertically from the peak point of a hemispherical tomb with a radius 3m. Find out the light energy intercepted by the tomb.

Q.5.125. A spherical body intercepts one-fourth of the total radiation emitted from a uniform point-source located a distance d from the centre. Prove that the radius of the sphere is $\sqrt{3}d/_2$.

Q.5.126. A spherical body with a radius 'R' intercepts x fraction of the total radiation energy emitted by a point-source in the space. Prove that the distance (d) of point-source from the centre of spherical body is given as

$$d = \frac{R}{2\sqrt{x - x^2}} \quad \forall\, x < \frac{1}{2}.$$

Q.5.127. There are two spherical bodies A & B with radii R_A&R_B respectively,radiating energy uniformly in all the directions.x_A&$x_B$are the fractions of total energy intercepted by bodies A & Brespectively. Prove that

$$\frac{R_A}{R_B} = \sqrt{\frac{x_A - {x_A}^2}{x_B - {x_B}^2}} \quad \forall\, x_A, x_B < \frac{1}{2}.$$

Q.5.128. There are two spherical bodies A & B with same luminosity &radii R_A&R_B respectively,separated by a distance d between their centres are radiating energy uniformly in all the directions. If E_A&E_B arethe total radiation energiesintercepted by bodies A & B respectively then prove that

$$\frac{E_A}{E_B} = \frac{d - \sqrt{d^2 - {R_A}^2}}{d - \sqrt{d^2 - {R_B}^2}} \quad \forall\, d > R_A + R_B\,.$$

Q.5.129. There are two spherical bodies A & B with luminosities L_A&L_B&radii R_A&R_B respectively,separated by a distance d between their centres are radiating energy uniformly in all the directions. $E_A \& E_B$ are the total radiation energy intercepted by bodies A & B respectively. Prove that

$$\frac{E_A}{E_B} = \frac{L_A}{L_B}\left(\frac{d - \sqrt{d^2 - {R_A}^2}}{d - \sqrt{d^2 - {R_B}^2}}\right) \quad \forall\ d > R_A + R_B\ .$$

Q.5.130. There are two spherical bodies A & B with emissivity $\epsilon_A \& \epsilon_B$, effective temperatures $T_A \& T_B \&$ radii $R_A \& R_B$ respectively,separated by a distance d between their centres are radiating energy uniformly in all the directions. $E_A \& E_B$ are the total radiation energy intercepted by bodies A & B respectively. Prove that

$$\frac{E_A}{E_B} = \frac{\epsilon_A}{\epsilon_B}\left(\frac{R_A}{R_B}\right)^2\left(\frac{T_A}{T_B}\right)^4\left(\frac{d - \sqrt{d^2 - {R_A}^2}}{d - \sqrt{d^2 - {R_B}^2}}\right) \quad \forall\ d > R_A + R_B\ .$$

*** Celestial bodies:**

Q.5.131. Find out the radiation energy intercepted by the Earth. The luminosity of the Sun is $3.84 \times 10^{26}\ W$, mean radius of the Earth is $6.37 \times 10^{6}\ m$& the average distance between the centres of Earth & Sun can be taken $1.49 \times 10^{11}\ m$. If the energy is uniformly distributed over the intercepted surface then find out the energy density.

Ans. $1.75 \times 10^{17} j/sec., 686.43\ W/m^2$.

Q.5.132. A star with luminosity $1.55 \times 10^{30}\ W$is at an approximate distance 1.2ly from the centre of the Earth. Find out the total energy intercepted by the Earth. (Take radius of the Earth = $6.37 \times 10^{6}\ m$)

Q.5.133. A star has brightness $1645.5\ W/m^2$ at the surface of the Earth. If an average density of radiation of specific wavelength emitted from the star, is $203.67\ W/m^2$ at the surface. Find out approximate distance of the star from the centre of the Earth. (Take radius of the Earth = $6.37 \times 10^{6}\ m$)

Q.5.134. A star has brightness $3028.34\ W/m^2$ at the surface of the Earth. If an average density of radiation of specific wavelength emitted from the star, is $434.87\ W/m^2$ at the surface. Find out approximate distance of the star from the centre of the Earth. (Take radius of the Earth = $6.37 \times 10^{6}\ m$)

*** Radiometry:**

Q.5.135. A uniform point-source of 200 W is located at a distance 2m from the centre of a spherical surface with a radius 25cm. Find out the total radiation energy intercepted by the surface.

Q.5.136. Total radiation energy intercepted by a spherical-radiometer with a radius 10cm, is $8.67\ j/sec$. A red hot sphere is located at a distance 3m between the centres of spheres. Find out the luminosity of red hot sphere.

(Take emissivity of material of red hot sphere 0.78).

Q.5.137. If radiation energy intercepted by a spherical-radiometer with a radius 10cm is $7.5\ j/sec$.A red hot sphere with radius 7cm &emissivity 0.8 is located at a distance 1.5m between their centres. Find out the approximate temperature of red hot sphere. Also find out the brightness of red hot sphere at a distance 2m from its centre.

Chapter 6
Graphical Analysis of Solid Angle

6.1 AXIOM OF POLYGON

For a given point in the space, each of the polygons can be divided internally or externally or both (w.r.t. F.O.P.)into a certain number of the elementary triangles all having a common vertex at the foot of perpendicular (F.O.P.) drawn from the given point to the plane of polygon,by joining all the vertices of polygon to the F.O.P. by extended straight lines.

1. Polygon is externally dividedif the F.O.P. lies outside the boundary
2. Polygon is divided internally or externallyor both if the F.O.P. lies inside or on the boundary depending on the geometrical shape of polygon (i.e. angles & sides)

   (See the figures (21), (22), (23)& (24) below)

6.2 AXIOM OF RIGHT TRIANGLE

If the perpendicular, drawn from a given point in the space to the plane of a given triangle, passes through one of the verticesthen the triangle can be dividedinternally or externally(w.r.t. F.O.P.) into two elementary right triangles having common vertex at the foot of perpendicular, by drawing a normal from the common vertex (i.e. F.O.P.) to theopposite side of given triangle.

1. An acute angled triangle is internally divided w.r.t. F.O.P. (i.e. common vertex)
2. A right angled triangle is internally divided w.r.t. F.O.P.(i.e. common vertex)
3. An obtuse angled triangle is divided
   a) Internally if and only if the angle of common vertex (F.O.P.) is obtuse
   b) Externally if and only if the angle of common vertex (F.O.P.) is acute

   (As shown in the figures (19) & (20) below)

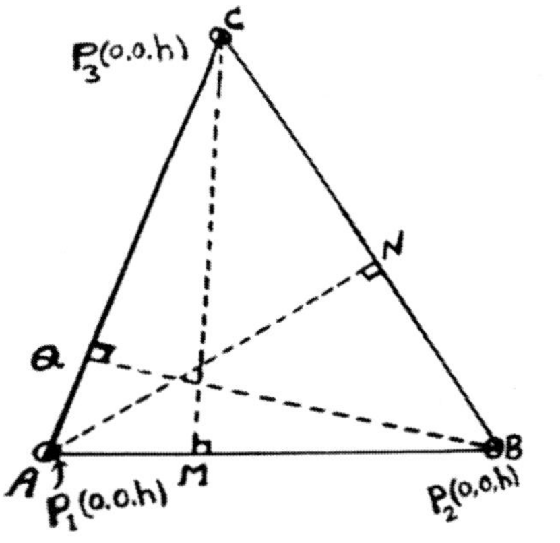

Fig (19)

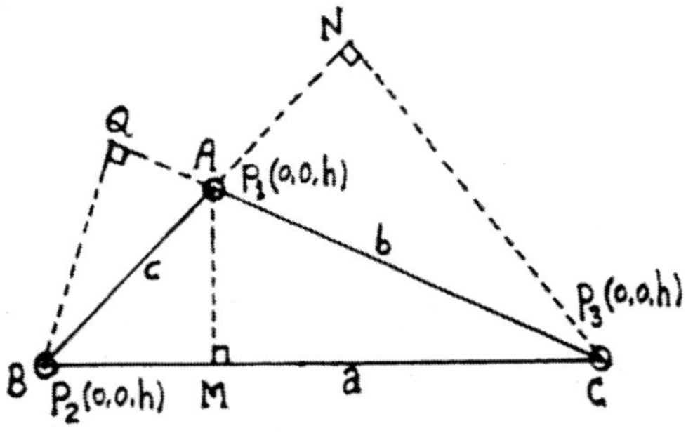

Fig (20)

6.3 GRAPHICAL METHOD: (FOR POLYGONAL-PLANES)

This method is based on above two axioms (proposed by the author) for analysis of polygonal planes. It is the most versatile method for finding out the solid angle subtended by a polygonal plane at any point in the space. This method overcomes all the limitations of Approximation formula used for symmetrical planes. Although, it necessitates the drawing of a given polygonal plane but it involves theoretically zero error. Still due to computational error some numerical error may be involved in the error.

Graphical method eliminatesa lot of calculations in finding out the solid angle subtended by a given plane if the location of foot of perpendicular, drawn from a given point in the space, is known w.r.t. the given plane.

Working Steps:

STEP 1: Trace/draw the diagram a given polygon (plane) with known geometrical dimensions.

STEP 2: Specify the location of foot of perpendicular drawn from a given point to the plane of polygon.

STEP 3: Join all the vertices of polygon to the foot of perpendicular by the extended straight lines. Thus the polygon is divided into a number of elementary triangles, all having a common vertex at the foot of perpendicular.

STEP 4: Now, consider each elementary triangle & divide it into two sub-elementary right triangles to find the solid angle subtended by each elementary triangle.

STEP 5: Solid angle subtended by each individual sub-elementary right triangle is determined using the formula of right triangle

$$\omega = \sin^{-1}\left\{\frac{b}{\sqrt{b^2+p^2}}\right\} - \sin^{-1}\left\{\left(\frac{b}{\sqrt{b^2+p^2}}\right)\left(\frac{h}{\sqrt{h^2+p^2}}\right)\right\}$$

STEP 6: Now, solid angle, subtended by polygonal plane at the given point, will be the sum of all its individual elementary planes as given (By Element-Method)

$$\omega_o = \omega_1 + \omega_2 + \omega_3 + \omega_4 + \omega_5 + \dots\dots\dots\dots$$

Let's us consider a polygonal plane (with vertices)123456& a given point say $P(0,0,h)$ at a normal height h from the given plane in the space.

Now, specify the location of foot of perpendicular say 'O' on the plane of polygon which may lie

1. Outside the boundary
2. Inside the boundary
3. On the boundary
   a) At one of the vertices or b. At any point of one of the vertices

Let's consider the above cases one by one as follows

1. F.O.P. outside the boundary:

Let the foot of perpendicular 'O' lie outside the boundary of polygon. Join all the vertices of polygon (plane) 123456 to the foot of perpendicular 'O' by the extended straight lines (as shown in the fig (21) below)

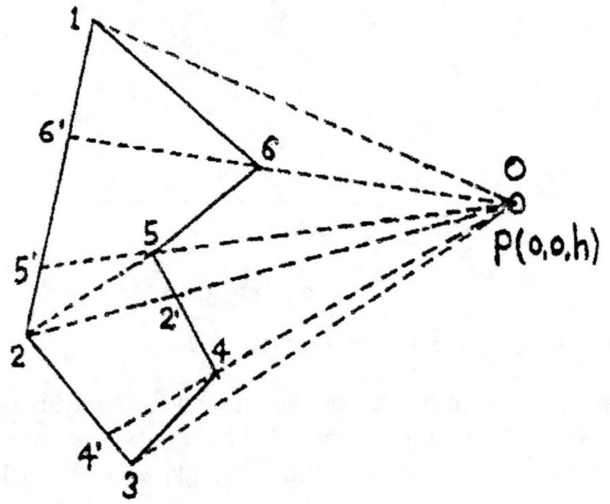

Fig (21)

Thus the polygon is divided into elementary triangles (obtained by the extension lines), all having common vertex at the foot of perpendicular 'O'. Now the solid angle subtended by the polygonal plane at the given point 'P' in the space is given

By Element-Method

$$\omega_{123456} = \omega_{616'} + \omega_{566'5'} + \omega_{22'55'} + \omega_{4'42'2} + \omega_{344'} \quad \dots\dots\dots\dots\dots\dots\dots.(I)$$

Where,

$$\omega_{616'} = \omega_{6'01} - \omega_{601}$$

$$\omega_{566'5'} = \omega_{5'06'} - \omega_{506}\omega_{22'55'} = \omega_{205'} - \omega_{2'05}$$

$$\omega_{4'42'2} = \omega_{4'02} - \omega_{402'}$$

$$\omega_{344'} = \omega_{304'} - \omega_{304}$$

Thus, the value of solid angle subtended by the polygon at the given point is obtained by setting these values in the eq. (I) as follows

$$\Rightarrow \omega_{123456} = (\omega_{6'01} - \omega_{601}) + (\omega_{5'06'} - \omega_{506}) + (\omega_{205'} - \omega_{2'05}) + (\omega_{4'02} - \omega_{402'})$$

Further, each of the individual elementary triangles can be divided into two sub-elementary right triangles for which the values of solid angle are determined by using formula of right triangle.

2. F.O.P. inside the boundary:

Let the foot of perpendicular 'O' lie inside the boundary of polygon. Join all the vertices of polygon (plane) 123456 to the foot of perpendicular 'O' by the extended straight lines (as shown in the fig (22) below)

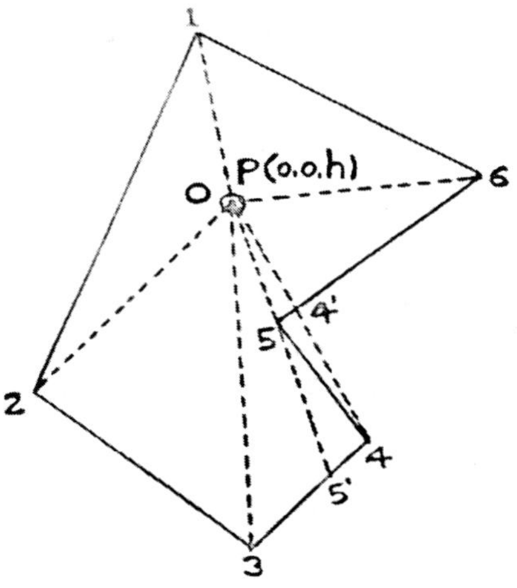

Fig (22)

Thus the polygon is divided into elementary triangles (obtained by the extension lines), all having common vertex at the foot of perpendicular 'O'. Now the solid angle subtended by the polygonal plane at the given point 'P' in the space is given

By Element-Method

$$\omega_{123456} = \omega_{102} + \omega_{203} + \omega_{305'} + \omega_{5'54} + \omega_{504'} \quad \dots\dots\dots\dots\dots\dots\dots.(II)$$

Where,

$$\omega_{5'54} = \omega_{5'04} - \omega_{504}$$

Thus, the value of solid angle subtended by the polygon at the given point is obtained by setting these values in the eq. (II) as follows

$$\Rightarrow \omega_{123456} = \omega_{102} + \omega_{203} + \omega_{305'} + (\omega_{5'04} - \omega_{504}) + \omega_{504'}$$

Further, each of the individual elementary triangles can be divided into two sub-elementary right triangles for which the values of solid angle are determined by using formula of right triangle.

3. F.O.P. ON THE BOUNDARY: FURTHER TWO CASES ARE POSSIBLE

3.1 F.O.P. lying at one of the sides:

Let the foot of perpendicular 'O' lie on one of the sides say '12' of polygon. Join all the vertices of polygon (plane) 123456 to the foot of perpendicular 'O' by the straight lines (as shown in the fig (23) below)

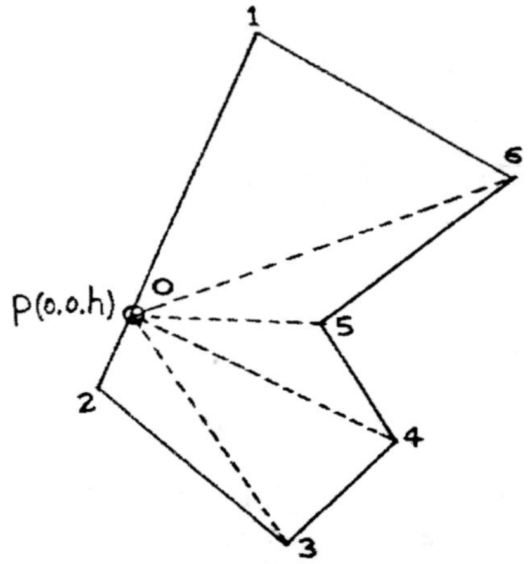

Fig (23)

Thus the polygon is divided into elementary triangles (obtained by the straight lines), all having common vertex at the foot of perpendicular 'O'. Now the solid angle subtended by the polygonal plane at the given point 'P' in the space is given

By Element-Method

$$\omega_{123456} = \omega_{106} + \omega_{605} + \omega_{504} + \omega_{403} + \omega_{302} \qquad \ldots\ldots\ldots\ldots\ldots\ldots\ldots (III)$$

Further, each of the individual elementary triangles can be divided into two sub-elementary right triangles for which the values of solid angle are determined by using formula of right triangle.

3.2 F.O.P. lying at one of the vertices:

Let the foot of perpendicular lie on one of the vertices say '5' of polygon. Join all the vertices of polygon (plane) 123456 to the foot of perpendicular (i.e. common vertex '5') by the straight lines (as shown in the fig (24) below)

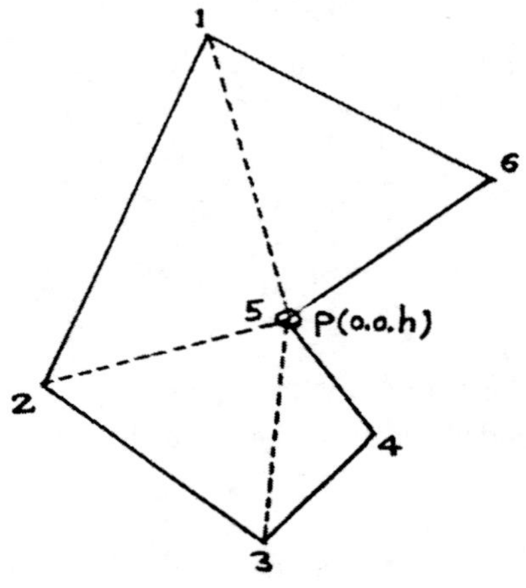

Fig (24)

Thus the polygon is divided into elementary triangles (obtained by the straight lines), all having common vertex at the foot of perpendicular '5'. Now the solid angle subtended by the polygonal plane at the given point 'P' in the space is given

By Element-Method

$$\omega_{123456} = \omega_{152} + \omega_{253} + \omega_{354} + \omega_{156} \quad \ldots\ldots\ldots\ldots\ldots (IV)$$

Further, each of the individual elementary triangles can be divided into two sub-elementary right triangles for which the values of solid angle are determined by using formula of right triangle.

6.4 COMPARISON B/T APPROXIMATION METHOD & GRAPHICAL METHOD

Approximation Method	Graphical Method
1. It is applicable only for symmetrical plane-figures. 2. It is not useful for finding the precise/exact value of solid angle i.e. a certain error is included in the results. 3. It depends on the geometrical shape of symmetrical plane & its configuration (factor of circularity). 4. It requires mere mathematical calculations which are easier as compared to that of graphical method.	1. It is applicable only for polygonal plane-figures. 2. It gives theoretical zero error in the values if computations/calculations are done carefully. 3. It necessitates exact location of foot perpendicular (F.O.P.) drawn from given point to the plane which is not easy to specify. 4. It necessitates drawing since mathematical calculations are easy to find all the dimensions of elementary planes.

Solved Examples

Example 6.1: Find out solid angle subtended by a triangle ABC having sides $AB = 8.6cm, BC = 4cm$ & $AC = 5.5cm$ at a point lying at a normal height 3cm from the vertex 'A'.

Solution

Step 1: Draw the triangle ABC with known values of the sides & specify the location of given point $P(0,0,3)$& F.O.P. (i.e. vertex 'A') (as shown in the figure (25) below)

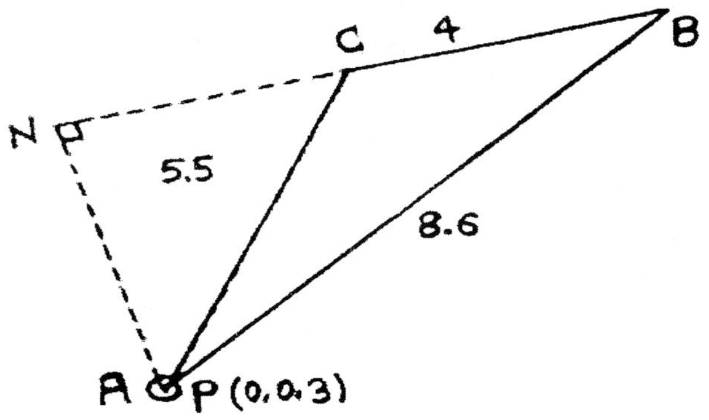

Fig (25)

Step 2: Draw the perpendicular AN from vertex 'A' to the opposite side BC thus ΔABC is externally divided into two right triangles ANB & ANC

Step 3: Measure necessary dimensions then perform the following calculations

$$AN = 4.4cm, \quad CN = 3.6cm \,\& \qquad (from\ the\ diagram)$$

$$BN = BC + CN = 4 + 3.6 = 7.6cm$$

Now, solid angle subtended by right ΔANB at the given point 'P'

On setting the corresponding values in formula of right triangle

$$\Rightarrow \omega_{\Delta ANB} = \sin^{-1}\left\{\frac{7.6}{\sqrt{7.6^2 + 4.4^2}}\right\} - \sin^{-1}\left\{\left(\frac{7.6}{\sqrt{7.6^2 + 4.4^2}}\right)\left(\frac{3}{\sqrt{3^2 + 4.4^2}}\right)\right\}$$

$$= \sin^{-1}\{0.865\} - \sin^{-1}\{0.487\} = 1.045 - 0.509 = 0.536\ sr$$

Similarly, solid angle subtended by right ΔANC at the given point 'P'

$$\Rightarrow \omega_{\Delta ANC} = \sin^{-1}\left\{\frac{3.6}{\sqrt{3.6^2 + 4.4^2}}\right\} - \sin^{-1}\left\{\left(\frac{3.6}{\sqrt{3.6^2 + 4.4^2}}\right)\left(\frac{3}{\sqrt{3^2 + 4.4^2}}\right)\right\}$$

$$= \sin^{-1}\{0.633\} - \sin^{-1}\{0.357\} = 0.685 - 0.365 = 0.32\ sr$$

Hence, solid angle subtended by ΔABC at the point 'P' (by Element Method)

$$\Rightarrow \omega_{\Delta ABC} = \omega_{\Delta ANB} - \omega_{\Delta ANC} = 0.536 - 0.32 = 0.216\ sr \qquad Ans.$$

Example 6.2: Find out solid angle subtended by a parallelogram ABCD having sides $AB = CD = 8cm$, $BC = AD = 5cm$ & $BAD = 70^o$ at a point lying at a normal height 5cm from the vertex 'A'.

Solution

Step 1: Draw the parallelogram ABCD with known values of the sides& angles& specify the location of given point $P(0, 0, 5)$& F.O.P. (i.e. vertex 'A') (as shown in the fig (26) below)

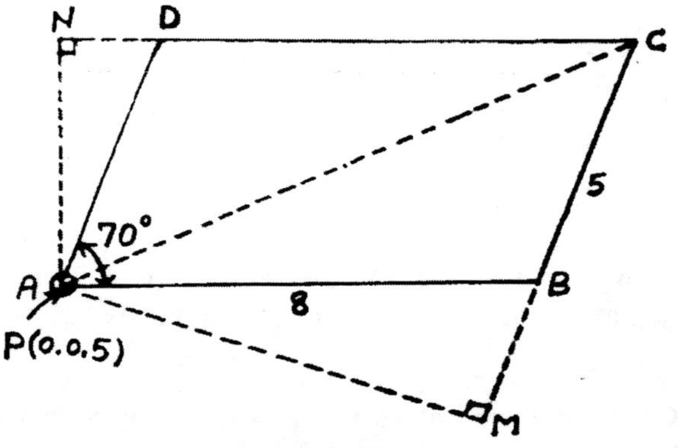

Fig (26)

Step 2: Join vertices 'A' & 'C' thus parallelogram ABCD is divided into two elementary triangles ABC & ACD

Step 3: Draw the perpendiculars AN & AM from vertex 'A' to the opposite sides CD & BC respectively thusΔACD is divided into two right triangles ANC & AND. $\Delta ABC$is divided into two right triangle AMC & AMB

Step 4: Measure necessary dimensions then perform the following calculations

$$AN = 4.7cm, \quad ND = 1.6cm, \quad AM = 7.5cm, \qquad (from\ the\ diagram)$$

$$BM = 2.8cm \Rightarrow CN = CD + ND = 8 + 1.6 = 9.6cm \ \&$$

$$CM = BC + BM = 5 + 2.8 = 7.8cm$$

Now, solid angle subtended by right ΔANC at the given point 'P'

On setting the corresponding values in formula of right triangle

$$\Rightarrow \omega_{\Delta ANC} = \sin^{-1}\left\{\frac{9.6}{\sqrt{9.6^2 + 4.7^2}}\right\} - \sin^{-1}\left\{\left(\frac{9.6}{\sqrt{9.6^2 + 4.7^2}}\right)\left(\frac{5}{\sqrt{5^2 + 4.7^2}}\right)\right\}$$

$$= \sin^{-1}\{0.898\} - \sin^{-1}\{0.654\} = 1.115 - 0.713 = 0.402\ sr$$

Similarly, solid angle subtended by right ΔAND at the given point 'P'

$$\Rightarrow \omega_{\Delta AND} = \sin^{-1}\left\{\frac{1.6}{\sqrt{1.6^2 + 4.7^2}}\right\} - \sin^{-1}\left\{\left(\frac{1.6}{\sqrt{1.6^2 + 4.7^2}}\right)\left(\frac{5}{\sqrt{5^2 + 4.7^2}}\right)\right\}$$

$$= \sin^{-1}\{0.322\} - \sin^{-1}\{0.235\} = 0.328 - 0.237 = 0.091\ sr$$

Similarly, solid angle subtended by right ΔAMC at the given point 'P'

$$\Rightarrow \omega_{\Delta AMC} = \sin^{-1}\left\{\frac{7.8}{\sqrt{7.8^2 + 7.5^2}}\right\} - \sin^{-1}\left\{\left(\frac{7.8}{\sqrt{7.8^2 + 7.5^2}}\right)\left(\frac{5}{\sqrt{5^2 + 7.5^2}}\right)\right\}$$

$$= \sin^{-1}\{0.721\} - \sin^{-1}\{0.4\} = 0.805 - 0.411 = 0.394\ sr$$

Similarly, solid angle subtended by right ΔAMB at the given point 'P'

$$\Rightarrow \omega_{\Delta AMB} = \sin^{-1}\left\{\frac{2.8}{\sqrt{2.8^2 + 7.5^2}}\right\} - \sin^{-1}\left\{\left(\frac{2.8}{\sqrt{2.8^2 + 7.5^2}}\right)\left(\frac{5}{\sqrt{5^2 + 7.5^2}}\right)\right\}$$

$$= \sin^{-1}\{0.3497\} - \sin^{-1}\{0.194\} = 0.357 - 0.195 = 0.162\ sr$$

Hence, solid angle subtended by $parallelogram\ ABCD$ at the point 'P' (by Element Method)

$$\Rightarrow \omega_{ABCD} = \omega_{\Delta ABC} + \omega_{\Delta ACD} = (\omega_{\Delta AMC} - \omega_{\Delta AMB}) + (\omega_{\Delta ANC} - \omega_{\Delta AND})$$

$$= 0.394 - 0.162 + 0.402 - 0.091 = 0.543\ sr \qquad Ans.$$

Example 6.3: Find out solid angle subtended by a quadrilateral ABCD having sides $AB = 6cm, BC = 8cm,$ $CD = 7cm, AD = 4cm$ & $BAD = 110^o$ at a point lying at a normal height 2cm from the vertex 'A'.

Solution

Step 1: Draw the quadrilateral ABCD with known values of the sides& angle& specify the location of given point $P(0, 0, 2)$& F.O.P. (i.e. vertex 'A') (as shown in the fig (27) below)

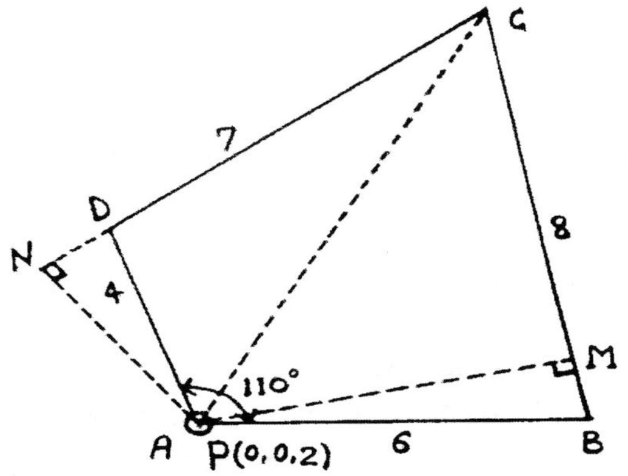

Fig (27)

Step 2: Join vertices 'A' & 'C' thus quadrilateral ABCD is divided into two elementary triangles ABC & ACD

Step 3: Draw the perpendiculars AN & AM from vertex 'A' to the opposite sides CD & BC respectively thus ΔACD is divided into two right triangles ANC & AND. ΔABC is divided into two right triangle AMC & AMB

Step 4: Measure necessary dimensions then perform the following calculations

$$AN = 3.8cm, \quad ND = 1.2cm, \quad AM = 5.9cm, \quad (from\ the\ diagram)$$

$$BM = 1.2cm \Rightarrow CN = CD + ND = 7 + 1.2 = 8.2cm\ \&$$

$$CM = BC - BM = 8 - 1.2 = 6.8cm$$

Now, solid angle subtended by right ΔANC at the given point 'P'

On setting the corresponding values in formula of right triangle

$$\Rightarrow \omega_{\Delta ANC} = \sin^{-1}\left\{\frac{8.2}{\sqrt{8.2^2 + 3.8^2}}\right\} - \sin^{-1}\left\{\left(\frac{8.2}{\sqrt{8.2^2 + 3.8^2}}\right)\left(\frac{2}{\sqrt{2^2 + 3.8^2}}\right)\right\}$$

$$= \sin^{-1}\{0.907\} - \sin^{-1}\{0.422\} = 1.136 - 0.436 = 0.7\ sr$$

Similarly, solid angle subtended by right ΔAND at the given point 'P'

$$\Rightarrow \omega_{\Delta AND} = \sin^{-1}\left\{\frac{1.2}{\sqrt{1.2^2 + 3.8^2}}\right\} - \sin^{-1}\left\{\left(\frac{1.2}{\sqrt{1.2^2 + 3.8^2}}\right)\left(\frac{2}{\sqrt{2^2 + 3.8^2}}\right)\right\}$$

$$= \sin^{-1}\{0.301\} - \sin^{-1}\{0.14\} = 0.306 - 0.14 = 0.166\ sr$$

Similarly, solid angle subtended by right ΔAMC at the given point 'P'

$$\Rightarrow \omega_{\Delta AMC} = \sin^{-1}\left\{\frac{6.8}{\sqrt{6.8^2 + 5.9^2}}\right\} - \sin^{-1}\left\{\left(\frac{6.8}{\sqrt{6.8^2 + 5.9^2}}\right)\left(\frac{2}{\sqrt{2^2 + 5.9^2}}\right)\right\}$$

$$= \sin^{-1}\{0.755\} - \sin^{-1}\{0.242\} = 0.856 - 0.244 = 0.612\ sr$$

Similarly, solid angle subtended by right ΔAMB at the given point 'P'

$$\Rightarrow \omega_{\Delta AMB} = \sin^{-1}\left\{\frac{1.2}{\sqrt{1.2^2 + 5.9^2}}\right\} - \sin^{-1}\left\{\left(\frac{1.2}{\sqrt{1.2^2 + 5.9^2}}\right)\left(\frac{2}{\sqrt{2^2 + 5.9^2}}\right)\right\}$$

$$= \sin^{-1}\{0.199\} - \sin^{-1}\{0.064\} = 0.2 - 0.064 = 0.136\ sr$$

Hence, solid angle subtended by *quadrilateral ABCD* at the point 'P' (by Element Method)

$$\Rightarrow \omega_{ABCD} = \omega_{\Delta ABC} + \omega_{\Delta ACD} = (\omega_{\Delta AMC} + \omega_{\Delta AMB}) + (\omega_{\Delta ANC} - \omega_{\Delta AND})$$

$$= 0.612 + 0.136 + 0.7 - 0.166 = 1.282\ sr \qquad Ans.$$

Exercise - 6

Q.6.1. Find out solid angle subtended by a triangle ABC having sides $AB = 10cm, BC = 5cm$ & $AC = 8cm$ at a point lying at a normal height 6cm from the vertex 'A'.

Q.6.2. Find out solid angle subtended by a triangle ABC having sides $AB = 10cm, BC = 5cm$ & $AC = 8cm$ at a point lying at a normal height 6cm from a point 'D' internally dividing the side AC in a ratio AD:DC = 1:3.

Q.6.3. Find out solid angle subtended by a triangle ABC having sides $AB = 15cm, BC = 25cm$ & $AC = 18cm$ at a point lying at a normal height 10cm from the mid-point of side AB.

Q.6.4. Find out solid angle subtended by a triangle ABC having sides $AB = 12cm, BC = 7cm$ & $AC = 9cm$ at a point lying at a normal height 3cm from a point 'O' externally dividing the side BC in a ratio OC: OB = 2:3.

Q.6.5. Find out the total flux intercepted by a parallelogram ABCD having sides $AB = CD = 14cm,\ BC = AD = 8cm$ & $BAD = 60^o$ if a uniform point-source of 100 cd is located at a normal height 4cm from the vertex 'A'.

Q.6.6. Find out solid angle subtended by a quadrilateral ABCD having sides $AB = 16cm, BC = 12cm, CD = 8cm, AD = 10cm$ & $BAD = 120^o$ at a point lying at a normal height 2cm from the vertex 'A'.

Q.6.7. A uniform point-source of 200 W is located at a normal height 5cm from the vertex 'B' of a quadrilateral ABCD having sides $AB = 13cm, BC = 8cm,\ CD = 9cm, AD = 15cm$ & $BAD = 100^o$.Find out the flux intercepted by the plane ABCD.

Unit 2: Miscellaneous Analysis

This unit is principally meant for additional information of geometry & trigonometry which carries two chapters. Each of the chapters deals with the articles related to the analysis & estimation of the different elementsof the solid-figures & plane-figures. Most of the articles of this unit are advanced & rest are modified. This unit has been included for the additional information of the Geometry. These chapters are as follows

Chapter 1.Analysis of frustums

a) sphere b) right cone & c) circular cylinder.

Chapter 2.Axiomatic analysis

a) axiom of two concurrent vectors. b) axiom of three concurrent vectors. c) axiom of triangle.

d) spiked regular-polygon* e) right pyramid.

Chapter 1
Analysis of frustums

This chapter deals with the analysis of frustums of the solid bodies such as sphere, right cone & circular cylinder. Here, we will study the frustums having only circular & elliptical sections obtained by cutting normally or obliquely with the axis from sphere, cone &cylinder.

1.1 SPHERE

Let's analyse the frustums of the sphere with one circular-section & two parallel circular-sections cut from the solid sphere with a smooth plane in order to find out the surface area & the volume of the respective frustum.

When a solid sphere is cut with a plane then the sphere is divided into two frustums with identical circular-sections.

1.1.1 Frustum of the sphere with one circular-section

Let there be a solid sphere with a radius 'R' & centre 'O' placed in the horizontal plane (reference plane) normal to the plane of paper. Now let it be cut with a smooth plane at a height 'h' parallel to the reference plane. (as shown in the figure(159) below)

*** Radius of the circular-section**

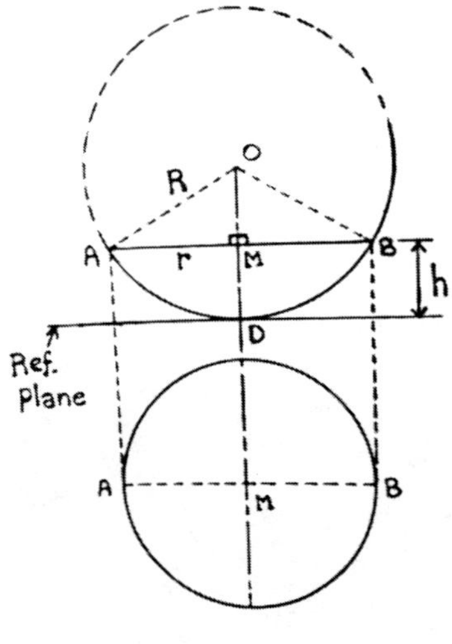

Fig (159)

The section of the frustum will be circular with a radius AM = BM = r

In right ΔAMO

$$\Rightarrow AM^2 = OA^2 - OM^2 = R^2 - (R-h)^2 = R^2 - R^2 - h^2 + 2Rh = 2Rh - h^2$$

$$\Rightarrow r^2 = 2Rh - h^2 \text{ or } r = \sqrt{2Rh - h^2}$$

$$\Rightarrow r = \sqrt{2Rh - h^2} \qquad \forall (0 < h \leq 2R) \qquad \dots\dots\dots\dots\dots\dots (87)$$

The above eq(87) is used to find the radius of the frustum (minor or major) cut from the sphere.

*** Surface area of the frustum**

Consider an elementary strip with a width '$Rd\theta$' & thickness '$Rsin\theta d\theta$' on the surface of the frustum at an angle 'θ' with X-axis.Let $AOM = \alpha$ (As shown in the figure (160) below)

Now, the area of the elementary strip

$$\Rightarrow dA = (circumference)(width)$$

$$\Rightarrow dA = (2\pi R sin\theta)(R d\theta) = 2\pi R^2 sin\theta d\theta$$

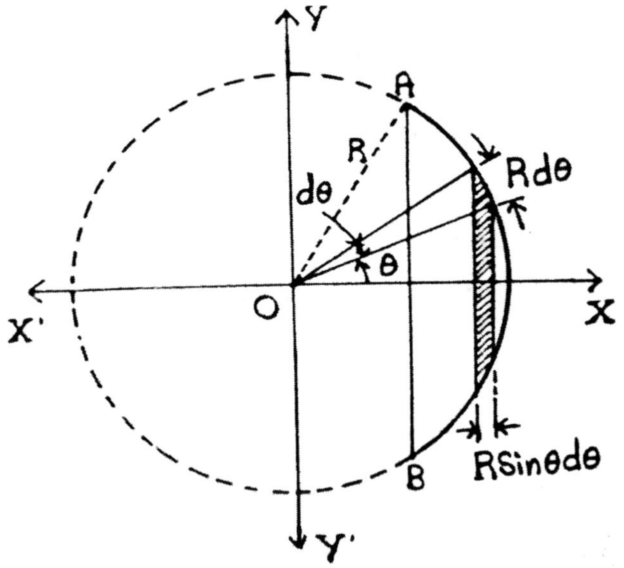

Fig (160)

Hence, by integrating the elementary from$\theta_1 = 0\ to\ \theta_2 = \alpha$,

The total surface of the frustum

$$\Rightarrow A = \int_{\theta_1=0}^{\theta_2=\alpha} 2\pi R^2 sin\theta d\theta = 2\pi R^2 \int_0^{\alpha} sin\theta d\theta = 2\pi R^2(-cos\theta)_0^{\alpha} = 2\pi R^2(-cos\alpha + cos0)$$

$$= 2\pi R^2(1 - cos\alpha)$$

In right ΔOMA

$$\Rightarrow cos\alpha = cos\ \ AOM = \frac{OM}{OA} = \frac{R-h}{R} = 1 - \frac{h}{R}$$

$$\therefore A = 2\pi R^2\left(1 - 1 + \frac{h}{R}\right) = 2\pi R^2 \frac{h}{R} = 2\pi Rh$$

$$\Rightarrow A = 2\pi Rh \qquad \forall(0 < h \le 2R) \qquad \ldots\ldots\ldots\ldots\ldots\ldots\ldots\ldots(88)$$

*** Volume of the frustum**

Volume of the elementary strip

$$\Rightarrow dV = (area\ of\ circular\ section\ of\ strip)(thickness)$$

$$\Rightarrow dV = [\pi(R sin\theta)^2][(R d\theta)sin\theta] = \pi R^3 sin^3\theta d\theta$$

Hence, by integrating the elementary volume from$\theta_1 = 0\ to\ \theta_2 = \alpha$,

The total volume of the frustum

$$\Rightarrow V = \int_{\theta_1=0}^{\theta_2=\alpha} \pi R^3 sin^3\theta d\theta = \pi R^3 \int_0^{\alpha} \left(\frac{3sin\theta - sin3\theta}{4}\right) d\theta$$

$$= \frac{\pi R^3}{4}\int_0^{\alpha}(3sin\theta - sin3\theta)d\theta = \frac{\pi R^3}{4}\left[-3cos\theta + \frac{1}{3}cos3\theta\right]_0^{\alpha}$$

$$= \frac{\pi R^3}{4}\left[-3cos\alpha + \frac{1}{3}cos3\alpha + 3cos0 - \frac{1}{3}cos0\right] = \frac{\pi R^3}{12}(cos3\alpha - 9cos\alpha + 8)$$

$$= \frac{\pi R^3}{12}(4cos^3\alpha - 3cos\alpha - 9cos\alpha + 8) = \frac{\pi R^3}{12}(4cos^3\alpha - 12cos\alpha + 8)$$

Now, on setting the value of $cos\alpha$, we have

$$\Rightarrow V = \frac{\pi R^3}{12}\left(4\left(\frac{R-h}{R}\right)^3 - 12\left(\frac{R-h}{R}\right) + 8\right)$$

$$= \frac{\pi R^3}{12}\left(\frac{4R^3 - 4h^3 + 12Rh^2 - 12R^2h - 12R^3 + 12R^2h + 8R^3}{R^3}\right) = \frac{\pi}{12}(12Rh^2 - 4h^3)$$

$$\Rightarrow V = \frac{\pi h^2}{3}(3R - h) \quad \forall (0 < h \leq 2R) \qquad \text{……………………… (89)}$$

Important deduction: if the frustum is the hemi-sphere then on setting h = R in the eq(87), (88) & (89), we get

1. Radius of the circular-section (r),

$$\Rightarrow r = \sqrt{2R(R) - R^2} = R = radius\ of\ the\ parent\ sphere$$

2. Surface area (A),

$$\Rightarrow A = 2\pi R(R) = 2\pi R^2$$

3. Volume (V),

$$\Rightarrow V = \frac{\pi R^2}{3}(3R - R) = \frac{\pi R^2}{3}(2R) = \frac{2}{3}\pi R^3$$

Thus, the above equations are applicable for any frustum of sphere having single circular-section.

1.1.2 Frustum of the sphere with two circular-sections

Let there be a solid sphere with a radius 'R' & centre 'O'. Now let it be cut with two smooth & parallel planes such that the angles subtended by the circular-sections at the centre 'O' of the sphere are $'2\theta_1'\&'2\theta_2'$ $\forall\ (\theta_1 < \theta_2 \Rightarrow \theta_1 \in [0,\pi)\&\theta_2 \in (0,\pi]\ )$ with the X-axis in clockwise sense. (as shown in the figure(161) below)

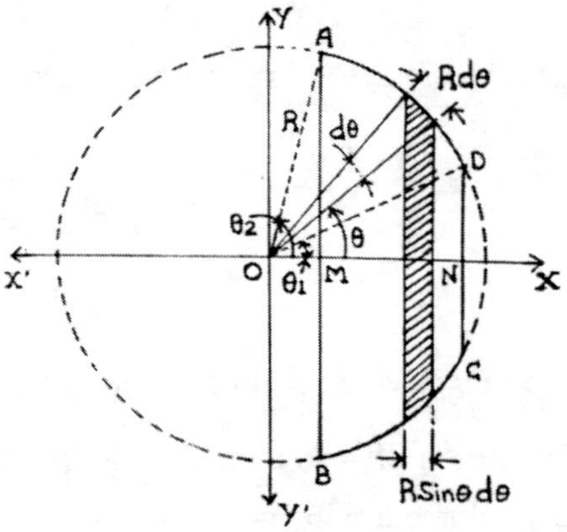

Fig (161)

* Radii of the circular-section (r_1& r_2)

The radii of the circular-sections are given as

1. Radius of section-1

$$\Rightarrow r_1 = DN = CN = ODsin\theta_1 = Rsin\theta_1 \&$$

2. Radius of section-2

$$\Rightarrow r_2 = AM = BM = OAsin\theta_2 = Rsin\theta_2$$

$$\forall \left(\theta_1 < \theta_2 \Rightarrow \theta_1 \in [0,\pi) \& \theta_2 \in (0,\pi]\right) \quad \ldots\ldots\ldots\ldots\ldots\ldots (90)$$

*** Surface area of the frustum**

Consider an elementary strip with a width '$Rd\theta$' & thickness '$Rsin\theta d\theta$' on the surface of the frustum at an angle 'θ' with X-axis. Given that $DOM = \theta_1 \&\ AOM = \theta_2$

Now, the area of the elementary strip

$$\Rightarrow dA = (circumference)(width)$$

$$\Rightarrow dA = (2\pi Rsin\theta)(Rd\theta) = 2\pi R^2 sin\theta d\theta$$

Hence, by integrating the elementary fromθ_1 *to* θ_2,

The total surface of the frustum

$$\Rightarrow A = \int_{\theta_1}^{\theta_2} 2\pi R^2 sin\theta d\theta = 2\pi R^2 \int_{\theta_1}^{\theta_2} sin\theta d\theta = 2\pi R^2(-cos\theta)_{\theta_1}^{\theta_2} = 2\pi R^2(-cos\theta_2 + cos\theta_1)$$

$$\Rightarrow A = 2\pi R^2(cos\theta_1 - cos\theta_2) \ \forall \left(\theta_1 < \theta_2 \Rightarrow \theta_1 \in [0,\pi) \& \theta_2 \in (0,\pi]\right) \quad \ldots\ldots (91)$$

*** Volume of the frustum**

Volume of the elementary strip

$$\Rightarrow dV = (area\ of\ circular\ section\ of\ strip)(thickness)$$

$$\Rightarrow dV = [\pi(Rsin\theta)^2][(Rd\theta)sin\theta] = \pi R^3 sin^3\theta d\theta$$

Hence, by integrating the elementary volume fromθ_1 *to* θ_2,

The total volume of the frustum

$$\Rightarrow V = \int_{\theta_1}^{\theta_2} \pi R^3 sin^3\theta d\theta = \pi R^3 \int_{\theta_1}^{\theta_2} \left(\frac{3sin\theta - sin3\theta}{4}\right) d\theta$$

$$= \frac{\pi R^3}{4}\int_{\theta_1}^{\theta_2} (3sin\theta - sin3\theta)d\theta = \frac{\pi R^3}{4}\left[-3cos\theta + \frac{1}{3}cos3\theta\right]_{\theta_1}^{\theta_2}$$

$$= \frac{\pi R^3}{4}\left[-3cos\theta_2 + \frac{1}{3}cos3\theta_2 + 3cos\theta_1 - \frac{1}{3}cos3\theta_1\right]$$

$$= \frac{\pi R^3}{12}[-9cos\theta_2 + cos3\theta_2 + 9cos\theta_1 - cos3\theta_1]$$

$$\Rightarrow V = \frac{\pi R^3}{12}[9(cos\theta_1 - cos\theta_2) - (cos3\theta_1 - cos3\theta_2)]$$

$$\forall \left(\theta_1 < \theta_2 \Rightarrow \theta_1 \in [0,\pi) \& \theta_2 \in (0,\pi]\right) \quad \ldots\ldots\ldots\ldots (92)$$

Important deduction 1: if the frustum has single circular-section then by setting $\theta_1 = 0\ \&\theta_2 = \theta$ in the eq(90), (91) & (92), we get

1. Radius of the circular-section (r),

$$\Rightarrow r = Rsin\theta$$

2. Surface area (A),

$$\Rightarrow A = 2\pi R^2(\cos 0 - \cos\theta) = 2\pi R^2(1 - \cos\theta)$$

3. Volume (V),

$$\Rightarrow V = \frac{\pi R^3}{12}[9(\cos 0 - \cos\theta) - (\cos 3(0) - \cos 3\theta)] = \frac{\pi R^3}{12}[\cos 3\theta - 9\cos\theta + 8]$$
$$\forall (0 < \theta \leq \pi)$$

Thus, the above equations are applicable for any frustum of sphere having single circular-section.

Important deduction 2: If the frustum is the hemi-sphere then on setting '$\theta_1 = 0$ &$\theta_2 = \frac{\pi}{2}$in the eq(90), (91) & (92), we get

1. Radius of the circular-section (r),

$$\Rightarrow r = R\sin\frac{\pi}{2} = R = Radius\ of\ the\ parent\ sphere$$

2. Surface area (A),

$$\Rightarrow A = 2\pi R^2\left(\cos 0 - \cos\frac{\pi}{2}\right) = 2\pi R^2(1 - 0) = 2\pi R^2$$

3. Volume (V),

$$\Rightarrow V = \frac{\pi R^3}{12}\left[9\left(\cos 0 - \cos\frac{\pi}{2}\right) - \left(\cos 3(0) - \cos 3\frac{\pi}{2}\right)\right] = \frac{\pi R^3}{12}[9(1 - 0) - (1 - 0)]$$
$$= \frac{8\pi R^3}{12} = \frac{2\pi R^3}{3}$$

1.1.3 Frustum of the sphere with two parallel & identical circular-sections

Let there be a frustum of the sphere with a radius 'R' having two identical & parallel circular-sections with a radius 'r' & angle subtended by each of the sections is 2θ at the centre 'O' of parent sphere. (as shown in the figure (162) below)

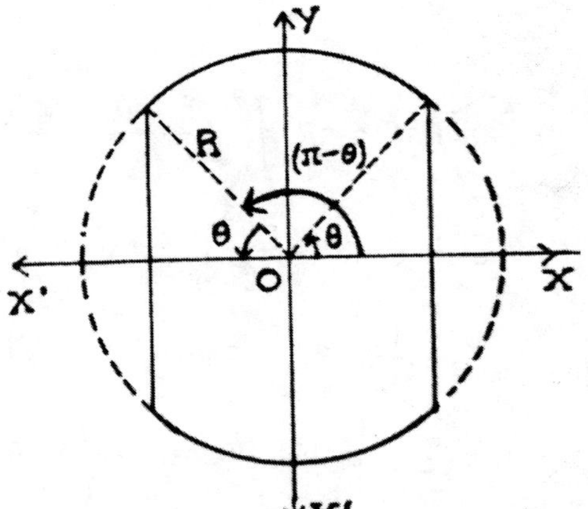

Fig (162)

In this case, the angle subtended by the respective sections will be $\theta_1 = \theta$ &$\theta_2 = \pi - \theta$ by the symmetry of the frustum. Hence, we obtain the results by setting $\theta_1 = \theta$ &$\theta_2 = \pi - \theta$in the eq(90), (91) & (92) as follows

1. Radius of the circular-sections (r),

$$\Rightarrow r_1 = r_2 = r = R\sin\theta$$

2. Surface area (A),

$$\Rightarrow A = 2\pi R^2(\cos\theta - \cos(\pi - \theta)) = 2\pi R^2(\cos\theta + \cos\theta) = 4\pi R^2 cos\theta \quad \ldots\ldots\ldots\ldots\ldots (93)$$

3. Volume (V),

$$\Rightarrow V = \frac{\pi R^3}{12}[9(cos\theta - \cos(\pi - \theta)) - (cos3\theta - cos3(\pi - \theta))]$$

$$= \frac{\pi R^3}{12}[9(cos\theta + cos\theta) - (cos3\theta + cos3\theta)] = \frac{\pi R^3}{12}[9cos\theta - cos3\theta]$$

$$\Rightarrow V = \frac{\pi R^3}{12}[9cos\theta - cos3\theta] \qquad \forall\left(0 \le \theta < \frac{\pi}{2}\right) \quad \ldots\ldots\ldots\ldots\ldots\ldots\ldots\ldots (94)$$

1.2 RIGHT CONE

When the right cone is thoroughly cut with a smooth plane then two types of the sections are obtained that may be circular or elliptical depending merely on the inclination of the cutting plane with the axis of right cone i.e. circular-section is obtained if and only if the cutting plane is normal to the axis of the right cone while in all other cases an elliptical-section is obtained.While their dimensions like radius of circular-section or minor & major axes of elliptical-section depend on the normal distance of the cutting plane from the apex of the right cone. (as shown in the figure (163) below)

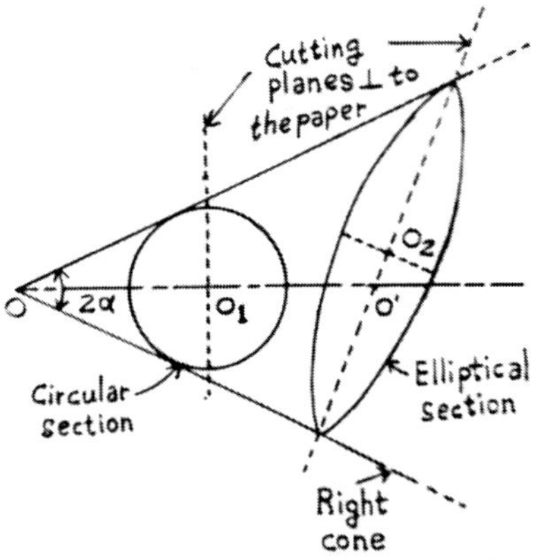

Fig (163)

Let's analyse the frustums of the right cone with elliptical & circular-section in order to find out the eccentricity of the section generated by cutting a right cone thoroughly.

1.2.1 Frustum of aright cone with elliptical-section

Let there be a right cone with apex 'O' & apex angle '2α'. Now let it be thoroughly cut with a smooth plane inclined at an angle 'θ' with the axis & at a normal distance 'h' from the apex 'O'. In this case, we obtain an elliptical-section. (as shown in the figure (164) below line AB shows the section normal to the plane of paper)

If '2a' is the major-axis, '2b' is the minor-axis & 'e' is the eccentricity of the elliptical-section then

In right ΔOMB

$$\Rightarrow sin(\theta + \alpha) = \frac{OM}{OB} = \frac{h}{OB} \Rightarrow OB = \frac{h}{sin(\theta + \alpha)} \quad \ldots\ldots\ldots\ldots\ldots (I)$$

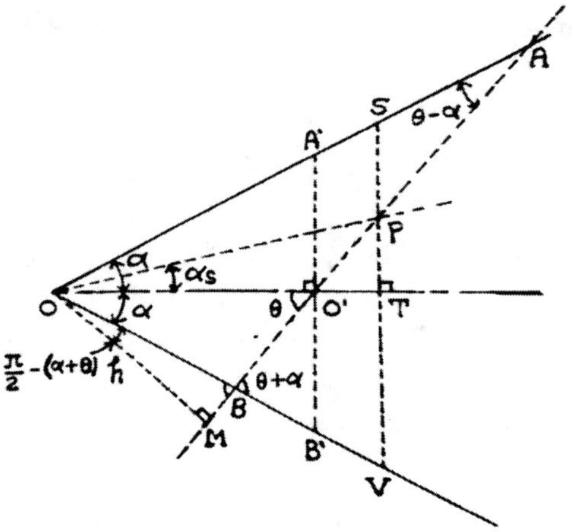

Fig (164)

in ΔOBA, applying Sine rule

$$\Rightarrow \frac{sin\ \ OAB}{OB} = \frac{sin\ \ AOB}{AB} \Rightarrow \frac{sin(\theta - \alpha)}{\left(\frac{h}{sin(\theta + \alpha)}\right)} = \frac{sin2\alpha}{AB} \quad (set\ value\ of\ OB\ from\ eq(I))$$

$$\Rightarrow AB = \left(\frac{h}{sin(\theta + \alpha)}\right)\frac{sin2\alpha}{sin(\theta - \alpha)} = \frac{hsin2\alpha}{sin(\theta + \alpha)sin(\theta - \alpha)} = \frac{hsin2\alpha}{sin^2\theta - sin^2\alpha}$$

But, the point 'P' is the centre of the elliptical-section (as shown in the figure(165) below)

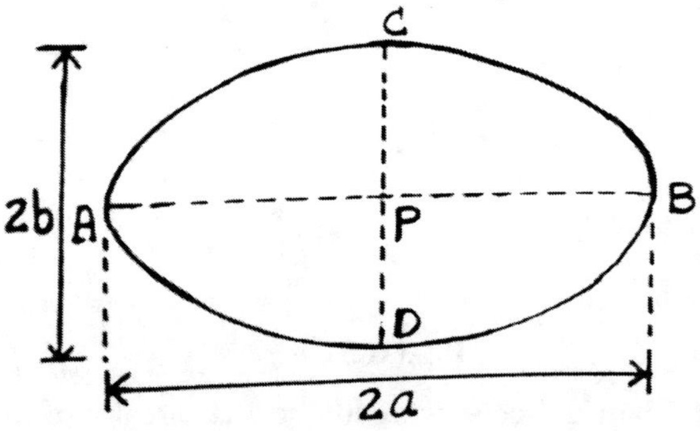

Fig (165)

Thus, we have

$$\Rightarrow AP = BP = \frac{AB}{2} = \frac{2a}{2} = a\ \ where\ AB\ is\ the\ majoraxis$$

Hence, the major axis of the elliptical-section

$$\Rightarrow 2a = AB = \frac{hsin2\alpha}{sin^2\theta - sin^2\alpha} \quad \ldots\ldots\ldots\ldots\ldots\ldots\ldots\ldots\ldots..(II)$$

In right $\Delta OMO'$

$$\Rightarrow sin\theta = \frac{h}{OO'} \Rightarrow OO' = hcosec\theta \quad \ldots\ldots\ldots\ldots\ldots\ldots\ldots\ldots.(III)$$

$$\Rightarrow cot\theta = \frac{MO'}{OM} = \frac{MO'}{h} \Rightarrow MO' = hcosec\theta \quad \ldots\ldots\ldots\ldots\ldots\ldots\ldots\ldots.(IV)$$

In right $\Delta OO'A'$

$$\Rightarrow tan\alpha = \frac{A'O'}{OO'} \Rightarrow A'O' = OO'tan\alpha = htan\alpha cosec\theta \quad (from\ eq(III)) \quad \dots\dots\dots\dots\dots (V)$$

In right ΔOMB

$$\Rightarrow cot(\theta+\alpha) = \frac{MB}{OM} = \frac{MB}{h} \Rightarrow MB = hcot(\theta+\alpha) \quad \dots\dots\dots\dots\dots\dots\dots\dots\dots\dots..(VI)$$

$$\therefore BO' = MO' - MB = hcot\theta - hcot(\theta+\alpha) \quad from\ the\ eq(IV)\&\ (VI)$$

Now,

$$O'P = BP - BO' = \frac{AB}{2} - \left(hcot\theta - hcot(\theta+\alpha)\right)$$

$$= \frac{1}{2}\frac{hsin2\alpha}{sin^2\theta - sin^2\alpha} - h\left(\frac{cos\theta}{sin\theta} - \frac{\cos(\theta+\alpha)}{\sin(\theta+\alpha)}\right)$$

$$= \frac{h(2sin\alpha cos\alpha)}{2\sin(\theta+\alpha)\sin(\theta-\alpha)} - h\left(\frac{\sin(\theta+\alpha)\,cos\theta - \cos(\theta+\alpha)sin\theta}{sin\theta\sin(\theta+\alpha)}\right)$$

$$[since, sin^2A - sin^2B = \sin(A+B).\sin(A-B)]$$

$$= \frac{hsin\alpha cos\alpha sin\theta - hsin(\theta+\alpha-\theta)\sin(\theta-\alpha)}{sin\theta\sin(\theta+\alpha)\sin(\theta-\alpha)}$$

$$[by\ trigonometrical\ relation, \quad sinAcosB - cosAsinB = \sin(A-B)]$$

$$= \frac{hsin\alpha cos\alpha sin\theta - hsin\alpha\sin(\theta-\alpha)}{sin\theta\sin(\theta+\alpha)\sin(\theta-\alpha)} = \frac{hsin\alpha(cos\alpha sin\theta - \sin(\theta-\alpha))}{sin\theta(sin^2\theta - sin^2\alpha)}$$

$$= \frac{hsin\alpha(cos\alpha sin\theta - \sin\theta cos\alpha + cos\theta sin\alpha)}{sin\theta(sin^2\theta - sin^2\alpha)} = \frac{hsin\alpha(cos\theta sin\alpha)}{sin\theta(sin^2\theta - sin^2\alpha)}$$

$$\Rightarrow O'P = \frac{hsin^2\alpha cot\theta}{sin^2\theta - sin^2\alpha} \quad \dots\dots\dots\dots\dots\dots\dots\dots\dots (VII)$$

In right $\Delta O'TP$

$$\Rightarrow sin\theta = \frac{PT}{O'P} \Rightarrow PT = O'Psin\theta = \left(\frac{hsin^2\alpha cot\theta}{sin^2\theta - sin^2\alpha}\right)sin\theta \quad from\ the\ eq(VII)$$

$$\Rightarrow PT = \frac{hsin^2\alpha cos\theta}{sin^2\theta - sin^2\alpha} \quad \dots\dots\dots\dots\dots\dots\dots\dots\dots..(VIII)$$

$$\Rightarrow cos\theta = \frac{O'T}{O'P} \Rightarrow O'T = O'Pcos\theta = \left(\frac{hsin^2\alpha cot\theta}{sin^2\theta - sin^2\alpha}\right)cos\theta \quad from\ the\ eq(VII)$$

$$\Rightarrow O'T = \frac{hsin^2\alpha cos^2\theta cosec\theta}{sin^2\theta - sin^2\alpha} \quad \dots\dots\dots\dots\dots\dots..(IX)$$

Now, we have

$$\Rightarrow OT = OO' + O'T = \frac{h}{sin\theta} + \frac{hsin^2\alpha cos^2\theta cosec\theta}{sin^2\theta - sin^2\alpha} \quad from\ eq(III)\&\ (IX)$$

$$= \frac{h(sin^2\theta - sin^2\alpha + sin^2\alpha cos^2\theta cosec\theta sin\theta)}{sin\theta(sin^2\theta - sin^2\alpha)} = \frac{h\left(sin^2\theta - sin^2\alpha(1 - cos^2\theta)\right)}{sin\theta(sin^2\theta - sin^2\alpha)}$$

$$OT = \frac{h(sin^2\theta - sin^2\alpha sin^2\theta)}{sin\theta(sin^2\theta - sin^2\alpha)} = \frac{hsin^2\theta(1 - sin^2\alpha)}{sin\theta(sin^2\theta - sin^2\alpha)} = \frac{hsin\theta cos^2\alpha}{sin^2\theta - sin^2\alpha} \quad \dots\dots\dots.(X)$$

Now, in right ΔOTS

$$\Rightarrow tan\alpha = \frac{ST}{OT} \Rightarrow ST = OTtan\alpha = \left(\frac{hsin\theta cos^2\alpha}{sin^2\theta - sin^2\alpha}\right)tan\alpha \quad from\ the\ eq(X)$$

$$\Rightarrow ST = \frac{hsin\theta sin\alpha cos\alpha}{sin^2\theta - sin^2\alpha} \quad \ldots\ldots\ldots\ldots\ldots (XI)$$

Now, consider a circular section with diameter SV intersecting the elliptical-section at the points 'C' & 'D'. (as shown in the figure(166) below)

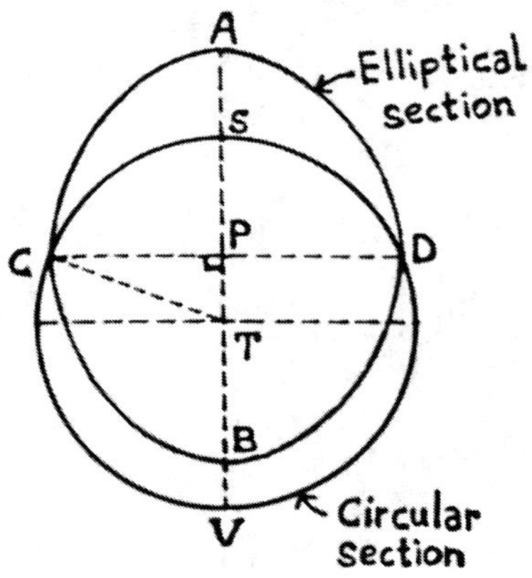

Fig (166)

In right ΔTPC

$$\Rightarrow CP = \sqrt{CT^2 - PT^2} = \sqrt{ST^2 - PT^2} \quad (since, CT = ST = radius\ of\ circular - section)$$

$$= \sqrt{\left(\frac{hsin\theta sin\alpha cos\alpha}{sin^2\theta - sin^2\alpha}\right)^2 - \left(\frac{hsin^2\alpha cos\theta}{sin^2\theta - sin^2\alpha}\right)^2} = \frac{h\sqrt{sin^2\theta cos^2\alpha sin^2\alpha - sin^4\alpha cos^2\theta}}{sin^2\theta - sin^2\alpha}$$

(On setting the values from the eq(VIII) & (XI))

$$= \frac{h\sqrt{sin^2\theta sin^2\alpha(1 - sin^2\alpha) - sin^4\alpha(1 - sin^2\theta)}}{sin^2\theta - sin^2\alpha}$$

$$= \frac{h\sqrt{sin^2\theta sin^2\alpha - sin^4\alpha sin^2\theta - sin^4\alpha + sin^4\alpha sin^2\theta}}{sin^2\theta - sin^2\alpha} = \frac{h\sqrt{sin^2\theta sin^2\alpha - sin^4\alpha}}{sin^2\theta - sin^2\alpha}$$

$$\Rightarrow CP = \frac{hsin\alpha\sqrt{sin^2\theta - sin^2\alpha}}{sin^2\theta - sin^2\alpha} = \frac{hsin\alpha}{\sqrt{sin^2\theta - sin^2\alpha}}$$

$$\Rightarrow CD = 2CP = \frac{2hsin\alpha}{\sqrt{sin^2\theta - sin^2\alpha}}$$

Hence, the minor axis of the elliptical-section

$$\Rightarrow 2b = CD = \frac{2hsin\alpha}{\sqrt{sin^2\theta - sin^2\alpha}} \quad \ldots\ldots\ldots\ldots\ldots (XII)$$

Hence, the eccentricity (e) of the elliptical-section is obtained by using formula as follows

$$\Rightarrow e = \sqrt{1 - \frac{b^2}{a^2}} \quad \forall(b \le a)$$

$$\Rightarrow e = \sqrt{1 - \frac{\left(\frac{hsin\alpha}{\sqrt{sin^2\theta - sin^2\alpha}}\right)^2}{\left(\frac{hsin2\alpha}{2(sin^2\theta - sin^2\alpha)}\right)^2}} \quad (on\ setting\ values\ from\ eq(II)\&(XII))$$

$$= \sqrt{1 - \frac{h^2 sin^2\alpha(sin^2\theta - sin^2\alpha)}{h^2 sin^2\alpha cos^2\alpha}} = \sqrt{\frac{cos^2\alpha - sin^2\theta + sin^2\alpha}{cos^2\alpha}} = \frac{\sqrt{1 - sin^2\theta}}{cos\alpha} = \frac{\sqrt{cos^2\theta}}{cos\alpha}$$

$$\Rightarrow e = \frac{cos\theta}{cos\alpha} \quad \ldots\ldots\ldots\ldots\ldots\ldots (XIII)$$

Thus, we get the following parameters of the elliptical-section generated by cutting the right cone with a smooth plane as follows

1. Major-axis

$$\Rightarrow 2a = \frac{hsin2\alpha}{sin^2\theta - sin^2\alpha} \quad \ldots\ldots\ldots\ldots\ldots\ldots (95)$$

2. Minor-axis

$$\Rightarrow 2b = \frac{2hsin\alpha}{\sqrt{sin^2\theta - sin^2\alpha}} \quad \ldots\ldots\ldots\ldots\ldots\ldots (96)$$

3. Eccentricity

$$\Rightarrow e = \frac{cos\theta}{cos\alpha} \qquad \forall\left(0 < \theta \le \frac{\pi}{2} \;\&\; 0 < \alpha < \frac{\pi}{2} \Rightarrow \theta > \alpha\right) \quad \ldots\ldots\ldots\ldots\ldots\ldots (97)$$

Thus, we find that the major & minor axes depend on the normal distance 'h' of the cutting plane from the apex while the eccentricity of the elliptical-section doesn't depends on the same but merely depend on the apex angle (2α) & the angle of inclination (θ) of the cutting plane with the axis of right cone.

* **Angular-shift (α_s):** The angle between axis of the right cone & the line joining the centre of elliptical-section to the apex is called the angular shift. Further, we are to find the angular shift (α_s) for the elliptical-section. (as shown in the figure (164) above)

In right ΔOMB

$$\Rightarrow \cot(\theta + \alpha) = \frac{MB}{OM} = \frac{MB}{h} \Rightarrow MB = h\cot(\theta + \alpha)$$

In right ΔOMP

$$\Rightarrow \tan\ MOP = \frac{MP}{OM} \Rightarrow tan\left(\frac{\pi}{2} - (\theta - \alpha_s)\right) = \frac{MP}{h} \Rightarrow MP = hcot(\theta - \alpha_s)\&$$

$$\Rightarrow \cos\ MOP = \frac{OM}{OP} \Rightarrow cos\left(\frac{\pi}{2} - (\theta - \alpha_s)\right) = \frac{h}{OP} \Rightarrow OP = \frac{h}{sin(\theta - \alpha_s)}$$

$$\Rightarrow OP = hcosec(\theta - \alpha_s)$$

$$\Rightarrow AP = BP = MP - MB = hcot(\theta - \alpha_s) - h\cot(\theta + \alpha) = h\big(cot(\theta - \alpha_s) - \cot(\theta + \alpha)\big)$$

Now, applying Sine-rule in ΔAOP as follows

$$\Rightarrow \frac{sin\ AOP}{AP} = \frac{sin\ OAP}{OP} \Rightarrow \frac{\sin(\alpha - \alpha_s)}{h\big(cot(\theta - \alpha_s) - \cot(\theta + \alpha)\big)} = \frac{sin(\theta - \alpha)}{hcosec(\theta - \alpha_s)}$$

$$\Rightarrow \frac{\sin(\alpha - \alpha_s)}{\left(\frac{sin(\theta + \alpha)cos(\theta - \alpha_s) - cos(\theta + \alpha)sin(\theta - \alpha_s)}{sin(\theta + \alpha)sin(\theta - \alpha_s)}\right)} = sin(\theta - \alpha)sin(\theta - \alpha_s)$$

$$\Rightarrow \frac{\sin(\alpha - \alpha_s)sin(\theta + \alpha)sin(\theta - \alpha_s)}{sin(\theta + \alpha - \theta + \alpha_s)} = sin(\theta - \alpha)sin(\theta - \alpha_s)$$

$$\Rightarrow \frac{\sin(\alpha - \alpha_s)}{sin(\alpha + \alpha_s)} = \frac{sin(\theta - \alpha)}{sin(\theta + \alpha)} \Rightarrow \frac{sin\alpha cos\alpha_s - cos\alpha sin\alpha_s}{sin\alpha cos\alpha_s + cos\alpha sin\alpha_s} = \frac{sin(\theta - \alpha)}{sin(\theta + \alpha)}$$

On dividing the numerator & denominator of LHS by $cos\alpha cos\alpha_s$and then adding 1 both the sides as follows

$$\Rightarrow \frac{\left(\frac{sin\alpha cos\alpha_s - cos\alpha sin\alpha_s}{cos\alpha cos\alpha_s}\right)}{\left(\frac{sin\alpha cos\alpha_s + cos\alpha sin\alpha_s}{cos\alpha cos\alpha_s}\right)} = \frac{sin(\theta-\alpha)}{sin(\theta+\alpha)} \Rightarrow \frac{tan\alpha - tan\alpha_s}{tan\alpha + tan\alpha_s} + 1 = \frac{sin(\theta-\alpha)}{sin(\theta+\alpha)} + 1$$

$$\Rightarrow \frac{tan\alpha - tan\alpha_s + tan\alpha + tan\alpha_s}{tan\alpha + tan\alpha_s} = \frac{sin(\theta-\alpha) + sin(\theta+\alpha)}{sin(\theta+\alpha)}$$

$$\Rightarrow \frac{2tan\alpha}{tan\alpha + tan\alpha_s} = \frac{sin(\theta-\alpha) + sin(\theta+\alpha)}{sin(\theta+\alpha)} \Rightarrow tan\alpha_s = \frac{2tan\alpha sin(\theta+\alpha)}{sin(\theta-\alpha) + sin(\theta+\alpha)} - tan\alpha$$

$$\Rightarrow tan\alpha_s = tan\alpha\left(\frac{2sin(\theta+\alpha) - sin(\theta-\alpha) - sin(\theta+\alpha)}{sin(\theta-\alpha) + sin(\theta+\alpha)}\right)$$

$$= tan\alpha\left(\frac{sin(\theta+\alpha) - sin(\theta-\alpha)}{sin(\theta+\alpha) + sin(\theta-\alpha)}\right)$$

$$= tan\alpha\left(\frac{2\cos\left(\frac{\theta+\alpha+\theta-\alpha}{2}\right)\sin\left(\frac{\theta+\alpha-\theta+\alpha}{2}\right)}{2\sin\left(\frac{\theta+\alpha+\theta-\alpha}{2}\right)\cos\left(\frac{\theta+\alpha-\theta+\alpha}{2}\right)}\right)$$

$$\left(sinA + sinB = 2sin\left(\frac{A+B}{2}\right)cos\left(\frac{A-B}{2}\right) \& \; sinA - sinB = 2cos\left(\frac{A+B}{2}\right)sin\left(\frac{A-B}{2}\right)\right)$$

$$\Rightarrow tan\alpha_s = tan\alpha\left(\frac{2\cos\theta\sin\alpha}{2\sin\theta\cos\alpha}\right) = tan\alpha\left(\frac{\tan\alpha}{\tan\theta}\right) = tan^2\alpha cot\theta$$

$$\Rightarrow \alpha_s = \tan^{-1}(tan^2\alpha cot\theta) \quad \forall\left(0 < \theta \le \frac{\pi}{2} \& \, 0 < \alpha < \frac{\pi}{2}\right) \quad \ldots\ldots\ldots\ldots\ldots\ldots (98)$$

It is obvious from the above eq(98) that the angular shift (α_s) merely depends on the apex angle (2α) & the angle of inclination (θ) of the cutting plane with the axis of right cone.

1.2.2 Frustum of aright cone with circular-section

It is obvious that the circular-section is generated if and only if the right cone is cut normally with the axis I.e. when the angle of inclination (θ) of the cutting plane with the axis of right cone is right angle i.e. $\theta = \frac{\pi}{2}$

Hence, by setting $\theta = \frac{\pi}{2}$ in the eq(95), (96), (97), & (98) we have

$$\Rightarrow major\ axis, 2a = \frac{hsin2\alpha}{sin^2\frac{\pi}{2} - sin^2\alpha} = \frac{2hsin\alpha cos\alpha}{1 - sin^2\alpha} = \frac{2hsin\alpha cos\alpha}{cos^2\alpha} = 2htan\alpha$$

$$\Rightarrow minor\ axis,\ \ 2b = \frac{2hsin\alpha}{\sqrt{sin^2\frac{\pi}{2} - sin^2\alpha}} = \frac{2hsin\alpha}{\sqrt{cos^2\alpha}} = \frac{2hsin\alpha}{cos\alpha} = 2htan\alpha$$

$$\Rightarrow eccentricity, e = \frac{cos\frac{\pi}{2}}{cos\alpha} = \frac{0}{cos\alpha} = 0 \ \ \&$$

$$\Rightarrow angular\ shift, \alpha_s = \tan^{-1}\left(tan^2\alpha cot\frac{\pi}{2}\right) = \tan^{-1}(tan^2\alpha(0)) = \tan^{-1}(0) = 0$$

Thus, we have

$$\Rightarrow diameter\ of\ the\ circular\ section = 2htan\alpha \quad \forall\left(0 < \alpha < \frac{\pi}{2}\right) \quad \ldots\ldots\ldots\ldots\ldots\ldots (99)$$

Where 'h' is the normal height of the cutting plane from apex of the right cone & '2α' is the apex-angle.

1.2.3 Practical-experiment to find out parameters of the elliptical-section of a right cone.

Let there a right cone with an apex angle '2α'. Now, let it be cut thoroughly with a smooth plane to generate an elliptical-section. (as shown in the figure (167) below)

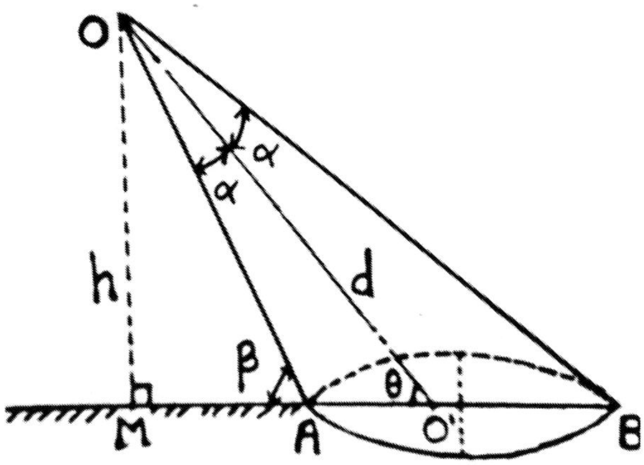

Fig (167)

Step 1: Take a frustum of the right cone with apex-point& put it with the elliptical-section on the plane-surface.

Step 2: Measure normal height 'h' of the apex-point from the plane-surface.

Step 3: Measure the minimum angle of inclination 'β' of lateral surface with the plane-surface. (see the above fig.)

Step 4: Find the angle of inclination (θ) of the cutting plane with the axis of right cone.

In $\Delta AOO'$

$$\Rightarrow \beta = \theta + \alpha \qquad (since, angle\ '\beta' is\ the\ exteior\ angle)$$

$$\Rightarrow \theta = \beta - \alpha$$

Step 5: Now put the required values of h, θ & α in the eq(95), (96), (97) & (98) to find the required parameters of elliptical-section.

Important deduction: If the a right cone with a certain apex-angle is cut with a number of parallel smooth planes inclined at a certain acute-angle with the axis then all the elliptical-sections generated have equal eccentricity but minor & axesare unequal it's because the value of eccentricity (e) of the elliptical-section merely depends on the angle of inclination of the plane with the axis of right cone while the values of minor & major axes depend on the normal distances of the plane from the apex-point.

The centres of all the elliptical-sections generated from a right cone thoroughly cut with the parallel planes always lie on the straight line (called shifted-axis) passing through the apex-point of right cone it's because the angular-shift (α_s) merely depends on the angle of inclination (θ)of the plane with the axis.

1.3 CIRCULAR CYLINDER

When a circular cylinder is thoroughly cut with a smooth plane then two types of the sections are obtained that may be circular or elliptical depending merely on the inclination of the cutting plane with the axis of cylinder i.e. circular-section is obtained if and only if the cutting plane is normal to the axis of the cylinder while in all other cases an elliptical-section is obtained. While their dimensions like radius of circular-section or minor & major axes of elliptical-section depend on the diameter of the cylinder.

1.3.1 Frustum of a circular cylinder with elliptical-section

Let there be a circular cylinder with a diameter 'D' & let it be thoroughly cut with a smooth plane inclined at an angle 'θ' with the axis of cylinder. (as shown in the figure (168) below)

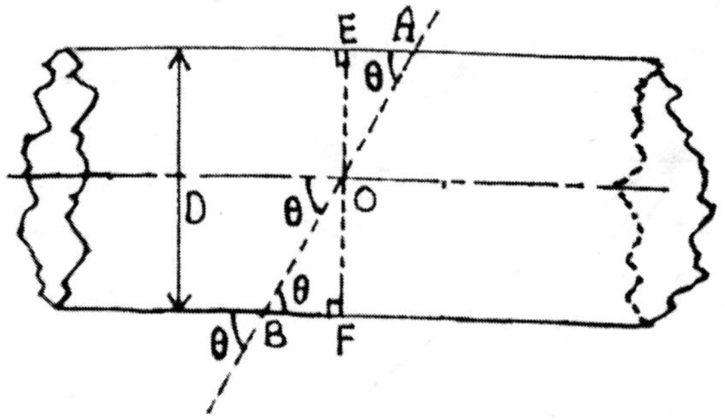

Fig (168)

In this case, the elliptical-section is obtained with major axis 'AB' & minor axis minor axis 'PQ' normal to the plane of paper. (as shown in the figure (169) below)

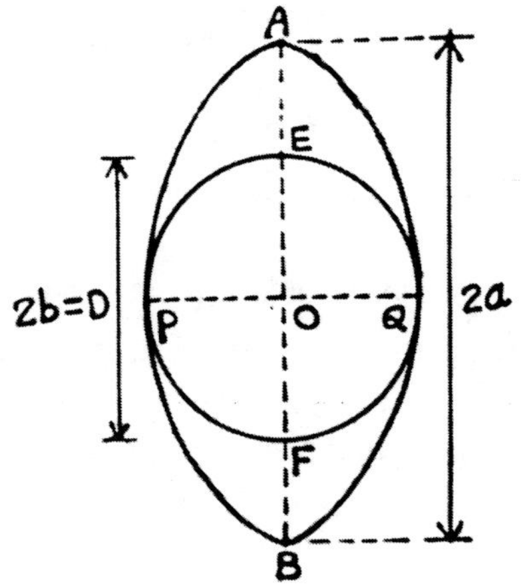

Fig (169)

In right ΔOFB

$$\Rightarrow sin\theta = \frac{OF}{OB} = \frac{\left(\frac{D}{2}\right)}{OB} \Rightarrow OB = \frac{D}{2}cosec\theta$$

$$\Rightarrow AB = 2a = major\ axis = 2OB = 2\frac{D}{2}cosec\theta = Dcosec\theta \quad \&$$

$$minor\ axis = 2b = Diameter\ of\ the\ cylinder = D$$

Eccentricity of the elliptical section (e) is obtained by using formula as follows

$$\Rightarrow e = \sqrt{1-\frac{b^2}{a^2}} = \sqrt{1-\frac{\left(\frac{D}{2}\right)^2}{\left(\frac{Dcosec\theta}{2}\right)^2}} = \sqrt{1-sin^2\theta} = cos\theta$$

Thus, we have

1. Major-axis

$$\Rightarrow 2a = Dcosec\theta \qquad \ldots\ldots\ldots\ldots\ldots\ldots\ldots (99)$$

2. Minor-axis

$$\Rightarrow 2b = D \qquad \ldots\ldots\ldots\ldots\ldots\ldots\ldots (99.1)$$

3. Eccentricity

$$\Rightarrow e = cos\theta \qquad \forall\left(0 < \theta \le \frac{\pi}{2}\right) \qquad \dots\dots\dots\dots\dots\dots\dots\dots\dots (99.2)$$

Eccentricity of the elliptical-section can also be obtained by setting $\alpha = 0$ in the eq(97) as follows (since the apex angle of the cylinder is zero.).

$$\Rightarrow e = \frac{cos\theta}{cos0} = cos\theta$$

1.3.2 Frustum of a circular cylinder with circular-section

It is obvious that the circular section of the cylinder will be obtained if and only if it is cut with a smooth plane normal with longitudinal axis.

Hence, by setting $\theta = \frac{\pi}{2}$ we get the parameters of the circular-section as follows

$$\Rightarrow 2a = Dcosec\frac{\pi}{2} = D$$

$$\Rightarrow 2b = D \quad \&$$

$$\Rightarrow e = cos\frac{\pi}{2} = 0 = eccentricity\ of\ the\ circle$$

1.3.3 Practical-experiment to find out parameters of the elliptical-section of a circular cylinder.

Let there be a circular cylinder with a diameter 'D'. Now, let it be cut thoroughly with a smooth plane to generate an elliptical-section. (as shown in the figure (170) below)

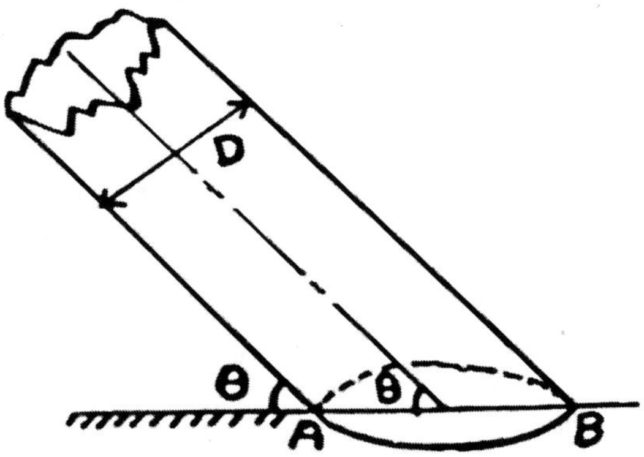

Fig (170)

Step 1: Take one of the frustums of the circular cylinder with elliptical& put it with the elliptical-section on the plane-surface.

Step 2: Measure the minimum angle of inclination 'θ' of the lateral surface with the plane-surface. (see the above fig.)

Step 3: Now put the required values of D & θ in the eq(99), (99.1) & (99.2) to find the required parameters of elliptical-section.

Important deduction: If a circular cylinder with a certain diameter is cut with a number of parallel smooth planes inclined at a certain acute-angle with the axis then identical elliptical-sections are obtained i.e. having equal major-axis, minor-axis& eccentricity and their centres always lie on the longitudinal axis of circular cylinder.

Solved Examples

Example 7.1: Find out radius of circular section, surface area & volume of a frustum of a sphere with a radius 14cm such that radial thickness of the frustum is 2.5cm.

Sol. given that

$$h = 2.5cm \ \& \ radius\ of\ parent\ sphere, R = 14cm$$

We know that the radius of circular section of frustum is given as

$$Radius, r = \sqrt{2Rh - h^2}$$

$$= \sqrt{2 \times 14 \times 2.5 - 2.5^2} = 7.98cm \qquad Ans.$$

Surface area of a frustum is given as

$$surface\ area, A = 2\pi Rh$$

$$= 2\pi \times 14 \times 2.5 = 219.911cm^2 \qquad Ans.$$

Now, the volume of frustum is given as

$$Volume, V = \frac{\pi h^2}{3}(3R - h)$$

$$= \frac{\pi(2.5)^2}{3}(3 \times 14 - 2.5)$$

$$= 6.545 \times 39.5 = 258.527cm^3 \qquad Ans.$$

Example 7.2: Find out surface area & volume of a minor frustum of a sphere with a radius 10cm such that radius of circular section of the frustum is 5.6cm.

Sol. given that

$$radius\ of\ circular\ section = 5.6cm \ \& \ radius\ of\ parent\ sphere, R = 10cm$$

We know that the radius of circular section of frustum is given as

$$Radius, r = \sqrt{2Rh - h^2}$$

$$h^2 - 2Rh + r^2 = 0$$

$$h^2 - 2 \times 10h + 5.6^2 = 0 \quad or \quad h^2 - 20h + 31.36 = 0$$

$$h = \frac{20 \pm \sqrt{20^2 - 4 \times 1 \times 31.36}}{2 \times 1} = \frac{20 \pm 16.57}{2}$$

$$\Rightarrow h = 18.285cm \ \ or \ 1.715cm$$

But the frustum is minor ($i.e.\, h < R$) hence, acceptable value of radial thickness is 1.715cm

$$\therefore surface\ area, A = 2\pi Rh$$

$$= 2\pi \times 10 \times 1.715 = 107.757cm^2 \qquad Ans.$$

Now, the volume of frustum is given as

$$Volume, V = \frac{\pi h^2}{3}(3R - h)$$

$$= \frac{\pi(1.715)^2}{3}(3 \times 10 - 1.715)$$

$$= 3.08 \times 28.285 = 87.118cm^3 \qquad Ans.$$

Example 7.3: Find out the radius of circular section, surface area & volume of a frustum of a sphere with a radius 11cm such that the angle subtended by it at the centre of parent sphere is 90°.

Sol. given that

$$2\theta = 90^o \,\&\, radius\ of\ parent\ sphere, R = 11cm$$

$$\therefore\ Radius\ of\ circular\ section, r = Rsin\theta$$

$$= 11 \times sin45^o = \frac{11}{\sqrt{2}} = 7.778cm \qquad Ans.$$

We know that the surface area of a frustum is given as

$$surface\ area, A = 2\pi R^2(1 - cos\theta)$$

$$= 2\pi(10)^2(1 - cos45^o) = 200\pi \times \left(1 - \frac{1}{\sqrt{2}}\right)$$

$$= 628.318 \times 0.293 = 184.097cm^2 \qquad Ans.$$

$$\therefore\ Volume, V = \frac{\pi R^3}{12}(cos3\theta - 9cos\theta + 8)$$

$$= \frac{\pi(11)^3}{12}(cos135^o - 9cos45^o + 8) = 345.455 \times 0.929 = 320.927cm^3 \qquad Ans.$$

Example 7.4: Find out volume of the minor frustum of a sphere with a radius 35cm such that surface area of the frustum is $3518.584cm^2$. Also, find out the angle subtended by the circular section of frustum at the centre of parent sphere.

Sol. given that

$$radius\ of\ parent\ sphere, R = 35cm\ \&$$

$$surface\ area\ of\ frustum, A = 3518.584cm^2$$

Let 'h' be the radial thickness of frustum then

$$surface\ area, A = 2\pi Rh$$

$$h = \frac{A}{2\pi R} = \frac{3518.584}{2\pi \times 35} = 16cm$$

$$\therefore volume\ of\ the\ minor\ frustum, V = \frac{\pi h^2}{3}(3R - h)$$

$$= \frac{\pi(16)^2}{3}(3 \times 35 - 16)$$

$$= 268.082 \times 89 = 23859.298cm^3 \qquad Ans.$$

Radius of circular section is given as

$$r = \sqrt{2Rh - h^2}$$

$$= \sqrt{2 \times 35 \times 16 - 16^2} = 29.394cm$$

If 2θ is the angle subtended by the circular section at the centre of parent sphere then

$$\Rightarrow r = Rsin\theta = 35sin\theta$$

$$or\ sin\theta = \frac{29.394}{35}\ or\ \theta = \sin^{-1} 0.84 = 57.14^o$$

$$\therefore \quad 2\theta = 2 \times 57.14^o = 114.28^o \qquad Ans.$$

Example 7.5: Find out surface area of the minor frustum of a sphere with a radius 15cm such that volume of the frustum is $350cm^3$. Also find out the angle subtended by the circular section of frustum at the centre of parent sphere.

Sol. given that

$$radius\ of\ parent\ sphere, R = 15cm\ \&$$
$$volume\ of\ frustum, V = 350cm^3$$

Let 'h' be the radial thickness of frustum then

$$volume\ of\ the\ minor\ frustum, V = \frac{\pi h^2}{3}(3R - h)$$
$$V = 350 = \frac{\pi h^2}{3}(3 \times 15 - h)\ or\ 334.225 = 45h^2 - h^3$$
$$h^3 - 45h^2 + 334.225 = 0$$

On solving the cubic equation in form of variable 'h', we have

$$\Rightarrow h = 44.834, -2.648\ \&\ 2.815$$

Since the frustum is minor ($i.e.\,h < R$) hence, radial thickness of frustum, h = 2.815

$$\therefore\ surface\ area, A = 2\pi Rh$$
$$= 2\pi \times 15 \times 2.815 = 265.307cm^2 \qquad Ans.$$

We know that the radius of circular section is given as

$$r = \sqrt{2Rh - h^2}$$
$$= \sqrt{2 \times 15 \times 2.815 - 2.815^2} = 8.748cm$$

If 2θ is the angle subtended by the circular section at the centre of parent sphere then

$$\Rightarrow r = R sin\theta = 35 sin\theta$$
$$or\ sin\theta = \frac{8.748}{15}\ or\ \theta = \sin^{-1} 0.583 = 35.66^o$$
$$\therefore \quad 2\theta = 2 \times 35.66^o = 71.32^o \qquad Ans.$$

Example 7.6: A frustum has two parallel circular sections which subtend the angles $36^o \& 76^o$ respectively at the centre of parent sphere with a radius 22cm. Find out surface area & volume of the frustum.

Sol. given that

$$radius\ of\ parent\ sphere,\ R = 22cm\ \&$$
$$2\theta_1 = 36^o \& 2\theta_2 = 76^o$$
$$\therefore\ Area\ of\ the\ surface,\ A = 2\pi R^2(cos\theta_1 - cos\theta_2)$$
$$= 2\pi \times 22^2(cos18^o - cos38^o)$$
$$= 3041.062 \times 0.163 = 495.693cm^2 \qquad Ans.$$
$$\therefore Volume, V = \frac{\pi R^3}{12}[9(cos\theta_1 - cos\theta_2) - (cos3\theta_1 - cos3\theta_2)]$$
$$= \frac{\pi(22)^3}{12}[9(cos18^o - cos38^o) - (cos54^o - cos114^o)]$$
$$= 2787.64 \times 0.473 = 1318.553cm^3 \qquad Ans.$$

Example 7.7: A frustum has two identical circular sections each, with radius 6cm, subtending an angle 36° at the centre of parent sphere. Whatis surface area & volume of the frustum?

Sol. given that

$$radius\ circular\ section,\ r = 6cm\ \&\ 2\theta = 36^o$$

If the radius of parent sphere is R then we have

$$Radius,\ r = R sin\theta$$

$$or\ R = \frac{r}{sin\theta} = \frac{6}{sin18^o} = 19.416cm$$

$$\therefore\ Area\ of\ the\ surface\ of\ the\ frustum,\ A = 4\pi R^2 cos\theta$$

$$= 4\pi(19.416)^2 cos18^o = 4505.424cm^2 \qquad Ans.$$

$$\therefore\ Volume, V = \frac{\pi R^3}{12}(9cos\theta - cos3\theta)$$

$$= \frac{\pi(19.416)^3}{12}(9cos18^o - cos54^o) = 15275.665cm^3 \qquad Ans.$$

Example 7.8: A frustum, of the sphere with a radius 17cm, with two parallel & identical circular sections, has surface area $1815.84cm^2$. Find out volume of the frustum & radius of the circular sections.

Sol. given that

$$surface\ area, A = 1815.84cm^2 \&\ radius\ of\ parent\ sphere, R = 17cm$$

If the angle subtended by the circular sections at the centre of parent sphere is 2θ then

$$Area\ of\ the\ surface,\ A = 4\pi R^2 cos\theta$$

$$or\ \ cos\theta = \frac{A}{4\pi R^2} = \frac{1815.84}{4\pi(17)^2} = 0.5$$

$$\theta = \cos^{-1} 0.5 = 60^o$$

$$\therefore\ 2\theta = 2 \times 60^o = 120^o \qquad Ans.$$

$$\therefore\ Volume, V = \frac{\pi R^3}{12}(9cos\theta - cos3\theta)$$

$$= \frac{\pi(17)^3}{12}(9cos60^o - cos180^o) = 7074.212cm^3 \qquad Ans.$$

Example 7.9: Find out the value of minor-axis, major-axis & eccentricity of an elliptical section generated by cutting a right cone having apex angle 80°, with a smooth plane inclined at an angle 65° with the axis of cone & having a normal distance 10cm from the apex-point.

Sol. given that

$$apex\ angle,\ 2\alpha = 80^o, \qquad angle\ of\ inclintion,\ \theta = 65^o \&$$

$$normal\ height,\ h = 10cm$$

$$\therefore\ major\ axis, 2a = \frac{hsin2\alpha}{sin^2\theta - sin^2\alpha}$$

$$= \frac{10sin80^o}{sin^2 65^o - sin^2 40^o} = \frac{9.848}{0.408} = 24.137cm \qquad Ans.$$

$$minor\ axis, 2b = \frac{2hsin\alpha}{\sqrt{sin^2\theta - sin^2\alpha}}$$

$$= \frac{2 \times 10sin40^o}{\sqrt{sin^2 65^o - sin^2 40^o}} = \frac{12.856}{0.639} = 20.119cm \qquad Ans.$$

$$\&\ eccentricity, e = \frac{cos\theta}{cos\alpha}$$

$$= \frac{cos65^o}{cos40^o} = 0.552 \qquad Ans.$$

Example 7.10: Determine the value of minor-axis, major-axis & eccentricity of an elliptical section generated by cutting a right cone having apex angle 60°, with a smooth plane inclined at an angle 56° with the axis of cone & having a normal distance 14cm from the apex-point.

Sol. given that

$$apex\ angle,\ \ 2\alpha = 60^o, \qquad angle\ of\ inclintion,\ \ \theta = 56^o \&$$

$$normal\ height,\ \ h = 14cm$$

$$\therefore\ major\ axis, 2a = \frac{hsin2\alpha}{sin^2\theta - sin^2\alpha}$$

$$= \frac{14sin60^o}{sin^2 56^o - sin^2 30^o} = \frac{12.124}{0.437} = 27.744cm \qquad Ans.$$

$$minor\ axis, 2b = \frac{2hsin\alpha}{\sqrt{sin^2\theta - sin^2\alpha}}$$

$$= \frac{2 \times 14sin30^o}{\sqrt{sin^2 56^o - sin^2 30^o}} = \frac{14}{0.661} = 21.18cm \qquad Ans.$$

$$\&\ eccentricity, e = \frac{cos\theta}{cos\alpha}$$

$$= \frac{cos56^o}{cos30^o} = 0.646 \qquad Ans.$$

Example 7.11: A truncated right cone having apex angle 45° is placed with elliptical section on a plane such that the normal height of apex point from the plane is 12cm & minimum angle of inclination of lateral edge with the plane is 60°.Determine the value of minor-axis, major-axis & eccentricity of elliptical section. Also calculate the angular shift.

Sol. given that

$$apex\ angle,\ \ 2\alpha = 45^o,\ \ minimum\ angle\ of\ inclintion,\ \ \beta = 60^o \&$$

$$normal\ height,\ \ h = 12cm$$

We know that angle of inclination (θ) of cutting plane with the axis of right cone is given as

$$\theta = \beta - \alpha = 60^o - 22.5^o = 37.5^o$$

$$\therefore\ major\ axis, 2a = \frac{hsin2\alpha}{sin^2\theta - sin^2\alpha}$$

$$= \frac{12sin45^o}{sin^2 37.5^o - sin^2 22.5^o} = \frac{8.485}{0.224} = 37.879cm \qquad Ans.$$

$$minor\ axis, 2b = \frac{2hsin\alpha}{\sqrt{sin^2\theta - sin^2\alpha}}$$

$$= \frac{2 \times 14sin22.5^o}{\sqrt{sin^2 37.5^o - sin^2 22.5^o}} = \frac{10.715}{0.473} = 22.653cm \qquad Ans.$$

$$\&\ eccentricity, e = \frac{cos\theta}{cos\alpha}$$

$$= \frac{cos37.5^o}{cos22.5^o} = 0.859 \qquad Ans.$$

Now, the angular shift (α_s)

$$\alpha_s = \tan^{-1}(tan^2\alpha cot\theta)$$

$$= \tan^{-1}(tan^2 22.5^o cot37.5^o)$$

$$= \tan^{-1}(0.2236) = 12.6^o\ or\ 12^o 36' \qquad Ans.$$

Example 7.12: A truncated right cone having apex angle 36° is placed with elliptical section on a plane such that the normal height of apex point from the plane is 18cm & minimum angle of inclination of lateral edge with the plane is 90°.Determine the value of minor-axis, major-axis, eccentricity & angular shift of the section generated.

Sol. given that

$$apex\ angle,\ \ 2\alpha = 36^o,\ \ minimum\ angle\ of\ inclintion,\ \ \beta = 90^o\&$$
$$normal\ height,\ \ h = 18cm$$

We know that angle of inclination (θ) of cutting plane with the axis of right cone is given as

$$\theta = \beta - \alpha = 90^o - 18^o = 72^o$$

$$\therefore\ major\ axis, 2a = \frac{hsin2\alpha}{sin^2\theta - sin^2\alpha}$$

$$= \frac{18sin36^o}{sin^2 72^o - sin^2 18^o} = \frac{10.58}{0.809} = 13.078cm \qquad Ans.$$

$$minor\ axis, 2b = \frac{2hsin\alpha}{\sqrt{sin^2\theta - sin^2\alpha}}$$

$$= \frac{2 \times 18sin18^o}{\sqrt{sin^2 72^o - sin^2 18^o}} = \frac{11.125}{0.899} = 12.375cm \qquad Ans.$$

$$\&\ eccentricity, e = \frac{cos\theta}{cos\alpha}$$

$$= \frac{cos72^o}{cos18^o} = 0.325 \qquad Ans.$$

Now, angular shift (α_s)

$$\alpha_s = \tan^{-1}(tan^2\alpha cot\theta)$$
$$= \tan^{-1}(tan^2 18^o cot72^o)$$
$$= \tan^{-1}(0.0343) = 1.96^o\ or\ 1^o57'36'' \qquad Ans.$$

Example 7.13: A truncated right cone having apex angle 66° is placed with elliptical section on a plane such that the normal height of apex point from the plane is 11cm & minimum angle of inclination of lateral edge with the plane is 123°.Determine the value of minor-axis, major-axis, eccentricity, angular shift& type of the section generated.

Sol. given that

$$apex\ angle,\ \ 2\alpha = 66^o,\ \ minimum\ angle\ of\ inclintion,\ \ \beta = 123^o\&$$
$$normal\ height,\ \ h = 11cm$$

We know that angle of inclination (θ) of cutting plane with the axis of right cone is given as

$$\theta = \beta - \alpha = 123^o - 33^o = 90^o$$

$$\therefore\ major\ axis, 2a = \frac{hsin2\alpha}{sin^2\theta - sin^2\alpha}$$

$$= \frac{11sin66^o}{sin^2 90^o - sin^2 33^o} = \frac{10.049}{0.703} = 14.294cm \qquad Ans.$$

$$minor\ axis, 2b = \frac{2hsin\alpha}{\sqrt{sin^2\theta - sin^2\alpha}}$$

$$= \frac{2 \times 11sin33^o}{\sqrt{sin^2 90^o - sin^2 33^o}} = \frac{11.982}{0.839} = 14.281cm \qquad Ans.$$

$$\&\ eccentricity, e = \frac{cos\theta}{cos\alpha}$$

$$= \frac{cos90^o}{cos33^o} = 0 \qquad Ans.$$

Now, angular shift (α_s)

$$\alpha_s = \tan^{-1}(tan^2\alpha cot\theta)$$

$$= \tan^{-1}(tan^2 33^o cot 90^o)$$

$$= \tan^{-1}(0) = 0 \qquad Ans.$$

Since, eccentricity (e) of the section is zero thus the section is circular.

Example 7.14: Find out eccentricity & angular shift of an elliptical section generated by cutting a right cone with apex angle 70° with a smooth plane inclined at an angle 80° with the axis of right cone.

Sol. given that

$$apex\ angle,\ 2\alpha = 70^o,\ angle\ of\ inclintion,\ \theta = 80^o$$

$$\therefore\ eccentricity, e = \frac{cos\theta}{cos\alpha}$$

$$= \frac{cos80^o}{cos35^o} = 0.212 \qquad Ans.$$

Angular shift (α_s)

$$\alpha_s = \tan^{-1}(tan^2\alpha cot\theta)$$

$$= \tan^{-1}(tan^2 35^o cot 80^o)$$

$$= \tan^{-1}(0.0864) = 4.9^o \qquad Ans.$$

Example 7.15: Find out eccentricity & angular shift of an elliptical section generated by cutting a right cone with apex angle 40° such that when it is placed with section on a plane, the minimum angle of inclination of its lateral edge with the plane is 100°.

Sol. given that

$$apex\ angle,\ 2\alpha = 40^o,\qquad minimum\ angle\ of\ inclintion,\ \beta = 100^o$$

We know that angle of inclination (θ) of cutting plane with the axis of right cone is given as

$$\theta = \beta - \alpha = 100^o - 20^o = 80^o$$

$$\therefore\ eccentricity, e = \frac{cos\theta}{cos\alpha}$$

$$= \frac{cos80^o}{cos20^o} = 0.185 \qquad Ans.$$

Angular shift (α_s)

$$\alpha_s = \tan^{-1}(tan^2\alpha cot\theta)$$

$$= \tan^{-1}(tan^2 20^o cot 80^o)$$

$$= \tan^{-1}(0.0233) = 1.33^o\ or\ 1^o 19' 48'' \qquad Ans.$$

Example 7.16: A right cone with apex angle 60° is cut with a smooth plane such that the elliptical section generated has eccentricity 0.354& minor axis is 8cm. Determine the normal distance of cutting plane from the apex-point of right cone.

Sol. given that

$$apex\ angle,\ 2\alpha = 60^o,\ eccentricity, e = 0.354\ \&$$

$$minor\ axis, 2b = 8cm$$

We know that eccentricity of the section is given as

$$e = \frac{cos\theta}{cos\alpha}$$

$$\Rightarrow cos\theta = ecos\alpha = 0.354cos30^o = 0.306$$

$$\theta = \cos^{-1}(0.306) = 72.18^o$$

$$minor\ axis, 2b = \frac{2hsin\alpha}{\sqrt{sin^2\theta - sin^2\alpha}}$$

$$or\ h = \frac{b\sqrt{sin^2\theta - sin^2\alpha}}{sin\alpha} = \frac{4\sqrt{sin^2 72.18^o - sin^2 30^o}}{sin30^o}$$

$$h = \frac{4 \times 0.81}{0.5} = 6.48cm \qquad Ans.$$

Example 7.17: A truncated right cone is placed with elliptical section on a plane such that the normal height of apex point from the plane is 13.5cm & minimum angle of inclination of lateral edge with the plane is 90°. Determine the value of minor-axis, major-axis& angular shift if eccentricity of section is 0.268.

Sol. given that

$$minimum\ angle\ of\ inclination,\ \beta = \alpha + \theta = 90^o,\ eccenticity,\ e = 0.268\ \&$$

$$normal\ height, h = 13.5cm$$

We know that eccentricity of the section is given as

$$e = \frac{cos\theta}{cos\alpha} = \frac{cos(\beta - \alpha)}{cos\alpha} (since, \qquad \beta = \alpha + \theta = 90^o)$$

$$\Rightarrow e = \frac{cos(90^o - \alpha)}{cos\alpha} = \frac{sin\alpha}{cos\alpha} = tan\alpha$$

$$\Rightarrow \alpha = \tan^{-1} e = \tan^{-1}(0.268) = 15^o \&$$

$$\theta = 90^o - 15^o = 75^o$$

Hence, we get

$$major\ axis, 2a = \frac{hsin2\alpha}{sin^2\theta - sin^2\alpha}$$

$$= \frac{13.5sin30^o}{sin^2 75^o - sin^2 15^o} = \frac{6.75}{0.866} = 7.794cm \qquad Ans.$$

$$minor\ axis, 2b = \frac{2hsin\alpha}{\sqrt{sin^2\theta - sin^2\alpha}}$$

$$= \frac{2 \times 13.5sin15^o}{\sqrt{sin^2 75^o - sin^2 15^o}} = \frac{6.988}{0.931} = 7.506cm \qquad Ans.$$

Angular shift (α_s)

$$\alpha_s = \tan^{-1}(tan^2\alpha cot\theta)$$

$$= \tan^{-1}(tan^2 15^o cot75^o)$$

$$= \tan^{-1}(0.0192) = 1.1^o\ or\ 1^o 6' \qquad Ans.$$

Example 7.18: A truncated right cone is placed with elliptical section on a plane such that the normal height of apex point from the plane is 9cm & minimum angle of inclination of lateral edge with the plane is 75°. Determine the value of minor-axis, major-axis & angular shift if eccentricity of section is 0.816.

Sol. given that

$$minimum\ angle\ of\ inclination,\ \beta = \alpha + \theta = 75^o,\ eccenticity,\ e = 0.816\ \&$$

$$normal\ height, h = 9cm$$

We know that eccentricity of the section is given as

$$e = \frac{cos\theta}{cos\alpha} = \frac{cos(\beta - \alpha)}{cos\alpha} (since, \qquad \beta = \alpha + \theta)$$

$$\Rightarrow e = 0.816 = \frac{cos(75^o - \alpha)}{cos\alpha} = \frac{cos75^o cos\alpha + sin75^o sin\alpha}{cos\alpha}$$

$$or\ 0.816cos\alpha = 0.259cos\alpha + 0.966sin\alpha$$

$$0.557cos\alpha = 0.966sin\alpha$$

$$or\ tan\alpha = \frac{0.557}{0.966} = 0.559 \ \Rightarrow \alpha = \tan^{-1}(0.577) = 30^o$$

$$\theta = 75^o - 30^o = 45^o$$

Hence, we get

$$major\ axis, 2a = \frac{hsin2\alpha}{sin^2\theta - sin^2\alpha}$$

$$= \frac{9sin60^o}{sin^2 45^o - sin^2 30^o} = \frac{7.794}{0.25} = 31.176cm \qquad Ans.$$

$$minor\ axis, 2b = \frac{2hsin\alpha}{\sqrt{sin^2\theta - sin^2\alpha}}$$

$$= \frac{2 \times 9sin30^o}{\sqrt{sin^2 45^o - sin^2 30}} = \frac{9}{0.5} = 18cm \qquad Ans.$$

Angular shift (α_s)

$$\alpha_s = \tan^{-1}(tan^2\alpha cot\theta)$$

$$= \tan^{-1}(tan^2 30^o cot45^o)$$

$$= \tan^{-1}(0.333) = 18.43^o\ or\ 18^o 25' 48'' \qquad Ans.$$

Example 7.19: A right cone is cut with a smooth plane at a distance 12cm from the apex point-point & inclined at an angle 45° with the axis of cone such that the elliptical section generated has minor axis 5cm. Find out

1) Apex angle of right cone 2) eccentricity 3) major axis of the elliptical section & 4) angular-shift (i.e. angle between the axis of cone & line joining the centre of elliptical section to the apex-point)

Sol. given that

$$normal\ distance, h = 12cm, angle\ of\ inclination, \theta = 45^o \&$$

$$minor\ axis, 2b = 5cm$$

Let the apex angle of right cone be α then

$$minor\ axis, 2b = \frac{2hsin\alpha}{\sqrt{sin^2\theta - sin^2\alpha}}$$

$$\Rightarrow 5 = \frac{2 \times 12sin\alpha}{\sqrt{sin^2 45^o - sin^2\alpha}}$$

$$sin^2 45^o - sin^2\alpha = \left(\frac{24sin\alpha}{5}\right)^2 = 23.04sin^2\alpha$$

$$sin^2\alpha = \frac{0.5}{24.04} \Rightarrow \alpha = \sin^{-1}(\sqrt{0.0208}) = 8.29^o$$

1. Apex angle of the right cone,

$$\alpha = 2 \times 8.29^o = 16.58^o \qquad Ans.$$

2. Eccentricity of the section,

$$e = \frac{cos\theta}{cos\alpha} = \frac{cos45^o}{cos8.29^o} = 0.714 \qquad Ans.$$

3. Major axis

$$2a = \frac{hsin2\alpha}{sin^2\theta - sin^2\alpha}$$

$$= \frac{12sin16.58^o}{sin^2 45^o - sin^2 8.29^o} = \frac{3.424}{0.479} = 7.148cm \qquad Ans.$$

4. Angular shift,

$$\alpha_s = \tan^{-1}(tan^2\alpha cot\theta)$$
$$= \tan^{-1}(tan^2 8.29^o cot45^o)$$
$$= \tan^{-1}(0.0212) = 11.97^o \ or\ 18^o58'12'' \qquad Ans.$$

Example 7.20: A truncated right cone with apex is placed with elliptical section on a plane such that the minimum angle of inclination of lateral edge with the plane is β. If eccentricity of the section is e then prove that angle of inclination of cutting plane with axis of right cone is given as

$$\theta = \cot^{-1}\left(\frac{esin\beta}{1 - ecos\beta}\right) \ \forall \ \left(0 < \beta < \pi \Rightarrow 0 < \theta \le \frac{\pi}{2}\right)$$

Sol. given that

$$eccentricity = e \ \& \ minimum\ angle\ of\ inclination = \beta$$

Let the angle of inclination of cutting plane with the axis of right cone be θ& apex angle of right cone be 2α then we have

$$eccenrisity,\ e = \frac{cos\theta}{cos\alpha} = \frac{cos\theta}{cos(\beta - \theta)} (\ since,\ \ \beta = \alpha + \theta)$$

$$\Rightarrow e = \frac{cos\theta}{cos\beta cos\theta + sin\beta sin\theta}$$

$$ecos\beta cos\theta + esin\beta sin\theta = cos\theta$$

$$\Rightarrow esin\beta sin\theta = cos\theta(1 - ecos\beta)$$

$$or\ \frac{cos\theta}{sin\theta} = \frac{esin\beta}{1 - ecos\beta} \Rightarrow cot\theta = \frac{esin\beta}{1 - ecos\beta}$$

Now we have

$$\beta = \alpha + \theta \qquad \forall\ 0 < \theta \le \frac{\pi}{2} \&\ 0 < \alpha < \frac{\pi}{2}$$

$$\Rightarrow 0 < \beta < \left(\frac{\pi}{2} + \frac{\pi}{2}\right)\ or\ 0 < \beta < \pi$$

$$\Rightarrow \qquad sin\beta > 0\ \&(1 - ecos\beta) > 0 \qquad (since,\ \ 0 \le e < 1\ \ \& -1 < cos\beta < 1)$$

$$\Rightarrow\ 0 \le cot\theta < \infty \quad or\ \ 0 < \theta \le \frac{\pi}{2}$$

We find that the value of $cot\theta$ varies from zero to infinity.Thus the value of θ will always be positive within proper limits.

Hence, the angle of inclination of cutting plane

$$\theta = \cot^{-1}\left(\frac{esin\beta}{1 - ecos\beta}\right) \quad \forall\ \left(0 < \beta < \pi \Rightarrow 0 < \theta \le \frac{\pi}{2}\right) \qquad proved.$$

Example 7.21: A truncated right cone with apex is placed with elliptical section on a plane such that the minimum angle of inclination of lateral edge with the plane is 60°. If eccentricity of the section is 0.815 then find out the apex angle& angular shift of right cone.

Sol. given that

$$eccentricity, e = 0.815\ \&\ minimum\ angle\ of\ inclination, \beta = 60^o$$

We know that eccentricity of the section is given as

$$e = \frac{cos\theta}{cos\alpha} = \frac{cos(\beta - \alpha)}{cos\alpha} (\ since,\ \ \beta = \alpha + \theta)$$

$$\Rightarrow e = \frac{cos\beta cos\alpha + sin\beta sin\alpha}{cos\alpha}$$

$$cos\beta cos\alpha + sin\beta sin\alpha = ecos\alpha$$

$$\Rightarrow sin\beta sin\alpha = cos\alpha(e - cos\beta) \ or \ \frac{cos\alpha}{sin\alpha} = \frac{sin\beta}{e - cos\beta}$$

$$\alpha = \cot^{-1}\left(\frac{sin\beta}{e - cos\beta}\right)$$

Now, on setting the values in the above expression, we get

$$\alpha = \cot^{-1}\left(\frac{sin60^o}{0.815 - cos60^o}\right) = \cot^{-1}(2.749) = 20^o$$

Hence, the apex angle of right cone

$$2\alpha = 2 \times 20^o = 40^o \qquad Ans.$$

Now, angle of inclination

$$\Rightarrow \theta = \beta - \alpha = 60^o - 20^o = 40^o$$

Hence, angular shift,

$$\alpha_s = \tan^{-1}(tan^2\alpha cot\theta)$$
$$= \tan^{-1}(tan^2 20^o cot40^o)$$
$$= \tan^{-1}(0.158) = 8.97^o \ or \ 8^o58'17.8'' \qquad Ans.$$

Example 7.22: Determine the values of minor-axis, major-axis & eccentricity of an elliptical section generated by cutting, a circular cylinder having a diameter 12cm, with a plane inclined at an angle 30° with the longitudinal axis.

Sol. given that

$$diameter\ of\ cylinder, D = 8cm\ \&\ angle\ of\ inclination, \theta = 30^o$$
$$\therefore major\ axis, 2a = Dcosec\theta$$
$$= Dcosec\theta = 12 \times cosec30^o = 24cm \qquad Ans.$$
$$minor\ axis, 2b = D = 12cm \qquad Ans.$$
$$eccentricity, e = cos\theta$$
$$= cos30^o = \frac{\sqrt{3}}{2} = 0.808 \qquad Ans.$$

Example 7.23: Find out the diameter of a circular cylinder if the elliptical section, generated by cutting the cylinder with a plane inclined at an angle 40° with the axis, has major axis 7cm

Sol. given that

$$major\ axis\ of\ section, 2a = 7cm\ \&\ angle\ of\ inclination, \theta = 40^o$$

We know that major axis of elliptical section

$$2a = Dcosec\theta$$

$$hence,\ diameter,\ D = 2asin\theta = 7 \times sin40^o = 4.5cm \qquad Ans.$$

Example 7.24: A circular cylinder is cut by a smooth plane such that the elliptical section generated has eccentricity 0.559 & major axis 6.8cm. Find out the

1. Diameter of the cylinder & 2) angle of inclination of the plane with the axis of the cylinder.

Sol. given that

$$eccentricity\ of\ section, e = 0.559\ \&\ major\ axis, 2a = 6.8cm$$

Now, eccentricity of the section

$$e = cos\theta \Rightarrow \theta = \cos^{-1}(e)$$
$$= \cos^{-1}(0.559) = 56^o$$

We know that major axis of elliptical section

$$2a = D cosec\theta$$

$$hence,\ diameter,\ D = 2a sin\theta = 6.8 \times sin56^o = 5.637cm \qquad Ans.$$

Hence, angle of inclination, $\theta = 56^o$ *Ans.*

Example 7.25: A right cone & a circular cylinder are thoroughly cut with a smooth plane keeping their axes parallel to each other. Now the truncated right cone with apex is placed with its elliptical section on a plane such that the minimum angle of inclination of its lateral edge with the plane is β. If area of cross section of circular cylinder is A then prove that area of elliptical section generated of cylinder is equal to

$$\frac{A\sqrt{1+e^2-2e cos\beta}}{1-e cos\beta} \qquad \forall\ (e > cos\beta\ \&\ 0 < \beta < \pi\ )$$

Where, e is the eccentricity of elliptical section of right cone.

Sol. let the apex angle & angle of inclination of plane with their axes are $2\alpha\ \&\ \theta$ respectively

Now the eccentricity of section of cone

$$e = \frac{cos\theta}{cos\alpha} = \frac{cos\theta}{cos(\beta-\theta)} (\ since,\ \beta = \alpha + \theta)$$

$$\Rightarrow\ e = \frac{cos\theta}{cos\beta cos\theta + sin\beta sin\theta}$$

$$e cos\beta cos\theta + e sin\beta sin\theta = cos\theta$$

$$\Rightarrow e sin\beta sin\theta = cos\theta(1 - e cos\beta)$$

$$or\ \frac{cos\theta}{sin\theta} = \frac{e sin\beta}{1 - e cos\beta} \Rightarrow cot\theta = \frac{e sin\beta}{1 - e cos\beta}$$

We know that

$$cosec^2\theta = 1 + cot^2\theta \quad or\ cosec\theta = \sqrt{1 + cot^2\theta}$$

$$\Rightarrow cosec\theta = \sqrt{1 + \left(\frac{e sin\beta}{1 - e cos\beta}\right)^2} = \sqrt{\frac{1 + (1 - e cos\beta)^2 + e^2 sin^2\beta}{(1 - e cos\beta)^2}}$$

$$= \sqrt{\frac{1 + (1 - e cos\beta)^2 + e^2 sin^2\beta}{(1 - e cos\beta)^2}} = \frac{\sqrt{1 + e^2 cos^2\beta - 2e cos\beta + e^2 sin^2\beta}}{1 - e cos\beta}$$

$$= \frac{\sqrt{1 + e^2(sin^2\beta + cos^2\beta) - 2e cos\beta}}{1 - e cos\beta} = \frac{\sqrt{1 + e^2 - 2e cos\beta}}{1 - e cos\beta}$$

$$cosec\theta = \frac{\sqrt{1 + e^2 - 2e cos\beta}}{1 - e cos\beta} \qquad \ldots\ldots\ldots\ldots\ldots\ldots (I)$$

Now we have

$$\beta = \alpha + \theta \qquad \forall\ 0 < \theta \le \frac{\pi}{2} \&\ 0 < \alpha < \frac{\pi}{2}$$

$$\Rightarrow 0 < \beta < \left(\frac{\pi}{2} + \frac{\pi}{2}\right)\ or\ 0 < \beta < \pi$$

$$\Rightarrow (1 - e cos\beta) > 0 \quad (since,\ 0 \le e < 1\ \&\ -1 < cos\beta < 1)$$

Now, we have

$$1 + e^2 - 2e cos\beta > 0$$

$$\Rightarrow \{(1 - ecos\beta) + e(e - cos\beta)\} > 0$$

$$\Rightarrow \quad e(e - cos\beta) > 0 \qquad since, \ (1 - ecos\beta) > 0$$

$$\Rightarrow \quad e - cos\beta > 0 \ \ or \qquad e > cos\beta$$

Also, we have

$$\Rightarrow 1 \le cosec\theta < \infty \ \ or \ \ 0 < \theta \le \frac{\pi}{2}$$

Let the diameter of cylinder be 'D' then we have

$$major\ axis, 2a = Dcosec\theta \ \&$$

$$minor\ axis, 2b = D$$

$$\therefore Area\ of\ elliptical\ section\ of\ cylinder = \frac{\pi}{4}(major\ axis)(minor\ axis)$$

$$= \frac{\pi}{4}(Dcosec\theta)(D) = \frac{\pi}{4}D^2 cosec\theta$$

$$= Acosec\theta \qquad \left(since,\ \frac{\pi}{4}D^2 = area\ of\ cross\ section\ of\ clinder\right)$$

$$= \frac{A\sqrt{1 + e^2 - 2ecos\beta}}{1 - ecos\beta}(from\ eq(I))$$

Now, applying the proper limits, we area of elliptical section of circular cylinder

$$= \frac{A\sqrt{1 + e^2 - 2ecos\beta}}{1 - ecos\beta} \qquad \forall\ (e > cos\beta\ \&\ 0 < \beta < \pi\ )$$

Proved

Summary

Frustum of sphere:

1. A frustum, of the sphere with a radius 'R', has asingle circular-section then the radius (r) of circular-section, surface area (A)& volume (V) are given as

$$Radius, r = \sqrt{2Rh - h^2}$$

$$surface\ area, A = 2\pi Rh\ \&$$

$$Volume, V = \frac{\pi h^2}{3}(3R - h) \quad \forall\ (0 < h \le 2R)$$

Where, 'h' is the radial thickness of the frustum.

If the angle subtended by the circular-section of the frustum at the centre of parent sphere with radius 'R' is 2θ then we have

$$Radius, r = Rsin\theta$$

$$surface\ area, A = 2\pi R^2(1 - cos\theta)\&$$

$$Volume, V = \frac{\pi R^3}{12}(cos3\theta - 9cos\theta + 8)[\ \forall\ 0 < \theta \le \pi\ ]$$

2. A frustum, of the sphere with a radius 'R', has two parallel circular-sections with radii $r_1 \& r_2$ subtending angles $\theta_1 \& \theta_2$ respectively at the centre of parent sphere then radii $(r_1 \& r_2)$ of circular-sections, surface area (A) & volume (V) of the frustum are given as

$$Radii, r_1 = Rsin\theta_1 \& r_2 = Rsin\theta_2$$

$$Area\ of\ the\ surface,\ A = 2\pi R^2(cos\theta_1 - cos\theta_2)\&$$

$$Volume, V = \frac{\pi R^3}{12}[9(cos\theta_1 - cos\theta_2) - (cos3\theta_1 - cos3\theta_2)]$$

$$\forall\ (\theta_1 < \theta_2 \Rightarrow \theta_1 \in [0,\pi)\&\theta_2 \in (0,\pi]\ )$$

3. A frustum of the sphere with a radius 'R' has two parallel & identical circular-sections each subtending an angle 2θ at the centre of parent sphere then radius 'r' of circular-sections, surface area (A)& volume of the frustum are given as

$$Radius,\ \ r = Rsin\theta$$

$$Area\ of\ the\ surface,\ \ A = 4\pi R^2 cos\theta\ \&$$

$$Volume, V = \frac{\pi R^3}{12}(9cos\theta - cos3\theta)\left[\forall\ 0 \le \theta < \frac{\pi}{2}\right]$$

Frustum of right cone:

4. Major axis $(2a)$, minor axis (2b) & eccentricity (e) of an elliptical section, generated by cutting a right cone with apex angle 2α with a smooth plane inclined at an angle θ with the axis & at a normal distance 'h' from the apex point, are given as

$$major\ axis, 2a = \frac{hsin2\alpha}{sin^2\theta - sin^2\alpha}$$

$$minor\ axis, 2b = \frac{2hsin\alpha}{\sqrt{sin^2\theta - sin^2\alpha}}\&$$

$$eccentricity, e = \frac{cos\theta}{cos\alpha}\quad \forall\left(0 < \theta \le \frac{\pi}{2}\&\ 0 < \alpha < \frac{\pi}{2} \Rightarrow \theta > \alpha\right)$$

5. **Angular shift (α_s):** Angle between axis of the right cone & line joining the centre of elliptical section (generated by cutting right cone obliquely with the axis) to the apex point. It is given as

$$\alpha_s = \tan^{-1}(tan^2\alpha cot\theta)\quad \forall\left(0 < \theta \le \frac{\pi}{2}\&\ 0 < \alpha < \frac{\pi}{2} \Rightarrow \theta > \alpha\right)$$

6. Radius of the circular section, generated by cutting a right cone with apex angle 2α with a smooth plane normal to the axis & at a normal distance 'h' from the apex point, is given as

$$R = 2htan\alpha\quad \forall\left(0 < \alpha < \frac{\pi}{2}\right)$$

Frustum of circular cylinder:

7. Major axis $(2a)$, minor axis (2b) & eccentricity (e) of an elliptical section, generated by cutting a circular cylinder with a diameter 'D' with a smooth plane inclined at an angle θ with the axis, are given as

$$major\ axis, 2a = Dcosec\theta$$

$$minor\ axis, 2b = D\ \&$$

$$eccentricity, e = cos\theta\qquad \forall\left(0 < \theta \le \frac{\pi}{2}\right)$$

Exercise 7

Q.7.1. Find out the surface area of a frustum of a sphere with a radius 12cm such that the angle subtended by it at the centre of parent sphere is 120°.

Q.7.2. Determine surface area of the frustum of a sphere with a radius 250mm if the frustum subtends an angle 60° at the centre of parent sphere.

Q.7.3. Find out the radius of the sphere if one of its frustums having surface area 512 cm^2 subtends an angle $\frac{\pi}{2}$ rad. at the centre.

Q.7.4. Surface area of the frustum of a sphere with a radius 15cm is 426 cm^2. Determine the angle subtended by it at the centre of parent sphere.

Q.7.5. A frustum has two parallel circular sectionswhich subtend the angles 30°&90° respectively at the centre of parent sphere with a radius 20cm. What will the surface area of frustum?

Q.7.6. A frustum has two identical circular sections each, with radius 4cm, subtending an angle 30° at the centre of parent sphere. What will be surface area of the frustum?

Q.7.7. What will the flux density on a sphere with a radius 10cm when a uniform point-source emitting a total flux 1400 lm is located at a distance 30cm from the centre of the sphere? (Assume that the flux is uniformly distributed over the intercepted surface).

Q.7.8. The total flux intercepted by a sphere is 20 lm when a uniform point-source emitting a total flux 2800 lm is located at a distance 40cm from the centre. What will be the flux density over the intercepted surface if the flux is uniformly distributed over the entire surface of interception?

Q.7.9. Find out the values of minor-axis, major-axis & eccentricity of an elliptical section generated by cutting a right cone having apex angle 90°, with a smooth plane inclined at an angle 60° with the axis of cone & having a normal distance 10cm from the apex-point.

Q.7.10. A truncated right cone having apex angle 60° is placed with elliptical section on a plane such that the normal height of apex point from the plane is 14cm & minimum angle of inclination of lateral edge with the plane is45^o.Determine the values of minor-axis, major-axis & eccentricity of elliptical section.

Q.7.11. A right cone with apex angle 80° is cut with a smooth plane at a normal distance 16cmfrom the apex point such that the elliptical section generated has eccentricity 0.4. Determine the

1) Major axis 2) minor axis & 3) angle of inclination of the plane with the axis of cone

Q.7.12. Find out the eccentricity of the elliptical section generated by cutting a right cone with apex angle 90° with a smooth plane inclined at an angle 60° with the axis of right cone.

Q.7.13. A right cone with apex angle 70° is cut by a smooth plane such that elliptical section generated has eccentricity 0.25& major axis 14.5cm. Determine normal distance of cutting plane from the apex-point of right cone.

Q.7.14. A right cone is cut with a smooth plane at a distance 12cm from the apex point-point &inclined at an angle 45° with the axis of cone such that the elliptical section generated has minor axis 5cm. Find out

1) Apex angle of right cone 2) eccentricity 3) major axis of the elliptical section & 4) angular-shift (i.e. angle between the axis of cone & line joining the centre of elliptical section to the apex-point)

Q.7.15. A truncated right cone with apex is placed with elliptical section on a plane such that the minimum angle of inclination of lateral edge with the plane is 100°. If eccentricity of the section is 0.653 then find out angle of inclination of cutting plane with axis of right cone.

Q.7.16. A right cone with apex angle 50° is cut by a smooth plane such that the elliptical section generated has eccentricity 0.845 & minor axis is 7.6cm. Determine the normal distance of cutting plane from the apex-point of right cone.

Q.7.17. A truncated right cone with apex is placed with elliptical section on a plane such that the minimum angle of inclination of lateral edge with the plane is 96°. If eccentricity of the section is 0.618 then find out the apex angle of right cone.

Q.7.18. Find out eccentricity & angular shift of an elliptical section generated by cutting a right cone with apex angle 50° such that when it is placed with section on a plane, the minimum angle of inclination of its lateral edge with the plane is 85°.

Q.7.19. A truncated right cone is placed with elliptical section on a plane such that the normal height of apex point from the plane is 14cm & minimum angle of inclination of lateral edge with the plane is 70°. Determine the value of minor-axis, major-axis & angular shift if eccentricity of section is 0.884.

Q.7.20. A truncated right cone with apex is placed with elliptical section on a plane such that the minimum angle of inclination of lateral edge with the plane is β. If eccentricity of the section is e then prove that the apex angle of right cone is given as

$$\alpha = \cot^{-1}\left(\frac{sin\beta}{e - cos\beta}\right) \quad \forall\ (e > cos\beta)$$

Q.7.21. A truncated right cone is placed with elliptical section on a plane such that the normal height of apex point from the plane is 9cm & minimum angle of inclination of lateral edge with the plane is 65°. Determine the value of minor-axis, major-axis & angular shift if eccentricity of section is 0.845.

Q.7.22. A truncated right cone having apex angle 66° is placed with elliptical section on a plane such that the normal height of apex point from the plane is 11cm & minimum angle of inclination of lateral edge with the plane is 123°. Determine the value of minor-axis, major-axis, eccentricity, angular shift & type of the section generated.

Q.7.23. Determine the values of minor-axis, major-axis & eccentricity of an elliptical section generated by cutting, a circular cylinder having a diameter 8cm, with a plane inclined at an angle 30° with the longitudinal axis.

Q.7.24. Find out the diameter of a circular cylinder if the elliptical section, generated by cutting the cylinder with a plane inclined at an angle 45° with the axis, has major axis 8cm

Q.7.25. A circular cylinder is cut by a smooth plane such that the elliptical section generated has eccentricity 0.45 & major axis 10cm. Find out the 1) diameter of the cylinder & 2) angle of inclination of the plane with the axis of the cylinder.

Chapter 2
Axiomatic analysis

This chapter deals with the analysis of axioms of the concurrent-vectors and the triangle and analysis of spiked regular-polygon.

***Concurrent vectors**: Hereby, the vectors having the same tail-point are called concurrent vectors.

2.1 AXIOM-1

If 'α' is the angle between two concurrent vectorsand one of them, keeping other stationary, is rotated by an angle 'β' about the point of concurrency in a plane inclined at an angle 'θ' with the plane containing both the vectors in initial position then the angle (α') between them in final position is given by the following expression

$$\Rightarrow \alpha' = \cos^{-1}(cos\alpha cos\beta + sin\alpha sin\beta cos\theta) \ \forall \ (0 \leq \alpha, \theta \leq \pi)$$

Let's call the above formula as HCR's or Vectors' cosine-formula.

Proof: Let there be two concurrent vectors$\overrightarrow{OA}$&$\overrightarrow{OB}$inclined at an angle 'α' and having point of concurrency as 'O'. Now, one of them say vector $\overrightarrow{OB}$ be rotated by an angle 'β' about the point of concurrency 'O' in a plane inclined at an angle 'θ' with the plane containing both the vectors in initial position& $MON' = \alpha'$ be the angle between the vectors $\overrightarrow{OA}$&$\overrightarrow{OB'}$ in the shifted position.

(as shown in the figure (171) below)

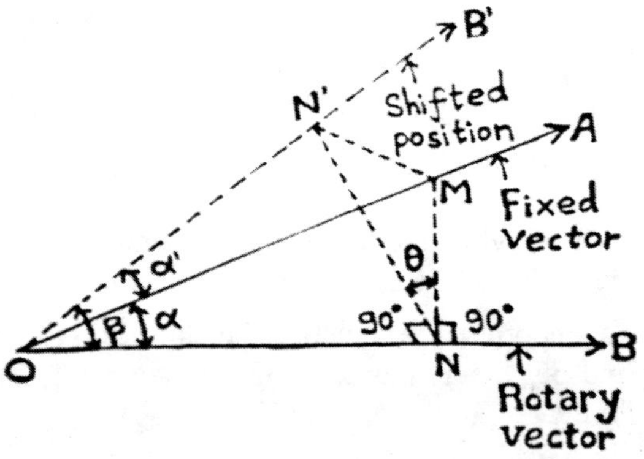

Fig (171)

Now, consider a point 'N' on the rotary vector $\overrightarrow{OB}$ and draw perpendicular NM & NN' from the point 'N' intersecting fixed vector $\overrightarrow{OA}$& rotated (shifted) vector $\overrightarrow{OB'}$ at the points 'M' and 'N'' respectively.

Let ON = a

In right ΔONM

$$\Rightarrow cos\alpha = \frac{ON}{OM} = \frac{a}{OM} \Rightarrow OM = asec\alpha \ \&$$

$$\Rightarrow tan\alpha = \frac{MN}{ON} = \frac{MN}{a} \Rightarrow MN = atan\alpha$$

Similarly, in right $\Delta ONN'$

$$\Rightarrow cos\beta = \frac{ON}{ON'} = \frac{a}{ON'} \Rightarrow ON' = asec\beta \ \&$$

$$\Rightarrow tan\beta = \frac{NN'}{ON} = \frac{NN'}{a} \Rightarrow NN' = atan\beta$$

Applying Cosine-rule in $\Delta MNN'$

$$\Rightarrow MN'^2 = MN^2 + NN'^2 - 2(MN)(NN')cos\ \ MNN'$$

$$\Rightarrow MN'^2 = (atan\alpha)^2 + (atan\beta)^2 - 2(atan\alpha)(atan\beta)cos\theta$$

$$\Rightarrow MN'^2 = a^2tan^2\alpha + a^2tan^2\beta - 2a^2tan\alpha tan\beta cos\theta \qquad \ldots\ldots\ldots\ldots\ldots\ldots\ldots..(I)$$

Again, applying Cosine-rule in $\Delta MON'$

$$\Rightarrow cos\alpha' = \frac{OM^2 + ON'^2 - MN'^2}{2(OM)(ON')}$$

$$= \frac{(asec\alpha)^2 + (asec\beta)^2 - (a^2tan^2\alpha + a^2tan^2\beta - 2a^2tan\alpha tan\beta cos\theta)}{2a^2sec\alpha sec\beta} \qquad from\ eq(I)$$

$$= \frac{a^2sec^2\alpha + a^2sec^2\beta - a^2tan^2\alpha - a^2tan^2\beta + 2a^2tan\alpha tan\beta cos\theta}{2a^2sec\alpha sec\beta}$$

$$= \frac{a^2(sec^2\alpha - tan^2\alpha) + a^2(sec^2\beta - tan^2\beta) + 2a^2tan\alpha tan\beta cos\theta}{2a^2sec\alpha sec\beta}$$

$$= \frac{a^2(1) + a^2(1) + 2a^2tan\alpha tan\beta cos\theta}{2a^2sec\alpha sec\beta} = \frac{2a^2 + 2a^2tan\alpha tan\beta cos\theta}{2a^2sec\alpha sec\beta}$$

$$= \frac{1 + tan\alpha tan\beta cos\theta}{sec\alpha sec\beta} = \frac{1}{sec\alpha sec\beta} + \frac{tan\alpha tan\beta cos\theta}{sec\alpha sec\beta} = cos\alpha cos\beta + \frac{sin\alpha sin\beta cos\theta}{cos\alpha cos\beta sec\alpha sec\beta}$$

$$\Rightarrow cos\alpha' = cos\alpha cos\beta + sin\alpha sin\beta cos\theta$$

$$\Rightarrow \alpha' = \cos^{-1}(cos\alpha cos\beta + sin\alpha sin\beta cos\theta)\ \forall\ (0 \le (\alpha, \theta) \le \pi) \qquad \ldots\ldots\ldots\ldots\ldots..\ (100)$$

Important deduction 1: If the angle α' between the vectors in the shifted position is known then the angle of inclination 'θ' of plane of rotation & plane of vectors in initial position is given from the eq(100) as follows

$$\Rightarrow \alpha' = \cos^{-1}(cos\alpha cos\beta + sin\alpha sin\beta cos\theta)\ \ or\ cos\alpha' = cos\alpha cos\beta + sin\alpha sin\beta cos\theta$$

$$\Rightarrow sin\alpha sin\beta cos\theta = cos\alpha' - cos\alpha cos\beta\ \ or\ cos\theta = \frac{cos\alpha' - cos\alpha cos\beta}{sin\alpha sin\beta}$$

$$\Rightarrow \theta = \cos^{-1}\left(\frac{cos\alpha' - cos\alpha cos\beta}{sin\alpha sin\beta}\right)\ \ \forall\ (0 \le (\alpha, \theta) \le \pi) \qquad \ldots\ldots\ldots\ldots\ldots\ldots\ldots\ldots (101)$$

Let's call the above formula as HCR's Cosine-formula.

The above formula is very important to analyse the three intersecting planes as well as three intersecting straight lines in the space. With the help of this formula, we can co-relate internal angles with the angles of lateral faces meeting at each vertex of a tetrahedron. (see fig (172))

Important deduction 2: If three planes are intersecting one another at a single point (such as at the vertex of a tetrahedron) in the space such that the angle between the lines of intersection of the planes are'α', 'β' & 'γ'respectively.

(as shown in the figure (172) below)

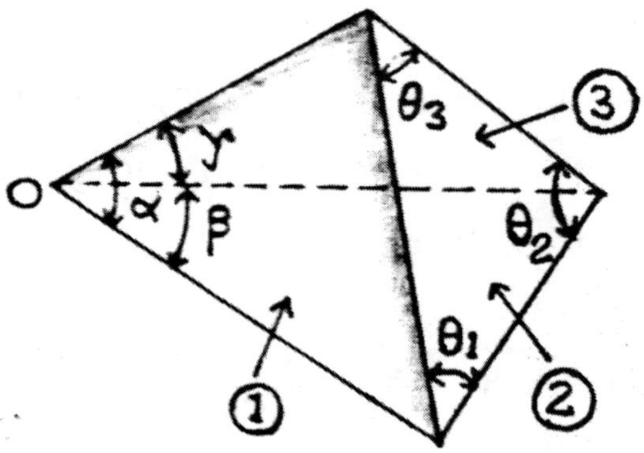

Fig (172)

This case is similar to the case of the two concurrent vectors i.e. the any two lines of intersection of the planes can be treated as two concurrent vectors and third one as the rotated vector.

Hence, using the eq(101) angles between the planes are given as follows

1. The angle between the plane-1 & 2

   By setting $\alpha' = angle\ between\ the\ lines\ of\ intersection\ opposite\ to\ the\ angle\ '\theta_1'$ $= \gamma$ in the eq(101)

$$\Rightarrow \theta_1 = \cos^{-1}\left(\frac{cos\gamma - cos\alpha cos\beta}{sin\alpha sin\beta}\right)$$

2. The angle between the plane-2 & 3

   By setting $\alpha' = angle\ between\ the\ lines\ of\ intersection\ opposite\ to\ the\ angle\ '\theta_2'$ $= \alpha$ in the eq(101)

$$\Rightarrow \theta_2 = \cos^{-1}\left(\frac{cos\alpha - cos\beta cos\gamma}{sin\beta sin\gamma}\right)$$

3. The angle between the plane-3 & 1

   By setting $\alpha' = \beta$ in the eq(101)

$$\Rightarrow \theta_3 = \cos^{-1}\left(\frac{cos\beta - cos\alpha cos\gamma}{sin\alpha sin\gamma}\right)$$

Important deduction 3: On setting $\theta = 0$ in the eq(100) as follows

$$\Rightarrow \alpha' = \cos^{-1}(cos\alpha cos\beta + sin\alpha sin\beta cos0) = \cos^{-1}(cos\alpha cos\beta + sin\alpha sin\beta)$$
$$= \cos^{-1}(\cos(\alpha - \beta)) = \alpha - \beta$$

It is obvious from the above result that whenone of the vectors is rotated in the same plane containing both the vectors towards other then the angle between them in shifted position is $\alpha' = \alpha - \beta$may be positive or negative depending on the values of $'\alpha'$ & $'\beta'$. If $\alpha > \beta \Rightarrow \alpha'$is positive (i.e. anticlockwise sense) & If $\alpha < \beta \Rightarrow \alpha'$ is negative (i.e. clockwise sense).

(as shown in the figure (173) below)

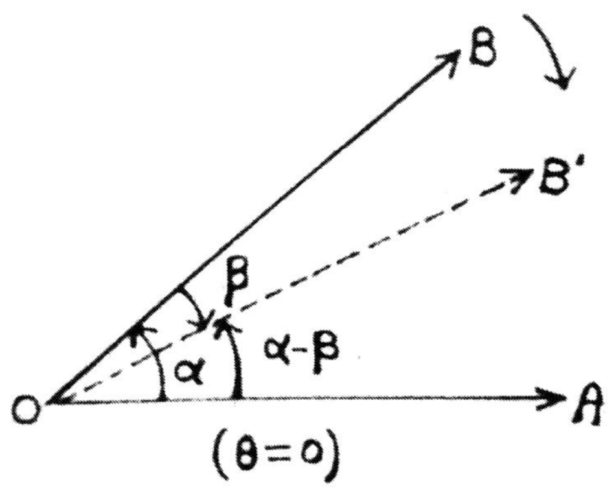

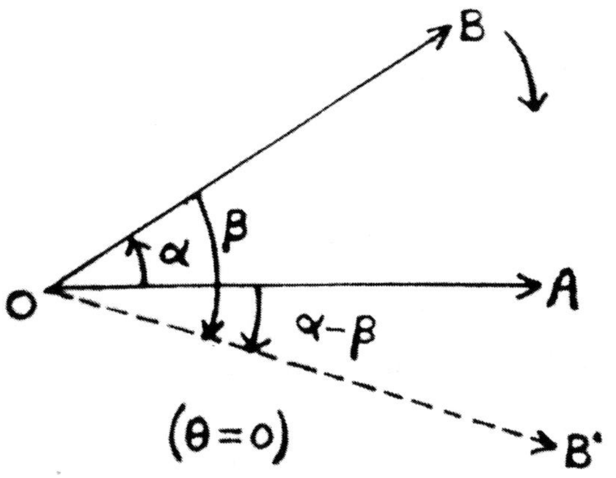

Fig (173)

Important deduction 4: On setting $\theta = \frac{\pi}{2}$ in the eq(100) as follows

$$\Rightarrow \alpha' = \cos^{-1}\left(cos\alpha cos\beta + sin\alpha sin\beta cos\frac{\pi}{2}\right) = \cos^{-1}(cos\alpha cos\beta - sin\alpha sin\beta(0))$$

$$\Rightarrow \alpha' = \cos^{-1}(cos\alpha cos\beta) \ \forall \ (0 \leq \alpha \leq \pi) \qquad \ldots\ldots\ldots\ldots\ldots\ldots..(102)$$

It is obvious from the above result that when one of the vectors is rotated in the plane normal to the plane containing both the vectors away from other then the angle between them in shifted position α' depends on the initial angle 'α' between them& the angle of rotation 'β'.

Case: If the one of the vectors is rotated by an angle $\beta = \frac{\pi}{2}$ about the point of concurrency in a plane normal to the plane containing both the vectors in initial position then the angle between them in shifted position

$$\Rightarrow \alpha' = \cos^{-1}\left(cos\alpha cos\frac{\pi}{2}\right) = \cos^{-1}(0) = \frac{\pi}{2} \qquad from\ the\ eq(102)$$

In this case, the rotated vector and the original vectors are normal to each other thus showing orthogonal axes X, Y & Z axes.

Important deduction 5: On setting $\theta = \pi$ in the eq(100) as follows

$$\Rightarrow \alpha' = \cos^{-1}(cos\alpha cos\beta + sin\alpha sin\beta cos\pi) = \cos^{-1}(cos\alpha cos\beta - sin\alpha sin\beta)$$

$$= \cos^{-1}(\cos(\alpha + \beta)) = |\alpha + \beta| = (\alpha + \beta)$$

It is obvious from the above result that when one of the vectors is rotated in the same plane containing both the vectors away from other then the angle between them in shifted position is $\alpha' = (\alpha + \beta)$is measured in the same sense i.e. anticlockwise. (as shown in the figure (174) below)

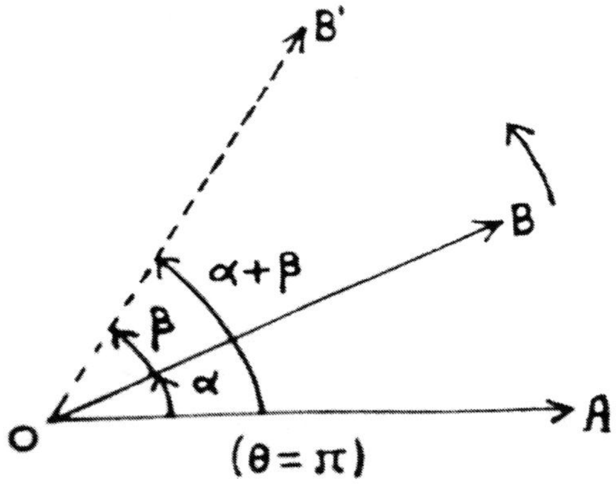

Fig (174)

2.2 AXIOM OF TETRAHEDRON

If α, β & γ are the angles between consecutive lateral edges meeting at a vertex of a tetrahedron then the tetrahedron will exist only and if only the sum of any two angles out of α, β & γ is always greater than third one i.e.

$$(\alpha+\beta)>\gamma,\ \ (\beta+\gamma)>\alpha\ \&(\alpha+\gamma)>\beta \qquad\qquad \forall\ 0<(\alpha+\beta+\gamma)<2\pi$$

In case of equality of the angles i.e. $(\alpha+\beta)=\gamma,\ \ (\beta+\gamma)=\alpha\ or\ \ (\alpha+\gamma)=\beta$

The lateral faces meeting at a vertex of the tetrahedron will coincide with each other in a single plane

Similarly, if $(\alpha+\beta+\gamma)=2\pi$

i.e. if sum of all the angles is 2π then all the lateral faces of tetrahedron will coincide in a single plane.

Hence for existence of a tetrahedron the above conditions should be satisfied.

2.3 AXIOM-2

"The perpendicular, drawn from the point of concurrency of three equal vectors to the plane of a triangle generated by joining the heads of all the vectors,always passes through the circumscribed-centre of the triangle"

Proof: Let there be three vectors$\overrightarrow{OP},\ \overrightarrow{OQ}\&\overrightarrow{OR}$ of equal length inclined at certain angle with one another & having point of concurrency 'O'. (as shown in the figure (175) below)

Now, join the heads to generate a triangle PQR having its circumscribed centre 'C'. Thus we have

$$\Rightarrow OP=OQ=OR$$

Assume that point C' is the foot of perpendicular drawn from the point of concurrency 'O' to the plane of ΔPQR. Thus we have

$$OC'P=\ \ OC'Q=\ \ OC'R=\frac{\pi}{2}\quad (since, OC' is\ normal\ to\ the\ plane)$$

Hence, $\Delta OC'P, \Delta OC'Q$ & $\Delta OC'R$ are right triangles. Thus we have

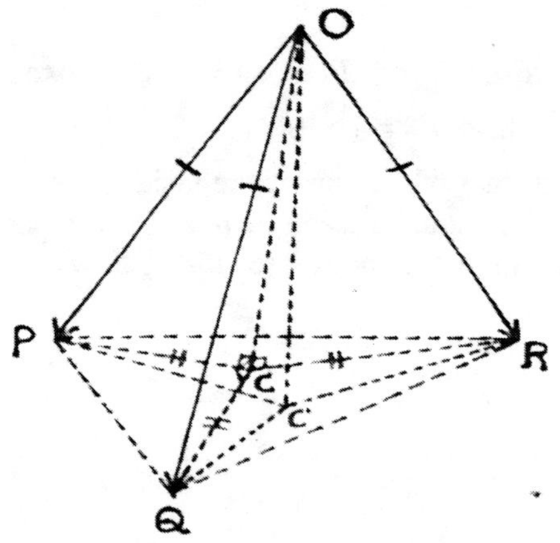

Fig (175)

$$in\,right\ \Delta OC'P \quad \Rightarrow C'P=\sqrt{OP^2-OC'^2}$$

$$in\,right\ \Delta OC'Q \quad \Rightarrow C'Q=\sqrt{OQ^2-OC'^2}=\sqrt{OP^2-OC'^2} \quad (given, OQ=OP)$$

$$in\,right\ \Delta OC'R \quad \Rightarrow C'R=\sqrt{OR^2-OC'^2}=\sqrt{OP^2-OC'^2} \quad (given, OR=OP)$$

Hence, from the above three expressions we get

$$C'P = C'Q = C'R = \sqrt{OP^2 - OC'^2} \qquad \ldots\ldots\ldots\ldots\ldots\ldots\ldots\ldots.(I)$$

From the above eq(I), it is clear that the foot of perpendicular C' is at an equal distance from all the vertices of ΔPQR but this is possible only & only when the foot of perpendicular C' is the circumscribed centre of ΔPQRbut circumscribed centre is the point 'C' thus the point C' must be the circumscribed centre 'C' of ΔPQR. Proved.

Mathematically, if 'p' is the length of perpendicular drawn from the point of concurrency of the equal concurrent vectors each of length 'a' to the plane of triangle generated by joining heads of all the vectors then the length of each vector is given as

In right triangle OCP, we have

$$OP^2 = OC^2 + CP^2 \Rightarrow a^2 = p^2 + R^2$$

$$or\ a = \sqrt{p^2 + R^2}$$

Where, R is the circumscribed radius of the triangle.

2.4 AXIOM-3

The distance between circumscribed & inscribed centres of a triangle is given by the following expression

$$\Rightarrow D_{CI} = \sqrt{\left\{(r - Rcos\theta_{min})^2 + (R + r - Rcos\theta_{min})^2 cot^2\left(\frac{2\theta + \theta_{min}}{2}\right)\right\}}$$

Where, $R \rightarrow radius\ of\ circumscribed\ circle$

$r \rightarrow radius\ of\ insscribed\ circle$

$\theta_{min} \rightarrow smallest\ angle\ \&$

$\theta \rightarrow any\ of\ the\ angles\ except\ \theta_{min}\ of\ the\ given\ triangle$

Proof: Let there be a triangle ABC having circumscribed centre C', inscribed centre 'I', circumscribed & inscribed radii 'R' & 'r' respectively and smallest angle B such that

B≤ A, C.

(as shown in the figure(176) below)

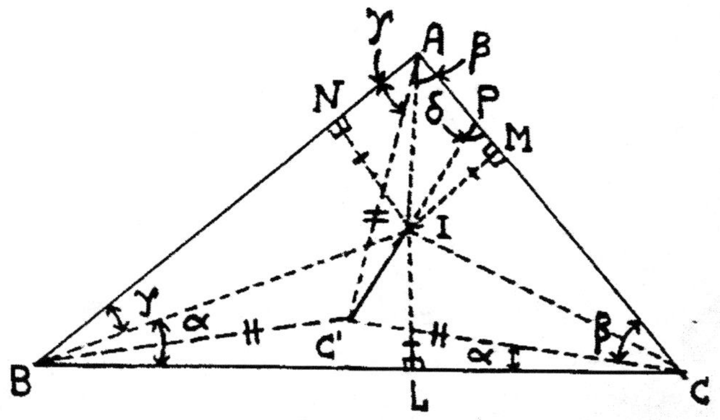

Fig (176)

Now, specify the circumscribed centre C'perpendicular bisectors of all the sides & inscribed centres'I' by drawing angle bisectors of ΔABC, draw the perpendiculars from the centre 'I' to all the sides and join centresby the line C'I which, on extending, intersects the side AC at the point 'P' such that $IPM = \delta$

We know that the sum of all the angles in a triangle is 180°

$$\Rightarrow A + B + C = \pi$$

Also, let

$$\Rightarrow \quad IBC = \quad ICB = \alpha, \quad ICA = \quad IAC = \beta, \quad IAB = \quad IBA = \gamma \ \& \ IPM = \delta$$

$$\Rightarrow A = \beta + \gamma, \qquad B = \alpha + \gamma \quad \& \qquad C = \alpha + \beta$$

Thus, on solving the above expressions, we have

$$\Rightarrow \alpha = \frac{B + C - A}{2} = \frac{(\pi - A) - A}{2} = \frac{\pi - 2A}{2} = \frac{\pi}{2} - A$$

$$\Rightarrow \beta = \frac{A + C - B}{2} = \frac{(\pi - B) - B}{2} = \frac{\pi - 2B}{2} = \frac{\pi}{2} - B \ \&$$

$$\Rightarrow \gamma = \frac{A + B - C}{2} = \frac{(\pi - C) - C}{2} = \frac{\pi - 2C}{2} = \frac{\pi}{2} - C$$

Now, in right ΔIMP

$$\Rightarrow sin\delta = \frac{IM}{IP} = \frac{r}{IP} (since, IM = IN = IL = inscribed\ radius = r)$$

Applying Sine-rule in $\Delta C'AP$

$$\Rightarrow \frac{sin\beta}{C'P} = \frac{sin\delta}{C'A} \Rightarrow \frac{sin\beta}{C'I + IP} = \frac{sin\delta}{R} \quad (since, C'A = C'B = C'C = circumradius = R)$$

$$\Rightarrow \frac{sin\beta}{x + IP} = \frac{\left(\frac{r}{IP}\right)}{R} = \frac{r}{R(IP)} \qquad (where, C'I = x\ (unknown))$$

$$\Rightarrow Rsin\beta(IP) = rx + r(IP) \Rightarrow IP = \frac{rx}{(Rsin\beta - r)}$$

Thus, on setting the value of 'IP', we get

$$\Rightarrow sin\delta = \frac{r}{IP} = \frac{r}{\left(\frac{rx}{Rsin\beta - r}\right)} = \frac{Rsin\beta - r}{x} \qquad \dots\dots\dots\dots\dots\dots\dots (I)$$

$$now \Rightarrow cos\delta = \sqrt{1 - sin^2\delta} = \sqrt{1 - \left(\frac{Rsin\beta - r}{x}\right)^2} = \frac{\sqrt{x^2 - (Rsin\beta - r)^2}}{x} \qquad \dots\dots\dots (II)$$

Since, line BI is bisecting$B = \alpha + \gamma$, hence

$$\Rightarrow \quad IBC = \frac{\alpha + \gamma}{2} \Rightarrow \quad C'BI = \quad C'BC - \quad IBC = \alpha - \frac{\alpha + \gamma}{2}$$

$$\Rightarrow \quad C'BI = \frac{\alpha - \gamma}{2} \qquad \dots\dots\dots\dots\dots\dots (III)$$

In right ΔILB

$$\Rightarrow \quad BIL = \frac{\pi}{2} - \quad IBL = \frac{\pi}{2} - \frac{\alpha + \gamma}{2} \left(since, \quad IBL = \quad IBC = \frac{\alpha + \gamma}{2}\right)$$

Similarly, we get

$$\Rightarrow \quad ICL = \frac{\alpha + \beta}{2} \Rightarrow \quad CIM = \quad LIC = \frac{\pi}{2} - \frac{\alpha + \beta}{2} \quad (since, line\ IC\ bisects \quad C)$$

$$CIP = \quad CIM - \quad PIM = \frac{\pi}{2} - \frac{\alpha + \beta}{2} - \left(\frac{\pi}{2} - \delta\right) = \delta - \frac{\alpha + \beta}{2} \ (since, \quad IPM = \delta)$$

$$\Rightarrow \quad BIC' = \pi - (\quad BIL + \quad LIC + CIP) \qquad (from\ the\ fig. (176))$$

$$= \pi - \left\{\left(\frac{\pi}{2} - \frac{\alpha + \gamma}{2}\right) + \left(\frac{\pi}{2} - \frac{\alpha + \beta}{2}\right) + \left(\delta - \frac{\alpha + \beta}{2}\right)\right\} = \left(\frac{3\alpha + 2\beta + \gamma}{2} - \delta\right)$$

Now, applying Sine-rule in ΔBIC

$$\Rightarrow \frac{sin\ \ C'BI}{CI} = \frac{sin\ \ BIC'}{C'B}$$

$$\Rightarrow \frac{sin\left(\frac{\alpha-\gamma}{2}\right)}{x} = \frac{sin\left(\frac{3\alpha+2\beta+\gamma}{2}-\delta\right)}{R} \qquad (set\ the\ value\ of\ \ C'BI\ from\ eq(III))$$

$$\Rightarrow x sin\left(\frac{3\alpha+2\beta+\gamma}{2}-\delta\right) = Rsin\left(\frac{\alpha-\gamma}{2}\right)$$

$$\Rightarrow x\left[sin\left(\frac{3\alpha+2\beta+\gamma}{2}\right)cos\delta - cos\left(\frac{3\alpha+2\beta+\gamma}{2}\right)sin\delta\right] = Rsin\left(\frac{\alpha-\gamma}{2}\right)$$

Now, on setting the values of $sin\delta$ & $cos\delta$ from the eq(I) & (II), we get

$$\Rightarrow x\left[sin\left(\frac{3\alpha+2\beta+\gamma}{2}\right)\left(\frac{\sqrt{x^2-(Rsin\beta-r)^2}}{x}\right) - cos\left(\frac{3\alpha+2\beta+\gamma}{2}\right)\left(\frac{Rsin\beta-r}{x}\right)\right] = Rsin\left(\frac{\alpha-\gamma}{2}\right)$$

$$\Rightarrow \sqrt{x^2-(Rsin\beta-r)^2} = \frac{Rsin\left(\frac{\alpha-\gamma}{2}\right) + (Rsin\beta-r)cos\left(\frac{3\alpha+2\beta+\gamma}{2}\right)}{sin\left(\frac{3\alpha+2\beta+\gamma}{2}\right)}$$

Now, on setting the values of α, β & γ in the above expression, we get

$$\Rightarrow \sqrt{x^2-\left(Rsin\left(\frac{\pi}{2}-B\right)-r\right)^2} = \frac{Rsin\left(\frac{C-A}{2}\right) + \left(Rsin\left(\frac{\pi}{2}-B\right)-r\right)cos\left(\frac{3\pi-(3A+2B+C)}{2}\right)}{sin\left(\frac{3\pi-(3A+2B+C)}{2}\right)}$$

$$\Rightarrow \sqrt{x^2-(RcosB-r)^2} = \frac{Rsin\left(\frac{C-A}{2}\right) + (RcosB-r)cos\left(\frac{2\pi-(2A+B)}{2}\right)}{sin\left(\frac{2\pi-(2A+B)}{2}\right)}$$

$$since, A+B+C=\pi$$

$$= \frac{Rsin\left(\frac{C-A}{2}\right) + (RcosB-r)cos\left(\pi-\frac{2A+B}{2}\right)}{sin\left(\pi-\frac{2A+B}{2}\right)}$$

$$= \frac{Rsin\left(\frac{C-A}{2}\right) - (RcosB-r)cos\left(\frac{2A+B}{2}\right)}{sin\left(\frac{2A+B}{2}\right)}$$

$$= Rsin\left(\frac{C-A}{2}\right)cosec\left(\frac{2A+B}{2}\right) - (RcosB-r)cot\left(\frac{2A+B}{2}\right)$$

On setting $C = \pi - A - B$, we get

$$\sqrt{x^2-(RcosB-r)^2} = Rsin\left(\frac{\pi}{2}-\frac{2A+B}{2}\right)cosec\left(\frac{2A+B}{2}\right) - (RcosB-r)cot\left(\frac{2A+B}{2}\right)$$

$$= R\frac{cos\left(\frac{2A+B}{2}\right)}{sin\left(\frac{2A+B}{2}\right)} - (RcosB-r)cot\left(\frac{2A+B}{2}\right)$$

$$= Rcot\left(\frac{2A+B}{2}\right) - (RcosB-r)cot\left(\frac{2A+B}{2}\right) = (R-RcosB+r)cot\left(\frac{2A+B}{2}\right)$$

On squaring both the sides, we get

$$\Rightarrow x^2-(RcosB-r)^2 = (R-RcosB+r)^2cot^2\left(\frac{2A+B}{2}\right)$$

$$x = \sqrt{(R + r - RcosB)^2 cot^2\left(\frac{2A + B}{2}\right) + (RcosB - r)^2}$$

Thus, we get

$$\Rightarrow D_{CI} = \sqrt{(RcosB - r)^2 + (R + r - RcosB)^2 cot^2\left(\frac{2A + B}{2}\right)}$$

But $B = \theta_{min}$

$$\Rightarrow D_{CI} = \sqrt{(Rcos\theta_{min} - r)^2 + (R + r - Rcos\theta_{min})^2 cot^2\left(\frac{2A + \theta_{min}}{2}\right)} \quad \ldots\ldots\ldots\ldots\ldots (IV)$$

Now, if we set $A = \pi - B - C = \pi - \theta_{min} - C$ in the above expression, we get

$$D_{CI} = \sqrt{(Rcos\theta_{min} - r)^2 + (R + r - Rcos\theta_{min})^2 cot^2\left(\frac{2(\pi - \theta_{min} - C) + \theta_{min}}{2}\right)}$$

$$= \sqrt{(Rcos\theta_{min} - r)^2 + (R + r - Rcos\theta_{min})^2 cot^2\left(\frac{2\pi - 2\theta_{min} - 2C + \theta_{min}}{2}\right)}$$

$$= \sqrt{(Rcos\theta_{min} - r)^2 + (R + r - Rcos\theta_{min})^2 cot^2\left(\frac{2\pi - (2C + \theta_{min})}{2}\right)}$$

$$= \sqrt{(Rcos\theta_{min} - r)^2 + (R + r - Rcos\theta_{min})^2 cot^2\left(\pi - \frac{2C + \theta_{min}}{2}\right)}$$

$$\Rightarrow D_{CI} = \sqrt{(Rcos\theta_{min} - r)^2 + (R + r - Rcos\theta_{min})^2 cot^2\left(\frac{2C + \theta_{min}}{2}\right)} \quad \ldots\ldots\ldots\ldots\ldots (V)$$

It is clear from the above eq(IV) & (V) that the distance between the circumscribed & inscribed centres depends on any one of the two angles except θ_{min} i.e. value of D_{CI} is not affected by the replacement of remaining two angles 'A' & 'C'

Hence, on setting $\theta = any\ of\ the\ two\ angles\ 'A' or\ 'C'$ in the eq(IV) or (V), hence we get

$$\Rightarrow D_{CI} = \sqrt{(Rcos\theta_{min} - r)^2 + (R + r - Rcos\theta_{min})^2 cot^2\left(\frac{2\theta + \theta_{min}}{2}\right)}$$

$$\Rightarrow D_{CI} = \sqrt{(r - Rcos\theta_{min})^2 + (R + r - Rcos\theta_{min})^2 cot^2\left(\frac{2\theta + \theta_{min}}{2}\right)} \quad \ldots\ldots\ldots\ldots\ldots (103)$$

$$\forall\ (\theta_{min} \leq \theta)$$

The value of the circumscribed radius of $\Delta ABC$is given as follows

$$\Rightarrow R = \frac{abc}{4\Delta} = \frac{a}{2sinA} = \frac{b}{2sinB} = \frac{c}{2sinC} \&$$

Inscribed radius is given as follows

$$\Rightarrow r = \frac{\Delta}{s} = 4Rsin\left(\frac{A}{2}\right)sin\left(\frac{B}{2}\right)sin\left(\frac{C}{2}\right) = \frac{a}{2}tan\left(\frac{A}{2}\right) = \frac{b}{2}tan\left(\frac{B}{2}\right) = \frac{c}{2}tan\left(\frac{C}{2}\right)$$

Where, a, b, c are the sides & A, B, C, are the angles of ΔABC and 'Δ' is the area of triangle which is given as follows

$$\Rightarrow \Delta = \sqrt{s(s-a)(s-b)(s-c)} = \frac{1}{2}absinC = \frac{1}{2}bcsinA = \frac{1}{2}acsinB$$

$$where, \qquad s = semiperimeter = \frac{a+b+c}{2}$$

Important deduction 1: Isosceles triangle

The distance 'D_{CI}', between the circumscribed & inscribed centres in an isosceles triangle having each of the equal angles θ, is given as follows

1. If $\theta < (\pi - 2\theta)$ thenby setting $\theta_{min} = \theta$ in the eq(103), we get

$$\Rightarrow D_{CI} = \sqrt{(r - Rcos\theta)^2 + (R + r - Rcos\theta)^2 cot^2\left(\frac{2\theta + \theta}{2}\right)}$$

$$= \sqrt{(r - Rcos\theta)^2 + (R + r - Rcos\theta)^2 cot^2\left(\frac{3\theta}{2}\right)} \quad \forall\,(\theta < (\pi - 2\theta))$$

It is applicable for an isosceles triangle having equal angles ($each\ equal\ to\ \theta$) smaller than third one $(\pi - 2\theta)$.

2. If $\theta > (\pi - 2\theta)$then by setting $\theta_{min} = \pi - 2\theta$ in the eq(103), we get

$$\Rightarrow D_{CI} = \sqrt{(r - Rcos(\pi - 2\theta))^2 + (R + r - Rcos(\pi - 2\theta))^2 cot^2\left(\frac{2\theta + \pi - 2\theta}{2}\right)}$$

$$= \sqrt{(r + Rcos2\theta)^2 + (R + r + Rcos2\theta)^2 cot^2\left(\frac{\pi}{2}\right)}$$

$$= \sqrt{(r + Rcos2\theta)^2 + (R + r + Rcos2\theta)^2(0)^2} = \sqrt{(r + Rcos2\theta)^2} = r + Rcos2\theta$$

$$\Rightarrow D_{CI} = r + Rcos2\theta \quad \forall\,(\theta > (\pi - 2\theta)) \qquad \ldots\ldots\ldots\ldots\ldots\ldots\ldots\ldots (104)$$

The above expression is applicable for an isosceles triangle having both the equal angles ($each\ equal\ to\ \theta$) greater than third one $(\pi - 2\theta)$.

Important deduction 2: Equilateral triangle

The distance 'D_{CI}' between the circumscribed & inscribed centres in an equilateral triangle is given as follows

We know that all the sides are equal & each of the angles of an equilateral triangle is $\frac{\pi}{3}$, hence we get

$$\Rightarrow circumscribed\ radius = R = \frac{a}{2sinA} = \frac{a}{2sin\frac{\pi}{3}} = \frac{a}{2\left(\frac{\sqrt{3}}{2}\right)} = \frac{a}{\sqrt{3}} \;\&\Rightarrow inscribed\ radius = r$$

$$= \frac{a}{2}tan\left(\frac{A}{2}\right) = \frac{a}{2}tan\left(\frac{\pi}{3}\right) = \frac{a}{2}\left(\frac{1}{\sqrt{3}}\right) = \frac{a}{2\sqrt{3}}$$

On setting $\theta_{min} = \theta = \dfrac{\pi}{3}$ in the eq(103) as follows

$$\Rightarrow D_{CI} = \sqrt{\left(\frac{a}{2\sqrt{3}} - \frac{a}{\sqrt{3}}cos\frac{\pi}{3}\right)^2 + \left(\frac{a}{\sqrt{3}} + \frac{a}{2\sqrt{3}} - \frac{a}{\sqrt{3}}cos\frac{\pi}{3}\right)^2 cot^2\left(\frac{2\frac{\pi}{3} + \frac{\pi}{3}}{2}\right)}$$

$$= \sqrt{\left(\frac{a}{2\sqrt{3}} - \frac{a}{\sqrt{3}}\frac{1}{2}\right)^2 + \left(\frac{a}{\sqrt{3}} + \frac{a}{2\sqrt{3}} - \frac{a}{\sqrt{3}}\frac{1}{2}\right)^2 cot^2\left(\frac{\pi}{2}\right)}$$

$$= \sqrt{(0)^2 + \left(\frac{a}{\sqrt{3}}\right)^2 (0)^2} = 0$$

It is clear that the distance 'D_{CI}' between the circumscribed & inscribed centres in an equilateral triangle is zero i.e. both the centres coincide with each other.

2.5 SPIKED/STAR-LIKE REGULAR-POLYGON:

"Regular-polygon obtained by the straight lines joining each vertex to the end-points of its opposite side in a regular-polygon having odd no. of the sides is called a spiked regular-polygon or a star-like regular-polygon" (as shown in the figure (177) below)

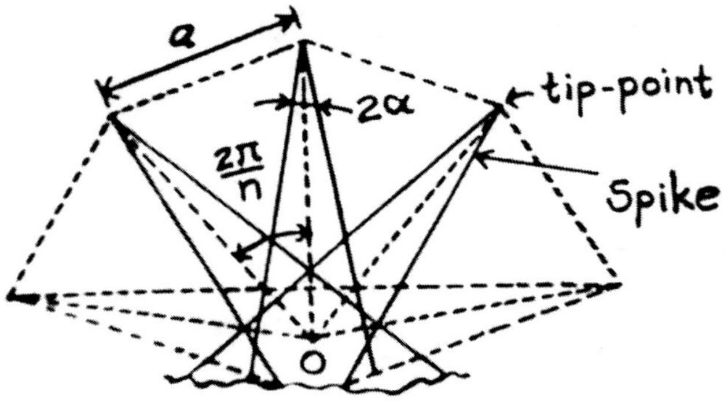

Fig (177)

If 'n' shows the no. of the sides of a spiked regular-polygon then

$$\Rightarrow n = (2m+1) \quad \forall (m \in N, \quad \textit{where } N \textit{ is the set of natural numbers})$$

* **Span**: "The distance between tip-points of any two consecutive spikes is called span."

It can be denoted by 'a'. (see the above fig(177))

2.5.1 Spike-angle: "Angle between the lines joining any of the vertices to the end-points of its opposite side in a regular polygon having odd no. of the sides is called spike angle" or

"Angle subtended at the tip-point of spike is called spike-angleof a spiked regular-polygon"

It is denoted by '2α'. It is given by the following formula

$$\Rightarrow 2\alpha = \frac{\pi}{n} \quad \forall \left(n = 2m+1 \Rightarrow m \in N\right)$$

Derivation: Let there be a spiked regular-polygon $A_1A_2A_3 \ldots\ldots A_{\left(\frac{n+3}{2}\right)} \ldots\ldots A_n$ having 'n' no. of the spikes such that $n = 2m+1 \Rightarrow m \in N$& 'a' is the span (i.e. distance between the tip-points of any twoconsecutive spikes).

(as shown in the figure (178) below showing two consecutive spikes & their opposite spike)

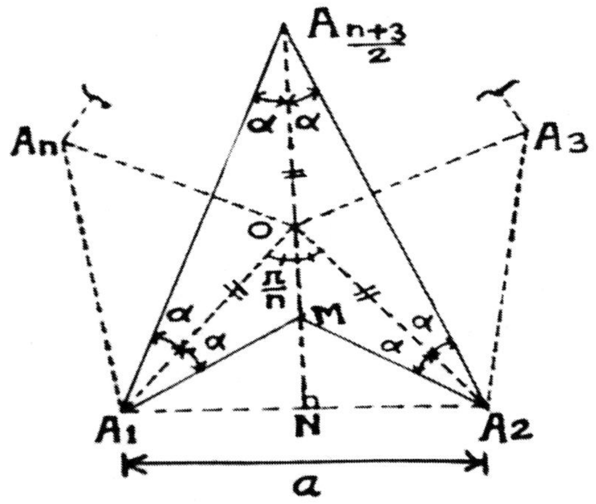

Fig (178)

Now, consider an isosceles $\Delta A_1A_2A_{\left(\frac{n+3}{2}\right)}$ obtained by joining two consecutive tip-points to the opposite tip-point. We are to find the spike angle $A_1A_{\left(\frac{n+3}{2}\right)}A_2 = 2\alpha$. We have

$$\Rightarrow \ A_1A_{\left(\frac{n+3}{2}\right)}N = \ A_2A_{\left(\frac{n+3}{2}\right)}N = \frac{A_1A_{\left(\frac{n+3}{2}\right)}A_2}{2} = \frac{2\alpha}{2} = \alpha \ \&$$

$$A_1OA_2 = \frac{2\pi}{n} \Rightarrow \ A_1ON = \ A_2ON = \frac{A_1OA_2}{2} = \frac{\pi}{n}A_1A_2 = a$$

In right ΔONA_1

$$\Rightarrow sin\frac{\pi}{n} = \frac{A_1N}{OA_1} = \frac{\left(\frac{a}{2}\right)}{OA_1} \Rightarrow OA_1 = \frac{a}{2}cosec\frac{\pi}{n}\ \&$$

$$\Rightarrow tan\frac{\pi}{n} = \frac{A_1N}{ON} = \frac{\left(\frac{a}{2}\right)}{ON} \Rightarrow ON = \frac{a}{2}cot\frac{\pi}{n}$$

Now, we have

$$\Rightarrow A_{\left(\frac{n+3}{2}\right)}N = OA_{\left(\frac{n+3}{2}\right)} + ON = OA_1 + ON = \frac{a}{2}cosec\frac{\pi}{n} + \frac{a}{2}cot\frac{\pi}{n}\left(OA_{\left(\frac{n+3}{2}\right)} = OA_1\right)$$

$$= \frac{a}{2}\left(cosec\frac{\pi}{n} + cot\frac{\pi}{n}\right)$$

In right $\Delta A_1NA_{\left(\frac{n+3}{2}\right)}$

$$\Rightarrow tan\ \ A_1A_{\left(\frac{n+3}{2}\right)}N = \frac{A_1N}{A_{\left(\frac{n+3}{2}\right)}N} = \frac{\left(\frac{a}{2}\right)}{\frac{a}{2}\left(cosec\frac{\pi}{n} + cot\frac{\pi}{n}\right)} = \frac{1}{cosec\frac{\pi}{n} + cot\frac{\pi}{n}}$$

$$\Rightarrow tan\alpha = \frac{1}{cosec\frac{\pi}{n} + cot\frac{\pi}{n}} \Rightarrow cot\alpha = cosec\frac{\pi}{n} + cot\frac{\pi}{n}$$

$$\Rightarrow cot\alpha = \frac{1}{sin\frac{\pi}{n}} + \frac{cos\frac{\pi}{n}}{sin\frac{\pi}{n}} = \frac{1 + cos\frac{\pi}{n}}{sin\frac{\pi}{n}} = \frac{2cos^2\frac{\pi}{2n}}{2sin\frac{\pi}{2n}cos\frac{\pi}{2n}} = \frac{cos\frac{\pi}{2n}}{sin\frac{\pi}{2n}} = cot\frac{\pi}{2n}$$

$$\Rightarrow cot\alpha = cot\frac{\pi}{2n}\ \ or\ \ \Rightarrow \alpha = \frac{\pi}{2n}\ \ or\ 2\alpha = \frac{\pi}{n}$$

Hence, the spike angle

$$\Rightarrow \left[2\alpha = \frac{\pi}{n}\right] \quad \forall\left(n = 2m + 1 \Rightarrow m \in N\right) \qquad \ldots\ldots\ldots\ldots\ldots (105)$$

2.5.2 Area (Δ) of the spiked regular-polygon

See the above figure (178).

The area of the isoscelesΔA_1OA_2

$$\Rightarrow \Delta_1 = \frac{1}{2}(A_1A_2)(ON) = \frac{1}{2}a \times \frac{a}{2}cot\frac{\pi}{n} = \frac{1}{4}a^2cot\frac{\pi}{n}$$

Since, the regular-polygon can be divided into 'n' no. of the congruent isosceles triangles as $\Delta A_1OA_2, \Delta A_2OA_3, \ldots\ldots\ldots\ldots \;\&\; \Delta A_nOA_1$.

Hence, area of the regular-polygon

$$\Rightarrow \Delta_2 = n \times (area\ of\ isosceles\ \Delta A_1OA_2) = n \times \frac{1}{4}a^2cot\frac{\pi}{n} = \frac{1}{4}na^2cot\frac{\pi}{n}$$

In right ΔMNA_1

$$\Rightarrow tan\ \ MA_1N = \frac{MN}{A_1N} \Rightarrow tan\left(\frac{\pi}{2} - \frac{\pi}{n} - \frac{\pi}{2n}\right) = \frac{MN}{\left(\frac{a}{2}\right)}$$

$$\Rightarrow tan\left(\frac{\pi}{2} - \left(\frac{3\pi}{2n}\right)\right) = \frac{2MN}{a} \Rightarrow cot\left(\frac{3\pi}{2n}\right) = \frac{2MN}{a}$$

$$\Rightarrow MN = \frac{a}{2}cot\frac{3\pi}{2n}$$

Hence, the area of the isosceles ΔA_1MA_2

$$\Rightarrow \Delta_3 = \frac{1}{2}(A_1A_2)(MN) = \frac{1}{2}a \times \frac{a}{2}cot\frac{3\pi}{2n} = \frac{1}{4}a^2cot\frac{3\pi}{2n}$$

But, the area of isosceles ΔA_1MA_2 is not included in the area of spiked regular-polygon. Hence, the area of the spiked regular-polygon

$$\Rightarrow \Delta = area\ of\ regular\ polygon - n \times area\ of\ \text{isosceles}\ \Delta A_1MA_2 = \Delta_2 - n\Delta_3$$

$$\Rightarrow \Delta = \frac{1}{4}na^2cot\frac{\pi}{n} - n \times \frac{1}{4}a^2cot\frac{3\pi}{2n} = \frac{1}{4}na^2\left[cot\frac{\pi}{n} - cot\frac{3\pi}{2n}\right]$$

$$= \frac{1}{4}na^2\left[\frac{cos\frac{\pi}{n}}{sin\frac{\pi}{n}} - \frac{cos\frac{3\pi}{2n}}{sin\frac{3\pi}{2n}}\right] = \frac{1}{4}na^2\left[\frac{sin\frac{3\pi}{2n}cos\frac{\pi}{n} - cos\frac{3\pi}{2n}sin\frac{\pi}{n}}{sin\frac{3\pi}{2n}in\frac{\pi}{n}}\right]$$

$$= \frac{1}{4}na^2\left[\frac{sin\left(\frac{3\pi}{2n} - \frac{\pi}{n}\right)}{sin\frac{3\pi}{2n}sin\frac{\pi}{n}}\right] = \frac{na^2sin\frac{\pi}{2n}}{4sin\frac{3\pi}{2n}sin\frac{\pi}{n}} = \frac{na^2sin\frac{\pi}{2n}}{4sin\frac{3\pi}{2n}\left(2sin\frac{\pi}{2n}cos\frac{\pi}{2n}\right)}$$

$$= \frac{na^2}{4\left(2sin\frac{3\pi}{2n}cos\frac{\pi}{2n}\right)} = \frac{na^2}{4\left(sin\left(\frac{3\pi}{2n} + \frac{\pi}{2n}\right) + sin\left(\frac{3\pi}{2n} - \frac{\pi}{2n}\right)\right)}$$

$$(since, \quad 2sinAcosB = \sin(A + B) + \sin(A - B))$$

$$= \frac{na^2}{4\left(sin\left(\frac{3\pi}{2n} + \frac{\pi}{2n}\right) + sin\left(\frac{3\pi}{2n} - \frac{\pi}{2n}\right)\right)} = \frac{na^2}{4\left(sin\frac{2\pi}{n} + sin\frac{\pi}{n}\right)}$$

$$\Rightarrow \Delta = \frac{na^2}{4\left(sin\frac{\pi}{n} + sin\frac{2\pi}{n}\right)} \quad \forall\left(n = 2m + 1 \Rightarrow m \in N\right) \qquad \ldots\ldots\ldots\ldots\ldots\ldots (106)$$

The above equation is in the generalised form depending on the no. of the spikes 'n' & span 'a' of the spiked regular-polygon

Since, spike-angle (2α) & area (Δ) of spiked regular-polygon depend on the no. of the spikes (n). Hence by setting the different values of 'n' we can obtain different expression for finding the values of '2α' & 'Δ'.

2.5.3 Regular-polygon with three spikes (n = 3)

Let there be a regular polygon with three spikes & span 'a'. (as shown in the figure(179) below)

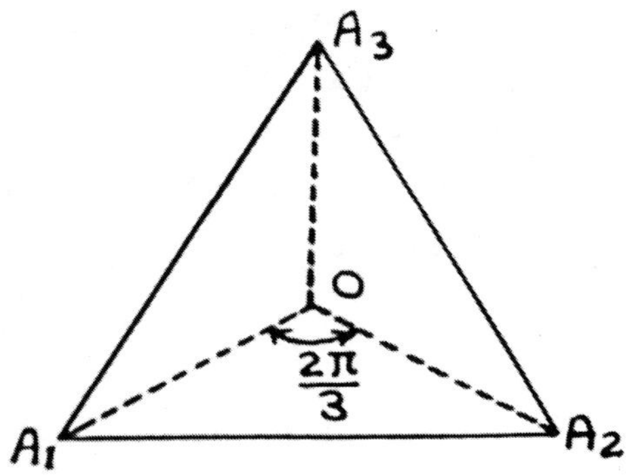

Fig (179)

On setting n = 3 in the eq(105) & (106) we have

1. Spike-angle

$$\Rightarrow 2\alpha = \frac{\pi}{3} = 60^o$$

2. Area (Δ)

$$\Rightarrow \Delta = \frac{3a^2}{4\left(sin\frac{\pi}{3} + sin\frac{2\pi}{3}\right)} = \frac{3a^2}{4\left(\frac{\sqrt{3}}{2} + \frac{\sqrt{3}}{2}\right)} = \frac{3a^2}{4(\sqrt{3})} = \frac{\sqrt{3}a^2}{4}$$

2.5.4 Regular-polygon with five spikes (n = 5)

Let there be a regular polygon with five spikes & span 'a'. (as shown in the figure(180) below)

On setting n = 5 in the eq(105) & (106) we have

1. Spike-angle

$$\Rightarrow 2\alpha = \frac{\pi}{5} = 36^o$$

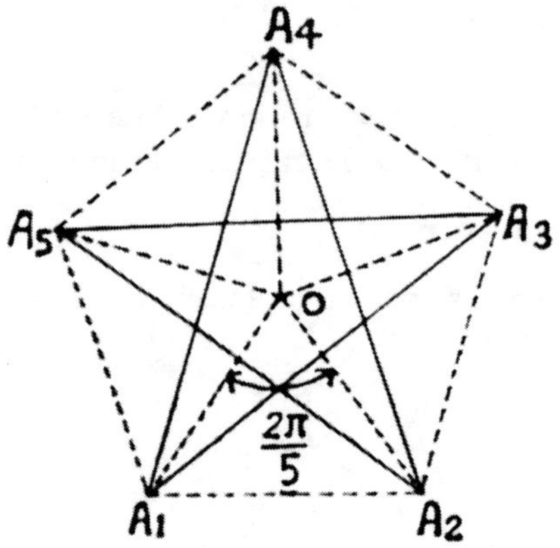

Fig (180)

2. Area (Δ)

$$\Rightarrow \Delta = \frac{5a^2}{4\left(\sin\frac{\pi}{5} + \sin\frac{2\pi}{5}\right)} = \frac{5a^2}{4(0.588 + 0.951)} = \frac{5a^2}{4(1.539)} = 0.812a^2$$

2.5.5 Regular-polygon with seven spikes (n = 7)

Let there be a regular polygon with seven spikes & span 'a'. (as shown in the figure(181) below)

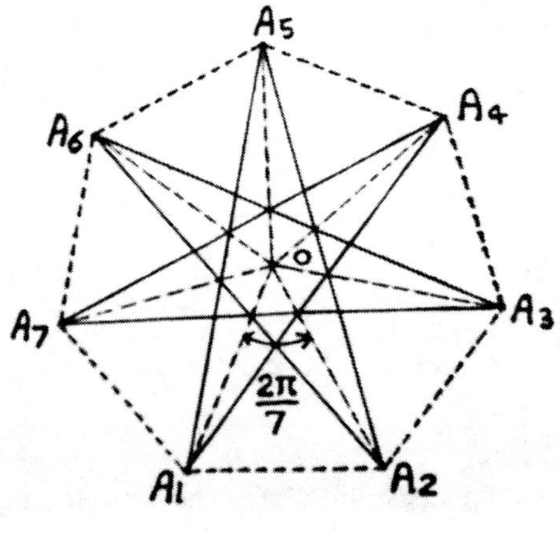

Fig (181)

On setting n = 7 in the eq(105) & (106) we have

1. Spike-angle

$$\Rightarrow 2\alpha = \frac{\pi}{7} = 25^o 42' 51.4''$$

2. Area (Δ)

$$\Rightarrow \Delta = \frac{7a^2}{4\left(\sin\frac{\pi}{7} + \sin\frac{2\pi}{7}\right)} = \frac{7a^2}{4(0.433 + 0.782)} = \frac{7a^2}{4(1.216)} = 1.439a^2$$

2.5.6 Regular-polygon with nine spikes (n = 9)

Let there be a regular polygon with nine spikes & span 'a'. (as shown in the figure(182) below)

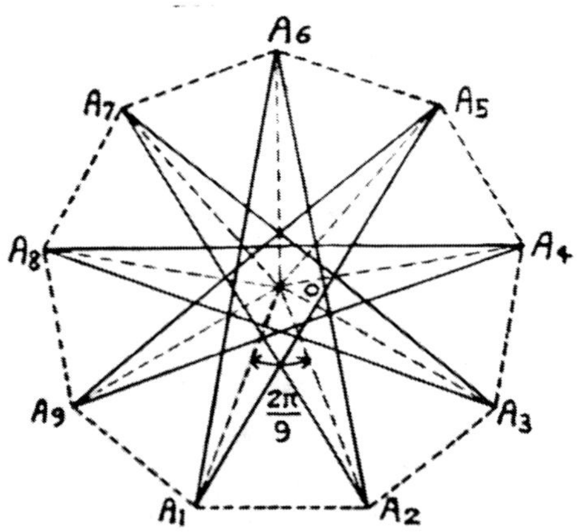

Fig (182)

On setting n = 9 in the eq(105) & (106) we have

1. Spike-angle

$$\Rightarrow 2\alpha = \frac{\pi}{9} = 20^o$$

2. Area (Δ)

$$\Rightarrow \Delta = \frac{9a^2}{4\left(\sin\frac{\pi}{9} + \sin\frac{2\pi}{9}\right)} = \frac{9a^2}{4(0.342 + 0.643)} = \frac{9a^2}{4(0.985)} = 2.285a^2$$

Similarly, further by setting n = 11, 13, 15, 17 .... upto $n = (2m + 1)$ $\forall\, m \in N$

We may obtain a number of expressions for the spiked regular-polygons having higher number of the sides.

2.6 ANALYSIS OF RIGHT PYRAMID:

In this section, we are interested to find out the surface area, volume &relation between the angles in a right pyramid with base as a regular polygon.

2.6.1 Area of lateral surface:

Let there be a right pyramid with base as a regular polygon $A_1A_2A_3 \dots A_n$ having 'n' number of the sides each of length 'a', angle between the consecutive lateral edges 'α'& normal height'H'. (as shown in the figure (183) below)

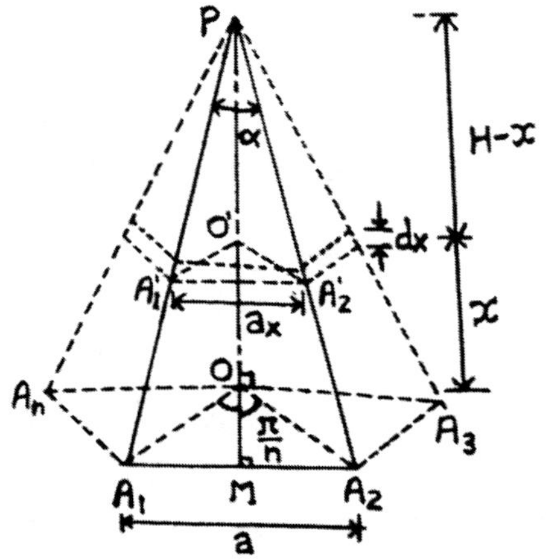

Fig (183)

Now, join all the vertices of the base to the centre 'O' thus we obtain 'n' number of isosceles triangles.

In right ΔOMA_2

$$\Rightarrow \tan \ A_2OM = \frac{MA_2}{OM}$$

$$or\ tan\frac{\pi}{n} = \frac{\left(\frac{a}{2}\right)}{OM} \Rightarrow\ OM = \frac{a}{2}cot\frac{\pi}{n}\left(since,\ A_1OA_2 = \frac{2\pi}{n}\right)$$

In right ΔPOM

$$\Rightarrow PM^2 = OP^2 + OM^2\ \ or\ PM = \sqrt{H^2 + \left(\frac{a}{2}cot\frac{\pi}{n}\right)^2}$$

$$\Rightarrow\ PM = \frac{1}{2}\sqrt{4H^2 + a^2cot^2\frac{\pi}{n}} \qquad \ldots\ldots\ldots\ldots\ldots..(I)$$

In right ΔPMA_2

$$\Rightarrow \tan \ A_2PM = \frac{MA_2}{PM}$$

$$\Rightarrow\ tan\frac{\alpha}{2} = \frac{\left(\frac{a}{2}\right)}{PM}\ \ or\ \ PM = \frac{a}{2}cot\frac{\alpha}{2} \qquad \ldots\ldots\ldots\ldots\ldots\ldots.(II)$$

From equation (I) & (II)

$$\Rightarrow \frac{a}{2}cot\frac{\alpha}{2} = \frac{1}{2}\sqrt{4H^2 + a^2cot^2\frac{\pi}{n}}$$

On squaring both the sides, we get

$$a^2cot^2\frac{\alpha}{2} = 4H^2 + a^2cot^2\frac{\pi}{n}$$

$$\Rightarrow\ 4H^2 = a^2\left(cot^2\frac{\alpha}{2} - cot^2\frac{\pi}{n}\right)$$

$$H = \frac{a}{2}\sqrt{cot^2\frac{\alpha}{2} - cot^2\frac{\pi}{n}} \qquad \ldots\ldots\ldots\ldots\ldots\ldots\ldots\ldots.(107)$$

Now, area of isosceles ΔA_1PA_2

$$\Rightarrow A' = \frac{1}{2} \times (A_1A_2)(PM)$$

$$= \frac{1}{2} \times (a)\left(\frac{1}{2}\sqrt{4H^2 + a^2cot^2\frac{\pi}{n}}\right) = \frac{1}{4}a\sqrt{4H^2 + a^2cot^2\frac{\pi}{n}}$$

Since, lateral surface of right pyramid consists of 'n' number of isosceles triangle congruent to ΔA_1PA_2 hence, area (A_l) of the lateral surface

$$\Rightarrow A_l = n \times (area\ of\ \Delta A_1PA_2) = n \times \frac{1}{4}a\sqrt{4H^2 + a^2cot^2\frac{\pi}{n}}$$

$$A_l = \frac{1}{4}na\sqrt{4H^2 + a^2cot^2\frac{\pi}{n}} \qquad \ldots\ldots\ldots\ldots..(III)$$

On setting the value of H from the eq(107) as follows

$$A_l = \frac{1}{4}na\sqrt{a^2\left(cot^2\frac{\alpha}{2} - cot^2\frac{\pi}{n}\right) + a^2cot^2\frac{\pi}{n}} = \frac{1}{4}na\sqrt{a^2cot^2\frac{\alpha}{2}}$$

$$\Rightarrow A_l = \frac{1}{4}na^2cot\frac{\alpha}{2} \qquad \ldots\ldots\ldots\ldots\ldots\ldots..(IV)$$

Hence, area (A_l) of the lateral surface of right pyramid is given as

$$A_l = \frac{1}{4}na\sqrt{4H^2 + a^2cot^2\frac{\pi}{n}} = \frac{1}{4}na^2cot\frac{\alpha}{2} \quad \dots\dots\dots\dots\dots\dots (108)$$

2.6.2 Area of regular-polygonal base:

Area of isosceles ΔA_1OA_2

$$\Rightarrow A'' = \frac{1}{2} \times (A_1A_2)(OM)$$

$$= \frac{1}{2} \times (a)\left(\frac{a}{2}cot\frac{\pi}{n}\right) = \frac{1}{4}a^2cot\frac{\pi}{n}$$

Since, base of the right pyramid consists of 'n' number of isosceles triangle congruent to ΔA_1OA_2 hence the area of the base (A_b)

$$\Rightarrow A_b = n \times (area\ of\ \Delta A_1OA_2) = n \times \frac{1}{4}a^2cot\frac{\pi}{n}$$

$$\Rightarrow A_b = \frac{1}{4}na^2cot\frac{\pi}{n} \quad \dots\dots\dots\dots\dots\dots (109)$$

2.6.3 Volume:

Consider an elementary regular polygonal strip having thickness 'dx' & 'n' number of the sides each of length a_x (as shown in the fig(183) above)

Area of the elementary strip (from the eq(109))

$$\Rightarrow A_x = \frac{1}{4}n{a_x}^2cot\frac{\pi}{n}$$

$$\therefore volume\ of\ the\ elementary\ strip, dV_x = A_xdx$$

$$\Rightarrow dV_x = \left(\frac{1}{4}ncot\frac{\pi}{n}\right){a_x}^2dx \quad \dots\dots\dots\dots\dots\dots (V)$$

Since, $\Delta PO'A_1'$ & ΔPOA_1 are similar triangles hence using similarity, we have

$$\frac{PA_1'}{PA_1} = \frac{O'P}{OP} = \frac{H-x}{H} \quad \dots\dots\dots\dots\dots\dots (VI)$$

Again using similarity in $\Delta PA_1'A_2'$ & ΔPA_1A_2

$$\frac{PA_1'}{PA_1} = \frac{A_1'A_2'}{A_1A_2} = \frac{a_x}{a}$$

From the above eq(VI), we have

$$\frac{PA_1'}{PA_1} = \frac{a_x}{a} = \frac{H-x}{H}$$

$$\Rightarrow a_x = \frac{a(H-x)}{H}$$

On setting the value of a_x in the eq(VI), we get

$$dV_x = \left(\frac{1}{4}ncot\frac{\pi}{n}\right)\left(\frac{a(H-x)}{H}\right)^2dx = \left(\frac{na}{4H}cot\frac{\pi}{n}\right)(H-x)^2dx$$

By integrating above expression, the volume (V) of the right pyramid

$$\Rightarrow V = \int_{x=0}^{x=H}\left(\frac{na}{4H}cot\frac{\pi}{n}\right)(H-x)^2dx$$

$$= \frac{na^2}{4H^2}cot\frac{\pi}{n}\int_{x=0}^{x=H}(H-x)^2dx = \frac{na^2}{4H^2}cot\frac{\pi}{n}\left[\frac{-(H-x)^3}{3}\right]_{x=0}^{x=H}$$

$$= \frac{na^2}{4H^2} cot\frac{\pi}{n}\left[\frac{-(H-H)^3}{3} + \frac{(H-0)^3}{3}\right] = \frac{na^2H}{12} cot\frac{\pi}{n}$$

$$\Rightarrow V = \frac{1}{12}na^2Hcot\frac{\pi}{n} \quad \forall\,(n \in N \,\&\, n \geq 3) \qquad \dots\dots\dots\dots\dots\dots\dots. (110)$$

Note: It is obvious from the above eq(110) that the volume of a right pyramid has symmetry (analogy) with that of a right cone

$$V = \frac{1}{12}na^2Hcot\frac{\pi}{n}$$

$$= \frac{1}{3}\left(\frac{1}{4}na^2cot\frac{\pi}{n}\right)(H) = \frac{1}{3}(area\ of\ base)(normal\ height)$$

Similarly, volume of a right cone

$$= \frac{1}{3}(area\ of\ base)(normal\ height) = \frac{1}{3}(\pi R^2)(H)$$

$$= \frac{1}{3}\pi R^2 H$$

2.6.4 Theorem (Analogy):(proposed by H.C.R.)

"Volume of a right conical body with cross-section varying linearly with its normal height (length) is given as

$$\boldsymbol{V = \frac{1}{3}(area\ of\ base)(normal\ height)}.$$

Note: Geometrical shape of the cross-section of a right conical body must be same throughout its normal height while geometrical dimensions of the cross-section (like perimeter of a curved cross-section or sides of a polygonal cross-section) must vary linearly with thenormal height i.e. when a right conical body is cut thoroughly with parallel planes, identical cross-sections are obtained.

Ex: Right-cone with cross-section as a rectangle, scalene-triangle, trapezium, parallelogram, rhombus, ellipse or polygon etc. just to name few of them.

2.6.5 Relation between angles (α & β) in a right pyramid:

Let there be a right pyramid with base as a regular polygon $A_1A_2A_3 \dots . A_n$ having 'n' number of the sides each of length 'a', angle between the consecutive lateral edges $'\alpha'$, normal height 'H' &acute angle between lateral edge & axis is $'\beta'$.(as shown in the fig(184) below)

Now, join all the vertices of the base to the centre 'O' thus we obtain 'n' number of isosceles triangles.

In right ΔOMA_2

$$\Rightarrow \sin\ A_2OM = \frac{MA_2}{OA_2}$$

$$or\ sin\frac{\pi}{n} = \frac{\left(\frac{a}{2}\right)}{OA_2} \Rightarrow OA_2 = \frac{a}{2}cosec\frac{\pi}{n}\left(since,\ A_1OA_2 = \frac{2\pi}{n}\right)$$

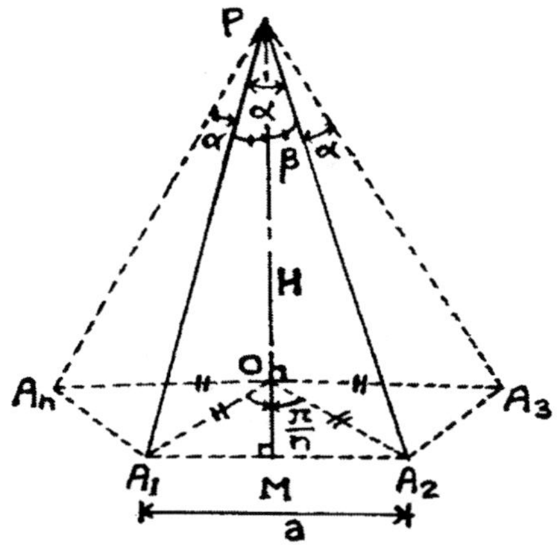

Fig (184)

In right ΔA_2OP

$$\Rightarrow \tan \quad OPA_2 = \frac{OA_2}{OP}$$

$$\Rightarrow \tan\beta = \frac{\left(\frac{a}{2}cosec\frac{\pi}{n}\right)}{H}$$

On setting the value of H from the eq(107) in the above expression as follows

$$\tan\beta = \frac{\left(\frac{a}{2}cosec\frac{\pi}{n}\right)}{H = \frac{a}{2}\sqrt{cot^2\frac{\alpha}{2} - cot^2\frac{\pi}{n}}} = \frac{cosec\frac{\pi}{n}}{\sqrt{cot^2\frac{\alpha}{2} - cot^2\frac{\pi}{n}}}$$

$$\Rightarrow cot\beta = sin\frac{\pi}{n}\sqrt{cot^2\frac{\alpha}{2} - cot^2\frac{\pi}{n}} \quad \forall\left(\alpha \le \frac{2\pi}{n} \;\&\; \beta \le \frac{\pi}{2}\right) \quad \ldots\ldots\ldots\ldots\ldots..(111)$$

2.6.6 Relation between angles (α & β) of equally inclined 'n' number of concurrent straight-lines

Let there be 'n' number of the concurrentstraight-lines inclined at an angle $'\alpha'$ consecutively such that the angle between any of the lines & axis, passing through point of concurrency, is $'\beta'$. (as shown in the figure (185) below)

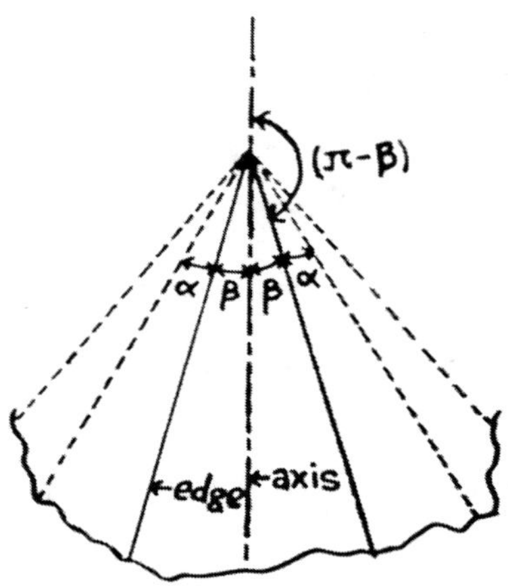

Fig (185)

This case is similar to that of a right pyramid with base as regular polygon having 'n' number of the sides.

Thus, in this case α & β can also be related as follows

$$cot\beta = sin\frac{\pi}{n}\sqrt{cot^2\frac{\alpha}{2} - cot^2\frac{\pi}{n}} (\,from\ the\ eq(111))$$

On squaring both the sides, we get

$$\Rightarrow cot^2\beta = sin^2\frac{\pi}{n}\left(cot^2\frac{\alpha}{2} - cot^2\frac{\pi}{n}\right) \qquad \forall\left(\alpha \leq \frac{2\pi}{n}\right) \qquad \dots\dots\dots\dots\dots..(112)$$

Note: The above formula is applicable for all the values of angle $'\beta'$. It is very useful in 3D analysis such as angles between the bonds in a molecule having regular orientation.

2.6.7 Angle between σ-bonds in a moleculehaving regular tetrahedral structure (Methane molecule (CH_4))

For ease of understanding consider a molecule having regular tetrahedral structure such as CH_4 molecule. There are four σ-bonds of hydrogen atoms with a carbon atom which are equally inclined with each other in 3D space. Let's calculate the angle between any two σ –bonds in CH_4 molecule (as shown in the figure (186) below)

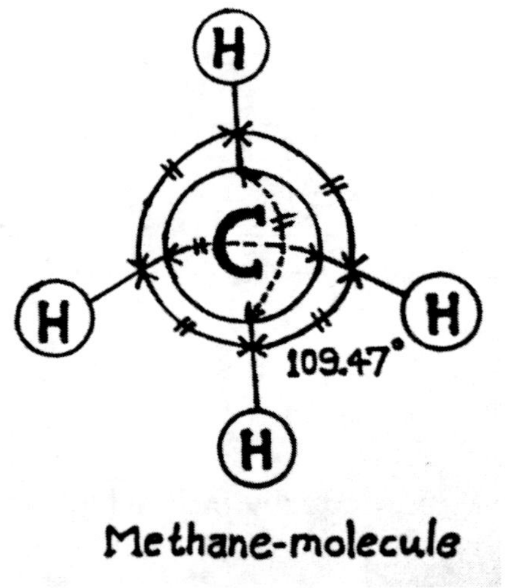

Fig (186)

Assume any three of four σ –bonds as concurrent straight lines & fourthσ –bond as axis passing through the point of concurrency (i.e. carbon atom)

Let $'\alpha'$ be the angle between any two σ –bonds in CH_4 molecule

We know the relation between α & β from the eq(112) as follows

$$cot^2\beta = sin^2\frac{\pi}{n}\left(cot^2\frac{\alpha}{2} - cot^2\frac{\pi}{n}\right)$$

On setting $\pi - \beta = \alpha\ or\ \beta = \pi - \alpha\ \&\ n = 3$ in the above expression, we get

$$cot^2(\pi - \alpha) = sin^2\frac{\pi}{3}\left(cot^2\frac{\alpha}{2} - cot^2\frac{\pi}{3}\right)$$

$$\Rightarrow cot^2\alpha = \left(\frac{\sqrt{3}}{2}\right)^2\left(cot^2\frac{\alpha}{2} - \left(\frac{1}{\sqrt{3}}\right)^2\right)$$

$$\Rightarrow \frac{cos^2\alpha}{sin^2\alpha} = \frac{3}{4}\left(\frac{cos^2\frac{\alpha}{2}}{sin^2\frac{\alpha}{2}} - \frac{1}{3}\right)$$

$$\Rightarrow \frac{cos^2\alpha}{sin^2\alpha} = \frac{3}{4}\left(\frac{3cos^2\frac{\alpha}{2} - sin^2\frac{\alpha}{2}}{3sin^2\frac{\alpha}{2}}\right)$$

$$\Rightarrow 4cos^2\alpha sin^2\frac{\alpha}{2} = sin^2\alpha\left(3cos^2\frac{\alpha}{2} - sin^2\frac{\alpha}{2}\right)$$

$$\Rightarrow 4\left(2cos^2\frac{\alpha}{2} - 1\right)^2\left(1 - cos^2\frac{\alpha}{2}\right) = (1 - cos^2\alpha)\left(3cos^2\frac{\alpha}{2} - \left(1 - cos^2\frac{\alpha}{2}\right)\right)$$

$$\left(since,\ \ cosA = 2cos^2\frac{A}{2} - 1\right)$$

$$\Rightarrow 4\left(4cos^4\frac{\alpha}{2} + 1 - 4cos^2\frac{\alpha}{2}\right)\left(1 - cos^2\frac{\alpha}{2}\right) = \left(1 - \left(2cos^2\frac{\alpha}{2} - 1\right)^2\right)\left(4cos^2\frac{\alpha}{2} - 1\right)$$

$$\Rightarrow 16cos^4\frac{\alpha}{2} - 16cos^2\frac{\alpha}{2} + 4 - 16cos^6\frac{\alpha}{2} + 16cos^4\frac{\alpha}{2} - 4cos^2\frac{\alpha}{2}$$
$$= \left(1 - 4cos^4\frac{\alpha}{2} - 1 + 4cos^2\frac{\alpha}{2}\right)\left(4cos^2\frac{\alpha}{2} - 1\right)$$

$$\Rightarrow -16cos^6\frac{\alpha}{2} + 32cos^4\frac{\alpha}{2} - 20cos^2\frac{\alpha}{2} + 4 = -16cos^6\frac{\alpha}{2} + 16cos^4\frac{\alpha}{2} + 4cos^4\frac{\alpha}{2} - 4cos^2\frac{\alpha}{2}$$

$$\Rightarrow 12cos^4\frac{\alpha}{2} - 16cos^2\frac{\alpha}{2} + 4 = 0$$

$$or\ 3cos^4\frac{\alpha}{2} - 4cos^2\frac{\alpha}{2} + 1 = 0$$

Above equation is quadratic equation in terms of $cos^2\,{}^{\alpha}/_{2}$. On solving the above equation, we have

$$cos^2\frac{\alpha}{2} = \frac{-(-4) \pm \sqrt{(-4)^2 - 4(3)(1)}}{2(3)}$$

$$\Rightarrow cos^2\frac{\alpha}{2} = \frac{4 \pm 2}{6} = \frac{4+2}{6} = 1\ or\ cos^2\frac{\alpha}{2} = \frac{4-2}{6} = \frac{1}{3}$$

But $\alpha \neq 0$ hence, we have

$$cos^2\frac{\alpha}{2} = \frac{1}{3}$$

$$or\ cos\frac{\alpha}{2} = \frac{1}{\sqrt{3}} \Rightarrow \frac{\alpha}{2} = \cos^{-1}\frac{1}{\sqrt{3}}$$

$$\Rightarrow \alpha = 2\cos^{-1}\frac{1}{\sqrt{3}} \approx 109.47^o \qquad \ldots\ldots\ldots\ldots\ldots\ldots\ldots\ldots (113)$$

Note: The above eq(113) also represents the value of angle between any two straight lines joining the centre of a tetrahedral void to the centres of spheres forming the void.

(See section 5.1.4 for analysis of a tetrahedral void)

Solved Examples

Example 8.1: If the angle between two concurrent vectors is 30°. Now one of them, keeping other stationary, is rotated by an angle 40° about the point of concurrency in a plane inclined at an angle 50° with the plane containing both the vectors in initial position. What will bethe angle between them in the final position?

Sol. given that

$$\alpha = 30^o,\ \ \beta = 40^o,\ \ \theta = 50^o \,\&\, \alpha' = ?$$

On setting the values in Vectors' cosine formula, we get

$$\alpha' = \cos^{-1}(cos\alpha cos\beta + sin\alpha sin\beta cos\theta)$$

$$= \cos^{-1}(cos30^o cos40^o + sin30^o sin40^o cos50^o)$$

$$= \cos^{-1}\left(\frac{\sqrt{3}}{2} \times 0.766 + \frac{1}{2} \times 0.643 \times 0.643\right)$$

$$= \cos^{-1}(0.87) = 29.54^o \ or \ 29^o31'55'' \qquad Ans.$$

Example 8.2: The angle between two concurrent vectors is 40°. If a vector, keeping other stationary is rotated by an angle 60° about the point of concurrency such that the angle between them in the final position is 55°. Determine the angle between plane of rotation of the vector & plane of concurrency.

Sol. given that

$$\alpha = 40^o, \ \beta = 60^o, \ \alpha' = 55^o \& \ \theta = ?$$

On setting the values in HCR's cosine formula, we get the value of angle between the planes

$$\theta = \cos^{-1}\left(\frac{cos\alpha' - cos\alpha cos\beta}{sin\alpha sin\beta}\right)$$

$$= \cos^{-1}\left(\frac{cos55^o - cos40^o cos60^o}{sin40^o sin60^o}\right)$$

$$= \cos^{-1}\left(\frac{0.573 - 0.766 \times 0.5}{0.643 \times 0.866}\right)$$

$$= \cos^{-1}(0.341) = 70.06^o \ or \ 70^o2'57'' \qquad Ans.$$

Example 8.3: Three planes are intersecting each other at a single in the space such that the angles between two consecutive lines of intersection are successively $30^o, 45^o \& 70^o$. Using HCR's cosine-formula find out the angles between the intersecting planes successively.

Sol. given that

$$\alpha = 30^o, \ \beta = 45^o, \ \gamma = 70^o \& \theta_1, \theta_2, \theta_3 \ = ?$$

On setting the values in HCR's cosine formula, we get the values of angles between the planes successively

$$\Rightarrow \theta_1 = \cos^{-1}\left(\frac{cos\alpha - cos\beta cos\gamma}{sin\beta sin\gamma}\right)$$

$$= \cos^{-1}\left(\frac{cos30^o - cos45^o cos70^o}{sin45^o sin70^o}\right)$$

$$= \cos^{-1}\left(\frac{0.866 - 0.707 \times 0.342}{0.707 \times 0.94}\right)$$

$$= \cos^{-1}(0.939) = 20.07^o \ or \ 20^o4'27'' \qquad Ans.$$

$$\Rightarrow \theta_2 = \cos^{-1}\left(\frac{cos\beta - cos\alpha cos\gamma}{sin\alpha sin\gamma}\right)$$

$$= \cos^{-1}\left(\frac{cos45^o - cos30^o cos70^o}{sin30^o sin70^o}\right)$$

$$= \cos^{-1}\left(\frac{0.707 - 0.866 \times 0.342}{0.5 \times 0.94}\right)$$

$$= \cos^{-1}(0.874) = 29.06^o \ or \ 29^o3'40'' \qquad Ans.$$

$$\Rightarrow \theta_3 = \cos^{-1}\left(\frac{cos\gamma - cos\alpha cos\beta}{sin\alpha sin\beta}\right)$$

$$= \cos^{-1}\left(\frac{cos70^o - cos30^o cos45^o}{sin30^o sin45^o}\right)$$

$$= \cos^{-1}\left(\frac{0.342 - 0.866 \times 0.707}{0.5 \times 0.707}\right)$$

$$= \cos^{-1}(-0.764) = 139.86^o \; or \; 139^o51'55'' \qquad Ans.$$

Example 8.4: The angles between the successive lateral edges meeting at a vertex of a tetrahedron are $40^o, 70^o \& \, 90^o$. Using HCR's Cosine-formula find out the internal angles between the consecutive lateral faces of the tetrahedron.

Sol. given that

$$\alpha = 40^o, \;\; \beta = 70^o, \;\; \gamma = 90^o \& \theta_1, \theta_2, \theta_3 \;=?$$

On setting the values in HCR's cosine formula, we get the values of internal angles between the consecutive lateral faces successively

$$\Rightarrow \theta_1 = \cos^{-1}\left(\frac{cos\alpha - cos\beta cos\gamma}{sin\beta sin\gamma}\right)$$

$$= \cos^{-1}\left(\frac{cos40^o - cos70^o cos90^o}{sin70^o sin90^o}\right)$$

$$= \cos^{-1}\left(\frac{0.766 - 0.342 \times 0}{0.94 \times 1}\right)$$

$$= \cos^{-1}(0.815) = 35.42^o \; or \; 35^o25'23'' \qquad Ans.$$

$$\Rightarrow \theta_2 = \cos^{-1}\left(\frac{cos\beta - cos\alpha cos\gamma}{sin\alpha sin\gamma}\right)$$

$$= \cos^{-1}\left(\frac{cos70^o - cos40^o cos90^o}{sin40^o sin90^o}\right)$$

$$= \cos^{-1}\left(\frac{0.342 - 0.766 \times 0}{0.643 \times 1}\right)$$

$$= \cos^{-1}(0.532) = 57.87^o \; or \; 57^o52'2'' \qquad Ans.$$

$$\Rightarrow \theta_3 = \cos^{-1}\left(\frac{cos\gamma - cos\alpha cos\beta}{sin\alpha sin\beta}\right)$$

$$= \cos^{-1}\left(\frac{cos90^o - cos40^o cos70^o}{sin40^o sin70^o}\right)$$

$$= \cos^{-1}\left(\frac{0 - 0.766 \times 0.342}{0.643 \times 0.94}\right)$$

$$= \cos^{-1}(-0.433) = 115.66^o \; or \; 115^o39'29'' \qquad Ans.$$

Example 8.5: Circumscribed-radius, of the triangle generated by joining the heads of three equal concurrent vectors each of length 15cm, is 8cm. Find out length of perpendicular drawn from the point of concurrency to the plane of triangle.

Sol. given that

Circumscribed radius, R = 8cm & length of equal vectors, a = 15cm

Let the length of perpendicular be 'p' drawn from the point of concurrency to the plane of triangle

From axiom, we know

$$a = \sqrt{p^2 + R^2} \Rightarrow \; 15 = \sqrt{p^2 + 8^2}$$

$$p = \sqrt{225 - 64} = 12.69cm \qquad Ans.$$

Example 8.6: The length of perpendicular drawn from the point of concurrency of three equal vectors each of length 20cm to the plane of a triangle generated by joining the heads of all the vectors is 12cm. What will be circumscribed-radius of the triangle (generated).

Sol. given that

Length of perpendicular, p = 12cm & length of equal vectors, a = 20cm

Let the circumscribed radius of the trianglebe R then

From axiom, we know

$$a = \sqrt{p^2 + R^2} \Rightarrow 20 = \sqrt{12^2 + R^2}$$

$$R = \sqrt{400 - 144} = 16\ cm \qquad Ans.$$

Example 8.7: A triangle has circumscribed radius 7cm, smallest angle 36°& another angle 70°. Find out the distance between the circumscribed & inscribed centres.

Sol. given that

$$\theta_{min} = 36^o,\ \theta = 70^o \&\ circumscribed\ radius, R = 7cm$$

We know the relation between circumscribed & inscribed radii as follows

$$r = 4Rsin\frac{A}{2}sin\frac{B}{2}sin\frac{C}{2}$$

$$\Rightarrow r = 4 \times 7sin\frac{36^o}{2}sin\frac{70^o}{2}sin\frac{74^o}{2} \qquad since, C = 180^o - A - B$$

$$= 28 \times 0.309 \times 0.573 \times 0.602 = 2.98cm$$

Hence, the distance between the circumscribed & inscribed centres

$$D_{CI} = \sqrt{\left\{(r - Rcos\theta_{min})^2 + (R + r - Rcos\theta_{min})^2 cot^2\left(\frac{2\theta + \theta_{min}}{2}\right)\right\}}$$

$$= \sqrt{\left\{(2.98 - 7cos36^o)^2 + (7 + 2.98 - 7cos36^o)^2 cot^2\left(\frac{2 \times 70^o + 36^o}{2}\right)\right\}}$$

$$= \sqrt{\{7.2 + 18.63 \times 0.00122\}} = 2.687\ cm \qquad Ans.$$

Example 8.8: A triangle has inscribed radius 5cm, smallest angle 30°& another angle 80°.Determine the distance between the circumscribed & inscribed centres.

Sol. given that

$$\theta_{min} = 30^o,\ \theta = 80^o \&\ inscribed\ radius, r = 5cm$$

We know the relation between circumscribed & inscribed radii as follows

$$r = 4Rsin\frac{A}{2}sin\frac{B}{2}sin\frac{C}{2}$$

$$\Rightarrow 5 = 4 \times Rsin\frac{30^o}{2}sin\frac{80^o}{2}sin\frac{70^o}{2} \qquad since, C = 180^o - A - B$$

$$or\ R = \frac{5}{4 \times 0.259 \times 0.643 \times 0.573} = 13.1cm$$

Hence, the distance between the circumscribed & inscribed centres

$$D_{CI} = \sqrt{\left\{(r - Rcos\theta_{min})^2 + (R + r - Rcos\theta_{min})^2 cot^2\left(\frac{2\theta + \theta_{min}}{2}\right)\right\}}$$

$$= \sqrt{\left\{(5 - 13.1cos30^o)^2 + (13.1 + 5 - 13.1cos30^o)^2 cot^2\left(\frac{2 \times 80^o + 30^o}{2}\right)\right\}}$$

$$= \sqrt{\{40.258 + 45.631 \times 0.00765\}} = 6.372\ cm \qquad Ans.$$

Example 8.9: A scalene triangle has circumscribed & inscribed radii 10cm & 3.6cm respectively and smallest angle 20°. If the distance between the circumscribed & inscribed centres is 7.8 cm find out the angles of the triangle.

Sol. given that

$$\theta_{min} = 20^o,\ circuscribed\ radius, R = 10cm,$$
$$inscribed\ radius, r = 3.6cm\ \&D_{CI} = 7.8cm$$

We know that, the distance between the circumscribed & inscribed centres

$$D_{CI} = \sqrt{\left\{(r - Rcos\theta_{min})^2 + (R + r - Rcos\theta_{min})^2 cot^2\left(\frac{2\theta + \theta_{min}}{2}\right)\right\}}$$

$$\Rightarrow 7.8 = \sqrt{\left\{(3.6 - 10cos20^o)^2 + (10 + 3.6 - 10cos20^o)^2 cot^2\left(\frac{2\theta + 20^o}{2}\right)\right\}}$$

$$cot^2\left(\frac{2\theta + 20^o}{2}\right) = \frac{(7.8)^2 - (-5.797)^2}{(4.203)^2} = 1.542\ or\ cot(\theta + 10^o) = 1.242$$

$$\theta + 10^o = 38.84^o\ or\ \theta = 28.84^o$$

Thus, we get value of third angle

$$= 180^o - \theta_{min} - \theta = 180^o - 20^o - 28.84^o = 131.16^o$$

Hence, the angles of triangle are $20^o,\ 28.84^o \& 131.16^o$ *Ans.*

Example 8.10: A scalene triangle has circumscribed & inscribed radii 15cm & 6.5cm respectively and smallest angle 30°. If the distance between circumscribed & inscribed centres is 8.3cm find out all the sides of triangle. Also find out area of the triangle.

Sol. given that

$$\theta_{min} = 30^o = A\ (let),\ circuscribed\ radius, R = 15cm,$$
$$inscribed\ radius, r = 6.5cm\ \&D_{CI} = 8.3cm$$

Let the sides of triangle be as

$a, b\ \&\ c$ *opposite to the angles* $A = \theta_{min}, B\ \&\ C$ *respectively* .

We know that, the distance between the circumscribed & inscribed centres

$$D_{CI} = \sqrt{\left\{(r - Rcos\theta_{min})^2 + (R + r - Rcos\theta_{min})^2 cot^2\left(\frac{2\theta + \theta_{min}}{2}\right)\right\}}$$

$$\Rightarrow 8.3 = \sqrt{\left\{(6.5 - 15cos30^o)^2 + (15 + 6.5 - 15cos30^o)^2 cot^2\left(\frac{2B + 30^o}{2}\right)\right\}}$$

$$cot^2\left(\frac{2B + 30^o}{2}\right) = \frac{(8.3)^2 - (-6.49)^2}{(8.51)^2} = 0.369\ or\ cot(\theta + 15^o) = 0.608$$

$$B + 15^o = 58.7^o\ or\ \theta = 43.7^o$$

Thus, we get value of third angle (C)

$$C = 180^o - A - B = 180^o - 30^o - 43.7^o = 106.3^o$$

Hence, the angles of triangle are $30^o,\ 43.7^o \& 106.3^o$

We know that

$$R = \frac{a}{2sinA}\ or\ a = 2RsinA$$

$$a = 2 \times 15sin30^o = 15cm$$

$$b = 2RsinB = 2 \times 15sin43.7^o = 20.73cm \ \&$$

$$c = 2RsinC = 2 \times 15sin106.3^o = 28.79cm$$

Thus, we have the sides of triangle as $15cm,\ 20.73cm\ \&\ 28.79cm$ *Ans.*

Now, semi perimeter of the triangle,

$$s = \frac{a+b+c}{2} = \frac{15+20.73+28.79}{2} = 32.26cm$$

Hence, the area of the triangle (Using Hero's formula)

$$\Delta = \sqrt{s(s-a)(s-b)(s-c)}$$

$$= \sqrt{32.26(32.26-15)(32.26-20.73)(32.26-28.79)}$$

$$= 149.256\ cm^2$$ *Ans.*

Example 8.11: Find out area of a right triangle having smallest angle 40°& the distance between circumscribed & inscribed centres is 4.2cm.

Sol. given that

$$\theta_{min} = 40^o = A\ (let),\ \ D_{CI} = 4.2cm$$

Since, the triangle is right angled triangle hence

$$other\ two\ angles, \quad B = 90^o\&$$

$$C = 180^o - 40^o - 90^o = 50^o \qquad (since, A+B+C = 180^o)$$

We know that the circumscribed & inscribed radii are related as

$$r = 4Rsin\frac{A}{2}sin\frac{B}{2}sin\frac{C}{2}$$

On setting the values of angles, we get

$$r = 4Rsin\frac{40^o}{2}sin\frac{50^o}{2}sin\frac{90^o}{2} = 0.409R$$

But, the distance between circumscribed & inscribed centres is given as

$$D_{CI} = \sqrt{\left\{(r - Rcos\theta_{min})^2 + (R + r - Rcos\theta_{min})^2cot^2\left(\frac{2\theta+\theta_{min}}{2}\right)\right\}}$$

$$\Rightarrow D_{CI} = \sqrt{\left\{(0.409R - Rcos40^o)^2 + (R + 0.409R - Rcos40^o)^2cot^2\left(\frac{2\times 50^o + 40^o}{2}\right)\right\}}$$

$$= R\sqrt{\{(0.409 - cos40^o)^2 + (1 + 0.409 - cos40^o)^2cot^2(70^o)\}}$$

$$= R\sqrt{\{(-0.357)^2 + (0.643)^2(0.132)^2\}} = 0.367R$$

$$\Rightarrow D_{CI} = 4.2 = 0.367R \qquad (given\ D_{CI} = 4.2)$$

$$R = \frac{4.2}{0.367} = 11.44cm$$

Now, orthogonal sides say a & c opposite to angles A & C of right triangle can be found out using formula

$$R = \frac{a}{2sinA}\ or\ a = 2RsinA$$

$$a = 2 \times 11.44sin40^o = 14.71cm\ \&$$

$$c = 2RsinC = 2 \times 11.44sin50^o = 17.53cm$$

Hence, the area of right triangle (using formula)

$$\Rightarrow \Delta = \frac{1}{2}(base)(perpendicular)$$

$$= \frac{1}{2}(14.71)(17.53) = 128.933cm^2 \qquad Ans.$$

Example 8.12: A triangle has sides 7cm, 13cm & 18cm. What is the distance between circumscribed & inscribed centre.

Sol. given that

$$a = 7cm, b = 13cm \;\&\; c = 18cm$$

Let the angles of triangles A, B & C opposite to the sides $a, b \;\&\; c\; respectively.$

We know that angle opposite to the smallest side is the smallest angle in a triangle

Thus, the angle A is the smallest angle opposite to the smallest side a,

Now, using Cosine formula

$$cosA = \frac{b^2 + c^2 - a^2}{2bc}$$

$$cosA = \frac{13^2 + 18^2 - 7^2}{2(13)(18)} = 0.949$$

$$A = \cos^{-1} 0.949 = 18.38^o$$

Similarly, we have

$$cosB = \frac{c^2 + a^2 - b^2}{2ca}$$

$$cosA = \frac{18^2 + 7^2 - 13^2}{2(18)(7)} = 0.809$$

$$A = \cos^{-1} 0.809 = 36^o$$

Now, semi perimeter (s) of the triangle

$$\Rightarrow s = \frac{a + b + c}{2} = \frac{7 + 13 + 18}{2} = 19cm$$

Area of the triangle (Using Hero's formula)

$$\Delta = \sqrt{s(s-a)(s-b)(s-c)}$$

$$= \sqrt{19(19-7)(19-13)(19-18)}$$

$$= 36.986\ cm^2$$

Hence, circumscribed radius of triangle (using formula)

$$R = \frac{abc}{4\Delta} = \frac{7 \times 13 \times 18}{4 \times 36.986} = 11.07cm \;\&$$

$$\Rightarrow inscribed\ radius,\ r = \frac{\Delta}{s} = \frac{36.986}{19} = 1.96cm$$

Hence, the distance between the circumscribed & inscribed centres

$$D_{CI} = \sqrt{\left\{(r - Rcos\theta_{min})^2 + (R + r - Rcos\theta_{min})^2 cot^2\left(\frac{2\theta + \theta_{min}}{2}\right)\right\}}$$

On setting the corresponding values as follows

$$D_{CI} = \sqrt{\left\{(1.96 - 11.07cos18.38^o)^2 + (11.07 + 1.96 - 11.07cos18.38^o)^2 cot^2\left(\frac{2 \times 36^o + 18.38^o}{2}\right)\right\}}$$

$$= \sqrt{\{(-8.545)^2 + (2.525)^2(0.993)^2\}} = 8.9cm \qquad Ans.$$

Example 8.13: Find out spike-angle & area of a spiked regular polygon having 9 spikes & span 10cm.

Sol. given that

$$n = 9 \ \& \ span, a = 10cm$$

$$\therefore \quad spike\ angle,\ 2\alpha = \frac{\pi}{n} = \frac{\pi}{9} = 20^o \qquad Ans.$$

$$area\ of\ spiked\ regular\ polygon,\ \Delta = \frac{na^2}{4\left(sin\frac{\pi}{n} + sin\frac{2\pi}{n}\right)}$$

$$\therefore\ \Delta = \frac{9 \times 10^2}{4\left(sin\frac{\pi}{9} + sin\frac{2\pi}{9}\right)} = \frac{225}{0.342 + 0.643} = 228.426cm^2 \qquad Ans.$$

Example 8.14: Find out spike-angle & area of a spiked regular polygon having 51 spikes & span 5cm.

Sol. given that

$$n = 51 \ \& \ span, a = 5cm$$

$$\therefore \quad spike\ angle,\ 2\alpha = \frac{\pi}{n} = \frac{\pi}{51} = 3^o31'46'' \qquad Ans.$$

$$area\ of\ spiked\ regular\ polygon,\ \Delta = \frac{na^2}{4\left(sin\frac{\pi}{n} + sin\frac{2\pi}{n}\right)}$$

$$\therefore\ \Delta = \frac{51 \times 5^2}{4\left(sin\frac{\pi}{51} + sin\frac{2\pi}{51}\right)} = \frac{1275}{4(0.184)} = 1732.337cm^2 \qquad Ans.$$

Example 8.15: A spiked regular polygon has spike angle 6.21^o& span 25cm. Calculate no. of the spikes & the area of spiked polygon.

Sol. given that

$$2\alpha = 6.21^o \& \ span, a = 25cm$$

$$spike\ angle,\ 2\alpha = \frac{\pi}{n} = \frac{180^o}{n}$$

$$or \quad n = \frac{180^o}{2\alpha} = \frac{180^o}{6.21^o} = 28.985 \approx 29 \qquad Ans.$$

$$awe\ know\ that\ area\ of\ spiked\ regular\ polygon,\ \Delta = \frac{na^2}{4\left(sin\frac{\pi}{n} + sin\frac{2\pi}{n}\right)}$$

$$\therefore\ \Delta = \frac{29 \times 25^2}{4\left(sin\frac{\pi}{29} + sin\frac{2\pi}{29}\right)} = \frac{18125}{4(0.503)} = 9008.449cm^2 \qquad Ans.$$

Example 8.16: A spiked regular polygon has spike angle $2^o20'15.6''$& span 16cm. Calculate no. of the spikes & the area of spiked polygon.

Sol. given that

$$2\alpha = 2^o20'15.6'' \& \ span, a = 16cm$$

$$spike\ angle,\ \ 2\alpha = 2^o20'15.6'' = \left(2+\frac{20}{60}+\frac{15.6}{3600}\right)^o = 2.34^o$$

$$\therefore\ \ 2\alpha = \frac{\pi}{n} = \frac{180^o}{n}$$

$$\Rightarrow\ n = \frac{180^o}{2\alpha} = \frac{180^o}{2.34^o} = 76.923 \approx 77 \qquad Ans.$$

$$we\ know\ that\ area\ of\ spiked\ regular\ polygon,\ \ \Delta = \frac{na^2}{4\left(sin\frac{\pi}{n}+sin\frac{2\pi}{n}\right)}$$

$$\therefore\ \Delta = \frac{77\times 16^2}{4\left(sin\frac{\pi}{77}+sin\frac{2\pi}{77}\right)} = \frac{19712}{4(0.122)} = 40393.442cm^2\ or\ 4.039\ m^2 \qquad Ans.$$

Example 8.17: Find out the radius of circle passing through the tip-points of all the spikes of a spiked regular polygon having area $1248.455cm^2$& spike-angle $5^o27'16.4''$.

Sol. given that

$$2\alpha = 5^o27'16.4''\ \ \&\ \ Area,\ \ \Delta = 1248.455cm^2$$

$$spike\ angle,\ \ 2\alpha = 5^o27'16.4'' = \left(5+\frac{27}{60}+\frac{16.4}{3600}\right)^o = 5.45^o$$

$$\therefore\ \ 2\alpha = \frac{\pi}{n} = \frac{180^o}{n}$$

$$\Rightarrow\ n = \frac{180^o}{2\alpha} = \frac{180^o}{5.45^o} = 33.02 \approx 33$$

$$we\ know\ that\ area\ of\ spiked\ regular\ polygon,\ \ \Delta = \frac{na^2}{4\left(sin\frac{\pi}{n}+sin\frac{2\pi}{n}\right)}$$

$$or\ \ a^2 = \frac{4\Delta\left(sin\frac{\pi}{n}+sin\frac{2\pi}{n}\right)}{n}$$

$$\Rightarrow\ \ a = \sqrt{\frac{4\Delta}{n}\left(sin\frac{\pi}{n}+sin\frac{2\pi}{n}\right)}$$

On setting the corresponding values in the above expression, we have

$$\therefore\ a = \sqrt{\frac{4\times 1248.455}{33}\left(sin\frac{\pi}{33}+sin\frac{2\pi}{33}\right)} = \sqrt{151.328\times 0.284} = 6.55cm$$

We know that the circumscribed radius (R) of a regular polygon with 'n' no. of the sides each of length 'a' is given as

$$R = \frac{a}{2}cosec\frac{\pi}{n}$$

Now, on setting $a = span = 6.55cm$, we get

$$R = \frac{6.55}{2}cosec\frac{\pi}{33} = 113.64cm\ \ or\ 1.136m \qquad Ans.$$

Example 8.18: Find out area of a spiked regular polygon with a span 7.6cm such that the radius of circle passing through the tip-points of all the spikes is 30.32cm. What is the area of spiked polygon?

Sol. given that

$$span, a = 7.6cm\ \&\ circumscribed\ radius,\ \ R = 30.32cm$$

We know that the circumscribed radius (R) of a regular polygon with 'n' no. of the sides each of length 'a' is given as

$$R=\frac{a}{2}cosec\frac{\pi}{n}$$

$$or\ sin\frac{\pi}{n}=\frac{a}{2R}\Rightarrow\frac{\pi}{n}=\sin^{-1}\left(\frac{a}{2R}\right)\ or\ n=\frac{\pi}{\sin^{-1}\left(\frac{a}{2R}\right)}$$

Now, on setting $a=span=7.6cm\ \&\ R=30.32cm$, we get

$$n=\frac{\pi}{\sin^{-1}\left(\frac{7.6}{2\times 30.32}\right)}=\frac{180^o}{7.2^o}=25$$

$$we\ know\ that\ area\ of\ spiked\ regular\ polygon,\ \Delta=\frac{na^2}{4\left(sin\frac{\pi}{n}+sin\frac{2\pi}{n}\right)}$$

$$\therefore\ \Delta=\frac{25\times 7.6^2}{4\left(sin\frac{\pi}{25}+sin\frac{2\pi}{25}\right)}=\frac{1444}{4(0.374)}=965.241cm^2$$ *Ans.*

Example 8.19: A solid right pyramid with base as an equilateral triangle having each side 12cm & normal height 18cm. Find out

1) Total area of surface 2) volume 3) angle between lateral edge & axis

Sol. given that

$$number\ of\ the\ sides, n=3, length, a=12cm\ \&\ normal\ height, H=18cm$$

1. Total surface area

Area of lateral surface,

$$A_l=\frac{1}{4}na\sqrt{4H^2+a^2cot^2\frac{\pi}{n}}$$

On setting the values, we get

$$A_l=\frac{1}{4}\times 3\times 12\sqrt{4(18)^2+(12)^2cot^2\frac{\pi}{3}}=329.945\ cm^2$$

Area of regular polygonal base,

$$A_b=\frac{1}{4}na^2cot\frac{\pi}{n}$$

On setting the values, we get

$$A_b=\frac{1}{4}\times 3(12)^2cot\frac{\pi}{3}=62.354\ cm^2$$

$$\therefore\ total\ surface\ area=A_l+A_b$$

$$=329.945+62.354=392.299\ cm^2$$ *Ans.*

2. Volume,

$$V=\frac{1}{12}na^2Hcot\frac{\pi}{n}$$

On setting the values,

$$V=\frac{1}{12}\times 3(12)^2(18)cot\frac{\pi}{3}=374.122cm^3$$ *Ans.*

3. Angle (β) between lateral edge & axis,

We know that the relation between side 'a' & normal height 'H'

$$H = \frac{a}{2}\sqrt{cot^2\frac{\alpha}{2} - cot^2\frac{\pi}{n}}$$

$$\Rightarrow 4H^2 = a^2\left(cot^2\frac{\alpha}{2} - cot^2\frac{\pi}{n}\right)$$

$$\Rightarrow cot^2\frac{\alpha}{2} = \frac{4H^2}{a^2} + cot^2\frac{\pi}{n}$$

On setting the values as follows

$$\Rightarrow cot^2\frac{\alpha}{2} = \frac{4(18)^2}{(12)^2} + cot^2\frac{\pi}{3} = 9.333$$

$$\Rightarrow cot\frac{\alpha}{2} = 3.055 \; or \; \frac{\alpha}{2} = \cot^{-1} 3.055 = \tan^{-1}\frac{1}{3.055} = 18.12^o$$

$$\Rightarrow \alpha = 2(18.12^o) = 36.24^o$$

$$cot\beta = sin\frac{\pi}{n}\sqrt{cot^2\frac{\alpha}{2} - cot^2\frac{\pi}{n}}$$

On setting the value, we get

$$cot\beta = sin\frac{\pi}{3}\sqrt{cot^2\frac{36.24^o}{2} - cot^2\frac{\pi}{3}} = 2.6 \; or \; \beta = \cot^{-1} 2.6$$

$$or \; \beta = \tan^{-1}\frac{1}{2.6} = 21.04^o$$ *Ans.*

Example 8.20: A solid right pyramid with base as a regular hexagon having each side 15cm & normal height 28cm. Find out

1) Total area of surface 2) volume 3) angle between consecutive lateral edges

Sol. given that

$$number\ of\ the\ sides, n = 6, length, a = 15cm\ \&\ normal\ height, H = 28cm$$

1. Total surface area

Area of lateral surface,

$$A_l = \frac{1}{4}na\sqrt{4H^2 + a^2cot^2\frac{\pi}{n}}$$

On setting the values, we get

$$A_l = \frac{1}{4} \times 6 \times 15\sqrt{4(28)^2 + (15)^2cot^2\frac{\pi}{6}} = 1389\ cm^2$$

Area of regular polygonal base,

$$A_b = \frac{1}{4}na^2cot\frac{\pi}{n}$$

On setting the values, we get

$$A_b = \frac{1}{4} \times 6(15)^2cot\frac{\pi}{6} = 584.567\ cm^2$$

$$\therefore total\ surface\ area = A_l + A_b$$

$$= 1389 + 584.567 = 1973.567\ cm^2$$ *Ans.*

2. Volume,

$$V = \frac{1}{12}na^2Hcot\frac{\pi}{n}$$

On setting the values,

$$V = \frac{1}{12} \times 6(15)^2(28)cot\frac{\pi}{6} = 5455.96cm^3 \qquad Ans.$$

3. Angle (α) between consecutive lateral edges,

We know that the relation between side 'a' & normal height 'H'

$$H = \frac{a}{2}\sqrt{cot^2\frac{\alpha}{2} - cot^2\frac{\pi}{n}}$$

$$\Rightarrow 4H^2 = a^2\left(cot^2\frac{\alpha}{2} - cot^2\frac{\pi}{n}\right)$$

$$\Rightarrow cot^2\frac{\alpha}{2} = \frac{4H^2}{a^2} + cot^2\frac{\pi}{n}$$

On setting the values as follows

$$\Rightarrow cot^2\frac{\alpha}{2} = \frac{4(28)^2}{(15)^2} + cot^2\frac{\pi}{6} = 16.938$$

$$\Rightarrow cot\frac{\alpha}{2} = 4.115 \; or \; \frac{\alpha}{2} = \cot^{-1} 4.115 = \tan^{-1}\frac{1}{4.115} = 13.66^o$$

$$\Rightarrow \alpha = 2(13.66^o) = 27.32^o \qquad Ans.$$

Summary

1. **Vectors' Cosine-formula:**(Derived by author of this book)

   If 'α' is the angle between two concurrent vectors and one of them, keeping other stationary, is rotated by an angle 'β' about the point of concurrency in a plane inclined at an angle 'θ' with the plane containing both the vectors in initial position then the angle (α') between them in final position is given by the following expression

   $$\Rightarrow \alpha' = \cos^{-1}(cos\alpha cos\beta + sin\alpha sin\beta cos\theta) \;\; \forall\, (0 \le (\alpha, \theta) \le \pi)$$

2. HCR's Cosine-formula:

   If the angle α' between two concurrent vectors in the shifted position is known then the angle θ between the plane of rotation & the plane containing the vectors in initial position is given as

   $$\theta = \cos^{-1}\left(\frac{cos\alpha' - cos\alpha cos\beta}{sin\alpha sin\beta}\right) \;\; \forall\, (0 \le (\alpha, \theta) \le \pi)$$

3. **Axiom of tetrahedron:**If α, β & γ are the angles between consecutive lateral edges meeting at a vertex of a tetrahedron then the tetrahedron will exist only and if only the sum of any two angles out of α, β & γ is always greater than third one i.e.

   $$(\alpha + \beta) > \gamma, \;\; (\beta + \gamma) > \alpha \;\&(\alpha + \gamma) > \beta \qquad \forall\; 0 < (\alpha + \beta + \gamma) < 2\pi$$

4. Internal angles $\theta_1, \theta_2 \& \theta_3$, between lateral faces respectively opposite to the angles α, β & γ between consecutive lateral edges meeting at a vertex of a tetrahedron, are given as

   $$\theta_1 = \cos^{-1}\left(\frac{cos\alpha - cos\beta cos\gamma}{sin\beta sin\gamma}\right), \;\; \theta_2 = \cos^{-1}\left(\frac{cos\beta - cos\alpha cos\gamma}{sin\alpha sin\gamma}\right) \;\&$$

   $$\theta_3 = \cos^{-1}\left(\frac{cos\gamma - cos\alpha cos\beta}{sin\alpha sin\beta}\right)$$

5. **Axiom of three concurrent-vectors:**The perpendicular, drawn from the point of concurrency of three equal vectors to the plane of a triangle generated by joining the heads of all the vectors, always passes through the circumscribed-centre of the triangle.
6. **Axiom of triangle:**The distance between circumscribed & inscribed centres in a triangle is given as

$$\Rightarrow D_{CI} = \sqrt{\left\{(r - Rcos\theta_{min})^2 + (R + r - Rcos\theta_{min})^2 cot^2\left(\frac{2\theta + \theta_{min}}{2}\right)\right\}}$$

Where, $R \rightarrow radius\ of\ circumscribed\ circle$

$r \rightarrow radius\ of\ insscribed\ circle$

$\theta_{min} \rightarrow smallest\ angle$ &

$\theta \rightarrow any\ of\ the\ angles\ except\ \theta_{min}\ in\ a\ given\ triangle$

7. The distance, between the circumscribed & inscribed centres of an isosceles triangle having each of the equal angles θ, is given as

$$D_{CI} = \sqrt{(r - Rcos\theta)^2 + (R + r - Rcos\theta)^2 cot^2\left(\frac{3\theta}{2}\right)} \quad if\ (\theta < (\pi - 2\theta))$$

$$D_{CI} = r + Rcos2\theta \quad if\ (\theta > (\pi - 2\theta))$$

8. **Spiked/star-like regular-polygon:** Regular-polygon obtained by the straight lines joining each vertex to the end-points of its opposite side in a regular-polygon having odd no. of the sides is called a spiked regular-polygon or a star-like regular-polygon.
9. Spike angle (2α) of spiked regular polygon is given as

$$2\alpha = \frac{\pi}{n}$$

10. Area (A) of a spiked regular polygon is given as

$$\Delta = \frac{na^2}{4\left(sin\frac{\pi}{n} + sin\frac{2\pi}{n}\right)} \quad \forall\left(n = 2m + 1 \Rightarrow m \in N\right)$$

11. Radius of the circle passing through all the spikes of a spiked regular polygon with 'n' number of the spikes & span 'a', is given as

$$R = \frac{a}{2} cosec\frac{\pi}{n} \quad \forall\left(n = 2m + 1 \Rightarrow m \in N\right)$$

12. **Right pyramid with base as a regular polygon:** Right pyramid with base as a regular polygon having 'n' number of the sides each of length 'a', angle between consecutive lateral edges $'\alpha'$& normal height 'H'

Area of lateral surface,

$$A_l = \frac{1}{4} na\sqrt{4H^2 + a^2 cot^2\frac{\pi}{n}} = \frac{1}{4} na^2 cot\frac{\alpha}{2}$$

Area of regular polygonal base,

$$A_b = \frac{1}{4} na^2 cot\frac{\pi}{n}$$

Volume,

$$V = \frac{1}{12} na^2 Hcot\frac{\pi}{n} \quad \forall\ (n \in N\ \&\ n \geq 3)$$

Relation between the angle α (between consecutive lateral edges) &acute angle β (between any of lateral edges& axis of the right pyramid)

$$cot\beta = sin\frac{\pi}{n}\sqrt{cot^2\frac{\alpha}{2} - cot^2\frac{\pi}{n}} \qquad \forall\left(\alpha \le \frac{2\pi}{n} \& \ \beta \le \frac{\pi}{2}\right)$$

Relation between normal height 'H' & length 'a' of side of the base

$$H = \frac{a}{2}\sqrt{cot^2\frac{\alpha}{2} - cot^2\frac{\pi}{n}}$$

13. Analogy:

Volume of a right conical body with cross-section varying linearly with its normal height (length) is given as

$$V = \frac{1}{3}(area\ of\ base)(normal\ height).$$

14. Angle (α) between the between any two bonds in a molecule having regular tetrahedral structure (like CH_4 molecule) is given as

$$\alpha = 2\cos^{-1}\left(\frac{1}{\sqrt{3}}\right) \approx 109.47^o$$

15. Angle between any two lines joining the centre of a tetrahedral void to the centres of spheres (forming the void) is equal to $2\cos^{-1}\left({}^{1}/_{\sqrt{3}}\right) \approx 109.47^o$

Note: In most of the books even in Chemistry it is given equal to 109.28^o which is very closer to the most correct value 109.47^o

Exercise - 8

Q.8.1. If the angle between two concurrent vectors is 30°. Now one of them, keeping other stationary, is rotated by an angle 45° about the point of concurrency in a plane inclined at an angle 60° with the plane containing both the vectors in initial position. Find out the angle between them in the final position.

Q.8.2. The angle between two concurrent vectors is 45°. If a vector, keeping other stationary is rotated by an angle 20° about the point of concurrency such that the angle between them in the final position is 60°. Determine angle between plane of rotation of the vector & plane of concurrency.

Q.8.3. Three planes are intersecting each other at a single in the space such that the angles between two consecutive lines of intersection are successively $30^o, 60^o \& 75^o$. Using HCR's Cosine-formula find out the angles between the intersecting planes successively.

Q.8.4. The angles between the successive lateral edges meeting at a vertex of a tetrahedron are $45^o, 50^o \&\ 70^o$. Using HCR's Cosine-formula find out the internal angles between the consecutive lateral faces of the tetrahedron.

Q.8.5. If the length of perpendicular drawn from the point of concurrency of three equal vectors to the plane of a triangle generated by joining the heads of all the vectors is 16cm & circumscribed-radius of the triangle generatedis 9cm.

Find out the length of the vectors.

Q.8.6. Circumscribed-radius, of the triangle generated by joining the heads of three equal concurrent vectors each of length 12cm, is 7cm. Find out length of perpendicular drawn from the point of concurrency to the plane of triangle.

Q.8.7. The length of perpendicular drawn from the point of concurrency of three equal vectors each of length 15cm to the plane of a triangle generated by joining the heads of all the vectors is 11cm. What will be circumscribed-radius of the triangle (generated).

Q.8.8. If the length of perpendicular drawn from the point of concurrency of three equal vectors each of length 5cm to the plane of a triangle generated by joining the heads of all the vectors is 3cm. What will be circumscribed-radius of the triangle (generated).

Q.8.9. The circumscribed-radius, of the triangle generated by joining the heads of three equal concurrent vectors each of length 30cm, is 17cm. Find out length of perpendicular drawn from point of concurrency to the plane of triangle.

Q.8.10. A triangle has circumscribed radius 9cm, smallest angle 30°& another angle 70°. Find out the distance between the circumscribed & inscribed centres.

Q.8.11. A triangle has inscribed radius 4cm, smallest angle 20°& another angle 75°.Determine distance between the circumscribed & inscribed centres.

Q.8.12. What is the distance between the circumscribed & inscribed centres of a triangle having inscribed radius 4cm, smallest angle 20°& another angle 75°.

Q.8.13. A triangle has sides 6cm, 8cm & 13cm. What is the distance between circumscribed & inscribed centre.

Q.8.14. An isosceles triangle has smallest angle 28°& circumscribed radius 10cm. Determine the distance between the circumscribed & inscribed centres.

Q.8.15. An isosceles triangle has greatest angle 100°& inscribed radius 4cm. Determine the distance between the circumscribed & inscribed centres.

Q.8.16. An isosceles triangle has greatest angle 80°& circumscribed radius 12cm. Determine the distance between the circumscribed & inscribed centres.

Q.8.17. An isosceles triangle has smallest angle 36°& inscribed radius 6cm. Determine the distance between the circumscribed & inscribed centres.

Q.8.18. Find out the distance between the circumscribed & inscribed centres of a right triangle has orthogonal sides 9cm & 12cm.

Q.8.19. Find out the distance between the circumscribed & inscribed centres of a right triangle has orthogonal sides 9cm & 12cm.

Q.8.20. A scalene triangle has circumscribed & inscribed radii 9cm & 3cm respectively and smallest angle 35°. If the distance between the circumscribed & inscribed centres is 4.8 cm, find out angles of the triangle.

Q.8.21. A scalene triangle has circumscribed & inscribed radii 16cm & 5cm respectively and smallest angle 30°. If the distance between the circumscribed & inscribed centres is 11.3cm, find out all the sides of triangle.

Q.8.22. Circumscribed & inscribed radii of a scalene triangle are 13cm & 3.6cm respectively and smallest angle 22°. If the distance between the circumscribed & inscribed centres is 10.2cm, find out all the sides of triangle.

Q.8.23. Find out area of a right triangle having smallest angle 30°& the distance between the circumscribed & inscribed centres is 5.6cm.

Q.8.24. Calculate the area of a scalene triangle having circumscribed & inscribed radii 12cm & 5.4cm respectively and smallest angle 24°. If the distance between circumscribed & inscribed centres is 7.8cm.

Q.8.25. Find out the area of a scalene triangle having circumscribed & inscribed radii 30cm & 14cm respectively and smallest angle 30° if the distance between circumscribed & inscribed centres is 15.2cm.

Q.8.26. Find outspike-angle & area of a spiked regular polygon having 7 spikes& span 5cm.

Q.8.27. Determine spike-angle & area of a spiked regular polygon having 15 spikes& span 12cm

Q.8.28. A spiked regular polygon has spike angle $13^o50'46''$& span 4cm. Find out the area of spiked polygon.

Q.8.29. Find out the area of a spiked regular polygon having spike angle $10^o35'18''$& span 15cm.

Q.8.30. A spiked regular polygon has spike angle 12°& span 8cm. Calculate the area of spiked polygon.

Q.8.31. Find out the radius of circle passing through the tip-points of spikes of a spiked regular polygon having area $100cm^2$& spike-angle 9°.

Q.8.32. A spiked regular polygon has spike angle 6.67^o& span 25cm. Calculate no. of the spikes &the area of spiked polygon.

Q.8.33. Find out the radius of circle passing through the tip-points of spikes of a spiked regular polygon having area $236cm^2$& spike-angle 5.45^o.

Appendix

1. The Solid Angle $(\boldsymbol{\omega})$ subtended by a regular pentagonal plane with each side $'\boldsymbol{a}'$ at any point lying at a normal height $'\boldsymbol{h}'$ from one of the vertices is given as

$$\omega = 2\left[\frac{3\pi}{10} + \sin^{-1}\left(\frac{hsin18^o}{\sqrt{h^2 + a^2cos^2 18^o}}\right) - \sin^{-1}\left(\frac{hcos36^o}{\sqrt{h^2 + a^2cos^2 18^o}}\right) - \sin^{-1}\left(\frac{2hsin18^o}{\sqrt{4h^2 + a^2cot^2 18^o}}\right)\right]$$

$$\forall\, h \to 0 \Rightarrow \omega \to \frac{3\pi}{5}$$

$$\forall\, a \to 0 \;\; or \;\; h \to \infty \Rightarrow \omega \to 0$$

2. The Solid Angle $(\boldsymbol{\omega})$ subtended by a regular hexagonal plane with each side $'\boldsymbol{a}'$ at any point lying at a normal height $'\boldsymbol{h}'$ from one of the vertices is given as

$$\omega = 2\left[\frac{\pi}{3} + \sin^{-1}\left(\frac{h}{\sqrt{4h^2 + 3a^2}}\right) - \sin^{-1}\left(\frac{h\sqrt{3}}{\sqrt{4h^2 + 3a^2}}\right) - \sin^{-1}\left(\frac{h}{2\sqrt{h^2 + 3a^2}}\right)\right]$$

$$\forall\, h \to 0 \Rightarrow \omega \to \frac{2\pi}{3}$$

$$\forall\, a \to 0 \;\; or \;\; h \to \infty \Rightarrow \omega \to 0$$

Competitive Questions

Q.1. An electric bulb of 200 W is mounted onto the centre of one of the walls of a room of $2.6m \times 2.6m \times 2.6m$. Find out the total radiation energy striking the floor area if the bulb emits radiation uniformly in all the directions.

(Neglect height of bulb from the centre of wall) Ans. 21.788 J/sec

Q.2. A uniform point-source of light is located at the centre of a torus having inner & outer radii 10cm & 18cm respectively. What will be the fraction of total radiation intercepted by the torus?

Ans. $^2/_7$

Q.3. A torus intercepts 60% of the total radiation emitted by a uniform point-source located at the centre of torus. Find out ratio of outer to inner radii of torus.

Ans. $^8/_3$

Q.4. A solar-panel of $30 \times 30\ unit^2$ is illuminated by a uniform point-source of light located at a normal height $15\sqrt{\sqrt{2}-1}\ unit$ from the centre of panel. Find out percentage of total radiation incident on panel.

Ans. 25%

Q.5. Find out the solid angle subtended by an infinitely long vertical cylindrical pole with a diameter D at a point lying in the plane of base at a normal distance 2D from the longitudinal axis.

Ans. $^{\pi}/_3\ sr$

Q.6. A torus having diameter 8cm intercepts 40% of the total radiation emitted by a uniform point-source located at the centre of torus. Find out inner & outer radii of torus.

Ans. 14cm & 6cm

Q.7. A uniform point-source of light is located at a normal height 8cm from the centre of an infinitely long rectangular plane having width 16cm. What will be the percentage of total radiation intercepted by the plane?

Ans. 25%

Q.8. Some identical spheres are arranged, touching one another, in a complete circular fashion such that each of them intercepts 25% of total radiation emitted by a uniform point-source located at the centre of polygonal plane obtained by joining the centres of spheres. Find out number of spheres.

Ans. 3

Q.9. A uniform point-source of 100 cd is located at the centre of a simple cubic cell. What will be total flux directly escaping the cell?

Ans. 334.24 lm

Q.10. Solid angle (ω) subtended by an ellipsoid having major and minor axes $2a$ & $2b$ respectively at any point lying on the major axis at a distance $d[\forall\ (d \geq a \geq b)]$ from the centre is given by

a) $2\pi\left(1-\sqrt{\frac{(d^2-a^2)}{(d^2-a^2)+b^2}}\right)$

b) $\pi\left(1+\sqrt{\frac{(d^2-a^2)}{(d^2-a^2)+b^2}}\right)$

c) $2\pi\left(1-\sqrt{\frac{(d^2-a^2)}{(d^2+a^2)-b^2}}\right)$

d) $None$

Q.11. An ellipsoid with major axis $2a$ unit & eccentricity $1/\sqrt{2}$ subtends a solid angle $\pi\ sr$ at a point lying on the major axis at a distance d unit from the centre. The value of h/a

a) $\sqrt{\frac{8}{7}}$ b) $\sqrt{\frac{7}{6}}$ c) $\sqrt{\frac{6}{5}}$ d) none.

Q.12. A sphere with radius R intercepts 30% of total radiation emitted by a uniform point-source located at a distance d from the centre. Then the value of d/R is equal to

a) $\frac{7}{\sqrt{21}}$ b) $\frac{5}{\sqrt{21}}$ c) $\frac{6}{\sqrt{21}}$ d) none.

Q.13. Solid angle, subtended by an infinitely long circular cylinder with a radius 14cm at a point lying in the plane of one of the circular-ends at a normal distance 84cm from the longitudinal axis, is equal to

a) $\sin^{-1}\left(\frac{1}{6}\right)\ sr$ b) $\sin^{-1}\left(\frac{1}{8}\right)\ sr$ c) $2\sin^{-1}\left(\frac{1}{6}\right)\ sr$ d) $2\sin^{-1}\left(\frac{1}{8}\right)\ sr$

Q.14. A uniform point-source of light is mounted at a normal height 'h' from the centre of an infinitely long rectangular plane having width 'b'. If plane intercepts 25% of total radiation emitted by the source then h/b equal to

a) 1 b) $\frac{1}{2}$ c) $\frac{1}{3}$ d) $\frac{1}{4}$

Q.15. A circular plane intercepts 40% of total radiation emitted by a uniform point-source located at a normal height h from the centre. Then the area of circular plane is equal to

a) $16\pi h^2$ b) $20\pi h^2$ c) $24\pi h^2$ d) none.

Q.16. A square plane intercepts x fraction of total radiation emitted by a uniform point-source located at a normal height h from one of the vertices. Then the area of square plane is equal to

a) $\frac{h^2}{cosec4\pi x - 1}$ b) $\frac{h^2}{cosec4\pi x + 1}$ c) $\frac{h^2}{cosec2\pi x - 1}$ d) $\frac{h^2}{cosec2\pi x + 1}$

Q.17. The angle between any two σ-bonds in CH_4 molecule is given as

a) $\cos^{-1}\left(\frac{1}{\sqrt{3}}\right)$ b) $2\cos^{-1}\left(\frac{1}{\sqrt{3}}\right)$ c) $\sin^{-1}\left(\sqrt{\frac{2}{3}}\right)$ d) $2\sin^{-1}\left(\sqrt{\frac{2}{3}}\right)$

Q.18. Solid angle (in sr.), subtended by a right pyramid having base as an equilateral triangle & angle between any two consecutive lateral edges 30°, at the vertex is equal to

a) $2\pi - 3\sin^{-1}\left(\sqrt{\frac{2}{3}}\right)$ b) $2\pi - 4\sin^{-1}\left(\sqrt{\frac{2}{3}}\right)$

c) $2\pi - 5\sin^{-1}\left(\sqrt{\frac{2}{3}}\right)$ d) $2\pi - 6\sin^{-1}\left(\sqrt{\frac{2}{3}}\right)$

Q.19. A uniform point-source of light is located at a point $(0, 0, -a)$ in the space. Fraction of total radiation, intercepted by a plane bounded by the straight lines $x = 0, x = a, y = 0\ \&\ y = a\ [\forall\ a > 0]$, is

a) $\frac{1}{6}$ b) $\frac{1}{12}$ c) $\frac{1}{18}$ d) $\frac{1}{24}$

Q.20. A uniform point-source of 400cd is located at a point $(0, 0, k)$ in the space. Total flux, intercepted by a plane bounded by the straight lines $x = 0, x = \infty, y = 0 \; \& \; y = k$ $[\forall \, k > 0]$, is

$a)$ $80\pi \; lm$ $b)$ $100\pi \; lm$ $c)$$120\pi \; lm$ $d)$ $none.$

Q.21. A right circular cone with apex angle 90° is thoroughly cut with a plane inclined at an angle 60° with the longitudinal axis of cone. Eccentricity of generated elliptical-section is

$a)$ $\frac{1}{\sqrt{2}}$ $b)$ $\frac{1}{\sqrt{3}}$ $c)$ $\frac{1}{\sqrt{5}}$ $d)$ $\frac{1}{\sqrt{7}}$

Q.22. A right circular cone with apex angle 60° is thoroughly cut with a plane inclined at an angle 45° with the longitudinal axis of cone. Angle, between the longitudinal axis & the line joining centre of elliptical-section to the apex-point of cone, is equal to

$a)$ $\tan^{-1}\left(\frac{1}{2}\right)$ $b)$ $\tan^{-1}\left(\frac{1}{3}\right)$ $c)$ $\tan^{-1}\left(\frac{1}{4}\right)$ $d)$ $\tan^{-1}\left(\frac{1}{5}\right)$

Q.23. A uniform point-source of light is located at the centre of a torus. If ratio of inner to outer radii is $^2/_3$ then the fraction of total radiation intercepted by the torus is equal to

$a)$ $\frac{1}{2}$ $b)$ $\frac{1}{3}$ $c)$ $\frac{1}{4}$ $d)$ $\frac{1}{5}$

Q.24. Total solid angle subtended by the identical spheres forming a tetrahedral void at the centre is equal to

$a)$ $4\pi\left(1-\frac{1}{\sqrt{3}}\right)$ sr $b)$ $6\pi\left(1-\frac{1}{\sqrt{3}}\right)$ sr $c)$ $8\pi\left(1-\frac{1}{\sqrt{3}}\right)$ sr $d)$ $none$

Q.25. Total solid angle subtended by the identical spheres forming an octahedral void at the centre is

$a)$ $8\pi\left(1-\frac{1}{\sqrt{2}}\right)$ sr $b)$ $12\pi\left(1-\frac{1}{\sqrt{2}}\right)$ sr $c)$ $16\pi\left(1-\frac{1}{\sqrt{2}}\right)$ sr $d)$ $none$

Q.26. Total solid angle subtended by the identical spheres forming a simple cubic unit-cell at the centre is

$a)$ $16\pi\left(1-\sqrt{\frac{2}{3}}\right)$ sr $b)$ $8\pi\left(1-\sqrt{\frac{2}{3}}\right)$ sr $c)$ $4\pi\left(1-\sqrt{\frac{2}{3}}\right)$ sr $d)$ $none$

Q.27. A uniform point-source of luminous intensity I is located at the centre of hollow cylindrical shell with a radius r& length l. Total flux intercepted by the shell is given by

$a)$ $4\pi I\left(\frac{l}{\sqrt{l^2+2r^2}}\right)$ $b)$ $4\pi I\left(\frac{l}{\sqrt{l^2+4r^2}}\right)$ $c)$ $4\pi I\left(\frac{l}{\sqrt{l^2+6r^2}}\right)$ $d)$ $none$

Q.28. Solid angle (in sr.), subtended by the largest sphere completely trapped in a tetrahedral void, at the centre of each of the identical spheres forming void is

$a)$ $\frac{\pi}{2}\left(1+\sqrt{\sqrt{\frac{4}{3}}-\frac{2}{3}}\right)$ $b)$ $\frac{\pi}{2}\left(1+\sqrt{\sqrt{\frac{5}{3}}-\frac{2}{3}}\right)$

$c)$ $2\pi\left(1-\sqrt{\sqrt{\frac{7}{3}}-\frac{2}{3}}\right)$ $d)$ $2\pi\left(1-\sqrt{\sqrt{\frac{8}{3}}-\frac{2}{3}}\right)$

Q.29. Solid angle (in sr.), subtended by the largest sphere completely trapped in an octahedral void, at the centre of each of the identical spheres forming void is

$a)$ $\pi\left(1+\frac{\sqrt{4\sqrt{3}-2}}{2}\right)$ $b)$ $\pi\left(1-\frac{\sqrt{4\sqrt{2}-2}}{2}\right)$

c) $2\pi\left(1-\frac{\sqrt{4\sqrt{2}-2}}{2}\right)$ d) $2\pi\left(1-\frac{\sqrt{4\sqrt{3}-2}}{2}\right)$

Q.30. A uniform point-source of luminous intensity I is located at the centre of hollow conical shell with a radius r& normal height h. Total flux intercepted by the shell is given by

a) $2\pi I\left(1+\frac{h}{\sqrt{h^2+9r^2}}\right)$ b) $2\pi I\left(1+\frac{h}{\sqrt{h^2+3r^2}}\right)$

c) $4\pi I\left(1-\frac{h}{\sqrt{h^2+9r^2}}\right)$ d) $4\pi I\left(1-\frac{h}{\sqrt{h^2+3r^2}}\right)$

Q.31. A hollow cylindrical shell intercepts $2/3$ of total radiation emitted by a uniform point-source of light located at the centre of shell. Ratio of length to diameter of the shell is

a) $\frac{1}{\sqrt{5}}$ b) $\frac{2}{\sqrt{5}}$ c) $\frac{1}{\sqrt{7}}$ d) $\frac{2}{\sqrt{7}}$

Q.32. A hollow conical shell intercepts 75% of total radiation emitted by a uniform point-source of light located at the centre of mass of shell. Ratio of radius to normal height of the shell is

a) $\frac{2}{\sqrt{5}}$ b) $\frac{1}{\sqrt{5}}$ c) $\frac{2}{\sqrt{3}}$ d) $\frac{1}{\sqrt{3}}$

Q.33. A hollow cylindrical shell intercepts $7/8$ of total radiation emitted by a uniform point-source of light located at the centre of shell. Ratio of diameter to length of the shell is

a) $\frac{\sqrt{15}}{8}$ b) $\frac{\sqrt{15}}{7}$ c) $\frac{8}{\sqrt{15}}$ d) $\frac{7}{\sqrt{15}}$

Q.34. Solid angle (in sr.), subtended by the largest sphere completely trapped in a simple cubic cell, at the centre of each of the identical spheres forming S.C.C. is

a) $\pi\left(1+\frac{\sqrt{2\sqrt{3}-1}}{\sqrt{3}}\right)$ b) $\pi\left(1+\frac{\sqrt{3\sqrt{2}-2}}{\sqrt{3}}\right)$

c) $2\pi\left(1-\frac{\sqrt{3\sqrt{2}-2}}{\sqrt{3}}\right)$ d) $2\pi\left(1-\frac{\sqrt{2\sqrt{3}-1}}{\sqrt{3}}\right)$

Q.35. A square plane intercepts $1/24$ of total radiation emitted by a uniform point-source located at a normal height h from one of the vertices then the area of plane is

a) $3h^2$ b) $2h^2$ c) h^2 d) *none.*

Q.36. A square plane intercepts $1/6$ of total radiation emitted by a uniform point-source located at a normal height h from the centre then the area of plane is

a) $4h^2$ b) $8h^2$ c) $12\,h^2$ d) *none.*

37. Solid angle, subtended by a rectangular plane having length l& width b at a point lying at a normal height h from one of the vertices of plane $[\forall \ l:b:h = 3:2:1]$, is

a) $\cos^{-1}\left(\frac{6}{\sqrt{50}}\right)$ b) $\cos^{-1}\left(\frac{7}{\sqrt{50}}\right)$ c) $\sin^{-1}\left(\frac{6}{\sqrt{50}}\right)$ d) $\sin^{-1}\left(\frac{7}{\sqrt{50}}\right)$

Q.38. An equilateral-triangular plane intercepts 25% of total radiation emitted by a uniform point-source located at a normal height h from the centre then the area of plane is

a) $4\sqrt{3}h^2$ b) $6\sqrt{3}h^2$ c) $8\sqrt{3}h^2$ d) *none.*

Q.39. Solid angle (in sr.) subtended by the solid generated by rotating the parabola, $y^2 = 4ax$ about x-axis at a point $(-a, 0)$ is

$a)\ \pi\left(1+\frac{1}{\sqrt{3}}\right)$ $b)\ \pi\left(1+\frac{1}{\sqrt{2}}\right)$ $c)\ 2\pi\left(1-\frac{1}{\sqrt{3}}\right)$ $d)\ 2\pi\left(1-\frac{1}{\sqrt{2}}\right)$

Q.40. Solid angle subtended by the plane bounded by the straight lines $x = 0, y = 0$ & $5x + 12y = 60$ at a point $(0, 5, -7)$ is

$a)\ 0.326\ sr$ $b)\ 0.426\ sr$ $c)\ 0.526\ sr$ $d)\ none$

Q.41. Solid angle subtended by the plane bounded by the straight lines $x = -5, x = 7, y = -2$ & $y = 10$ at a point $(7, -2, 4)$ is

$a)\ \cos^{-1}\left(\frac{9}{10}\right)$ $b)\ 2\cos^{-1}\left(\frac{7}{10}\right)$ $c)\ \sin^{-1}\left(\frac{9}{10}\right)$ $d)\ 2\sin^{-1}\left(\frac{7}{10}\right)$

Q.42. Solid angle subtended by the plane bounded by the straight lines $x = 2, x = 12, y = -3$ & $y = 5$ at a point $(2, 5, -4)$ is

$a)\ 2\cos^{-1}\left(\sqrt{\frac{19}{29}}\right)$ $b)\ \cos^{-1}\left(\sqrt{\frac{20}{29}}\right)$ $c)\ 2\sin^{-1}\left(\sqrt{\frac{19}{29}}\right)$ $d)\ \sin^{-1}\left(\sqrt{\frac{20}{29}}\right)$

Q.43. Solid angle subtended by the plane bounded by the straight lines $x = 3, x = \infty, y = 3$ & $y = \infty$ at a point $(0, 0, 4)$ is

$a)\ \sin^{-1}\left(\frac{9}{25}\right) - \cos^{-1}\left(\frac{24}{25}\right)$ $b)\ 2\sin^{-1}\left(\frac{9}{25}\right) - \cos^{-1}\left(\frac{24}{25}\right)$

$c)\ \sin^{-1}\left(\frac{9}{25}\right) + \cos^{-1}\left(\frac{24}{25}\right)$ $d)\ 2\sin^{-1}\left(\frac{9}{25}\right) + \cos^{-1}\left(\frac{24}{25}\right)$

Q.44. Solid angle subtended by XY-plane,in the first quadrant, at a point $(-5, -5, 12)$ is

$a)\ \sin^{-1}\left(\frac{25}{169}\right) + \cos^{-1}\left(\frac{100}{169}\right)$ $b)\ \sin^{-1}\left(\frac{25}{169}\right) + \cos^{-1}\left(\frac{120}{169}\right)$

$c)\ \sin^{-1}\left(\frac{25}{169}\right) + \cos^{-1}\left(\frac{140}{169}\right)$ $d)\ \sin^{-1}\left(\frac{25}{169}\right) + \cos^{-1}\left(\frac{160}{169}\right)$

Answers:

10. *a*) 11. *b*) 12. *b*) 13. *c*) 14. *b*) 15. *c*) 16. *a*) 17. *b*) & *d*)
18. *d*) 19. *d*) 20. *b*) 21. *a*) 22. *b*) 23. *d*) 24. *c*) 25. *b*) 26. *a*) 27. *b*) 28. *d*)
29 *c*) 30. *a*) 31. *b*) 32. *d*) 33. *b*) 34. *d*) 35. *c*) 36. *a*) 37. *c*) 38. *b*)
39. *d*) 40. *a*) 41. *c*) 42. *d*) 43. *c*) 44. *b*)

OBJECTIVE QUESTIONS

Q.1. Solid angle, subtended by an infinitely long circular cylinder with a radius 14cm at a point lying in the plane of one of the circular-ends at a normal distance 84cm from the longitudinal axis, is equal to

$a)\ \sin^{-1}\left(\frac{1}{4}\right)\ sr$ $b)\ \sin^{-1}\left(\frac{1}{6}\right)\ sr$ $c)\ 2\sin^{-1}\left(\frac{1}{4}\right)\ sr$ $d)\ 2\sin^{-1}\left(\frac{1}{6}\right)\ sr$

Q.2. A uniform point-source of light is mounted at a normal height 'h' from the centre of an infinitely long rectangular plane having width 'b'. If plane intercepts 25% of total radiation emitted by the source then $^h/_b$ equal to

$a)\ \frac{1}{4}$ $b)\ \frac{1}{3}$ $c)\ \frac{1}{2}$ $d)\ 1$

Q.3. A circular plane intercepts 40% of total radiation emitted by a uniform point-source located at a normal height h from the centre. Then the area of circular plane is equal to

$a)\ 24\pi h^2$ $b)\ 25\pi h^2$ $c) 26\pi h^2 d)\ 27\pi h^2.$

Q.4. A square plane intercepts x fraction of total radiation emitted by a uniform point-source located at a normal height h from one of the vertices. Then the area of square plane is equal to

$a)\ \frac{h^2}{cosec4\pi x+1}$ $b)\ \frac{h^2}{cosec2\pi x+1}$ $c)\ \frac{h^2}{cosec4\pi x-1}$ $d)\ \frac{h^2}{cosec2\pi x-1}$

Q.5. The angle between any two σ-bonds in CH_4 molecule is given as

$a)\ 2\cos^{-1}\left(\frac{1}{\sqrt{3}}\right)$ $b)\ \cos^{-1}\left(\frac{1}{\sqrt{3}}\right)$ $c)\ 2\sin^{-1}\left(\sqrt{\frac{2}{3}}\right)$ $d)\ \sin^{-1}\left(\sqrt{\frac{2}{3}}\right)$

Q.6. Solid angle (in sr.), subtended by a right pyramid having base as an equilateral triangle & angle between any two consecutive lateral edges 30°, at the vertex is equal to

$a)\ 2\pi-4\sin^{-1}\left(\sqrt{\frac{2}{3}}\right)$ $b)\ 2\pi-5\sin^{-1}\left(\sqrt{\frac{2}{3}}\right)$ $c)\ 2\pi-6\sin^{-1}\left(\sqrt{\frac{2}{3}}\right)$

$d)\ 2\pi-7\sin^{-1}\left(\sqrt{\frac{2}{3}}\right)$

Q.7. A uniform point-source of light is located at a point $(0,0,-a)$ in the space. Fraction of total radiation, intercepted by a plane bounded by the straight lines $x=0, x=a, y=0\ \&\ y=a\ [\forall\ a>0]$, is

$a)\ \frac{1}{24}$ $b)\ \frac{1}{25}$ $c)\ \frac{1}{26}$ $d)\ \frac{1}{27}$

Q.8. A uniform point-source of 400cd is located at a point $(0,0,k)$ in the space. Total flux, intercepted by a plane bounded by the straight lines $x=0, x=\infty, y=0\ \&\ y=k\ [\forall\ k>0]$, is

$a)\ 80\pi\ lm$ $b)\ 90\pi\ lm$ $c) 100\pi\ lm$ $d)\ none.$

Q.9. A right circular cone with apex angle 90° is thoroughly cut with a plane inclined at an angle 60° with the longitudinal axis of cone. Eccentricity of generated elliptical-section is

$a)\ \frac{1}{\sqrt{7}}$ $b)\ \frac{1}{\sqrt{5}}$ $c)\ \frac{1}{\sqrt{3}}$ $d)\ \frac{1}{\sqrt{2}}$

Q.10. A right circular cone with apex angle 60° is thoroughly cut with a plane inclined at an angle 45° with the longitudinal axis of cone. Angle, between the longitudinal axis & the line joining centre of elliptical-section to the apex-point of cone, is equal to

a) $\tan^{-1}\left(\frac{1}{2}\right)$ b) $\tan^{-1}\left(\frac{1}{3}\right)$ c) $\tan^{-1}\left(\frac{1}{4}\right)$ d) $\tan^{-1}\left(\frac{1}{5}\right)$

Q.11. A uniform point-source of light is located at the centre of a torus. If ratio of inner to outer radii is $2/3$ then the fraction of total radiation intercepted by the torus is equal to

a) $\frac{1}{5}$ b) $\frac{1}{6}$ c) $\frac{1}{7}$ d) $\frac{1}{8}$

Q.12. Total solid angle subtended by the identical spheres forming a tetrahedral void at the centre is

a) $6\pi\left(1-\frac{1}{\sqrt{3}}\right)$ *sr* b) $8\pi\left(1-\frac{1}{\sqrt{3}}\right)$ *sr* c) $10\pi\left(1-\frac{1}{\sqrt{3}}\right)$ *sr* d) *none*

Q.13. Total solid angle subtended by the identical spheres forming an octahedral void at the centre is

a) $8\pi\left(1-\frac{1}{\sqrt{2}}\right)$ *sr* b) $12\pi\left(1-\frac{1}{\sqrt{2}}\right)$ *sr* c) $16\pi\left(1-\frac{1}{\sqrt{2}}\right)$ *sr* d) *none*

Q.14. Total solid angle subtended by the identical spheres forming a simple cubic unit-cellat the centre is

a) $16\pi\left(1-\sqrt{\frac{2}{3}}\right)$ *sr* b) $8\pi\left(1-\sqrt{\frac{2}{3}}\right)$ *sr* c) $4\pi\left(1-\sqrt{\frac{2}{3}}\right)$ *sr* d) *none*

Q.15. A uniform point-source of luminous intensity I is located at the centre of hollow cylindrical shell with a radius r& length l. Total flux intercepted by the shell is given by

a) $4\pi I\left(\frac{l}{\sqrt{l^2+2r^2}}\right)$ b) $4\pi I\left(\frac{l}{\sqrt{l^2+4r^2}}\right)$ c) $4\pi I\left(\frac{l}{\sqrt{l^2+6r^2}}\right)$ d) *none*

Answers: 1. d) 2. c) 3. a) 4. c) 5. a)& c) 6. c) 7. a) 8. c) 9. d) 10. b) 11. a) 12. b) 13. b) 14. a) 15. b)

Miscellaneous Review Questions

These numerical questions are relevant to all the articles from this book.

Q.1. Derive expression for solid angle subtended by a right triangular lamina having orthogonal sides 'a' & 'b' at a point lying at a normal height 'h' from the right angled vertex.

Q.2. Using Integral theorem, derive expression for solid angle subtended by a rectangular lamina having sides 'a' & 'b' at a point lying at a normal height 'h' from the centre.

Q.3. Find out the solid angle subtended by a right triangular plane with orthogonal sides 6cm & 8cm at a point lying at a normal height 3cm from the centre of hypotenuse.

Q.4. What will be the total flux incident on a square plane with each side 20cm if four uniform point-sources each of 200cd is located at the same normal height 25cm from the vertex?

Q.5. Find out the solid angle subtended by a plane bounded by the straight lines $x = 0, y = 0$ & $2x + 3y = 12$ at a point $(6, 0, 5)$.

Q.6. Find out the solid angle subtended by a plane bounded by the straight lines $x = 4, x = 16, y = 2$ & $y = 12$ at a point $(0, 0, 4)$.

Q.7. Find out the solid angle subtended by a right triangular plane with orthogonal sides 15cm & 20cm at a point lying at a normal height 8cm from a point which divides the hypotenuse in an internal ratio 2:3measured from vertex of smallest angle.

Q.8. A uniform point-source of 200 cd is located at a point lying at a normal height 6cm from a point which internally divides the hypotenuse of a right triangular lamina in a ratio 6:7 measured from vertex of larger acute angle. Calculate the total flux incident on the plane.

(Given orthogonal sides of triangle are 15cm & 36cm)

Q.9. Find out the total radiation energy incident on a triangular lamina having sides 14cm, 16cm & 20cm if a uniform point source of 100W is located at a normal height 4cm from the inscribed centre.

Q.10. A square panel of $16cm \times 16cm$, consisting of square photo-sensitive cells, is illuminated by a uniform point-source (S) of 2kW located at point 'S' lying at a normal height 6cm from the vertex of panel. Find out total radiation energy incident on the shaded cells (as indicated in the fig (1) below)

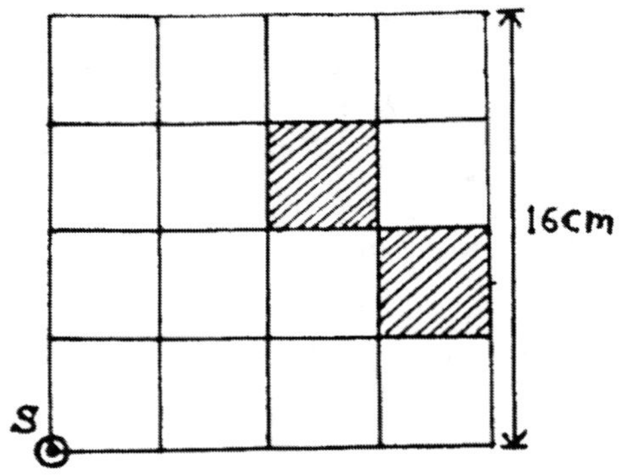

Fig (1)

Q.11. A rectangular panel of $54cm \times 27cm$, consisting of square photo-sensitive cells, is illuminated by a uniform point-source (S) of 300cd located at point lying at a normal height 10cm from the centre of panel. Find out total radiation energy incident on the shaded cells (as indicated in the fig (2) below)

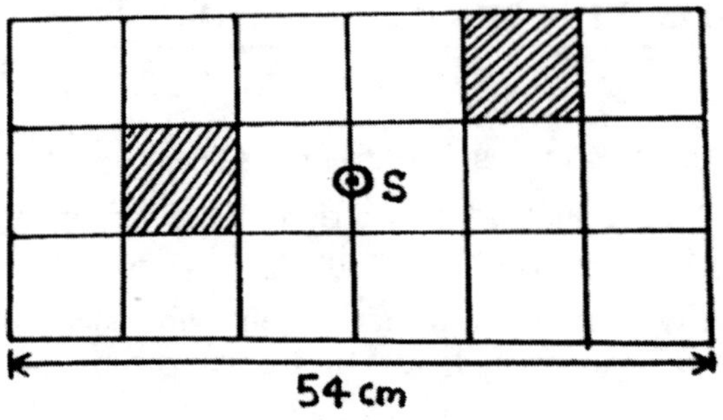

Fig(2)

Q.12. Find out total radiation energy incident on a large rectangular solar-panel $4m \times 6m$ consisting of photo-sensitive cells, installed on a roof. If at any instant, angles elevation & deviation of the Sun (i.e. straight line joining the centre of panel to the centre of the Sun) are 70°&33.7^o respectively (as shown in the fig(3) below) (Neglect the effect of absorption by intermediate medium)

(Take luminosity of the Sun as $3.84 \times 10^{26}\ W$ & its average distance from the centre of panel as $1.48 \times 10^{11}\ m$)

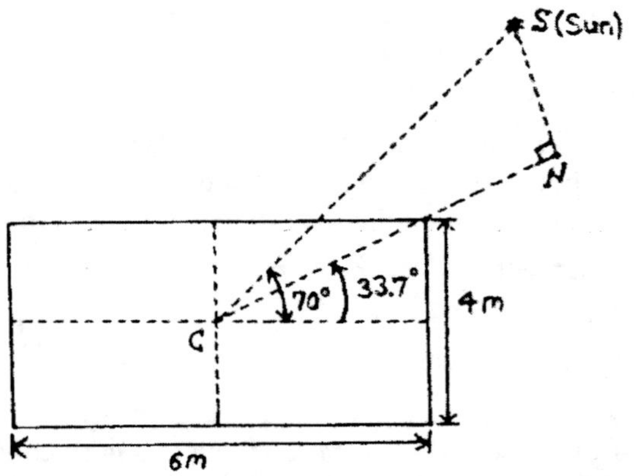

Fig (3)

Q.13. A uniform point-source of 110cd is located at a normal distance 1.5m from the centre of a photo-sensitive circular plate having radius 28cm & a square hole with each side 5cm at its centre. What will be flux density if intercepted flux is uniformly distributed over the entire plate?

Q.14. Find out the solid angle subtended by a regular polygonal plane with 15 no. of the sides each of 10cm at a point lying at a normal height 80cm from the centre.

Q.15. Find out the total radiation incident on an equilateral triangular lamina having each side 25cm & a square hole of $4cm \times 4cm$at the centre if a uniform point-source (S) of 200 W is located at a normal height 12cm from the centre.

(as shown in fig(4) below)

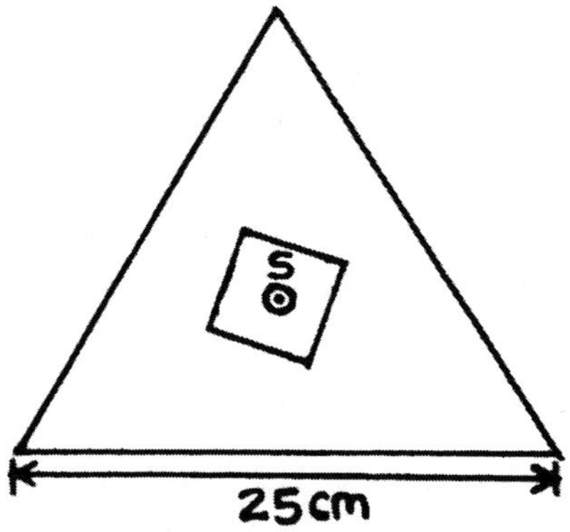

Fig (4)

Q.16. Calculate approximate value of solid angle subtended by an elliptical plane with axes 12cm & 16cm at a point lying at a normal height 20cm from centre. What will be total flux intercepted by the plane if a uniform point-source of 200cd is located at the same point?

Q.17. Find out the solid angle subtended by a 12m long conical tower at a point lying in the plane of base at a distance 40cm from its centre.

Q.18. If $40^o, 60^o \& 100^o$ are the angles between consecutive lateral edges meeting at one of the vertices of a tetrahedron find out the internal angles between consecutive lateral faces meeting at the same vertex.

Q.19. A uniform point-source (S) of 1kW is located at a normal height 8cm from the centre of a rectangular lamina of $16cm \times 30cm$ having a circular hole at the centre & two identical square holes (as shown in the fig(5) below). Find out the total radiation energy incident on the lamina

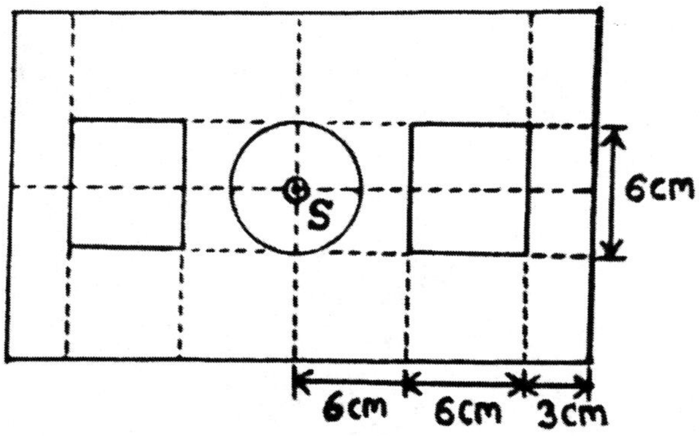

Fig (5)

Q.20. A uniform point-source (S) of 100cd is located at a normal height 5cm from the centre of a semi-regular hexagonal plane having a semi-circular hole at the centre (as shown in the fig(6) below). What will be total flux striking plane (as shaded)

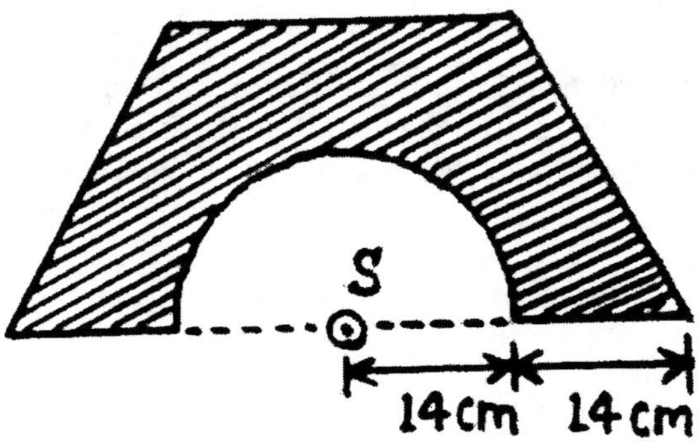

Fig (6)

Q.21. Seven identical spheres are arranged, touching one another, in a complete circular-fashion. If a uniform point-source of 100 W is located at the centre of regular heptagonal plane obtained by joining the centres of all the spheres calculate the radiation incident of the spheres.

Q.22. A panel consisting of identical square photo-sensitive shells, is illuminated by a uniform point-source (S) of 5kW (as shown in the fig(7) below) Calculate the total flux striking the panel. (Take normal height = 10cm)

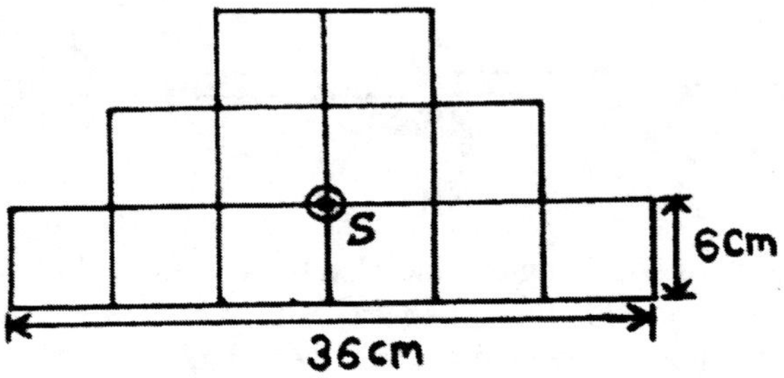

Fig (7)

Q.23. Find out the total flux directly escaping a tetrahedral void if a uniform point-source of 100 W is located at the centre of void.

Q.24 Find out the total flux directly escaping an octahedral void if a uniform point-source of 50 cd is located at the centre of void.

Q.25. Prove that angle between any two consecutive σ-bondsin a tetrahedral structure is $2\cos^{-1}\left(1/\sqrt{3}\right) \approx 109.47^o$.

Q.26. Find out the total radiation energy striking hatched equilateral triangular plane if a uniform point-source (S) of 200cd is located at a normal height 4cm from the centre (as shown in the fig(8) below)

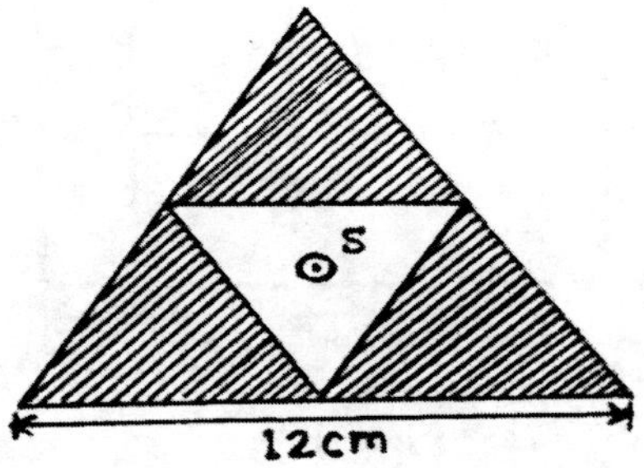

Fig (8)

Q.27. A uniform point source of 100 cd is located at the centre of a closed cube having each side 45cm. If a small rectangular hole of $4cm \times 8cm$ is made at the centre of one of the faces what will be flux escaping the cube?

(Neglect the effect of reflection)

Q.28. Two circular discs of equal width 6cm having radii 12cm & 21cm are co-axially placed. (Smaller on larger, as shown in the fig(9) below). Find out the total radiation incident on the lower one if a uniform point-source (S) of 0.5 kW is located at a normal height 24cm from the centre of upper disc.

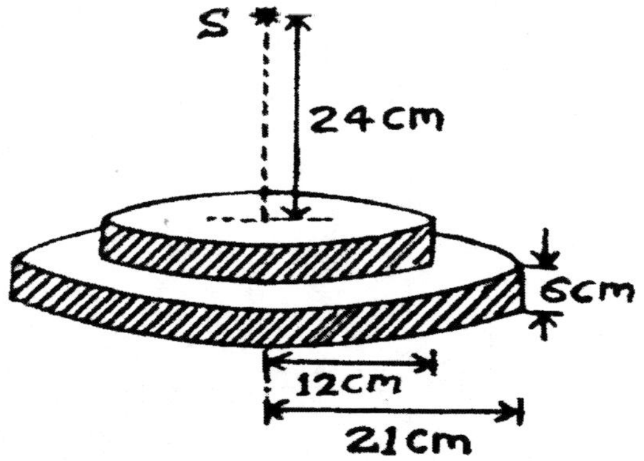

Fig (9)

Q.29. Two spheres A & B with radii 10cm & 28cm respectively & a uniform point-source (S) of 2kW are located linearly (as shown in the fig(9) below). Find out the total radiation energy striking sphere-B.

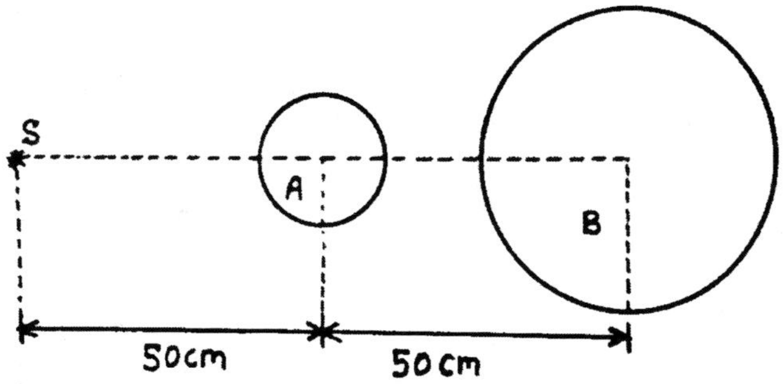

Fig (10)

Q.30. Two spheres A & B with radii 12cm & 32cm and luminosity 80 W & 55W respectively are placed in non-absorbing medium (see fig(10) below)

Find out the radiation energy striking each emitted by other one.

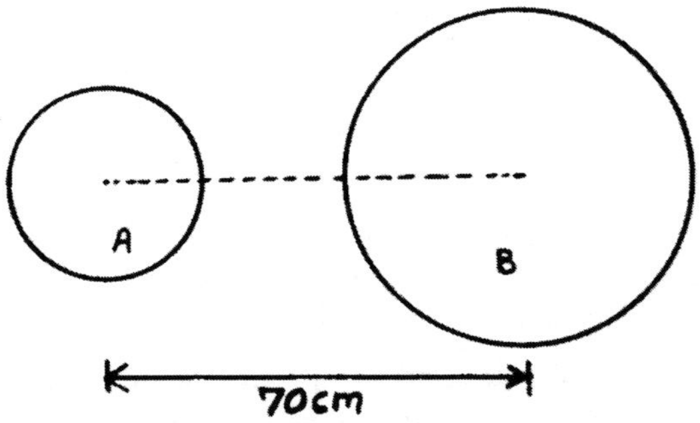

Fig (11)

Q.31. Determine the flux density on a square lamina having each side 18cm & a square hole with vertices at the mid-points of sides of larger one. If a uniform point-source (S) of 100 cd is located at a normal height 10cm from the centre. (As shown in the fig(12) below) (Assume that intercepted flux is uniformly distributed over the entire plane)

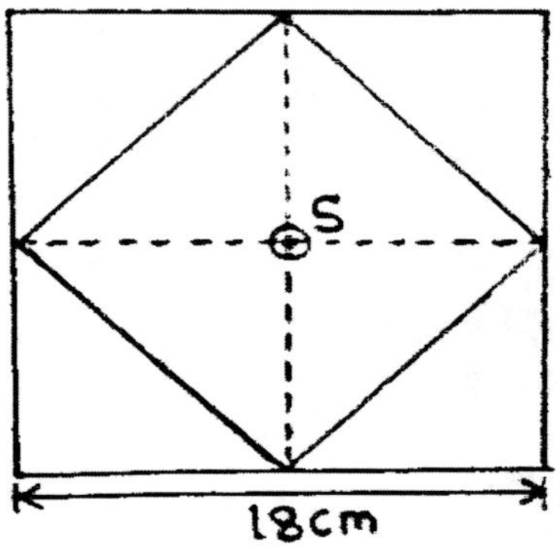

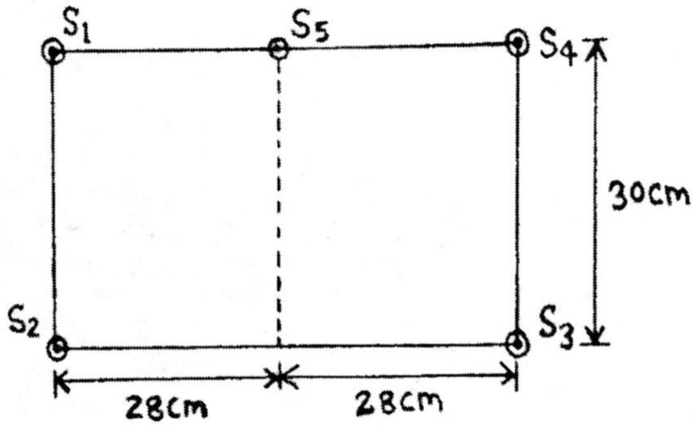

Fig (12) Fig (13)

Q.32. Find out the total radiation energyintercepted by a rectangular plane of $30cm \times 56cm$ if five uniform point-sources $S_1, S_2, S_3, S_4 \& S_5$ each of 200 W are located at an equal normal height 25cm perpendicular to the plane of paper (as shown in the fig(13) above).

Q.33. Determine total flux incident on a right triangular plane with orthogonal sides 18cm & 25cm if two uniform point-sources $S_1, \& S_2$ each of 110 cd are located at an equal normal height 14cm from the respective acute angled vertices (as indicated in the fig(14) below)

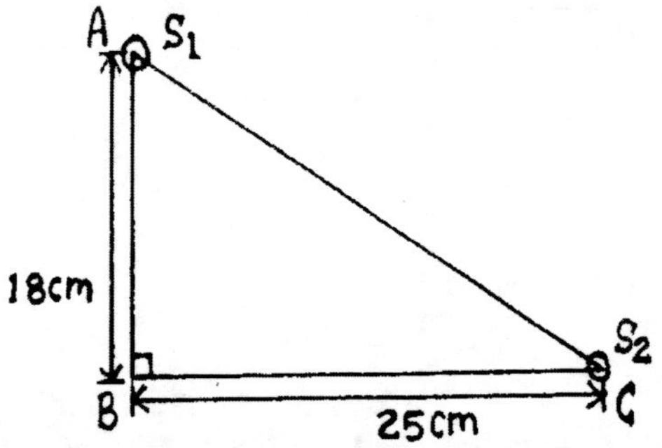

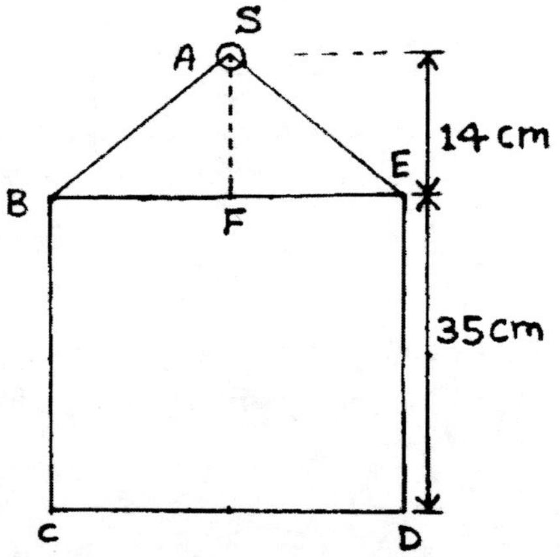

Fig (14) Fig (15)

Q.34. Determine total radiation energy incident on a plane lamina ABCDE consisting of square BCDE with each side 35cm & an isosceles ΔABE if a uniform point-source (S) of 200 W is located at a normal height 8cm from the vertex 'A' (as indicated in the fig(15) above)

Q.35. A uniform point-source (S) of 200 cd is located at a normal height 10cm from the centre of a circular disc with a radius 14cm if a small rectangular plane is removed from the centre (see the fig(16) below) what will be change in the flux striking the disc?

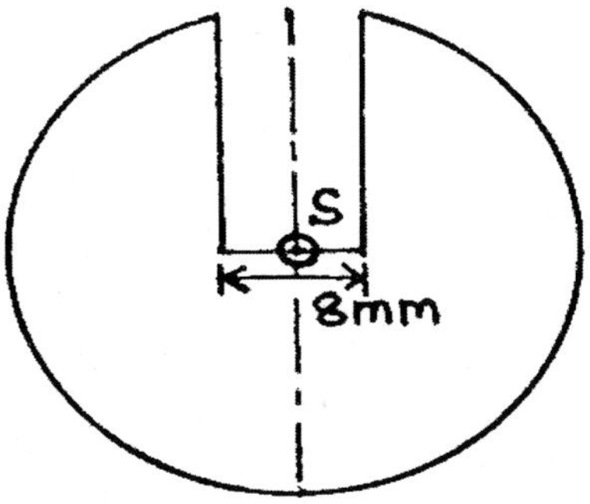

Fig (16)

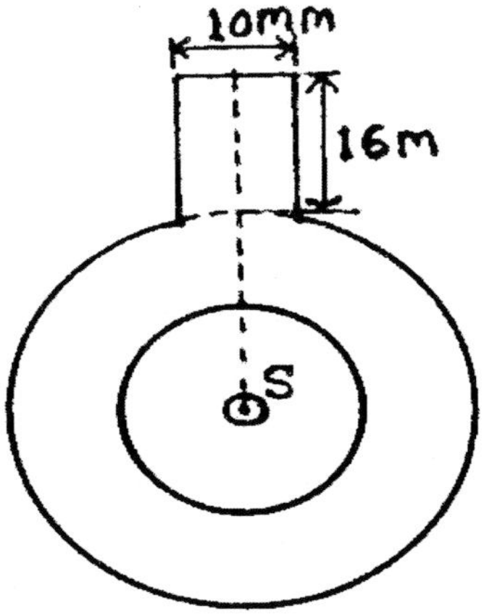

Fig (17)

Q.36. A uniform point-source (S) of 2kW is located at a normal height 6cm from the centre of eye-section of an eye-bolt with inner & outer radii 20mm& 30mm respectively (as indicated in the fig(17) above). Calculate the total radiation striking the entire section?

Q.37. Figure (18) below shows cross-section of a coupling-sleeve having inner & outer radii 25mm & 40mm and a square key-way of $5mm \times 5mm$ what will be the total radiation striking the section if a uniform point-source (S) of 1kW is located at a normal height 10cm from the centre?

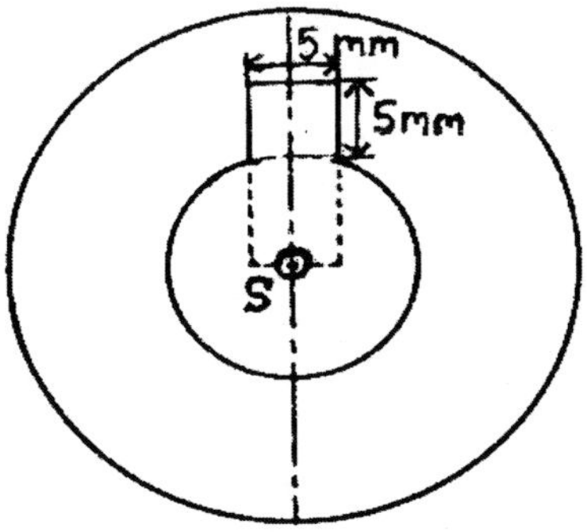

Fig (18)

Note: So far, mostly the answers of numerical questions have not been provided although all of them can be solved by concepts & formulae in this book. Next edition will be improved for the same.